# MITOCHONDRIA IN OBESITY AND TYPE 2 DIABETES

# MITOCHONDRIA IN OBESITY AND TYPE 2 DIABETES

## Comprehensive Review on Mitochondrial Functioning and Involvement in Metabolic Diseases

*Edited By*

Béatrice Morio

Luc Pénicaud

Michel Rigoulet

Academic Press is an imprint of Elsevier
125 London Wall, London EC2Y 5AS, United Kingdom
525 B Street, Suite 1650, San Diego, CA 92101, United States
50 Hampshire Street, 5th Floor, Cambridge, MA 02139, United States
The Boulevard, Langford Lane, Kidlington, Oxford OX5 1GB, United Kingdom



**Library of Congress Cataloging-in-Publication Data**
A catalog record for this book is available from the Library of Congress

**British Library Cataloguing-in-Publication Data**
A catalogue record for this book is available from the British Library

ISBN 978-0-12-811752-1

For information on all Academic Press publications
visit our website at https://www.elsevier.com/books-and-journals

*Publisher:* Stacy Masucci
*Acquisition Editor:* Tari K. Broderick
*Editorial Project Manager:* Timothy Bennett
*Production Project Manager:* Maria Bernard
*Cover Designer:* Mark Rogers

Typeset by SPi Global, India

# Contents

# II

# TISSUES INVOLVED IN THE PROGRESSION OF THE PATHOLOGIES

## 6. Role of Mitochondria in the Skeletal Muscle Metabolism in Obesity and Type 2 Diabetes

PAULA M. MIOTTO, GRAHAM P. HOLLOWAY



## 7. Role of Mitochondria in Adipose Tissues Metabolism and Plasticity

AUDREY CARRIÈRE, LOUIS CASTEILLA



## 8. Role of Mitochondria in the Liver Metabolism in Obesity and Type 2 Diabetes

HISAYUKI KATSUYAMA, JULIANE K. CZECZOR, MICHAEL RODEN



## 9. Contributions of Mitochondrial Dysfunction to ß Cell Failure in Diabetes Mellitus

JULIA PARNIS, GUY A. RUTTER

## 10. Role of Mitochondria in Brain Nutrient Sensing: Control of Energy Balance and Dysregulation in Obesity and Type 2 Diabetes

CORINNE LELOUP, LUC PÉNICAUD



# III

# TISSUES SUFFERING CONSEQUENCES FROM THE PATHOLOGIES

## 11. Role of Mitochondria in Cardiovascular Comorbidities Associated with Obesity and Type 2 Diabetes

SIHEM BOUDINA



## 12. Role of Mitochondria in the Regulation of Kidney Function and Metabolism in Type 2 Diabetes

XIANLIN HAN, YUGUANG SHI, MAGGIE DIAMOND-STANIC, KUMAR SHARMA



## 13. Role of Mitochondria in Neurodegeneration in Obesity and Type 2 Diabetes

SUSANA CARDOSO, RAQUEL M. SEIÇA, PAULA I. MOREIRA



# IV

# FACTORS THAT MAY TRIGGER OR AGGRAVATE THE PATHOLOGIES

## 14. Impact of Fetal Programming on Mitochondrial Function and Susceptibility to Obesity and Type 2 Diabetes

AMITA BANSAL, CETEWAYO RASHID, REBECCA A. SIMMONS

## 15. Mitochondrial Dysfunction Induced by Xenobiotics: Involvement in Steatosis and Steatohepatitis

KARIMA BEGRICHE, JULIE MASSART, BERNARD FROMENTY



# V PATHOLOGY PREVENTION AND CARE

## 16. Impact of Lifestyle and Clinical Interventions on Mitochondrial Function in Obesity and Type 2 Diabetes

BRENNA OSBORNE, AMANDA E. BRANDON, GREG C. SMITH, NIGEL TURNER



## 17. State of Knowledge and Recent Advances in Prevention and Treatment of Mitochondrial Dysfunction in Obesity and Type 2 Diabetes

CARLES CANTÓ



# VI SYNTHESIS

## 18. Mitochondria in Obesity and Type 2 Diabetes: Concluding Review and Research Perspectives

BÉATRICE MORIO, LUC PÉNICAUD, MICHEL RIGOULET

# Contributors

**Nicole Averet** Institut de Biochimie et Génétique Cellulaires, UMR 5095, CNRS, Université de Bordeaux, Bordeaux, France

**Amita Bansal** Center for Research on Reproduction and Women's Health; Center of Excellence in Environmental Toxicology, Perelman School of Medicine, University of Pennsylvania, Philadelphia, PA, United States

**Karima Begriche** INSERM, Univ Rennes, INRA, Institut NUMECAN (Nutrition Metabolisms and Cancer) UMR_A 1341, UMR_S 1241, Rennes, France

**Cyrielle Bouchez** Institut de Biochimie et Génétique Cellulaires, UMR 5095, CNRS, Université de Bordeaux, Bordeaux, France

**Sihem Boudina** Department of Nutrition and Integrative Physiology, College of Health, University of Utah, Salt Lake City, UT, United States

**Amanda E. Brandon** Sydney Medical School, Charles Perkins Centre, University of Sydney, Sydney, Australia

**Yan Burelle** Interdisciplinary School of Health Sciences, Faculty of Health Sciences; Department of Cellular and Molecular Medicine, Faculty of Medicine, University of Ottawa, Ottawa, ON, Canada

**Nadine Camougrand** Institut de Biochimie et Génétique Cellulaires, UMR 5095, CNRS, Université de Bordeaux, Bordeaux, France

**Carles Cantó** Nestlé Institute of Health Sciences S.A; École Polytechnique Fédérale de Lausanne, Lausanne, Switzerland

**Susana Cardoso** CNC-Center for Neuroscience and Cell Biology; Institute for Interdisciplinary Research, University of Coimbra, Coimbra, Portugal

**Audrey Carrière** STROMALab, Université de Toulouse, EFS, ENVT, Inserm U1031, ERL CNRS 5311, CHU Rangueil, Toulouse, France

**François Casas** DMEM – Dynamique Musculaire et Métabolisme, UMR 866, Montpellier Cedex, France

**Louis Casteilla** STROMALab, Université de Toulouse, EFS, ENVT, Inserm U1031, ERL CNRS 5311, CHU Rangueil, Toulouse, France

**Juliane K. Czeczor** Institute for Clinical Diabetology, German Diabetes Center, Leibniz Center for Diabetes Research at Heinrich-Heine University; German Center for Diabetes Research, Düsseldorf, Germany

**Claudine David** Institut de Biochimie et Génétique Cellulaires, UMR 5095, CNRS, Université de Bordeaux, Bordeaux, France

**Anne Devin** Institut de Biochimie et Génétique Cellulaires, UMR 5095, CNRS, Université de Bordeaux, Bordeaux, France

**Maggie Diamond-Stanic** Center for Renal Precision Medicine, Division of Nephrology, Department of Medicine, University of Texas Health San Antonio, San Antonio, TX, United States

**Stéphane Duvezin-Caubet** Institut de Biochimie et Génétique Cellulaires, UMR 5095, CNRS, Université de Bordeaux, Bordeaux, France

**Bernard Fromenty** INSERM, Univ Rennes, INRA, Institut NUMECAN (Nutrition Metabolisms and Cancer) UMR_A 1341, UMR_S 1241, Rennes, France

**Xianlin Han** Barshop Institute for Longevity and Aging Studies; Center for Renal Precision Medicine, Division of Nephrology, Department of Medicine; Department of Biochemistry, University of Texas Health San Antonio, San Antonio, TX, United States

**Graham P. Holloway** Department of Human Health and Nutritional Sciences, University of Guelph, Guelph, ON, Canada

**Hisayuki Katsuyama** Institute for Clinical Diabetology, German Diabetes Center, Leibniz Center for Diabetes Research at Heinrich-Heine University; German Center for Diabetes Research, Düsseldorf, Germany

**Albert M. Kroon** Department of Clinical and Movement Neurosciences, Institute of Neurology, University College London, London, United Kingdom

**Corinne Leloup** Center of Taste Sciences and Feeding Behaviour (CSGA), UMR CNRS 6265, INRA 1324, Bourgogne Franche-Comté University, AgroSup Dijon, Dijon, France

**Stéphen Manon** Institut de Biochimie et Génétique Cellulaires, UMR 5095, CNRS, Université de Bordeaux, Bordeaux, France

**Julie Massart** Department of Molecular Medicine and Surgery, Karolinska University Hospital, Karolinska Institutet, Stockholm, Sweden

**Keir J. Menzies** Interdisciplinary School of Health Sciences, Faculty of Health Sciences; Department of Biochemistry, Microbiology and Immunology, University of Ottawa Brain and Mind Research Institute and Centre for Neuromuscular Disease, University of Ottawa, Ottawa, ON, Canada

**Paula M. Miotto** Department of Human Health and Nutritional Sciences, University of Guelph, Guelph, ON, Canada

**Thibaut Molinié** Université Bordeaux, IBGC, UMR 5095; Institut de Biochimie et Génétique Cellulaires, UMR 5095, CNRS, Université de Bordeaux, Bordeaux, France

**Paula I. Moreira** CNC-Center for Neuroscience and Cell Biology; Laboratory of Physiology—Faculty of Medicine, University of Coimbra, Coimbra, Portugal

**Béatrice Morio** CarMeN Laboratory, Research Unit U. 1060 Inserm/Lyon1 University/INRA 1397, Lyon Sud Medical School, Oullins, France

**Arnaud Mourier** Institut de Biochimie et Génétique Cellulaires, UMR 5095, CNRS, Université de Bordeaux, Bordeaux, France

**Brenna Osborne** Department of Pharmacology, School of Medical Sciences, The University of New South Wales, Sydney, Australia; Department of Cellular and Molecular Medicine, University of Copenhagen, Copenhagen, Denmark

**Julia Parnis** Section of Cell Biology and Functional Genomics, Imperial College London, Hammersmith Hospital, London; Division of Cardiovascular Medicine, Radcliffe Department of Medicine; Wellcome Trust Centre for Human Genetics, University of Oxford; Oxford Centre for Diabetes, Endocrinology and Metabolism, University of Oxford, Churchill Hospital, Oxford, United Kingdom

**Luc Pénicaud** STROMALab, Université de Toulouse, EFS, ENVT, Inserm U1031, ERL CNRS 5311, CHU Rangueil, Toulouse, France

**Cetewayo Rashid** Center for Research on Reproduction and Women's Health; Center of Excellence in Environmental Toxicology, Perelman School of Medicine, University of Pennsylvania, Philadelphia, PA, United States

**Kimberly Reid** Interdisciplinary School of Health Sciences, Faculty of Health Sciences; Department of Biology, Faculty of Science, University of Ottawa, Ottawa, ON, Canada

**Michel Rigoulet** Institut de Biochimie et Génétique Cellulaires, UMR 5095, CNRS, Université de Bordeaux, Bordeaux, France

**Michael Roden** Institute for Clinical Diabetology, German Diabetes Center, Leibniz Center for Diabetes Research at Heinrich-Heine University; German Center for Diabetes Research; Division of Endocrinology and Diabetology, Medical Faculty, Heinrich-Heine University, Düsseldorf, Germany

**Manuel Rojo** Institut de Biochimie et Génétique Cellulaires, UMR 5095, CNRS, Université de Bordeaux, Bordeaux, France

**Guy A. Rutter** Section of Cell Biology and Functional Genomics, Imperial College London, Hammersmith Hospital, London, United Kingdom

**Raquel M. Seiça** Laboratory of Physiology; iCBR - Coimbra Institute for Clinical and Biomedical Research, Faculty of Medicine, University of Coimbra, Coimbra, Portugal

**Kumar Sharma** Center for Renal Precision Medicine, Division of Nephrology, Department of Medicine, University of Texas Health San Antonio; Audie L. Murphy Memorial VA Hospital, South Texas Veterans Health Care System (STVHCS), San Antonio, TX, United States

**Yuguang Shi** Barshop Institute for Longevity and Aging Studies; Department of Pharmacology, University of Texas Health San Antonio, San Antonio, TX, United States

**Rebecca A. Simmons** Center for Research on Reproduction and Women's Health; Center of Excellence in Environmental Toxicology, Perelman School of Medicine, University of Pennsylvania; Division of Neonatology, The Children's Hospital of Philadelphia, Philadelphia, PA, United States

**Greg C. Smith** Department of Pharmacology, School of Medical Sciences, The University of New South Wales, Sydney, Australia

**Jan-Willem Taanman** Department of Clinical and Movement Neurosciences, Institute of Neurology, University College London, London, United Kingdom

**Nigel Turner** Department of Pharmacology, School of Medical Sciences, The University of New South Wales, Sydney, Australia

**Goutham Vasam** Interdisciplinary School of Health Sciences, Faculty of Health Sciences, University of Ottawa, Ottawa, ON, Canada

**Edgar Djaha Yoboué** Protein Transport and Secretion Unit, Division of Genetics and Cell Biology, IRCCS Ospedale San Raffaele, Milan, Italy

# Preface

The World Health Organization has highlighted that, since 1975–80 (around 40 years), the global prevalence of overweight and obesity has tripled, and that of diabetes has nearly doubled in the adult population. This almost parallel increase in both diseases can be explained by the fact that overweight or obesity has been shown to represent the strongest risk factors for type 2 diabetes. During the past decade, obesity and diabetes prevalence have risen faster in low- and middle-income countries than in high-income countries. When diabetes is not well managed, complications develop and can damage the heart, blood vessels, eyes, kidneys, central nervous system, and nerves, leading to disability and premature death.

Obesity and type 2 diabetes are nutritional and metabolic pathologies mainly explained by an energy imbalance between calories consumed and calories expended. This leads to changes in metabolic fluxes at the level of different tissues or organs and, as a consequence, at the level of the whole organism. Mitochondria are at the heart of the cellular energy metabolism and contribute for about 90% of the energy conversion necessary for a smooth running of eukaryote cells. During the past two decades, research about mitochondria has revolutionized our understanding of its structure, dynamics, function, and regulation. In particular, these works have contributed to the understanding of the place that the mitochondria take in the metabolic adaptations and disorders observed in the previously mentioned pathologies. Throughout the world, several laboratories have contributed significantly to the advancement of our knowledge about the tissue-specific composition and functioning of mitochondria, their regulation, and their significance in the metabolic regulation of main tissues and organs implicated to most of the defects observed in obesity and type 2 diabetes. They have led to substantial progress in the comprehension of the involvement of mitochondria in the etiology and in the prevention of these nutritional and metabolic diseases.

The aim of this book is to provide an educational tool that synthesized the actual knowledge on mitochondria as a major player in the etiology, prevention, and treatment of obesity and type 2 diabetes. The book is primarily dedicated to students and educational teams in biology, physiology and medicine, researchers in basic and applied sciences in biology, physiology and medicine, and R&D researchers from food and pharmaceutical companies.

The first part is composed of five chapters that present the basic notions on mitochondria. They outline mitochondrial structure, dynamics and function, underlying its role in metabolism and as a signaling platform; their regulation by hormonal and nutritional factors; and the involvement of the mitochondrial genome and heteroplasmy.

In Chapters 6–10, the role of the mitochondria in the main tissues (skeletal muscle and adipose tissues) and organs (liver, pancreas) involved in the regulation of metabolism is described underlying the changes observed in obesity and/or type 2 diabetes. We then address the consequences of these metabolic disorders on other tissues, such as the cardiovascular system (Chapter 11), the kidney (Chapter 12), and the brain (Chapter 13). Chapters 14 and 15 focus on fetal programming and environmental factors that can play a role in activating or aggravating metabolic disorders through their effects on mitochondrial functions. Chapters 16 and 17 address the effects on mitochondrial function of current lifestyle and clinical intervention strategies for preventing obesity and type 2 diabetes, and how these impacts on mitochondria can help to improve health outcomes, and the state of knowledge and recent advances in prevention and treatment of mitochondrial dysfunction in obesity and type 2 diabetes.

In the last chapter (Chapter 18), we propose a comprehensive review with an integrated view of the role played by mitochondria in obesity and type 2 diabetes and identification of future research perspectives.

The editors would like to thank all the authors for having agreed to participate and for their valuable contributions. Many thanks to the reviewers for their kind support and precious contribution: Philippe Beaune, Karim Bouzakri, Olga Corti, Eric Fontaine, Emmanuel Letavernier, Anne Lombes, Anne-Marie Madec, Christophe Magnan, Cédric Moro, Mélanie Paillard, Baptiste Panthu, Bernard Portha, and Francesc Villaroya. They also acknowledge the help of Timothy Bennett, Megan Ball, Joslyn T. Chaiprasert-Paguio, and Tari K. Broderick from Elsevier Life Science editorial team. Finally, our most profound thanks go to Dusacq, a generous artist, whose work illuminates our cover page.

**Béatrice Morio**
**Luc Pénicaud**
**Michel Rigoulet**

# PART I

# EDUCATIONAL CHAPTERS

CHAPTER

1

# Mitochondria: Ultrastructure, Dynamics, Biogenesis and Main Functions

*Anne Devin**, *Cyrielle Bouchez**, *Thibaut Molinié**,†, *Claudine David**, *Stéphane Duvezin-Caubet**, *Manuel Rojo**, *Arnaud Mourier**, *Nicole Averet**, *Michel Rigoulet**

***Institut de Biochimie et Génétique Cellulaires, UMR 5095, CNRS, Université de Bordeaux, Bordeaux, France †Université Bordeaux, IBGC, UMR 5095, Bordeaux, France**

## 1 INTRODUCTION

Mitochondria are double-membrane-enclosed organelles that ensure a number of pivotal functions in the vast majority of eukaryotic cells. These include key steps in lipid and amino acid metabolism, biosynthesis of iron-sulfur clusters, heme, and further prosthetic groups, regulation of programmed cell death, and, most prominently, the final transformation of proteins, fats, and sugars into the cellular energy currency adenosine-5′-triphosphate (ATP) by oxidative phosphorylation (OXPHOS). Severe mitochondrial dysfunctions are associated with so-called mitochondrial diseases that appear linked to specific or broad OXPHOS defects. In addition, alterations of mitochondrial properties have been associated with further diseases linked to other metabolic and/or physiologic alterations. This chapter gives a general introduction on mitochondrial properties and functions that will be developed further in subsequent specialized chapters.

## 2 MITOCHONDRIAL ULTRASTRUCTURE

Mitochondria display a unique and highly characteristic ultrastructure that allows their unambiguous identification in electron microscopy sections. Mitochondria are enveloped by an

*Mitochondria in Obesity and Type 2 Diabetes*
https://doi.org/10.1016/B978-0-12-811752-1.00001-8

outer and an inner membrane enclosing a narrow intermembrane space. The outer membrane represents the border to the cytosol, and the inner membrane separates the intermembrane space from the matrix. The inner membrane is not limited to the mitochondrial boundary, but forms protrusions into the matrix——known as cristae——that provide a tremendous increase in surface.[1] The direct contacts between outer and inner membranes, as well as the transition between inner boundary and inner cristae membranes, occur at specific structures known as contact sites and cristae junctions, respectively (Fig. 1A). These structures are made and regulated by specific protein complexes known as MICOS (mitochondrial contact site and cristae organizing system) that are essential for several mitochondrial functions, including the maintenance of mitochondrial ultrastructure and the transport and exchange of proteins and lipids (reviewed in Ref. 2).

These ultrastructural characteristics are common to the mitochondria of all eukaryotes. The topology, size, and relative abundance of the different membranes and subcompartments, however, vary significantly among species and tissues and as a function of the metabolic or physiological state.[3] This is in accordance with the fact that each mitochondrial membrane and compartment hosts specific functions and activities.

Electron microscopy analysis often depicts close proximity or even direct contacts between outer mitochondrial membranes and the endoplasmic reticulum (ER) (Fig. 1A) that are termed MAMs (mitochondrial associated membranes) or ERMES (ER-mitochondria encounter structure). The molecular composition of yeast ERMES complex is well established,[4] but that of mammalian MAMs remain highly debated.[5] These contacts mediate the interorganellar exchange of calcium and lipids, are involved in mitochondrial fission, and play a role in insulin signaling, autophagy, and apoptosis (see also Chapter 2). Consequently, alterations of MAMs properties and of MAM-dependent functions have been associated with human diseases, including diabetic syndromes.[6,7]

## 3 MITOCHONDRIAL MORPHOLOGY, DISTRIBUTION, AND DYNAMICS

Mitochondria are mobile structures that do not distribute homogeneously throughout cells, but localize to intracellular sites (e.g., apical membranes of secretary cells, synapses of neurons) where their function (e.g., ATP-synthesis by OXPHOS) is required. Mitochondrial distribution and mobility are ensured by adaptors and motors that mediate interactions with the microtubule and actin cytoskeletons.[8,9]

Electron microscopy of ultrathin sections often has depicted mitochondria as isolated punctate structures that resemble bacteria (their phylogenetic ancestors) in form and in size. This appearance and the resemblance to that of their prokaryotic ancestors led to the longstanding concept that the mitochondrial compartment is composed of a collection of numerous small and independent organelles. Work in the last decades, however, has revealed that mitochondria continuously fuse and divide and that the equilibrium between antagonizing fusion and fission determine their morphology: Mitochondria appear as a network of interconnected filaments at high fusion/fission ratios and as a collection of small punctate structures when the fusion/fission ratio shifts toward fission (Fig. 1B). Fusion enables molecular exchanges between mitochondria (proteins, RNA, DNA) and leads to the existence of a single mitochondrial compartment within cells. Division enables severing of mitochondrial filaments and

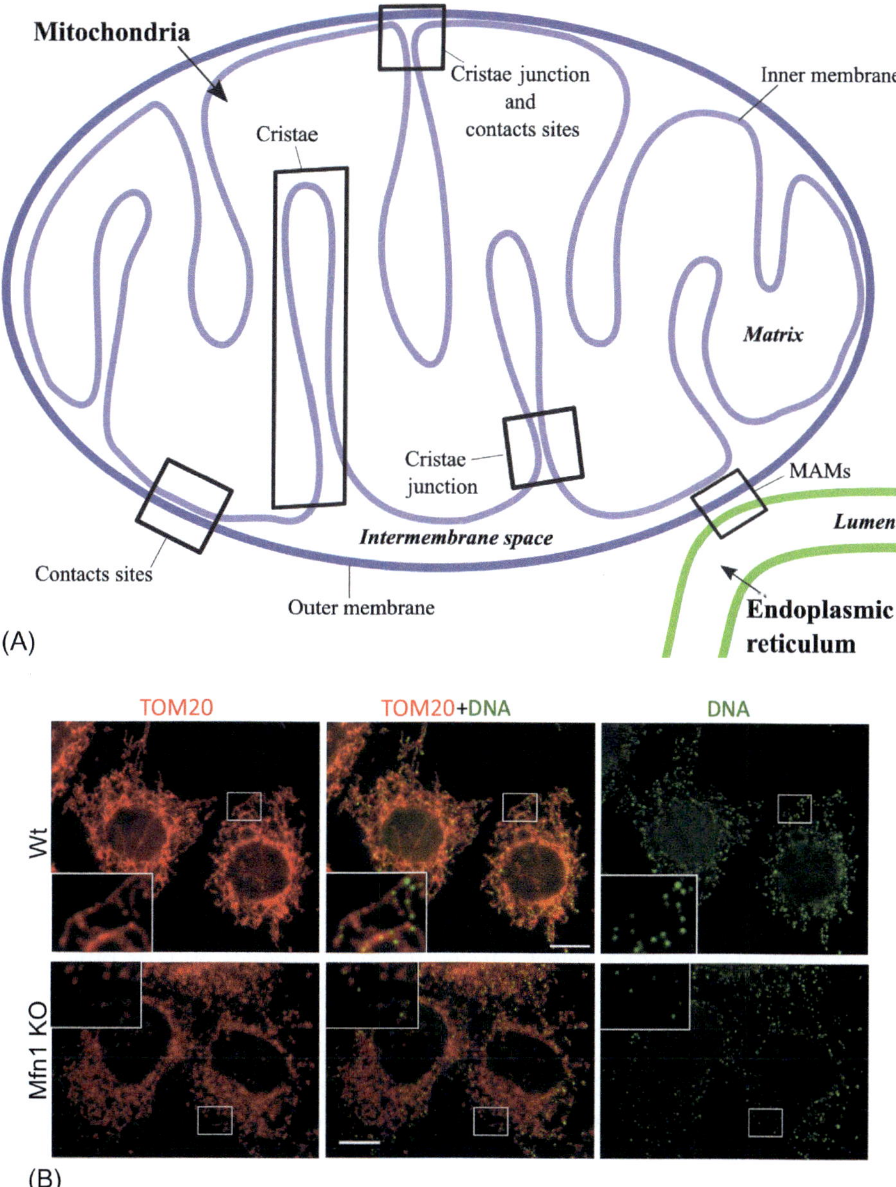


FIG. 1 Mitochondrial ultrastructure. (A) Mitochondria are enveloped by outer and inner membranes enclosing a narrow intermembrane space. The inner membrane encloses the mitochondrial matrix and forms invaginations called cristae. The direct contact between outer and inner membranes, as well as the transition between inner boundary and inner cristae membranes, occur at specific structures known as contact sites and cristae junctions, respectively. Outer mitochondrial membranes and the endoplasmic reticulum are in close or direct contact at sites termed MAMs (mitochondria associated membranes). (B) Mitochondria form filaments containing numerous mtDNA nucleoids in wild-type cells (WT) and display fragmented mitochondria in fusion-defective cells lacking fusion factor Mfn1 (MFN1 KO). Morphology and mitochondrial DNA distribution in cultured cells was visualized with antibodies against outer membrane Tom20 *(red)* and DNA *(green)*, respectively. The inset represents an enlargement of the indicated area. Bar = 20 μm.

segregation of functional and defective mitochondria, a requisite for the selective degradation of individual mitochondria by autophagy or mitophagy (see Chapter 2).

Fusion and fission are energy consuming processes that depend on the hydrolysis of nucleotides by specific GTPases of the dynamin-superfamily. In addition, the fusion of inner membranes (but not of outer membranes) also depends on the inner membrane electrical potential difference.[10] It is now well established that mitochondrial dynamics and bioenergetics are linked: The fusion/fission steady state is modulated by the bioenergetic status of cells and mitochondria and reciprocally, fusion/fission dynamics can modulate mitochondrial bioenergetics.[11, 12] Accordingly, conditions leading to mitochondrial dysfunction and/or depolarization, such as the permeabilization of the outer membrane at the onset of apoptosis, are accompanied by mitochondrial fragmentation.[13] Numerous pathological situations have been associated with alterations of the fusion/fission steady state, and mitochondrial fragmentation has become a hallmark of mitochondrial or cellular dysfunctions,[14] including diabetes and related metabolic disorders.[15]

Several components of the mitochondrial fusion and fission machineries have been identified and many of them, notably some GTP-binding proteins of the dynamin superfamily, are conserved between animals and fungi.[16] In mammals, outer membrane fusion is governed by mitofusins MFN1 and MFN2, two GTPases anchored to the outer mitochondrial membrane. Further reports reveal that mitofusin MFN2 can modulate contacts between ER and mitochondria, but the precise role of MFN2 in MAM biogenesis/modulation remains controversial.[17, 18] Inner membrane fusion depends on OPA1, a GTPase that localizes to the intermembrane space. The size, membrane association, and activity of OPA1 is regulated by mitochondrial metalloproteases that generate isoforms of different lengths. Proteolytic processing of OPA1 is modulated by bioenergetics[19] and altered by mitochondrial dysfunctions.[20] Mitochondrial fission is modulated by a machinery that relies on two cytosolic GTPases (DRP1/DNM1L and dynamin 2) and their complex interactions with outer membrane receptors, ER, MAMs, and the actin cytoskeleton.[21–23] To date, no specific machinery for inner membrane fission has been identified.

Mitochondrial fusion and fission are essential processes that are required for embryonic development of mice.[24] Further work has revealed that alterations of mitochondrial fusion/fission dynamics provoke defects in cellular bioenergetics and/or physiology[25] and that mutations of genes encoding fusion/fission proteins provoke hereditary neuropathies with highly variable symptoms.[24] While mutations of a fusion factor of the outer membrane (MFN2) are associated with a peripheral sensory and motor neuropathy (Charcot–Marie-Tooth disease of type 2A, CMT2A), that of an inner membrane fusion factor (OPA1) or of a fission factor (DNM1L) have been linked to optic nerve atrophy (dominant optic atrophy, DOA).[26] Alteration of fusion/fission dynamics, however, also has been associated with metabolic diseases and/or diabetes-related syndromes, including the response to high fat diet and/or the function of pancreatic beta cells.[27–29]

## 4 BIOGENESIS, MAINTENANCE, AND QUALITY-CONTROL

Present-day mitochondria are the descendants of prokaryotic endosymbionts that were internalized by the ancestor of today's eukaryotic cells; they carry their own genome that represents a relic of this evolutionary past.[30] The mitochondrial genome encodes a few proteins

(13 in humans) that are essential components of the OXPHOS machinery, as well as ribosomal and transfer RNAs (rRNAs, tRNAs) essential for their intramitochondrial translation. All human mtDNA-encoded genes give rise to hydrophobic, transmembrane subunits of the respiratory complexes and the ATP synthase. Besides complex II, the respiratory complexes and the ATP synthase that catalyze the OXPHOS reactions in the inner mitochondrial membrane, therefore, have a double genetic origin: although the majority of their components are encoded by the nuclear genome, a few essential subunits are encoded by a relatively small mitochondrial genome (16.5 kb in humans) that is present in thousands of copies per cell and localizes to hundreds of nucleoprotein complexes termed nucleoids.[31, 32] The mitochondrial proteome comprises more than 1000 different proteins.[33] The vast majority of these proteins (>99%) is encoded by the nuclear genome, synthesized on cytosolic ribosomes, and imported into the organelle. The proper biogenesis of mitochondria requires a number of processes ensuring, for example, the maintenance of the mitochondrial genome, the coordination of the two genomes' expression, the import of nuclear-encoded components, as well as the maturation, folding, membrane insertion, and the final assembly in functional molecular complexes. The biogenesis of mitochondria, therefore, is a challenge for the cell with the coordination of two genomes, the membrane compartmentalization, the elaboration of multisubunit complexes containing co-factors/prosthetic groups. Furthermore, mitochondria are the main source of reactive oxygen species (ROS) and their components and thus are the first targets of ROS-induced damage. An array of proteins with chaperone and protease functions are required to ensure the protein quality control and proteostasis of mitochondria. For example, no less than 40 proteases are identified in mitochondria; 20 are resident and 20 are transient proteases translocated to mitochondria only in certain conditions performing additional proteolytic activities essentially related to apoptosis or autophagy.[34, 35]

## 4.1 Mitochondrial Protein Import and Sorting (Fig. 2)

The vast majority of mitochondrial proteins are translated on cytosolic ribosomes and subsequently targeted to and imported into mitochondria. Targeting to the matrix, the inner membrane and the intermembrane space mainly relies on the presence of cleavable targeting signals: 60% of all mitochondrial proteins have a N-terminal mitochondrial targeting signal (MTS) on their precursor that is cleaved upon/after mitochondrial import. Other proteins of the inner membrane and intermembrane space have internal targeting signals that, in some cases, have not yet been identified or characterized. The targeting of outer membrane protein follows somewhat different paths. Highly hydrophobic channels/beta-barrels are imported/inserted by a dedicated machinery, but the machineries allowing for the targeting of proteins with a single membrane anchor remain elusive.[36] It is important to note that targeting, folding, processing, and assembly appear exquisitely interlinked and thus, that translocases, proteases, and chaperones fulfill complementary, linked, and partially overlapping functions that are difficult to dissociate/describe separately. In addition, several of these protein import machineries interact with each other and with other complexes such as MICOS or ERMES. Therefore, all these machineries must be considered as a dynamic network rather than individual complexes, and interorganellar interactions have to be considered. Most of the genes encoding these proteins are essential and/or linked to a variety of diseases, showing that these functions are not dispensable. In the following sections, we present a succinct view

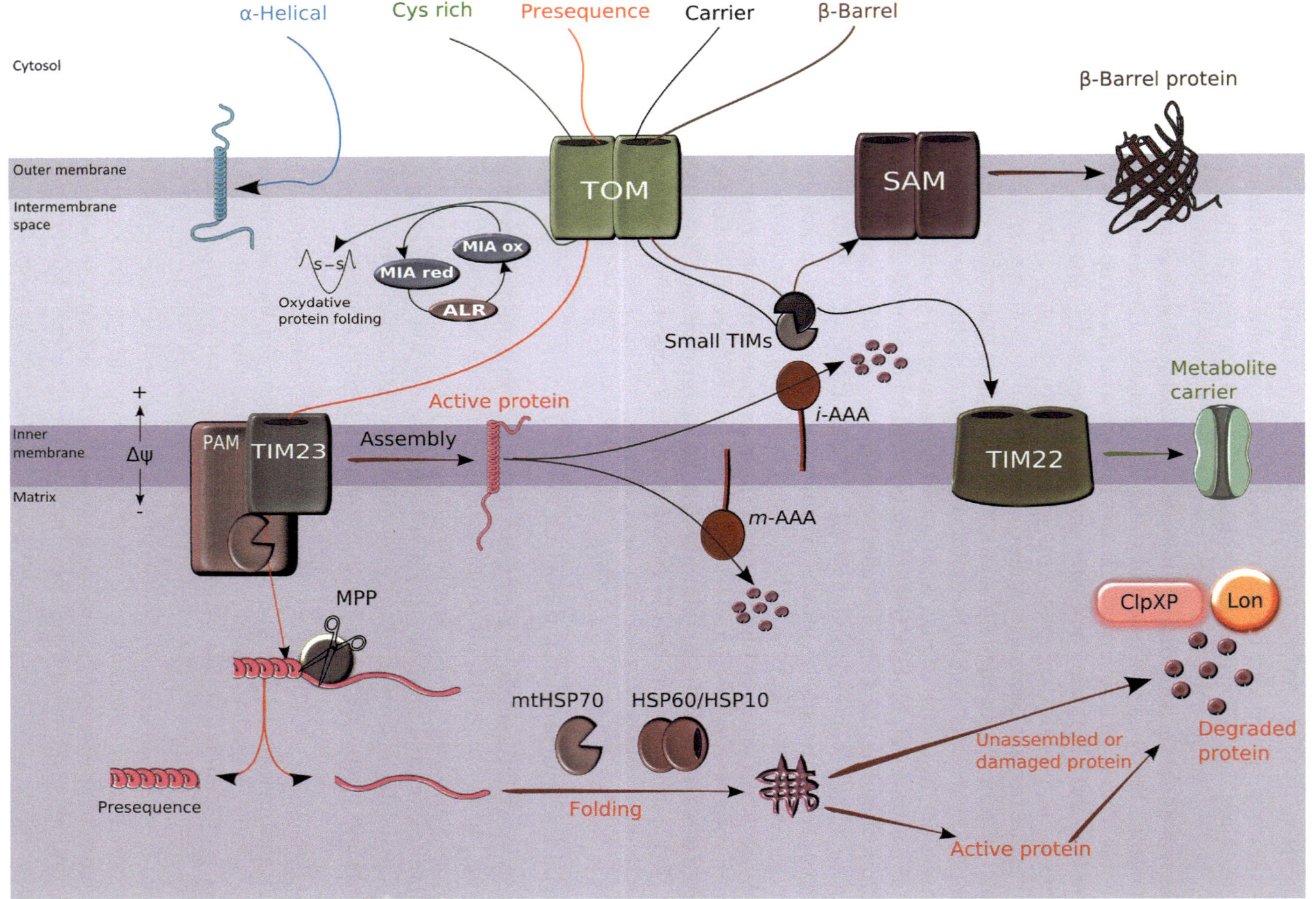


FIG. 2 Overview of the main machineries for protein import and quality control in mitochondria. Most mitochondrial proteins are nuclear-encoded and imported into mitochondria. Several types of targeting signals determine the routes and final destination of precursor proteins. Main protein import machineries have been represented on this scheme: TOM, the translocase of inner membrane 23 (TIM23) coupled to presequence-associated motor (PAM), the sorting and assembly machinery complex (SAM), mitochondrial import and assembly machinery (MIA) composed by Mia40 and the sulfhydryl oxidase augmenter of liver regeneration (ALR), small TIMs and the carrier translocase TIM22. A subset of precursor proteins is subjected to proteolytic maturation, for example by the mitochondrial processing peptidase (MPP) for matrix proteins. To acquire a proper folding to become functional, mitochondrial proteins can be assisted by molecular chaperones, such as HSP60/HSP10 and mtHSP70. At last, damaged, unfolded, or unassembled proteins are degraded by quality-control proteases such as the Lon protease homolog (Lon) and the ATP-dependent Clp protease (ClpXP) in the matrix, and the intermembrane-oriented (i-AAA) or the matrix-oriented (m-AAA) proteases of the AAA family in the inner membrane.

about the main machineries mediating import and sorting of mitochondrial proteins (see Refs. 37, 38 for more extensive reviews).

### 4.1.1 *Translocase of Outer Mitochondrial Membrane (TOM) Complex*

The translocase of outer membrane represents the main entry gate for proteins bearing mitochondrial targeting signals. Precursors of mitochondrial proteins are chaperoned by combinations of cytosolic factors that are determined by the sequence/nature of their targeting signal, hydrophobicity, and many other factors. These precursors are recognized by the receptor proteins of the TOM complex. When exiting the TOM complex, precursor proteins follow different routes, depending on their targeting signals and final destination within mitochondria.

### 4.1.2 *TIM23 Presequence Translocase*

TIM23 is the presequence pathway. It resides in the inner membrane (TIM: translocase of inner membrane) and recognizes precursor proteins with cleavable amphipathic N-terminal targeting sequences emerging from the TOM complex. The inner membrane electrical potential difference is required to insert the positively charged N-terminal presequence precursors into the TIM23 pore, but it is not sufficient for a complete translocation across the IM. The mitochondrial HSP70 chaperone protein (mtHSP70), together with its regulating factors, form the presequence- associated motor (PAM) that interacts with the TIM23 core components and hydrolyses ATP as a secondary driving force to pull the proteins entirely inside the matrix. In most cases, the presequence is processed in the matrix by the mitochondrial processing peptidase (MPP) and/or additional matrix proteases. Alternatively, the TIM23 ensures the lateral insertion of precursors into the inner membrane. In this process, named stop-transfer, the translocation is interrupted when a hydrophobic stretch reaches the membrane pore and the protein is released in the inner membrane. Some of these precursors are processed further by the inner membrane protease IMP to release the mature protein in the intermembrane space. Consequently, TIM23 allows proteins to be directed either to the matrix, the inner membrane, or the intermembrane space.

### 4.1.3 *The Carrier Pathway Into the Inner Membrane*

The TIM22 complex (translocase of the inner mitochondrial membrane 22) is required for the import of a subset of multispanning hydrophobic proteins of the inner membrane that do not exhibit any cleavable presequence but rather several internal targeting elements. It also is called the carrier translocase as it mediates the import of a large family of carriers that contain six transmembrane domains and catalyzes the exchange of metabolites across the inner membrane.[39] TIM22 substrates first are taken up at the TOM exit by soluble chaperones of the small TIM chaperone family in the intermembrane space. The precursor proteins then are delivered to the TIM22 complex from the intermembrane space side and inserted into the inner membrane.

### 4.1.4 *The Intermembrane Space Mitochondrial Import and Assembly Machinery (MIA)*

The intermembrane space contains a number of proteins with characteristic cysteine motifs. The MIA complex couples oxidative protein folding to the biogenesis of intermembrane space proteins. It mediates the import and oxidative folding of substrates with conserved cysteine motifs (CX3C, CX9C). Key components are the oxidoreductase Mia40 (CHCHD4) and the sulfhydryl oxidase ALR (augmenter of liver regeneration). Mia40 engages in an intermolecular disulfide bond with the incoming substrate. Following oxidation of the substrate, MIA40 is reoxidized by ALR. In yeast, it has been shown that reduced FAD in Erv1 (yeast ALR) is reoxidized either by transferring the electrons directly to molecular oxygen, or to cytochrome c.[40–43]

#### 4.1.5 *Import of Mitochondrial Outer Membrane Proteins*

Integral outer mitochondrial membrane proteins are of two types: beta-barrel proteins that are anchored in the outer membrane by multiple transmembrane beta strands and that are evolutionary conserved from Gram-negative bacteria (mitochondria's ancestor); and proteins that are anchored in the outer membrane by one or more alpha-helical segments. A large variety of import pathways has been reported over the last years for mitochondrial outer membrane proteins and the best described system is the sorting and assembly machinery complex (SAM) for beta-barrel proteins. Several beta-barrel proteins such as VDAC (voltage dependent anion channel), TOM40 or SAM50 (channel forming proteins) are in the outer mitochondrial membrane. The members of this family contain an array of multiple beta strands arranged as transmembrane segments in the outer membrane. The mitochondrial targeting signal in these proteins consists of the last beta strands and the loop in between arranged as a hairpin. This signal is recognized by the TOM complex, which translocates the precursors. The interaction between the TOM complex and the SAM complex promotes the release of beta-barrel proteins into the membrane.

### 4.2 Protein Quality-Control (Fig. 2)

#### 4.2.1 *Protein Trafficking, Processing, and Activation: Processing Peptidases*

Most mitochondrial proteins are imported from the cytosol thanks to targeting signals. The majority of these proteins require further processing of this targeting signal once they reach their final location in order to be active. Processing peptidases in the different mitochondrial compartments perform this maturation. In the matrix, MPP (mitochondrial processing peptidase) is in charge of removing the N-terminal targeting peptides from precursor proteins. Likewise, IMMPL1/2 (mitochondrial inner membrane protease) removes hydrophobic sorting signals from proteins sorted to the intermembrane space. Several mitochondrial proteins undergo additional proteolytic cleavages in order to regulate their activities (ribosome assembly and consequently synthesis of mitochondrially encoded subunits of the respiratory chain (AFG3L2), regulation of import (YME1L1), control of mitochondrial membrane composition (YME1L1), regulation of apoptotic resistance (YME1L1), mtDNA maintenance, and gene expression (LONP1, HTRA2), mitochondrial dynamics (YME1L1 and OMA1).[34]

#### 4.2.2 *Chaperones Involved in the Correct Folding of Proteins Within Mitochondria*

During import, precursor proteins must be maintained soluble and later must be folded and assembled (reviewed in Ref. 44). As stated previously, MIA40 and small TIM in the intermembrane space and mtHSP70 (HSPA9) together with HSP60 (HSPD1) and HSP10 (HSPE1) in the mitochondrial matrix fulfill chaperone functions. MIA40 and small TIM function in an ATP- independent manner, whereas in the mitochondrial matrix, chaperones are ATP-dependent. The mtHSP70 has an essential role in protein import by acting in the import motor to translocate proteins in the matrix. It also has a concomitant role in maintaining the preproteins unfolded during import to stabilize them and to avoid irregular associations. The mtHSP70 has a further role in the correct folding of polypeptides in the matrix together with HSP60, not only for incoming polypeptides but also for the biosynthesis and assembly of mtDNA-encoded subunits of the OXPHOS system. HSP60 is a member of the chaperonins/GroEL family also located in the matrix. It forms large complexes that interact with HSP10. ATP hydrolysis driven conformational changes allow the folding of the substrate proteins. Both mtHSP70 and HSP60 are essential proteins highlighting their unique and important roles in cellular metabolism.

### 4.2.3 *Additional Factors Required for the Biosynthesis of OXPHOS Complexes*

In addition to classical chaperones with general functions in mitochondria, numerous substrate-specific factors are required to obtain functional complexes of the respiratory chain or functional ATP synthase. Indeed, the biogenesis of large hetero-oligomeric complexes such as OXPHOS complexes is a challenge for the cells because of the large number of subunits added to their dual genetic origin and the presence of metal cofactors. Therefore, these processes usually are assisted by dedicated chaperones and multiple quality-control checkpoints. A striking example is the biogenesis of the cytochrome-c-oxidase, the copper-heme a terminal oxidase of the respiratory chain (complex IV). This transmembrane complex consists of 14 subunits of which three are mtDNA-encoded. To date, 30 COX assembly factors already have been described, with roles ranging from mitochondrial mRNA stability and translation activation to metalation and assembly of hemes and of the different subunits in functional complexes and supercomplexes.[45]

### 4.2.4 *Protein Quality Control (PQC) (Table 1)*

Protein quality control refers to the mechanisms and factors required for the maintenance of protein homeostasis in mitochondria in normal or under stress conditions. These factors include chaperones to take care of unfolded or misfolded proteins, as well as proteases to degrade proteins that are terminally damaged (oxidized, aggregated) and/or do not reach their final destination (not assembled properly). The mtHSP70 and HSP60 have important roles in preventing aggregation, refolding of misfolded proteins to an active conformation, or stabilization in a soluble state awaiting for degradation by proteases. They are strongly induced in stress conditions where damaged/unfolded proteins accumulate (elevated temperatures, exposure to toxic compounds, elevated ROS, mitochondrial unfolded protein response), and were accordingly initially recognized as HSPs (heat shock proteins).[44]

Proteases involved in PQC are mostly chambered proteases of the AAA family (ATPases associated with diverse cellular activities) that hold a chaperone-like activity allowing for the necessary unfolding of substrate for their ATP-driven translocation inside the protease cavity and proteolytic cleavage: the Lon protease homolog (LONP1) and the ATP-dependent ClpP protease in the matrix, intermembrane-oriented i-AAA (YME1L1) and matrix-oriented m-AAA both anchored in the inner mitochondrial membrane.[34, 35, 44] They degrade membrane spanning and membrane-associated subunits of the electron transport chain that are

TABLE 1 Main Proteins Involved in Mitochondrial PQC

| Gene Name | Localization/Complex | Function | Associated Disorders/Phenotypes of Interest |
|---|---|---|---|
| HSPA9<br>Mortalin<br>GRP75 | Matrix, mtHSP70 | ATP-dependent molecular chaperone: mitochondrial biogenesis and PQC | Mutations associated with EVEN-Plus syndrome (MIM600548) |
| HSPD1<br>HSP60<br>CHA PERONIN | Matrix, 60kDa heat shock protein | ATP-dependent molecular chaperone: mitochondrial biogenesis and PQC | Mutations associated with spastic paraplegia and hypomyelinating leukodystrophy (MIM118190). The expression levels of HSP60 are decreased in rats under high fat diet and accompanied by impaired glucose tolerance.[46] Similar results are found in adipose tissue of ob/ob[47] and db/db[48] mice. Treatment of ob/ob mice with rosiglitazone reverses these effects[47] |
| LONP1 | Matrix, ATP-dependent Lon protease homolog ($AAA^+$ protein family) | Serine protease and ABC ATPase: mitochondrial biogenesis and PQC | Mutations associated with multisystem developmental CODAS syndrome (MIM600373). Downregulation of Lon can experimentally provoke hepatic insulin resistance[49] |
| CLPP | Matrix, proteolytic subunit (caseinolytic peptidase P, ClpP) of the ATP- dependent clp protease (ClpXP complex) | Serine protease: mitochondrial PQC | Mutations associated with Perrault syndrome (MIM601119). The knock-out mouse shows protection from diet-induced obesity and insulin resistance[50] |
| YME1L1 | Inner membrane, IMS-oriented, subunit of *i*-AAA (IMS-facing ATP- dependent $AAA^+$ metalloprotease | Metalloprotease and ABC ATPase: mitochondrial biogenesis and PQC | Mutations associated with infantile-onset developmental delay, muscle weakness, ataxia, and optic nerve atrophy (MIM607472) |
| AFG3L2 | Inner membrane, matrix-oriented, subunit of *m*-AAA (matrix-facing ATP-dependent $AAA^+$ metalloprotease) | Metalloprotease and ABC ATPase: mitochondrial biogenesis and PQC | Mutations associated with spinocerebellar ataxia and spastic-ataxia (MIM604581) |
| OMA1 | Inner membrane, IMS-oriented ATP-independent OMA1 metalloendopeptidase | Stress-activated zinc metalloprotease: mitochondrial dynamics, mitophagy and apoptosis | The knock-out mouse shows obesity, defective thermogenesis and reduced energy expenditure[51] |

An array of mitochondrial proteins (molecular chaperones and proteases) are involved in the maintenance of protein homeostasis in mitochondria. The main components have been listed in this table together with their known implications in disease and in metabolism related to diabetes. *IMS*, inter-membrane space; *AAA+*, ATPases associated with diverse cellular activities.

damaged, oxidized, or nonassembled, and have regulatory functions (mitochondrial protein synthesis, dynamics, calcium homeostasis). LONP1 substrates are aconitase, succinate dehydrogenase subunit 5, and transcription factor A (TFAM), and several other OXPHOS complex subunits, such as heme-related enzymes.

Mitoproteases are also essential components of the mitochondrial stress response (for instance heat stress, elevated ROS levels, low inner membrane potential difference). UPRmt (mitochondrial unfolded protein response) is a good example of recently described mitochondrial stress response. It is triggered by mitochondrial proteotoxic stress (accumulation of unfolded/unassembled proteins) that induces the expression of nuclear genes, notably nuclear- encoded mitochondrial quality-control components (protease and chaperones) in order to restore mitochondrial protein homeostasis. There is evidence for the implication of CLPP and LONP1 in this process probably together with YME1L1. Another example is the inactivation of mitochondrial fusion through the proteolytic processing of OPA1 by the inner membrane metalloendopeptidase OMA1 in stress conditions.[52]

Intricate links between mitochondrial quality control and energetic metabolism have been reported in human metabolic diseases or mouse models.[53–55] For example, a knockout mouse model of the stress-induced metalloprotease OMA1 alters the metabolic function. Oma1-deficient mice show an obesity phenotype, characterized by an increase in body weight and hepatic steatosis, in addition to defective thermogenesis because of reduced energy expenditure. These alterations were especially significant under stress conditions (high-fat diet or cold-shock).[51,56] Surprisingly, however, a recent report showed that loss of another mitochondrial protease participating in quality-control (ClpP) in mice leads to reduced adiposity and improved insulin sensitivity, increased whole-body energy expenditure, and upregulated mitochondrial biogenesis in white adipose tissue. Moreover, the absence of ClpP protects mice from diet-induced obesity and insulin resistance.[50]

# 5 MITOCHONDRIAL FUNCTIONS

## 5.1 Mitochondrial Oxidative Phosphorylations and Mitochondrial Carriers

Mitochondrial outer membrane has a high permeability, thanks to porine (VDAC) which, on isolated mitochondria, allows the free diffusion of molecules up to 5000 Da. It has been shown that in situ, however, permeability of this channel could be regulated,[57] leading to subtle regulations of mitochondrial oxidative phosphorylations and regulation of metabolic channeling processes.[58] In eukaryotic cells, oxidative phosphorylations take place in the inner membrane of mitochondria. According to the chemiosmotic theory initially formulated by Peter Mitchell,[59] this process converts a redox energy——from food conversion processes——into a proton osmotic gradient (electrochemical proton potential difference that has an electric component $\Delta\Psi$ and a chemical component $\Delta pH$) that is itself converted into another kind of chemical energy usable for various cell functions, that is, the phosphate potential (ATP/ADP.Pi). Since Mitchell's proposal of the chemiosmotic theory, numerous studies have validated this hypothesis, which is now well accepted. Oxidative phosphorylations require the mitochondrial respiratory chain, the ATPsynthase, ATP/ADP antiporter, and the $Pi^-/H^+$ symporter. In contrast to the outer membrane, the mitochondrial inner membrane is impermeable,

TABLE 2 Main Mitochondrial Carriers and the Carried Compounds Under Physiological Conditions

| | Carried Compounds | | | |
|---|---|---|---|---|
| Name | Cytoplasm | Matrix | Vectorial Aspect | Energetic Mechanism |
| ANT | $ADP^{-3}$ | $ATP^{-4}$ | Antiport | Electrogenic |
| AGC | Glutamate $H^+$ | Aspartate | Antiport | Electrogenic |
| MCU | $Ca^{2+}$ | | Uniport | Electrogenic |
| PiC | $Pi^-$<br>$H^+$ | | Symport | Electroneutral |
| OGC | Malate | Oxoglutarate | Antiport | Electroneutral |
| CIC | malate | citrate | Antiport | Electroneutral |
| DIC | Succinate<br>Malate | $Pi^{2-}$ | Antiport | Electroneutral |
| Tricarboxylate carrier | Citrate | Malate<br>Succinate | Antiport | Electroneutral |
| MPC | $Pyruvate^-H^+$ | | Uniport | Electroneutral |

Mitochondrial carriers allow an exchange of varied compounds between the inter-membrane space and the mitochondrial matrix. These carriers can be separated in two categories: the electrogenic carriers, such as adenine nucleotide translocator (ANT), aspartate/glutamate carrier (AGC) and mitochondrial $Ca^{2+}$ uniporter (MCU), and the electroneutral carriers, such as phosphate inorganic carrier (PiC), oxoglutarate carrier (OGC), citrate carrier (CIC), dicarboxylate carrier (DIC), tricarboxylate carrier and mitochondrial pyruvate carrier (MPC).

particularly for molecules that harbor a positive or negative charge. Metabolites exchanges between the intermembrane space and mitochondrial matrix are ensured by a family of proteins called mitochondrial carriers. The main carriers are listed in Table 2.[60, 61] These carriers can be classified in regards to their dependence to the components of the inner membrane potential difference ($\Delta\Psi$ and $\Delta$pH) and the transport mechanism. Of interest for the purpose of this chapter is the ATP/ADP carrier (ANT) that catalyzes the electrogenic exchange (dependent on $\Delta\Psi$) of $ATP^{4-}$ with $ADP^{3-}$; another electrogenic exchanger is the glutamate/aspartate exchanger that catalyzes the electrogenic exchange of aspartate with glutamate+$H^+$. This mechanism is responsible for the vectorial aspect of the exchange (i.e., glutamate import and aspartate export), which is mandatory for the transfer of reduced equivalents in the mitochondrial matrix by the malate aspartate shuttle. The $Pi^-/H^+$ and pyruvate/$H^+$ symporters are electroneutral but dependent on the $\Delta$pH, which favors Pi and pyruvate import. Di- and tricarboxylate carriers are electroneutral. Calcium import/export to/from mitochondrial matrix involves three distinct carriers. Calcium import goes through mitochondrial calcium uniport, which is electrogenic, and calcium release from the matrix requires an electrogenic exchanger ($Ca^{2+}/3$–$4Na^+$). It also has been proposed a $Ca^{2+}/2K^+$ and a $Ca^{2+}/2H^+$. Their relevance, however, is still controversial.[61]

The respiratory chain comprises four complexes: I–IV. Complexes I, III, and IV couple electron transfer to proton extrusion whereas complex II only transfers electrons to the quinone pool. Complex I (NADH-CoQ oxidoreductase) catalyzes the electron transfer from NADH to quinone. Complex II (succinate-CoQ oxidoreductase) couples succinate oxidation to quinone reduction. Complex III (quinol-cytochrome c oxidoreductase) couples quinol oxidation to cytochrome c reduction. Complex IV (cytochrome c oxidase) catalyzes the last step of electron transfer from reduced cytochrome c to oxygen. Several other enzymes exist in the inner mitochondrial membrane, which can feed the respiratory chain with electrons such as glycerol-3-phosphate (G3P) dehydrogenase, electron transfer flavoprotein dehydrogenase, dihydroorotate dehydrogenase, and choline dehydrogenase. Complexes I, III, and IV use the redox span along the respiratory chain (Gibbs energy of the redox reaction) to pump out protons against the electrochemical proton gradient (see Fig. 3). The inner membrane electrical potential difference is used by some mitochondrial carriers and mainly by the ATPsynthase, which is able to convert passive proton transfer into ATP synthesis from ADP and Pi (Fig. 3). The structure of the respiratory chain as a succession of complexes relies on both functional studies and biochemical isolation.[62] Electron transfer in between the complexes depends on both the quinone pool and cytochrome c, which are electrons mobile carriers. Use of nonionic detergents allowed the isolation of more organized structures, called supercomplexes, in which some of the previously described complexes are associated with specific stoechiometries.[63] The composition of these supercomplexes has been shown to be variable and to depend on physiological conditions. It has been proposed that these supercomplexes would be an advantage to the functioning of the respiratory chain, such as electrons channeling, kinetic efficiency, decrease in ROS production. These proposals, however, are still debated to this day.[64–66]

## 5.2 Thermokinetic Control and Yield of Oxidative Phosphorylations

According to the chemiosmotic hypothesis,[67,68] the energy transduction between redox free energy and phosphate potential is allowed by inner mitochondrial membrane proton pumps structurally independent but functionally connected by an inner membrane electrical potential difference between two bulk phases: intermembranal space and mitochondrial matrix ($\Delta\mu H^+$). By analogy with the functioning of an electric battery, Mitchell introduced the term *protonmotive force* ($\Delta p = \Delta\mu H^+/F$), which is used largely in bioenergetics. Two main questions to be taken into account are the yield of energy conversion (ATP/O) and the relative fluxes involved in this process where both ATP synthesis and NADH oxidation are important. It is necessary, therefore, to understand how the values of the different coupled fluxes are determined in an integrated system, such as oxidative phosphorylations. Obviously, the coupling of fluxes is mediated by forces, but, at first sight, the quantitative relationships between fluxes might depend on the properties of the whole system. In a complex metabolic network like mitochondrial oxidative phosphorylations, a very simple quantitative analysis is the determination of the yield of the overall reaction, as is performed by measuring ATP synthesis flux over oxygen consumption flux (ATP/O). With this approach, the yield can vary and several mechanisms possibly involved have been proposed. The first mechanism decreasing the coupling efficiency, the proton leak, is a direct consequence of the nature of the energetic intermediary, the protonmotive force. Biological membranes always present some proton conductance (LH), and the resulting proton flux is strictly dependent on protonmotive force ($JH = LH \cdot \Delta p$). This membrane conductance is a specific property of the membrane itself, but is not entirely independent from the protonmotive force. The size of this proton leak can modulate the yield

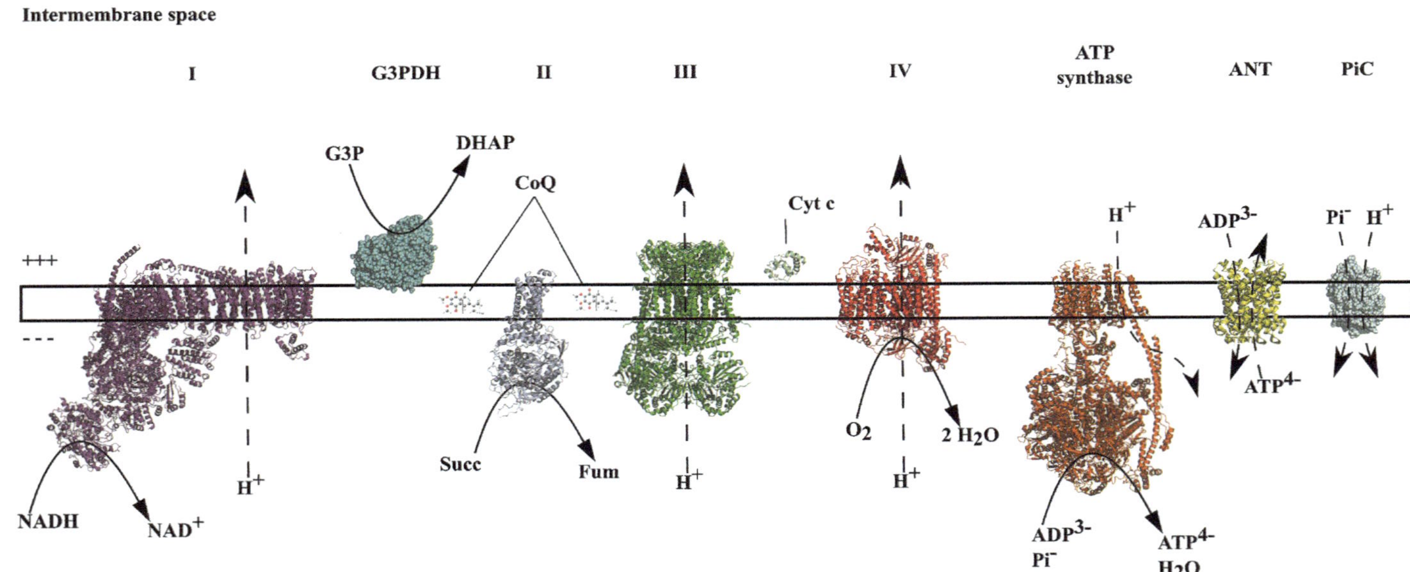


FIG. 3 Scheme of the mammalian oxidative phosphorylation (OXPHOS) system. The OXPHOS system can be described as two functional entities: the respiratory chain (RC) and the phosphorylating system. The RC is composed by holoenzymes as the NADH dehydrogenase (complex I), succinate dehydrogenase (complex II), cytochrome-c reductase (complex III), and cytochrome-c oxidase (complex IV). Other dehydrogenases, such as the glycerol-3-phosphate dehydrogenase (G3PDH), can be connected to the RC. The RC also is composed by mobile electrons carrier such as coenzyme Q and cytochrome-c. The RC is functionally coupled to the phosphorylating system composed by the ATP synthase, adenine nucleotide transporter (ANT), and phosphate inorganic carrier (PiC).

of oxidative phosphorylations (ATP/O), and it has been shown that protonophores uncouple oxidative phosphorylations by increasing proton membrane conductance.[69–75] In this uncoupling process, the first event is a dissipation of protonmotive force, which leads to an increase in respiratory rate and a decrease in the ATP synthesis rate, thus leading to a decrease in ATP/O ratio. Even if much experimental work has established strong evidence that some protonophoric action can quantitatively account for the uncoupling of oxidative phosphorylations,[76] no definitive proof that the uncoupling is exclusively and quantitatively because of an increase in cation membranal conductance has been obtained. In fact, from a growing amount of data, it is evident that the question of the multiplicity of uncoupling mechanisms is still open. The first member of the uncoupling protein (UCP) family, brown adipose tissue uncoupling protein 1 (UCP1), was identified in 1976.[77] Its uncoupling activity has been demonstrated largely, as well as the regulators of this activity. It is activated by fatty acids and inhibited by GDP. Its activation ensures thermogenesis. Based on sequence homology, closely related proteins were described in mammals. The actual function of these proteins (i.e., UCP2 and UCP3) is still a matter of debate.[78, 79] The definition of the function of these proteins is crucial in terms of bioenergetics in the sense that a slight uncoupling would lead to a decrease in protonmotive force, in ATP synthesis and in mitochondrial ROS production.

Among the multiplicity of uncoupling mechanisms, two kinds of experimental evidence can be noted: It has been shown that some uncoupling effects are not linked to a significant decrease in protonmotive force, which indicates that the decrease in oxidative phosphorylations efficiency is not, in this case, the consequence of an increase in membranal proton conductance[80–86]; and direct or indirect estimations of the coupled flow through different proton pumps indicate that their intrinsic stoichiometry, that is, the $H^+/2e^-$ stoichiometry of the respiratory chain and the $H^+$/ATP stoichiometry of the ATPsynthase, might vary as a function of many physical parameters or some drug addition.[83, 84, 86–97] This led Azzone et al.[98, 99] to propose another mechanism causing a loss of oxidative phosphorylations yield. Such a new possibility called *slip* is a decrease in the efficiency of a proton pump because of partial and variable decoupling of chemical reaction and proton transport. We have described a new energy wastage mechanism of interest. In isolated yeast mitochondria, the membrane proton conductance is shown to be strictly dependent on external dehydrogenases activity. An increase in their activity leads to an increase in the membrane proton conductance. This proton permeability is independent of the respiratory chain and ATP synthase proton pumps. This mechanism is an active proton leak. In such a process, the high cellular redox constraints can be alleviated without affecting much the ATP synthesis flux and allows a decrease in mitochondrial ROS production.[100]

## 5.3 Mitochondrial ROS and Their Detoxification

### *5.3.1 Dioxygen: A Poison for Life*

Dioxygen, owning two unpaired electrons each located in a different antibonding orbital, is a biradical. These two electrons have the same spin quantum number or, as often is written, they have a parallel spin. This constitutes the most stable state, that is, ground state, of dioxygen. If a diatomic oxygen molecule attempts to oxidize another atom or molecule by accepting a pair of electrons, both of these electrons must have an antiparallel spin, so as to fit the vacant space in the orbitals. In most cases, electrons forming a pair in an orbital have opposite spins in accordance with Pauli's principle. This imposes a restriction on electron transfer, which leads dioxygen to accept electrons one at a time. Consequently, dioxygen reacts only very slowly with all nonradicals. Even if the redox potential of the $O_2/H_2O$ couple is extremely positive, oxygen in the air cannot immediately burn living organisms because of the structure of their organic compounds. This spin restriction of dioxygen reactivity allowed the complex evolution of organisms and a metabolic adaptation, that is, the use of dioxygen as last electron acceptor during catabolism leading to an efficient energy transfer in cells.[101]

When a unique electron is accepted by the ground state dioxygen, it enters one of the antibonding orbitals and forms the superoxide radical, $O^{2-}$. This compound is a very active electron donor, potentially able to generate a toxic cascade of electron transfer reactions. At physiological pH, however, the main reason for superoxide disappearance in aqueous solution is its dismutation catalyzed by the superoxide dismutase enzyme. It often is said that spontaneous dismutation is very rapid, but it is worth noting that such a reaction becomes slower as the pH rises and yeast mutant devoid of superoxide dismutases (sod 1 and sod 2) cannot survive when cells switch from fermentation to growth fueled by respiration.[102, 103] If dioxygen accepts two electrons, it generates the peroxide ion, that is, $O_2^{2-}$ which, in biological media, is found as hydrogen peroxide ($H_2O_2$). Hydrogen peroxide is not a radical but is considered a reactive oxygen species because, in the presence of a transition metal, it is able to form the hydroxyl radical, $OH^-$ (i.e., the Fenton reaction). In biological systems, this product can induce a cascade of radical reactions potentially involving any cell compound.

### 5.3.2 *ROS Production by the Respiratory Chain*

It generally is believed that the main superoxide producer in the cell is the respiratory chain (Fig. 4). Two of the respiratory chain complexes (I and III) have been, for a long time, recognized as involved in superoxide production.[104] Under physiological conditions, predictions estimate that the superoxide production is about 0.1% of the respiratory rate.[105, 106] One should keep in mind, however, that such a production is highly dependent on the respiratory state/rate of the mitochondria. At the level of complex III, it has been shown that the Q-cycle can produce superoxides on the inner and outer surfaces of the inner mitochondrial membrane.[107, 108] Because superoxides do not cross this membrane, it is important to know the size of the relative fluxes in the intermembranal space and into the matrix.

Another site in the respiratory chain involved in superoxide formation is complex I. There is now general agreement that, in this large multisubunit complex, the electron transfer is working at near equilibrium and, therefore, superoxide production might be linked to both forward electron transport (FET) and reverse electron transport (RET).[104, 109–115] It is likely that different sites of superoxide production exist in complex I and that the sites involved are different for FET or in RET.[116, 117] The sites of complex I associated superoxide generation are still controversial. Three main sites have been proposed: the flavine mononucleotide (FMN),[118–120] the Fe-S clusters (more likely N1a or N2)[112, 121, 122] and a ubiquinone specifically linked to complex I (iQ site).[123] Rotenone inhibits electron transfer right upstream the quinone binding site. Consequently, superoxides that are produced by complex I in the presence of NADH are most likely because of the electron carriers (flavin or Fe/S clusters). ROS generation flux, however, seems mainly linked to reverse electron flux, and the iQ site could be a major player in this flux. It should be stressed that, unlike ROS production for FET, ROS production from RET can originate either at the actual RET site or at the dehydrogenase level.

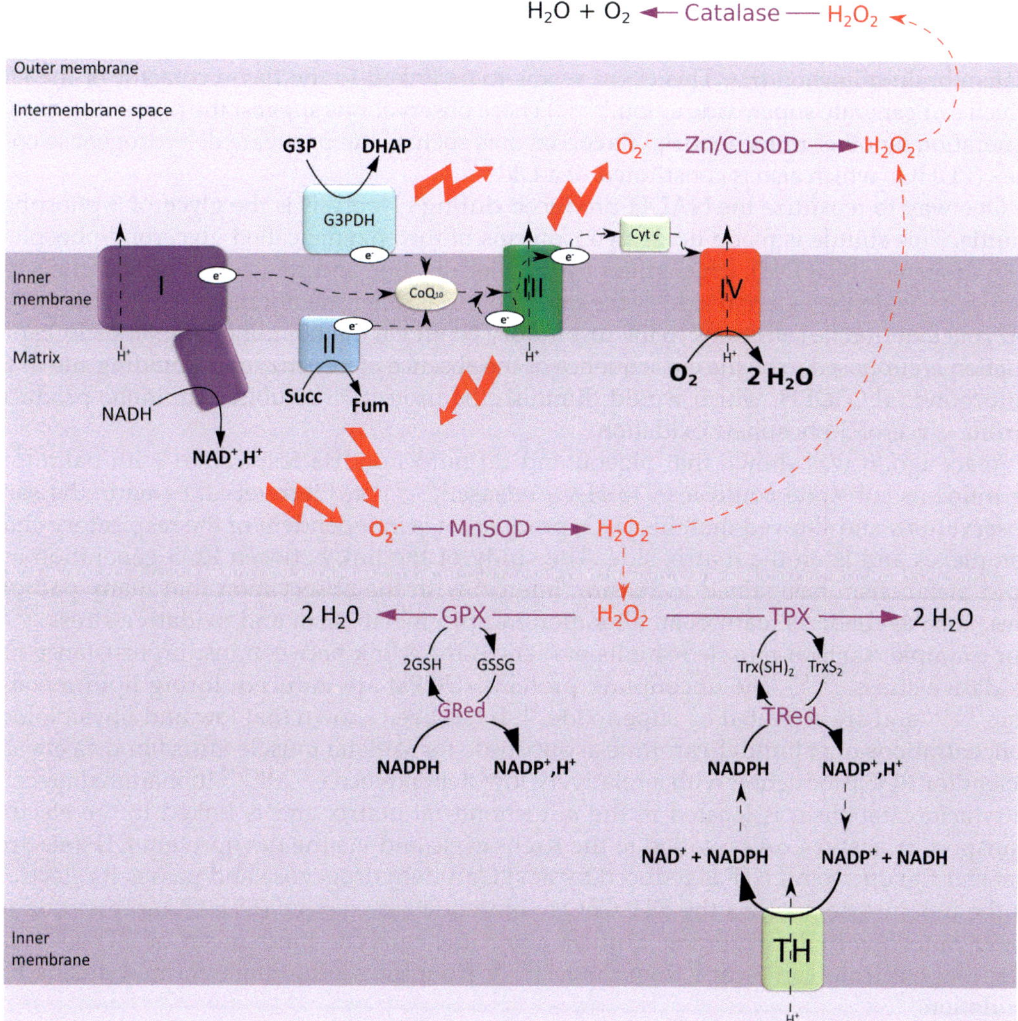


FIG. 4 Main enzymatic systems of detoxification of ROS generated by the OXPHOS system. NADH dehydrogenase (I), glycerol-3-phosphate dehydrogenase (G3PDH), succinate dehydrogenase (II), cytochrome c reductase (III) generate superoxides; *CoQ*, coenzyme Q; *Cyt c*, cytochrome c; *GPX*, glutathione peroxidase; *Gred*, glutathione reductase; *GSH*, reductive form of glutathione; *GSSG*, oxidative form of glutathione; *TPX*, thiol peroxidase; *Tred*, thiol reductase; *Trx*, thioredoxin; $O_2^{\cdot -}$, superoxide; *SOD*, superoxide dismutase; $H_2O_2$, hydrogen peroxide; *TH*, transhydrogenase.

As previously mentioned, the respiratory chain complexes are the most studied elements regarding the contribution of mitochondria to ROS production. The mitochondrial dehydrogenases, however, also have been shown to be involved in the following process.

The α-ketoglutarate dehydrogenase complex (αKGDHC) catalyzes the oxidation of α- ketoglutarate in succinyl-coA with the production of NADH in the Krebs cycle. It is composed of multiple copies of three enzymes: α-ketoglutarate dehydrogenase (E1), dihydrolipoamide succinyl-transferase (E2), and dihydrolipoyl dehydrogenase (Dld or LADH, E3).[124–126] Starkov et al.[127] showed that this complex is able to produce ROS during its catalytic process in mammalian brain mitochondria. This event seems to be linked to the flavin cofactor of the Dld, which can generate superoxide anion.[128, 129] These observations suggest the possibility of ROS generation by other mitochondrial flavoenzymes such as the pyruvate dehydrogenase complex (PDHC), which also is constituted of a Dld.[127]

One way to reoxidize the NADH produced during glycolysis is the glycerol-3-phosphate shuttle. This shuttle is made up of two isoforms of the enzyme called glycerol-3-phosphate dehydrogenase (G3PDH), which differ by their localization and cofactors. The mitochondrial isoform[130] is on the external side of the mitochondrial inner membrane and is a FAD-linked enzyme that donates electrons to the respiratory chain via the ubiquinone pool. Its ROS production is supposed to be the consequence of the absence of a coenzyme Q binding site in the mitochondrial G3DPH, which would diminish the protection of ubisemiquinone produced during glycerol-3-phosphate oxidation.

Years ago, it was shown that pigeon and rat mitochondria respiration with palmitoyl-carnitine as substrate could lead to $H_2O_2$ release.[131, 132] St. Pierre et al.[133] made the same observations and showed that this $H_2O_2$ production is independent of the respiratory chain complexes and is on the matrix side. The study of the link between ROS generation and lipid metabolism has gained increasing interest with the observation that many pathologies, such as diabetes feature, an alteration in lipid metabolism and oxidative stress.[134–137] For example, skeletal muscle exhibits evidences for a link between insulin resistance and oxidative stress.[138, 139] The uncoupling proteins (UCPs) are induced during lipid metabolism[140, 141] and are activated by superoxide.[142] It has been shown that low and physiological concentrations of palmitoyl-carnitine as substrate for skeletal muscle mitochondria are sufficient for ROS production with a relatively low dependence on $\Delta\Psi$.[143] In mammalian cells, fatty acid catabolism is located in the mitochondrial matrix and is linked to the electron transport by acetyl-CoA provided to the Krebs cycle and via the flavoprotein ETF (electron transfer flavoprotein). ETF is reduced by acyl CoA dehydrogenase and passes its electrons to the ubiquinone pool via the ETF CoQ oxidoreductase enzyme (ETF-QO) present on the inner mitochondrial membrane. Studies about the catalytic mechanism of ETF-QO suggest that electrons can escape from the $ETF^-$ or from the semiquinone formed during ETF oxidation.[144–146]

Consequently, ETF and/or ETF-QO are supposed to generate ROS with palmitoyl-carnitine as substrate.

### 5.3.3 *Control of ROS Production Flux by Mitochondria*

The redox level of the respiratory chain electron carriers, including ubisemiquinone, is thermokinetically controlled. Thus, the forces directly associated with respiratory chain activity, that are the redox potential of the $NADH/NAD^+$ couple and the protonmotive force, are powerful regulators of the steady-state concentration of the free ubisemiquinone radical. An increase in electron supply to the respiratory chain or in protonmotive force must lead to a rise in ubisemiquinone radical content. Moreover, kinetic constraints exerted downstream of the quinone pool (e.g., at the level of cytochrome oxidase) could further increase the level of free radical ubisemiquinones. Moderate uncoupling, however, will lead to a decrease in force and kinetic constraints and, therefore, effectively decrease superoxide production by the respiratory chain.[147] The opposite effect can be observed when the respiratory chain is inhibited.[132, 148] Other mechanisms that induce a decrease of the redox proton pump efficiency (such as slipping) are expected to be less effective because they do not affect the protonmotive force significantly.[149–152] It is well known that the transition nonphosphorylating/phosphorylating or the uncoupling state decrease ROS generation in isolated mitochondria.[153] Free fatty acids exert different effects on mitochondria; they inhibit electron flux at the complex I and probably complex III levels inducing an increase in ROS production associated with forward electron transport in isolated rat heart or liver mitochondria.[154] They also induce a slight uncoupling effect that seems responsible for a large decrease in ROS formation linked to reverse electron transfer. This illustrates the subtle ROS production response to changes in electron flux through the respiratory chain.

One of the physiological functions of the uncoupling protein family (UCPs) could be the modulation of the protonmotive force in response to an increase in mitochondrial ROS production.[155–158]

### *5.3.4 ROS Enzymatic Detoxification*

ROS are involved in a number of physiological processes such as cell signaling,[159–161] mitochondrial biogenesis,[162, 163] and numerous physiopathological processes. ROS production and ROS level are tightly controlled within the cell through a number of processes. We will focus here on ROS enzymatic detoxification processes even though a number of antioxidant compounds could participate to the control of ROS levels. At the level of the mitochondrial inner membrane, ROS can be produced both on the matrix side and intermembrane space. As stated previously, electron transfer through both mitochondrial dehydrogenases and oxidative phosphorylations complexes can generate the superoxide ion. After being produced, this unstable molecule is dismuted thanks to matricial MnSOD and cytosolic/intermembrane space Zn/Cu SOD (Fig. 4). These enzymes will generate $H_2O_2$ and oxygen from the superoxide ion. $H_2O_2$ itself is not a radical species but is considered as a highly reactive oxygen species because it is the substrate of the Fenton reaction catalyzed by transition metals. This reaction generates $OH^-$, which is the most reactive oxygen radical. It should be stressed that both superoxide and $OH^-$ are highly reactive and poorly diffusible, whereas $H_2O_2$ is highly diffusible through biological membranes. This leads to compartmentation of radicals.

$H_2O_2$ detoxification relies on a number of enzymes. The only one that is specific for $H_2O_2$ is catalase, and it is not present in the mitochondrial matrix (Fig. 4). This enzyme will dismute $H_2O_2$ in $H_2O$ and $O_2$. Within the mitochondrial matrix, two main enzymatic systems are able to reduce $H_2O_2$ in $H_2O$: glutathione peroxidase and thioredoxin peroxidase. Both these enzymes are not specific to $H_2O_2$, but also will reduce peroxide species produced by ROS damage. This requires regeneration of the reduced forms of either glutathione or thioredoxin. This is achieved by glutathione reductase and thioredoxin reductase, respectively. Both these enzymes require NADPH. The efficiency of these systems is highly controlled by NADPH regeneration. In the mitochondrial matrix, the main system for NADPH regeneration is the transhydrogenase that catalyzes $NADP^+$ reduction from NADH using the inner membrane potential difference. In the cytosol, NADPH is produced mainly by the pentose phosphate pathway and, similarly to what occurs in the mitochondrial matrix, is involved in ROS detoxification.

# 6 TISSUE SPECIFIC MITOCHONDRIAL ENERGY METABOLISM

Multisystemic organisms present several hundreds of different types of cells, differing in shape, structure, function, division cycle, and metabolic activities. Likewise, mitochondrial content,[164–168] strongly differs in a tissue-specific manner. We will give a short overview of processes leading to mitochondrial tissue specificity, which can be driven by intrinsic modification of OXPHOS subunits composition or through the tissue-specific expression of enzymes modulating mitochondrial metabolic pathways.

## 6.1 Tissue-Specific Expression of OXPHOS Subunits

Historically, the cytochrome-c oxidase has been the first enzyme of the respiratory chain described as having tissue-specific subunits. The mammalian COX consists of 14 subunits per monomeric COX of which 3 are mitochondrial and 11 are nuclear-encoded.[169] In mammals, isoforms have been found for subunits IV, VIa, VIIa, and VIII, which are expressed in a tissue-specific manner.[169–171] The molecular diversity observed for subunit VIa isoforms (VIaL, liver type, vs. VIaH, heart type) was proposed to be responsible for tissue-specific regulation of the efficiency of COX.[172, 173] The largest nuclear encoded COX subunit, that is, subunit IV, also is known to exhibit two isoforms expressed in a tissue-specific way. In contrast to the IV-1 isoform ubiquitously expressed, IV-2 expression is activated at high oxygen tension[174] and high IV-2 levels are observed in adult lung, trachea, and placenta and lower expression in all other tissues investigated.[171] In contrast to COX, cytochrome bc1 complexes from bovine liver and heart do not exhibit tissue-specific differences.[175] Two independent groups have reported that the accessory complex I subunit NDUFV3 can exist under two isoforms generated by alternative splicing.[176, 177] Short (~10 kDa, NDUFV3-S) and long (~50 kDa, NDUFV3-L) isoforms are expressed in a tissue-specific manner, the long one being more abundant in brain and liver, whereas the high level of the short isoform is detected in heart and skeletal muscle. The presence of the long isoform seems to correlate with a higher affinity of the complex I for NADH, however, further studies will be required to elucidate the regulatory role of posttranslational modification of NDUFV3 on complex I activity in different tissues.

The phosphorylating system through the ATP synthase and ANT exhibit tissue-specific isoforms. The subunit c of the ATP synthase is encoded by three different genes, P1, P2 and P3, which encode identical mature protein.[178, 179] The P1 to P2 mRNA ratio is high in ATP synthase-rich tissues; P2 mRNA dominates in tissues harboring low ATP synthase level. The strong P1 mRNA downregulation observed in BAT under cold exposure could explain the low ATP synthase abundance in activated BAT. Beyond subunit c isoforms, the ATP synthase γ-subunit also presents tissue-specific isoforms generated by alternative splicing.[180]

In human tissues, the H type mRNA devoid of exon 9 is expressed specifically in the heart and skeletal muscle, whereas the L type mRNA is expressed in the brain, liver, and kidney. This tissue specificity of transcript heterogeneity suggests distinct functional or regulatory roles of the γ-subunit isoforms in the catalysis of mitochondrial ATP synthase. Other components of the phosphorylating system as the adenine nucleotide translocase (ANT) protein are present in multiple isoforms in mammals. Two ANT isoforms have been identified in rat and mouse. In contrast, three adenine nucleotide translocase isoforms encoded by different genes (ANT1, ANT2 and ANT3) are expressed in human tissue.[181] The three human ANT isoforms are co-expressed in all tissues revealing tissue-specific transcription patterns. The highest ANT1 mRNA proportions were found in skeletal muscle, heart, and brain, whereas ANT2 was expressed mainly in tissues capable of proliferation and regeneration, as in the kidneys, spleen, liver, fibroblasts, and lymphocytes. The ANT3 mRNA was rare in all of the tissues.

## 6.2 Tissue-Specific Enzymes Connected to the Respiratory Chain

Mitochondrial glycerol-3-phosphate dehydrogenase (mGPDH) is a flavin-linked respiratory chain dehydrogenase bound on the inter membrane space side of the inner mitochondrial membrane.[182, 183] The mGPDH oxidizes glycerol-3-phosphate (G3P) to dihydroxyacetone phosphate (DHAP) with concurrent reduction of flavin adenine dinucleotide (FAD) to FADH2 and transfers electrons to coenzyme Q (CoQ). The mGPDH is encoded by GPD2 gene, which is controlled by different promoters conferring a unique tissue-specific expression profile and placing the gene expression under the regulation of thyroid hormones. Despite their complete lack of homology, GPD2 often is co-expressed with the GPD1 enzyme encoding cytosolic glycerol-3-phosphate dehydrogenase, strongly supporting their mutual contribution in the glycerol phosphate redox shuttle. These two enzymes through the glycerol phosphate redox shuttle will regenerate $NAD^+$ from the NADH formed during glycolysis.[184, 185] As mGPDH is expressed mainly in brown adipose tissue, brain, skeletal muscle, and testes, it has been assumed that glycerol phosphate shuttle could be functionally important in tissues that rapidly oxidize glucose.[186]

As previously described, the electron transfer flavoprotein:ubiquinone oxidoreductase (ETF-QO) is an enzyme involved in mitochondrial β-oxidation, which can transfer electron from acyl-CoA dehydrogenase to the respiratory chain coenzyme Q.[187, 188] This enzyme and mitochondrial β-oxidation are key metabolic pathways to sustain the high energy requirement of tissues such as heart and skeletal muscle. Together with the Thiolase I,[189] the ETF:QO is almost absent in brain mitochondria in agreement with their almost complete lack of β-oxidation capacity.[190] The presence of Thiolase II, however, enables neuronal mitochondria to metabolize ketone bodies.

The fourth enzyme of the pyrimidine-de-novo synthesis pathway, i.e. the dihydroorotate dehydrogenase, catalyzes the conversion of dihydroorotate to orotate by transferring electrons to the coenzyme Q. This enzyme essential to ensure proper pyrimidine synthesis has been found highly expressed in tissues presenting important proliferative or regenerative activity and in tissues presenting absorptive or excretory activity.[191]

## 6.3 Tissue-Specific Mitochondrial Metabolic Pathways

The tissue-specific proteome, beyond modulating the mitochondrial metabolic pathway, can completely hijack the OXPHOS from its primary function of disconnecting the RC activity from the ATP synthesis. In contrast to cardiac mitochondria mostly devoted to ATP synthesis, mitochondria found in brown adipose tissues (BAT) can, under certain physiological conditions, be totally uncoupled through the activity of a tissue-specific uncoupling protein (UCP1). The dissipation of the protonmotive force by UCP1 mostly expressed in BAT regulates heat production, a process linked to nonshivering thermogenesis.

Key metabolic pathways, such as gluconeogenesis, are known to take place mainly in liver and kidney mitochondria. Another example is the urea synthesis, which occurs almost exclusively in liver mitochondria, as this tissue expresses a liver-specific ornithine-citruline antiporter, and the carbamoylphosphate synthetase and ornithine transcarbamylase enzymes.[192] Likewise, generation of ketone bodies mostly occurs in liver mitochondria because of the liver specific β- hydroxybutyrate dehydrogenase.[193] Mitochondria present in steroidogenic tissues, such as adrenal cortex, gonads, placenta, and brain, are essential to steroidogenesis because they contain the cholesterol side-chain cleavage enzyme P450ssc.[194, 195]

# 7 CONCLUSION

Mitochondria are dynamic organelles that continuously move, fuse, and divide. The relationships among the structural changes of the mitochondrial network and their eventual roles in regulation processes remain an open question that is being studied thoroughly. Mitochondrial biogenesis is a complex process that involves a great number of proteins and quality-control processes will play an important role in the development of pathologies. Within the most important mitochondrial functions, energy conversion (ATP synthesis) and coenzymes reoxidation play a key role. These two functions are tightly but indirectly linked through the oxidative phosphorylation process. This is a key point because in physio- and pathological situations, the need for either ATP production or NADH oxidation could be kinetically very different. Imbalance between ATP synthesis and NADH reoxidation often is linked to an increase in ROS production. A lot of evidence shows that cells better adjust to a deficit in ATP synthesis than to a deficit in coenzymes reoxidation. A number of correlations between mitochondrial network and mitochondrial functions alterations are found in the literature. There is little direct evidence, however, that these two processes are causal. It is now clear that ROS—produced at the mitochondrial level or by other cellular alternative pathways—are involved both in cell signaling and deleterious redox alterations. One of the remaining issues is to identify in pathophysiological situations the localization of ROS production and the ROS species involved.

## Acknowledgments

The authors received continuous support from the CNRS (Centre National de la Recherche Scientifique), the Comité de Dordogne & Gironde de la Ligue Nationale Contre le Cancer, The Fondation ARC pour la recherche sur le Cancer, the Plan cancer 2014-2019 No BIO 2014 06, the **ANR-16-CE14-0013**, and Bordeaux University. This work is dedicated to the memory of professor Xavier Leverve, whose exceptional contribution and friendly collaboration were continuous throughout the years.

## References

1. Frey TG, Mannella CA. The internal structure of mitochondria. *Trends Biochem Sci* 2000;**25**(7):319–24.
2. Van der Laan M, Horvath SE, Pfanner N. Mitochondrial contact site and cristae organizing system. *Curr Opin Cell Biol* 2016;**41**:33–42.
3. Mannella CA. Structural diversity of mitochondria: functional implications. *Ann N Y Acad Sci* 2008;**1147**:171–9.
4. Kornmann B, Walter P. ERMES-mediated ER-mitochondria contacts: molecular hubs for the regulation of mitochondrial biology. *J Cell Sci* 2010;**123**:1389–93. Pt 9.

5. Herrera-Cruz MS, Simmen T. Of yeast, mice and men: MAMs come in two flavors. *Biol Direct* 2017;**12**(1):3.
6. Vance JE. MAM (mitochondria-associated membranes) in mammalian cells: lipids and beyond. *Biochim Biophys Acta* 2014;**1841**(4):595–609.
7. Rieusset J. Role of endoplasmic reticulum-mitochondria communication in type 2 diabetes. *Adv Exp Med Biol* 2017;**997**:171–86.
8. Frederick RL, Shaw JM. Moving mitochondria: establishing distribution of an essential organelle. *Traffic* 2007;**8**(12):1668–75.
9. López-Doménech G, Covill-Cooke C, Ivankovic D, Halff EF, Sheehan DF, Norkett R, et al. Miro proteins coordinate microtubule- and actin-dependent mitochondrial transport and distribution. *EMBO J* 2018;**37**(3):321–36.
10. Malka F, Guillery O, Cifuentes-Diaz C, Guillou E, Belenguer P, Lombès A, et al. Separate fusion of outer and inner mitochondrial membranes. *EMBO Rep* 2005;**6**(9):853–9.
11. Sauvanet C, Duvezin-Caubet S, di Rago JP, Rojo M. Energetic requirements and bioenergetic modulation of mitochondrial morphology and dynamics. *Semin Cell Dev Biol* 2010;**21**(6):558–65.
12. Wai T, Langer T. Mitochondrial dynamics and metabolic regulation. *Trends Endocrinol Metab* 2016;**27**(2):105–17.
13. Arnoult D. Mitochondrial fragmentation in apoptosis. *Trends Cell Biol* 2007;**17**(1):6–12.
14. Knott AB, Perkins G, Schwarzenbacher R, Bossy-Wetzel E. Mitochondrial fragmentation in neurodegeneration. *Nat Rev Neurosci* 2008;**9**(7):505–18.
15. Yoon Y, Galloway CA, Jhun BS, Yu T. Mitochondrial dynamics in diabetes. *Antioxid Redox Signal* 2011;**14**(3):439–57.
16. Westermann B. Mitochondrial fusion and fission in cell life and death. *Nat Rev Mol Cell Biol* 2010;**11**(12):872–84.
17. De Brito OM, Scorrano L. Mitofusin 2 tethers endoplasmic reticulum to mitochondria. *Nature* 2008;**456**(7222):605–10.
18. Filadi R, Greotti E, Turacchio G, Luini A, Pozzan T, Pizzo P. Mitofusin 2 ablation increases endoplasmic reticulum-mitochondria coupling. *Proc Natl Acad Sci U S A* 2015;**112**(17):E2174–81.
19. Guillery O, Malka F, Landes T, Guillou E, Blackstone C, Lombès A, et al. Metalloprotease-mediated OPA1 processing is modulated by the mitochondrial membrane potential. *Biol Cell* 2008;**100**(5):315–25.
20. Duvezin-Caubet S, Jagasia R, Wagener J, Hofmann S, Trifunovic A, Hansson A, et al. Proteolytic processing of OPA1 links mitochondrial dysfunction to alterations in mitochondrial morphology. *J Biol Chem* 2006;**281**(49):37972–9.
21. Sesaki H, Adachi Y, Kageyama Y, Itoh K, Iijima M. In vivo functions of Drp1: lessons learned from yeast genetics and mouse knockouts. *Biochim Biophys Acta* 2014;**1842**(8):1179–85.
22. Manor U, Bartholomew S, Golani G, Christenson E, Kozlov M, Higgs H, et al. A mitochondria-anchored isoform of the actin-nucleating spire protein regulates mitochondrial division. *elife* 2015;**4**.
23. Lee JE, Westrate LM, Wu H, Page C, Voeltz GK. Multiple dynamin family members collaborate to drive mitochondrial division. *Nature* 2016;**540**(7631):139–43.
24. Bertholet AM, Delerue T, Millet AM, Moulis MF, David C, Daloyau M, et al. Mitochondrial fusion/fission dynamics in neurodegeneration and neuronal plasticity. *Neurobiol Dis* 2016;**90**:3–19.
25. Mourier A, Motori E, Brandt T, Lagouge M, Atanassov I, Galinier A, et al. Mitofusin 2 is required to maintain mitochondrial coenzyme Q levels. *J Cell Biol* 2015;**208**(4):429–42.
26. Gerber S, Charif M, Chevrollier A, Chaumette T, Angebault C, Kane MS, et al. Mutations in DNM1L, as in OPA1, result in dominant optic atrophy despite opposite effects on mitochondrial fusion and fission. *Brain* 2017;**140**(10):2586–96.
27. Zhang Z, Wakabayashi N, Wakabayashi J, Tamura Y, Song WJ, Sereda S, et al. The dynamin-related GTPase Opa1 is required for glucose-stimulated ATP production in pancreatic beta cells. *Mol Biol Cell* 2011;**22**(13):2235–45.
28. Wang L, Ishihara T, Ibayashi Y, Tatsushima K, Setoyama D, Hanada Y, et al. Disruption of mitochondrial fission in the liver protects mice from diet-induced obesity and metabolic deterioration. *Diabetologia* 2015;**58**(10):2371–80.
29. Kulkarni SS, Joffraud M, Boutant M, Ratajczak J, Gao AW, Maclachlan C, et al. Mfn1 deficiency in the liver protects against diet-induced insulin resistance and enhances the hypoglycemic effect of metformin. *Diabetes* 2016;**65**(12):3552–60.
30. Gray MW, Burger G, Lang BF. Mitochondrial evolution. *Science* 1999;**283**:1476–81.
31. Taylor RW, Turnbull DM. Mitochondrial DNA mutations in human disease. *Nat Rev Genet* 2005;**6**(5):389–402.
32. F1 M, Lombès A, Rojo M. Organization, dynamics and transmission of mitochondrial DNA: focus on vertebrate nucleoids. *Biochim Biophys Acta* 2006;**1763**(5–6):463–72.

33. Calvo SE, Clauser KR, Mootha VK. MitoCarta2.0: an updated inventory of mammalian mitochondrial proteins. *Nucleic Acids Res* 2016;**44**(D1):D1251–7.
34. Quirós PM, Langer T, López-Otín C. New roles for mitochondrial proteases in health, ageing and disease. *Nat Rev Mol Cell Biol* 2015;**16**(6):345–59.
35. Levytskyy RM, Bohovych I, Khalimonchuk O. Metalloproteases of the inner mitochondrial membrane. *Biochemistry* 2017;**56**(36):4737–46.
36. Costello JL, Castro IG, Camões F, Schrader TA, McNeall D, Yang J, et al. Predicting the targeting of tail-anchored proteins to subcellular compartments in mammalian cells. *J Cell Sci* 2017;**130**(9):1675–87.
37. Kang Y, Fielden LF, Stojanovski D. Mitochondrial protein transport in health and disease. *Semin Cell Dev Biol* 2017; S1084-9521(17)30021-6.
38. Wiedemann N, Pfanner N. Mitochondrial machineries for protein import and assembly. *Annu Rev Biochem* 2017;**86**:685–714.
39. Taylor EB. Functional properties of the mitochondrial carrier system. *Trends Cell Biol* 2017;**27**(9):633–44.
40. Allen S, Balabanidou V, Sideris DP, Lisowsky T, Tokatlidis K. Erv1 mediates the Mia40-dependent protein import pathway and provides a functional link to the respiratory chain by shuttling electrons to cytochrome c. *J Mol Biol* 2005;**353**(5):937–44.
41. Bihlmaier K, Mesecke N, Terziyska N, Bien M, Hell K, Herrmann JM. The disulfide relay system of mitochondria is connected to the respiratory chain. *J Cell Biol* 2007;**179**(3):389–95.
42. Dabir DV, Leverich EP, Kim SK, Tsai FD, Hirasawa M, Knaff DB, et al. A role for cytochrome c and cytochrome c peroxidase in electron shuttling from Erv1. *EMBO J* 2007;**26**(23):4801–11.
43. Banci L, Bertini I, Calderone V, Cefaro C, Ciofi-Baffoni S, Gallo A, et al. An electron-transfer path through an extended disulfide relay system: the case of the redox protein ALR. *J Am Chem Soc* 2012;**134**(3):1442–5.
44. Voos W. Chaperone-protease networks in mitochondrial protein homeostasis. *Biochim Biophys Acta* 2013;**1833**(2):388–99.
45. Timón-Gómez A, Nývltová E, Abriata LA, Vila AJ, Hosler J, Barrientos A. Mitochondrial cytochrome c oxidase biogenesis: recent developments. *Semin Cell Dev Biol* 2017; S1084-9521(17)30275-6.
46. Sutherland LN, Bomhof MR, Capozzi LC, Basaraba SA, Wright DC. Exercise and adrenaline increase PGC-1{alpha} mRNA expression in rat adipose tissue. *J Physiol* 2009;**587**:1607–17. Pt 7.
47. Wilson-Fritch L, Nicoloro S, Chouinard M, Lazar MA, Chui PC, Leszyk J, et al. Mitochondrial remodeling in adipose tissue associated with obesity and treatment with rosiglitazone. *J Clin Invest* 2004;**114**(9):1281–9.
48. Rong JX, Qiu Y, Hansen MK, Zhu L, Zhang V, Xie M, et al. Adipose mitochondrial biogenesis is suppressed in db/db and high-fat diet-fed mice and improved by rosiglitazone. *Diabetes* 2007;**56**(7):1751–60.
49. Bota DA, Davies KJ. Mitochondrial Lon protease in human disease and aging: Including an etiologic classification of Lon-related diseases and disorders. *Free Radic Biol Med* 2016;**100**:188–98.
50. Bhaskaran S, Pharaoh G, Ranjit R, Murphy A, Matsuzaki S, Nair BC, et al. Loss of mitochondrial protease ClpP protects mice from diet-induced obesity and insulin resistance. *EMBO Rep* 2018;e45009.
51. Quirós PM, Ramsay AJ, Sala D, Fernández-Vizarra E, Rodríguez F, Peinado JR, et al. Loss of mitochondrial protease OMA1 alters processing of the GTPase OPA1 and causes obesity and defective thermogenesis in mice. *EMBO J* 2012;**31**(9):2117–33.
52. Head B, Griparic L, Amiri M, Gandre-Babbe S, van der Bliek AM. Inducible proteolytic inactivation of OPA1 mediated by the OMA1 protease in mammalian cells. *J Cell Biol* 2009;**187**(7):959–66.
53. Civitarese AE, MacLean PS, Carling S, Kerr-Bayles L, McMillan RP, Pierce A, et al. Regulation of skeletal muscle oxidative capacity and insulin signaling by the mitochondrial rhomboid protease PARL. *Cell Metab* 2010;**11**(5):412–26.
54. Almontashiri NA, Chen HH, Mailloux RJ, Tatsuta T, Teng AC, Mahmoud AB, et al. SPG7 variant escapes phosphorylation-regulated processing by AFG3L2, elevates mitochondrial ROS, and is associated with multiple clinical phenotypes. *Cell Rep* 2014;**7**(3):834–47.
55. Cavalcanti DM, Castro LM, Rosa Neto JC, Seelaender M, Neves RX, Oliveira V, et al. Neurolysin knockout mice generation and initial phenotype characterization. *J Biol Chem* 2014;**289**(22):15426–40.
56. Quirós PM, Ramsay AJ, López-Otín C. New roles for OMA1 metalloprotease: From mitochondrial proteostasis to metabolic homeostasis. *Adipocytes* 2013;**2**(1):7–11.
57. Rostovtseva TK, Sheldon KL, Hassanzadeh E, Monge C, Saks V, Bezrukov SM, et al. Tubulin binding blocks mitochondrial voltage-dependent anion channel and regulates respiration. *Proc Natl Acad Sci U S A* 2008;**105**(48):18746–51.

58. Rigoulet M, Aguilaniu H, Avéret N, Bunoust O, Camougrand N, Grandier-Vazeille X, et al. Organization and regulation of the cytosolic NADH metabolism in the yeast *Saccharomyces cerevisiae*. *Mol Cell Biochem* 2004;**256-257**(1–2):73–81.
59. Mitchell P, Moyle J. Chemiosmotic hypothesis of oxidative phosphorylation. *Nature* 1967;**213**(5072):137–9.
60. Monné M, Palmieri F. Antiporters of the mitochondrial carrier family. *Curr Top Membr* 2014;**73**:289–320.
61. De Stefani D, Rizzuto R, Pozzan T. Enjoy the trip: calcium in mitochondria back and forth. *Annu Rev Biochem* 2016;**85**:161–92.
62. Hatefi Y. The mitochondrial electron transport and oxidative phosphorylation system. *Annu Rev Biochem* 1985;**54**:1015–69.
63. Schägger H, Pfeiffer K. Supercomplexes in the respiratory chains of yeast and mammalian mitochondria. *EMBO J* 2000;**19**(8):1777–83.
64. Trouillard M, Meunier B, Rappaport F. Questioning the functional relevance of mitochondrial supercomplexes by time-resolved analysis of the respiratory chain. *Proc Natl Acad Sci U S A* 2011;**108**(45):E1027–34.
65. Moreno-Loshuertos R, Enríquez JA. Respiratory supercomplexes and the functional segmentation of the CoQ pool. *Free Radic Biol Med* 2016;**100**:5–13.
66. Rigoulet M, Mourier A, Galinier A, Casteilla L, Devin A. Electron competition process in respiratory chain: regulatory mechanisms and physiological functions. *Biochim Biophys Acta* 2010;**1797**(6–7):671–7.
67. Mitchell P. Coupling of phosphorylation to electron and hydrogen transfer by a chemi-osmotic type of mechanism. *Nature* 1961;**191**:144–8.
68. Mitchell P. Vectorial chemistry and the molecular mechanics of chemiosmotic coupling: power transmission by proticity. *Biochem Soc Trans* 1976;**4**(3):399–430.
69. Bielawski J, Thompson TE, Lehninger AL. The effect of 2,4-dinitrophenol on the electrical resistance of phospholipid bilayer membranes. *Biochem Biophys Res Commun* 1966;**24**(6):948–54.
70. Mitchell P, Moyle J. Acid-base titration across the membrane system of rat-liver mitochondria. Catalysis by uncouplers. *Biochem J* 1967;**104**(2):588–600.
71. Mitchell P, Moyle J. Respiration-driven proton translocation in rat liver mitochondria. *Biochem J* 1967;**105**(3):1147–62.
72. Hopfer U, Lehninger AL, Thompson TE. Protonic conductance across phospholipid bilayer membranes induced by uncoupling agents for oxidative phosphorylation. *Proc Natl Acad Sci U S A* 1968;**59**(2):484–90.
73. Liberman EA, Topaly VP. Selective transport of ions through bimolecular phospholipid membranes. *Biochim Biophys Acta* 1968;**163**(2):125–36.
74. Leblanc Jr. OH. The effect of uncouplers of oxidative phosphorylation on lipid bilayer membranes: carbonylcyanidem-chlorophenylhydrazone. *J Membr Biol* 1971;**4**(1):227–51.
75. McLaughlin S. The mechanism of action of DNP on phospholipid bilayer membranes. *J Membr Biol* 1972;**9**(1):361–72.
76. Hanstein WG. Uncoupling of oxidative phosphorylation. *Biochim Biophys Acta* 1976;**456**(2):129–48.
77. Ricquier D, Bouillaud F. Mitochondrial uncoupling proteins: from mitochondria to the regulation of energy balance. *J Physiol* 2000;**529**(Pt 1):3–10.
78. Nègre-Salvayre A, Hirtz C, Carrera G, Cazenave R, Troly M, Salvayre R, et al. A role for uncoupling protein-2 as a regulator of mitochondrial hydrogen peroxide generation. *FASEB J* 1997;**11**(10):809–15.
79. Bouillaud F, Alves-Guerra MC, Ricquier D. UCPs, at the interface between bioenergetics and metabolism. *Biochim Biophys Acta* 2016;**1863**(10):2443–56.
80. Rottenberg H. Uncoupling of oxidative phosphorylation in rat liver mitochondria by general anesthetics. *Proc Natl Acad Sci U S A* 1983;**80**(11):3313–7.
81. Rottenberg H, Hashimoto K. Fatty acid uncoupling of oxidative phosphorylation in rat liver mitochondria. *Biochemistry* 1986;**25**(7):1747–55.
82. Pick U, Weiss M, Rottenberg H. Anomalous uncoupling of photophosphorylation by palmitic acid and by gramicidin D. *Biochemistry* 1987;**26**(25):8295–302.
83. Luvisetto S, Pietrobon D, Azzone GF. Uncoupling of oxidative phosphorylation. 1. Protonophoric effects account only partially for uncoupling. *Biochemistry* 1987;**26**(23):7332–8.
84. Luvisetto S, Azzone GF. Nature of proton cycling during gramicidin uncoupling of oxidative phosphorylation. *Biochemistry* 1989;**28**(3):1100–8.
85. Rottenberg H, Koeppe RE. Mechanism of uncoupling of oxidative phosphorylation by gramicidin. *Biochemistry* 1989;**28**(10):4355–60.

86. Rigoulet M, Ouhabi R, Leverve X, Putod-Paramelle F, Guérin B. Almitrine, a new kind of energy-transduction inhibitor acting on mitochondrial ATP synthase. *Biochim Biophys Acta* 1989;**975**(3):325–9.
87. Ouhabi R, Rigoulet M, Lavie JL, Guérin B. Respiration in non-phosphorylating yeast mitochondria. Roles of non-ohmic proton conductance and intrinsic uncoupling. *Biochim Biophys Acta* 1991;**1060**(3):293–8.
88. Krenn BE, Van Walraven HS, Scholts MJ, Kraayenhof R. Modulation of the proton-translocation stoichiometry of H(+)-ATP synthases in two phototrophic prokaryotes by external pH. *Biochem J* 1993;**294**(Pt 3):705–9.
89. Murphy MP, Brand MD. Variable stoichiometry of proton pumping by the mitochondrial respiratory chain. *Nature* 1987;**329**(6135):170–2.
90. Van Walraven HS, Scholts MJC, Koppenaal F, Bakels RHA, Krab K. Dependence of the proton translocation stoichiometry of cyanobacterial and chloroplat H+-ATP synthase on the membrane composition. *Biochim Biophys Acta* 1990;**1015**:425–34.
91. Luvisetto S, Conti E, Buso M, Azzone GF. Flux ratios and pump stoichiometries at sites II and III in liver mitochondria. Effect of slips and leaks. *J Biol Chem* 1991;**266**(2):1034–42.
92. Bechmann G, Weiss H. Regulation of the proton/electron stoichiometry of mitochondrial ubiquinol:cytochrome c reductase by the membrane potential. *Eur J Biochem* 1991;**195**(2):431–8.
93. Steverding D, Kadenbach B. Influence of N-ethoxycarbonyl-2-ethoxy-1,2-dihydroquinoline modification on proton translocation and membrane potential of reconstituted cytochrome-c oxidase support "proton slippage". *J Biol Chem* 1991;**266**(13):8097–101.
94. Capitanio N, Capitanio G, De Nitto E, Villani G, Papa S. H+/e- stoichiometry of mitochondrial cytochrome complexes reconstituted in liposomes. Rate-dependent changes of the stoichiometry in the cytochrome c oxidase vesicles. *FEBS Lett* 1991;**288**(1–2):179–82.
95. Papa S, Capitanio N, Capitanio G, De Nitto E, Minuto M. The cytochrome chain of mitochondria exhibits variable H+/e- stoichiometry. *FEBS Lett* 1991;**288**(1–2):183–6.
96. Groth G, Junge W. Proton slip of the chloroplast ATPase: its nucleotide dependence, energetic threshold, and relation to an alternating site mechanism of catalysis. *Biochemistry* 1993;**32**(32):8103–11.
97. Fitton V, Rigoulet M, Ouhabi R, Guérin B. Mechanistic stoichiometry of yeast mitochondrial oxidative phosphorylation. *Biochemistry* 1994;**33**(32):9692–8.
98. Pietrobon D, Azzone GF, Walz D. Effect of funiculosin and antimycin A on the redox- driven H+-pumps in mitochondria: on the nature of "leaks". *Eur J Biochem* 1981;**117**(2):389–94.
99. Pietrobon D, Zoratti M, Azzone GF. Molecular slipping in redox and ATPase H+ pumps. *Biochim Biophys Acta* 1983;**723**(2):317–21.
100. Mourier A, Devin A, Rigoulet M. Active proton leak in mitochondria: a new way to regulate substrate oxidation. *Biochim Biophys Acta* 2010;**1797**(2):255–61.
101. Halliwell B. Reactive species and antioxidants. Redox biology is a fundamental theme of aerobic life. *Plant Physiol* 2006;**141**(2):312–22.
102. Longo VD, Gralla EB, Valentine JS. Superoxide dismutase activity is essential for stationary phase survival in *Saccharomyces cerevisiae*. Mitochondrial production of toxic oxygen species in vivo. *J Biol Chem* 1996;**271**(21):12275–80.
103. Longo VD, Liou LL, Valentine JS, Gralla EB. Mitochondrial superoxide decreases yeast survival in stationary phase. *Arch Biochem Biophys* 1999;**365**(1):131–42. [Erratum in: Arch Biochem Biophys. 2000;377(1):213].
104. Chance B, Hollunger G. The interaction of energy and electron transfer reactions in mitochondria. I. General properties and nature of the products of succinate-linked reduction of pyridine nucleotide. *J Biol Chem* 1961;**236**:1534–43.
105. Hansford RG, Hogue BA, Mildaziene V. Dependence of $H_2O_2$ formation by rat heart mitochondria on substrate availability and donor age. *J Bioenerg Biomembr* 1997;**29**(1):89–95.
106. Tahara EB, Navarete FD, Kowaltowski AJ. Tissue-, substrate-, and site-specific characteristics of mitochondrial reactive oxygen species generation. *Free Radic Biol Med* 2009;**46**(9):1283–97.
107. Boveris A, Cadenas E, Stoppani AO. Role of ubiquinone in the mitochondrial generation of hydrogen peroxide. *Biochem J* 1976;**156**(2):435–44.
108. Cadenas E, Boveris A, Ragan CI, Stoppani AO. Production of superoxide radicals and hydrogen peroxide by NADH-ubiquinone reductase and ubiquinol-cytochrome c reductase from beef-heart mitochondria. *Arch Biochem Biophys* 1977;**180**(2):248–57.
109. Chance B, Hollunger G. The interaction of energy and electron transfer reactions in mitochondria. IV. The pathway of electron transfer. *J Biol Chem* 1961;**236**:1562–8.

110. Hinkle PC, Butow RA, Racker E, Chance B. Partial resolution of the enzymes catalyzing oxidative phosphorylation. XV. Reverse electron transfer in the flavin-cytochrome beta region of the respiratory chain of beef heart submitochondrial particles. *J Biol Chem* 1967;**242**(22):5169–73.
111. Korshunov SS, Skulachev VP, Starkov AA. High protonic potential actuates a mechanism of production of reactive oxygen species in mitochondria. *FEBS Lett* 1997;**416**(1):15–8.
112. Kushnareva Y, Murphy AN, Andreyev A. Complex I-mediated reactive oxygen species generation: modulation by cytochrome c and NAD(P)+ oxidation-reduction state. *Biochem J* 2002;**368**(Pt 2):545–53.
113. Kwong LK, Sohal RS. Substrate and site specificity of hydrogen peroxide generation in mouse mitochondria. *Arch Biochem Biophys* 1998;**350**(1):118–26.
114. Liu Y, Fiskum G, Schubert D. Generation of reactive oxygen species by the mitochondrial electron transport chain. *J Neurochem* 2002;**80**(5):780–7.
115. Votyakova TV, Reynolds IJ. DeltaPsi(m)-dependent and -independent production of reactive oxygen species by rat brain mitochondria. *J Neurochem* 2001;**79**(2):266–77.
116. Batandier C, Guigas B, Detaille D, El-Mir MY, Fontaine E, Rigoulet M, et al. The ROS production induced by a reverse-electron flux at respiratory-chain complex 1 is hampered by metformin. *J Bioenerg Biomembr* 2006;**38**(1):33–42.
117. Vinogradov AD, Grivennikova VG. The mitochondrial complex I: progress in understanding of catalytic properties. *IUBMB Life* 2001;**52**(3–5):129–34.
118. Johnson Jr. JE, Choksi K, Widger WR. NADH-Ubiquinone oxidoreductase: substrate-dependent oxygen turnover to superoxide anion as a function of flavin mononucleotide. *Mitochondrion* 2003;**3**(2):97–110.
119. Kudin AP, Bimpong-Buta NY, Vielhaber S, Elger CE, Kunz WS. Characterization of superoxide-producing sites in isolated brain mitochondria. *J Biol Chem* 2004;**279**(6):4127–35.
120. Kussmaul L, Hirst J. The mechanism of superoxide production by NADH:ubiquinone oxidoreductase (complex I) from bovine heart mitochondria. *Proc Natl Acad Sci U S A* 2006;**103**(20):7607–12.
121. Genova ML, Ventura B, Giuliano G, Bovina C, Formiggini G, Parenti Castelli G, et al. The site of production of superoxide radical in mitochondrial Complex I is not a bound ubisemiquinone but presumably iron-sulfur cluster N2. *FEBS Lett* 2001;**505**(3):364–8.
122. Herrero A, Barja G. Localization of the site of oxygen radical generation inside the complex I of heart and nonsynaptic brain mammalian mitochondria. *J Bioenerg Biomembr* 2000;**32**(6):609–15.
123. Lambert AJ, Brand MD. Inhibitors of the quinone-binding site allow rapid superoxide production from mitochondrial NADH:ubiquinone oxidoreductase (complex I). *J Biol Chem* 2004;**279**(38):39414–20.
124. Massey V. The composition of the ketoglutarate dehydrogenase complex. *Biochim Biophys Acta* 1960;**38**:447–60.
125. Sheu KF, Blass JP. The alpha-ketoglutarate dehydrogenase complex. *Ann N Y Acad Sci* 1999;**893**:61–78.
126. Wagenknecht T, Francis N, DeRosier DJ. alpha-Ketoglutarate dehydrogenase complex may be heterogeneous in quaternary structure. *J Mol Biol* 1983;**165**(3):523–39.
127. Starkov AA, Fiskum G, Chinopoulos C, Lorenzo BJ, Browne SE, Patel MS, et al. Mitochondrial alpha-ketoglutarate dehydrogenase complex generates reactive oxygen species. *J Neurosci* 2004;**24**(36):7779–88.
128. Bunik VI, Sievers C. Inactivation of the 2-oxo acid dehydrogenase complexes upon generation of intrinsic radical species. *Eur J Biochem* 2002;**269**(20):5004–15.
129. Massey V. Activation of molecular oxygen by flavins and flavoproteins. *J Biol Chem* 1994;**269**(36):22459–62.
130. Klingenberg M. Localization of the glycerol-phosphate dehydrogenase in the outer phase of the mitochondrial inner membrane. *Eur J Biochem* 1970;**13**(2):247–52.
131. Boveris A, Oshino N, Chance B. The cellular production of hydrogen peroxide. *Biochem J* 1972;**128**(3):617–30.
132. Boveris A, Chance B. The mitochondrial generation of hydrogen peroxide. General properties and effect of hyperbaric oxygen. *Biochem J* 1973;**134**(3):707–16.
133. St-Pierre J, Buckingham JA, Roebuck SJ, Brand MD. Topology of superoxide production from different sites in the mitochondrial electron transport chain. *J Biol Chem* 2002;**277**(47):44784–90.
134. Adams SH, Hoppel CL, Lok KH, Zhao L, Wong SW, Minkler PE, et al. Plasma acylcarnitine profiles suggest incomplete long-chain fatty acid beta-oxidation and altered tricarboxylic acid cycle activity in type 2 diabetic African-American women. *J Nutr* 2009;**139**(6):1073–81.
135. Hancock CR, Han DH, Chen M, Terada S, Yasuda T, Wright DC, et al. High-fat diets cause insulin resistance despite an increase in muscle mitochondria. *Proc Natl Acad Sci U S A* 2008;**105**(22):7815–20.
136. Koves TR, Ussher JR, Noland RC, Slentz D, Mosedale M, Ilkayeva O, et al. Mitochondrial overload and incomplete fatty acid oxidation contribute to skeletal muscle insulin resistance. *Cell Metab* 2008;**7**(1):45–56.

137. Turner N, Bruce CR, Beale SM, Hoehn KL, So T, Rolph MS, et al. Excess lipid availability increases mitochondrial fatty acid oxidative capacity in muscle: evidence against a role for reduced fatty acid oxidation in lipid-induced insulin resistance in rodents. *Diabetes* 2007;**56**(8):2085–92.
138. Anderson EJ, Lustig ME, Boyle KE, Woodlief TL, Kane DA, Lin CT, et al. Mitochondrial H2O2 emission and cellular redox state link excess fat intake to insulin resistance in both rodents and humans. *J Clin Invest* 2009;**119**(3):573–81.
139. Bonnard C, Durand A, Peyrol S, Chanseaume E, Chauvin MA, Morio B, et al. Mitochondrial dysfunction results from oxidative stress in the skeletal muscle of diet-induced insulin-resistant mice. *J Clin Invest* 2008;**118**(2):789–800.
140. Cadenas S, Buckingham JA, Samec S, Seydoux J, Din N, Dulloo AG, et al. UCP2 and UCP3 rise in starved rat skeletal muscle but mitochondrial proton conductance is unchanged. *FEBS Lett* 1999;**462**(3):257–60.
141. Samec S, Seydoux J, Dulloo AG. Role of UCP homologues in skeletal muscles and brown adipose tissue: mediators of thermogenesis or regulators of lipids as fuel substrate? *FASEB J* 1998;**12**(9):715–24.
142. Echtay KS, Roussel D, St-Pierre J, Jekabsons MB, Cadenas S, Stuart JA, et al. Superoxide activates mitochondrial uncoupling proteins. *Nature* 2002;**415**(6867):96–9.
143. Seifert EL, Estey C, Xuan JY, Harper ME. Electron transport chain-dependent and -independent mechanisms of mitochondrial $H_2O_2$ emission during long-chain fatty acid oxidation. *J Biol Chem* 2010;**285**(8):5748–58.
144. Beckmann JD, Frerman FE. Electron-transfer flavoprotein-ubiquinone oxidoreductase from pig liver: purification and molecular, redox, and catalytic properties. *Biochemistry* 1985;**24**(15):3913–21.
145. Beckmann JD, Frerman FE. Reaction of electron-transfer flavoprotein with electron-transfer flavoprotein-ubiquinone oxidoreductase. *Biochemistry* 1985;**24**(15):3922–5.
146. Ramsay RR, Steenkamp DJ, Husain M. Reactions of electron-transfer flavoprotein and electron-transfer flavoprotein:ubiquinone oxidoreductase. *Biochem J* 1987;**241**(3):883–92.
147. Skulachev VP. Uncoupling: new approaches to an old problem of bioenergetics. *Biochim Biophys Acta* 1998;**1363**(2):100–24.
148. Cadenas E, Boveris A. Enhancement of hydrogen peroxide formation by protophores and ionophores in antimycin-supplemented mitochondria. *Biochem J* 1980;**188**(1):31–7.
149. Ouhabi R, Rigoulet M, Lavie JL, Guérin B. Respiration in non-phosphorylating yeast mitochondria. Roles of non-ohmic proton conductance and intrinsic uncoupling. *Biochim Biophys Acta* 1991;**1060**(3):293–8.
150. Rigoulet M, Leverve X, Fontaine E, Ouhabi R, Guérin B. Quantitative analysis of some mechanisms affecting the yield of oxidative phosphorylation: dependence upon both fluxes and forces. *Mol Cell Biochem* 1998;**184**(1–2):35–52.
151. Rottenberg H. Uncoupling of oxidative phosphorylation in rat liver mitochondria by general anesthetics. *Proc Natl Acad Sci U S A* 1983;**80**(11):3313–7.
152. Zoratti M, Favaron M, Pietrobon D, Azzone GF. Intrinsic uncoupling of mitochondrial proton pumps. 1. Non-ohmic conductance cannot account for the nonlinear dependence of static head respiration on delta microH. *Biochemistry* 1986;**25**(4):760–7.
153. Korshunov SS, Skulachev VP, Starkov AA. High protonic potential actuates a mechanism of production of reactive oxygen species in mitochondria. *FEBS Lett* 1997;**416**(1):15–8.
154. Schönfeld P, Wojtczak L. Fatty acids decrease mitochondrial generation of reactive oxygen species at the reverse electron transport but increase it at the forward transport. *Biochim Biophys Acta* 2007;**1767**(8):1032–40.
155. Echtay KS, Winkler E, Klingenberg M. Coenzyme Q is an obligatory cofactor for uncoupling protein function. *Nature* 2000;**408**(6812):609–13.
156. Echtay KS, Winkler E, Frischmuth K, Klingenberg M. Uncoupling proteins 2 and 3 are highly active H(+) transporters and highly nucleotide sensitive when activated by coenzyme Q (ubiquinone). *Proc Natl Acad Sci U S A* 2001;**98**(4):1416–21.
157. Casteilla L, Rigoulet M, Pénicaud L. Mitochondrial ROS metabolism: modulation by uncoupling proteins. *IUBMB Life* 2001;**52**(3–5):181–8.
158. Echtay KS, Esteves TC, Pakay JL, Jekabsons MB, Lambert AJ, Portero-Otín M, et al. A signalling role for 4-hydroxy-2-nonenal in regulation of mitochondrial uncoupling. *EMBO J* 2003;**22**(16):4103–10.
159. Carrière A, Fernandez Y, Rigoulet M, Pénicaud L, Casteilla L. Inhibition of preadipocyte proliferation by mitochondrial reactive oxygen species. *FEBS Lett* 2003;**550**(1–3):163–7.
160. Carrière A, Carmona MC, Fernandez Y, Rigoulet M, Wenger RH, Pénicaud L, Casteilla L. Mitochondrial reactive oxygen species control the transcription factor CHOP-10/GADD153 and adipocyte differentiation: a mechanism for hypoxia-dependent effect. *J Biol Chem* 2004;**279**(39):40462–9.

161. Frezza C. Mitochondrial metabolites: undercover signalling molecules. *Interface Focus* 2017;**7**(2):20160100.
162. Chevtzoff C, Yoboue ED, Galinier A, Casteilla L, Daignan-Fornier B, Rigoulet M, Devin A. Reactive oxygen species-mediated regulation of mitochondrial biogenesis in the yeast *Saccharomyces cerevisiae*. *J Biol Chem* 2010;**285**(3):1733–42.
163. Yoboue ED, Augier E, Galinier A, Blancard C, Pinson B, Casteilla L, et al. cAMP- induced mitochondrial compartment biogenesis: role of glutathione redox state. *J Biol Chem* 2012;**287**(18):14569–78.
164. Fernández-Vizarra E, Enríquez JA, Pérez-Martos A, Montoya J, Fernández-Silva P. Tissue-specific differences in mitochondrial activity and biogenesis. *Mitochondrion* 2011;**11**(1):207–13.
165. D'Erchia AM, Atlante A, Gadaleta G, Pavesi G, Chiara M, De Virgilio C, et al. Tissue-specific mtDNA abundance from exome data and its correlation with mitochondrial transcription, mass and respiratory activity. *Mitochondrion* 2015;**20**:13–21.
166. Rossignol R, Malgat M, Mazat JP, Letellier T. Threshold effect and tissue specificity. Implication for mitochondrial cytopathies. *J Biol Chem* 1999;**274**(47):33426–32.
167. Shitara H, Shimanuki M, Hayashi J, Yonekawa H. Global imaging of mitochondrial morphology in tissues using transgenic mice expressing mitochondrially targeted enhanced green fluorescent protein. *Exp Anim* 2010;**59**(1):99–103.
168. Brandt T, Mourier A, Tain LS, Partridge L, Larsson NG, Kühlbrandt W. Changes of mitochondrial ultrastructure and function during ageing in mice and Drosophila. *elife* 2017;**6**. pii:e24662.
169. Grossman LI, Lomax MI. Nuclear genes for cytochrome c oxidase. *Biochim Biophys Acta* 1997;**1352**(2):174–92.
170. Kadenbach B, Kuhnnentwig L, Buge U. Evolution of a regulatory enzyme—cytochrome-C oxidase (complex-IV). *Curr Top Bioenerg* 1987;**15**:113–61.
171. Hüttemann M, Kadenbach B, Grossman LI. Mammalian subunit IV isoforms of cytochrome c oxidase. *Gene* 2001;**267**(1):111–23.
172. Anthony G, Reimann A, Kadenbach B. Tissue-specific regulation of bovine heart cytochrome-c oxidase activity by ADP via interaction with subunit VIa. *Proc Natl Acad Sci U S A* 1993;**90**(5):1652–6.
173. Frank V, Kadenbach B. Regulation of the H+/e-stoichiometry of cytochrome c oxidase from bovine heart by intramitochondrial ATP/ADP ratios. *FEBS Lett* 1996;**382**(1–2):121–4.
174. Aras S, Pak O, Sommer N, Finley Jr. R, Hüttemann M, Weissmann N, et al. Oxygen-dependent expression of cytochrome c oxidase subunit 4-2 gene expression is mediated by transcription factors RBPJ, CXXC5 and CHCHD2. *Nucleic Acids Res* 2013;**41**(4):2255–66.
175. Vázquez-Acevedo M, Antaramian A, Corona N, González-Halphen D. Subunit structures of purified beef mitochondrial cytochrome bc1 complex from liver and heart. *J Bioenerg Biomembr* 1993;**25**(4):401–10.
176. Bridges HR, Mohammed K, Harbour ME, Hirst J. Subunit NDUFV3 is present in two distinct isoforms in mammalian complex I. *Biochim Biophys Acta* 2017;**1858**(3):197–207.
177. Guerrero-Castillo S, Cabrera-Orefice A, Huynen MA, Arnold S. Identification and evolutionary analysis of tissue-specific isoforms of mitochondrial complex I subunit NDUFV3. *Biochim Biophys Acta* 2017;**1858**(3):208–17.
178. Farrell LB, Nagley P. Human liver cDNA clones encoding proteolipid subunit 9 of the mitochondrial ATPase complex. *Biochem Biophys Res Commun* 1987;**144**(3):1257–64.
179. Andersson U, Houstek J, Cannon B. ATP synthase subunit c expression: physiological regulation of the P1 and P2 genes. *Biochem J* 1997;**323**(Pt 2):379–85.
180. Matsuda C, Endo H, Ohta S, Kagawa Y. Gene structure of human mitochondrial ATP synthase gamma-subunit. Tissue specificity produced by alternative RNA splicing. *J Biol Chem* 1993;**268**(33):24950–8.
181. Dörner A, Olesch M, Giessen S, Pauschinger M, Schultheiss HP. Transcription of the adenine nucleotide translocase isoforms in various types of tissues in the rat. *Biochim Biophys Acta* 1999;**1417**(1):16–24.
182. Green DE. alpha-Glycerophosphate dehydrogenase. *Biochem J* 1936;**30**(4):629–44.
183. Ohkawa KI, Vogt MT, Farber E. Unusually high mitochondrial alpha glycerophosphate dehydrogenase activity in rat brown adipose tissue. *J Cell Biol* 1969;**41**(2):441–9.
184. Bücher TA, Klingenberg M. Wege des wasserstoffs in der lebendigen organisation. *Angew Chem* 1958;**70**(17/18):552–70.
185. Estabrook RW, Sacktor B. alpha-Glycerophosphate oxidase of flight muscle mitochondria. *J Biol Chem* 1958;**233**(4):1014–9.
186. Koza RA, Kozak UC, Brown LJ, Leiter EH, MacDonald MJ, Kozak LP. Sequence and tissue-dependent RNA expression of mouse FAD-linked glycerol-3-phosphate dehydrogenase. *Arch Biochem Biophys* 1996;**336**(1):97–104.

187. Crane FL, Beinert H. On the mechanism of dehydrogenation of fatty acyl derivatives of coenzyme A. II. The electron-transferring flavoprotein. *J Biol Chem* 1956;**218**(2):717–31.
188. Watmough NJ, Frerman FE. The electron transfer flavoprotein: ubiquinone oxidoreductases. *Biochim Biophys Acta* 2010;**1797**(12):1910–6.
189. Yang SY, He XY, Schulz H. Fatty acid oxidation in rat brain is limited by the low activity of 3-ketoacyl-coenzyme A thiolase. *J Biol Chem* 1987;**262**(27):13027–32.
190. Kunz WS, Gellerich FN. Quantification of the content of fluorescent flavoproteins in mitochondria from liver, kidney cortex, skeletal muscle, and brain. *Biochem Med Metab Biol* 1993;**50**(1):103–10.
191. Löffler M, Jöckel J, Schuster G, Becker C. Dihydroorotat-ubiquinone oxidoreductase links mitochondria in the biosynthesis of pyrimidine nucleotides. *Mol Cell Biochem* 1997;**174**(1–2):125–9.
192. Küster U, Letko G, Kunz W, Duszyńsky J, Bogucka K, Wojtczak L. Influence of different energy drains on the interrelationship between the rate of respiration, proton-motive force and adenine nucleotide patterns in isolated mitochondria. *Biochim Biophys Acta* 1981;**636**(1):32–8.
193. Williamson DH, Lund P, Krebs HA. The redox state of free nicotinamide-adenine dinucleotide in the cytoplasm and mitochondria of rat liver. *Biochem J* 1967;**103**(2):514–27.
194. Cherradi N, Chambaz EM, Defaye G. Organization of 3 beta-hydroxysteroid dehydrogenase/isomerase and cytochrome P450scc into a catalytically active molecular complex in bovine adrenocortical mitochondria. *J Steroid Biochem Mol Biol* 1995;**55**(5–6):507–14.
195. Miller WL. Steroid hormone synthesis in mitochondria. *Mol Cell Endocrinol* 2013;**379**(1–2):62–73.

CHAPTER

# 2

# Mitochondria as Signaling Platforms

*Edgar Djaha Yoboué**, *Stéphen Manon*[†], *Nadine Camougrand*[†]

***Protein Transport and Secretion Unit, Division of Genetics and Cell Biology, IRCCS Ospedale San Raffaele, Milan, Italy [†]Institut de Biochimie et Génétique Cellulaires, UMR 5095, CNRS, Université de Bordeaux, Bordeaux, France**

## 1 INTRODUCTION

Although mitochondria are known for energy conversion, their roles encompass a wider spectrum than their textbook nickname of powerhouses of the cell. It is now well-established that mitochondria are associated with numerous signaling processes. Furthermore, these signals are not only local events, but also are used by mitochondria for communicating with other intracellular compartments. Thus, diverse signals are coming in and out from mitochondria and can govern the fate of the cells and organs.

Until the 1990s, it was known that mitochondria were both a target and a player in $Ca^{2+}$-regulation. The molecular support and the mechanisms involved remained elusive, however, until new cellular approaches identified mitochondrial $Ca^{2+}$-transporters and their regulators. Mitochondria now are recognized as a central player of $Ca^{2+}$-signaling pathways.

ROS are side products of the normal function of the mitochondrial electron transport chain. Cells are well-equipped with enzymatic and nonenzymatic ROS-scavenging systems. For a long time, studies about ROS focused on conditions where they are over-produced, generally because of alterations of electron carriers, reflecting pathological situations. Recent studies, however, reconsidered ROS as signaling molecules under both physiological and pathological conditions.

Mitochondria produce a large variety of metabolites, fueling different anabolic pathways. Many intermediates also act as signaling factors that contribute both to the regulation of the whole cellular metabolism and to more specific aspects of cellular function.

Autophagy and apoptosis are among the most studied processes for the last 30 years. Their alterations have been reported in most noninfectious diseases, including cancer, neurodegenerative diseases, and metabolic diseases. They also contribute to virus-related diseases, and certain diseases caused by bacteria and protozoa. The role of mitochondria in apoptosis has emerged with the discovery that it was initiated by the cytosolic relocation of mitochondrial



https://doi.org/10.1016/B978-0-12-811752-1.00002-X

proteins. Mitochondria also were described early as a target of autophagy (mitophagy), but recently have been identified as a mediator of autophagy. Considering the large amount of research about the role of mitochondria in autophagy and apoptosis, many contradictory data have been reported, but several hypotheses and concepts now are accepted widely.

Mitochondria also have emerged as a player of innate immunity response, with signaling pathways that might connect partly to other processes described previously.

We will finish this short overview by addressing the functional and physical integration of mitochondria with other signaling contributing organelles. It has been well-established that some signaling processes through these chatty mitochondria required particular spatial organization, thereby creating well-defined physical contacts between mitochondria and other cellular constituents, such as the endoplasmic reticulum (ER).

# 2 MITOCHONDRIA AND CALCIUM

Since Sydney Ringer's observations in 1880s that $Ca^{2+}$-supplemented perfusion buffer triggered heart contractions, we have learned that $Ca^{2+}$-ions regulate other processes, such as synaptic transmission, gene expression, oocyte fertilization, cell death, metabolism, and protein folding.

$Ca^{2+}$-signaling requires dynamic and regulated concentration changes. The easiest way would be through synthesis and degradation, similarly to another second messenger, cyclic AMP (cAMP). But because cells cannot synthesize $Ca^{2+}$ de novo, how do they do it?

The free extracellular $Ca^{2+}$ concentration is within the millimolar (mM) range while around only 100 nM in the cytosol, under resting conditions. The low cytosolic $Ca^{2+}$ concentration ($[Ca^{2+}]$cyto) is because of both cytosolic $Ca^{2+}$-binding proteins and to pumps and exchangers located at the plasma membrane (PM), which constantly extrude $Ca^{2+}$. The low $[Ca^{2+}]$cyto is maintained by intracellular $Ca^{2+}$-storage compartments, primarily the ER. The ER harbors the sarcoendoplasmic reticulum calcium ATPase (SERCA) that pumps cytosolic $Ca^{2+}$ into the ER lumen, where free $Ca^{2+}$ concentration reaches 0.5–1 mM. Therefore, $Ca^{2+}$ signaling mainly relies on the mobilization of the extracellular media and the intracellular stores. In nonexcitable cells, $Ca^{2+}$ release from ER is mediated mainly by the inositol 1,4,5-triphosphate receptors (IP3Rs) located in the ER membrane. IP3Rs release $Ca^{2+}$ after having bound inositol 1,4,5-triphosphate (IP3) produced subsequently to agonists binding (histamine, ATP, bradykinin, etc.) to their PM receptors.

Mitochondria also are prominent players in $Ca^{2+}$-homeostasis and signaling. Links between mitochondria and calcium ($Ca^{2+}$) are contemporaneous to Peter Mitchell's chemiosmotic theory. In the mid-1950s, mitochondria were the first organelles associated with $Ca^{2+}$ handling. Mitochondrial $Ca^{2+}$ entry serves different roles ranging from cytosolic $Ca^{2+}$ buffering to the regulation of oxidative phosphorylation and cell death. In the following lines, we will describe the actors of mitochondrial $Ca^{2+}$ homeostasis and the physiological consequences of mitochondrial $Ca^{2+}$ concentration variation.

## 2.1 Mitochondrial $Ca^{2+}$-Transport

Mitochondrial $Ca^{2+}$ uptake has been known since the mid-1950s. The identification of the molecular actors and mechanisms, however, took a long time. One of the reasons has been the identification of ER as the main intracellular $Ca^{2+}$ store and agonist responder.[1] Furthermore, the low affinity of the putative mitochondrial $Ca^{2+}$-transporter (10 μM, far above [$Ca^{2+}$]cyto) argued against physiological transport. Breakthrough studies published in the 1990s, however, relaunched the interest for the study of mitochondrial $Ca^{2+}$ transport, culminating in the recent discovery of the molecular effectors. We will discuss some of these molecular $Ca^{2+}$ transporters and regulators.

## 2.2 The Mitochondrial Calcium Uniporter (MCU)

The development of genetically encoded $Ca^{2+}$ probes (aequorin, GFP-derivatives) has been crucial for the $Ca^{2+}$ research field. It is possible to target them to specific compartments, such as mitochondria, to monitor the local $Ca^{2+}$ concentration.[2] It was detected that agonist-induced ER $Ca^{2+}$ release was followed by a transient increase of mitochondrial $Ca^{2+}$, reaching up to 100 μM in some cell types.[3] Thus, mitochondria actually were able to uptake $Ca^{2+}$ in physiological conditions, most likely through an electrogenic process. Fluorescence microscopy studies further demonstrated that mitochondrial $Ca^{2+}$ uptake was localized mainly in specific areas where ER and mitochondria were extremely close.[4] Through analogy with the synaptic cleft, it was suggested that these regions of close proximity represented hotspots where local $Ca^{2+}$ concentration meets the affinity of the transporter. Together with the presence of a high mitochondrial electrical potential ($\Delta\Psi m$), this elevated local concentration might contribute to a highly efficient mitochondrial $Ca^{2+}$ uptake.

The physical nature of the transporter, however, remained unknown. Although $Ca^{2+}$ is small enough to cross the outer mitochondrial membrane (OMM) through the voltage dependent activated channel (VDAC, or mitochondrial porin), it cannot easily cross the ion-impermeable inner mitochondrial membrane (IMM). The thermodynamic transport mode and inhibitors had been identified long time ago, but it was only in 2011 that two teams independently identified what was named the MCU.[5,6]

The MCU is a 351 residues-protein located in the IMM, with two transmembrane spanning domains. Its topology is hairpin-like, with both N- and C-termini extremities localized in the mitochondrial matrix. A highly conserved loop (EYSWDIMEP) is exposed in the intermembrane space (IMS) and connects the transmembrane domains. The $D^5$ and $E^8$ in the loop interact with $Ca^{2+}$ and are required for transport. The active form of the MCU is presumably an oligomer of four or five subunits.[7,8] The MCU gene is ubiquitously expressed in human tissues, with higher levels detected in thyroid, skeletal muscle, and lung compared to liver, heart, or pancreas.[9]

MCU knockdown blunts mitochondrial $Ca^{2+}$ entry in numerous cell types, such as pancreatic and neuronal cells, and MCU KO mice have a dramatic decrease of mitochondrial $Ca^{2+}$.[10] The first experiments of general KO of MCU in mice, however, led to viable mice with no dramatic phenotypes.[11] Although these unexpected results suggest that a compensatory mechanism exists, it has not been identified yet. Furthermore, the genetic background of the mice used for KO experiments has an impact on the results. More recent experiments focused on the analysis of organ selective KO and KD and revealed MCU importance in different stresses.[12] The presence of additional mitochondrial $Ca^{2+}$ import systems, namely the protein Letm1, still is debated.[13,14] Besides $Ca^{2+}$ importing carriers, the IMM contains $Ca^{2+}$ efflux proteins. One well-identified is the protein NCLX, which mediates the exit of one $Ca^{2+}$ ion for the entry of three to four $Na^+$ ions.[9] Another protein is assumed to mediate $Ca^{2+}/H^+$ exchange, but its nature is still debated.

## 2.3 MCU Regulators or the Notion of MCU Complex

Additional biochemical and genetic analyses further demonstrated that what was first thought to be a homo-oligomer was a complex containing different proteins, with MCU as the core of the complex, and interacting proteins crucial for its regulation. We will discuss some of these bona fide additional components of what is now called the MCU complex.

### 2.3.1 *MCUb*

Bioinformatic analyses identified MCUb as a paralog of MCU; the primary sequences of both proteins share 50% homology and similar predicted topology. One striking difference is that MCUb lacks the $Ca^{2+}$-binding glutamic residue present in the IMS loop of MCU. Consequently, MCUb is a poor $Ca^{2+}$ transporter and has been established as an inhibitor of the MCU activity.[15] MCU and MCUb form heteromeric complexes that depend on their ratio. Accordingly, MCU activity is 30-fold higher in the heart than in the skeletal muscle where MCU:MCUb ratios are 40:1 and 3:1, respectively.[15, 16]

### 2.3.2 *The MICU Family: MICU1 and MICU2 (and MICU3)*

Before the molecular identification of MCU, other putative mitochondrial $Ca^{2+}$ transporters were proposed tentatively. One was MICU1, a 50 kDa IMS-located protein (first thought to be transmembrane), which contains an EF-hand motif that binds $Ca^{2+}$ ion.[17] A paralog of MICU1 was later identified and named MICU2.[18] MICU2 is a 45 kDa protein that also contains an EF-hand motif and is localized in the IMS. Both MICU1 and MICU2 are regulators of MCU with a still not fully understood model that could be summarized as follows: MICU1 is an activator of MCU and MICU2 is an inhibitor; MICU1 and MICU2 form a disulfide bond-linked hetero-complex that interacts with MCU and acts as a gatekeeper to avoid $Ca^{2+}$ transport when not required (low extracellular $Ca^{2+}$ concentration); and when external $Ca^{2+}$ concentration rises (e.g., opening of IP3R channels), $Ca^{2+}$ binding to the MICU1-MICU2 complex through their EF-hand motifs induces conformational changes that facilitate the opening of MCU and subsequent mitochondrial $Ca^{2+}$ import.[19, 20]

Another paralog, named MICU3, has been identified simultaneously with MICU2. Its expression seems specific to the central nervous system, and its function has not yet been identified. Among members of the MICU family, MICU1 has been found as the most frequently mutated in humans, associated with pathologies.[8]

### 2.3.3 *Essential MCU Regulator (EMRE)*

EMRE was identified by proteomic study as an interactor of MCU and MICU1/2.[21] EMRE is a small 10 kDa protein with one transmembrane helix and is specific to metazoans. The essential regulatory function of EMRE is illustrated by the fact that, in its absence, the over-expression of MCU is not sufficient to restore mitochondrial $Ca^{2+}$ uptake. Although not fully established yet, studies suggest that EMRE facilitates the interaction between the MICU1/2 complex and the MCU.[22]

## 2.4 Mitochondria, $Ca^{2+}$ and Modulation of Cellular Processes

### 2.4.1 *Microdomains of $Ca^{2+}$ Buffering and Signaling*

An obvious conceptual consequence of $Ca^{2+}$ import by mitochondria is (cytosolic) $Ca^{2+}$ buffering. Indeed, mitochondria contribute to the low [$Ca^{2+}$]cyto, especially when considering their proximity with ER release channels. [$Ca^{2+}$]cyto buffering might be important to prevent the excessive activation of $Ca^{2+}$ regulated proteins. For example, in polarized pancreatic acinar cells, mitochondrial $Ca^{2+}$ uptake close to the apical region is necessary to avoid $Ca^{2+}$ diffusion toward basolateral regions, thus restricting granules secretion.[23] Another consequence of this buffering lies in the regulation of $Ca^{2+}$ release channels, namely IP3R receptors, that are both negatively and positively regulated by $Ca^{2+}$. Data suggested that mitochondrial $Ca^{2+}$ uptake was important for modulating IP3R-dependent $Ca^{2+}$ waves.[24–26]

### 2.4.2 *Regulation of Mitochondrial Dehydrogenases*

One of the first functions attributed to mitochondrial $Ca^{2+}$ was the regulation of mitochondrial enzymes. The activity of dehydrogenases, such as the pyruvate dehydrogenase and enzymes of the tricarboxylic (TCA) cycle such as α-ketoglutarate dehydrogenase or isocitrate dehydrogenase, is regulated positively by $Ca^{2+}$. These dehydrogenases generate NADH that fuels the respiratory chain. Therefore, it was hypothesized that $Ca^{2+}$ influx led to a stimulation of the respiratory chain and an increase of ATP production.[27] The demonstration has been made in different cell types. For example, in pancreatic β cells, the increase of ATP synthesis following $Ca^{2+}$ stimulation of mitochondrial dehydrogenases is important for insulin secretion.[28, 29]

## 2.5 $Ca^{2+}$ Transport and Cell Death

Although [$Ca^{2+}$]mito can stimulate energetic metabolism, it is also a regulator of cell death. Mitochondria are key players in cell death processes such as apoptosis and necrosis (see Part 6). Mitochondrial $Ca^{2+}$ overload induces the opening of the mitochondrial permeability transition pore (mPTP). The opening of mPTP allows the passage of molecules below 1500 Da, inducing a dramatic loss of $\Delta\Psi m$ and favoring mitochondrial swelling.[30] This impairment of mitochondrial integrity and fitness paves the way for a series of events leading to cell death.[31, 32]

In accordance with this model, MCU overexpression increased cell death, while MCU silencing reduced it.[33–35] The level of MCU regulators, such as MICU1, also influences stress-induced cell death.[36, 37] Although the first results suggesting that $Ca^{2+}$ modulates mitochondria integrity dates back to the 1980s, the complete knowledge of all the actors and mechanisms has not yet been reached. Thus, [$Ca^{2+}$]cyto also seems to be an important player, but how its interplay with [$Ca^{2+}$]mito affects $Ca^{2+}$-mediated cell death is still under investigation.

# 3 MITOCHONDRIA, REDOX SIGNALING, ROS AND OXIDATIVE STRESS

## 3.1 ROS as Signaling Molecules

Among mitochondrial redox events, the production of reactive oxygen species (ROS) is probably the best known by investigators outside the mitochondria research field. Indeed, mitochondria are widely recognized as the primary intracellular ROS sources. Main ROS produced by mitochondria are superoxide anion ($O_2^-$) and hydrogen peroxide ($H_2O_2$). As detailed in Chapter 1, ROS production is a side effect of mitochondrial electron transport chain activity. Basal mitochondrial ROS production is typically low, but can increase dramatically depending on the physiopathological status of the cell.

ROS reaction with biomolecules induces a wide spectrum of modifications. ROS-induced protein modifications have long been considered to be destructive events, but were later recognized as signaling means, not unlike phosphorylation or ubiquitination. The concept of oxidative eustress versus oxidative stress is crucial in apprehending the ROS signaling field.[38, 39] Intracellular ROS outcomes often are described as consequences of oxidative stress that occurs under conditions where the ROS burden leads to irreversible and detrimental effects on cell survival. These conditions are reached through experiments involving the extracellular addition of elevated concentration of molecules such as $H_2O_2$ or drugs triggering $H_2O_2$ production. ROS signaling, however, occurs under conditions where ROS concentrations are bearable by the cell. This physiological ROS signaling has been defined as oxidative eustress and refers to states in which the cell deals with the stress by adapting permanently or temporarily. Thus, ROS signaling has been shown to regulate activity of transcription factors, phosphatases, receptors, and events such as cell proliferation or differentiation. Among ROS, $H_2O_2$ is considered the best suited for signaling, because of its low reactivity compared with free radicals and its capacity to be transported through membranes by aquaporins and peroxiporins.[40–42]

## 3.2 Mitochondrial ROS and Signaling: Examples

Although many studies link mitochondria to oxidative stress and cell death, only a few actually demonstrate the involvement of mitochondrial ROS in those processes. This is because of the lack of ROS probes trustworthy for whole cell measurements and the ability to discriminate mitochondrial ROS from other sources. Consequently, most comprehensive studies relied on the reversion of phenotypes by the overexpression of mitochondrial ROS scavenging enzymes.

Kamata et al. reported that c-Jun N-terminal kinase (JNK) inactivation relied on phosphatases inhibited by $H_2O_2$. Expression of the mitochondrial superoxide dismutase 2 (SOD2) was crucial for controlling the $H_2O_2$ level and, therefore, JNK and cell survival,[43] making $H_2O_2$ a negative regulator of JNK-induced cell death.

Many proteins containing structural disulfide bonds transit through the ER before being secreted or addressed to the PM. Yang et al. showed that mitochondrial ROS modulate the disulfide proteome (i.e., the formation of disulfide bonds in proteins).[44] They were important for the adequate PM localization of ligand receptors such as the insulin-like growth factor 1 receptor (IGF1R). Consequently, decrease of mitochondrial ROS reduced insulin-derived signaling.

Hyperglycemia increases intracellular ROS levels.[45, 46] The impairment of the regulatory capacity of the mitochondrial respiratory chain leads to IMM hyperpolarization, favoring ROS production.[47, 48] Diabetes is characterized by impairments in the control of glucose and insulin homeostasis. In type 2 diabetes, alterations in mitochondrial density and functions were observed in the skeletal muscles.[49–51] Transcriptomic analysis indicated a lower expression of genes encoding proteins of the OXPHOS system.[52] In hyperglycemic and hyperlipidemic mice, an increase in muscle ROS production, associated with mitochondrial alterations and a decrease in the expression of genes encoding mitochondrial proteins, have been shown, and the antioxidant N-acetylcystein (NAC) restored mitochondrial density and structure.[53] This suggested that mitochondrial dysfunction results from hyperglycemia and hyperlipidemia-induced mitochondrial ROS.

The emergence of $H_2O_2$ specific probes targetable to specific subcellular compartments provided useful tools for mitochondrial ROS signaling studies.[54, 55] For example, the $H_2O_2$ specific fluorescent probe HyPer was used in a study by Hajnoczky's team.[56] As indicated earlier, regions of ER-mitochondria close contacts are important for ER/mitochondria $Ca^{2+}$-transfer. By targeting HyPer to the ER-mitochondria interface, they observed that IP3R-dependent $Ca^{2+}$ release was accompanied by mitochondrial $H_2O_2$ release at the interface between both organelles. That spatiotemporal $H_2O_2$ efflux was the consequence of a $Ca^{2+}$-induced $K^+$ and water influx into the mitochondrial matrix, resulting in mitochondrial matrix swelling, thus leading to the release of their $H_2O_2$ content.

Although the physiological meanings of these $H_2O_2$ nanodomains still are being investigated, they are presumed to set up a positive feedback, potentiating $Ca^{2+}$ release through the oxidation of cytosolic cystein residues of IP3Rs.[57, 58] This local efflux of $H_2O_2$ also might modulate oxidative folding in the ER. Analyzing how these $H_2O_2$ nanodomains behave during $Ca^{2+}$ overload-induced cell death is also of interest.

This result illustrates the notion of microenvironment and ROS signaling. Cells contain many nonenzymatic and enzymatic antioxidants, limiting the ability of mitochondrial ROS to reach a distant target. ROS signaling processes likely occur locally in adequate microenvironments. They also might rely on oxidative relay, in which intermediates would be oxidized before the ultimate targets were oxidized.[59] Although illustrations of these different processes have been documented, the molecular supports of mitochondrial ROS signaling still remain largely unknown and require further identification.

## 4 MITOCHONDRIA METABOLITES AS REGULATING AGENTS

As metabolite hubs, mitochondria fuel numerous metabolic reactions with precursors and cosubstrates. A typical example is the TCA cycle, which generates reduced substrates for OXPHOS, and precursors for amino acids, fatty acids, and heme biosynthetic pathways. Besides this metabolic function, some intermediates of mitochondrial metabolism uncover signaling properties, both inside and outside mitochondria.

### 4.1 Acetyl-CoA (AcCoA)

AcCoA is generated from the oxidation of the glycolysis end-product pyruvate catalyzed by pyruvate dehydrogenase and by the β-oxidation of fatty acids. AcCoA is a substrate of citrate synthase in the TCA cycle, and the precursor of sterol and fatty acid biosynthesis.

Furthermore, AcCoA is the acetyl donor for protein acetylation. This posttranslational modification (PTM) consists in the attachment of an acetyl group to lysine residues or the N-termini of proteins, catalyzed by K and Nt-acetyl transferases, respectively.[60] Among the identified acetylation targets, there are histones and metabolic enzymes,[61] the tumor suppressor p53,[62, 63] or the pro-apoptotic protein Bax.[64] Contrary to Nt-acetylation, K-acetylation is reversible, as acetyl is removed by deacetylases. Some histone deacetylases (e.g., HDAC2, 5, and 6) have been described as regulators of insulin signaling,[65–67] making AcCoA a key player in diabetes and obesity.

## 4.2 Citrate

Citrate results from the reaction between AcCoA and oxaloacetate. Citrate is a well-known allosteric inhibitor of the glycolytic enzyme phosphofructokinase. It is also an inhibitor of PDH and succinate dehydrogenase (SDH), which both are involved in the TCA cycle.[68] Citrate also seems important for the stimulation of insulin secretion. This effect seems indirect, however, and would require subsequent metabolic transformation of citrate in the cytosol.[69]

## 4.3 2-Oxoglutarate (2OG; α-Ketoglutarate)

The 2OG is crucial in nitrogen metabolism, and also is important for epigenetic regulations. The 2OG is the cosubstrate of 2OG-dependent dioxygenases (2OGDD) that are involved in histone and DNA regulation, thus regulating genome expression in mammals.[70, 71] The 2OGDDs also are involved in protein hydroxylation (i.e., addition of hydroxyl group to residues such as histidine).[72] The regulation of 2OGDD by 2OG has been linked to the pluripotency of stem cells and collagen maturation.[73, 74]

## 4.4 Succinyl-coA (ScoA)

ScoA is produced via the enzyme 2OG dehydogenase using 2OG as substrate. ScoA is the precursor of heme biosynthesis and is also the donor of a recently discovered PTM named protein succinylation.[75, 76] Although the full consequences of succinylation still need to be understood, it is known that it is reversible, and members of the histone deacetylase sirtuin family (SIRT5, SIRT7) have been identified as desuccinylases.[77, 78]

## 4.5 Succinate

Succinate inhibits 2OGDD, counteracting the regulatory roles of 2OG described previously. Consequently, succinate accumulation leads to strong epigenetic modifications and degradation of some transcription factors under the regulation of 2OGDD.[79]

Succinate also has emerged as an extracellular regulator. Succinate can be detected in extracellular fluids and a succinate receptor (named GPR91 or SUCNR1) has been identified and characterized.[80] Succinate binding to GPR91 is associated with various processes such as retinal angiogenesis or platelets activation. Succinate accumulation in kidneys was observed in animal models of diabetes and metabolic diseases. Activation of GPR91 by succinate stimulates renin secretion, and so succinate could be a contributor to diabetes-induced hypertension and to nephropathy in metabolic complications linked to obesity.[81, 82]

### 4.6 Fumarate

Fumarate is the product of succinate oxidation in the TCA cycle, but is also produced through tyrosine breakdown and urea cycle. Like succinate, fumarate inhibits 2OGDD. Fumarate also is involved in a PTM named succination, which is the irreversible formation of S-2-succino-cysteine, resulting from the nonenzymatic reaction of fumarate with cysteine residues.[83] Because it modifies cysteins, succination often targets redox reactive proteins. Succination induced the aberrant activation of Keap1/Nrf2 antioxidant response pathway. Increased succination of many skeletal muscles and adipocyte proteins has been observed in diabetes mice models.[84] Recent works suggest that the succination of ER chaperones could be the link between mitochondrial and ER-stress observed in diabetes.[85]

## 5 MITOCHONDRIA AND QUALITY CONTROL

### 5.1 Ubiquitin: A Dual Role in the Mitochondria Quality Control

Accumulation of mitochondrial damages contributes to numerous diseases, and quality control of mitochondria is a critical issue to maintain cellular functions. The first line of defense operates at the molecular level. Cells have a large panel of chaperones and proteases dedicated to mitochondria functioning, located in the cytosol or in the different mitochondrial subcompartments. Some are involved in mitochondrial import regulation. The cytosolic chaperones Hsp70 or Hsp90 protect nascent precursor peptides before their mitochondrial translocation, where matrix chaperones take over. Other proteins have a role in protein maturation, respiratory chain complexes assembly, or misfolded protein degradation.[86] When mitochondria damages overstep the capacities of these controls, however, whole organelles might be degraded. For this, damaged mitochondria must be recognized as such by the presence of tags on both OMM and IMM: ubiquitin or mitophagy receptors.

### 5.2 Role of Ubiquitin-Proteasome System (UPS)

Ubiquitin is a 76 amino-acid polypeptide that generally tags other proteins, including mitochondrial proteins, for degradation, making ubiquitin a major player of mitochondria quality control. The proteasome is a large machinery responsible for the degradation of short-lived, misfolded, or damaged proteins after their ubiquitination. Ubiquitin is linked to protein as a monomer or as a polyubiquitin chain, orchestrated by ubiquitin ligases. They create an isopeptide bond between a lysine residue of a target protein and the C-terminal glycine of ubiquitin.[87] OMM ubiquitin ligases, such as MULAN, MARCHV/MITOL, and Mdm30 ubiquitinate proteins involved in mitochondrial dynamics, tagging them for proteasomal degradation, thus affecting mitochondria morphology.[88] In mammalian cells, the UPS regulates the quality of mitochondrial proteins located in other subcompartments such as oligomycin sensitivity-conferring protein (OSCP, mitochondrial matrix),[89] Endonuclease G (IMS),[90] and UCP2 and UCP3 (IMM).[91] In *Saccharomyces cerevisiae*, Lehmann et al. showed the presence of ubiquitinated proteins in mitochondrial matrix and characterized the Dma1p ubiquitin ligase.[92, 93]

## 5.3 Role of Ubiquitin in Mitophagy

Chaperones, proteases, and the UPS act together to maintain mitochondrial proteostasis. When mitochondrial damage is too important or becomes irreversible because fissioned mitochondria have lost their membrane potential, however, cells have developed a process called mitophagy, a specific form of autophagy. Autophagy is a conserved mechanism of intracellular degradation. It involves a specific molecular machinery and degradation organelles (vacuoles in fungi and plants, lysosomes in mammals). Autophagy was described for the first time around the mid-20th century.[94] Three types of autophagy have been described, depending on the mechanism by which intracellular material is delivered to lysosome/vacuole for degradation: macroautophagy, microautophagy, and chaperone-mediated autophagy (CMA). Macroautophagy is the most prevalent form (and is referred to as autophagy hereafter). It involves de novo formation of double phospholipid membranes called phagophores that elongate, close, and form autophagosome vesicles containing sequestered portions of the cytosol and organelles. Mature autophagosomes merge with vacuole/lysosomes, and the cargo is degraded by hydrolytic enzymes. Degradation products are re-exported to the cytosol and reused by the cell.[95] During microautophagy, cytosolic material is sequestered directly through the invagination of vacuolar/lysosomal membrane. During CMA, mammalian proteins flagged by a pentapeptide motif (KFERQ) are selectively degraded through direct translocation into lysosomes.[96]

The term mitophagy was coined to account for the nonrandom nature of mitochondrial autophagy,[97] coming from a work on *S. cerevisiae*, subjected to nitrogen starvation after being grown on a nonfermentable carbon source.[98] Numerous key proteins involved either in autophagy core machinery or regulation were characterized first in yeast, making it a powerful tool to study both selective and nonselective autophagy.[99] In mammals, Kim et al. observed sequestered mitochondria in rat hepatocytes during nutrient starvation or after laser-induced damage.[100] Since 2008, interest in mitophagy has increased constantly with an emphasis on its role in mitochondria quality control and its link with pathologies such as Parkinson's disease, Alzheimer's disease, and cancers.[101, 102]

In mammalian cells, the PINK/PARKIN pathway is a major mitophagy processes that involves ubiquitin-tagging of damaged mitochondria. PINK and PARKIN are mutated in some familial forms of Parkinson's disease.[103] In *Drosophila*, both proteins work together to mediate mitophagy.[104] PINK1 is a mitochondrially targeted serine/threonine kinase, but its turnover is rapid because of its constitutive degradation by IMM proteases, reflecting mitochondria fitness.[105] When mitochondria are damaged, PINK is stabilized at the OMM following its association with the TOM complex, which allows the recruitment of PARKIN, an E3 ubiquitin ligase that normally resides in the cytosol. This is the first step of the mitophagy cascade. PINK1 phosphorylates specific proteins including PARKIN and ubiquitin.[106] Phosphorylated PARKIN performs different types of lysine ubiquitination (K27, K48, K63) on several mitochondrial proteins, including VDAC1, Mfn1, Mfn2, and MIRO.[107] K48 ubiquitination is associated with proteasomal degradation; that of K63 and K27 is associated with mitophagy (Fig. 1).

These processes are interconnected and different scenarios have been proposed.[107–110] The presence of high levels of ubiquitin on the mitochondrial surface can induce mitophagy.[111] The following steps in the mitophagy process require the polyubiquitin chains recognition

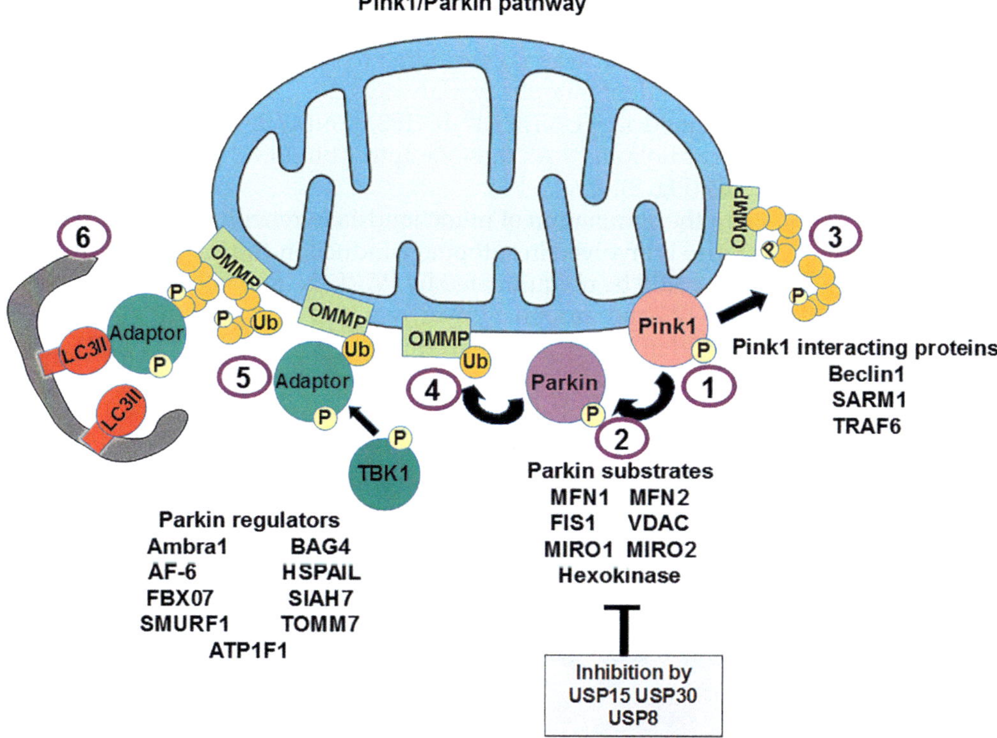


FIG. 1 PINK1/PARKIN dependent mitophagy. When mitophagy is induced, PINK1 is stabilized to the outer mitochondrial membrane and activated (1). PINK1 phosphorylates both PARKIN (2) and ubiquitin on conserved residue Ser65 (3), a residue conserved in both PARKIN and ubiquitin proteins (2). OMM proteins that already are ubiquitinated are phosphorylated by PINK1 on their ubiquitin chains (3), leading to PARKIN recruitment and activation. Alternatively, nonubiquitinated OMM proteins are ubiquinated by mitochondria-recruited and activated PARKIN (4). Adaptors then are recruited to these highly ubiquitinated mitochondrial areas (5). This recruitment could be activated by TBK1. Finally, the interaction between adaptor and LC3 allows the engulfment of mitochondria within the phagophore (6).

by different adaptor proteins. Optineurin (OPTN) and nuclear dot protein 52 kDa (NDP52) might be the two preferential adaptors required for the PINK1/PARKIN pathway.[112] In addition, the Tank-binding kinase (TBK1) could be a signal amplifier of mitophagy because it mediated phosphorylation of adaptors, expanding their binding capacity to diverse ubiquitin chains.[113] These adaptors possess a LIR domain (LC3 interacting region) allowing the interaction with the MAP1LC3B/LC3B (microtubule associated protein1 light chain 3beta) family proteins present in the autophagosomal membranes. This leads to the engulfment of damaged mitochondria by autophagosomes and subsequent lysosomal degradation. Other E3 ligases could substitute PARKIN in the proceedings of mitophagy: Levine's group has identified SMAD-specific E3 ubiquitin protein ligase 1 (SMURF1) as a mitophagy promoter.[114] ARIH1, belonging to the same E3 ligases family as PARKIN, is overexpressed in cancer cell lines in which it is the main regulator of mitophagy.[115]

## 5.4 Mitochondrial Receptors Involved in Mitophagy

Although the PINK1/PARKIN pathway is considered to be the central regulator of mitophagy, PINK1/PARKIN-independent mitophagy pathways exist, mediated by receptors constitutively localized to mitochondria: FUNDC1,[116] BNIP3,[117] NIX/BNIP3L,[118, 119] BCL2L13,[120] prohibitin 2 (PBH2),[121] and cardiolipin.[122] All these receptors bind LC3 proteins through their LC3-interacting region (LIR) (Fig. 2).

NIX/BNIPL3 is crucial in the elimination of mitochondria in reticulocytes during erythrocyte maturation.[123, 124] NIX also is involved in autophagy induction and PARKIN translocation to mitochondria.[125] NIX also could be ubiquitinated by PARKIN and induce mitophagy.[126]

The first link between BNIP3 and autophagy was reported in malignant glioma cells.[127] The BNIP3 receptor is involved in hypoxia-induced mitophagy,[128] such as FUNDC1, another MOM receptor.[116] Both BNIP3 and FUNDC1 are dephosphorylated and phosphorylated upon membrane potential loss and hypoxia respectively, resulting in an increase of LC3 affinity but in a different way. Phosphorylation of BNIP3 is required to interact with LC3. FUNDC1 is regulated by two different pathways. The kinase CK2 phosphorylates FUNDC1 on Ser13, and the dephosphorylation by the phosphatase PGAM5 increased LC3 interaction.[129] A second

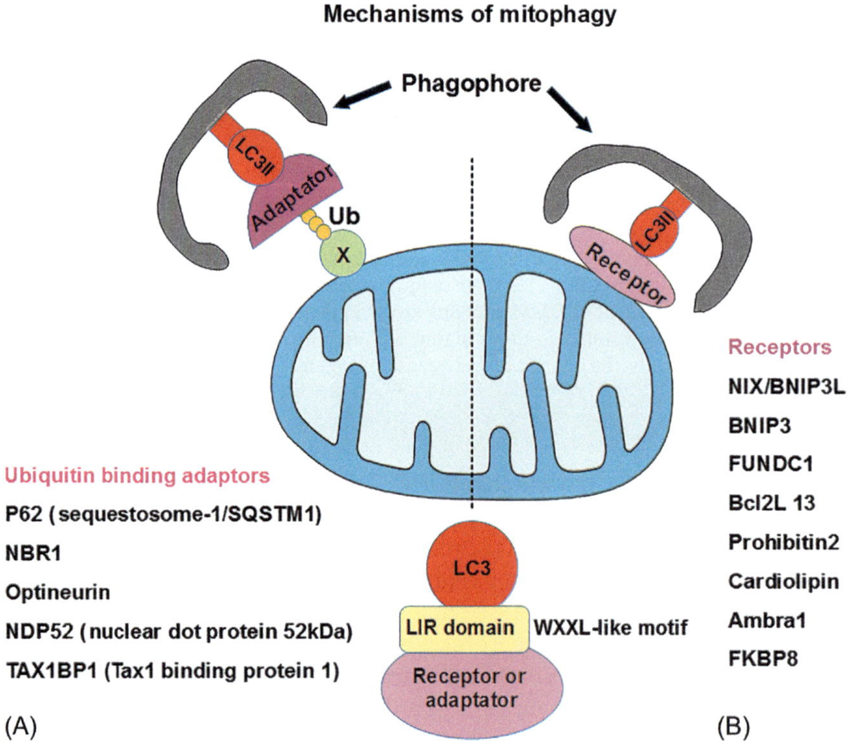

FIG. 2 Mitophagy mechanisms. Mitophagy occurs in two ways: (A) ubiquitinated outer mitochondrial membrane proteins are recognized by adaptor proteins, that interact with phagophore-inserted LC3; (B) receptors localized in the outer or inner mitochondrial membranes interact with LC3, allowing the recruitment of mitochondria in the phagophore.

pathway involves the kinase ULK1 that is recruited to fragmented mitochondria under hypoxia and phosphorylates FUNDC1 on Ser17, increasing the interaction with LC3.[130]

The role of BCL2L13 in mitophagy was highlighted recently. This Bcl-2 family member might be the mammalian ortholog of the yeast mitophagy receptor Atg32p. The ectopic expression of BCL2L13 in the atg32Δ yeast mutant rescued the mitophagy defect.[120]

Strappazzon et al. have shown that pro-autophagic protein AMBRA1 interacted with LC3 protein and promoted mitophagy in PARKIN-KD cell lines.[131] Wei et al. identified the IMM protein prohibitin 2 (PHB2) as a receptor involved in mitochondrial targeting for autophagic degradation.[121] PHB2 binds LC3 through its LIR domain upon mitochondrial depolarization and proteasome-dependent OMM rupture. PHB2 is required for PARKIN-induced mitophagy in mammalian cells and for the clearance of paternal mitochondria after embryonic fertilization in *Caenorhabditis elegans*.

The role of cardiolipin in mitophagy was characterized in neuronal cells.[122] Cardiolipin is a phospholipid located in mitochondria exclusively, mainly in the IMM. After treatments with mitophagy inducers, cardiolipin is translocated to the OMM and interacts with LC3.

The diversity of mitophagy receptors highlights the intricacy of mitochondrial degradation and shows that mitophagy is not a linear process. Possible cross-talks between ubiquitin and receptors-mediated mitophagy were observed. BNIP3 interacts with PINK1 to inhibit its proteolytic degradation.[132] NIX might provide an alternative mitophagy pathway that could restore mitochondrial function in the setting of PINK1/PARKIN deficiency.[133] Both BNIP3/NIX-mediated and PINK1/Parkin-mediated mitophagy are involved in the normal turnover of myocardium mitochondria.[134–136] It was suggested that BNIP3L/NIX and PARK2 function in independent pathways to mediate mitophagy in ischemic brains and that BNIP3L could be involved in a pathway that is induced after PARK2 is exhausted during ischemia-reperfusion.[137] A cross-talk between mitophagy and apoptosis also is proposed, because both NIX and BNIP3 also regulate cell death.[118, 119]

## 5.5 Mitophagy Inducers

The molecular cascade underlying mitophagy was studied with inducers of mitochondrial damages, mostly on mammalian cell cultures. Among them, the protonophoric uncouplers carbonyl cyanide m-chlorophenyl hydrazone (CCCP), carbonyl cyanide-p-(trifluoromethoxy)phenylhydrazone (FCCP), and 2,4-dinitrophenol (DNP) have been used extensively to decipher the PINK/PARKIN pathway. Potassium ionophores (valinomycin, salinomycin) also have been used. These toxins, however, are detrimental to the whole mitochondria population and mediate off-target effects through acting on other membranes. More specific OXPHOS inhibitors (rotenome, antimycin A, oligomycin), and iron chelators (deferiprone, 1,10′-phenathroline) also induce mitophagy.[138, 139]

The accumulation of misfolded proteins in the mitochondrial matrix promotes mitophagy through the PINK1/PARKIN pathway without mitochondria depolarization.[140] More physiologically, nutrient starvation stimulates mitophagy on hepatocytes,[141] and the depolarization of individual mitochondria selectively triggers them into degradation by mitophagy.[100, 142]

Considering the role of ROS in signaling pathways, and their detrimental effect on proteins, lipids, and DNA, their impact on autophagy has been investigated. ROS modify a cysteine residue near the catalytic site of Atg4, required for the core machinery of autophagy and mitophagy.[143] Chen et al. reported that superoxide ion was the major ROS regulating autophagy.[144] The selenite treatment of various glioma cells induces a nonapoptotic cell death accompanied by excessive mitophagy, in which superoxide plays a key role.[145] Xiao et al. highlighted the role of superoxide in driving the progression of PARKIN/PINK1-dependent mitophagy following translocation of PARKIN to mitochondria.[146]

# 6 MITOCHONDRIA AS A MEDIATOR IN APOPTOSIS

Apoptosis is the main form of regulated cell death in mammals. It is involved in development, morphogenesis, immune response, and elimination of potentially dangerous cells. The extrinsic pathway is activated by various PM death receptors such as FAS or TNFα receptors. These receptors activate caspases, a group of proteases responsible for the degradation of a range of substrates, initiating the apoptotic characteristics of the cell. The extrinsic pathway does not necessarily require mitochondria, but they might be involved through the caspase 8-dependent activation of the protein Bid.[147] The intrinsic pathway is initiated by intracellular signals, such as DNA alterations during anticancer therapies. It involves the activation of the transcription factor p53 that regulates genes involved in the apoptotic response, among which the Bcl-2 family members, that directly acts on mitochondria.[148] The intrinsic pathway, therefore, often is called the mitochondrial pathway.

The central role of mitochondria in the intrinsic pathway has been documented since the 1990s,[149, 150] culminating with the demonstration that they are the main target of the Bcl-2 family. Following apoptotic induction, pro-apoptotic proteins Bax and Bak oligomerize and form a large pore that permeabilizes the OMM.[151, 152] This pore promotes the release of several proteins from the IMS to the cytosol.[153] These proteins, together known as apoptogenic factors, include Cytochrome c, Smac/Diablo, HtrA2/omi, endonuclease G, and AIF. They participate in the caspases activation and to the degradation of nuclear DNA.[154] Anti-apoptotic proteins Bcl-2, Bcl-xL, or Mcl-1 inhibit the action of Bax/Bak by acting on the steps leading to their activation, including mitochondrial localization, insertion, and oligomerization[155] (Fig. 3).

Whether mitochondria play an active role in apoptosis signaling, or are a passive target of Bcl-2 family members, is an ongoing discussion.

## 6.1 Pore Formation

The most striking event of mitochondria-dependent apoptosis is the formation of the OMM pore. Bax oligomerization and pore formation are simultaneous.[156] There is now no doubt that Bax (or the closely related Bak) is a central component of the pore.[157] Its formation and arrangement, the involvement of other components, and the regulation of the whole system, however, remain debated.

Inactive Bax is a globular protein organized as 9 α-helices.[158] α9 is hydrophobic and resembles its counterparts in anti-apoptotic proteins Bcl-2/Bcl-xL, which are required for mitochondrial localization and insertion,[159] but the deletion of Bax-α9 does not preclude permeabilization of artificial membranes and mitochondria.[160] Switching C-terminal helices between Bax and Bcl-xL demonstrated that Bax-α9 does not drive membrane insertion of

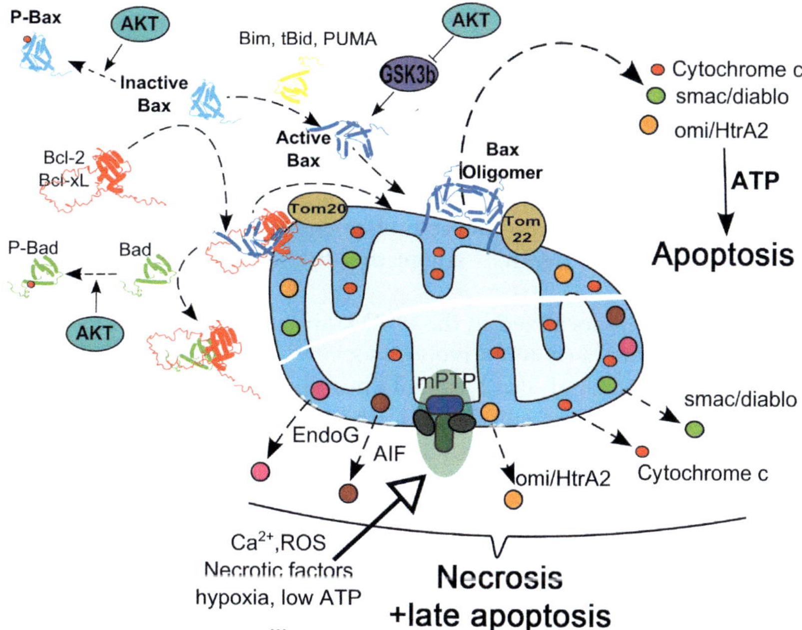


FIG. 3 Mitochondria in cell deaths. Following an apoptotic signal, Bax is activated directly through a transient interaction with BH3-only proteins Bim, tBid, and Puma. It then is relocated to the OMM, where it is oligomerized and forms a large-size pore that allows the release of apoptogenic factors such as cytochrome c, smac/diablo, and omi/HtrA2. This process is helped by the interaction of Bax with Tom22, and by its phosphorylation by GSK3β on Ser163. Alternatively, primed Bax is released from its interaction with anti-apoptotic proteins Bcl-2 and Bcl-xL by BH3-only protein Bad. This process is regulated by Tom20, which contributes to the mitochondrial localization of Bcl-2, and thus of the complex Bcl-2/Bax, and by AKT, which phosphorylates and inhibits Bad. Akt also regulates the whole apoptotic process by phosphorylating Bax and GSK3β. Following a necrotic event, mPTP opening induces a swelling of mitochondria. Subsequent alterations of the OMM promote the release of the same apoptogenic factors and additional ones such as AIF and endonuclease G. Note that a decrease of ATP concentration converts apoptosis to necrosis.

another Bcl-2 family member, contrary to its Bcl-xL counterpart.[161] Bax contains two amphipathic helices α5 and α6, forming a hairpin, reminiscent of membrane-permeabilizing bacterial toxins.[162] For a long time, it has been considered that membrane-inserted α5-α6 hairpin was the pore unit, with 10–12 units needed to reach a size large enough.[163] The structure resolution of an activated Bax dimer, however, revealed that α5-α6 helices were actually laid out top to tail with α5-α6 helices of a facing monomer.[164] Recent hypotheses propose that Bax dimers lie at the membrane surface and tilt through a still-unknown process.[165] Imaging data have confirmed that Bax forms a ring-shaped hole in both mitochondrial and artificial membranes.[166,167] Conductance recordings suggest that the pore is formed in a stepwise manner[163] and, once initiated, the whole process can be autocatalyzed.[168]

Because purified Bax is able to permeabilize artificial lipid bilayers, the possible role of resident mitochondrial proteins often has been overlooked, but additional regulations might take place.

## 6.2 OMM Components

The mPTP is a large $Ca^{2+}$-regulated pore, of which the opening has been associated with cell death.[169] For a long time, Bax-induced mPTP activation has been considered as causal in the release of apoptogenic factors. This relied on physical interactions between Bax and putative mPTP components, namely VDAC and the adenine nucleotides translocator. The deletions of these proteins, however, did not affect mPTP.[170, 171] The deletion of cyclophilin D, a regulator of mPTP, had no consequence on Bax-dependent cell death, but impaired ischemia/reperfusion-induced cell death,[172] suggesting that mPTP was involved in necrosis but not in apoptosis.

Bcl-2 family members are localized in the OMM, and possibly interact with proteins from this membrane. VDAC has been tentatively associated with apoptosis, both as a putative component of mPTP, and through its capacity to modulate OMM permeability. The maximal size of VDAC, however, remains smaller than the size of apoptogenic factors. Fully open VDAC can let molecules with a size of ~3 kDa go through,[59] while the smallest apoptogenic factor, cytochrome c, is 12.5 kDa.

Subunits of the mitochondrial proteins import machinery, namely TOM and SAM, are major components of the OMM.[173] Physical interactions have been measured between Bcl-2 and Tom20[174] and between Bax and Tom22.[175] Tom20 is directly involved in Bcl-2 sorting to mitochondria, through a pathway distinct from the general import pathway. Tom22 is not required for Bax mitochondrial localization, but could help Bax to acquire its active conformation.[176]

SAM is involved in the mitochondrial import of OMM proteins. Bak interacts with two SAM components, Metaxins 1 and 2,[177] and with VDAC2, an isoform of VDAC1.[178] VDAC2 might be a mitochondrial receptor for Bak, and the interaction with metaxin 1 might be involved in the formation of a Bak-formed pore.[179]

Besides Bcl-2 family members, other regulators, such as nuclear proteins p53 and Histone 1.2, are partly translocated to mitochondria during apoptosis. They might modulate Bax/Bak function, but mitochondrial p53 also might regulate mtDNA stability and transcription.[180]

## 6.3 ROS Signaling and $Ca^{2+}$ in Cell Death

The mPTP is a major player in necrosis and is regulated by both ROS and $Ca^{2+}$; ROS decreases the threshold $Ca^{2+}$ concentration triggering mPTP.[181] The fact that mPTP was considered to be a major apoptosis regulator led to the widely spread belief that ROS and $Ca^{2+}$ were apoptosis regulators. Demonstrations that mPTP was crucial for necrosis, rather than apoptosis, however, questioned their actual roles. Both could help the formation of tBid-regulated Bax/Bak-pores by acting on cardiolipins[182, 183] and/or ceramides.[184] Although the process is unclear, $Ca^{2+}$ could stabilize OMM/IMM contact sites[185] and/or stabilize the pore itself.[186] ROS might contribute to membrane disruption through their action on unsaturated acyl chains and subsequent formation of lysophospholipids that would help tBid action.[187] Lysophospholipid formation also would decrease the affinity of cytochrome c for IMM,[188] favoring its release.

Two apoptogenic factors, cytochrome c and AIF, are redox proteins that potentially could modulate cytosolic redox potential and affect the oxidation of components such as membrane lipids.[189, 190]

Bax has two cystein residues making it a possible target of ROS signaling. The oxidation of Cys62 is involved in the apoptotic response to $H_2O_2$[191]; and Cys126 is crucial to apoptosis activation by prostaglandin E2.[192]

## 6.4 Kinases-Dependent Signaling

Death and survival signaling pathways involve kinases and phosphatases. A typical example is provided by Bad, a BH3-only protein that binds to Bcl-2, preventing its inhibition of Bax.[193] Bad is phosphorylated by AKT, which is activated through different surviving signals.[194] Bad phosphorylation decreases its affinity to Bcl-2. In dying cells, AKT function is limited, Bad is less phosphorylated and can interact with Bcl-2, preventing its anti-apoptotic function.[195] Bad phosphorylation also regulates the balance between oxidative glycolysis and gluconeogenesis in hepatocytes through its association to a glucokinase-containing complex loosely bound to mitochondria.[196, 197]

AKT also regulates Bax directly. AKT-dependent phosphorylation of Bax-Ser184 prevents its mitochondrial relocalization[198] and modulates its regulation by Bcl-xL.[199] Ser184 is located in the hydrophobic helix α9; its phosphorylation opens Bax structure, making it more sensitive to proteolytic degradation, unless it is protected through its interaction with Bcl-xL. AKT also negatively regulates other protein-kinases such as GSK3β, involved in death pathways. Conversely, GSK3β activates apoptosis by phosphorylating Bax-S163, favoring Bax mitochondrial localization.[200]

Bax phosphorylation and mitochondrial localization are linked.[201] Beyond Ser163 and Ser184, other Bax residues contribute to regulate conformational changes leading to Bax activation. For example, inhibition of the $p38^{MAPK}$-dependent phosphorylation of an unknown residue is responsible for the absence of Bax oligomerization after its mitochondrial insertion during anoikis.[202] Anti-apoptotic proteins Bcl-2 and Bcl-xL also have potential phosphorylation sites by PKA and PKC-types kinases, although it has not been demonstrated univocally that they are responsible for significant regulation. The dephosphorylation of Bax and other anti-apoptotic proteins also might be a control point for the regulation of their function, such as Bax activation by phosphatase PP2A.[203]

Kinases and phosphatases regulating Bcl-2 family members have many substrates and, therefore, are not adequate targets to modulate their functions. Molecules that specifically interact with a Bax domain, including phosphorylatable Ser184, however, efficiently inhibit Bax phosphorylation, thereby helping Bax activation and inhibiting tumor growth.[204]

Both AKT and GSK3β have been partly localized in mitochondria under peculiar conditions.[205–207] Although the regulation of this localization is unknown, this opens interesting perspectives.

These kinases are involved in signaling pathways regulating cell metabolism. For example, AKT stimulates glucose transport and metabolism,[208] which might provide fuel to tumoral cells, but they are less able to convert energy through mitochondrial OXPHOS.

This leads us to ask if and how mitochondria metabolic status regulates cell death.

## 6.5 Mitochondria Bioenergetics and Apoptosis

Bioenergetic collapse associated with mPTP leads to necrosis, because it happens during ischemia/reperfusion. Apart from this extreme situation, apoptosis might be regulated by the metabolic status of mitochondria.

ATP is required for apoptotic cell death, because decreasing cellular ATP concentration while triggering apoptosis leads to the conversion to necrotic cell death.[209] It is unclear how ATP is involved. ATP/dATP stabilizes the tripartite APAF-1/Cytochrome c/Procaspase 9 complex leading to Caspase 9 activation.[210] ATP also is needed for the phosphorylation of several proteins such as Bax.

Physiological apoptosis usually occurs in post-mitotic cells that have a low bioenergetic activity, opposite to dividing or freshly confluent cells that are used in the literature. It is possible that the ATP-dependence of apoptosis in in vitro experiments does not reflect the in vivo status. Apoptosis, however, might be induced in dividing cells, such as tumoral cells targeted by chemotherapy. In this situation, the energetic status of the cell might be crucial for the efficiency of chemotherapy-induced apoptosis. Tumoral cells generally support a shift from oxidative to glycolytic metabolism; the understanding of how these alterations modulate apoptosis is a major issue. Although no exhaustive study has been done, several marginal observations are noteworthy. Rho0 cells deprived of mitochondrial DNA tended to be more resistant to chemotherapeutic agents-induced apoptosis than their Rho+ counterparts,[211] and exhibited necrotic rather than apoptotic characteristics,[212] which can be paralleled with the observation about ATP depletion.

Fragmentation of the mitochondrial network is an early event following apoptosis initiation.[213] The molecular mechanism is unknown, but it is prevented by a dominant-negative mutant of Drp1. Mitochondrial fission might help the release of apoptogenic factors; the inhibition of fission does not prevent Bax mitochondrial localization, but it does prevent cytochrome c release. The connection, however, remains unclear because cells still engage in apoptosis.[214]

Cytochrome c is largely in excess compared to respiratory complexes III and IV. Therefore, the release of a fraction of cytochrome c is not expected to alter electron transfer between these complexes. In addition, positively charged cytochrome c is partly bound to negatively charged phospholipids of IMM. The massive release of cytochrome c observed in cell cultures might affect electron transfer between complexes III and IV, but might not reflect the in vivo situation where the release of a small fraction of cytochrome c (and other apoptogenic factors) is sufficient to initiate apoptosis without affecting mitochondrial electron transfer. This parallels the idea that fully active OXPHOS is needed for the implementation of apoptosis.

## 7 MITOCHONDRIA AND THE INNATE IMMUNE RESPONSE

The innate immune response is an evolutionary conserved process that represents the first line of defense against pathogens or other harmful immune stimuli. Its initiation relies on pattern recognition receptors (PRRs), which recognized pathogen-associated molecular patterns (PAMPs). PRRs are expressed mainly by immune cells and are classified in different groups and subgroups. The obvious reason for PRR diversity lies in the diversity of their ligands; PAMPs might be lipoproteins, carbohydrates, or nucleic acids from protozoa, bacteria, fungi, or viruses. Furthermore, PRRs also differ by their subcellular localization, and they might initiate distinct signaling cascades.[215]

Mitochondria have emerged recently as major regulators of the innate immune response. Descriptions follow of two aspects of this feature illustrating how mitochondria detect and activate the innate immune response.

## 7.1 Mitochondria and the Antiviral Immunity

One family of PRRs is the retinoic acid-inducible gene I (RIG-I)-like receptors (RLRs). RLRs are cytosolic proteins expressed in immune and nonimmune cells and are able to bind cytosolic viral RNA and trigger signaling pathways. Four groups have independently identified the mitochondrial anti-viral signaling protein (MAVS, aka IPS1, CARDIF, VISA) and demonstrated that it is an important RLR adaptor molecule.[216–218] MAVS is a 540 amino-acid protein spanning OMM through its C-terminal transmembrane helix (amino acids 514–535), and the core of the protein is cytosolic. The N-terminal region contains a caspase recruiting domain (CARD). Two main RLRs family members (RIG-I, MDA5) also harbor a CARD through which they are able to interact with MAVS. This interaction is required for the induction of pro-inflammatory cytokine and type I interferons (IFNs) subsequent to the recognition of viral RNA.

As a component of antiviral defense, MAVS is the target of some viruses. Hepatitis C virus (HCV) encodes for a protease that cleaves MAVS, releasing it from the OMM, and allows HCV to evade RIG-I signaling, contributing to HCV pathogenicity.[219, 220] This emphasizes the importance of MAVS localization to mitochondria.[216] MAVS function requires its oligomerization and the recruitment of additional proteins, forming a MAVS signalosome that activates different responses, depending on the interactors.[221] In the case of HCV, an important MAVS signalosome has been described in the MAM.[222]

## 7.2 Mitochondrial DNA as Trigger of Innate Immunity

Innate immunity response also can result from damage associated molecular patterns (DAMPs). The main difference between DAMPs and PAMPs results from the fact that the former are endogenous molecules that are intracellular in healthy cells, but are released extracellularly in danger situations, such as cellular stress or injuries. In other words, while PAMPs correspond to the non-self, DAMPs can be seen like a hidden self of which the detection by PRRs set off the alarm.[223]

Many mitochondrial components have emerged as DAMPs. Among them, mitochondrial DNA (mtDNA) is one of the hottest topics. Recent studies have established its role in immunity. The intraarticular injection of mtDNA in mice leads to inflammation, contrary to nuclear DNA.[224] This inflammation occurs through binding to the PRRs toll-like receptor 9 (TLR9), which is expressed on the surface of macrophages and neutrophils.[225] Following damage, mitochondria might release mtDNA as a DAMP to trigger diverse extracellular and intracellular pathways, which alerts the immune system.[226]

This connection between mitochondria and the immune system is interesting when one considers the endosymbiotic theory, which claims that mitochondria are derived from a bacteria-like ancestor. Assuming that the immune system has evolved to recognize bacterial molecules, this could explain why some ancestral mitochondrial molecules act like DAMPs. Contrary to nuclear DNA, mtDNA contains hypomethylation motifs similar to those of bacterial DNA. Other mitochondrial DAMPs are N-formyl peptides, which is another common trait between mitochondria and bacteria.

## 8 THE EMERGENCE OF THE MITOCHONDRIA-CONTACT SITES AS SIGNALING DOMAINS

It would be impossible to conclude this chapter without emphasizing what has emerged as a hot topic for cell signaling: interorganelles contacts. During the 20th century, improvements of microscopy and biochemical techniques revealed a plethora of interorganelles and organelles/PM contacts.[227, 228] The understanding of their physiological importance is recent, however, and the diversity of these contacts and their functions is expanding.[229] Mitochondria/ER contact sites, called mitochondria associated membranes (MAM), probably are the most studied.

Investigators had noticed for a long time that some ER regions were in close proximity to mitochondria, and that mitochondria isolated by subfractionation were tightly associated with ER tubules. The term MAM first was used in early 1990s by Jean Vance, who separated MAM and pure mitochondria from a crude mitochondria fraction.[230, 231] Those studies assigned the first known function of MAM, as a phospholipid biosynthesis and transfer hub. Indeed, enzymes catalyzing the synthesis of phosphatidylserine (PS) and phosphatidylcholine (PC) were among the first identified MAM markers, and MAM-dependent transfer of PS toward mitochondria has been demonstrated.[232, 233] But, since these early experiments, the number of processes associated with MAM have expanded, including $Ca^{2+}$ signaling, cell death, antiviral response, and autophagy.[234–238]

We are still at the beginning of a long trip as advances should be made to better define MAM components and how their formation and stability are regulated. These questions will need to expand the proteomic characterization of MAM. Available fractionation techniques often lead to a substantial number of false-positives (e.g., proteins of the mitochondrial matrix), strongly hindering the interpretation of whole-scale proteomic analysis.[239] An interesting emerging concept is MAM heterogeneity. IIt is known that different types of ER-mitochondria contact sites exist, based not only on the distance separating both organelles but also on the presence or not of ribosomes.[240] Therefore, MAM could be a subclass of contact sites in which some type of signaling events (e.g., $Ca^{2+}$ exchange) could take place, while other signaling events could occur at other subclasses of contact sites that are yet to be defined.

## 9 CONCLUSION

This overview of mitochondria roles in signaling cannot be exhaustive. It aims at providing several clues showing that mitochondria are a major integration platform of signals coming from both inside and outside the organelle. It is noteworthy that mitochondria participate in signaling pathways in which they are not directly involved. An interesting evolutionary example is given by the comparison of the apoptotic process in *C. elegans* and mammals. The general process involving Bcl-2 family members and caspase activation is conserved. Evolution, however, has introduced the permeabilization of mitochondria by Bax/Bak as an additionnal step in mammals.[241] In *Drosophila melanogaster*, mitochondria serve as a platform for caspases activation, but are not permeabilized. The general evolutionary process of apoptosis, therefore, suggests that mitochondria get increasingly involved in life/death decisions of the cell. Considering the endosymbiotic theory of mitochondria origin, this would indicate that they were relatively independent within the primitive cells, but are becoming more involved in processes in which they are not the primary target.

The acceptance of mitochondria as a full player in signaling pathways has not always been a smooth process. In the 1980s, the first experiments showing that mitochondria were at least as important as ER for intracellular $Ca^{2+}$-fluxes regulation were received with skepticism. A similar reluctancy appeared in the 1990s, when mitochondria were proposed as a major player of apoptosis in mammals, and the hypothesis that mitochondria can regulate autophagic processes distinct from mitophagy still is not widely accepted. It is likely that new investigations about the role of mitochondria in processes to which they are not supposed to participate still will unveil unexpected discoveries.

## References


1. Carafoli E. Calcium signaling: a tale for all seasons. *Proc Natl Acad Sci U S A* 2002;**99**:1115–22.
2. Pendin D, Greotti E, Lefkimmiatis K, Pozzan T. Exploring cells with targeted biosensors. *J Gen Physiol* 2017;**149**:1–36.
3. Sheu SS, Jou MJ. Mitochondrial free $Ca^{2+}$ concentration in living cells. *J Bioenerg Biomembr* 1994;**26**:487–93.
4. Rizzuto R, Pinton P, Carrington W, Fay FS, Fogarty KE, Lifshitz LM, et al. Close contacts with the endoplasmic reticulum as determinants of mitochondrial $Ca^{2+}$ responses. *Science* 1998;**280**:1763–6.
5. Baughman JM, Perocchi F, Girgis HS, Plovanich M, Belcher-Timme CA, Sancak Y, et al. Integrative genomics identifies MCU as an essential component of the mitochondrial calcium uniporter. *Nature* 2011;**476**:341–5.
6. De Stefani D, Raffaello A, Teardo E, Szabò I, Rizzuto R. A forty-kilodalton protein of the inner membrane is the mitochondrial calcium uniporter. *Nature* 2011;**476**:336–40.
7. Raffaello A, De Stefani D, Sabbadin D, Teardo E, Merli G, Picard A, et al. The mitochondrial calcium uniporter is a multimer that can include a dominant-negative pore-forming subunit. *EMBO J* 2013;**32**:2362–76.
8. Marchi S, Pinton P. The mitochondrial calcium uniporter complex: molecular components, structure and physiopathological implications. *J Physiol* 2014;**592**:829–39.
9. De Stefani D, Rizzuto R, Pozzan T. Enjoy the trip: calcium in mitochondria back and forth. *Annu Rev Biochem* 2016;**85**:161–92.
10. Pan X, Liu J, Nguyen T, Liu C, Sun J, Teng Y, et al. The physiological role of mitochondrial calcium revealed by mice lacking the mitochondrial calcium uniporter. *Nat Cell Biol* 2013;**15**:1464–72.
11. Murphy E, Pan X, Nguyen T, Liu J, Holmström KM, Finkel T. Unresolved questions from the analysis of mice lacking MCU expression. *Biochem Biophys Res Commun* 2014;**449**:384–5.
12. Wu Y, Rasmussen TP, Koval OM, Joiner ML, Hall DD, Chen B, et al. The mitochondrial uniporter controls fight or flight heart rate increases. *Nat Commun* 2015;**6**:6081.
13. Doonan PJ, Chandramoorthy HC, Hoffman NE, Zhang X, Cárdenas C, Shanmughapriya S, et al. LETM1-dependent mitochondrial $Ca^{2+}$ flux modulates cellular bioenergetics and proliferation. *FASEB J* 2014;**28**:4936–49.
14. Austin S, Tavakoli M, Pfeiffer C, Seifert J, Mattarei A, De Stefani D, et al. LETM1-mediated $K^+$ and $Na^+$ homeostasis regulates mitochondrial $Ca^{2+}$ efflux. *Front Physiol* 2017;**8**:839.
15. Raffaello A, De Stefani D, Sabbadin D, Teardo E, Merli G, Picard A, et al. The mitochondrial calcium uniporter is a multimer that can include a dominant-negative pore-forming subunit. *EMBO J* 2013;**32**:2362–76.
16. Murgia M, Rizzuto R. Molecular diversity and pleiotropic role of the mitochondrial calcium uniporter. *Cell Calcium* 2015;**58**:11–7.
17. Perocchi F, Gohil VM, Girgis HS, Bao XR, McCombs JE, Palmer AE, et al. MICU1 encodes a mitochondrial EF hand protein required for Ca(2+) uptake. *Nature* 2010;**467**:291–6.
18. Plovanich M, Bogorad RL, Sancak Y, Kamer KJ, Strittmatter L, Li AA, et al. MICU2, a paralog of MICU1 resides within the mitochondrial uniporter complex to regulate calcium handling. *PLoS ONE* 2013;**8**:e55785.
19. Patron M, Checchetto V, Raffaello A, Teardo E, Vecellio Reane D, Mantoan M, et al. MICU1 and MICU2 finely tune the mitochondrial $Ca^{2+}$ uniporter by exerting opposite effects on MCU activity. *Mol Cell* 2014;**53**:726–37.
20. Matesanz-Isabel J, Arias-del-Val J, Alvarez-Illera P, Fonteriz RI, Montero M, Alvarez J. Functional roles of MICU1 and MICU2 in mitochondrial Ca(2+) uptake. *Biochim Biophys Acta* 2016;**1858**:1110–7.

21. Sancak Y, Markhard AL, Kitami T, Kovács-Bogdán E, Kamer KJ, Udeshi ND, et al. EMRE is an essential component of the mitochondrial calcium uniporter complex. *Science* 2013;**342**:1379–82.
22. Vais H, Mallilankaraman K, Mak DD, Hoff H, Payne R, Tanis JE, et al. EMRE is a matrix Ca(2+) sensor that governs gatekeeping of the mitochondrial Ca(2+) uniporter. *Cell Rep* 2016;**14**:403–10.
23. Tinel H, Cancela JM, Mogami H, Gerasimenko JV, Gerasimenko OV, Tepikin AV, et al. Active mitochondria surrounding the pancreatic acinar granule region prevent spreading of inositol trisphosphate-evoked local cytosolic Ca(2+) signals. *EMBO J* 1999;**18**:4999–5008.
24. Hajnóczky G, Hager R, Thomas AP. Mitochondria suppress local feedback activation of inositol 1,4,5-trisphosphate receptors by $Ca^{2+}$. *J Biol Chem* 1999;**274**:14157–62.
25. Lin X, Várnai P, Csordás G, Balla A, Nagai T, Miyawaki A, et al. Control of calcium signal propagation to the mitochondria by inositol 1,4,5-trisphosphate-binding proteins. *J Biol Chem* 2005;**280**:12820–32.
26. Qi H, Li L, Shuai J. Optimal microdomain crosstalk between endoplasmic reticulum and mitochondria for $Ca^{2+}$ oscillations. *Sci Rep* 2015;**5**:7984.
27. Denton RM. Regulation of mitochondrial dehydrogenases by calcium ions. *Biochim Biophys Acta* 2009;**1787**:1309–16.
28. Kennedy ED, Wollheim CB. Role of mitochondrial calcium in metabolism-secretion coupling in nutrient-stimulated insulin release. *Diabetes Metab* 1998;**24**:15–24.
29. Maechler P, Carobbio S, Rubi B. In beta-cells, mitochondria integrate and generate metabolic signals controlling insulin secretion. *Int J Biochem Cell Biol* 2006;**38**:696–709.
30. Bernardi P, Broekemeier KM, Pfeiffer DR. Recent progress on regulation of the mitochondrial permeability transition pore; a cyclosporin-sensitive pore in the inner mitochondrial membrane. *J Bioenerg Biomembr* 1994;**26**:509–17.
31. Bernardi P. The permeability transition pore. Control points of a cyclosporin A-sensitive mitochondrial channel involved in cell death. *Biochim Biophys Acta* 1996;**1275**:5–9.
32. Rasola A, Bernardi P. Mitochondrial permeability transition in Ca(2+)-dependent apoptosis and necrosis. *Cell Calcium* 2011;**50**:222–33.
33. Liao Y, Hao Y, Chen H, He Q, Yuan Z, Cheng J. Mitochondrial calcium uniporter protein MCU is involved in oxidative stress-induced cell death. *Protein Cell* 2015;**6**:434–42.
34. Xie N, Wu C, Wang C, Cheng X, Zhang L, Zhang H, et al. Inhibition of the mitochondrial calcium uniporter inhibits Aβ-induced apoptosis by reducing reactive oxygen species-mediated endoplasmic reticulum stress in cultured microglia. *Brain Res* 2017;**1676**:100–6.
35. Oropeza-Almazán Y, Vázquez-Garza E, Chapoy-Villanueva H, Torre-Amione G, García-Rivas G. Small Interfering RNA Targeting Mitochondrial Calcium Uniporter Improves Cardiomyocyte Cell Viability in Hypoxia/Reoxygenation Injury by Reducing Calcium Overload. *Oxidative Med Cell Longev* 2017;**2017**:5750897.
36. Mallilankaraman K, Doonan P, Cárdenas C, Chandramoorthy HC, Müller M, Miller R, et al. MICU1 is an essential gatekeeper for MCU-mediated mitochondrial Ca(2+) uptake that regulates cell survival. *Cell* 2012;**151**:630–44.
37. Xue Q, Pei H, Liu Q, Zhao M, Sun J, Gao E, et al. MICU1 protects against myocardial ischemia/reperfusion injury and its control by the importer receptor Tom70. *Cell Death Dis* 2017;**8**:e2923.
38. Niki E. Oxidative stress and antioxidants: distress or eustress? *Arch Biochem Biophys* 2016;**595**:19–24.
39. Sies H, Berndt C, Jones DP. Oxidative Stress. *Annu Rev Biochem* 2017;**86**:715–48.
40. Fisher AB. Redox signaling across cell membranes. *Antioxid Redox Signal* 2009;**11**:1349–56.
41. Hachez C, Chaumont F. Aquaporins: a family of highly regulated multifunctional channels. *Adv Exp Med Biol* 2010;**679**:1–17.
42. Marinelli RA, Marchissio MJ. Mitochondrial aquaporin-8: a functional peroxiporin? *Antioxid Redox Signal* 2013;**19**:896.
43. Kamata H, Honda S, Maeda S, Chang L, Hirata H, Karin M. Reactive Oxygen Species Promote TNFα-Induced Death and Sustained JNK Activation by Inhibiting MAP Kinase Phosphatases. *Cell* 2005;**120**:649–61.
44. Yang Y, Song Y, Loscalzo J. Regulation of the Protein disulfide Proteome by mitochondria in mammalian Cell. *Proc Natl Acad Sci U S A* 2007;**104**:10813–7. [Erratum in: Proc Natl Acad Sci U S A 106, 14734.
45. Kassab A, Piwowar A. Cell oxidant stress delivery and cell dysfunction onset in type 2 diabetes. *Biochimie* 2012;**94**:1837–48.
46. Sifuentes-Franco S, Pacheco-Moisés FP, Rodríguez-Carrizalez AD, Miranda-Díaz AG. The Role of Oxidative Stress, Mitochondrial Function, and Autophagy in Diabetic Polyneuropathy. *J Diabetes Res* 2017;**2017**:1673081.

47. Lin Y, Berg AH, Iyengar P, Lam TK, Giacca A, Combs TP, et al. The hyperglycemia-induced inflammatory response in adipocytes: the role of reactive oxygen species. *J Biol Chem* 2005;**280**:4617–26.
48. Munusamy S, MacMillan-Crow LA. Mitochondrial superoxide plays a crucial role in the development of mitochondrial dysfunction during high glucose exposure in rat renal proximal tubular cells. *Free Radic Biol Med* 2009;**46**:1149–57.
49. Morino K, Petersen KF, Dufour S, Befroy D, Frattini J, Shatzkes N, et al. Reduced mitochondrial density and increased IRS-1 serine phosphorylation in muscle of insulin-resistant offspring of type 2 diabetic parents. *J Clin Invest* 2005;**115**:3587–93.
50. Befroy DE, Petersen KF, Dufour S, Mason GF, de Graaf RA, Rothman DL, et al. Impaired mitochondrial substrate oxidation in muscle of insulin-resistant offspring of type 2 diabetic patients. *Diabetes* 2007;**56**:1376–81.
51. Abdul-Ghani MA, DeFronzo RA. Mitochondrial dysfunction, insulin resistance, and type 2 diabetes mellitus. *Curr Diab Rep* 2008;**8**:173–8.
52. Nilsson E, Jansson PA, Perfilyev A, Volkov P, Pedersen M, Svensson MK, et al. Altered DNA methylation and differential expression of genes influencing metabolism and inflammation in adipose tissue from subjects with type 2 diabetes. *Diabetes* 2014;**63**:2962–76.
53. Bonnard C, Durand A, Peyrol S, Chanseaume E, Chauvin MA, Morio B, et al. Mitochondrial dysfunction results from oxidative stress in the skeletal muscle of diet-induced insulin-resistant mice. *J Clin Invest* 2008;**118**:789–800.
54. Lukyanov KA, Belousov VV. Genetically encoded fluorescent redox sensors. *Biochim Biophys Acta* 2014;**1840**:745–56.
55. Bilan DS, Belousov VV. New tools for redox biology: from imaging to manipulation. *Free Radic Biol Med* 2017;**109**:167–88.
56. Booth DM, Joseph SK, Hajnóczky G. Subcellular ROS imaging methods: relevance for the study of calcium signaling. *Cell Calcium* 2016;**60**:65–73.
57. Bánsághi S, Golenár T, Madesh M, Csordás G, RamachandraRao S, Sharma K, et al. Isoform- and species-specific control of inositol 1,4,5-trisphosphate (IP3) receptors by reactive oxygen species. *J Biol Chem* 2014;**289**:8170–81.
58. Joseph SK, Nakao SK, Sukumvanich S. Reactivity of free thiol groups in type-I inositol trisphosphate receptors. *Biochem J* 2006;**393**:575–82.
59. Stöcker S, Maurer M, Ruppert T, Dick TP. A role for 2-Cys peroxiredoxins in facilitating cytosolic protein thiol oxidation. *Nat Chem Biol* 2017;https://doi.org/10.1038/nchembio.2536.
60. Drazic A, Myklebust LM, Ree R, Arnesen T. The world of protein acetylation. *Biochim Biophys Acta* 2016;**1864**:1372–401.
61. Van Damme P, Arnesen T, Gevaert K. Protein alpha-N-acetylation studied by N-terminomics. *FEBS J* 2011;**278**:3822–34.
62. Gu W, Roeder RG. Activation of p53 sequence-specific DNA binding by acetylation of the p53 C-terminal domain. *Cell* 1997;**90**:595–606.
63. Tang Y, Zhao W, Chen Y, Zhao Y, Gu W. Acetylation Is Indispensable for p53 Activation. *Cell* 2008;**133**:612–26.
64. Alves S, Neiri L, Chaves SR, Vieira S, Trindade D, Manon S, et al. N-terminal acetylation modulates Bax targeting to mitochondria. *Int J Biochem Cell Biol* 2017;**95**:35–42.
65. Yao ZG, Liu Y, Zhang L, Huang L, Ma CM, Xu YF, et al. Co-location of HDAC2 and insulin signaling components in the adult mouse hippocampus. *Cell Mol Neurobiol* 2012;**32**:1337–42.
66. Raichur S, Teh SH, Ohwaki K, Gaur V, Long YC, Hargreaves M, et al. Histone deacetylase 5 regulates glucose uptake and insulin action in muscle cells. *J Mol Endocrinol* 2012;**49**:203–11.
67. Bricambert J, Favre D, Brajkovic S, Bonnefond A, Boutry R, Salvi R, et al. Impaired histone deacetylases 5 and 6 expression mimics the effects of obesity and hypoxia on adipocyte function. *Mol Metab* 2016;**5**:1200–7.
68. Iacobazzi V, Infantino V. Citrate—new functions for an old metabolite. *Biol Chem* 2014;**395**:387–99.
69. Ronnebaum SM, Ilkayeva O, Burgess SC, Joseph JW, Lu D, Stevens RD, et al. A pyruvate cycling pathway involving cytosolic NADP-dependent isocitrate dehydrogenase regulates glucose-stimulated insulin secretion. *J Biol Chem* 2006;**281**:30593–602.
70. Tsukada Y, Fang J, Erdjument-Bromage H, Warren ME, Borchers CH, Tempst P, et al. Histone demethylation by a family of JmjC domain-containing proteins. *Nature* 2006;**439**:811–6.
71. Murn J, Shi Y. The winding path of protein methylation research: milestones and new frontiers. *Nat Rev Mol Cell Biol* 2017;**18**:517.

72. Markolovic S, Wilkins SE, Schofield CJ. Protein Hydroxylation Catalyzed by 2-Oxoglutarate-dependent Oxygenases. *J Biol Chem* 2015;**290**:20712–22.
73. Son ED, Choi GH, Kim H, Lee B, Chang IS, Hwang JS. Alpha-ketoglutarate stimulates procollagen production in cultured human dermal fibroblasts, and decreases UVB-induced wrinkle formation following topical application on the dorsal skin of hairless mice. *Biol Pharm Bull* 2007;**30**:1395–9.
74. Carey BW, Finley LWS, Cross JR, Allis CD, Thompson CB. Intracellular α-ketoglutarate maintains the pluripotency of embryonic stem cells. *Nature* 2015;**518**:413–6.
75. Zhang Z, Tan M, Xie Z, Dai L, Chen Y, Zhao Y. Identification of lysine succinylation as a new post-translational modification. *Nat Chem Biol* 2011;**7**:58–63.
76. Alleyn M, Breitzig M, Lockey R, Kolliputi N. The dawn of succinylation: a posttranslational modification. *Am J Phys Cell Physiol* 2017;. ajpcell.00148.2017.
77. Du J, Zhou Y, Su X, Yu JJ, Khan S, Jiang H, et al. Sirt5 is a NAD-dependent protein lysine demalonylase and desuccinylase. *Science* 2011;**334**:806–9.
78. Li L, Shi L, Yang S, Yan R, Zhang D, Yang J, et al. SIRT7 is a histone desuccinylase that functionally links to chromatin compaction and genome stability. *Nat Commun* 2016;**7**:12235.
79. Selak MA, Armour SM, MacKenzie ED, Boulahbel H, Watson DG, Mansfield KD, et al. Succinate links TCA cycle dysfunction to oncogenesis by inhibiting HIF-alpha prolyl hydroxylase. *Cancer Cell* 2005;**7**:77–85.
80. He W, Miao FJ, Lin DC, Schwandner RT, Wang Z, Gao J, et al. Citric acid cycle intermediates as ligands for orphan G-protein-coupled receptors. *Nature* 2004;**429**:188–93.
81. de Castro Fonseca M, Aguiar CJ, da Rocha Franco JA, Gingold RN, Leite MF. GPR91: expanding the frontiers of Krebs cycle intermediates. *Cell Commun Signal* 2016;**14**:3.
82. Gilissen J, Jouret F, Pirotte B, Hanson J. Insight into SUCNR1 (GPR91) structure and function. *Pharmacol Ther* 2016;**159**:56–65.
83. Frizzell N, Lima M, Baynes JW. Succination of proteins in diabetes. *Free Radic Res* 2011;**45**:101–9.
84. Merkley ED, Metz TO RD, Smith, Baynes JW, Frizzell N. The succinated proteome. *Mass Spectrom Rev* 2014;**33**:98–109.
85. Manuel AM, Walla MD, Faccenda A, Martin SL, Tanis RM, Piroli GG, et al. Succination of Protein Disulfide Isomerase Links Mitochondrial Stress and Endoplasmic Reticulum Stress in the Adipocyte During Diabetes. *Antioxid Redox Signal* 2017;**27**:1281–96.
86. Gumeni S, Trougakos IP. Cross talk of proteostasis and mitostasis in cellular homeodynamics, ageing and disease. *Oxidative Med Cell Longev* 2016;**2016**:4587691.
87. Ciachanover A. The ubiquitin-proteasome proteolytic pathway. *Cell* 1994;**79**:13–21.
88. Yonashiro R, Ishido S, Kyo S, Fukuda T, Goto E, Matsuki Y, et al. A novel mitochondrial ubiquitin ligase plays a critical role in mitochondrial dynamics. *EMBO J* 2006;**25**:3618–26.
89. Margineantu DH, Emerson CB, Diaz D, Hockenbery DM. Hsp90 inhibition decreases mitochondrial protein turnover. *PLoS ONE* 2007;**2**:e1066.
90. Radke S, Chander H, Schäfer P, Meiss G, Krüger R, Schulz JB, et al. Mitochondrial Protein Quality Control by the Proteasome Involves Ubiquitination and the Protease Omi. *J Biol Chem* 2008;**283**:12681–5.
91. Azzu V, Mookerjee SA, Brand MD. Rapid turnover of mitochondrial uncoupling protein 3. *Biochem J* 2010;**426**:13–7.
92. Lehmann G, Ziv T, Braten O, Admon A, Udasin RG, Ciechanover A. Ubiquitination of specific mitochondrial matrix proteins. *Biochem Biophys Res Commun* 2016;**475**:13–8.
93. Lehmann G, Udasin RG, Ciechanover A. On the linkage between the ubiquitin-proteasome system and the mitochondria. *Biochem Biophys Res Commun* 2016;**473**:80–6.
94. de Duve C. In: De Reuck A, Cameron MP, editors. *Foundation C, Ciba foundation symposium: lysosome*. Little, Brown; 1963.
95. Xie Z, Klionsky DJ. Autophagosome formation: core machinery and adaptations. *Nat Cell Biol* 2007;**9**:1102–9.
96. Cuervo AM, Wong E. Chaperone-mediated autophagy: roles in disease and aging. *Cell Res* 2014;**24**:92–104.
97. Lemasters JJ. Selective mitochondrial autophagy, or mitophagy, as a targeted defense against oxidative stress, mitochondrial dysfunction, and aging. *Rejuvenation Res* 2005;**8**:3–5.
98. Kissová I, Deffieu M, Manon S, Camougrand N. Uth1p is involved in the autophagic degradation of mitochondria. *J Biol Chem* 2004;**279**:39068–74.
99. Tsukada M, Ohsumi Y. Isolation and characterization of autophagy-defective mutants of *Saccharomyces cerevisiae*. *FEBS Lett* 1993;**333**:169–74.

100. Kim I, Rodriguez-Enriquez S, Lemasters JJ. Selective degradation of mitochondria by mitophagy. *Arch Biochem Biophys* 2007;**462**:245–53.
101. Gallagher LE, Williamson LE, Chan EY. Advances in autophagy regulatory mechanisms. *Cell* 2016;**5**:. pii:E24.
102. Levine B, Packer M, Codogno P. Development of autophagy inducers in clinical medicine. *J Clin Invest* 2015;**125**:14–24.
103. Gasser T. Molecular pathogenesis of Parkinson disease: insights from genetics studies. *Expert Rev Mol Med* 2009;**11**:e22.
104. Clark IE, Dodson MW, Jiang C, Cao JH, Huh JR, Scol JH, et al. Drosophila pink1 is required for mitochondrial function and interactsgenetically with parkin. *Nature* 2006;**441**:1162–6.
105. Jin DM, Lazarou M, Wang C, Kane LA, Narendra DP, Youle RJ. Mitochondrial membrane potential regulates Pink import and proteolytic destabilization by PARL. *J Cell Biol* 2010;**191**:933–42.
106. Kazlauskaite A, Martinez-Torres RJ, Wilkie S, Kumar A, Peltier J, Gonzalez A, et al. Binding to serine 65-phosphorylated ubiquitin promes Parkin for optimal PINK1-dependent phosphorylation and activation. *EMBO Rep* 2015;**16**:939–54.
107. Chan NC, Salazar AM, Phamm AH, Sweredoski MJ, Kolawa NJ, Graham RL, et al. Broad activation of the ubiquitin-proteasome system by Parkin is critical for mitophagy. *Hum Mol Genet* 2011;**20**:1726–37.
108. Geisser S, Holmstroem KM, Skujat D, Fiessel FC, Rothfuss OC, Kahle PJ, et al. PINK/Parkin-mediated mitophagy is dependent on VDCA1 and p62/SQSTM1. *Nat Cell Biol* 2010;**12**:119–31.
109. Youle RJ, Narendra DP. Mechanisms of mitophagy. *Nat Rev Mol Cell Biol* 2011;**12**:9–14.
110. Tanaka A, Cleland MM, Xu S, Narendra DP, Suen DF, Karbowski M, et al. Proteasome and p97 mediate mitophagy and degradation of mitofusins induced by Parkin. *J Cell Biol* 2010;**191**:1367–80.
111. Ashrafi G, Schwartz TL. The pathways of mitophagy for quality control and clearence of mitochondria. *Cell Death Differ* 2013;**20**:31–42.
112. Lazarou M, Sliter DA, Kane LA, Sarraf SA, Wang C, Burman JL, et al. The ubiquitin Kinase PINK1 recruits autophagy receptors to induce mitophagy. *Nature* 2015;**524**:309–14.
113. Ritcher B, Sliter DA, Herhaus L, Stolz A, Wang C, Beli P, et al. Phosphorylation of OPTN by TBK1 enhances its binding to Ub chains and promotes selective autophagy of damaged mitochondria. *Proc Natl Acad Sci U S A* 2016;**113**:4039–44.
114. Orvedahl A, Sumpter Jr. R, Xiao G, Ng A, Zou Z, Tang Y, et al. Image-based genome-wide siRNA screen identifies selective autophagy factors. *Nature* 2011;**480**:113–7.
115. Villa E, Proïcs E, Rubio-Patino C, Obba S, Zunino B, Bossowski JP, et al. Parkin-independent mitophagy controls chemotherapeutic response in cancer cells. *Cell Rep* 2017;**20**:2846–59.
116. Liu L, Feng D, Chen G, Chen M, Zheng Q, Song P, et al. Mitochondrial outer-membrane protein FUNDC1 mediates hypoxia-induced mitophagy in mammalian cells. *Nat Cell Biol* 2012;**14**:177–85.
117. Chinnadurai G, Vijayalingam S, Gibson SB. BNIP3 subfamily BH3-only proteins: mitochondrial stress sensors in normal and pathological functions. *Oncogene* 2009;**27**:5114–27.
118. Novak I, Kirkin V, McEwan DG, Zhang J, Wild P, Rozenknop A, et al. Nix is a selective autophagy receptor for mitochondrial clearance. *EMBO Rep* 2010;**11**:45–51.
119. Zhu Y, Massen S, Terenzio M, Lang V, Chen-Lindner S, Eils R, et al. Modulation of serines 17 and 24 in the LC3-interacting region of Bnip3 determines pro-survival mitophagy versus apoptosis. *J Biol Chem* 2013;**288**:1099–113.
120. Murakawa T, Yamaguchi O, Hashimoto A, Hikoso S, Takeda T, Oka T, et al. Bcl-2-like protein 13 is a mammalian Atg32 homologue that mediates mitophagy and mitochondrial fragmentation. *Nat Commun* 2015;**6**:7527–41.
121. Wei Y, Chiang WC, Sumpter Jr. R, Mishra P, Levine B. Prohibitin 2 Is an Inner Mitochondrial Membrane Mitophagy Receptor. *Cell* 2017;**168**:224–38.
122. Chu CT, Ji J, Dagda RK, Jiang JF, Tyurina YY, Kapralov AA, et al. Cardiolipin externalization to the outer mitochondrial membrane acts as an elimination signal for mitophagy in neuronal cells. *Nat Cell Biol* 2013;**15**:1197–205.
123. Schweers RL, Zhang J, Randall MS, Loyd MR, Li W, Dorsey FC, et al. NIX is required for programmed mitochondrial clearance during reticulocyte maturation. *Proc Natl Acad Sci U S A* 2007;**104**:19500–5.
124. Sandoval H, Thiagarajan P, Dasgupta SK, Schumacher A, Prchal JT, Chen M, et al. Essential role for NIX in autophagic maturation of erythroid cells. *Nature* 2008;**454**:232–5.
125. Ding WX, Ni HM, Li M, Liao Y, Chen X, Stolz DB, et al. Two distinct phases of mitophagy, reactive oxygen species-mediated autophagy induction and Parkin-Ubiquitin-p62-mediated mitochondrial priming. *J Biol Chem* 2010;**285**:27879–90.

126. Gao F, Chen D, Si J, Hu Q, Qin Z, Fang M, et al. The mitochondrial protein BNIP3L is the substrate of PARK2 and mediates mitophagy in PINK1/PARK2 pathway. *Hum Mol Genet* 2015;**24**:2528–38.
127. Daido S, Kanzawa T, Yamamoto A, Takeuchi H, Kondo Y, Kondo S. Pivotal role of the cell death factor BNIP3 in ceramide-induced autophagic cell death in malignant glioma cells. *Cancer Res* 2004;**64**:4286–93.
128. Quinsay MN, Thomas RL, Lee Y, Gustafsson AB. Bnip3-mediated mitochondrial autophagy is independent of the mitochondrial permeability transition pore. *Autophagy* 2010;**6**:855–62.
129. Chen G, Han Z, Feng D, Chen Y, Chen L, Wu H, et al. A regulatory signaling loop comprising the PCAM5 phosphatase and CK2 controls receptor-mediated mitophagy. *Mol Cell* 2014;**54**:362–77.
130. Wu W, Tian W, Hu Z, Chen G, Huang L, Li W, et al. ULK1 translocates to mitochondria and phosphorylates FUNDC1 to regulate mitophagy. *EMBO Rep* 2014;**15**:566–75.
131. Strappazzon F, Nazio F, Corrado M, Cianfanelli V, Romagnoli A, Fimia GM, et al. AMBRA1 is able to induce mitophagy via LC3 binding, regardless of PARKIN and p62/SQSTM1. *Cell Death Differ* 2015;**22**:419–32.
132. Zhang T, Xue L, Li L, Tang C, Wan Z, Wang R, et al. BNIP3 protein suppresses PINK1 kinase proteolytic cleavage to promote mitophagy. *J Biol Chem* 2016;**291**:21616–29.
133. Koentjoro B, Park JS, Sue CM. Nix restores mitophagy and mitochondrial function to protect against PINK1/Parkin-related Parkinson's disease. *Sci Rep* 2017;**7**:44373.
134. Hoshino A, Mita Y, Okawa Y, Ariyoshi M, Iwai-Kanai E, Ueyama T, et al. Cytosolic p53 inhibits Parkin-mediated mitophagy and promotes mitochondrial dysfunction in the mouse heart. *Nat Commun* 2013;**4**:2308.
135. Dorn 2nd GW. Mitochondrial pruning by Nix and BNip3: an essential function for cardiac-expressed death factors. *J Cardiovasc Transl Res* 2010;**3**:374–83.
136. Kubli DA, Quinsay MN, Gustafsson AB. Parkin deficiency results in accumulation of abnormal mitochondria in aging myocytes. *Commun Integr Biol* 2013;**6**:e24511.
137. Yuan Y, Zheng Y, Zhang X, Chen Y, Wu X, Wu J, et al. BNIP3L/NIX-mediated mitophagy protects against ischemic brain injury independent of PARK2. *Autophagy* 2017;**3**:1754–66.
138. Allen GFG, Toth R, James J, Ganley IG. Loss of iron triggers PINK/Parkin independent mitophagy. *EMBO Rep* 2013;**14**:1127–35.
139. Park SJ, Shin JH, Kim ES, Jo YK, Kim JH, Hwang JJ, et al. Mitochondrial fragmentation caused by phenanthroline promotes mitophagy. *FEBS Lett* 2012;**586**:4303–10.
140. Jin SM, Youle RJ. The accumulation of misfolded proteins in the mitochondrial matrix is sensed by PINK1 to induce PARK2/Parkin-mediated mitophagy of polarized mitochondria. *Autophagy* 2013;**9**:1750–7.
141. Kim I, Rodriguez-Enriquez S, Mizushima N, Ohsumi Y, Lemasters JJ. Autophagic degradation of mitochondria in GFP-LC3 transgenic mouse hepatocytes after nutrient deprivation. *Hepatology* 2004;**40**(Suppl 1):291A.
142. Kim I, Mizushima N, Lemasters JJ. Selective removal of damaged mitochondria by autophagy (mitophagy). *Hepatology* 2006;**44**(Suppl 1):241A.
143. Scherz-Shouval R, Shvets E, Fass E, Shorer H, Gil L, Elazar Z. Reactive oxygen species are essential for autophagy and specifically regulate the activity of Atg4. *EMBO J* 2007;**26**:1749–60.
144. Chen Y, Azad MB, Gibson SB. Superoxide is the major reactive oxygen species regulating autophagy. *Cell Death Differ* 2009;**16**:1040–52.
145. Kim EH, Choi KS. A critical role of superoxide anion in selenite-induced mitophagic cell death. *Autophagy* 2008;**4**:76–8.
146. Xiao B, Deng X, Lim GGY, Xie S, Zhou ZD, Lim KL, et al. Superoxide drives progression of Parkin/PINK1-dependent mitophagy following translocation of Parkin to mitochondria. *Cell Death Dis* 2017;**8**:e3097.
147. Li H, Zhu H, Xu CJ, Yuan J. Cleavage of BID by caspase 8 mediates the mitochondrial damage in the Fas pathway of apoptosis. *Cell* 1998;**94**:491–501.
148. Wyllie A. Apoptosis. Clues in the p53 murder mystery. *Nature* 1997;**389**:237–8.
149. Vayssiere JL, Petit PX, Risler Y, Mignotte B. Commitment to apoptosis is associated with changes in mitochondrial biogenesis and activity in cell lines conditionally immortalized with simian virus 40. *Proc Natl Acad Sci U S A* 1994;**91**:11752–6.
150. Zamzami N, Susin SA, Marchetti P, Hirsch T, Gómez-Monterrey I, Castedo M, et al. Mitochondrial control of nuclear apoptosis. *J Exp Med* 1996;**183**:1533–44.
151. Pavlov EV, Priault M, Pietkiewicz D, Cheng EH, Antonsson B, Manon S, et al. A novel, high conductance channel of mitochondria linked to apoptosis in mammalian cells and Bax expression in yeast. *J Cell Biol* 2001;**155**:725–31.

152. Mikhailov V, Mikhailova M, Degenhardt K, Venkatachalam MA, White E, Saikumar P. Association of Bax and Bak homo-oligomers in mitochondria. Bax requirement for Bak reorganization and cytochrome c release. *J Biol Chem* 2003;**278**:5367–76.
153. Dejean LM, Martinez-Caballero S, Guo L, Hughes C, Teijido O, Ducret T, et al. Oligomeric Bax is a component of the putative cytochrome c release channel MAC, mitochondrial apoptosis-induced channel. *Mol Biol Cell* 2005;**16**:2424–32.
154. Vaux DL. Apoptogenic factors released from mitochondria. *Biochim Biophys Acta* 2011;**1813**:546–50.
155. Mignotte B, Vayssiere JL. Mitochondria and apoptosis. *Eur J Biochem* 1998;**252**:1–15.
156. Korsmeyer SJ, Wei MC, Saito M, Weiler S, Oh KJ, Schlesinger PH. Pro-apoptotic cascade activates BID, which oligomerizes BAK or BAX into pores that result in the release of cytochrome c. *Cell Death Differ* 2000;**7**:1166–73.
157. Annis MG, Soucie EL, Dlugosz PJ, Cruz-Aguado JA, Penn LZ, Leber B, et al. Bax forms multispanning monomers that oligomerize to permeabilize membranes during apoptosis. *EMBO J* 2005;**24**:2096–103.
158. Suzuki M, Youle RJ, Tjandra N. Structure of Bax: coregulation of dimer formation and intracellular localization. *Cell* 2000;**103**:645–54.
159. Kaufmann T, Schlipf S, Sanz J, Neubert K, Stein R, Borner C. Characterization of the signal that directs Bcl-x(L), but not Bcl-2, to the mitochondrial outer membrane. *J Cell Biol* 2003;**160**:53–64.
160. Wieckowski MR, Vyssokikh M, Dymkowska D, Antonsson B, Brdiczka D, Wojtczak L. Oligomeric C-terminal truncated Bax preferentially releases cytochrome c but not adenylate kinase from mitochondria, outer membrane vesicles and proteoliposomes. *FEBS Lett* 2001;**505**:453–9.
161. Tremblais K, Oliver L, Juin P, Le Cabellec TM, Meflah K, Vallette FM. The C-terminus of bax is not a membrane addressing/anchoring signal. *Biochem Biophys Res Commun* 1999;**260**:582–91.
162. Nouraini S, Six E, Matsuyama S, Krajewski S, Reed JC. The putative pore-forming domain of Bax regulates mitochondrial localization and interaction with Bcl-X(L). *Mol Cell Biol* 2000;**20**:1604–15.
163. Martinez-Caballero S, Dejean LM, Kinnally MS, Oh KJ, Mannella CA, Kinnally KW. Assembly of the mitochondrial apoptosis-induced channel, MAC. *J Biol Chem* 2009;**284**:12235–45.
164. Czabotar PE, Westphal D, Dewson G, Ma S, Hockings C, Fairlie WD, et al. Bax crystal structures reveal how BH3 domains activate Bax and nucleate its oligomerization to induce apoptosis. *Cell* 2013;**152**:519–31.
165. Westphal D, Dewson G, Menard M, Frederick P, Iyer S, Bartolo R, et al. Apoptotic pore formation is associated with in-plane insertion of Bak or Bax central helices into the mitochondrial outer membrane. *Proc Natl Acad Sci U S A* 2014;**111**:E4076–85.
166. Salvador-Gallego R, Mund M, Cosentino K, Schneider J, Unsay J, Schraermeyer U, et al. Bax assembly into rings and arcs in apoptotic mitochondria is linked to membrane pores. *EMBO J* 2016;**35**:389–401.
167. Große L, Wurm CA, Brüser C, Neumann D, Jans DC, Jakobs S. Bax assembles into large ring-like structures remodeling the mitochondrial outer membrane in apoptosis. *EMBO J* 2016;**35**:402–13.
168. Cartron PF, Bellot G, Oliver L, Grandier-Vazeille X, Manon S, Vallette FM. Bax inserts into the mitochondrial outer membrane by different mechanisms. *FEBS Lett* 2008;**582**:3045–51.
169. Marzo I, Brenner C, Zamzami N, Susin SA, Beutner G, Brdiczka D, et al. The permeability transition pore complex: a target for apoptosis regulation by caspases and bcl-2-related proteins. *J Exp Med* 1998;**187**:1261–71.
170. Kokoszka JE, Waymire KG, Levy SE, Sligh JE, Cai J, Jones DP, et al. The ADP/ATP translocator is not essential for the mitochondrial permeability transition pore. *Nature* 2004;**427**:461–5.
171. Baines CP, Kaiser RA, Sheiko T, Craigen WJ, Molkentin JD. Voltage-dependent anion channels are dispensable for mitochondrial-dependent cell death. *Nat Cell Biol* 2007;**9**:550–5.
172. Nakagawa T, Shimizu S, Watanabe T, Yamaguchi O, Otsu K, Yamagata H, et al. Cyclophilin D-dependent mitochondrial permeability transition regulates some necrotic but not apoptotic cell death. *Nature* 2005;**434**:652–8.
173. Pfanner N, Wiedemann N, Meisinger C, Lithgow T. Assembling the mitochondrial outer membrane. *Nat Struct Mol Biol* 2004;**11**:1044–8.
174. Motz C, Martin H, Krimmer T, Rassow J. Bcl-2 and porin follow different pathways of TOM-dependent insertion into the mitochondrial outer membrane. *J Mol Biol* 2002;**323**:729–38.
175. Bellot G, Cartron PF, Er E, Oliver L, Juin P, Armstrong LC, et al. TOM22, a core component of the mitochondria outer membrane protein translocation pore, is a mitochondrial receptor for the proapoptotic protein Bax. *Cell Death Differ* 2007;**14**:785–94.
176. Renault TT, Grandier-Vazeille X, Arokium H, Velours G, Camougrand N, Priault M, et al. The cytosolic domain of human Tom22 modulates human Bax mitochondrial translocation and conformation in yeast. *FEBS Lett* 2012;**586**:116–21.

177. Cartron PF, Petit E, Bellot G, Oliver L, Vallette FM. Metaxins 1 and 2, two proteins of the mitochondrial protein sorting and assembly machinery, are essential for Bak activation during TNF alpha triggered apoptosis. *Cell Signal* 2014;**26**:1928–34.
178. Cheng EH, Sheiko TV, Fisher JK, Craigen WJ, Korsmeyer SJ. VDAC2 inhibits BAK activation and mitochondrial apoptosis. *Science* 2003;**301**:513–7.
179. Petit E, Cartron PF, Oliver L, Vallette FM. The phosphorylation of Metaxin 1 controls Bak activation during TNFα induced cell death. *Cell Signal* 2017;**30**:171–8.
180. Lindenboim L, Borner C, Stein R. Nuclear proteins acting on mitochondria. *Biochim Biophys Acta* 2011;**1813**:584–96.
181. Brookes PS, Yoon Y, Robotham JL, Anders MW, Sheu SS. Calcium, ATP, and ROS: a mitochondrial love-hate triangle. *Am J Phys Cell Physiol* 2004;**287**:C817–33.
182. Gonzalvez F, Pariselli F, Dupaigne P, Budihardjo I, Lutter M, Antonsson B, et al. tBid interaction with cardiolipin primarily orchestrates mitochondrial dysfunctions and subsequently activates Bax and Bak. *Cell Death Differ* 2005;**12**:614–26.
183. Dingeldein APG, Pokorná Š, Lidman M, Sparrman T, Šachl R, Hof M, et al. Apoptotic Bax at oxidatively stressed mitochondrial membranes: lipid dynamics and permeabilization. *Biophys J* 2017;**112**:2147–58.
184. Perera MN, Lin SH, Peterson YK, Bielawska A, Szulc ZM, Bittman R, et al. Bax and Bcl-xL exert their regulation on different sites of the ceramide channel. *Biochem J* 2012;**445**:81–91.
185. Ortiz A, Killian JA, Verkleij AJ, Wilschut J. Membrane fusion and the lamellar-to-inverted-hexagonal phase transition in cardiolipin vesicle systems induced by divalent cations. *Biophys J* 1999;**77**:2003–14.
186. Van Mau N, Kajava AV, Bonfils C, Martinou JC, Harricane MC. Interactions of Bax and tBid with lipid monolayers. *J Membr Biol* 2005;**207**:1–9.
187. Degli Esposti M, Ferry G, Masdehors P, Boutin JA, Hickman JA, Dive C. Post-translational modification of Bid has differential effects on its susceptibility to cleavage by caspase 8 or caspase 3. *J Biol Chem* 2003;**278**:15749–57.
188. Rytömaa M, Kinnunen PK. Reversibility of the binding of cytochrome c to liposomes. Implications for lipid-protein interactions. *J Biol Chem* 1995;**270**:3197–202.
189. Bayir H, Fadeel B, Palladino MJ, Witasp E, Kurnikov IV, Tyurina YY, et al. Apoptotic interactions of cytochrome c: redox flirting with anionic phospholipids within and outside of mitochondria. *Biochim Biophys Acta* 2006;**1757**:648–59.
190. Joza N, Pospisilik JA, Hangen E, Hanada T, Modjtahedi N, Penninger JM, et al. AIF: not just an apoptosis-inducing factor. *Ann N Y Acad Sci* 2009;**1171**:2–11.
191. Nie C, Tian C, Zhao L, Petit PX, Mehrpour M, Chen Q. Cysteine 62 of Bax is critical for its conformational activation and its proapoptotic activity in response to H2O2-induced apoptosis. *J Biol Chem* 2008;**283**:15359–69.
192. Lalier L, Cartron PF, Olivier C, Logé C, Bougras G, Robert JM, et al. Prostaglandins antagonistically control Bax activation during apoptosis. *Cell Death Differ* 2011;**18**:528–37.
193. Yang E, Zha J, Jockel J, Boise LH, Thompson CB, Korsmeyer SJ. Bad, a heterodimeric partner for Bcl-XL and Bcl-2, displaces Bax and promotes cell death. *Cell* 1995;**80**:285–91.
194. del Peso L, González-García M, Page C, Herrera R, Nuñez G. Interleukin-3-induced phosphorylation of BAD through the protein kinase Akt. *Science* 1997;**278**:687–9.
195. Datta SR, Dudek H, Tao X, Masters S, Fu H, Gotoh Y, et al. Akt phosphorylation of BAD couples survival signals to the cell-intrinsic death machinery. *Cell* 1997;**91**:231–41.
196. Giménez-Cassina A, Garcia-Haro L, Choi CS, Osundiji MA, Lane EA, Huang H, et al. Regulation of hepatic energy metabolism and gluconeogenesis by BAD. *Cell Metab* 2014;**19**:272–84.
197. Giménez-Cassina A, Danial NN. Regulation of mitochondrial nutrient and energy metabolism by BCL-2 family proteins. *Trends Endocrinol Metab* 2015;**26**:165–75.
198. Gardai SJ, Hildeman DA, Frankel SK, Whitlock BB, Frasch SC, Borregaard N, et al. Phosphorylation of Bax Ser184 by Akt regulates its activity and apoptosis in neutrophils. *J Biol Chem* 2004;**279**:21085–95.
199. Simonyan L, Renault TT, Novais MJ, Sousa MJ, Côrte-Real M, Camougrand N, et al. Regulation of Bax/mitochondria interaction by AKT. *FEBS Lett* 2016;**590**:13–21.
200. Linseman DA, Butts BD, Precht TA, Phelps RA, Le SS, Laessig TA, et al. Glycogen synthase kinase-3beta phosphorylates Bax and promotes its mitochondrial localization during neuronal apoptosis. *J Neurosci* 2004;**24**:9993–10002.

201. Arokium H, Ouerfelli H, Velours G, Camougrand N, Vallette FM, Manon S. Substitutions of potentially phosphorylatable serine residues of Bax reveal how they may regulate its interaction with mitochondria. *J Biol Chem* 2007;**282**:35104–12.
202. Owens TW, Valentijn AJ, Upton JP, Keeble J, Zhang L, Lindsay J, et al. Apoptosis commitment and activation of mitochondrial Bax during anoikis is regulated by p38MAPK. *Cell Death Differ* 2009;**16**:1551–62.
203. Xin M, Deng X. Protein phosphatase 2A enhances the proapoptotic function of Bax through dephosphorylation. *J Biol Chem* 2006;**281**:18859–67.
204. Xin M, Li R, Xie M, Park D, Owonikoko TK, Sica GL, et al. Small-molecule Bax agonists for cancer therapy. *Nat Commun* 2014;**5**:4935.
205. Sasaki K, Sato M, Umezawa Y. Fluorescent indicators for Akt/protein kinase B and dynamics of Akt activity visualized in living cells. *J Biol Chem* 2003;**278**:30945–51.
206. Santi SA, Lee H. The Akt isoforms are present at distinct subcellular locations. *Am J Phys Cell Physiol* 2010;**298**:C580–91.
207. Hoshi M, Sato M, Kondo S, Takashima A, Noguchi K, Takahashi M, et al. Different localization of tau protein kinase I/glycogen synthase kinase-3 beta from glycogen synthase kinase-3 alpha in cerebellum mitochondria. *J Biochem* 1995;**118**:683–5.
208. Whiteman EL, Cho H, Birnbaum MJ. Role of Akt/protein kinase B in metabolism. *Trends Endocrinol Metab* 2002;**13**:444–51.
209. Leist M, Single B, Castoldi AF, Kühnle S, Nicotera P. Intracellular adenosine triphosphate (ATP) concentration: a switch in the decision between apoptosis and necrosis. *J Exp Med* 1997;**185**:1481–6.
210. Hu Y, Benedict MA, Ding L, Núñez G. Role of cytochrome c and dATP/ATP hydrolysis in Apaf-1-mediated caspase-9 activation and apoptosis. *EMBO J* 1999;**18**:3586–95.
211. Singh KK, Russell J, Sigala B, Zhang Y, Williams J, Keshav KF. Mitochondrial DNA determines the cellular response to cancer therapeutic agents. *Oncogene* 1999;**18**:6641–6.
212. Wochna A, Niemczyk E, Kurono C, Masaoka M, Majczak A, Kedzior J, et al. Role of mitochondria in the switch mechanism of the cell death mode from apoptosis to necrosis—studies on rho0 cells. *J Electron Microsc* 2005;**54**:127–38.
213. Frank S, Gaume B, Bergmann-Leitner ES, Leitner WW, Robert EG, Catez F, et al. The role of dynamin-related protein 1, a mediator of mitochondrial fission, in apoptosis. *Dev Cell* 2001;**1**:515–25.
214. Parone PA, Martinou JC. Mitochondrial fission and apoptosis: an ongoing trial. *Biochim Biophys Acta* 2006;**1763**:522–30.
215. Pandey S, Kawai T, Akira S. Microbial Sensing by Toll-Like Receptors and Intracellular Nucleic Acid Sensors. *Cold Spring Harb Perspect Biol* 2015;**7**:a016246.
216. Seth RB, Sun L, Ea CK, Chen ZJ. Identification and characterization of MAVS, a mitochondrial antiviral signaling protein that activates NF-kappaB and IRF 3. *Cell* 2005;**122**:669–82.
217. Meylan E, Curran J, Hofmann K, Moradpour D, Binder M, Bartenschlager R, et al. Cardif is an adaptor protein in the RIG-I antiviral pathway and is targeted by hepatitis C virus. *Nature* 2005;**437**:1167–72.
218. Johnson CL, Gale Jr. M. CARD games between virus and host get a new player. *Trends Immunol* 2006;**27**:1–4.
219. Li XD, Sun L, Seth RB, Pineda G, Chen ZJ. Hepatitis C virus protease NS3/4A cleaves mitochondrial antiviral signaling protein off the mitochondria to evade innate immunity. *Proc Natl Acad Sci U S A* 2005;**102**:17717–22.
220. Lin R, Lacoste J, Nakhaei P, Sun Q, Yang L, Paz S, et al. Dissociation of a MAVS/IPS-1/VISA/Cardif-IKKepsilon molecular complex from the mitochondrial outer membrane by hepatitis C virus NS3-4A proteolytic cleavage. *J Virol* 2006;**80**:6072–83.
221. Horner SM, Wilkins C, Badil S, Iskarpatyoti J, Gale Jr. M. Proteomic analysis of mitochondrial-associated ER membranes (MAM) during RNA virus infection reveals dynamic changes in protein and organelle trafficking. *PLoS ONE* 2015;**10**:e0117963.
222. Horner SM, Liu HM, Park HS, Briley J, Gale Jr. M. Mitochondrial-associated endoplasmic reticulum membranes (MAM) form innate immune synapses and are targeted by hepatitis C virus. *Proc Natl Acad Sci U S A* 2011;**108**:14590–5.
223. Vénéreau E, Ceriotti C, Bianchi ME. DAMPs from Cell Death to New Life. *Front Immunol* 2015;**6**:422.
224. Collins LV, Hajizadeh S, Holme E, Jonsson IM, Tarkowski A. Endogenously oxidized mitochondrial DNA induces in vivo and in vitro inflammatory responses. *J Leukoc Biol* 2004;**75**:995–1000.
225. Zhang Q, Raoof M, Chen Y, Sumi Y, Sursal T, Junger W, Brohi K, Itagaki K, Hauser CJ. Circulating mitochondrial DAMPs cause inflammatory responses to injury. *Nature* 2010;**464**:104–7.

226. West AP, Shadel GS. Mitochondrial DNA in innate immune responses and inflammatory pathology. *Nat Rev Immunol* 2017;**17**:363–75.
227. Bernhard W, Rouiller C. Close topographical relationship between mitochondria and ergastoplasm of liver cells in a definite phase of cellular activity. *J Biophys Biochem Cytol* 1956;**2**:73–8.
228. Porter KR, Palade GE. Studies on the endoplasmic reticulum. III. Its form and distribution in striated muscle cells. *J Biophys Biochem Cytol* 1957;**3**:269–300.
229. Hoffmann PC, Kukulski W. Perspective on architecture and assembly of membrane contact sites. *Biol Cell* 2017;**109**:400–8.
230. Vance JE. Phospholipid synthesis in a membrane fraction associated with mitochondria. *J Biol Chem* 1990;**265**:7248–56.
231. Rusiñol AE, Cui Z, Chen MH, Vance JE. A unique mitochondria-associated membrane fraction from rat liver has a high capacity for lipid synthesis and contains pre-Golgi secretory proteins including nascent lipoproteins. *J Biol Chem* 1994;**269**:27494–502.
232. Shiao YJ, Lupo G, Vance JE. Evidence that phosphatidylserine is imported into mitochondria via a mitochondria- associated membrane and that the majority of mitochondrial phosphatidylethanolamine is derived from decarboxylation of phosphatidylserine. *J Biol Chem* 1995;**270**:11190–8.
233. Stone SJ, Vance JE. Phosphatidylserine synthase-1 and -2 are localized to mitochondria-associated membranes. *J Biol Chem* 2000;**275**:34534–40.
234. Grimm S. The ER-mitochondria interface: the social network of cell death. *Biochim Biophys Acta* 2012;**1823**:327–34.
235. Vance JE. MAM (mitochondria-associated membranes) in mammalian cells: lipids and beyond. *Biochim Biophys Acta* 2014;**1841**:595–609.
236. van Vliet AR, Verfaillie T, Agostinis P. New functions of mitochondria associated membranes in cellular signaling. *Biochim Biophys Acta* 2014;**1843**:2253–62.
237. Herrera-Cruz MS, Simmen T. Of yeast, mice and men: MAMs come in two flavors. *Biol Direct* 2017;**12**:3.
238. Marchi S, Patergnani S, Missiroli S, Morciano G, Rimessi A, Wieckowski MR, et al. Mitochondrial and endoplasmic reticulum calcium homeostasis and cell death. *Cell Calcium* 2018;**69**:62–72.
239. Hung V, Lam SS, Udeshi ND, Svinkina T, Guzman G, Mootha VK, et al. Proteomic mapping of cytosol-facing outer mitochondrial and ER membranes in living human cells by proximity biotinylation. *elife* 2017;**6**:. pii:e24463.
240. Giacomello M, Pellegrini L. The coming of age of the mitochondria-ER contact: a matter of thickness. *Cell Death Differ* 2016;**23**:1417–27.
241. Igaki T, Miura M. Role of Bcl-2 family members in invertebrates. *Biochim Biophys Acta* 2004;**1644**:73–81.

CHAPTER

# 3

# Overview of the Cross-Talk Between Hormones and Mitochondria

*Béatrice Morio**, *François Casas*[†], *Luc Pénicaud*[‡]

*CarMeN Laboratory, Research Unit U. 1060 Inserm/Lyon1 University/INRA 1397, Lyon Sud Medical School, Oullins, France [†]DMEM – Dynamique Musculaire et Métabolisme, UMR 866, Montpellier Cedex, France [‡]STROMALab, Université de Toulouse, EFS, ENVT, Inserm U1031, ERL CNRS 5311, CHU Rangueil, Toulouse, France

## 1 INTRODUCTION

Mitochondria play highly specialized, indispensable roles in the synthesis and secretion of hormones, and are involved in many physiological functions in vertebrates including humans. In turn, these hormones act on tissues' metabolism and functions via different mechanisms including all aspects of mitochondria biology. This chapter will review some of these important cross-talks between hormones and mitochondria by underlying striking examples.

## 2 INVOLVEMENT OF MITOCHONDRIA IN HORMONE SYNTHESIS, AN EXAMPLE: STEROID HORMONES

Mitochondria are involved in steroid hormone (glucocorticoids, mineralocorticoids, estrogens, progesterons, androgens, and neurosteroids) biosynthesis of which cholesterol is the precursor. For the synthesis of steroid hormones (Fig. 1), cholesterol must be transported from the cytosol to the inner membrane of the mitochondria via a series of protein-protein interactions involving cytosolic and mitochondrial proteins located at both the outer and inner mitochondrial membranes.[1,2] This transfer is provided mainly by the StAR (steroidogenic acute regulatory) proteins later renamed STARD1. Full-length StAR is a 37 kDa



https://doi.org/10.1016/B978-0-12-811752-1.00003-1

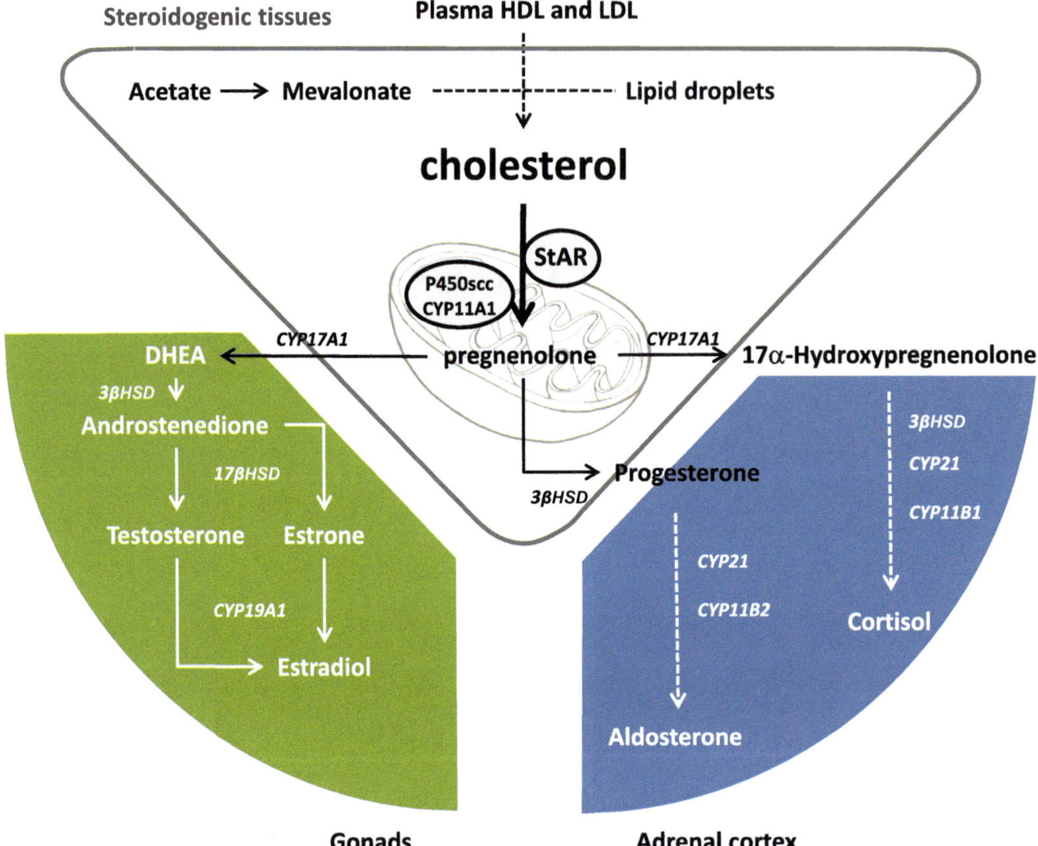


FIG. 1 De novo synthesis of steroids from cholesterol. The synthesis of steroid hormones requires the contribution of several enzymes from the family of the cytochrome P450 oxidase, the main one being coded by the CYP11A1 gene and located at the matrix side of the inner mitochondrial membrane. The conversion of cholesterol to pregnenolone by mitochondrial P450scc is the first rate-limiting step in the synthesis of all steroid hormones. It is made of three stages—the 22-hydroxylation of cholesterol, the 20-hydroxylation of 22-hydroxycholesterol, and the oxidative scission of the C20–22 bond of the subsequent 20,22-dihydroxycholesterol—to yield pregnenolone and isocaproaldehyde. Pregnenolone, the precursor of all steroids, is then converted to tissue-specific steroid hormones through steroidogenic enzymes located in mitochondria or the endoplasmic reticulum (ER). For mitochondrial P450, the carriers of electrons are the adrenodoxine and the adrenodoxine reductase. NADPH is the donor of electrons in each case.

protein with a C-terminal cholesterol-binding domain and an N-terminal mitochondrial targeting sequence common to matrix proteins. StAR is a big cholesterol transporter that is synthesized just after hormonal stimulation and contributes to steroidogenesis. StAR, however, does not act in every tissue, such as in the nervous system and the placenta, involved in steroid synthesis. Thus, in these steroidogenic tissues, the transfer of cholesterol from the cytosol to the mitochondria is done by other means.

## 3 MITOCHONDRIA AND HORMONE SECRETION

Mitochondria have been described as being involved in the synthesis and secretion of other hormones, including pancreatic hormones (insulin, glucagon, and probably somatostatin) and other hormones or peptides from the gastrointestinal tract (GLP-1 for example). It is in pancreatic β-cells, however, that it has been largely studied. In these cells, mitochondrial activity translates glucose metabolism into signals that control the rate of insulin exocytosis. The concept of metabolism-secretion coupling that characterizes the β-cell is tightly controlled by on and off signals, most of them requiring mitochondrial function. These mechanisms are described in detail in Chapter 9.

## 4 MITOCHONDRIA AS TARGETS OF HORMONAL REGULATION

We focus here on major hormones known to regulate mitochondrial functions (thyroid hormones, steroid hormones, insulin and growth factors, adipokines, ghrelin, glucagon-like peptide 1, natriuretic peptides, cytokines, and myokines), but other peptidic factors or neurotransmitters also can regulate mitochondrial functions (see Chapter 7).

### 4.1 Thyroid Hormones

Thyroid hormones (TH) have a multiplicity of effects and influence developmental processes, thermogenesis, and metabolism. TH also are considered to be major regulators of mitochondrial function. This part focuses on physiological influences of TH mediated by the mitochondrial T3 receptor (p43).

Thyroxine (T4) and triiodothyronine (T3) are the main iodothyronines produced by thyroid gland. T3 is the active form of TH, and most of the systemic T3 is generated by deiodination of T4 by deiodinase within peripheral tissues. Because TH action takes place intracellularly, a specific transport of the hormone across the plasma membrane is required. Uptake of the hormone into peripheral tissues is mediated through the monocarboxylate transporter (MCT) MCT8 and MCT10, which act as specific TH transporters.[3]

At the cellular level, TH actions can be initiated within the nucleus, at the plasma membrane and in mitochondria. In the nucleus, T3 regulates genes expression through binding to the nuclear receptors α (TRα) and β (TRβ), which function as transcription factors. The transcriptional activity of TRs is regulated at multiple levels by: the type of thyroid hormone response elements located on the promoters of T3 target gene; the tissue-dependent expression of TR isoforms; the interaction with heterodimeric partners; and nuclear coregulatory proteins that modulate the transcription activity of TRs in a T3-dependent manner (Fig. 2).

Actions of thyroid hormone that are not initiated in the nucleus by TRs are called nongenomic. At the plasma membrane, the work of Davis's team demonstrates that the interaction of TH to integrin αvβ3 activates ERK1/2 pathways. More recently, Kalyanaraman et al.[4] reported that a 30 kDa TRα1 membrane protein activates protein kinase GII, Src, ERK, and Akt signaling after T3 binding.

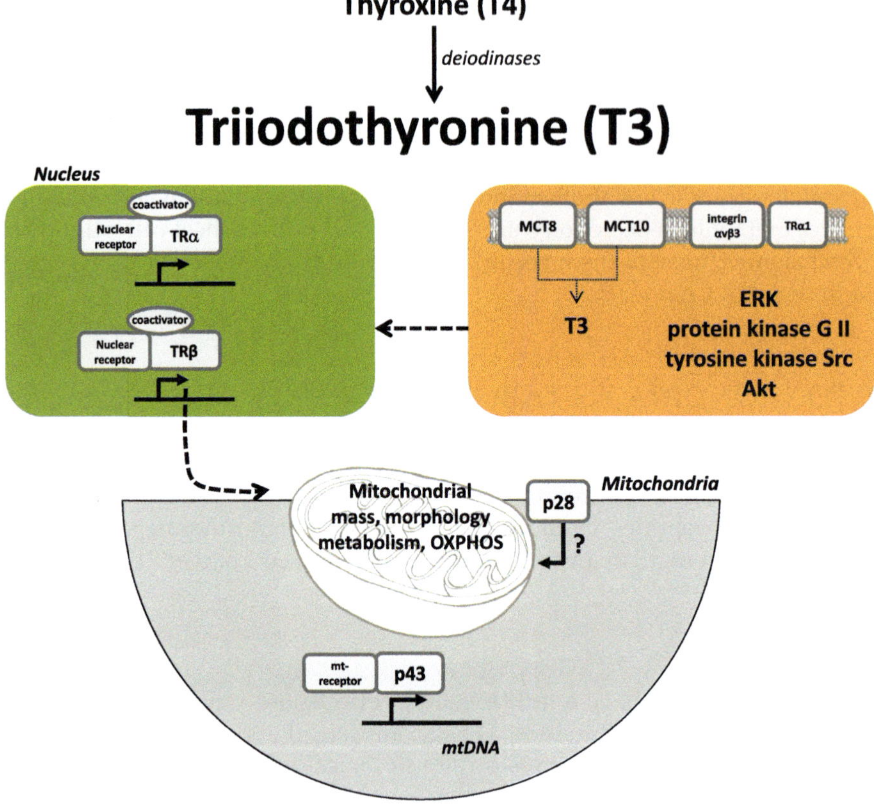


FIG. 2 Cellular targets of and triiodothyronine (T3) in the regulation of mitochondrial function.

At the mitochondrial level, TH are well known to regulate energy use and mitochondrial mass. Short-term and long-term effects of TH actions, however, have been reported in mitochondria.[5] In few minutes, hormone treatment induces a rapid stimulation of mitochondrial respiration and mitochondrial transcription. These effects occur in hypothyroid rats or in isolated mitochondria, and most probably involve a direct action of T3 at mitochondrial level. In several days, hormone treatment leads to a stimulation of mitochondrial biogenesis, suggesting coordination between nuclear and mitochondrial transcription.

### 4.1.1 Direct Action of T3

*Identification of a mitochondrial TRα1 receptors:* Numerous studies reported that the thyroid hormone stimulates mitochondrial genome expression and that important amounts of the hormone are internalized rapidly in the organelle,[6] thus raising the hypothesis that T3 could affect mitochondrial activity directly.[7] Using photoaffinity labeling of highly purified mitochondrial proteins with PAL-T3, two binding proteins with molecular masses of 43 (p43) and 28 kDa (p28) were identified, respectively located in the mitochondrial matrix and in the inner membrane. These two proteins are generated by the use of internal AUGs identified in the TRα1 mRNA.[8] In addition, p43 specifically binds to mitochondrial T3 responsive element (TRE) sequences and displays an affinity for T3 similar to that of the T3 nuclear receptor ($K_a$ = 109 $M^{-1}$). On isolated mitochondria, p43 stimulates mitochondrial transcription and protein synthesis in the presence of T3.[9] All these data demonstrate that p43 acts as a mitochondrial transcription factor. We found that this receptor is ubiquitously expressed.[8]

Concerning the 28 kDa protein, the picture is less clear. This protein, devoid of the ligand binding domain, displays an affinity for T3 higher than the nuclear receptor ($K_a$ = 3.3 1010 $M^{-1}$). Mice overexpressing p28 using the human a-skeletal actin promoter, present an early embryonic lethality probably linked to a transient expression of p28 in trophoblast giant cells. This could be explained partly by the observation that overexpression of p28 in human fibroblasts induced alterations of mitochondrial physiology.[10] Since these initial studies, several other truncated nuclear receptors, such as glucocorticoid, estrogen, vitamin D, RXRα or PPARγ receptors, have been identified in the organelle.[11–14] In agreement with the occurrence of dimerization domains in all these proteins, we reported that p43 interacts with mt-RXR or mt-PPAR to regulate mitochondrial transcription.[14]

*p43 influences at the cell level:* In order to assess influence of p43 at the cell level, in vitro experiments using cell cultures were performed. In avian QM7 myoblasts or murine C2C12 cells, p43 overexpression stimulates mitochondrial activity and potentiates terminal differentiation, whereas an inhibition of this pathway induces exactly the opposite changes.[15,16] Several target genes involve in this myogenic influence have been identified. In particular, in myoblasts, p43 overexpression stimulates terminal differentiation, by downregulating c-Myc expression and upregulating myogenin expression.[15,16] Thereafter, it also induces a preferential expression of the myosin heavy chain I (MHC-I), a slow twitch contractile myosin, through increasing calcineurin levels.[17] In parallel, overexpression of p43 in human dermal fibroblasts induces a strong stimulation of ROS production and a potent oxidative stress leading to cell transformation.[18] This transforming phenotype could be explained by an increase expression of c-Jun and c-Fos proto-oncogenes and an inhibition of tumor suppressor genes expression, such as p53, p21WAF1, and Rb.[18] Saelim et al.[19] reported that the ability of T3 treatment to inhibit apoptosis in CV1 cells and Xenopus oocytes was dependent on the expression of its TR mitochondrial receptor.

*Physiological influences of p43:* To further investigate the physiological importance of p43, mice specifically depleted of this receptor or overexpressing p43 in skeletal muscle were generated.[20,21] The physiological involvement of p43 in glucose homeostasis, thermogenesis and Sertoli cells proliferation was performed only in p43−/− mice (Fig. 3).

## MUSCLE PHENOTYPE

In line with our in vitro studies, in 2-month-old mice, p43 overexpression in skeletal muscle increases mt-DNA content, mitochondrial transcription, and respiratory chain activity leading to a shift in metabolic and contractile features of muscle fibers toward a slower and more oxidative phenotype, in particular, through the regulation of the expression of PGC-1α and PPARδ.[20] During aging, however, p43 overexpression induces an oxidative stress characterized by a strong increase of lipid peroxidation and protein oxidation. This leads to a progressive degradation of mitochondrial activity and to muscle atrophy.[22] Conversely, p43 depletion in mice (p43−/−) reduces mt-DNA content, mitochondrial respiratory chain

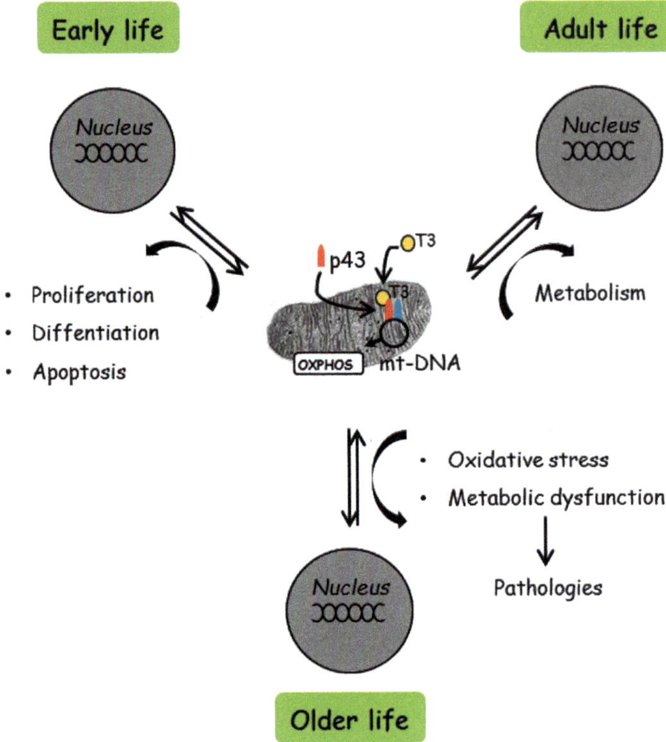


FIG. 3 p43 actions take place during development, adult life, and aging.

activity and induces a shift toward a faster and more glycolytic muscle fiber phenotype.[21] In addition, the absence of p43 leads to an increase of muscle mass. These sets of data demonstrate that the mitochondrial thyroid hormone pathway, through its p43 receptor and the modulation of mitochondrial activity, regulates the metabolic and contractile features of myofibers as well as muscle mass.

## GLUCOSE HOMEOSTASIS

Both in vivo and in isolated pancreatic islets, the absence of p43 induces loss of glucose-stimulated insulin secretion.[23] In addition, the pancreatic islet density is severely decreased in p43−/− mice. If plasma glucose levels are not altered in p43−/− mice fed with standard diet, however, high-fat/high-sucrose diet elicited more severe diabetes and glucose intolerance than that recorded in normal animals. Interestingly, p43−/− mice evidenced a decrease in the activity of complexes of the respiratory chain in isolated pancreatic islets. These dysfunctions were associated with a downregulation of the expression of the glucose transporter Glut2 and of Kir6.2, a key component of the KATP channel (see Chapter 9). During aging, the depletion of p43 in mice progressively induces an increased glycemia in the fasted state, glucose intolerance, and several features of insulin-resistance in type-2 diabetes.[24] These observations suggest that the absence of p43 does not affect glycemia regulation in normal conditions in young animals, but impairs the adaptability of this regulation to adverse conditions such as high caloric diets intake or aging. Therefore, all these findings establish that p43 is an important regulator of glucose homeostasis and β-cells function, and they support the hypothesis that deletion or mutation in p43 could be a factor in patients with type 2 diabetes.

#### THERMOGENESIS AND THYROID HORMONE LEVELS

Whereas body temperature is increased in p43 overexpressing mice, it is decreased in p43 −/− animals in a normal environment.[20,21] In contrast to whole TRα invalidation,[25] however, p43 −/− mice do not present cold intolerance or defect of facultative thermogenesis.[26] In line with these observations, the mitochondrial function of brown adipose tissue (BAT) is slightly affected in the absence of p43. Therefore, p43 could be involved in the determination of the set point in temperature regulation, whereas the nuclear pathway could regulate adaptive thermogenesis.[26]

## 4.2 Steroid Hormones

Sex hormones (estrogen, progesterone, and testosterone) can influence the expression of mitochondrial genes directly or indirectly by binding to their nuclear receptors [estrogen receptors (ERα and ERβ), progesterone receptors (PR-A and PR-B), and androgen receptors (AR1 and AR2)]. Estrogens, unlike testosterone and progesterone, are particularly involved in the regulation of mitochondrial function and biogenesis.

### *4.2.1 Estrogen*

Estrogens coordinate and integrate cellular metabolism and mitochondrial activities through direct and indirect mechanisms involving ER expression and localization (Fig. 4). Estrogens regulate nuclear gene transcription by binding and activating ERα and ERβ receptors, and by activating intracellular signaling pathways (e.g. ERK MAPK, PI3K, CREB) through interaction with the membrane-bound G protein-coupled estrogen receptor 1 (GPER1) or other G protein-coupled receptors (e.g. GPR30). Beneficial effects of estrogen on mitochondrial function are particularly important in tissues with high energy demands, such as the central nervous system, cardiac and skeletal muscles, or brown adipose tissue. ERα and ERβ both are able to induce mitochondrial function, but ERβ activation often induces a greater increase in the functional activity of mitochondria.[27]

At the nuclear level, estrogens can directly regulate the transcription of the nuclear respiratory factor-1 (NRF1), a key transcription factor necessary to regulate the expression of most proteins of the mitochondrial respiratory chain. The NRF1 promoter contains putative estrogen receptor response elements capable of binding the two ERs.[28] ERα and ERβ also can interact with PGC-1α and PGC-1β cofactors, and participate in the regulation of NRF1 expression and mitochondrial target proteins, such as cytochrome *c*, ATP synthase beta,[29] and mitochondrial transcription factor A (TFAM).[30,31] Therefore, estrogens' impacts on mitochondrial function are mainly mediated through enhancement of NRF1 and PGC-1 expression.[32,33] Finally, ERα with PGC-1α activates gene expression of mitofusin 2 (Mfn2), an integrated GTPase in the mitochondrial outer membrane and essential for mitochondrial fusion.[34]

The localization of ERα and ERβ receptors in mitochondria suggests that estrogens modulate mitochondrial function by directly affecting the transcriptional activity of mitochondrial

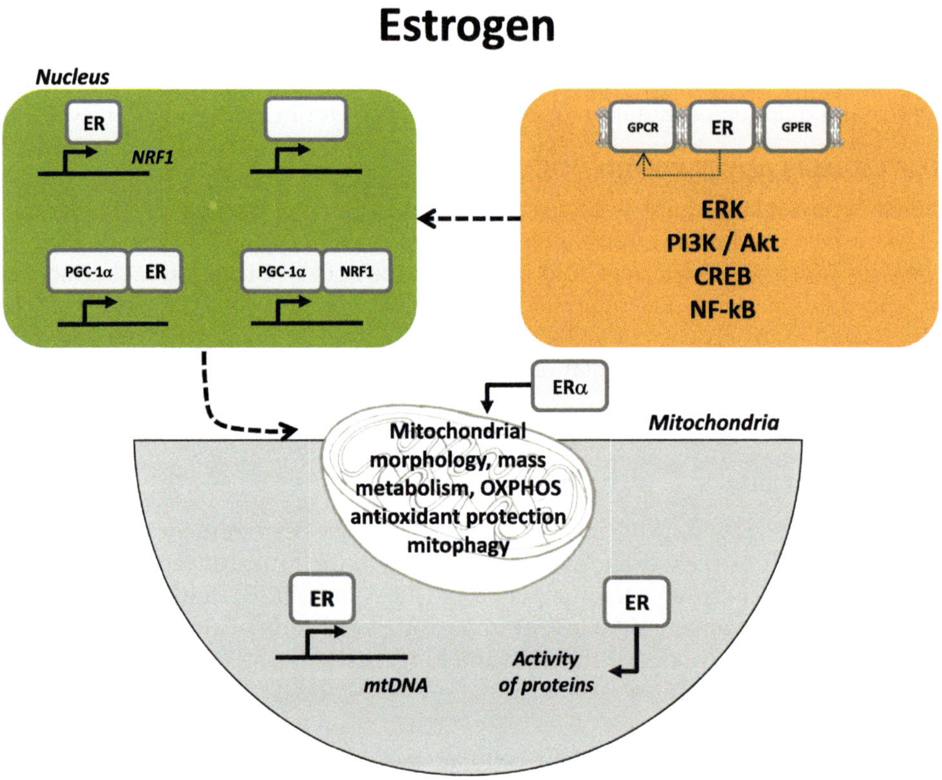

FIG. 4 Cellular targets of estrogen in the regulation of mitochondrial function.

DNA (mtDNA). Mitochondrial ERβ has been described as binding to estrogen response element (ERE)-like sequences in mtDNA.[35,36] This could help promote the expression of mtDNA genes associated with the electron transport chain and oxidative capacity, antioxidant defenses, and calcium buffering capacity.[37–39] Interactions between ERs and mitochondrial proteins also have been suggested. For example, ERβ was found to co-immunoprecipitate with the mitochondrial ATP synthase protein.[40] Whether this interaction is functional, however, remains to be demonstrated. Moreover, ERα has been shown recently to induce mitochondrial fission through Ser616 phosphorylation of DRP1.[41]

It is acknowledged that estrogens have an antioxidant effect-protection in particular mitochondrial functions. ERβ has been identified as the most effective receptor against oxidative stress in several types of tissues and cells, but some studies also have supported a protective role for ERα (see review[42]). Early work has shown that estrogens also act indirectly through intracellular phosphorylation cascades that involve the Akt, ERK MAP kinases, and NF-kB pathways, and result in the induction of the superoxide dismutase (SOD) activity and glutathione pathway (see review[32]). More recently, estrogens have been implicated in the mitochondrial unfolded protein response (UPRmt) that orchestrates several protective pathways, including protein quality control, antioxidant defenses, oxidative phosphorylation, mitophagy, and mitochondrial biogenesis. Thus, Papa and Germain[43] showed that under mitochondrial stress (located in the intermembrane space, but not in the matrix), ERα was specifically activated by phosphorylation by Akt. This resulted in induction of NRF1 and mitochondrial unfolded protein response (UPRmt), characterized by enhanced OME protease and 26S proteasome activity, which help preventing accumulation of unfolded proteins within the inter-membrane mitochondrial space.[44] The authors concluded that the ERα/NRF1/proteasome axis of the UPRmt is a cytoprotective response that contributes to maintaining the integrity of mitochondria.

Estrogens, however, can have opposite effects on cell death. The mechanisms vary according to the tissue involved, and past and present exposure to hormones and, as for the response to oxidative stress, involve different pathways. Under some conditions, estrogens are potentially pro-apoptotic via the involvement of death receptors as well as mitochondrial pro-apoptotic pathways (see review[32]).

### 4.2.2 *Progesterone*

Progesterone regulates gene expression via two nuclear progesterone receptors (PR-B and PR-A). Yet, actions of progesterone independent of gene regulation have been observed for decades, although mechanisms are less deciphered than for estrogen. Progesterone was shown to induce whole brain mitochondria oxidative capacity but not biogenesis. This was associated with enhanced gene expression of COX subunits, resulting in enhanced COX activity.[45] Furthermore, neuroprotective properties of progesterone were attributed to its ability to preserve mitochondrial function in mice brain[46] and spinal motoneurons.[47] Recently discovered mitochondrial progesterone receptor (PR-M) has been involved in modulation of mitochondrial membrane potential in hTERT-infected myometrial cells[48] and human spermatozoa.[49]

### 4.2.3 *Testosterone*

Circulating testosterone concentrations are positively correlated with Vo2max and with muscle mRNA content of oxidative phosphorylation genes in 60-year-old men,[50] but 6 months of testosterone treatment did not alter the expression of genes involved in mitochondrial biogenesis, OXPHOS or lipid metabolism, although whole body lipid oxidation was enhanced after treatment.[51]

Studies have reported that testosterone deprivation aggravated the impairment of mitochondrial function, mitochondrial respiratory complex, mitochondrial dynamics proteins, and apoptosis, leading to cardiac dysfunction[52] or neurodegenerative disorders.[53] In male rodents, removal of the testes decreased expression of genes involved in energy metabolism (OXPHOS, ubiquinone pathway) and testosterone treatment restored these parameters to baseline control levels.[54] Importantly, testes ablation was associated with decrease in muscle mRNA content in insulin-like growth factor 1 (IGF-1), IGF-1 receptor (IGF-1R), and androgen receptor (AR) suggesting the possible involvement of the growth hormone pathway in mediating testosterone effects. In another study, testosterone treatment induces heat production and gene expression of PGC-1α, ATP5B, and Cox4 in skeletal muscle, but not in the brown adipose tissue and liver of male mice.[55] These adaptations were absent in androgen receptor deficient mice. Similarly, testosterone treatment potentiates the effect of low-intensity exercise training in older male mice by increasing spontaneous physical activity, muscle mass, and strength, as well as mitochondrial biogenesis and respiration.[56] Finally, overexpression of the androgen receptor in cultured myocytes increases mitochondrial enzymatic activities and oxygen consumption.[57,58] However, potentially because of a toxic effect, testosterone overload in rats reduces mitochondrial function and increases lipid peroxidation in skeletal muscle.[59]

#### 4.2.4 *Glucocorticoids*

Glucocorticoids are important metabolic regulators, which increase hepatic glucose production and decrease peripheral glucose uptake by skeletal muscles. The action of glucocorticoids involves the cytoplasmic glucocorticoid receptor II (GR), the activity of which can be influenced by GR isoforms, ligand binding, and several regulatory compounds and processes (see review[60]). Glucocorticoids regulate mitochondrial biogenesis and bioenergetics through the regulation of gene expression both at the mitochondrial and nuclear levels (see review[61]). Data evidenced that GRs are present in mitochondria and that glucocorticoid response elements (GREs) are present in the mitochondrial genome (see review[60]). Furthermore, when bound to ligands, GRs can translocate from cytoplasm into the mitochondria, allowing an acute response to stress. In the short term, therefore, glucocorticoids can increase mitochondrial mass and induce OXPHOS activity to support enhanced energy needs. In contrast, in the long term, chronic exposure to glucocorticoids can have a detrimental impact on mitochondrial functions. It is associated with mitochondrial fission (glucocorticoids induce the expression of the fission protein Drp1[62]), alteration of mitochondrial membrane potential, enhanced ROS production and increased sensitivity to apoptosis (see review[61]).

### 4.3 Insulin and Growth Factors

Growth hormone (GH), IGF-1, and insulin are three intrinsically linked hormones. GH regulates the production of IGF-1 in liver and nonliver tissues. In turn, IGF-1 inhibits GH production in a negative feedback process. The liver is the primary source of IGF-1 ($\approx$70%), but IGF-1 also is produced in other tissues, including the brain, testes, skeletal muscle, bone, and cartilage. Secretion of IGF-1 and insulin is enhanced in postprandial situations and decreased during fasting, and GH and IGF-1 are involved in the regulation of insulin response. Furthermore, IGF-1 type I receptor and insulin receptor have 60% homology and the intracellular signaling pathways induced by insulin are shared with IGF-1.

#### 4.3.1 *Insulin*

Impacts of insulin on mitochondrial oxidative capacity have been especially studied in human skeletal muscle by Shree Nair laboratory. The authors reported an insulin-induced increase in the expression of genes encoded by mitochondria (ND1, COX1) in human muscle biopsies carried out before and after an insulin clamp.[63] Furthermore, Stump et al.[64] supported a stimulating action of insulin on mitochondrial protein fractional synthesis rate in human muscle and showed in parallel a positive impact on mitochondrial function, for example, increases in COX; citrate synthase activity; and ATP production. This was associated with an increased mRNA content of mitochondrial (NADH dehydrogenase subunit N) and nuclear (cytochrome *c* oxidase subunit IV)-encoded mitochondrial proteins. The same laboratory showed in type 1 diabetic patients that insulin deprivation decreased muscle mitochondrial ATP production and altered the expression of many mitochondrial genes, particularly those involved in OXPHOS.[65] Insulin-induced stimulation of mitochondrial protein synthesis rate was confirmed in skeletal muscle of insulin-infused pigs (but not in the liver or heart).[66] More recently, insulin stimulatory effect on mitochondrial protein synthesis rate was shown to be dependent on amino acid availability.[67,68] Insulin also stimulated mitochondrial OXPHOS activity and gene expression of mitochondrial proteins in the skeletal muscle of insulin-sensitive human subjects but not of insulin-resistant type 2 diabetic patients.[69] For further details regarding involvement of mitochondria in insulin resistance, see Chapters 6 and 8.

### 4.3.2 *Growth Hormone*

GH infusion induced mitochondrial ATP production and citrate synthase activity in skeletal muscle.[70] GH also induced gene expression of IGF-I, Tfam, and nuclear (cytochrome *c* oxidase subunit IV) and mitochondrial (cytochrome *c* oxidase subunit III) subunits of mitochondrial proteins. No changes in mitochondrial protein synthesis, however, were observed.[70] Because GH infusion stimulates plasma concentrations of IGF-1 and insulin, it is questionable whether the effects of GH on muscle mitochondrial activity are direct or indirect. Contrasting with these findings, results obtained in GH mutant mice evidenced upregulated mitochondrial metabolism in the liver during aging compared to wild-type animals or mutants treated with GH.[71] In particular, dwarf mice treated with GH evidenced a 45% decrease in complex II-driven mitochondrial respiration in the liver compared to dwarf mice treated with saline. GH treatment also decreased by 25% COX activity as well as gene expression and/or protein content of mitochondrial OXPHOS components, suggesting that GH could decrease mitochondrial functions.[71]

### 4.3.3 *IGF-1*

Low dose administration of IGF-1 in aging rats (characterized by low level of circulating IGF-1) was shown to improve glucose and lipid metabolism (insulin sensitivity, circulating levels of cholesterol, and triglycerides). This was associated with enhanced plasma testosterone levels and antioxidant capacity, and reduced oxidative damages.[72] In the liver, low-dose treatment with IGF-1 reversed mitochondrial dysfunctions associated with aging (decreased membrane potential, increased proton leakage rates, increased intramitochondrial production of free radicals [ROS], and reduced activity cytochrome *c* oxidase and ATPase) and reduced caspases 3 and 9 activation. The authors concluded, therefore, that the cytoprotective effects of IGF-1 could be related closely to mitochondrial protection, resulting in a reduction in ROS production, oxidative damage and apoptosis, and an increase in ATP production.[72] Similar results were obtained in experimental cirrhosis in rats. IGF-1 treatment was able to restore all parameters of mitochondrial dysfunction observed in untreated cirrhotic rats.[73,74] Mice with partial deficiency of IGF-1 (igf +/−) are characterized by enhanced oxidative damages in the liver compared to wild-type (igf +/+) and IGF-1 treated igf +/− mice.[75] These data are consistent with a study of human HUVEC cells showing that IGF-1 is preventive against oxidative stress-induced apoptosis.[76]

## 4.4 Adipokines

Adipose tissue is an active organ, with endocrine functions that regulate whole-body energy homeostasis (for review see 77). Among the hormones secreted by the adipose tissue, named adipokines, leptin, adiponectin, apelin, and CTRP3 (complement-C1q/tumor necrosis factor-related protein 3) are involved in the regulation of mitochondrial function, particularly in skeletal muscle and in the liver, two tissues that will be taken as examples in the following sections.

### 4.4.1 *Adiponectin*

Adiponectin (Acrp30) is the most abundant peptide secreted by adipocytes. It also is synthesized by other cell types, such as skeletal and cardiac myocytes as well as endothelial cells. In the plasma, adiponectin is multimerized with variable molecular weights, high molecular weight (HMW) complexes having the predominant action in metabolic tissues. Adiponectin regulates metabolism in the liver, skeletal muscle, cardiovascular system, and other tissues such as white adipose tissue, brain, and kidneys. Its secretion rate decreases in obesity and type 2 diabetes,[78] and this is known to play a central role in diseases related to obesity, including insulin resistance, type 2 diabetes and nonalcoholic fatty liver diseases. Adiponectin effects are mediated by two adiponectin receptors, AdipoR1 and AdipoR2, and T-cadherin receptor (see reviews[79,80]). AdipoR1 and AdipoR2 are expressed mainly in the skeletal muscle and in the liver, respectively. In contrast, T-cadherin receptor is expressed primarily in the cardiovascular system where it promotes cardioprotective effects. The exact physiological consequence of this ligand-receptor relationship, however, is not known.[80]

After binding to its receptors, adiponectin impacts on energy metabolism and mitochondrial function are mediated through activation of 5′-AMP-activated protein kinase (AMPK) in almost all major target tissues.[81–85] AMPK activation causes phosphorylation and deactivation of acetyl-CoA carboxylase (ACC), thus blocking the synthesis of malonyl-CoA, the result being an activation of fatty acid beta-oxidation. In skeletal muscle and liver,[86,87] some AMPK effects also are mediated by the activation of p38 mitogen-activated protein kinase (p38MAPK) (see review 79). Activation of AMPK and p38MAPK induces gene expression of PPARα, a nuclear receptor that regulates gene expression of enzymes regulating fatty acid oxidation (see review[88,89]). In skeletal muscle (see review[88]) and potentially in the liver,[86] adiponectin stimulates the expression of mitochondrial genes, such as PGC1α and PGC1β, suggesting that adiponectin might be involved in the regulation of mitochondrial biogenesis. These effects might be mediated through two independent pathways, one involving AMPK and the other the $Ca^{2+}$/calmodulin-dependent protein kinase (CaMK) pathway. Iwabu et al.[87] showed that adiponectin via the AdipoR1 receptor activates extracellular $Ca^{2+}$ influx, which activates calmodulin-dependent β-protein kinase kinase (CaMKKβ) and therefore CaMK.

### 4.4.2 *Leptin*

Leptin is a satiety signal, acting mainly at the level of the hypothalamus to reduce appetite.[90] Leptin receptors (Ob-R or LEPR), however, are present in virtually all tissues, giving leptin a role in the regulation of energy metabolism. Unlike adiponectin, leptin secretion is associated positively with body fat, and leptin resistance appears in obesity with a sustained increase in circulating leptin.[91] A long form (ObR-L or LEPRb) and a short form (ObR-S) of leptin receptors have been described. The effect of leptin on mitochondrial energetics[92] is somewhat similar to that of adiponectin and involves activation of AMPK and induction of fatty acid beta-oxidation.[93]

In the liver,[94] binding of leptin to ObR-L activates JAK2, allowing the recruitment of downstream effectors such as STAT3 (signal transducer and activator of transcription 3), which leads to the activation of AMPK by the LKB1 kinase. In addition, by activating JAK-2, leptin also can suppress the repression of PPARα activity by its physical sequestration by the activated nonphosphorylated form of FoxO1 by activating protein kinase B (Akt).[95] Finally, it has been shown that, especially in the liver and in the heart, STAT3 activation by phosphorylation on the serine residue (S727) promotes its recruitment in the mitochondria, where it is suggested to modulate the oxidative metabolism by interacting with complexes I and II of the respiratory chain.[96] Data on mitochondrial functions in the liver of leptin-deficient Ob/Ob mice, however, are inconsistent.[97–99] In particular, Singh et al.[99] found that, in Ob/Ob mice treated with leptin for 8 days, hepatic metabolism paradoxically decreased after treatment, resulting in a reduction in mitochondrial volume density and altered mitochondrial respiration. This was associated with decreased levels of the subunits IV and VIa of cytochrome *c* oxidase of the mitochondrial respiratory chain.

In vitro studies on C2C12 and mouse primary myotubes have shown that the effects of leptin on muscle fatty acid oxidation are time-dependent; an initial increase in fatty acid oxidation was observed within the first 6 h of treatment with leptin and a secondary increase was observed after 24 h.[100] Although ObR-L long has been considered the only functional form mediating the effects of leptin, Akasaka et al.[101] showed that fatty acid oxidation induction upon chronic leptin treatment resulted from the activation of the ObR-S/JAK2/p38 MAPK/STAT3 pathway, which was associated with enhanced gene expression of Mcad (medium-chain acyl-coenzyme A dehydrogenase), Cpt1 (carnitine palmitoyltransferase 1) and Ucp2 (uncoupling protein 2).

### 4.4.3 Apelin

Apelin (APJ endogenous ligand) is a peptide hormone and an endogenous ligand for a G protein-coupled receptor called APJ.[102] The presence of apelin in adipose tissue first was described in 2005.[103,104] Apelin and its receptor APJ are expressed in many organs, including hypothalamus, heart, lungs, kidneys, adipose tissue, and muscles.[105] In adipose tissue, apelin expression is stimulated by insulin, and plasma apelin levels markedly increase in obesity associated with insulin resistance and hyperinsulinemia.[106] Contrary to leptin, however, apelin treatments still are effective in obesity and type 2 diabetes.[107] Apelin initially is synthesized as a precursor of 77 amino acids, which then is cleaved into several mature forms of different lengths.[108] Apelin-17 and [pyr-1] -apelin-13 are the major circulating forms (see review[105]).

APJ receptor activates different G proteins depending on the cell type and, therefore, stimulates several key intracellular effectors such as ERK, phosphatidylinositol-3-kinase (PI3K)/Akt or p70S6 kinase, which might explain its multiple intracellular effects (see review[105]). AMPK is also a key intracellular effector of the APJ receptor in skeletal muscle cells, adipocytes, and enterocytes. In these cells, AMPK activation induces fatty acid oxidation and promotes mitochondrial biogenesis and function through mechanisms similar to those activated by adiponectin and leptin.[107,109,110] Recent data evidenced that apelin also induces PPARα protein expression in the liver by a mechanism involving activation of

AMPK and MAP kinases (p38 MAPK and ERK).[111] Dray and collaborators[112] demonstrated that correcting the age-associated decrease in apelin signaling in skeletal muscle enhanced the tissue function by triggering in myofibers mitochondrial biogenesis, autophagy, and anti-inflammatory pathways and enhanced the muscle regenerative capacity by targeting muscle stem cells.

### 4.4.4 CTRP3

CTRP3 is a newly identified adipokine that belongs to a widely expressed and highly conserved family of adiponectin paralogs, named C1q/tumor necrosis factor-alpha-related proteins (CTRPs).[113] CTRP3 is expressed predominantly in adipose tissue, but also in various tissues and cell types.[114] Like adiponectin, all secreted CTRP proteins form trimers as basic structural units that are assembled further into higher order oligomeric complexes.[114] The receptors to CTRP3 are not yet clearly identified.[115] Although some studies have shown a potent biological effect of CTRP3 on the liver,[116] only two in vitro studies are available on the impact of CTRP3 on mitochondrial functions. In cultured vascular smooth muscle cells (VSMCs), CTRP3 enhanced ATP synthesis and expression of key proteins, including PGC-1α, sirtuin-3 (SIRT3), and subunits of the complexes I, II, III, and V of the respiratory chain. The mechanisms might imply CTRP3-induced increase in mitochondrial membrane potential and mitochondrial reactive oxygen species.[117] In cultured cardiomyocytes, CTRP3 increased gene expression of PGC-1α, nuclear respiratory factors 1 and 2 (NRF-1, NRF-2), TFAM, cytochrome B, and subunits of the complexes III and V of the respiratory chain. In addition, CTRP3 increased the number of mitochondrial DNA copies and intracellular ATP content.[118] The authors found in this model that the mechanisms might involve AMPK and sirtuin 1 (SIRT1) activation, both involved in the activation of PGC-1α cofactor.[118]

### 4.4.5 Remarks

As illustrated in Fig. 5, although adiponectin, leptin, apelin, and CTRP3 interact with distinct receptors, they share a common pathway involving AMPK activation to regulation mitochondrial fatty acid oxidative capacity. AMPK activation inhibits ACC by phosphorylation, thus blocking the synthesis of malonyl-CoA and rising the inhibition of CPT1, the result being an activation of fatty acid beta-oxidation. Combined activation of AMPK and of alternative intracellular signaling pathways (e.g., p38MAP, Akt/FOXO1), however, induces gene expression of PPARα, a nuclear receptor that regulates gene expression of enzymes regulating fatty acid beta-oxidation.

## 4.5 Ghrelin

Ghrelin is a peptide of 28 amino acids released by the cells of the stomach during fasting. It acts as a signal of hunger to promote food intake. Since its discovery in 1999, numerous studies have shown that ghrelin has multifaceted peripheral and central biological activities (see review[119]). The enzyme Ghrelin O-acyltransferase (GOAT) specifically acylates ghrelin at Ser3 with octanoic acid, forming an acylated ghrelin recognized by the GHS-R1a

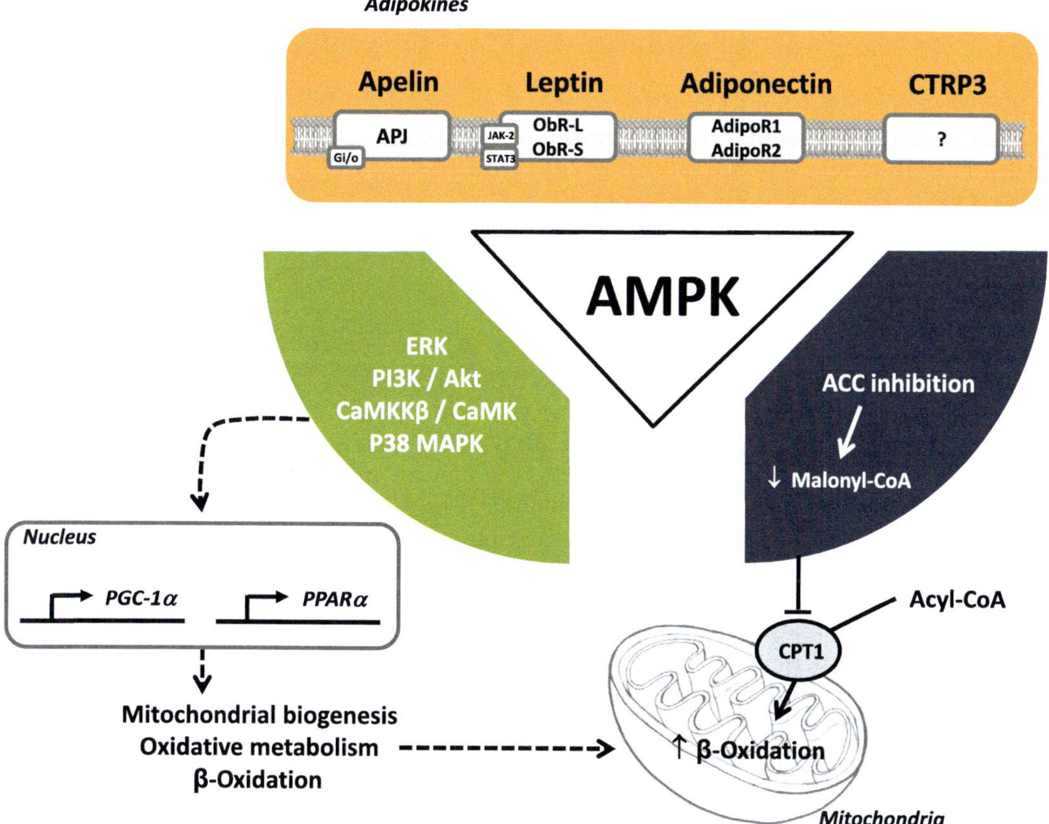


FIG. 5 Cellular pathways involved in the regulation of mitochondrial function by adipokines.

receptor, a G protein-coupled receptor (GPCR). GHS-R1a is expressed in the hypothalamic and extra-hypothalamic regions, as well as in metabolic organs such as liver, kidney, gut, adipose tissue, and myocardium (see review[119]). The various roles of ghrelin in the control of energy metabolism might arise from the heterodimerization of the GHS-R1a receptor with other G protein-coupled receptors. The deacylated form of ghrelin is the most abundant circulating form (70%–90%). Several lines of evidence have shown that the biological functions of deacyl ghrelin are mediated independently of the GHS-R1a receptor, suggesting the existence of another as-yet unidentified receptor (see review[119]). The ratio between deacylated ghrelin (decreased during obesity) and acylated ghrelin (unchanged during obesity) might have an impact on the regulation of tissue energy homeostasis.[120]

Ghrelin's impact on food intake involves modification of mitochondrial functions in the ventromedians of the hypothalamus (VMH), ultimately resulting in the activation of NPY/AgRP neurons (for details through SIRT1-p53 and CaMKKβ pathways, see[121,122]). In this tissue, cannabinoids are necessary for the stimulating effects of ghrelin on AMPK activity.[123] As previously described, AMPK activation and subsequent inhibition of ACC cause a reduction in malonyl-CoA concentration favoring mitochondrial β-oxidation, enhanced mitochondrial ROS production, thus potentiating the activation of NPY/AgRP neurons, resulting in food intake (for details see Chapter 10).

In animal models of chronic kidney disease, physical decline, or chronic heart failure, ghrelin treatment prevented mitochondrial dysfunction in skeletal muscle[124–127] and kidneys.[128] Similarly, ghrelin treatment prevented mitochondrial dysfunction in hepatic injury caused by ischemia/reperfusion[129] and in the brain of a mouse model with Parkinson's disease.[130] The mechanisms might involve SIRT1 induction[131] and mitophagy.[132,133]

## 4.6 Glucagon-Like Peptide 1

The incretin glucagon-like peptide 1 (GLP-1) is secreted by the enteroendocrine L cells in the distal gut in response to a meal. The biologically active forms of GLP-1 are GLP-1 (7–37) and GLP-1 (7–36) NH2, which both bind to the specific receptor GLP-1R expressed in many tissues. After it is in the circulation, the lifetime of GLP-1 is a few minutes before being inactivated by dipeptidyl peptidase 4 (DPP-IV). GLP-1 is a satiety hormone and is well known for its effects on glucose homeostasis, insulin secretion, and action. GLP-1 analogues or DPP-IV inhibitors, therefore, have been developed for treating type 2 diabetes. Experimental models suggest that these GLP-1 analogues exert several extrapancreatic effects and regulate several cellular pathways, including mitochondrial function (see review[134]). GLP-1R activation induces a reduction in mitochondrial ROS production; an increased expression of anti-apoptotic proteins (Bcl-2 and Bcl-XL), thus inhibiting the mitochondrial apoptotic pathway; and an increased mitochondrial biogenesis and respiration potentially through induction of SIRT1 and PGC-1α (see review[134]). Recent studies also suggest that GLP-1 agonists might improve mitochondrial function possibly through Akt/FoxO3a and AMPK/FoxO3a signal pathways.[135] Furthermore, in VSMC cells, GLP-1 was shown to increase expression of mitofusin-2 (Mfn2), to enhance the number of contact points between mitochondria and endoplasmic reticulum (ER), and to improve ER-mitochondrial calcium exchanges.[136]

## 4.7 Natriuretic Peptides

Since the discovery of the atrial natriuretic peptide (NAP) in 1980,[137] numerous studies have highlighted the multidimensional role of natriuretic peptides (NPs) in the regulation of blood volume and pressure. NPs are members of a family of structurally similar but genetically distinct endogenous peptide hormones. The family of mammalian NPs comprises at least three peptides: atrial, cerebral, and C-type (ANP, BNP, and CNP, respectively). ANP and BNP preferentially bind to the specific receptor for guanylyl cyclase (GC) called GC-A, whereas CNP is the physiological ligand of GC-B.[138] Miyashita et al.[139] and Engeli et al.[140] demonstrated that the NP-cGMP system also plays a role in the regulation of mitochondrial biogenesis, OXPHOS functions, and fatty acid oxidation in skeletal muscle via a mechanism involving cGMP, in agreement with previous results showing that the NO-sGC-cGMP pathway induces the expression of PGC-1α.[141]

## 4.8 Cytokines and Myokines

It is acknowledged that obesity is associated with a state of chronic low-grade inflammation (increased expression and circulating levels of pro-inflammatory cytokines; see review[142]). Tumor necrosis alpha (TNF-α) and the interleukin 1β (IL-1 β) are produced by many tissue and immune cells and upregulated in obesity.[142] Recent evidence in cultured neuronal cells showed that TNF-α acutely altered mitochondrial functioning through its receptor 1 (TNF-R1) and caused reduced mitochondrial respiration and membrane potential, increased caspase 8 activity and cytochrome *c* release from the mitochondria into the cytosol.[143] TNF-α also was shown to reduce complexes I, III and IV activity in hepatocytes, whereas in vivo inhibition of TNF-α restored complex III and ATP synthase activity in a model of experimental heart failure (for review see 144).

Other pathways also might be involved. Inflammation-induced oxidative stress can potently alter mitochondrial function, through inhibition of pyruvate and α-ketoglutarate dehydrogenases (for review see 144). In addition, in vitro and in vivo data have shown that TNF-α suppresses AMPK activity via transcriptional upregulation of protein phosphatase 2C (PP2C),[145] thus suppressing fatty-acid beta-oxidation. TNF-α also has been shown to increase the expression of the inducible isoform of the nitric oxide synthase (iNOS)[146] and to downregulate that of the endothelial isoform (eNOS).[147] NO generated by eNOS, however, was shown to promote mitochondrial biogenesis, oxidative metabolism, and ATP levels in several cell types, including white and brown adipocytes and muscle satellite cells,[147] through NO-dependent activation of soluble guanylate cyclase-cGMP-protein kinase G pathway which results in increased expression of PGC-1α, NRF-1 and TFAM.[148]

Pro-inflammatory cytokines were found to alter mitochondrial dynamics, through induction of mitochondrial fission (for review see 144). IL-6 was involved in cachexia-mediated downregulation of Mfn1/Mfn2 and FIS1 proteins.[149] IL1-β was shown to induce the profission protein Drp1 in astrocytes.[150] TNF-α was found to induce the expression of FIS1 and to decrease that of OPA1 in 3T3-L1 adipocytes.[151] In rat intestinal epithelial cells, TNF-α induced mitochondrial dysfunctions, potentially enhanced mitophagy.[152]

Mitochondria, the central regulators of the inflammatory response to stress, have been shown to promote the formation and activation of inflammatory signaling platform known as inflammasomes, among which NLRP3 is the best-described (see review[153]). Mitochondrial ROS, $Ca^{2+}$ overload, reduced $NAD^+$, cardiolipin, mitofusin, and mitochondrial DNA all have been shown to promote NLRP3 assembly and activation. This induces autocleavage and activation of the caspase-1, which then processes the proinflammatory cytokines pro-IL-1β and pro-IL-18 into their bioactive mature forms. Caspase-1 activation also provokes an inflammatory pyroptotic cell death (see review[153]).

By contrast to the pro-inflammatory cytokines expressed in response to nutritional stresses, skeletal muscle was found to secrete cytokines, referred to as myokines, in response to exercise. Myokines have been hypothesized to contribute to the acute and long-term benefits of exercise. Among those myokines, IL-15 induces fatty acid beta-oxidation in skeletal muscle, adipose tissue, and possibly liver.[154] IL-15 shares high degree of functional similarity with IL-2. It binds to and signals through a complex composed of IL-2/IL-15 receptor beta chain (CD122) and the common gamma chain (gamma-C, CD132). In vitro studies on primary myotubes showed that IL-15 is able to directly stimulate transcriptional expression of PGC-1α and PPARδ, enhancing mitochondrial density in the cultures myotubes.[155] Similarly, IL-15 transgenic mice (IL-15tg) evidence elevated mitochondrial activity and mass in adipose tissue compared to controls.[156] Few studies investigated IL-15 expression in obesity and did not evidence any significant alterations.[157]

Irisin was discovered in 2012.[158] The authors showed in mice that PGC1-α expression in muscle stimulates an increase in expression of FNDC5, a membrane protein that is cleaved and secreted as irisin. FNDC5 expression in skeletal muscle was found to be increased in obesity and T2D although hyperglycemia was suspected to inhibit it.[159] Recent evidence also suggests that FNDC5/irisin expression might be inhibited by myostatin in a miR-34a-dependent manner.[160] Irisin first was shown to act on white adipose cells in culture and in vivo to stimulate UCP1 expression and a broad program of brown-fat-like development.[158] On cultured C2C12 myotubes, irisin enhances oxidative metabolism and mitochondrial biogenesis, and induced gene expression of PGC-1α, NRF1, Tfam, and UCP3.[161] Irisin also plays a pivotal role in cardio- and pulmonary-protection because it prevents mitochondrial apoptosis in ischemia and reperfusion injury.[162,163] In those contexts, irisin enters cells via lipid raft-mediated endocytosis and targets mitochondria. Interaction between irisin and mitochondrial UCP2 helps to prevent ischemia-reperfusion induced oxidative stress and to preserve mitochondrial function. This protective effect is lost in UCP2-Ko animals.[162]

## 5 NONCODING NCRNA, NEW AUTOCRINE REGULATORY PATHWAYS FOR MITOCHONDRIAL FUNCTIONS

Small and long noncoding RNAs (ncRNAs) have regulatory roles in the modulation of gene expression. Small and long ncRNAs are not translated into proteins. They include the well-known tRNAs and rRNAs, which are key to production of functional proteins. Several classes of small regulatory ncRNAs (<200 nt) have been described, including small interfering (siRNAs), piwi-associated RNAs (piRNAs), microRNAs (miRNAs), small nucleolar RNAs (snoRNAs), and small nuclear RNAs (snRNAs) (see review[164]). Long ncRNAs (>200 nt, lncRNA) can be single- or double-stranded[165] and might be polyadenylated, determining their stability.[166] Both small and long ncRNAs play a role in modulating gene expression or in the treatment of other RNA species. Ambros[167] evidenced that regulatory ncRNAs, particularly miRNAs and lncRNAs, localize within the mitochondria across diverse physiological and pathological states. It was not until the 2010s that the concept was validated that human and murine mitogenome potentially encodes thousands of small and long ncRNAs, which can selectively localize in different cell compartments, including the nucleus, cytoplasm, and mitochondria (see review[164]).

### 5.1 miRNAs

MiRNAs are encoded by the nuclear genome and are tightly regulated from biogenesis to maturation (see review[82,164]). Although speculative, however, the mapping of a series of pre-miRNA and mature miRNA sequences on the human mitogenome suggest that some miRNAs might originate from the mitochondrial DNA (see review[164]). Pre- and mature miRNAs and proteins involved in their biogenesis are not confined to the cytosol and have been observed enriched in/associated with mitochondria (see review[82]). It has been suggested that mitochondria potentially might be an important site for miRNA-mediated posttranscriptional regulation, its outer-membrane providing a platform to assemble novel signaling complexes.

MiRNAs play an essential role in modulating gene expression by binding to complementary sites on the target mRNA both in the cytosol and in the mitochondria, thereby blocking translation or degrading the target mRNA (see review[164]). Several miRNAs have been found in isolated mitochondria in a tissue specific manner, for example, in rat liver (15 miRNAs), mouse liver (20 miRNAs), myotubes (20 miRNAs), HeLa (6 miRNAs), 143B (3 miRNAs), HEK293 (6 miRNAs), and human muscles (46 mature miRNAs and 2 pre-miRNAs) (see review[82,164]). As mentioned by Silver et al.,[164] many authors have speculated on miRNAs' potential gene and protein targets using target prediction tools, but these predictions should be considered with extreme caution. Among the notable examples that demonstrated miRNA impact on mitochondrial function, Zhang et al.[168] corroborated that miR-1 was localized in the mitochondria of C2C12 cells and that its mitochondrial content increased with myotube differentiation. Furthermore, they showed that mi-R1 formed a functional complex with AGO2, a protein necessary to catalyze the cleavage of the target transcript. This complex was bound to the mitochondrial transcripts COX1, ND1, Cytb, COX3, and Atp8 and surprisingly enhanced their translation, providing the first evidence for miRNA-mediated mitochondrial gene regulation in skeletal muscle cells.[168] The authors showed that miR-1 had a dual impact that involved the repression of the expression of cytoplasmic target genes, as well as the promotion of the expression of mitochondrial target genes.[168] This was particularly relevant to skeletal muscle, which needs to adapt quickly to changes in environmental conditions and energy demands. In human HepG2 cells, Ji et al.[169] reported that miR-141-3p might modulate mitochondrial ATP production and ROS production by targeting the phosphatase and tensin homolog (PTEN). In other respects, miR-27a and miR-27b, which were suggested to target prohibitin and impair adipocyte differentiation and mitochondrial function in human adipose-derived stem cells,[30] have been identified as suppressors of PINK1 expression at the translational level. The resulting blockage of the autophagic clearance of damaged mitochondria was suspected to participate to the autosomal recessive Parkinson's disease.[170]

## 5.2 Long ncRNAs

It has been estimated that the human nuclear genome might encode about 17,000 lncRNAs.[165] LncRNAs modulate gene and protein expression through a variety of molecular processes, including promotion or repression of gene transcription, constitution of support through allosteric modifications to regulatory proteins (facilitating histone methylation and regulating protein activities), and regulation of other ncRNA species (see review[164]). Like miRNAs, lncRNAs regulate mitochondrial function and are dysregulated in pathological states. For example, lncRNA Tug1 binds to and positively regulates Pgc-1α, favoring mitochondrial biogenesis.[171] Conversely, decreased Tug1 expression reduces Pgc-1α transcripts and contributes to alteration of mitochondrial function in diabetic nephropathy.[171] LncRNAs also are encoded by the mitochondrial genome. Recently, two new human mitogenome encoded lncRNAs, MDL1, and MDL1AS, have been discovered in the MCF-7 breast cancer cell line and were suggested to engage new mechanisms to regulate gene expression.[172] Their putative role within the mitochondria, however, still remains to be determined.

## 6 CONCLUSIONS

Mitochondria play an important role in the endocrine/hormonal regulation of numerous physiological functions contributing to whole-body homeostasis (Fig. 6).

Although, apart from steroid synthesis and insulin secretion, data are needed for a better description of the involvement of the organelle in hormone synthesis and secretion. Numerous data are available for the involvement of mitochondria in hormonal regulation of cells and tissues functions. Mitochondria also can be targets of endocrine-disrupting chemicals that can mimic, block, or interfere with the action of hormones, thus contributing to the onset or to the aggravation of metabolic disorders. This issue is addressed in the liver in Chapter 15 and is reviewed in more detail in Marroqui et al.[173]

To summarize, we will highlight some points that have been reviewed in this chapter.

**(1)** The identification of thyroid hormone receptors in mitochondria brings new insights to better understand the pleiotropic influence of thyroid hormones. Recent works demonstrate that thyroid hormones regulate skeletal muscle phenotype, testes development, glucose homeostasis, basal metabolism, and body temperature through complementary action between the mitochondrial and the nuclear thyroid receptors

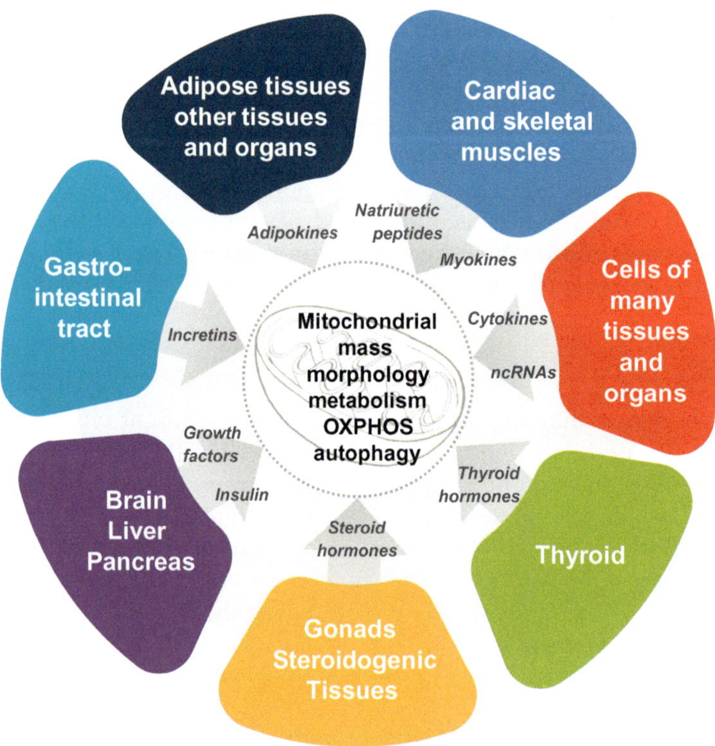


FIG. 6 Main hormones targeting mitochondrial function depicted in Chapter 4. *OXPHOS*, oxidative phosphorylation; *ncRNAs*, noncoding ribonucleic acids.

encoded by the Thra gene. These actions of p43 are diverse and take place during development, adult life, and aging. Because the mitochondrial genome encodes only 13 proteins, it is important to note that p43 actions mainly are indirect. Therefore, through changes in the organelle activity, p43 influences the mitochondrial retrograde signaling and modulates the expression of numerous nuclear genes. Such a regulation enlarges the number of thyroid hormone target genes. After the characterization of p43 and p28, different teams also have identified other members of the nuclear receptor superfamily inside mitochondria, indicating that other hormones are able to regulate nuclear and mitochondrial transcription in a coordinate way through the expression of receptors addressed to different cell compartments.

**(2)** Involvement of estrogen in the regulation of mitochondrial function and biogenesis is well documented, highlighting the crucial role of estrogens in the regulation of mitochondrial function, particularly during female aging. Surprisingly, this is less the case with progesterone and testosterone, although these hormones seem to play key direct or indirect roles in the preservation of mitochondrial functions, particularly in the central nervous system, in the heart, and/or in reproductive organs.

**(3)** Steroid hormones are intrinsically linked to growth factors, which also can have an impact on mitochondrial functions. Because there are also strong interactions between growth factors and insulin, whether the respective effect of each hormone is direct or indirect and whether these effects engage only one specific player, remains to be determined.

**(4)** Coordination between subcellular compartments—the nucleus, cytosol, and mitochondria—is essential for ensuring effective mitochondrial function in physiology and pathologies. Research about ncRNAs entering mitochondria and targeting mitochondrial functions is still in the early stages. In vitro studies suggest that interactions between ncRNAs and mitochondrial mRNAs might directly affect mitochondrial metabolism, but the question of the role that ncRNAs play in mitochondrial translation in vivo remains open. In addition, the involvement of ncRNAs in the mitochondrial-nucleus communication pathway remains to be understood. A possibility is that the distribution of relative amounts of some ncRNAs in the cytoplasm and mitochondria regulates their activity on specific targets to coordinate mitochondrial activities with other cellular functions. Such regulation might extend to simultaneous changes in ncRNA expression in several subcellular compartments. Much remains to do in this field.

## References


1. Fan J, Papadopoulos V. Evolutionary origin of the mitochondrial cholesterol transport machinery reveals a universal mechanism of steroid hormone biosynthesis in animals. *PLoS ONE* 2013;**8**:e76701.
2. Papadopoulos V, Miller WL. Role of mitochondria in steroidogenesis. *Best Pract Res Clin Endocrinol Metab* 2012;**26**:771–90.
3. Visser WE, van Mullem AA, Jansen J, Visser TJ. The thyroid hormone transporters MCT8 and MCT10 transport the affinity-label N-bromoacetyl-[(125)I]T3 but are not modified by it. *Mol Cell Endocrinol* 2011;**337**:96–100.
4. Kalyanaraman H, Schwappacher R, Joshua J, Zhuang S, Scott BT, Klos M, Casteel DE, Frangos JA, Dillmann W, Boss GR, Pilz RB. Nongenomic thyroid hormone signaling occurs through a plasma membrane-localized receptor. *Sci Signal* 2014;**7**:ra48.

5. Wrutniak-Cabello C, Casas F, Cabello G. Thyroid hormone action in mitochondria. *J Mol Endocrinol* 2001;**26**:67–77.
6. Sterling K, Campbell GA, Taliadouros GS, Nunez EA. Mitochondrial binding of triiodothyronine (T3). Demonstration by electron-microscopic radioautography of dispersed liver cells. *Cell Tissue Res* 1984;**236**:321–5.
7. Sterling K, Milch PO, Brenner MA, Lazarus JH. Thyroid hormone action: the mitochondrial pathway. *Science* 1977;**197**:996–9.
8. Wrutniak C, Cassar-Malek I, Marchal S, Rascle A, Heusser S, Keller JM, Fléchon J, Dauça M, Samarut J, Ghysdael J, et al. A 43-kDa protein related to c-Erb A alpha 1 is located in the mitochondrial matrix of rat liver. *J Biol Chem* 1995;**270**:16347–54.
9. Casas F, Rochard P, Rodier A, Cassar-Malek I, Marchal-Victorion S, Wiesner RJ, Cabello G, Wrutniak C. A variant form of the nuclear triiodothyronine receptor c-ErbAalpha1 plays a direct role in regulation of mitochondrial RNA synthesis. *Mol Cell Biol* 1999;**19**:7913–24.
10. Pessemesse L, Lepourry L, Bouton K, Levin J, Cabello G, Wrutniak-Cabello C. Casas F. p28, a truncated form of TRα1 regulates mitochondrial physiology. *FEBS Lett* 2014;**588**:4037–43.
11. Scheller K, Sekeris CE, Krohne G, Hock R, Hansen IA, Scheer U. Localization of glucocorticoid hormone receptors in mitochondria of human cells. *Eur J Cell Biol* 2000;**79**:299–307.
12. Yager JD, Chen JQ. Mitochondrial estrogen receptors--new insights into specific functions. *Trends Endocrinol Metab* 2007;**18**:89–91.
13. Silvagno F, De Vivo E, Attanasio A, Gallo V, Mazzucco G, Pescarmona G. Mitochondrial localization of vitamin D receptor in human platelets and differentiated megakaryocytes. *PLoS One* 2010;**5**:e8670.
14. Casas F, Daury L, Grandemange S, Busson M, Seyer P, Hatier R, Carazo A, Cabello G, Wrutniak-Cabello C. Endocrine regulation of mitochondrial activity: involvement of truncated RXRalpha and c-Erb Aalpha1 proteins. *FASEB J* 2003;**17**:426–36.
15. Rochard P, Rodier A, Casas F, Cassar-Malek I, Marchal-Victorion S, Daury L, Wrutniak C, Cabello G. Mitochondrial activity is involved in the regulation of myoblast differentiation through myogenin expression and activity of myogenic factors. *J Biol Chem* 2000;**275**:2733–44.
16. Seyer P, Grandemange S, Busson M, Carazo A, Gamaléri F, Pessemesse L, Casas F, Cabello G, Wrutniak-Cabello C. Mitochondrial activity regulates myoblast differentiation by control of c-Myc expression. *J Cell Physiol* 2006;**207**:75–86.
17. Seyer P, Grandemange S, Rochard P, Busson M, Pessemesse L, Casas F, Cabello G, Wrutniak-Cabello C. P43-dependent mitochondrial activity regulates myoblast differentiation and slow myosin isoform expression by control of Calcineurin expression. *Exp Cell Res* 2011;**317**:2059–71.
18. Grandemange S, Seyer P, Carazo A, Bécuwe P, Pessemesse L, Busson M, Marsac C, Roger P, Casas F, Cabello G, Wrutniak-Cabello C. Stimulation of mitochondrial activity by p43 overexpression induces human dermal fibroblast transformation. *Cancer Res* 2005;**65**:4282–91.
19. Saelim N, Holstein D, Chocron ES, Camacho P, Lechleiter JD. Inhibition of apoptotic potency by ligand stimulated thyroid hormone receptors located in mitochondria. *Apoptosis* 2007;**12**:1781–94.
20. Casas F, Pessemesse L, Grandemange S, Seyer P, Gueguen N, Baris O, Lepourry L, Cabello G, Wrutniak-Cabello C. Overexpression of the mitochondrial T3 receptor p43 induces a shift in skeletal muscle fiber types. *PLoS One* 2008;**3**:e2501.
21. Pessemesse L, Schlernitzauer A, Sar C, Levin J, Grandemange S, Seyer P, Favier FB, Kaminski S, Cabello G, Wrutniak-Cabello C, Casas F. Depletion of the p43 mitochondrial T3 receptor in mice affects skeletal muscle development and activity. *FASEB J* 2012;**26**:748–56.
22. Casas F, Pessemesse L, Grandemange S, Seyer P, Baris O, Gueguen N, Ramonatxo C, Perrin F, Fouret G, Lepourry L, Cabello G, Wrutniak-Cabello C. Overexpression of the mitochondrial T3 receptor induces skeletal muscle atrophy during aging. *PLoS One* 2009;**4**:e5631.
23. Blanchet E, Bertrand C, Annicotte JS, Schlernitzauer A, Pessemesse L, Levin J, Fouret G, Feillet-Coudray C, Bonafos B, Fajas L, Cabello G, Wrutniak-Cabello C, Casas F. Mitochondrial T3 receptor p43 regulates insulin secretion and glucose homeostasis. *FASEB J* 2012;**26**:40–50.
24. Bertrand C, Valet P, Castan-Laurell I. Apelin and energy metabolism. *Front Physiol* 2015;**6**:115.
25. Pelletier P, Gauthier K, Sideleva O, Samarut J, Silva JE. Mice lacking the thyroid hormone receptor-alpha gene spend more energy in thermogenesis, burn more fat, and are less sensitive to high-fat diet-induced obesity. *Endocrinology* 2008;**149**:6471–86.

26. Bertrand-Gaday C, Pessemesse L, Cabello G, Wrutniak-Cabello C, Casas F. Temperature homeostasis in mice lacking the p43 mitochondrial T3 receptor. *FEBS Lett* 2016;**590**:982–91.
27. Irwin RW, Yao J, To J, Hamilton RT, Cadenas E, Brinton RD. Selective oestrogen receptor modulators differentially potentiate brain mitochondrial function. *J Neuroendocrinol* 2012;**24**:236–48.
28. Ivanova MM, Radde BN, Son J, Mehta FF, Chung SH, Klinge CM. Estradiol and tamoxifen regulate NRF-1 and mitochondrial function in mouse mammary gland and uterus. *J Mol Endocrinol* 2013;**51**:233–46.
29. Kemper MF, Zhao Y, Duckles SP, Krause DN. Endogenous ovarian hormones affect mitochondrial efficiency in cerebral endothelium via distinct regulation of PGC-1 isoforms. *J Cereb Blood Flow Metab* 2013;**33**:122–8.
30. Kang T, Lu W, Xu W, Anderson L, Bacanamwo M, Thompson W, Chen YE, Liu D. MicroRNA-27 (miR-27) targets prohibitin and impairs adipocyte differentiation and mitochondrial function in human adipose-derived stem cells. *J Biol Chem* 2013;**288**:34394–402.
31. Virbasius JV, Scarpulla RC. Activation of the human mitochondrial transcription factor A gene by nuclear respiratory factors: a potential regulatory link between nuclear and mitochondrial gene expression in organelle biogenesis. *Proc Natl Acad Sci U S A* 1994;**91**:1309–13.
32. Klinge CM. Estrogens regulate life and death in mitochondria. *J Bioenerg Biomembr* 2017;**49**:307–24.
33. Scarpulla RC. Nuclear control of respiratory chain expression by nuclear respiratory factors and PGC-1-related coactivator. *Ann N Y Acad Sci* 2008;**1147**:321–34.
34. Soriano FX, Liesa M, Bach D, Chan DC, Palacín M, Zorzano A. Evidence for a mitochondrial regulatory pathway defined by peroxisome proliferator-activated receptor-gamma coactivator-1 alpha, estrogen-related receptor-alpha, and mitofusin 2. *Diabetes* 2006;**55**:1783–91.
35. Demonacos CV, Karayanni N, Hatzoglou E, Tsiriyiotis C, Spandidos DA, Sekeris CE. Mitochondrial genes as sites of primary action of steroid hormones. *Steroids* 1996;**61**:226–32.
36. Chen JQ, Eshete M, Alworth WL, Yager JD. Binding of MCF-7 cell mitochondrial proteins and recombinant human estrogen receptors alpha and beta to human mitochondrial DNA estrogen response elements. *J Cell Biochem* 2004;**93**:358–73.
37. Nilsen J, Diaz Brinton R. Mechanism of estrogen-mediated neuroprotection: regulation of mitochondrial calcium and Bcl-2 expression. *Proc Natl Acad Sci U S A* 2003;**100**:2842–7.
38. Simpkins JW, Yang SH, Sarkar SN, Pearce V. Estrogen actions on mitochondria--physiological and pathological implications. *Mol Cell Endocrinol* 2008;**290**:51–9.
39. Rettberg JR, Yao J, Brinton RD. Estrogen: a master regulator of bioenergetics systems in the brain and body. *Front Neuroendocrinol* 2014;**35**:8–30.
40. Alvarez-Delgado C, Mendoza-Rodríguez CA, Picazo O, Cerbón M. Different expression of alpha and beta mitochondrial estrogen receptors in the aging rat brain: interaction with respiratory complex V. *Exp Gerontol* 2010;**45**:580–5.
41. Oo PS, Yamaguchi Y, Sawaguchi A, Tin Htwe Kyaw M, Choijookhuu N, Noor Ali M, Srisowanna N, Hino SI, Hishikawa Y. Estrogen regulates mitochondrial morphology through phosphorylation of dynamin-related protein 1 in MCF7 human breast cancer cells. *Acta Histochem Cytochem* 2018;**51**:21–31.
42. Gollapudi L, Oblinger MM. Estrogen and NGF synergistically protect terminally differentiated, ERalpha-transfected PC12 cells from apoptosis. *J Neurosci Res* 1999;**56**:471–81.
43. Papa L, Germain D. Estrogen receptor mediates a distinct mitochondrial unfolded protein response. *J Cell Sci* 2011;**124**:1396–402.
44. Radke S, Chander H, Schäfer P, Meiss G, Krüger R, Schulz JB, Germain D. Mitochondrial protein quality control by the proteasome involves ubiquitination and the protease Omi. *J Biol Chem* 2008;**283**:12681–5.
45. Irwin RW, Yao J, Hamilton RT, Cadenas E, Brinton RD, Nilsen J. Progesterone and estrogen regulate oxidative metabolism in brain mitochondria. *Endocrinology* 2008;**149**:3167–75.
46. Gaignard P, Fréchou M, Schumacher M, Thérond P, Mattern C, Slama A, Guennoun R. Progesterone reduces brain mitochondrial dysfunction after transient focal ischemia in male and female mice. *J Cereb Blood Flow Metab* 2016;**36**:562–8.
47. Gonzalez Deniselle MC, Carreras MC, Garay L, Gargiulo-Monachelli G, Meyer M, Poderoso JJ, De Nicola AF. Progesterone prevents mitochondrial dysfunction in the spinal cord of wobbler mice. *J Neurochem* 2012;**122**:185–95.
48. Feng Q, Crochet JR, Dai Q, Leppert PC, Price TM. Expression of a mitochondrial progesterone receptor (PR-M) in leiomyomata and association with increased mitochondrial membrane potential. *J Clin Endocrinol Metab* 2014;**99**:E390–9.

49. Tantibhedhyangkul J, Hawkins KC, Dai Q, Mu K, Dunn CN, Miller SE, Price TM. Expression of a mitochondrial progesterone receptor in human spermatozoa correlates with a progestin-dependent increase in mitochondrial membrane potential. *Andrology* 2014;**2**:875–83.
50. Pitteloud N, Mootha VK, Dwyer AA, Hardin M, Lee H, Eriksson KF, Tripathy D, Yialamas M, Groop L, Elahi D, Hayes FJ. Relationship between testosterone levels, insulin sensitivity, and mitochondrial function in men. *Diabetes Care* 2005;**28**:1636–42.
51. Petersson SJ, Christensen LL, Kristensen JM, Kruse R, Andersen M, Højlund K. Effect of testosterone on markers of mitochondrial oxidative phosphorylation and lipid metabolism in muscle of aging men with subnormal bioavailable testosterone. *Eur J Endocrinol* 2014;**171**:77–88.
52. Apaiajai N, Chunchai T, Jaiwongkam T, Kerdphoo S, Chattipakorn SC, Chattipakorn N. Testosterone deprivation aggravates left-ventricular dysfunction in male obese insulin-resistant rats via impairing cardiac mitochondrial function and dynamics proteins. *Gerontology* 2018;**64**:333–43.
53. Yan W, Kang Y, Ji X, Li S, Li Y, Zhang G, Cui H, Shi G. Testosterone upregulates the expression of mitochondrial ND1 and ND4 and alleviates the oxidative damage to the nigrostriatal dopaminergic system in Orchiectomized rats. *Oxidative Med Cell Longev* 2017;**2017**:1202459.
54. Ibebunjo C, Eash JK, Li C, Ma Q, Glass DJ. Voluntary running, skeletal muscle gene expression, and signaling inversely regulated by orchidectomy and testosterone replacement. *Am J Physiol Endocrinol Metab* 2011;**300**:E327–40.
55. Usui T, Kajita K, Kajita T, Mori I, Hanamoto T, Ikeda T, Okada H, Taguchi K, Kitada Y, Morita H, Sasaki T, Kitamura T, Sato T, Kojima I, Ishizuka T. Elevated mitochondrial biogenesis in skeletal muscle is associated with testosterone-induced body weight loss in male mice. *FEBS Lett* 2014;**588**:1935–41.
56. Guo W, Wong S, Li M, Liang W, Liesa M, Serra C, Jasuja R, Bartke A, Kirkland JL, Shirihai O, Bhasin S. Testosterone plus low-intensity physical training in late life improves functional performance, skeletal muscle mitochondrial biogenesis, and mitochondrial quality control in male mice. *PLoS One* 2012;**7**:e51180.
57. Fernando SM, Rao P, Niel L, Chatterjee D, Stagljar M, Monks DA. Myocyte androgen receptors increase metabolic rate and improve body composition by reducing fat mass. *Endocrinology* 2010;**151**:3125–32.
58. Musa M, Fernando SM, Chatterjee D, Monks DA. Subcellular effects of myocyte-specific androgen receptor overexpression in mice. *J Endocrinol* 2011;**210**:93–104.
59. Pansarasa O, D'Antona G, Gualea MR, Marzani B, Pellegrino MA, Marzatico F. "Oxidative stress": effects of mild endurance training and testosterone treatment on rat gastrocnemius muscle. *Eur J Appl Physiol* 2002;**87**:550–5.
60. Lee SR, Kim HK, Song IS, Youm J, Dizon LA, Jeong SH, Ko TH, Heo HJ, Ko KS, Rhee BD, Kim N, Han J. Glucocorticoids and their receptors: insights into specific roles in mitochondria. *Prog Biophys Mol Biol* 2013;**112**:44–54.
61. Picard M, Juster RP, McEwen BS. Mitochondrial allostatic load puts the 'gluc' back in glucocorticoids. *Nat Rev Endocrinol* 2014;**10**:303–10.
62. Hernández-Alvarez MI, Paz JC, Sebastián D, Muñoz JP, Liesa M, Segalés J, Palacín M, Zorzano A. Glucocorticoid modulation of mitochondrial function in hepatoma cells requires the mitochondrial fission protein Drp1. *Antioxid Redox Signal* 2013;**19**:366–78.
63. Huang X, Eriksson KF, Vaag A, Lehtovirta M, Hansson M, Laurila E, Kanninen T, Olesen BT, Kurucz I, Koranyi L, Groop L. Insulin-regulated mitochondrial gene expression is associated with glucose flux in human skeletal muscle. *Diabetes* 1999;**48**:1508–14.
64. Stump CS, Short KR, Bigelow ML, Schimke JM, Nair KS. Effect of insulin on human skeletal muscle mitochondrial ATP production, protein synthesis, and mRNA transcripts. *Proc Natl Acad Sci U S A* 2003;**100**:7996–8001.
65. Karakelides H, Asmann YW, Bigelow ML, Short KR, Dhatariya K, Coenen-Schimke J, Kahl J, Mukhopadhyay D, Nair KS. Effect of insulin deprivation on muscle mitochondrial ATP production and gene transcript levels in type 1 diabetic subjects. *Diabetes* 2007;**56**:2683–9.
66. Boirie Y, Short KR, Ahlman B, Charlton M, Nair KS. Tissue-specific regulation of mitochondrial and cytoplasmic protein synthesis rates by insulin. *Diabetes* 2001;**50**:2652–8.
67. Barazzoni R, Short KR, Asmann Y, Coenen-Schimke JM, Robinson MM, Nair KS. Insulin fails to enhance mTOR phosphorylation, mitochondrial protein synthesis, and ATP production in human skeletal muscle without amino acid replacement. *Am J Physiol Endocrinol Metab* 2012;**303**:E1117–25.
68. Robinson MM, Soop M, Sohn TS, Morse DM, Schimke JM, Klaus KA, Nair KS. High insulin combined with essential amino acids stimulates skeletal muscle mitochondrial protein synthesis while decreasing insulin sensitivity in healthy humans. *J Clin Endocrinol Metab* 2014;**99**:E2574–83.

69. Asmann YW, Stump CS, Short KR, Coenen-Schimke JM, Guo Z, Bigelow ML, Nair KS. Skeletal muscle mitochondrial functions, mitochondrial DNA copy numbers, and gene transcript profiles in type 2 diabetic and nondiabetic subjects at equal levels of low or high insulin and euglycemia. *Diabetes* 2006;**55**:3309–19.
70. Short KR, Moller N, Bigelow ML, Coenen-Schimke J, Nair KS. Enhancement of muscle mitochondrial function by growth hormone. *J Clin Endocrinol Metab* 2008;**93**:597–604.
71. Brown-Borg HM, Bartke A. GH and IGF1: roles in energy metabolism of long-living GH mutant mice. *J Gerontol A Biol Sci Med Sci* 2012;**67**:652–60.
72. Puche JE, García-Fernández M, Muntané J, Rioja J, González-Barón S, Castilla CI. Low doses of insulin-like growth factor-I induce mitochondrial protection in aging rats. *Endocrinology* 2008;**149**:2620–7.
73. Castilla-Cortazar I, Garcia M, Muguerza B, Quiroga J, Perez R, Santidrian S, Prieto J. Hepatoprotective effects of insulin-like growth factor I in rats with carbon tetrachloride-induced cirrhosis. *Gastroenterology* 1997;**113**:1682–91.
74. Pérez R, García-Fernández M, Díaz-Sánchez M, Puche JE, Delgado G, Conchillo M, Muntané J, Castilla-Cortázar I. Mitochondrial protection by low doses of insulin-like growth factor-I in experimental cirrhosis. *World J Gastroenterol* 2008;**14**:2731–9.
75. De Ita JR, Castilla-Cortázar I, Aguirre GA, Sánchez-Yago C, Santos-Ruiz MO, Guerra-Menéndez L, Martín-Estal I, García-Magariño M, Lara-Díaz VJ, Puche JE, Muñoz U. Altered liver expression of genes involved in lipid and glucose metabolism in mice with partial IGF-1 deficiency: an experimental approach to metabolic syndrome. *J Transl Med* 2015;**13**:326.
76. Hao CN, Geng YJ, Li F, Yang T, Su DF, Duan JL, Li Y. Insulin-like growth factor-1 receptor activation prevents hydrogen peroxide-induced oxidative stress, mitochondrial dysfunction and apoptosis. *Apoptosis* 2011;**16**:1118–27.
77. Galic S, Oakhill JS, Steinberg GR. Adipose tissue as an endocrine organ. *Mol Cell Endocrinol* 2010;**316**:129–39.
78. Fisman EZ, Tenenbaum A. Adiponectin: a manifold therapeutic target for metabolic syndrome, diabetes, and coronary disease? *Cardiovasc Diabetol* 2014;**13**:103.
79. Achari AE, Jain SK. Adiponectin, a therapeutic target for obesity, diabetes, and endothelial dysfunction. *Int J Mol Sci* 2017;**18**(6).
80. Clark JL, Taylor CG, Zahradka P. Exploring the cardio-metabolic relevance of T-cadherin: a pleiotropic adiponectin receptor. *Endocr Metab Immune Disord Drug Targets* 2017;**17**:200–6.
81. Cheng KK, Lam KS, Wang Y, Huang Y, Carling D, Wu D, et al. Adiponectin-induced endothelial nitric oxide synthase activation and nitric oxide production are mediated by APPL1 in endothelial cells. *Diabetes* 2007;**56**:1387–94.
82. Sripada L, Tomar D, Singh R. Mitochondria: one of the destinations of miRNAs. *Mitochondrion* 2012;**12**:593–9.
83. Tsao TS, Tomas E, Murrey HE, Hug C, Lee DH, Ruderman NB, et al. Role of disulfide bonds in Acrp30/adiponectin structure and signaling specificity. Different oligomers activate different signal transduction pathways. *J Biol Chem* 2003;**278**:50810–7.
84. Wang Y, Lam KS, Chan L, Chan KW, Lam JB, Lam MC, et al. Posttranslational modifications of the four conserved lysine residues within the collagenous domain of adiponectin are required for the formation of its high molecular weight oligomeric complex. *J Biol Chem* 2006;**281**:16391–400.
85. Yamauchi T, Kamon J, Minokoshi Y, Ito Y, Waki H, Uchida S, et al. Adiponectin stimulates glucose utilization and fatty-acid oxidation by activating AMP-activated protein kinase. *Nat Med* 2002;**8**:1288–95.
86. Handa P, Maliken BD, Nelson JE, Morgan-Stevenson V, Messner DJ, Dhillon BK, Klintworth HM, Beauchamp M, Yeh MM, Elfers CT, Roth CL, Kowdley KV. Reduced adiponectin signaling due to weight gain results in nonalcoholic steatohepatitis through impaired mitochondrial biogenesis. *Hepatology* 2014;**60**:133–45.
87. Iwabu M, Yamauchi T, Okada-Iwabu M, Sato K, Nakagawa T, Funata M, Yamaguchi M, Namiki S, Nakayama R, Tabata M, Ogata H, Kubota N, Takamoto I, Hayashi YK, Yamauchi N, Waki H, Fukayama M, Nishino I, Tokuyama K, Ueki K, Oike Y, Ishii S, Hirose K, Shimizu T, Touhara K, Kadowaki T. Adiponectin and AdipoR1 regulate PGC-1alpha and mitochondria by Ca(2+) and AMPK/SIRT1. *Nature* 2010;**464**:1313–9.
88. Aguer C, Harper ME. Skeletal muscle mitochondrial energetics in obesity and type 2 diabetes mellitus: endocrine aspects. *Best Pract Res Clin Endocrinol Metab* 2012;**26**:805–19.
89. Gong Z, Tas E, Yakar S, Muzumdar R. Hepatic lipid metabolism and non-alcoholic fatty liver disease in aging. *Mol Cell Endocrinol* 2017;**455**:115–30.
90. Friedman JM, Halaas JL. Leptin and the regulation of body weight in mammals. *Nature* 1998;**395**:763–70.
91. Frederich RC, Hamann A, Anderson S, Löllmann B, Lowell BB, Flier JS. Leptin levels reflect body lipid content in mice: evidence for diet-induced resistance to leptin action. *Nat Med* 1995;**1**:1311–4.

92. Wein S, Ukropec J, Gasperíková D, Klimes I, Seböková E. Concerted action of leptin in regulation of fatty acid oxidation in skeletal muscle and liver. *Exp Clin Endocrinol Diabetes* 2007;**115**:244–51.
93. Minokoshi Y, Kim YB, Peroni OD, et al. Leptin stimulates fatty-acid oxidation by activating AMP-activated protein kinase. *Nature* 2002;**415**:339–43.
94. Vilà L, Roglans N, Alegret M, Sánchez RM, Vázquez-Carrera M, Laguna JC. Suppressor of cytokine signaling-3 (SOCS-3) and a deficit of serine/threonine (Ser/Thr) phosphoproteins involved in leptin transduction mediate the effect of fructose on rat liver lipid metabolism. *Hepatology* 2008;**48**:1506–16.
95. Roglans N, Vilà L, Alegret M, Sánchez RM, Vázquez-Carrera M, Laguna JC. Impairment of hepatic STAT-3 activation and reduction of PPARα activity in fructose-fed rats. *Hepatology* 2007;**45**:778–88.
96. Wegrzyn J, Potla R, Chwae YJ, Sepuri NB, Zhang Q, Koeck T, Derecka M, Szczepanek K, Szelag M, Gornicka A, Moh A, Moghaddas S, Chen Q, Bobbili S, Cichy J, Dulak J, Baker DP, Wolfman A, Stuehr D, Hassan MO, Fu XY, Avadhani N, Drake JI, Fawcett P, Lesnefsky EJ, Larner AC. Function of mitochondrial Stat3 in cellular respiration. *Science* 2009;**323**:793–7.
97. Holmström MH, Tom RZ, Björnholm M, Garcia-Roves PM, Zierath JR. Effect of leptin treatment on mitochondrial function in obese leptin-deficient ob/ob mice. *Metabolism* 2013;**62**:1258–67.
98. Perfield 2nd JW, Ortinau LC, Pickering RT, Ruebel ML, Meers GM, Rector RS. Altered hepatic lipid metabolism contributes to nonalcoholic fatty liver disease in leptin-deficient Ob/Ob mice. *J Obes* 2013;**2013**:296537.
99. Singh A, Wirtz M, Parker N, Hogan M, Strahler J, Michailidis G, Schmidt S, Vidal-Puig A, Diano S, Andrews P, Brand MD, Friedman J. Leptin-mediated changes in hepatic mitochondrial metabolism, structure, and protein levels. *Proc Natl Acad Sci U S A* 2009;**106**:13100–5.
100. Akasaka Y, Tsunoda M, Ide T, et al. Chronic leptin treatment stimulates lipid oxidation in immortalized and primary mouse skeletal muscle cells. *Biochim Biophys Acta* 2009;**1791**:103–9.
101. Akasaka Y, Tsunoda M, Ogata T, et al. Direct evidence for leptin-induced lipid oxidation independent of long-form leptin receptor. *Biochim Biophys Acta* 2010;**1801**:1115–22.
102. Tatemoto K, Hosoya M, Habata Y, Fujii R, Kakegawa T, Zou MX, Kawamata Y, Fukusumi S, Hinuma S, Kitada C, Kurokawa T, Onda H, Fujino M. Isolation and characterization of a novel endogenous peptide ligand for the human APJ receptor. *Biochem Biophys Res Commun* 1998;**251**:471–6.
103. Boucher J, Masri B, Daviaud D, Gesta S, Guigné C, Mazzucotelli A, Castan-Laurell I, Tack I, Knibiehler B, Carpéné C, Audigier Y, Saulnier-Blache JS, Valet P. Apelin, a newly identified adipokine up-regulated by insulin and obesity. *Endocrinology* 2005;**146**:1764–71.
104. Wei L, Hou X, Tatemoto K. Regulation of apelin mRNA expression by insulin and glucocorticoids in mouse 3T3-L1 adipocytes. *Regul Pept* 2005;**132**:27–32.
105. Chaves-Almagro C, Castan-Laurell I, Dray C, Knauf C, Valet P, Masri B. Apelin receptors: from signaling to antidiabetic strategy. *Eur J Pharmacol* 2015;**763**:149–59.
106. Bełtowski J. Apelin and visfatin: unique "beneficial" adipokines upregulated in obesity? *Med Sci Monit* 2006;**12**:RA112–9.
107. Attané C, Foussal C, Le Gonidec S, Benani A, Daviaud D, Wanecq E, Guzmán-Ruiz R, Dray C, Bezaire V, Rancoule C, Kuba K, Ruiz-Gayo M, Levade T, Penninger J, Burcelin R, Pénicaud L, Valet P, Castan-Laurell I. Apelin treatment increases complete fatty acid oxidation, mitochondrial oxidative capacity, and biogenesis in muscle of insulin-resistant mice. *Diabetes* 2012;**61**:310–20.
108. Masri B, Knibiehler B, Audigier Y. Apelin signalling: a promising pathway from cloning to pharmacology. *Cell Signal* 2005;**17**:415–26.
109. Frier BC, Williams DB, Wright DC. The effects of apelin treatment on skeletal muscle mitochondrial content. *Am J Phys Regul Integr Comp Phys* 2009;**297**:R1761–8.
110. Yamamoto T, Habata Y, Matsumoto Y, Yasuhara Y, Hashimoto T, Hamajyo H, Anayama H, Fujii R, Fuse H, Shintani Y, Mori M. Apelin-transgenic mice exhibit a resistance against diet-induced obesity by increasing vascular mass and mitochondrial biogenesis in skeletal muscle. *Biochim Biophys Acta* 2011;**1810**:853–62.
111. Huang J, Kang S, Park SJ, Im DS. Apelin protects against liver X receptor-mediated steatosis through AMPK and PPARα in human and mouse hepatocytes. *Cell Signal* 2017;**39**:84–94.
112. Vinel C, Lukjanenko L, Batut A, Deleruyelle S, Pradère JP, Le Gonidec S, Dortignac A, Geoffre N, Pereira O, Karaz S, Lee U, Camus M, Chaoui K, Mouisel E, Bigot A, Mouly V, Vigneau M, Pagano AF, Chopard A, Pillard F, Guyonnet S, Cesari M, Burlet-Schiltz O, Pahor M, Feige JN, Vellas B, Valet P, Dray C. The exerkine apelin reverses age-associated sarcopenia. *Nat Med* 2018;**24**:1360–71.
113. Wong GW, Wang J, Hug C, Tsao TS, Lodish HF. A family of Acrp30/adiponectin structural and functional paralogs. *Proc Natl Acad Sci U S A* 2004;**101**:10302–7.

114. Wong GW, Krawczyk SA, Kitidis-Mitrokostas C, Revett T, Gimeno R, Lodish HF. Molecular, biochemical and functional characterizations of C1q/TNF family members: adipose-tissue-selective expression patterns, regulation by PPAR-gamma agonist, cysteine-mediated oligomerizations, combinatorial associations and metabolic functions. *Biochem J* 2008;**416**:161–77.
115. Li Y, Ozment T, Wright GL, Peterson JM. Identification of putative receptors for the novel adipokine CTRP3 using ligand-receptor capture technology. *PLoS ONE* 2016;**11**:e0164593.
116. Peterson JM, Wei Z, Wong GW. C1q/TNF-related protein-3 (CTRP3), a novel adipokine that regulates hepatic glucose output. *J Biol Chem* 2010;**285**:39691–701.
117. Feng H, Wang JY, Zheng M, Zhang CL, An YM, Li L, Wu LL. CTRP3 promotes energy production by inducing mitochondrial ROS and up-expression of PGC-1α in vascular smooth muscle cells. *Exp Cell Res* 2016;**341**:177–86.
118. Zhang CL, Feng H, Li L, Wang JY, Wu D, Hao YT, Wang Z, Zhang Y, Wu LL. Globular CTRP3 promotes mitochondrial biogenesis in cardiomyocytes through AMPK/PGC-1α pathway. *Biochim Biophys Acta* 2017;**1861**:3085–94.
119. Yanagi S, Sato T, Kangawa K, Nakazato M. The homeostatic force of ghrelin. *Cell Metab* 2018;**27**:786–804.
120. Ezquerro S, Méndez-Giménez L, Becerril S, Moncada R, Valentí V, Catalán V, Gómez-Ambrosi J, Frühbeck G, Rodríguez A. Acylated and desacyl ghrelin are associated with hepatic lipogenesis, β-oxidation and autophagy: role in NAFLD amelioration after sleeve gastrectomy in obese rats. *Sci Rep* 2016;**6**:39942.
121. Anderson KA, Ribar TJ, Lin F, Noeldner PK, Green MF, Muehlbauer MJ, Witters LA, Kemp BE, Means AR. Hypothalamic CaMKK2 contributes to the regulation of energy balance. *Cell Metab* 2008;**7**:377–88.
122. Hawley SA, Pan DA, Mustard KJ, Ross L, Bain J, Edelman AM, Frenguelli BG, Hardie DG. Calmodulin-dependent protein kinase kinase-beta is an alternative upstream kinase for AMP-activated protein kinase. *Cell Metab* 2005;**2**:9–19.
123. Velásquez DA, Martínez G, Romero A, Vázquez MJ, Boit KD, Dopeso-Reyes IG, López M, Vidal A, Nogueiras R, Diéguez C. The central Sirtuin 1/p53 pathway is essential for the orexigenic action of ghrelin. *Diabetes* 2011;**60**:1177–85.
124. Barazzoni R, Gortan Cappellari G, Palus S, Vinci P, Ruozi G, Zanetti M, Semolic A, Ebner N, von Haehling S, Sinagra G, Giacca M, Springer J. Acylated ghrelin treatment normalizes skeletal muscle mitochondrial oxidative capacity and AKT phosphorylation in rat chronic heart failure. *J Cachexia Sarcopenia Muscle* 2017;**8**:991–8.
125. Barazzoni R, Zhu X, Deboer M, Datta R, Culler MD, Zanetti M, Guarnieri G, Marks DL. Combined effects of ghrelin and higher food intake enhance skeletal muscle mitochondrial oxidative capacity and AKT phosphorylation in rats with chronic kidney disease. *Kidney Int* 2010;**77**:23–8.
126. Tamaki M, Hagiwara A, Miyashita K, Wakino S, Inoue H, Fujii K, Fujii C, Sato M, Mitsuishi M, Muraki A, Hayashi K, Doi T, Itoh H. Improvement of physical decline through combined effects of muscle enhancement and mitochondrial activation by a gastric hormone ghrelin in male 5/6Nx CKD model mice. *Endocrinology* 2015;**156**:3638–48.
127. Tamaki M, Miyashita K, Hagiwara A, Wakino S, Inoue H, Fujii K, Fujii C, Endo S, Uto A, Mitsuishi M, Sato M, Doi T, Itoh H. Ghrelin treatment improves physical decline in sarcopenia model mice through muscular enhancement and mitochondrial activation. *Endocr J* 2017;**64**:S47–51.
128. Fujimura K, Wakino S, Minakuchi H, Hasegawa K, Hosoya K, Komatsu M, Kaneko Y, Shinozuka K, Washida N, Kanda T, Tokuyama H, Hayashi K, Itoh H. Ghrelin protects against renal damages induced by angiotensin-II via an antioxidative stress mechanism in mice. *PLoS ONE* 2014;**9**:e94373.
129. Rossetti A, Togliatto G, Rolo AP, Teodoro JS, Granata R, Ghigo E, Columbano A, Palmeira CM, Brizzi MF. Unacylated ghrelin prevents mitochondrial dysfunction in a model of ischemia/reperfusion liver injury. *Cell Death Dis* 2017;**3**:17077.
130. Andrews ZB, Erion D, Beiler R, Liu ZW, Abizaid A, Zigman J, Elsworth JD, Savitt JM, DiMarchi R, Tschoep M, Roth RH, Gao XB, Horvath TL. Ghrelin promotes and protects nigrostriatal dopamine function via a UCP2-dependent mitochondrial mechanism. *J Neurosci* 2009;**29**:14057–65.
131. Fujitsuka N, Asakawa A, Morinaga A, Amitani MS, Amitani H, Katsuura G, Sawada Y, Sudo Y, Uezono Y, Mochiki E, Sakata I, Sakai T, Hanazaki K, Yada T, Yakabi K, Sakuma E, Ueki T, Niijima A, Nakagawa K, Okubo N, Takeda H, Asaka M, Inui A. Increased ghrelin signaling prolongs survival in mouse models of human aging through activation of sirtuin1. *Mol Psychiatry* 2016;**21**:1613–23.
132. Ruozi G, Bortolotti F, Falcione A, Dal Ferro M, Ukovich L, Macedo A, Zentilin L, Filigheddu N, Gortan Cappellari G, Baldini G, Zweyer M, Barazzoni R, Graziani A, Zacchigna S, Giacca M. AAV-mediated in vivo functional selection of tissue-protective factors against ischaemia. *Nat Commun* 2015;**6**:7388.

133. Gortan Cappellari G, Semolic A, Ruozi G, Vinci P, Guarnieri G, Bortolotti F, Barbetta D, Zanetti M, Giacca M, Barazzoni R. Unacylated ghrelin normalizes skeletal muscle oxidative stress and prevents muscle catabolism by enhancing tissue mitophagy in experimental chronic kidney disease. *FASEB J* 2017;**31**:5159–71.
134. Athauda D, Foltynie T. The glucagon-like peptide 1 (GLP) receptor as a therapeutic target in Parkinson's disease: mechanisms of action. *Drug Discov Today* 2016;**21**:802–18.
135. Li S, Wu H, Han D, Zhang M, Li N, Yu W, Sun D, Sun Z, Ma S, Gao E, Li C, Shen M, Cao F. ZP2495 protects against myocardial ischemia/reperfusion injury in diabetic mice through improvement of cardiac metabolism and mitochondrial function: the possible involvement of AMPK-FoxO3a signal pathway. *Oxidative Med Cell Longev* 2018;**2018**:6451902.
136. Morales PE, Torres G, Sotomayor-Flores C, Peña-Oyarzún D, Rivera-Mejías P, Paredes F, Chiong M. GLP-1 promotes mitochondrial metabolism in vascular smooth muscle cells by enhancing endoplasmic reticulum-mitochondria coupling. *Biochem Biophys Res Commun* 2014;**446**:410–6.
137. de Bold AJ, Borenstein HB, Veress AT, Sonnenberg H. A rapid and potent natriuretic response to intravenous injection of atrial myocardial extract in rats. *Life Sci* 1981;**28**:89–94.
138. Kuhn M. Structure, regulation, and function of mammalian membrane guanylyl cyclase receptors, with a focus on guanylyl cyclase-A. *Circ Res* 2003;**93**:700–9.
139. Miyashita K, Itoh H, Tsujimoto H, Tamura N, Fukunaga Y, Sone M, Yamahara K, Taura D, Inuzuka M, Sonoyama T, Nakao K. Natriuretic peptides/cGMP/cGMP-dependent protein kinase cascades promote muscle mitochondrial biogenesis and prevent obesity. *Diabetes* 2009;**58**:2880–92.
140. Engeli S, Birkenfeld AL, Badin PM, Bourlier V, Louche K, Viguerie N, Thalamas C, Montastier E, Larrouy D, Harant I, de Glisezinski I, Lieske S, Reinke J, Beckmann B, Langin D, Jordan J, Moro C. Natriuretic peptides enhance the oxidative capacity of human skeletal muscle. *J Clin Invest* 2012;**122**:4675–9.
141. Nisoli E, Tonello C, Cardile A, Cozzi V, Bracale R, Tedesco L, Falcone S, Valerio A, Cantoni O, Clementi E, Moncada S, Carruba MO. Calorie restriction promotes mitochondrial biogenesis by inducing the expression of eNOS. *Science* 2005;**310**:314–7.
142. Chanséaume E, Morio B. Potential mechanisms of muscle mitochondrial dysfunction in aging and obesity and cellular consequences. *Int J Mol Sci* 2009;**10**:306–24.
143. Doll DN, Rellick SL, Barr TL, Ren X, Simpkins JW. Rapid mitochondrial dysfunction mediates TNF-alpha-induced neurotoxicity. *J Neurochem* 2015;**132**:443–51.
144. van Horssen J, van Schaik P, Witte M. Inflammation and mitochondrial dysfunction: a vicious circle in neurodegenerative disorders? *Neurosci Lett* 2017;pii:S0304-3940(17)30542-6.
145. Steinberg GR, Michell BJ, van Denderen BJ, Watt MJ, Carey AL, Fam BC, Andrikopoulos S, Proietto J, Görgün CZ, Carling D, Hotamisligil GS, Febbraio MA, Kay TW, Kemp BE. Tumor necrosis factor alpha-induced skeletal muscle insulin resistance involves suppression of AMP-kinase signaling. *Cell Metab* 2006;**4**:465–74.
146. Medeiros R, Prediger RD, Passos GF, Pandolfo P, Duarte FS, Franco JL, Dafre AL, Di Giunta G, Figueiredo CP, Takahashi RN, Campos MM, Calixto JB. Connecting TNF-alpha signaling pathways to iNOS expression in a mouse model of Alzheimer's disease: relevance for the behavioral and synaptic deficits induced by amyloid beta protein. *J Neurosci* 2007;**27**:5394–404.
147. Valerio A, Cardile A, Cozzi V, Bracale R, Tedesco L, Pisconti A, Palomba L, Cantoni O, Clementi E, Moncada S, Carruba MO, Nisoli E. TNF-alpha downregulates eNOS expression and mitochondrial biogenesis in fat and muscle of obese rodents. *J Clin Invest* 2006;**116**:2791–8.
148. Nisoli E, Clementi E, Paolucci C, Cozzi V, Tonello C, Sciorati C, Bracale R, Valerio A, Francolini M, Moncada S, Carruba MO. Mitochondrial biogenesis in mammals: the role of endogenous nitric oxide. *Science* 2003;**299**:896–9.
149. White JP, Puppa MJ, Sato S, Gao S, Price RL, Baynes JW, Kostek MC, Matesic LE, Carson JA. IL-6 regulation on skeletal muscle mitochondrial remodeling during cancer cachexia in the ApcMin/+ mouse. *Skelet Muscle* 2012;**2**:14.
150. Motori E, Puyal J, Toni N, Ghanem A, Angeloni C, Malaguti M, Cantelli-Forti G, Berninger B, Conzelmann KK, Götz M, Winklhofer KF, Hrelia S, Bergami M. Inflammation-induced alteration of astrocyte mitochondrial dynamics requires autophagy for mitochondrial network maintenance. *Cell Metab* 2013;**18**:844–59.
151. Anusree SS, Nisha VM, Priyanka A, Raghu KG. Insulin resistance by TNF-α is associated with mitochondrial dysfunction in 3T3-L1 adipocytes and is ameliorated by punicic acid, a PPARγ agonist. *Mol Cell Endocrinol* 2015;**413**:120–8.
152. Baregamian N, Song J, Bailey CE, Papaconstantinou J, Evers BM, Chung DH. Tumor necrosis factor-alpha and apoptosis signal-regulating kinase 1 control reactive oxygen species release, mitochondrial autophagy, and c-Jun N-terminal kinase/p38 phosphorylation during necrotizing enterocolitis. *Oxidative Med Cell Longev* 2009;**2**:297–306.

153. Gurung P, Lukens JR, Kanneganti TD. Mitochondria: diversity in the regulation of the NLRP3 inflammasome. *Trends Mol Med* 2015;**21**:193–201.
154. Almendro V, Busquets S, Ametller E, Carbó N, Figueras M, Fuster G, Argilés JM, López-Soriano FJ. Effects of interleukin-15 on lipid oxidation: disposal of an oral [(14)C]-triolein load. *Biochim Biophys Acta* 2006;**1761**:37–42.
155. O'Connell GC, Pistilli EE. Interleukin-15 directly stimulates pro-oxidative gene expression in skeletal muscle in-vitro via a mechanism that requires interleukin-15 receptor alpha. *Biochem Biophys Res Commun* 2015;**458**:614–9.
156. Barra NG, Palanivel R, Denou E, Chew MV, Gillgrass A, Walker TD, Kong J, Richards CD, Jordana M, Collins SM, Trigatti BL, Holloway AC, Raha S, Steinberg GR, Ashkar AA. Interleukin-15 modulates adipose tissue by altering mitochondrial mass and activity. *PLoS ONE* 2014;**9**:e114799.
157. Pierce JR, Maples JM, Hickner RC. IL-15 concentrations in skeletal muscle and subcutaneous adipose tissue in lean and obese humans: local effects of IL-15 on adipose tissue lipolysis. *Am J Physiol Endocrinol Metab* 2015;**308**:E1131–9.
158. Boström P, Wu J, Jedrychowski MP, Korde A, Ye L, Lo JC, Rasbach KA, Boström EA, Choi JH, Long JZ, Kajimura S, Zingaretti MC, Vind BF, Tu H, Cinti S, Højlund K, Gygi SP, Spiegelman BM. A PGC1-α-dependent myokine that drives brown-fat-like development of white fat and thermogenesis. *Nature* 2012;**481**:463–8.
159. Perakakis N, Triantafyllou GA, Fernández-Real JM, Huh JY, Park KH, Seufert J, Mantzoros CS. Physiology and role of irisin in glucose homeostasis. *Nat Rev Endocrinol* 2017;**13**:324–37.
160. Ge X, Sathiakumar D, Lua BJ, Kukreti H, Lee M, McFarlane C. Myostatin signals through miR-34a to regulate Fndc5 expression and browning of white adipocytes. *Int J Obes* 2017;**41**:137–48.
161. Vaughan RA, Gannon NP, Barberena MA, Garcia-Smith R, Bisoffi M, Mermier CM, Conn CA, Trujillo KA. Characterization of the metabolic effects of irisin on skeletal muscle in vitro. *Diabetes Obes Metab* 2014;**16**:711–8.
162. Chen K, Xu Z, Liu Y, Wang Z, Li Y, Xu X, Chen C, Xia T, Liao Q, Yao Y, Zeng C, He D, Yang Y, Tan T, Yi J, Zhou J, Zhu H, Ma J, Zeng C. Irisin protects mitochondria function during pulmonary ischemia/reperfusion injury. *Sci Transl Med* 2017;**9**:.
163. Wang H, Zhao YT, Zhang S, Dubielecka PM, Du J, Yano N, Chin YE, Zhuang S, Qin G, Zhao TC. Irisin plays a pivotal role to protect the heart against ischemia and reperfusion injury. *J Cell Physiol* 2017;**232**:3775–85.
164. Silver J, Wadley G, Lamon S. Mitochondrial regulation in skeletal muscle: a role for non-coding RNAs? *Exp Physiol* 2018;**103**:1132–44.
165. Wang X, Song X, Glass CK, Rosenfeld MG. The long arm of long noncoding RNAs: roles as sensors regulating gene transcriptional programs. *Cold Spring Harb Perspect Biol* 2011;**3**:a003756.
166. Wilusz JE, JnBaptiste CK, Lu LY, Kuhn CD, Joshua-Tor L, Sharp PA. A triple helix stabilizes the 3′ ends of long noncoding RNAs that lack poly(A) tails. *Genes Dev* 2012;**26**:2392–407.
167. Ambros V. microRNAs: tiny regulators with great potential. *Cell* 2001;**107**:823–6.
168. Zhang X, Zuo X, Yang B, Li Z, Xue Y, Zhou Y, Huang J, Zhao X, Zhou J, Yan Y, Zhang H, Guo P, Sun H, Guo L, Zhang Y, Fu XD. MicroRNA directly enhances mitochondrial translation during muscle differentiation. *Cell* 2014;**158**:607–19.
169. Ji J, Qin Y, Ren J, Lu C, Wang R, Dai X, Zhou R, Huang Z, Xu M, Chen M, Wu W, Song L, Shen H, Hu Z, Miao D, Xia Y, Wang X. Mitochondria-related miR-141-3p contributes to mitochondrial dysfunction in HFD-induced obesity by inhibiting PTEN. *Sci Rep* 2015;**5**:16262.
170. Kim J, Fiesel FC, Belmonte KC, Hudec R, Wang WX, Kim C, Nelson PT, Springer W, Kim J. miR-27a and miR-27b regulate autophagic clearance of damaged mitochondria by targeting PTEN-induced putative kinase 1 (PINK1). *Mol Neurodegener* 2016;**11**:55.
171. Long J, Badal SS, Ye Z, Wang Y, Ayanga BA, Galvan DL, Green NH, Chang BH, Overbeek PA, Danesh FR. Long noncoding RNA Tug1 regulates mitochondrial bioenergetics in diabetic nephropathy. *J Clin Invest* 2016;**126**:4205–18.
172. Gao S, Tian X, Chang H, Sun Y, Wu Z, Cheng Z, Dong P, Zhao Q, Ruan J, Bu W. Two novel lncRNAs discovered in human mitochondrial DNA using PacBio full-length transcriptome data. *Mitochondrion* 2018;**38**:41–7.
173. Marroqui L, Tuduri E, Alonso-Magdalena P, Quesada I, Nadal A, Dos Santos RS. Mitochondria as a target of endocrine-disrupting chemicals: implications for type 2 diabetes. *J Endocrinol* 2018;**2**, Aug.

CHAPTER

# 4

# Nutritional Regulation of Mitochondrial Function

*Goutham Vasam*[*,a], Kimberly Reid*[*,†,a], Yan Burelle*[*,‡], Keir J. Menzies*[*,§]*

*Interdisciplinary School of Health Sciences, Faculty of Health Sciences, University of Ottawa, Ottawa, ON, Canada †Department of Biology, Faculty of Science, University of Ottawa, Ottawa, ON, Canada ‡Department of Cellular and Molecular Medicine, Faculty of Medicine, University of Ottawa, Ottawa, ON, Canada §Department of Biochemistry, Microbiology and Immunology, University of Ottawa Brain and Mind Research Institute and Centre for Neuromuscular Disease, University of Ottawa, Ottawa, ON, Canada

The nutritive composition of a diet has important influence on mitochondrial health—nutrients provide the substrates to harvest energy in the form of adenosine triphosphate (ATP) to affect key cellular functions and processes. Principle macronutrients providing energy are carbohydrates and fats. Carbohydrates are processed through glycolysis and the tricarboxylic acid (TCA) cycle to generate reduced nicotinamide adenine dinucleotide (NADH) and flavin adenine dinucleotide ($FADH_2$), which feed into the electron transport chain (ETC) of mitochondria for ATP production. Fats are processed through β-oxidation to generate reduced NADH and $FADH_2$, along with acetyl-CoA, which feeds into the TCA cycle. Under certain metabolic conditions, β-oxidation of fatty acids (FAs) can generate high-energy byproducts called ketone bodies in the liver and kidneys, which are transported to other tissues via blood for further processing and entry into the TCA cycle. Tissues with impaired glucose oxidation can use ketone bodies as an alternate and valuable energy source. This can occur with ketogenic diets (KDs), rich in fat and low in carbohydrates, stimulating the synthesis of ketone bodies. In KD, ketone bodies replace carbohydrates as a primary energy source and are well-known to alter mitochondrial function. The composition of a diet can have an influence on mitochondrial-associated diseases. In a *Drosophila* model of mitochondrial defects, ETC dysfunction is exacerbated with a diet that has a high carbohydrate to protein ratio.[1]

[a] Co-first authors.


https://doi.org/10.1016/B978-0-12-811752-1.00004-3

Thus, it is presumed that diet can influence mitochondrial metabolism; therefore, a change in dietary regimen has the potential to mitigate certain mitochondrial phenotypes in a wide range of mitochondrial pathologies. There is increasing evidence that mitochondrial dysfunction is central to aging and various diseases, such as obesity, cardiovascular disease, diabetes, metabolic syndrome, Alzheimer's disease, and cardiomyopathies. Mitochondrial dysfunction can be characterized by an impaired ETC, mitochondrial enzyme impairment, and excessive oxidative damage. Beyond the macronutrient dietary regimen, key micronutrients (minerals and vitamins) can modulate mitochondrial efficiency or function during health or disease. Of these dietary constituents, those that protect the mitochondria from oxidative damage and improve mitochondrial function are referred to as "mitochondrial nutrients." Mitochondrial nutrients can be subdivided into antioxidants that scavenge reactive oxygen species (ROS); cofactors that are integral components of several enzymes required for mitochondrial metabolism; and energy enhancers or metabolic modifying agents, which repair mitochondria or enhance their biogenesis thus boosting metabolism. Often, mitochondrial nutrients can belong to more than one of these categories.[2] This chapter discusses the effects of macronutrients and micronutrients—classified based on their predominant function with respect to mitochondria in health and disease.

# 1 MACRONUTRIENTS

## 1.1 Protein

Unlike carbohydrates and fats, there is no way to store proteins in the body; therefore, excess and recycled proteins are broken down into their carbon skeletons and nitrogen-containing molecules, which enter the urea cycle. Except for those that have branched-chains, the catabolism of amino acids takes place primarily in the mitochondria and cytoplasm of liver hepatocytes. Branched-chain amino acid (leucine, isoleucine, and valine) catabolism occurs primarily in skeletal muscle cells and adipocytes.[3] There is minimal evidence suggesting that protein source impacts mitochondrial function, apart from amino acid deficiencies. The effects of dietary protein and its components on mitochondrial functions are depicted in Fig. 1.

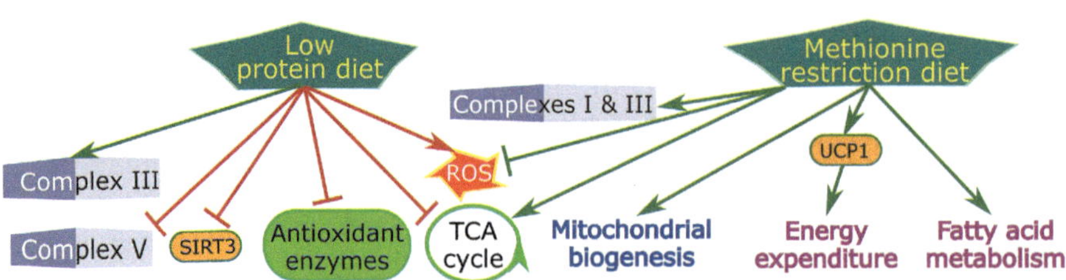


FIG. 1 The effects of a protein diet on the functions of mitochondria. Low protein diet impedes, and methionine restriction diet induces mitochondrial functions. Arrows: *green*—beneficial or neutral effects, *red*—deleterious effects; arrowheads: sharp—activation, and blunt—inhibition.

### 1.1.1 *Maternal Low-Protein Diet and Mitochondrial Function*

A poor maternal diet (particularly a low protein [LP] diet) can lead to intrauterine growth retardation and the early development of glucose intolerance in offspring[4] because of lower expression of malate dehydrogenase (a TCA cycle enzyme) and subunits of ATP synthase (complex V),[5] along with increased oxidative stress (increased β-cell xanthine oxidase) and decreased activity of antioxidant enzymes.[6] Male offspring-specific effects include reduced pancreatic β-cell mass, and increased ROS production accompanied by increased expression of NADH-ubiquinone oxidoreductase subunit 4L (subunit of ETC complex I) and over-expression of uncoupling protein (UCP) 2 and peroxisome proliferator-activated receptor-γ (PPARγ).[5] PPARγ, and UCP2 overexpression effectively reduces ATP production, which is thought to weaken insulin secretion. These results suggest that neonatal and early-life metabolic programming might contribute to increased risk of insulin dysregulation in adulthood.

In a rat model, a maternal LP diet shows remodeling effects within mitochondria isolated from male offspring skeletal muscle.[7] Adenosine diphosphate (ADP)-stimulated (state 3) respiration and ADP-limited (state 4) respiration (both signifying reduced mitochondrial function) are decreased in LP compared to control offspring. Expression of sirtuin protein-3 (SIRT3), a mitochondria-specific $NAD^+$-dependent deacetylase, also are reduced in LP offspring.[7] Associated with this, a reduction in SIRT3 deacetylase expression accelerates the development of metabolic syndrome.[8] These results suggest that a maternal LP diet can remodel offspring skeletal muscle mitochondria.

These data are congruent with findings in a mouse model showing that a maternal LP diet causes substantial changes in mitochondrial oxidative phosphorylation (OXPHOS) gene expression in both the liver and skeletal muscle (upregulation in liver and downregulation in muscle), when compared to control offspring. Interestingly, maternal taurine supplementation partially rescued the gene expression differences in both tissues.[9] Taurine is a sulfonic-acid found in meat and shellfish that has antioxidant properties[10] and is important for optimal function of mitochondrial tRNA during mitochondrial protein synthesis.[11] Given that taurine partially rescued the LP offspring phenotype, it likely plays an important role in mitochondrial gene expression and metabolic programming.[9]

### 1.1.2 *Methionine Restriction Diet and Mitochondrial Function*

Methionine is an essential amino acid found principally in meat and fish.[12] It is critical for protein synthesis and antioxidant function, particularly for the synthesis of homocysteine and, after transsulfuration, the generation of cysteine and glutathione (GSH). Similar to a 40% caloric restriction (CR) diet (but without restricting overall food intake), a methionine restriction (MR) diet will increase invertebrate[13] and rodent[14] life span as well as increase energy expenditure, improve insulin sensitivity, and reduce oxidative damage.[15] A potential mechanism underlying MR-mediated improved insulin sensitivity and lifespan extension might involve fibroblast growth factor-21 (FGF21), a hormone excreted by the liver that is known to stimulate glucose uptake in adipocytes. An MR diet can increase both hepatic expression and blood circulation levels of FGF21.[16] Increased FGF21 circulation also is observed with KD and CR diets and, like the MR diet, both of these diets can improve insulin sensitivity and extend lifespan.[17,18]

Oxidative stress is reduced in the brain, heart, liver, and kidneys of rodents fed a MR diet,[19] which can attenuate aging through reduced oxidative damage. Specifically, an MR diet can decrease the amount of ETC complexes I and III in rat liver mitochondria, reducing ROS production.[20] Within adipose tissue, an MR diet appears to have a different effect—lipogenesis and FA oxidation increase with a concomitant decrease in fat mass.[21, 22] This is matched by an increase in UCP1 expression, elevating energy expenditure, and the number of mitochondria, mitochondrial size, and cristae density in white adipose tissue.[23] Likewise, TCA cycle enzymes increase, as do the levels of complexes I and III.[23] Altogether, these findings appear to demonstrate that an MR diet increases adipocyte uncoupled respiration, which might explain the increased energy expenditure observed with this restrictive diet.

## 1.2 Fat

Fats are high-energy compounds that are an important source for ATP generation and important to structural and metabolic function. Excess energy is stored in mammals in the form of fat that can be mobilized when there is a demand for ATP.[24] Dietary or stored fats contain triglycerides that are hydrolyzed to FAs and glycerol. The fate of FAs includes incorporation into membrane lipids and β-oxidation to generate acetyl-CoA and subsequently, ATP. The $FADH_2$/NADH ratio is affected by β-oxidation of fats—lower for short chain FAs (SCFAs) and higher for long chain FAs (LCFAs), which will influence substrate entry into complex I of ETC.[25] In most mammals, including humans, liver acetyl-CoA can either feed into the TCA cycle or convert into ketone bodies, which include acetoacetate and D-β-hydroxybutyrate. Ketone bodies are exported to other tissues (most importantly the brain) to regenerate acetyl-CoA for energy production. A high fat (HF) diet can reduce ETC subunit formation, their assembly, and the activity of ETC complexes.[26] A discussion of the effects of dietary intake and composition of lipids on mitochondrial functions follows and is depicted in Fig. 2.

### *1.2.1 Dietary Lipids and Mitochondrial Membrane Composition*

The inner (IMM) and outer (OMM) mitochondrial membranes play an important function in the cell by partitioning mitochondrial processes, restricting or easing access to the

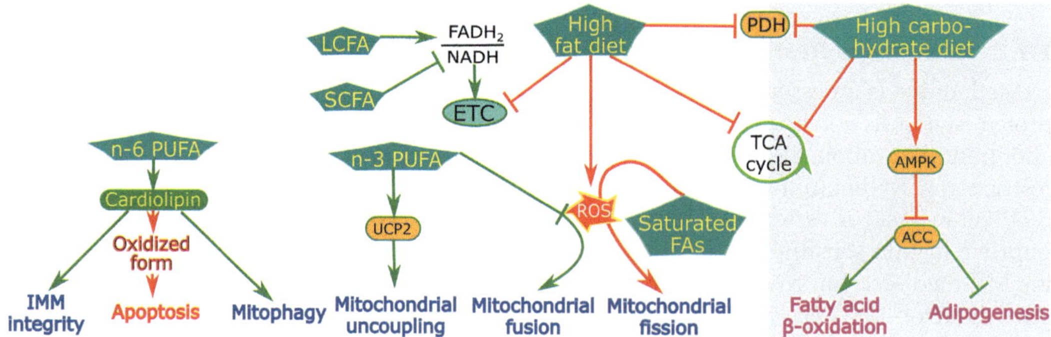


FIG. 2 The effects of a high carbohydrate diet, a high fat diet, and dietary fat constituents on the functions of mitochondria. High carbohydrate or high fat diet impedes mitochondrial functions; FA chain length (LCFA or SCFA) determines the substrate entry into complex I of the ETC; n-3 or n-6 PUFAs are essential for mitochondrial functions. Arrows: *green*—beneficial or neutral effects, *red*—deleterious effects; arrowheads: sharp—activation, blunt—inhibition.

mitochondria depending on cellular needs, and permitting the generation of an electrochemical gradient that drives ATP synthesis. Therefore, the composition of the mitochondrial membrane is essential to its permeability and flexibility, and its ability to harvest ATP and generate metabolic signals.

The source of dietary lipids can alter the FA makeup of mitochondrial membranes such that membrane composition reflects the omega-3 and omega-6 polyunsaturated FA (n-3 and n-6 PUFA) content of the primary lipid source. In mammals, saturated FAs (SFAs) and monounsaturated FAs (MUFAs) can be synthesized de novo from nonlipid sources. Both SFAs and MUFAs have a mitochondrial membrane profile that remains relatively constant over a wide dietary range; mitochondrial membrane SFA and MUFA composition appears to remain stable regardless of dietary levels.[27] In contrast, the n-3 and n-6 PUFAs cannot be synthesized de novo and their profile within the mitochondrial membranes is sensitive to dietary levels, in particular n-3 PUFA and to the ratio of n-3/n-6.

The FA composition of mitochondrial membranes affect function,[27] particularly proton leak (where dietary energy is not captured as ATP) and membrane potential, with evidence that dietary lipid source can affect mitochondrial function. In Chen et al. (2012),[28] three groups of mice on a CR diet with fat derived from sources that varied in their degree of unsaturation (from high to low saturation: lard, soybean oil, and fish oil). The CR high saturated fat group had the lowest proton leak, suggesting that the IMM had lower permeability to protons relative to the less saturated fat sources.[28] This was in keeping with other studies, in which the degree of mitochondrial membrane unsaturation mirrors the proportion of unsaturated FAs in the diet.[28–30]

### 1.2.2 *The Roles of Cardiolipin and n-3 and n-6 PUFAs on Mitochondrial Function*

Cardiolipin (CL) is an important phospholipid found in mitochondrial membranes, predominantly in the IMM. There are many forms of CL because it possesses four (rather than the usual two) fatty acyl side chains that confer various functions, such as to help stabilize protein complexes within the IMM.[31] In a healthy mammal, the most common FA incorporated in cardiac CL side chains is linoleic acid (C18:2), an n-6 PUFA. CL is known to interact with complex V,[32] the ETC complexes I, IV, and potentially III,[33] as well as supercomplexes $III_2$-IV and $III_2$-$IV_2$ (ETC complexes I, III, and IV are able to form various supercomplexes that are thought to increase aerobic respiration efficiency and decrease production of both mitochondrial ROS) and cytochrome *c* (cyt *c*). The interactions between CL and IMM proteins are thought to improve mitochondrial function.[34] In addition, translocation and oxidation of CL plays an important role in triggering apoptosis[35] and mitophagy (selective clearance of damaged mitochondria), the latter occurs after CL translocates to the OMM and binds microtubule-associated protein 1 light chain 3 (LC3), which is important for delivering damaged mitochondria to autophagosomes and lysosomes.[36]

A diet rich in SFA does not appear to affect the percentage of cardiac CL that have saturated fatty acyl side chains; CL with saturated fatty acyl side chains will remain approximately 8%–10% of the total CL, regardless of the amount of SFAs consumed.[37] In contrast, a diet rich in n-6 linoleic acid increases cardiac CL with tetralinoleoyl side chains until this n-6 PUFA version of CL reaches 80% of the total CL. Cortie and Else (2012) examined multiple FA diet rodent studies and determined that changes in CL fatty acyl side chain composition depends on the type of fats consumed and their relative dietary proportions.[38] In particular, when n-6 linoleic acid makes up to 20% of all consumed dietary FAs, incorporation of n-6 linoleic acid in CL sidechains will increase steadily. When dietary n-6 linoleic acid exceeds 20% of the total FAs consumed, however, incorporation of linoleic acid in cardiac CL levels off at approximately 70% of total CL. Docosahexanoic acid (DHA, C22:6) (an n-3 PUFA) is preferentially incorporated into the CL side chains of heart and liver mitochondria until the amount of n-3 DHA consumed reaches 10% and 20% of total dietary FAs, for the heart and liver, respectively.[38]

In obese and diabetic rodents and in diabetic humans, remodeling occurs within cardiac mitochondria such that n-6 linoleic acid is replaced by n-3 DHA. Dietary DHA is considered cardioprotective in that it improves the prognosis of patients with symptomatic heart failure or recent myocardial infarction.[39] Mice fed a Western diet (42% of calories from milk fat with or without 2% supplementation with DHA), however, demonstrated a DHA-related remodeling of cardiac CL and subsequent reduction in ETC complexes I, IV, V and supercomplex I+III activity.[40] DHA-associated cardiac impairment was rescued by introducing dietary n-6 linoleic acid, which suggests that balancing dietary n-6 linoleic acid and n-3 PUFAs (like DHA) is important for healthy cardiac function.

### 1.2.3 *Use of Long-Chain FAs as Fuel*

LCFAs are a preferred source of fuel for aerobic respiration within mitochondria; one molecule of palmitate (C16:0) can generate as much as 129 molecules of ATP compared to one molecule of glucose, which can generate up to 38 molecules of ATP. Mitochondrial membranes are impermeable to LCFAs (>12 carbon chain); therefore, to cross the IMM and OMM, LCFAs rely on the carnitine shuttle (Fig. 3.). Once inside the mitochondrial matrix, LCFAs are broken down via a four-step β-oxidation pathway and the products enter the TCA cycle. Both β-oxidation and the TCA cycle strip electrons from the catabolism of LCFAs and pass them to the electron carrier FAD, reducing it to $FADH_2$. The TCA cycle also passes electrons to the electron carrier $NAD^+$, reducing it to NADH. Both NADH and $FADH_2$ pass their electrons to the ETC complexes I and II, respectively, which generates the electropotential gradient that powers ATP synthase.

### 1.2.4 *High Fat Diets and ROS Production*

HF diets long have been associated with reduced mitochondrial respiration and ATP production. More specifically, a high SFA diet will increase liver, heart, and skeletal muscle mitochondrial oxidative stress in both rats and mice.[41,42] In contrast, a high n-3 PUFA diet (i.e., fish oil or menhaden oil) will decrease ROS production.[41,42] After clamping both ADP and membrane potential levels, Yu et al.[41] demonstrated that a high SFA diet increased ROS production at a lower mitochondrial membrane potential threshold, compared to a high n-3 PUFA diet.[41] This indicates that the biological cost (ROS production per unit of ATP) of oxidative phosphorylation is higher with a high SFA diet.

Tentative suggests evidence that increased expression of UCP3 in skeletal muscle mitochondria might mitigate the oxidative stress observed with HF diets.[43] Compared to wild-type control, UCP3 knockout mice fed a HF diet for 26-weeks had elevated ROS production and reduced mitochondrial function, as measured in isolated skeletal muscle mitochondria.[43]

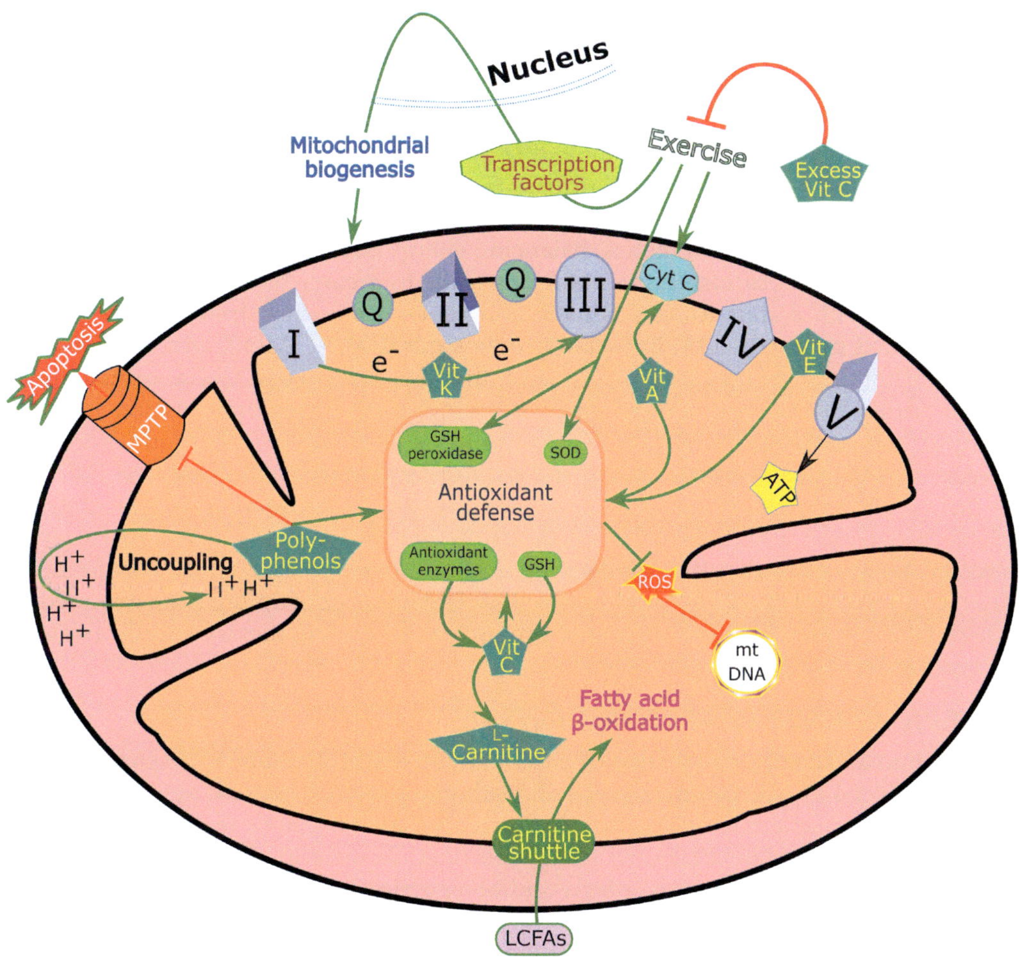


FIG. 3 The effects of various dietary constituents with antioxidant functions on mitochondria. Antioxidant nutrients boost mitochondrial antioxidant defense. Some of the alternate effects of these nutrients involve electron transfer (Vit K), L-carnitine synthesis (Vit C), inhibition of exercise-induced beneficial mitochondrial effects (excess Vit C) or inhibition of apoptosis (polyphenols). Arrows: *green*—beneficial or neutral effects and *red*—deleterious effects; arrowheads: sharp—activation, blunt—inhibition.

### *1.2.5 Long-Chain PUFAs Change With a Caloric Restriction Diet (Without Malnutrition)*

Across all ages, genders, ethnicities, and body weights, a CR diet can normalize blood pressure, reduce heart rate, improve arterial stiffness and endothelial dysfunction—all of which likely affect cardiovascular mortality and morbidity.[44] While the mechanisms behind the CR diet's effects are unknown, they likely involve mitochondrial adaptation, in which a CR diet might increase metabolic efficiency and reduce cellular damage.[45] A CR diet will lower the concentration of long-chain PUFAs in rat liver mitochondrial membranes,[46] lower proton leak, and lower ROS production.[47] Because SFAs are less prone to oxidation, it is thought that CR diet effects could be mediated by the alteration in PUFAs in mitochondrial membranes and through the modulation of proton leak.[28]

### 1.2.6 Mitochondrial Dynamics and Dietary FAs

The balance between mitochondrial fusion and fission (collectively termed mitochondrial dynamics) is thought to be essential to mitochondrial quality control and an important regulator of mitochondrial energetics, balancing energy demand with nutrient supply.[48, 49] A high SFA diet in obese insulin-resistant rats leads to greater tissue oxidative stress (increased ROS) with a shift toward hepatic and skeletal mitochondrial fission. This shift is accompanied by a reduction in fusion proteins mitofusin-2 (Mfn2, which is critical for outer-mitochondrial membrane fusion) and the dynamin-like 120 kDa protein (Opa1, which is critical for the regulation of the inner-mitochondrial membrane fusion and cristae structure).[42] In contrast, a high n-3 PUFA diet leads to reduced ROS production, an increase in hepatic mitochondrial fusion, and an increase in Mfn2 and Opa1. Overall, a high SFA diet results in reduced fusion and increased fission (mitochondrial fragmentation) in skeletal muscle and a high n-3 PUFA diet results in the opposite—increased fusion and decreased fission (mitochondrial elongation and networking). This study also found that a high n-3 PUFA diet causes skeletal muscle mitochondrial uncoupling (possibly mediated through UCP2 expression) and increased proton leak, diverting dietary energy away from ATP production.[42]

### 1.2.7 Fatty Acids and Apoptotic Signaling

In addition to reducing age-related diseases and increasing maximum lifespan,[50] a CR diet decreases skeletal muscle programmed cell death (apoptosis).[51] Because aging is associated with the loss of skeletal muscle and function (sarcopenia), it might be that the positive aging effects of a CR diet can be attributed to the observed decrease in apoptosis, leading to decreased sarcopenia.[51] Mitochondria play a central regulatory role in apoptosis, and mitochondrial dysfunction can trigger apoptosis following cyt *c* release from the mitochondria into the cytoplasm. A CR diet might decrease apoptosis levels through the mitochondria; Lopez-Dominguez et al. (2013) showed that a 6-month CR diet with either dietary fish oil (high in n-3 PUFAs) or lard (high in SFAs) decreased levels of cytosolic cyt *c*, an indicator of apoptosis, in skeletal muscle of old (21-month) mice.[52]

### 1.2.8 Ketogenic Diet

KD typically has high fat, moderate protein, and low carbohydrates (75%, 20%, and 5% of calories, respectively, in a classic KD) and must be balanced in overall micronutrient requirements.[53, 54] In response to a diet low in carbohydrates and high in fats, those of blood glucose levels are low and fat-derived ketone bodies are high (primarily β-hydroxybutyrate [β-OHB], but also acetoacetate and acetone, which is exhaled). These ketone bodies serve as a primary energy source that replaces glucose[55] and results in hyperketonemia (generally between 2 and 6 mM/L of β-OHB). In the presence of sufficient carbohydrates, pyruvate generated via glycolysis enters the mitochondria and is broken down in the TCA cycle to feed mitochondrial oxidative metabolism. In the absence of carbohydrates (as seen with KD or during prolonged fasting), there is a shift from glucose to lipids as the main source of energy, along with an increase in the production of ketone bodies. During ketosis, the TCA cycle is downregulated in the liver, and the acetyl-CoA produced during mitochondrial β-oxidation is converted to ketone bodies,[53] which are transported to other tissues and converted back to acetyl-CoA for energy.

Mitochondrial enzymes that are important for the generation of ketone bodies in the liver include succinyl-CoA:3-ketoacid CoA transferase (converts acetyl-CoA to acetoacetate) and β-OHB dehydrogenase (converts acetoacetate to β-OHB). In target tissues, important mitochondrial enzymes include β-OHB dehydrogenase (converts β-OHB back to acetoacetate) and thiolase (converts acetoacetate back to acetyl-CoA, which then enters the TCA cycle).[53]

The fat content within KD can be composed of triglycerides with different FA chain lengths: medium chain FAs (MCFAs), which are transported directly to the liver after ingestion via blood albumin transport and enter the mitochondria via diffusion; LCFAs; or a mixture of MCFAs and LCFAs. Compared to a KD high in MCFAs, LCFA-mediated ketosis is reached slowly because, after ingestion, LCFAs reach the liver indirectly via the thoracic duct and rely on an energetically expensive chylomicron transport. Furthermore, LCFAs require the carnitine shuttle (Fig. 3) to enter mitochondria, an overall slower and energetically more costly process compared to MCFA-diffused transport. For this reason, MCFA-based KDs can have a relatively lower fat content in relation to protein and carbohydrates, which increases palatability (important for diet compliance) while still achieving ketosis.[53]

The mitochondrial effects of KD include increased mitochondrial biogenesis (elevating energy production capacity), improved mitochondrial function, decreased oxidative stress, an increase in the reduced form of GSH (an antioxidant), and a reduced glycolytic rate (through an increase in lipid oxidation and an increase in mitochondrial respiration).[56] There are concerning adverse effects associated with long-term KD, however, primarily hyperlipidemia, hypercholesterolemia, nephrolithiasis (kidney stones), and cardiomyopathy.[53]

Interestingly, KD has been successfully employed to treat children with epilepsy.[53, 54] Glucose is an important energy source for the brain and is thought to lead to increased neuronal excitability, which leads to epileptic seizures. A low amount of carbohydrates in KD forces the brain to switch fuels from glucose to ketone bodies, resulting in antiepileptic and anticonvulsant effects. There are several proposed ways in which KD could have an antiepileptic effect: a ketone body-mediated decrease in glutamate release from synaptic vesicles and/or a decrease in mechanistic target of rapamycin (mTOR) pathway activation.[53] The mTOR pathway is a key regulator of metabolism and physiology and overactivation of mTOR is associated with epileptic seizures.[53] KD's antiepileptic effects also might involve the inhibition of the mitochondrial permeability transition pore (mPTP). In mice, the antiepileptic effect of ketone bodies can be mediated through their inhibition of mPTP formation in the mitochondrial IMM, which might help stabilize mitochondrial calcium levels and contribute to the observed KD-mediated reduction in oxidative stress.[53, 57]

There is also evidence that KD will increase longevity and health-span in mice. As with epilepsy, these positive health effects might be mediated through the mTOR pathway. KD decreases mTOR activity in adult mouse liver (possibly because of the reduced protein content in KD), which results in increased protein acetylation and decreased protein translation.[58] Decreased mTOR activity in rodents also is observed with MD and CR diets, and both diets also can increase lifespan.[58]

## 1.3 Carbohydrates

### *1.3.1 High Carbohydrate Diet and Mitochondria*

A high-carbohydrate (HC) diet (both eucaloric and hypercaloric, in which carbohydrate intake exceeds glycogen storage capacity) will trigger lipogenesis.[59, 60] Essentially, excess carbohydrates are converted to lipids through de novo lipogenesis (DNL) and stored primarily in white adipose tissue as triacylglycerides.[61] Within hepatocytes, excess carbohydrates leading to DNL will increase the synthesis and secretion of very low density lipoprotein (VLDL), which primarily target adipose and muscle tissues in which the lipids are either re-esterified and stored or used for energy.[61] Where there is excess carbohydrate capacity, a regulatory axis (sometimes called the AMPK/ACC/malonyl-CoA/CPT1 axis) is induced to shut down FA oxidation and turn on DNL.[62] AMP-activated protein kinase (AMPK) is a highly conserved regulatory molecule that senses the energy needs of the cell through the AMP/ATP ratio. When this ratio is large (high energy demand), then AMPK will inactivate acetyl-CoA carboxylase (ACC) and thereby promote FA oxidation (and the generation of ATP) and inhibit DNL at the same time. When the AMP/ATP ratio is low (low energy demand), then ACC promotes DNL (energy storage).

ACC converts acetyl-CoA to malonyl-CoA, which is the first step in FA synthesis and a critical regulator in the balance between DNL and FA oxidation. There are two isoforms of ACC (ACC1 and ACC2). ACC1 is found in the cytosol and is thought to be primarily responsible for generating malonyl CoA for FA synthesis. ACC1 is found predominantly in lipogenic tissues such as white fat and liver,[63–65] although it also is present in pancreatic islet and cardiac cells.[62] ACC2 is localized to the OMM and is thought to be responsible for regulating FA oxidation. In the OMM, ACC2 converts acetyl-CoA to malonyl-CoA, which is an inhibitor of carnitine/palmitoyl-transferase 1 (CPT1). CPT1 is the rate-limiting step for the transfer of acyl carnitine into the mitochondrial matrix, (the carnitine shuttle, which is a necessary step for FA oxidation; therefore, ACC2's regulation of CPT1 through malonyl-CoA production is a means of regulating FA oxidation. ACC2 is found predominantly in oxidative tissues such as skeletal muscle and the heart,[63–65] but it also is present in the liver, brown fat, and the mammary glands.[62] ACC1 and ACC2 appear to generate separate pools of malonyl-CoA, although there is some evidence suggesting that both isoforms cooperate to maintain cellular levels of malonyl-CoA and, therefore, the synthesis potential of FAs.[65]

In response to an HC diet, ACC1 will generate malonyl-CoA from acetyl-CoA in the cytosol for FA synthesis, and ACC2 will generate malonyl-CoA within the mitochondrial inner membrane space (IMS) for the downregulation of FA oxidation. The malonyl-CoA generated in the IMS will inhibit the carnitine shuttle system as previously described. In this way, FA oxidation is inhibited with a HC diet.[64] Thus, malonyl-CoA serves as a balancing regulatory molecule between FA synthesis and FA oxidation. The citrate carrier (CiC, or tricarboxylate carrier) is also an important gate-keeper protein. In the mitochondria, citrate can escape TCA cycle oxidation and is transported via the CiC to the cytosol, where it is cleaved into oxaloacetate (OAA) and acetyl-CoA. Acetyl-CoA then can be converted to malonyl-CoA by ACC1 and ACC2, as previously described.[66] Therefore, through the production of malonyl-CoA, glucose-derived acetyl-CoA from a HC diet will inhibit FA oxidation and favor FA synthesis.[67] In addition, OAA is reduced to malate and then $NAD^+$-dependent malic enzyme converts it to pyruvate while producing cytosolic NADPH, which is likely then used for FA synthesis. An HC diet will upregulate malic enzyme and increase cytosolic NADPH.[66] Citrate itself is an essential metabolic gatekeeper molecule, where its production by the TCA cycle provides negative feedback to both glycolysis and the TCA cycle (Pasteur effect). After it is in the cytosol, citrate will inhibit phosphofructokinase-1, (a glycolysis regulator enzyme), which will further activate ACC.[68] The effects of HC diet on mitochondrial functions is depicted in Fig. 2.

With a healthy diet, there is reciprocal regulation of glucose and FA oxidation that relies on mitochondrial metabolic flexibility (the Randle effect). When FAs are available then glucose oxidation decreases, which is in part mediated by FA oxidation itself. The products of FA oxidation (acetyl-CoA and NADH) inhibit pyruvate dehydrogenase (PDH), which prevents the transformation of pyruvate into acetyl-CoA.[67] ETC complex I appears to prefer NADH generated from FA oxidation and this might be because of substrate channeling (perhaps via compartmentalizing complex I). Complex I might become compartmentalized (i.e., isolated such that it preferentially receives electrons from NADH) through supercomplexing with other ETC protein complexes.[69] Compared to glucose utilization, FA oxidation results in more electrons passed to complex II than complex I, leading to less efficient oxidative phosphorylation.[67] The resultant decrease in proton electrochemical gradient ($\Delta p$) will favor electrons coming from FA oxidation because a low $\Delta p$ will lead to a decrease in influx of malate and glutamate via the malate/aspartate shuttle.[67] Overall, it is apparent that regulation of FA oxidation involves both malonyl-CoA levels and processes within the mitochondria.

Metabolic flexibility allows fuel switching depending on substrate availability, in which plasma FA promotes FA oxidation and suppresses glucose oxidation, and vice versa. Under chronic exposure to extreme diets, such as high sugar (HS) or HF (also called obesogenic diets), it is speculated that mitochondrial metabolic flexibility could be lost. This is observed in isolated skeletal muscle mitochondria from rats fed either a HS (70% sucrose) or HF (60% fat) diet for 12 weeks.[70] In this rat model, both extreme HS and HF diets interfere with the normally flexible interactions between pyruvate and FA oxidation within isolated mitochondria. An extreme HS and HF diet causes a decrease in TCA cycle flux within isolated skeletal muscle mitochondria, apparently mediated by a total reduction in PDH as well as reduced PDH flux.[70]

# 2 MICRONUTRIENTS

## 2.1 Antioxidants

Multiple oxidative reactions are involved in generating energy within the mitochondria, which also generate reactive free radicals. Unless neutralized the cumulative accumulation of these radicals can result in macromolecule damage and importantly the impairment of mitochondrial function, leading to cell senescence and potentially organismal aging. Mitochondria are extremely sensitive to oxidative stress, which is an integral part of OXPHOS defects.[71] Antioxidants play a key role in preventing these adverse effects by preserving mitochondrial function.[72] Functions of various dietary antioxidants in relation to mitochondria are depicted in Fig. 3.

### *2.1.1 Vitamin A*

Vitamin A, an antioxidant, usually consumed in the form of retinol or β-carotene, is taken up by mitochondria and protects cyt *c* and mitochondrial DNA from age-associated oxidative damage.[73,74] People need to exercise caution, however, because excessive vitamin A intake is toxic to liver, which could be an additive adverse effect during chronic alcohol consumption. Affected mitochondria undergo structural and functional changes leaking mitochondrial enzymes resulting in necrosis and fibrosis of liver.[75]

### 2.1.2 Vitamin C

Vitamin C or ascorbic acid, a water-soluble vitamin, is a reducing agent and prevents oxidative damage. Dehydroascorbate, the oxidized form of ascorbate, is recycled to a reduced form by GSH and antioxidant enzymes. Vitamin C also acts as a cofactor for hydroxylation of collagen and plays a role in biosynthesis of carnitine.[74] Owing to antioxidant properties, vitamin C has been used to help treat patients with diseases associated with OXPHOS dysfunction.[76] In patients with complex III deficiency, vitamin C combined with vitamin K (menadione), which transports electrons from NADH to CoQ or cyt *c*, act as a shunt to donate electrons more efficiently to cyt *c*, resulting in an improved phosphate metabolism and recovery from an exercise challenge.[77] This combination also improved clinical and biochemical parameters in complex I deficiency.[78] Benefits of vitamin C supplementation during exercise is questionable, however, because in the healthy population it might reduce exercise-enhanced expression of mitochondrial biogenesis-related transcription factors, such as nuclear respiratory factor 1, mitochondrial transcription factor A, and PPAR coactivator 1, along with cyt *c*, GSH peroxidase, and superoxide dismutase. This is because exercise generates low levels of ROS, which is needed as an important signaling molecule during exercise and stimulant of antioxidant enzyme synthesis. ROS is abolished by the antioxidant activity of vitamin C to cause a reduction in exercise endurance.[79]

### 2.1.3 Vitamin E

Vitamin E is a radical scavenging antioxidant that includes eight compounds among which α-tocopherol is biologically more active.[80] It is lipid-soluble, accumulates in biological membranes, and is abundant in mitochondrial membranes, where it scavenges ROS, inhibits lipid peroxidation and maintains mitochondrial integrity.[81] Like any other radical scavenger, it can act independently or synergistically with other antioxidants. Ubiquinol, ubisemiquinone, coenzyme $Q_{10}$, or ascorbate[82] are known to regenerate active vitamin E from tocopheroxyl radical. There is evidence to show that a supranutritional diet including all-rac-a-tocopheryl acetate will act as precursor to enrich vitamin E levels in the mitochondria of various animal models.[83] Mitochondria-targeted vitamin E also has been synthesized and experimentally used in various pathological conditions to boost its localization in mitochondria.[84]

### 2.1.4 Vitamin K

Though not a typical antioxidant in relation to mitochondrial function, vitamin K (phylloquinone or menadione) helps bridge the gap between complexes of the ETC in case of dysfunction of a complex. Based on this concept, vitamin K has been employed in combination with vitamin C in patients with OXPHOS disorders to help mediate electron transport (as previously described). In complex IV deficiency, vitamin K oral supplementation was shown to transfer electrons from complex I and II to $O_2$ to continue driving ATP production and improve mitochondrial dysfunction.[85] The type of vitamin K to be taken as oral supplementation for patients with various OXPHOS deficiencies, however, is debatable. Phylloquinone is rarely studied when compared to menadione, but it is lipid-soluble and biologically active. Water-soluble menadione, on the other hand, needs to be converted to menaquinone-4 after alkylation to be active.[86] Menadione supplementation was shown to have side effects in newborns, causing hemolytic anemia and hyperbilirubinemia, but no such reports exist with phylloquinone.[87]

### 2.1.5 *Polyphenols*

Polyphenols include a group of >500 compounds derived from plants with pleiotropic biological effects and are available in regular human diets.[88] They act as antioxidants and prevent oxidative damage generated by the mitochondria. In aged rats, wine polyphenols were shown to normalize ROS production and restore maximal muscle mitochondrial oxidative capacity.[89] Depending on the type of polyphenol, various mechanisms might mediate alterations in mitochondrial function, such as mitochondrial uncoupling, modulation of the mPTP[90] and modulation of energy uptake.[91]

## 2.2 Cofactors

Functions of various dietary constituents that serve as or are precursors of cofactors in mitochondria are depicted in Fig. 4.

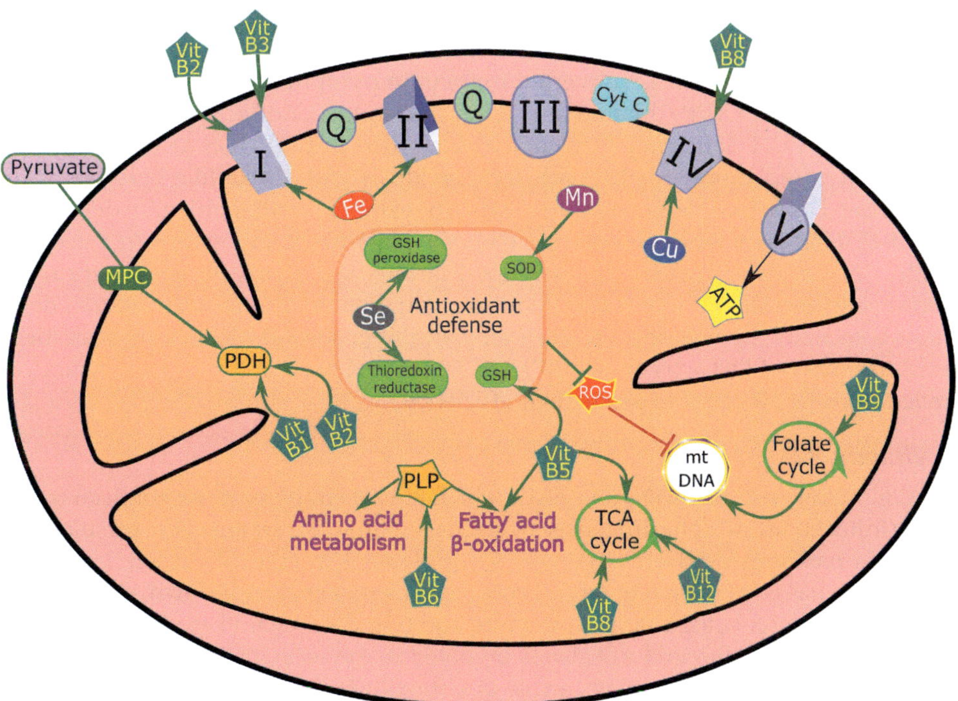


FIG. 4 The effects of various dietary constituents that directly or indirectly serve as cofactors in mitochondria. Cofactor nutrients are essential for the activities of mitochondrial complexes I (Vit B2, Vit B3 and Fe), II (Fe) and IV (Vit B8 and Cu), antioxidant enzymes (Vit B5, Se and Mn), energy metabolism (Vit B1, B2, B5, B6, B8 and B12), and mtDNA synthesis (Vit B9). Arrows: *green*—beneficial or neutral effects, *red*—deleterious effects; arrowheads: sharp—activation, blunt—inhibition.

### 2.2.1 Minerals

Trace minerals such as copper, zinc, and manganese are cofactors for superoxide dismutase (SOD). Deficiency of these minerals in a diet leads to decreased SOD activity, resulting in oxidative damage causing mitochondrial dysfunction.[92] Similarly, selenium (Se) is required for synthesis of selenoproteins such as GSH peroxidase and thioredoxin reductases, both having antioxidant properties, where its deficiency leads to peroxidative damage and dysfunctional mitochondria.[93] Often Se cellular levels are correlated to antioxidant potential in mitochondria and its function.[94] Copper is also a cofactor for a mitochondrial metabolic enzyme, cytochrome *c* oxidase (COX) of complex IV, therefore leads to mitochondrial dysfunction during its deficiency.[74] Iron (Fe) is essential for the formation of heme and iron-sulfur (Fe-S) clusters, important cofactors for assembly and activity of mitochondrial electron transfer complexes and aconitase enzyme of the TCA cycle. Impaired iron homeostasis is correlated with oxidative stress, disassembly of COX, loss of DNA integrity, and dysfunction in mitochondria.[74] Iron is an essential micronutrient, the deficiency of which will cause a decrease in mitochondrial respiratory chain proteins containing iron. Iron deficiency, however, does not affect non-iron containing proteins of the TCA cycle and FA oxidation pathway.[95]

### 2.2.2 Vitamin B1

Vitamin B1 or thiamin is a cofactor of the PDH complex, α-ketoglutarate dehydrogenase, branched-chain α-ketoacid dehydrogenase, transketolase (pentose phosphate pathway), and 2-hydroxyacyl-CoA lyase.[96] It is converted to thiamin pyrophosphate (TPP), which is transported rapidly through TPP/thiamin antiporter into mitochondria.[97] Thiamin is effective in the treatment of PDH deficiency[98] but not well studied in OXPHOS disorders. Some 3243 mitochondrial encephalomyopathy with lactic acidosis and stroke-like episodes (MELAS) mitochondrial DNA (mtDNA) mutation patients exhibit familial thiamin deficiency and decreased PDH activity that is associated with skeletal muscle myopathy. In these patients, myopathy and levels of creatine kinase, blood lactate, and pyruvate were improved with thiamin supplementation.[99] Currently, thiamin along with biotin (another vitamin described later in this section) supplements are used empirically in clinical practice because they have potential to enhance PDH activity or OXPHOS flux as reviewed in Ref. [100] but no randomized controlled trials (RCTs) exist to prove their efficiency.

### 2.2.3 Vitamin B2

Flavin mononucleotide (FMN) and FAD, which are cofactors for mitochondrial complexes I and II (mitochondrial dehydrogenases), needs vitamin B2 (riboflavin) for biosynthesis. This vitamin has been used effectively as a supplement in complex I deficient patients.[82] Riboflavin also has been shown to inhibit complex I proteolytic breakdown, enhancing enzymatic activity.[82] Alpha-ketoglutarate, pyruvate and branched-chain ketoacid dehydrogenase activities also need riboflavin. FAD also is involved in the recycling of oxidized glutathione (GSSG) to GSH and, therefore, plays a role in antioxidant defense.[101] As a result, riboflavin can both modulate antioxidant defenses and ETC function to enhance overall mitochondrial function. Riboflavin deficiency causes morphological changes in mitochondria, including the formation of giant mitochondria with alterations in cristae structure,[102] complex II, and amino acid metabolism-related enzymatic abnormalities in rats.[103] As expected, several of these studies demonstrated a reversal of these changes with riboflavin supplementation.[102]

In addition, complex I-deficient myopathic patients showed clinical improvement and increased exercise performance after riboflavin supplementation, although there was no correlation with complex I enzymatic activity.[104] With riboflavin supplementation, metabolic performance, and motor development improved in an infant with partial complex I defect,[105] and ATP synthesis enhanced in fibroblasts isolated from a patient with nuclear encoded complex I deficiency.[106] Combination therapies, including riboflavin and carnitine, also have been used to successfully treat myopathy involving complex I deficiency; clinical improvements include bicycle ergometry, measured muscular endurance strength and increased ETC complex activity in muscle biopsies.[82] Some of these studies showed that clinical improvements were exclusively because of riboflavin supplementation and not from the combination therapy of riboflavin and carnitine.[107] Adverse effects were not reported even with high doses of riboflavin.[108] Muscle strength in multiple acyl-CoA dehydrogenase deficiency patients with defects in the mitochondrial ETC is improved by riboflavin supplementation.[109] Riboflavin supplementation, but not the combination therapies (with other vitamins) in some open-label studies in complex I or II-deficient patients showed improvements in complex I and II enzyme activities, neurological exams, and muscle strength.[110, 111] The efficacy of riboflavin in primary mitochondrial disorders (PMDs), characterized by pathogenic mtDNA or nuclear DNA (nDNA) mutation, is yet to be explored in RCTs.

### 2.2.4 *Vitamin B5*

Oxidation of carbohydrates and FAs require vitamin B5 (pantothenic acid). Being a precursor for coenzyme A (CoA) and functioning as a carbonyl-activating group or an acyl group carrier, it is involved in mitochondrial enzymatic reactions including that of the TCA cycle's pyruvate, and α-ketoglutarate dehydrogenases and FA β-oxidation pathway. Enhancing the biosynthesis of GSH and CoA, and activity of GSH peroxidase enzyme prevents oxidative damage of mitochondrial constituents.[81]

### 2.2.5 *Vitamin B6*

Pyridoxal 5′-phosphate (PLP), a coenzyme formed from the precursor pyridoxine or vitamin B6, is essential for amino acid and lipid metabolism. PLP is imported passively into mitochondria,[112] which are the hosts for several PLP-dependent reactions. In addition, GSH biosynthesis from homocysteine require pyridoxine, therefore, it plays a role in regulating oxidative stress in mitochondria.[81]

### 2.2.6 *Biotin*

Carbohydrate, lipid, and protein metabolism require biotin (vitamin B8). Biotin is essential for synthesis of heme, which is a precursor for heme-a, a key component of complex IV in mitochondria. Therefore, its deficiency causes loss of mitochondrial complex IV, resulting in ROS-induced oxidative damage, which becomes more prominent in senescent cells.[113] Biotin also maintains the biochemical integrity of TCA cycle in other ways. Mitochondria hosts the coenzyme biotin-dependent pyruvate carboxylase (PC), propionyl-CoA carboxylase (PCC), and 3-methylcrotonyl-CoA carboxylase (MCC), which feeds intermediates (oxaloacetate, succinyl-CoA, and 3-methylglutalconyl CoA, respectively) into the TCA cycle and to acetyl-CoA carboxylase (ACC)-2, which is involved in FA metabolism.[114, 115] Therefore, biotin deficiency not only disrupts the TCA cycle by depleting intermediates, but it also decreases mitochondrial matrix glycine because of the accumulation of methylcrotonyl-CoA, because of low activity of MCC, which reacts with glycine. In situ mitochondrial heme synthesis, an essential step because dietary heme undergoes degradation by heme oxygenase before reaching the mitochondrial matrix, requires condensation of glycine and succinyl-CoA to form D-aminolevulinate. Moreover, in rodents, biotin deficiency is teratogenic and involves mitochondrial dysfunction along with disrupted heme metabolism.[113] Biotin also might play a role in enhancing mitochondrial biogenesis via the activation of guanylate cyclase through chemical interaction.[116, 117]

### 2.2.7 *Folic Acid*

Folinic acid, the activated form of folic acid or vitamin B9, is a cofactor involved in one-carbon-transfer reactions that are required for methylation reactions and synthesizing purines, such as methionine and thymidylate.[96] Half of the cellular folate levels reside in the mitochondria, which might play a role in the initiation of mitochondrial protein synthesis. Mitochondrial oxidative damage is aggravated by folate deficiency.[81] Some PMDs, such as Kearns-Sayre syndrome, is characterized by cerebral folate deficiency, although others do not involve folate. Empiric supplementation of folinic acid is recommended in PMDs with folate deficiency and neurological symptoms.[118, 119]

### 2.2.8 *Vitamin B12*

Like folinic acid, vitamin B12 or cobalamin also participates in one-carbon transfer reactions. Synthesis of mitochondrial succinyl-CoA of the TCA cycle requires adenosylcobalamin. Therefore, cobalamin deficiency leads to inhibition of TCA cycle causing mitochondrial dysfunction.[120]

## 2.3 Energy Enhancers or Metabolic Modifying Agents

Functions of various dietary constituents that boost metabolism or enhance repair and biogenesis of mitochondria are depicted in Fig. 5.

### 2.3.1 *L-Carnitine*

L-Carnitine, a quarternary ammonium compound, is an integral part of metabolism in several organisms. Its major function is shuttling long-chain FAs, for β-oxidation, from the cytosol to the mitochondrial matrix across the mitochondrial membrane via carnitine palmitoyltransferase enzymes by forming acetylcarnitine ester. As an acetyl-group donor and acceptor, it is involved in functions such as boosting CoA levels in mitochondria, oxidation of branched-chain α-ketoacid oxidation, directing peroxisomal β-oxidation acyl-CoA products to the liver mitochondrial matrix, neutralizing toxic acyl-CoA metabolites by esterification to maintain the TCA cycle,[121] storage of energy in the form of acetyl-carnitine in muscle,[122] and mitochondrial membrane stabilization.[123] Lysine and methionine are precursors of L-carnitine de novo biosynthesis in the liver and kidney, which contributes to 25% of the daily carnitine requirement. The remaining 75% is from dietary sources such as red meat and dairy.[121] Deficiency of L-carnitine can occur only in skeletal and heart muscle, as they are major users of this substance and are unable to synthesize the compound.[124] Plasma and muscle carnitine levels were reported to be depleted during mitochondrial diseases. The depletion, however,

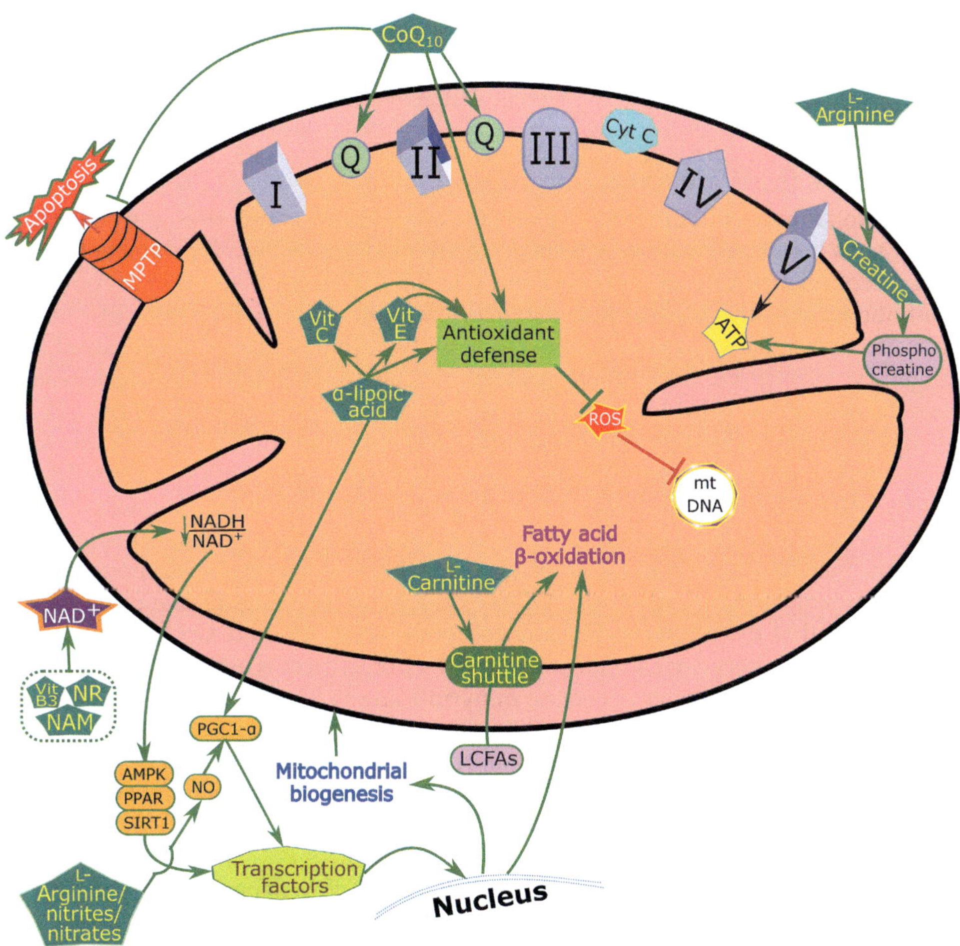


FIG. 5 The effects of various dietary constituents involved in repair and biogenesis of mitochondria, and enhancement of metabolism. These nutrients are involved in mitochondrial electron transfer and inhibition of apoptosis ($CoQ_{10}$), antioxidant defense ($CoQ_{10}$ and α-lipoic acid), ATP generation (L-arginine), fatty acid metabolism (L-carnitine), and biogenesis (α-lipoic acid, $NAD^+$ and NO precursors). Arrows: *green*—beneficial or neutral effects, *red*—deleterious effects; arrowheads: sharp—activation, blunt—inhibition.

appears to be more of a secondary effect than a cause, probably because acyl-CoA accumulation during mitochondrial dysfunction with mitochondrial disease results in increased esterification of carnitine depleting its free form.[125] Rising acyl-CoA intermediates might prevent exchange of ADP for ATP across the IMM by impairing adenine nucleotide translocase function. Carnitine supplementation in patients with its deficiency showed some improvements in muscle tone and strength, and in cardiomyopathic patients improved echocardiographic and clinical outcomes.[126] Primary carnitine deficiency in patients, such as in case of the carnitine transporter OCTN2 inherited defect, leads to hypoketotic hypoglycemia and cardiomyopathy. This could be because of mitochondrial dysfunction and can be attenuated by oral L-carnitine supplementation, which increases free L-carnitine levels in blood.[127] The potential benefit versus risk ratio of L-carnitine supplementation, however, needs to be further evaluated amidst concerns of increased risk of adverse cardiovascular events in specific intestinal microbiota containing mice and humans consuming red meat because of its metabolite trimethylamine-N-oxide.[128]

### 2.3.2 *Coenzyme $Q_{10}$ (Ubiquinone)*

$CoQ_{10}$ (also known as ubiquinone) is a lipid found in all membranes, particularly in the IMM. $CoQ_{10}$ is best known as an essential electron carrier within the mitochondrial ETC, where it accepts electrons from complex I and II in the ETC and passes them to complex III.[129] Protein-bound $CoQ_{10}$ also functions to stabilize yeast mitochondrial complex III,[130] therefore, it might affect protein complex structure and activity.[131] In addition, mitochondrial $CoQ_{10}$ is an obligatory cofactor for uncoupling proteins,[132] thereby facilitating induced proton leak between the matrix and IMM to reduce mitochondrial membrane potential. $CoQ_{10}$ likely also has an antiapoptotic function by preventing the opening of the mPTP.[133, 134] Finally, $CoQ_{10}$ is also a cosubstrate for dihydroorotate dehydrogenase (DHOD), which is responsible for the de novo synthesis of pyrimidine nucleotides (the building blocks of nucleic acids).[135] $CoQH_2$ (the fully reduced form, also known as ubiquinol) can act as a powerful fat-soluble antioxidant within the IMM by virtue of its location and high redox-status regeneration rate. As such, $CoQ_{10}$ inhibits protein, lipid, and DNA oxidation by scavenging free radicals.[131] It has been suggested that the homeostatic balance between $CoQH_2$ and $CoQ_{10}$ might influence mitochondrial aging, because an increased $CoQH_2/CoQ_{10}$ ratio is associated with decreased DHOD activity and a subsequent decrease in pyrimidine availability. A lower level of pyrimidine nucleotides can cause misincorporation of nucleotides during DNA replication, which might play a role in the increase in mitochondria DNA mutations observed with increasing $CoQH_2/CoQ_{10}$ ratio with age.[136] From the previous functions, it appears that $CoQH_2/CoQ_{10}$ homeostasis (i.e., the balance between reduced and oxidized forms) links bioenergetics, antioxidant protection, and cell death.

Sources of $CoQ_{10}$ include endogenous biosynthesis, which is dependent on the mevalonate pathway and uses tyrosine,[137] and exogenous dietary $CoQ_{10}$ (primarily meat and fish).[138] All cells can produce $CoQ_{10}$, although tissue concentrations vary considerably. Heart and kidney have the largest concentrations and lower concentrations are found in liver, skeletal muscle, and brain.[139, 140] It appears that exogenous $CoQ_{10}$ (dietary uptake) occurs when levels of endogenous $CoQ_{10}$ (biosynthesized) fall below a critical threshold.[141] With age and the onset of age-related disease, there appears to be a general decline in $CoQ_{10}$ levels in mammals.[131, 137]

#### DIETARY COENZYME Q, MITOCHONDRIAL FUNCTION, AND OXIDATIVE STRESS

There is evidence that supplemental dietary $CoQ_{10}$ can enhance its mitochondrial localization and have an antioxidant effect in skeletal muscle.[139] Rats fed 150 mg/kg/day of $CoQ_{10}$ for 13 weeks showed increased mitochondrial levels of homologues $CoQ_{10}$ and $CoQ_9$, as well as a reduction in skeletal muscle oxidative stress, specifically reduced mitochondrial protein oxidative damage.[139] Other studies confirm that mice fed supplemental $CoQ_{10}$ showed increased $CoQ_{10}$ in liver, heart, kidney, and skeletal muscle,[142] as well as the brain.[143] These and other findings led to the synthesis of the mitochondria-targeted ubiquinone (MitoQ), which selectively targets mitochondria to increase the accumulation of $CoQ_{10}$ and reduce mitochondrial oxidative stress.[142, 144] In a small double-blind randomized crossover control trial of older adult humans with impaired endothelial function, 6 weeks of oral supplementation with MitoQ mitigated mitochondrial ROS-related suppression of endothelial function and improved brachial arterial flow.[145]

#### COENZYME Q AND CALORIE RESTRICTION DIET

Following a long-term CR diet, which has been shown to increase lifespan and healthy aging in rodents,[146] an increase in total CoQ was observed in mouse and rat skeletal muscle, liver, heart, and kidney tissues.[52, 147, 148] Parrado-Fernandez et al. (2011) demonstrated that this increase in tissue-specific CoQ is in response to CR and a result of alterations in endogenous biosynthesis of CoQ.[140] The life-span effects of increased CoQ during CR, however, might be because of an activated plasma membrane redox system,[149] perhaps in addition to mitochondrial-specific effects. Specifically, plasma membrane NADH CoQ reductase (NCQR) is an important antioxidant enzyme that helps prevent lipid oxidation and stabilizes other antioxidants, such as α-tocopherol and ascorbate. NCQR is increased in rodents exposed to CR and might be partially responsible for observed CR life-span effects through its antioxidant functions.[150] It has been shown that activation of NQR1 (the yeast homologue of NCQR) prolongs lifespan in yeast by shifting cells to respiratory metabolism. NQR1 activation in the plasma membrane depletes cellular levels of NADH and increases $NAD^+$; therefore, the shift from fermentation to respiratory metabolism is thought to be facilitated by NQR1 regulation of NADH/$NAD^+$ pools.[151]

#### COENZYME Q AND HIGH PUFA DIET

A diet high in PUFAs is well-accepted as healthy with respect to prevention of cardiovascular disease. A diet high in PUFA, however, has been shown in rats to reduce mitochondrial function and increase ROS production. The increase in ROS is because of a higher degree of membrane polyunsaturation and subsequent risk of increased peroxidation.[152] To ameliorate these effects, $CoQ_{10}$ supplementation along with a high PUFA diet can increase ATP synthesis and decrease ROS production, where both are indicators of improved mitochondrial function.[153]

### 2.3.3 *Creatine*

Creatine, a guanidino compound, is biosynthesized from the amino acids glycine, arginine, and methionine in the liver, kidney, and pancreas. With meat products as the major source, it is also supplemented through diet.[154] Sodium-dependent creatine transporter actively transports it into mitochondria.[155] About 1.7% of the creatine pool is converted to creatinine and excreted daily. Being a mitochondrial nutrient, its main function as an energy-boosting compound is to prevent ATP depletion by increasing creatine/phosphocreatine stores, the ratio of which might cause mitochondrial dysfunction in muscle and brain.[82, 156] Mitochondrial creatine kinase converts creatine to phosphocreatine, which phosphorylates ADP to ATP, therefore acting as a high-energy buffer and protecting the efficiency of ATP synthesis. The other suggested roles for creatine are an indirect antioxidant and inhibitor of mitochondrial permeability transition,[157] a process that can lead to mitochondrial swelling and cell death. Creatine supplementation was shown to be effective only when muscle creatine levels are low in athletes and healthy control subjects,[158] and in more severely affected patients with mitochondrial disorders.[159] In mitochondrial disorder patients, high-intensity aerobic and anaerobic activities, but not lower-intensity aerobic activities, were improved by creatine supplementation in a short-term study.[159] More research is needed to determine the long-term effects of creatine supplementation in mitochondrial disease patients, whereas, $CoQ_{10}$ has been extensively studied as an energy-boosting compound in mitochondrial disorders. A placebo-controlled, double-blind, randomized cross-over study with a combination therapy of creatine, $CoQ_{10}$ and lipoic acid, each with different mechanisms of actions (increased ATP production from $CoQ_{10}$ and creatine; scavenging ROS from $CoQ_{10}$ and lipoic acid; and alternate energy source from creatine), showed improved muscle strength and blood lactate levels in mitochondrial-disease related myopathic patients.[160]

### 2.3.4 L-Arginine

L-Arginine, a precursor in the biosynthesis of nitric oxide (NO), creatine, proteins, polyamines, and α-ketoglutarate, is both synthesized endogenously and obtained from diet; the dietary intake is especially needed during some disease and stress conditions. Decarboxylation of L-arginine forms agmatine, which is an anti-inflammatory agent, antioxidant, and neuromodulator. NO formed from L-arginine activates peroxisome PPARα and PGC-1α, which affects gene expression and metabolism. Because of its beneficial effect of stroke-like episode prevention, arginine now is considered in frontline treatment for MELAS.[96] MELAS patients might experience vasoconstriction because of altered endothelial function in brain arterioles, and L-arginine presumably reverses it via NO synthesis.[127] L-Arginine supplementation in some MELAS siblings also showed improvements in maximum work at anaerobic threshold, $VO_2$ peak, and acidification along with an elevated phosphorous to phosphocreatine ratio and a recovery of phosphocreatine following moderate exercise.[161] Given these improvements in MELAS L-arginine should now be investigated in other PMDs.

### 2.3.5 $NAD^+$ Precursors

$NAD^+$ and $NAD^+$ phosphate ($NADP^+$) are well-known and crucial cofactors in cellular metabolism.[162] Naturally occurring precursors of $NAD^+$ include tryptophan, niacin (NA; vitamin B3 or nicotinic acid [FDA approved for hypertriglyceridemia]), nicotinamide (NAM), and nicotinamide riboside (NR). NR can enhance $NAD^+$ levels even in nonpathological conditions in mammalian cells and mitochondria and is tissue-dependent, such as in muscle and liver.[163] NR supplementation also was shown to induce mitochondrial biogenesis[164] and increases life span in yeast[165, 166] and mice.[167] $NAD^+$ biosynthesis intermediate, nicotinamide mononucleotide (NMN), also can be used as supplement for $NAD^+$ synthesis.[168] Among these, NMN and NR are popular supplements in NAD research, whereas clinical compliance of NA usage is low because of its skin flushing side effect.[169] $NAD^+$ plays an important role as a rate-limiting cofactor for the enzymatic activity of the sirtuin, poly (ADP-ribose) polymerase (PARP), and cluster of differentiation (CD) 38/157 families of enzymes.[168, 170] Reductions in the activities of $NAD^+$-dependent PARPs or CD38/157 can increase bioavailability of $NAD^+$ for SIRT1 activation and protect organisms from mitochondrial disease.[171] Moreover, calorie restriction (CR), fasting, and exercise boost $NAD^+$ levels, increasing the $NAD^+$/NADH ratio, thereby activating SIRT1, which in turn induces mitochondrial biogenesis and activates metabolic pathways simulating fasting response, FA oxidation, and ATP production.[168] In mice, $NAD^+$ levels were shown to deplete with age in muscle stem cells[167] or high-fat high-sucrose diet in liver.[172] Boosting $NAD^+$

levels through CR, pharmaceuticals directed at $NAD^+$ consumers or with $NAD^+$ precursor supplementation can improve mitochondrial function[102] and delay the onset of age-related disease and increase healthspan or maximum lifespan.[167, 173–175]

The ratio of $NAD^+$/NADH plays a significant role in PMDs. During ATP production NADH is converted to $NAD^+$ by mitochondrial complex I following transfer of electrons to ubiquinone, therefore, $NAD^+$ precursors have the potential to be used to correct the changes in the $NAD^+$/NADH ratio in OXPHOS disorders that specifically relate to complex I deficiencies. For example, NAM treatment was shown to increase blood $NAD^+$ levels, simultaneously decreasing serum lactate and pyruvate, in a 3243 MELAS PMD patient.[176] In another patient, clinical improvement was observed with NAM and riboflavin combination.[177]

Lower relative $NAD^+$ levels also can disrupt components of nutrient-sensing signaling network such as mTORC1 and AMPK, increase mitochondrial oxidant burden and decrease cellular respiratory capacity. Fibroblasts from Leigh syndrome patients with complex I subunit gene mutations have lower concentrations of NADH and $NAD^+$ with a low $NAD^+$/NADH ratio. Treatment with NA restores the $NAD^+$/NADH redox balance, and thereby the nutrient-sensing signaling network and cellular respiratory capacity, in these cells.[178] Similarly, NR supplementation enhanced $NAD^+$ content in NDUFS1 mutation containing patient fibroblasts[179] and in germline KO Ndufs4 mice.[180] Despite deleterious effects of disrupted NAD metabolism and potential benefits of $NAD^+$ precursors as supplements are known, they are not well studied in PMDs and RCTs are lacking, which warrants further investigations. Overall, NR and other $NAD^+$ precursors can be potential nutritional supplements to fight age-related and metabolic disorders characterized by mitochondrial dysfunction.

### *2.3.6 α-Lipoic Acid*

Antioxidant α-lipoic acid is a coenzyme for β-ketoacid dehydrogenases and PDH found in mitochondria and maintains energy homeostasis. It is a widely studied nutrient for examining changes in mitochondrial function both in vitro and in preclinical studies of neurodegeneration and brain aging. Diets containing meats, such as heart, kidney, and liver, and vegetables, such as broccoli, potatoes and spinach, contain α-lipoic acid. It is also synthesized de novo by mitochondrial lipoyl synthase[127] and in various tissues is reduced to dihydrolipoic acid by an NADH-dependent mitochondrial dihydrolipoamide dehydrogenase, which acts as an antioxidant, recycles other antioxidants, such as vitamins C and E and GSH (thus protects membranes), and chelates iron and copper (antioxidant effects).[127, 181] As mitochondria-associated antioxidants modulate cognitive function, α-lipoic acid has been shown to be effective in improving cognitive deficits in Alzheimer's disease in both animal models and patients.[181, 182] These effects are likely because of its oxidative stress-reducing and mitochondrial function improving effects.[82] There is also evidence of increases in phosphorylation potential, measured as [ATP]/[ADP][Pi], and brain phosphocreatine with α-lipoic acid supplementation in mtDNA-related mitochondrial disease patients.[183] Because of to the potential beneficial effects of α-lipoic acid in diseases such as diabetes and dyslipidemia that involve mitochondrial functional impairments,[96] extensive research is warranted in PMDs.

The α-lipoic acid partially recovers age-associated mitochondrial damage and increases ambulatory activity in aging rodent populations. Because of its antioxidant and chelation properties, α-lipoic acid plays a key role in restoring mitochondrial function following various forms of damage to neurons.[181] The α-lipoic acid also induces transcription factors and enzymes, which acts to enhance antioxidative defense and reduce cytosolic oxidative stress, altogether indirectly protecting mitochondria from oxidative stress.[184] Alone or in combination with acetyl L-carnitine, it also might stimulate mitochondrial biogenesis by enhancing PGC-1α activity, thus improving mitochondrial function, which presumably is beneficial in age-related neurodegenerative diseases. This combination with different mechanisms was shown to have synergistic action in improving cognitive and memory deficits in older rats by preventing the age-related decay of mitochondria as one of the mechanisms. Furthermore, in combination with other nutritional cofactors and antioxidants, the antioxidant recycling capacity of α-lipoic acid attenuated age-dependent cognitive decline in dogs.[181]

### *2.3.7 Nitrates and Nitrites (Nitric Oxide)*

Nitric oxide (NO) interacts with mitochondrial respiratory chain and TCA cycle[185] enzymes and controls cellular respiration. Vegetable diets contain inorganic anions called nitrates that are the source of NO in vivo. Nitrate is converted to nitrite by gut bacterial nitrate reductases and xanthine oxidase. Nitrite is spontaneously converted to NO by reducing or acidic environments that are accelerated by various proteins such as deoxyhaemoglobin, deoxymyoglobin, xanthine oxidase and mitochondrial respiratory chain enzymes. NO from dietary nitrate interacts with complex IV and IMM, therefore improving basal mitochondrial function (increases efficiency of mitochondrial oxidative phosphorylation), which was shown to improve oxygen use during physical exercise in healthy volunteers and increased oxidative phosphorylation efficiency in skeletal muscle mitochondria.[186] NO interaction with COX also might mediate signaling events that lead to improvements in cellular defense and adaptive immune responses. Recent evidence has shown that oxygen consumption can be reduced and muscle efficiency can be improved at submaximal exercise by dietary nitrate intake such as in the form of beetroot juice in young healthy volunteers without enhancing anaerobic metabolism.[187] Similar observations were made with sodium nitrate dietary intake along with improved mitochondrial coupling efficiency and ATP maximal rate production.[186, 188] Thus, nitrate supplementation appears to improve respiratory efficiency.

## 3 HIGH CALORIE AND CALORIE RESTRICTION DIETS ON MITOCHONDRIA

As reviewed by Wallace,[189] a high calorie diet leads to ROS production, OXPHOS defects, and reduced ATP synthesis. Excessive caloric intake overburdens the cell with reducing equivalents, hyperpolarizes mitochondria, and stalls the ETC, along with glucose use. The increased blood glucose levels stimulate pancreas to secrete more insulin, which is the first step towards the development of insulin resistance. Chronic insulin signaling suppresses transcription of PGC-1α, therefore inhibits OXPHOS and aggravates mitochondrial energy deficiency. The same vicious cycle occurs in β-cells of the pancreas, increases ROS, blocks ATP production, opens the mPTP, leading to apoptosis and eventually a reduction in insulin secretion. Thus, a high calorie diet has the propensity to cause insulin resistance and eventually diabetes.[189]

In contrast, a calorie restriction (CR) diet decreases free radical production from mitochondria, with lower ROS production from complex I, resulting in a lower incidence of mitochondrial DNA damage and reduced accumulation of mtDNA mutations.[190] Mitochondrial and TCA cycle activity increases, and the antioxidant defense system is enhanced during a CR diet.[191–193] Consequently, a CR diet has shown been shown to decelerate age-related phenotypes[189] in species ranging from worms to humans, while boosting longevity in yeast,[194] worms,[195] and rodents.[189] The nutrient-sensing network, involving AMPK activation and its inhibition of mTORC1, plays a key role in adaptation of mitochondria to CR diet.[102] Enhanced expression of PGC-1α, eNOS, and SIRT1 also might mediate these beneficial effects of a CR diet,[192] which also was confirmed in humans.[196] Beyond CR, inhibition of the mTOR pathway or boosting sirtuin activity, with rapamycin or NR, respectively, are examples of treatments that also increase lifespan of various organisms.[167, 173, 197]

# 4 CLINICAL TRIALS OF DIETARY CONSTITUENTS IN MITOCHONDRIAL DISORDERS

Mitochondrial disorders are characterized by progressive decline of mitochondrial capacity to respond to cellular energy needs. The origin of these disorders often can be heterogeneous, owing to biochemical, clinical, and genetic parameters, and can include myopathies, encephalopathies, neuropathies, and cardiomyopathies.[197] Genetic parameters of mitochondrial disorders can be inherited or through de novo mutations of mitochondrial and/or nuclear DNA that result in respiratory chain functional impairment.[198]

Although the causes of mitochondrial disorders are well elucidated, treatment options have not been well explored. Because diet can have a potential mitigatory or remedial effect on dysfunctional mitochondria, several researchers have attempted single or multi-nutrient diet clinical studies to treat mitochondrial disorders. The goal of such therapies is to delay the progression of the disease and enhance ATP production. As discussed in previous sections, several mitochondrial nutrients demonstrated therapeutic potential in OXPHOS disorders. In spite of the need for more research, however, a few dietary constituents have advanced to phase III and IV clinical trials, as shown in the Table 1. (retrieved from ClinicalTrials.gov). Phase II trials with reported results also are listed in the Table 1. Apart from these, dietary constituents in phase I and phase II (with unreported results) clinical trials for treating mitochondrial dysfunction-associated diseases are N-carbamylglutamate, epicatechin, short-chain quinones, resveratrol, citrulline, high protein or protein supplementation, sodium nitrite, urolithin A, L-glutamine, and medium-chain triglycerides (retrieved from ClinicalTrials.gov). A comprehensive list of other dietary constituent human trials are listed elsewhere.[82, 205] Based on this information, it is obvious that using dietary constituents as a therapeutic approach for mitochondrial dysfunction-associated diseases is still in the preliminary stages, but emphasizes their importance in regulating mitochondrial functions. Therefore, more research and extensive clinical trials are needed to confirm their efficacy.

TABLE 1 Advanced Clinical Trials of Dietary Supplements in Mitochondria-Related Diseases

| Dietary Agent(s) | Dosage | Diseases or Disorders With Mitochondrial Dysfunction | Study Details | Effects/Results (If Reported) | References or ClinicalTrials.gov Identifier |
|---|---|---|---|---|---|
| Nicotinamide | 2–8 g daily for 8 weeks | Friedreich's ataxia | Phase II: exploratory, open label, dose escalation study | Expression of the Frataxin protein (the deficiency of which causes the disease) increased with a reduction in heterochromatin modifications in its gene locus | 199 |
| α-Tocopheryl quinone (vitamin E precursor) | 0.5 g or 0.75 g b.i.d. for 28 days | Friedreich's ataxia | Phase II: randomized, double-blind, crossover study ($n=31$) | Neurological function improved on Friedreich Ataxia Rating Scale compared to the placebo | 200 |
| Coenzyme $Q_{10}$ | 2700 mg daily for 9 months | Amyotrophic lateral sclerosis (ALS) | Phase II: randomized, placebo controlled trial ($n=185$) | ALS Functional Rating Scale-revised score not better than placebo. Does not warrant Phase III testing. | 201 |
| Nitrate rich beetroot juice | 140 mL of 12.9 mmol nitrate | Heart failure with preserved ejection fraction | Phase II – randomized, double-blind, crossover study ($n=17$) | Increased exercise capacity with improved mitochondrial oxidative capacity. | 202 |
| Alpha-lipoic acid and L-acetyl carnitine | 600 mg and 1.5 g respectively for 6 months | Progressive supranuclear palsy | Phase II: open-label trial ($n=11$) | Changes of cerebral lactate and GSH levels measured (data not shown) | NCT01537549 |
| Ubiquinone | 400 mg daily for 12 weeks | Nonproliferative diabetic retinopathy | Phase II: double-blind, placebo-controlled trial ($n=61$) | Improves clinical outcomes and nerve conduction parameters of diabetic polyneuropathy with reduced oxidative stress | 203 |
| Uridine | Escalated doses t.i.d. | Mitochondrial (mt) toxicity of nucleoside analogue reverse transcriptase inhibitors in HIV patients | Phase II: randomized trial ($n=3$) | Deemed safe to use in HIV patients who cannot be switched to less-toxic drugs | 204 |
| Alpha-lipoic acid and acetyl-L-carnitine | 600–1800 mg/day and 1000–3000 mg/day respectively | Bipolar depression | Phase II: randomized, placebo controlled trial ($n=40$) | No statistical analysis reported but no or very slight improvement, if any, was observed in depressive and maniac symptoms compared to placebo | NCT00719706 |

| L-Arginine | 100 mg/kg t.i.d. for 6 weeks | Mitochondrial encephalomyopathy, lactic acidosis and stroke-like episodes (MELAS) syndrome | Phase II: case control study ($n = 3$) | Increase in mean % maximum work and heart rate at anaerobic threshold, slight increase in $VO_{2peak}$, less acidosis after exercise, increased work capacity, and improved mitochondrial activity were observed | 161 |
|---|---|---|---|---|---|
| Vitamin D | 60,000 units per week for 12 weeks | Statin therapy-induced myopathy in dyslipidemia patients | Phase III: randomized placebo controlled trial ($n = 33$) | No statistical analysis reported but no or very slight improvement, if any, in peak oxygen consumption and no change in skeletal muscle mitochondrial content were observed compared to the placebo | NCT02030041 |
| Coenzyme $Q_{10}$ | 10 mg/kg up to 400 mg daily for 6 months | Mitochondrial disease | Phase III: randomized placebo controlled trial ($n = 24$) | Accrual of patients was very limited, precluding interpretation of the data | NCT00432744 |
| Coenzyme $Q_{10}$ | 200 mg/day for 6 weeks | Statin therapy-induced myalgia in older athletes | Phase IV: randomized placebo controlled trial ($n = 20$) | | NCT01026311 |
| Cofactor supplementation (thiamine, riboflavin, L-carnitine) | 50, 100 and 1880 mg/day | Lactic acidosis in HIV patients | Phase IV: open label trial ($n = 30$) | | NCT00202228 |
| Tetrahydrobiopterin | 5 mg/kg once | Peripheral artery disease | Phase IV: randomized placebo controlled trial ($n = 10$) | | NCT03493412 |
| MitoQ (mitochondria targeted antioxidant) | 20 mg/day for 4 weeks | Chronic kidney disease | Phase IV: randomized placebo controlled trial ($n = 24$) | | NCT02364348 |
| L-Arginine | 2 g for 30 min infusion | Skeletal muscle ischemia | Phase IV: randomized placebo controlled trial ($n = 60$) | | NCT02117206 |

## 5 CONCLUSION

Macronutrients, carbohydrates, proteins and fats, not only serve as substrates for metabolic pathways and energy production, but they also regulate mitochondrial function. An excessive supply of macronutrients, occurring in high calorie, high carbohydrate, and high fat diets, or their deficiency, such as in a low protein diet, can be detrimental to mitochondria, whereas, caloric restriction without malnutrition can boost mitochondrial biogenesis and function. Mitochondrial nutrients, micronutrients that are essential for mitochondrial maintenance and function, often participate as antioxidants, cofactors or coenzymes, and integral components of mitochondrial biogenesis and/or metabolic functions. Deficiency of such nutrients would cause mitochondrial abnormalities. Clinical usage of mitochondrial nutrients, such as nicotinamide, coenzyme $Q_{10}$ and L-arginine, in diseases associated with mitochondrial abnormalities proved promising by mitigating functional defects and improving metabolic parameters. Overall, nutrient composition of a diet influences mitochondrial health, which in turn affects metabolic homeostasis, energy balance, survival and longevity of an organism.

## References


1. McInnes J. Mitochondrial-associated metabolic disorders: foundations, pathologies and recent progress. *Nutr Metab* 2013;**10**(1):63–75.
2. Shaum KM, Polotsky AJ. Nutrition and reproduction: is there evidence to support a "fertility diet" to improve mitochondrial function? *Maturitas* 2013;**74**(4):309–12.
3. Brosnan JT, Brosnan ME. Branched-chain amino acids: metabolism, physiological function, and application. *J Nutr* 2006;**136**:207S–11S.
4. Hales CN, Barker DJP. Type 2 (non-insulin-dependent) diabetes mellitus: the thrifty phenotype hypothesis. *Int J Epidemiol* 2013;**42**(5):1215–22.
5. Theys N, Bouckenooghe T, Ahn M-T, Remacle C, Reusens B. Maternal low-protein diet alters pancreatic islet mitochondrial function in a sex-specific manner in the adult rat. *Am J Phys Regul Integr Comp Phys* 2009;**297**(5):R1516–25.
6. Tarry-Adkins JL, Chen J-H, Jones RH, Smith NH, Ozanne SE. Poor maternal nutrition leads to alterations in oxidative stress, antioxidant defense capacity, and markers of fibrosis in rat islets: potential underlying mechanisms for development of the diabetic phenotype in later life. *FASEB J* 2010;**24**(8):2762–71.
7. Claycombe KJ, Roemmich JN, Johnson LK, Vomhof-Dekrey EE, Johnson WT. Skeletal muscle Sirt3 expression and mitochondrial respiration are regulated by a prenatal low-protein diet. *J Nutr Biochem* 2015;**26**(2):184–9.
8. Hirschey MD, Shimazu T, Jing E, Grueter CA, Collins AM, Aouizerat B, et al. SIRT3 deficiency and mitochondrial protein hyperacetylation accelerate the development of the metabolic syndrome. *Mol Cell* 2011;**44**(2):177–90.
9. Mortensen OH, Olsen HL, Frandsen L, Nielsen PE, Nielsen FC, Grunnet N, et al. A maternal low protein diet has pronounced effects on mitochondrial gene expression in offspring liver and skeletal muscle; protective effect of taurine. *J Biomed Sci* 2010;**17**(Suppl 1):S38.
10. Jong CJ, Azuma J, Schaffer S. Mechanism underlying the antioxidant activity of taurine: prevention of mitochondrial oxidant production. *Amino Acids* 2012;**42**(6):2223–32.
11. Suzuki T, Suzuki T, Wada T, Saigo K, Watanabe K. Novel taurine-containing uridine derivatives and mitochondrial human diseases. *Nucleic Acids Res Suppl* 2001;**1**:257–8.
12. Ables GP, Johnson JE. Pleiotropic responses to methionine restriction. *Exp Gerontol* 2017;**94**:83–8.
13. Lee BC, Kaya A, Ma S, Kim G, Gerashchenko MV, Yim SH, et al. Methionine restriction extends lifespan of *Drosophila melanogaster* under conditions of low amino-acid status. *Nat Commun* 2014;**5**:3592.
14. Richie JP, Leutzinger Y, Parthasarathy S, Malloy V, Orentreich N, Zimmerman JA. Methionine restriction increases blood glutathione and longevity in F344 rats. *FASEB J* 1994;**8**(15):1302–7.

15. Zhou X, He L, Wan D, Yang H, Yao K, Wu G, et al. Methionine restriction on lipid metabolism and its possible mechanisms. *Amino Acids* 2016;**48**(7):1533–40.
16. Lees EK, Król E, Grant L, Shearer K, Wyse C, Moncur E, et al. Methionine restriction restores a younger metabolic phenotype in adult mice with alterations in fibroblast growth factor 21. *Aging Cell* 2014;**13**(5):817–27.
17. Badman MK, Pissios P, Kennedy AR, Koukos G, Flier JS, Maratos-Flier E. Hepatic fibroblast growth factor 21 is regulated by PPARalpha and is a key mediator of hepatic lipid metabolism in ketotic states. *Cell Metab* 2007;**5**(6):426–37.
18. Mendelsohn AR, Larrick JW. Fibroblast growth factor-21 is a promising dietary restriction mimetic. *Rejuvenation Res* 2012;**15**(6):624–8.
19. Sanchez-Roman I, Barja G. Regulation of longevity and oxidative stress by nutritional interventions: role of methionine restriction. *Exp Gerontol* 2013;**48**(10):1030–42.
20. Sanchez-Roman I, Gómez A, Pérez I, Sanchez C, Suarez H, Naudí A, et al. Effects of aging and methionine restriction applied at old age on ROS generation and oxidative damage in rat liver mitochondria. *Biogerontology* 2012;**13**(4):399–411.
21. Perrone CE, Mattocks D a L, Jarvis-Morar M, Plummer JD, Orentreich N. Methionine restriction effects on mitochondrial biogenesis and aerobic capacity in white adipose tissue, liver, and skeletal muscle of F344 rats. *Metabolism* 2010;**59**(7):1000–11.
22. Hasek BE, Boudreau A, Shin J, Feng D, Hulver M, Van NT, et al. Remodeling the integration of lipid metabolism between liver and adipose tissue by dietary methionine restriction in rats. *Diabetes* 2013;**62**(10):3362–72.
23. Patil YN, Dille KN, Burk DH, Cortez CC, Gettys TW. Cellular and molecular remodeling of inguinal adipose tissue mitochondria by dietary methionine restriction. *J Nutr Biochem* 2015;**26**(11):1235–47.
24. Nakamura MT, Yudell BE, Loor JJ. Regulation of energy metabolism by long-chain fatty acids. *Prog Lipid Res* 2014;**53**:124–44.
25. Ballard JWO, Youngson NA. Review: can diet influence the selective advantage of mitochondrial DNA haplotypes? *Biosci Rep* 2015;**35**(6):e00277.
26. García-Ruiz I, Solís-Muñoz P, Fernández-Moreira D, Grau M, Colina F, Muñoz-Yagüe T, et al. High-fat diet decreases activity of the oxidative phosphorylation complexes and causes nonalcoholic steatohepatitis in mice. *Dis Model Mech* 2014;**7**(11):1287–96.
27. Hulbert AJ, Turner N, Storlien LH, Else PL. Dietary fats and membrane function: implications for metabolism and disease. *Biol Rev Camb Philos Soc* 2005;**80**(1):155–69.
28. Chen Y, Hagopian K, McDonald RB, Bibus D, López-Lluch G, Villalba JM, et al. The influence of dietary lipid composition on skeletal muscle mitochondria from mice following 1 month of calorie restriction. *J Gerontol A Biol Sci Med Sci* 2012;**67**(11):1121–31.
29. Ramsey JJ, Harper M-E, Humble SJ, Koomson EK, Ram JJ, Bevilacqua L, et al. Influence of mitochondrial membrane fatty acid composition on proton leak and $H_2O_2$ production in liver. *Comp Biochem Physiol Part B Biochem Mol Biol* 2005;**140**(1):99–108.
30. Quiles JL, Martínez E, Ibáñez S, Ochoa JJ, Martín Y, López-Frías M, et al. Ageing-related tissue-specific alterations in mitochondrial composition and function are modulated by dietary fat type in the rat. *J Bioenerg Biomembr* 2002;**34**(6):517–24.
31. Pfeiffer K, Gohil V, Stuart RA, Hunte C, Brandt U, Greenberg ML, et al. Cardiolipin stabilizes respiratory chain supercomplexes. *J Biol Chem* 2003;**278**(52):52873–80.
32. Acehan D, Malhotra A, Xu Y, Ren M, Stokes DL, Schlame M. Cardiolipin affects the supramolecular organization of ATP synthase in mitochondria. *Biophys J* 2011;**100**(9):2184–92.
33. Arnarez C, Marrink SJ, Periole X. Identification of cardiolipin binding sites on cytochrome c oxidase at the entrance of proton channels. *Sci Rep* 2013;**3**:1263.
34. Abramovitch DA, Marsh D, Powell GL. Activation of beef-heart cytochrome c oxidase by cardiolipin and analogues of cardiolipin. *BBA-Bioenergetics* 1990;**1020**(1):34–42.
35. Kagan VE, Bayir HA, Belikova NA, Kapralov O, Tyurina YY, Tyurin VA, et al. Cytochrome c/cardiolipin relations in mitochondria: a kiss of death. *Free Radic Biol Med* 2009;**46**(11):1439–53.
36. Chu CT, Bayir H, Kagan VE. LC3 binds externalized cardiolipin on injured mitochondria to signal mitophagy in neurons: implications for Parkinson disease. *Autophagy* 2014;**10**(2):376–8.
37. Aoun M, Feillet-Coudray C, Fouret G, Chabi B, Crouzier D, Ferreri C, et al. Rat liver mitochondrial membrane characteristics and mitochondrial functions are more profoundly altered by dietary lipid quantity than by dietary lipid quality: effect of different nutritional lipid patterns. *Br J Nutr* 2012;**107**(5):647–59.

38. Cortie CH, Else PL. Dietary docosahexaenoic acid (22:6) incorporates into cardiolipin at the expense of linoleic acid (18:2): analysis and potential implications. *Int J Mol Sci* 2012;**13**(11):15447–63.
39. Endo J, Arita M. Cardioprotective mechanism of omega-3 polyunsaturated fatty acids. *J Cardiol* 2016;**67**(1):22–7.
40. Madison Sullivan E, Pennington ER, Sparagna GC, Torres MJ, Darrell Neufer P, Harris M, et al. Docosahexaenoic acid lowers cardiac mitochondrial enzyme activity by replacing linoleic acid in the phospholipidome. *J Biol Chem* 2018;**293**(2):466–83.
41. Yu L, Fink BD, Herlein JA, Oltman CL, Lamping KG, Sivitz WI. Dietary fat, fatty acid saturation and mitochondrial bioenergetics. *J Bioenerg Biomembr* 2014;**46**(1):33–44.
42. Lionetti L, Mollica MP, Donizzetti I, Gifuni G, Sica R, Pignalosa A, et al. High-lard and high-fish-oil diets differ in their effects on function and dynamic behaviour of rat hepatic mitochondria. *PLoS ONE* 2014;**9**(3).
43. Nabben M, Hoeks J, Moonen-Kornips E, Van Beurden D, Briedé JJ, Hesselink MKC, et al. Significance of uncoupling protein 3 in mitochondrial function upon mid- and long-term dietary high-fat exposure. *FEBS Lett* 2011;**585**(24):4010–7.
44. Nicoll R, Henein MY. Caloric restriction and its effect on blood pressure, heart rate variability and arterial stiffness and dilatation: a review of the evidence. *Int J Mol Sci* 2018;**19**(3):E751.
45. Picca A, Pesce V, Lezza AMS. Does eating less make you live longer and better? An update on calorie restriction. *Clin Interv Aging* 2017;**12**:1887–902.
46. Laganiere S, Yu BP. Modulation of membrane phospholipid fatty acid composition by age and food restriction. *Gerontology* 1993;**39**(1):7–18.
47. Hagopian K, Harper M-E, Ram JJ, Humble SJ, Weindruch R, Ramsey JJ. Long-term calorie restriction reduces proton leak and hydrogen peroxide production in liver mitochondria. *Am J Physiol Metab* 2005;**288**(4):E674–84.
48. Putti R, Sica R, Migliaccio V, Lionetti L. Diet impact on mitochondrial bioenergetics and dynamics. *Front Physiol* 2015;**6**:109.
49. Liesa M, Shirihai OS. Mitochondrial dynamics in the regulation of nutrient utilization and energy expenditure. *Cell Metab* 2013;**17**(4):491–506.
50. Sohal RS, Weindruch R. Oxidative stress, caloric restriction, and aging. *Science* 1996;**273**(5271):59–63.
51. Wohlgemuth SE, Seo AY, Marzetti E, Lees HA, Leeuwenburgh C. Skeletal muscle autophagy and apoptosis during aging: effects of calorie restriction and life-long exercise. *Exp Gerontol* 2010;**45**(2):138–48.
52. López-Domínguez JA, Khraiwesh H, González-Reyes JA, López-Lluch G, Navas P, Ramsey JJ, et al. Dietary fat modifies mitochondrial and plasma membrane apoptotic signaling in skeletal muscle of calorie-restricted mice. *Age (Dordr)* 2013;**35**(6):2027–44.
53. Branco AF, Ferreira A, Simões RF, Magalhães-Novais S, Zehowski C, Cope E, et al. Ketogenic diets: from cancer to mitochondrial diseases and beyond. *Eur J Clin Investig* 2016;**46**(3):285–98.
54. Yuen AWC, Walcutt IA, Sander JW. An acidosis-sparing ketogenic (ASK) diet to improve efficacy and reduce adverse effects in the treatment of refractory epilepsy. *Epilepsy Behav* 2017;**74**:15–21.
55. Clanton RM, Wu G, Akabani G, Aramayo R. Control of seizures by ketogenic diet-induced modulation of metabolic pathways. *Amino Acids* 2017;**49**(1):1–20.
56. Rho JM. How does the ketogenic diet induce anti-seizure effects? *Neurosci Lett* 2017;**637**:4–10.
57. Kim DY, Simeone KA, Simeone TA, Pandya JD, Wilke JC, Ahn Y, et al. Ketone bodies mediate antiseizure effects through mitochondrial permeability transition. *Ann Neurol* 2015;**78**(1):77–87.
58. Roberts MN, Wallace MA, Tomilov AA, Zhou Z, Marcotte GR, Tran D, et al. A ketogenic diet extends longevity and healthspan in adult mice. *Cell Metab* 2017;**26**(3):539–546.e5.
59. Hudgins LC, Hellerstein M, Seidman C, Neese R, Diakun J, Hirsch J. Human fatty acid synthesis is stimulated by a eucaloric low fat, high carbohydrate diet. *J Clin Invest* 1996;**97**(9):2081–91.
60. Ferramosca A, Conte A, Damiano F, Siculella L, Zara V. Differential effects of high-carbohydrate and high-fat diets on hepatic lipogenesis in rats. *Eur J Nutr* 2014;**53**(4):1103–14.
61. Strable MS, Ntambi JM. Genetic control of de novo lipogenesis: role in diet-induced obesity. *Crit Rev Biochem Mol Biol* 2010;**45**(3):199–214.
62. Saggerson D. Malonyl-CoA, a key signaling molecule in mammalian cells. *Annu Rev Nutr* 2008;**28**(1):253–72.
63. Abu-Elheiga L, Oh W, Kordari P, Wakil SJ. Acetyl-CoA carboxylase 2 mutant mice are protected against obesity and diabetes induced by high-fat/high-carbohydrate diets. *Proc Natl Acad Sci* 2003;**100**(18):10207–12.
64. Abu-Elheiga L, Wu H, Gu Z, Bressler R, Wakil SJ. Acetyl-CoA carboxylase 2−/−mutant mice are protected against fatty liver under high-fat, high-carbohydrate dietary and *de novo* lipogenic conditions. *J Biol Chem* 2012;**287**(15):12578–88.

65. Olson DP, Pulinilkunnil T, Cline GW, Shulman GI, Lowell BB. Gene knockout of Acc2 has little effect on body weight, fat mass, or food intake. *Proc Natl Acad Sci* 2010;**107**(16):7598–603.
66. Palmieri F. The mitochondrial transporter family (SLC25): physiological and pathological implications. *Pflugers Arch* 2004;**447**(5):689–709.
67. Hue L, Taegtmeyer H. The Randle cycle revisited: a new head for an old hat. *AJP Endocrinol Metab* 2009;**297**(3):E578–91.
68. Icard P, Poulain L, Lincet H. Understanding the central role of citrate in the metabolism of cancer cells. *Biochim Biophys Acta* 2012;**1825**(1):111–6.
69. Acín-Pérez R, Fernández-Silva P, Peleato ML, Pérez-Martos A, Enriquez JA. Respiratory active mitochondrial supercomplexes. *Mol Cell* 2008;**32**(4):529–39.
70. Jorgensen W, Rud KA, Mortensen OH, Frandsen L, Grunnet N, Quistorff B. Your mitochondria are what you eat: a high-fat or a high-sucrose diet eliminates metabolic flexibility in isolated mitochondria from rat skeletal muscle. *Phys Rep* 2017;**5**(6) [pii:e13207].
71. Dassa EP, Dufour E, Goncalves S, Paupe V, Hakkaart GA, Jacobs HT, et al. Expression of the alternative oxidase complements cytochrome c oxidase deficiency in human cells. *EMBO MolMed* 2009;**1**(1):30–6.
72. Gershon D. The mitochondrial theory of aging: is the culprit a faulty disposal system rather than indigenous mitochondrial alterations? *Exp Gerontol* 1999;**34**(5):613–9.
73. Acin-Perez R, Hoyos B, Zhao F, Vinogradov V, Fischman DA, Harris RA, et al. Control of oxidative phosphorylation by vitamin A illuminates a fundamental role in mitochondrial energy homoeostasis. *FASEB J* 2010;**24**(2):627–36.
74. Liu J, Ames BN. Reducing mitochondrial decay with mitochondrial nutrients to delay and treat cognitive dysfunction, Alzheimer's disease, and Parkinson's disease. *Nutr Neurosci* 2005;**8**(2):67–89.
75. Lieber CS. Alcohol, liver, and nutrition. *J Am Coll Nutr* 1991;**10**(6):602–32.
76. Przyrembel H. Therapy of mitochondrial disorders. *J Inherit Metab Dis* 1987;**10**:129–46.
77. Eleff S, Kennaway NG, Buist NR, Darley-Usmar VM, Capaldi RA, Bank WJ, et al. 31P NMR study of improvement in oxidative phosphorylation by vitamins K3 and C in a patient with a defect in electron transport at complex III in skeletal muscle. *Proc Natl Acad Sci U S A* 1984;**81**(11):3529–33.
78. Wijburg FA, Barth PG, Ruitenbeek W, Wanders RJA, Vos GD, Plods van Amstel SLB, et al. Familial NADH: Q1 oxidoreductase (complex I) deficiency: variable expression and possible treatment. *J Inherit Metab Dis* 1989;**12**(Suppl. 2):349–51.
79. Gomez-Cabrera MC, Domenech E, Romagnoli M, Arduini A, Borras C, Pallardo FV, et al. Oral administration of vitamin C decreases muscle mitochondrial biogenesis and hampers training-induced adaptations in endurance performance. *Am J Clin Nutr* 2008;**87**(1):142–9.
80. Groff JL, Gropper SAS, Hunt SM. *Advanced nutrition and human metabolism*. St. Paul, MN: West Publishing Company; 1995.
81. Lapointe J. Mitochondria as promising targets for nutritional interventions aiming to improve performance and longevity of sows. *J Anim Physiol Anim Nutr* 2014;**98**(5):809–21.
82. Marriage B, Clandinin MT, Glerum DM. Nutritional cofactor treatment in mitochondrial disorders. *J Am Diet Assoc* 2003;**103**(8):1029–38.
83. Lauridsen C, Jensen SK. α-Tocopherol incorporation in mitochondria and microsomes upon supranutritional vitamin E supplementation. *Genes Nutr* 2012;**7**(4):475–82.
84. Ramis MR, Esteban S, Miralles A, Tan D-X, Reiter RJ. Protective effects of melatonin and mitochondria-targeted antioxidants against oxidative stress: a review. *Curr Med Chem* 2015;**22**(22):2690–711.
85. Aw TY, Jones DP. Nutrient supply and mitochondrial function. *Annu Rev Nutr* 1989;**9**:229–51.
86. Suttie JW. Vitamin K-dependent carboxylase. *Annu Rev Biochem* 1985;**54**:459–77.
87. Shoffner JM. *Oxidative phosphorylation diseases. The metabolic & molecular bases of inherited disease*. New York, NY: McGraw-Hill; 2001. p. 2367–423.
88. Serrano J, Cassanye A, Martín-Gari M, Granado-Serrano A, Portero-Otín M. Effect of dietary bioactive compounds on mitochondrial and metabolic flexibility. *Diseases* 2016;**4**(1):14.
89. Charles AL, Meyer A, Dal-Ros S, Auger C, Keller N, Ramamoorthy TG, et al. Polyphenols prevent ageing-related impairment in skeletal muscle mitochondrial function through decreased reactive oxygen species production. *Exp Physiol* 2013;**98**(2):536–45.
90. Sandoval-Acuña C, Ferreira J, Speisky H. Polyphenols and mitochondria: an update on their increasingly emerging ROS-scavenging independent actions. *Arch Biochem Biophys* 2014;**559**:75–90.

91. Duckles SP, Krause DN. Mechanisms of cerebrovascular protection: oestrogen, inflammation and mitochondria. *Acta Physiol (Oxford)* 2011;**203**(1):149–54.
92. Aruoma OI. Free radicals, oxidative stress, and antioxidants in human health and disease. *J Am Oil Chem Soc* 1998;**75**(2):199–212.
93. Xia Y, Hill KE, Burk RF. Effect of selenium deficiency on hydroperoxide-induced glutathione release from the isolated perfused rat heart. *J Nutr* 1985;**115**(6):733–42.
94. Dursun N, Taşkin E, Yerer Aycan MB, Şahin L. Selenium-mediated cardioprotection against adriamycin-induced mitochondrial damage. *Drug Chem Toxicol* 2011;**34**(2):199–207.
95. Cartier LJ, Ohira Y, Chen M, Cuddihee RW, Holloszy JO. Perturbation of mitochondrial composition in muscle by iron deficiency. implications regarding regulation of mitochondrial assembly. *J Biol Chem* 1986;**261**(29):13827–32.
96. Russell JO, Monga SP. Nutritional interventions for mitochondrial OXPHOS deficiencies: mechanisms and model systems. *Annu Rev Pathol Mech Dis* 2018;**1314**:1–829.
97. Barile M, Passarella S, Quagliariello E. Thiamine pyrophosphate uptake into isolated rat liver mitochondria. *Arch Biochem Biophys* 1990;**280**(2):352–7.
98. Naito E, Ito M, Takeda E, Yokota I, Yoshijima S, Kuroda Y. Molecular analysis of abnormal pyruvate dehydrogenase in a patient with thiamine-responsive congenital lactic acidemia. *Pediatr Res* 1994;**36**(3):340–6.
99. Sato Y, Nakagawa M, Higuchi I, Osame M, Naito E, Oizumi K. Mitochondrial myopathy and familial thiamine deficiency. *Muscle Nerve* 2000;**23**(7):1069–75.
100. Parikh S, Goldstein A, Koenig MK, Scaglia F, Enns GM, Saneto R, et al. Diagnosis and management of mitochondrial disease: a consensus statement from the Mitochondrial Medicine Society. *Genet Med* 2015;**17**(9):689–701.
101. Depeint F, Bruce WR, Shangari N, Mehta R, O'Brien PJ. Mitochondrial function and toxicity: role of the B vitamin family on mitochondrial energy metabolism. Chem Biol Interact 2006;163(1–2):94–112.
102. Theurey P, Rieusset J. Mitochondria-associated membranes response to nutrient availability and role in metabolic diseases. *Trends Endocrinol Metab* 2016;**28**(1):32–45.
103. Addison R, McCormick DB. Biogenesis of flavoprotein and cytochrome components in hepatic mitochondria from riboflavin-deficient rats. *Biochem Biophys Res Commun* 1978;**81**(1):133–8.
104. Bernsen PL, Gabreëls FJ, Ruitenbeek W, Hamburger HL. Treatment of complex I deficiency with riboflavin. *J Neurol Sci* 1993;**118**(2):181–7.
105. Griebel V, Krägeloh-Mann I, Ruitenbeek W, Trijbels JMF, Paulus W. A mitochondrial myopathy in an infant with lactic acidosis. *Dev Med Child Neurol* 1990;**32**(6):528–31.
106. Bar-Meir M, Elpeleg ON, Saada A. Effect of various agents on adenosine triphosphate synthesis in mitochondrial complex I deficiency. *J Pediatr* 2001;**139**(6):868–70.
107. Ogle RF, Christodoulou J, Fagan E, Blok RB, Kirby DM, Seller KL, et al. Mitochondrial myopathy with tRNA(Leu(UUR)) mutation and complex I deficiency responsive to riboflavin. *J Pediatr* 1997;**130**(1):138–45.
108. Institute of Medicine (US), Compounds I of M (US) P on DA and R. *Dietary reference intakes for vitamin C, vitamin E, selenium, and carotenoids*. 2000. p. 1–529.
109. Olsen RKJ, Olpin SE, Andresen BS, Miedzybrodzka ZH, Pourfarzam M, Merinero B, et al. ETFDH mutations as a major cause of riboflavin-responsive multiple acyl-CoA dehydrogenation deficiency. *Brain* 2007;**130**(8):2045–54.
110. Artuch R, Vilaseca MA, Pineda M. Biochemical monitoring of the treatment in paediatric patients with mitochondrial disease. *J Inherit Metab Dis* 1998;**21**(8):837–45.
111. Matthews PM, Ford B, Dandurand RJ, Eidelman DH, Oconnor D, Sherwin A, et al. Coenzyme Q10 with multiple vitamins is generally ineffective in treatment of mitochondrial disease. *Neurology* 1993;**43**:884–90.
112. Lui A, Lumeng L, Li TK. Transport of pyridoxine and pyridoxal 5′-phosphate in isolated rat liver mitochondria. *J Biol Chem* 1982;**257**(24):14903–6.
113. Atamna H, Newberry J, Erlitzki R, Schultz CS, Ames BN. Biotin deficiency inhibits heme synthesis and impairs mitochondria in human lung fibroblasts. *J Nutr* 2007;**137**(1):25–30.
114. Owen OE, Kalhan SC, Hanson RW. The key role of anaplerosis and cataplerosis for citric acid cycle function. *J Biol Chem* 2002;**277**(34):30409–12.
115. Mock DM. *Biotin*. 7th ed. Washington: International Life Sciences Institute Press; 1996.
116. Vesely DL. Biotin enhances guanylate cyclase activity. *Science* 1982;**216**(4552):1329–30.
117. Nisoli E, Clementi E, Paolucci C, Cozzi V, Tonello C, Sciorati C, et al. Mitochondrial biogenesis in mammals: the role of endogenous nitric oxide. *Science* 2003;**299**(5608):896–9.

118. Serrano M, Pérez-Dueñas B, Montoya J, Ormazabal A, Artuch R. Genetic causes of cerebral folate deficiency: clinical, biochemical and therapeutic aspects. *Drug Discov Today* 2012;**17**(23–24):1299–306.
119. Pineda M, Ormazabal A, López-Gallardo E, Nascimento A, Solano A, Herrero MD, et al. Cerebral folate deficiency and leukoencephalopathy caused by a mitochondrial DNA deletion. *Ann Neurol* 2006;**59**(2):394–8.
120. Toyoshima S, Watanabe F, Saido H, Pezacka EH, Jacobsen DW, Miyatake K, et al. Accumulation of methylmalonic acid caused by vitamin B12-deficiency disrupts normal cellular metabolism in rat liver. *Br J Nutr* 1996;**75**:929–38.
121. De Vivo DC, Tein I. Primary and secondary disorders of carnitine metabolism. *Int Pediatr* 1990;**5**:134–41.
122. El-Hattab AW, Scaglia F. Disorders of carnitine biosynthesis and transport. *Mol Genet Metab* 2015;**116**(3):107–12.
123. Carter AL, Abney TO, Lapp DF. Biosynthesis and metabolism of carnitine. *J Child Neurol* 1995;**10**:3–7.
124. Angelini C, Vergani L, Martinuzzi A. Clinical and biochemical aspects of carnitine deficiency and insufficiency: transport defects and inborn errors of β-oxidation. *Crit Rev Clin Lab Sci* 1992;**29**(3–4):217–42.
125. Arenas J, Gonzalez-Crespo MR, Campos Y, Martin MA, Cabello A, Gomez-Reino JJ. Abnormal carnitine distribution in the muscles of patients with idiopathic inflammatory myopathy. *Arthritis Rheum* 1996;**39**(11):1869–74.
126. Campos Y, Huertas R, Lorenzo G, Bautista J, Gutierrez E, Aparicio M, et al. Plasma carnitine insufficiency and effectiveness of L-carnitine therapy in patients with mitochondril myopathy. *Muscle Nerve* 1993;**16**(2):150–3.
127. Schiff M, Bénit P, Coulibaly A, Loublier S, El-Khoury R, Rustin P. Mitochondrial response to controlled nutrition in health and disease. *Nutr Rev* 2011;**69**(2):65–75.
128. Koeth RA, Wang Z, Levison BS, Buffa JA, Org E, Sheehy BT, et al. Intestinal microbiota metabolism of l-carnitine, a nutrient in red meat, promotes atherosclerosis. *Nat Med* 2013;**19**(5):576–85.
129. Mitchell P. Protonmotive redox mechanisms of cytochrome b-cl complex in the respiratory chain: protonmotive ubiquinone cycl. *FEBS Lett* 1975;**56**(1):1–6.
130. Santos-Ocaña C, Do TQ, Padilla S, Navas P, Clarke CF. Uptake of exogenous coenzyme Q and transport to mitochondria is required for bc1 complex stability in yeast coq mutants. *J Biol Chem* 2002;**277**(13):10973–81.
131. López-Lluch G, Rodríguez-Aguilera JC, Santos-Ocaña C, Navas P. Is coenzyme Q a key factor in aging? *Mech Ageing Dev* 2010;**131**(4):225–35.
132. Echtay KS, Winkler E, Frischmuth K, Klingenberg M. Uncoupling proteins 2 and 3 are highly active H(+) transporters and highly nucleotide sensitive when activated by coenzyme Q (ubiquinone). *Proc Natl Acad Sci U S A* 2001;**98**(4):1416–21.
133. Genova ML, Lenaz G. New developments on the functions of coenzyme Q in mitochondria. *Biofactors* 2011;**37**(5):330–54.
134. Papucci L, Schiavone N, Witort E, Donnini M, Lapucci A, Tempestini A, et al. Coenzyme Q10 prevents apoptosis by inhibiting mitochondrial depolarization independently of its free radical scavenging property. *J Biol Chem* 2003;**278**(30):28220–8.
135. Löffler M, Jöckel J, Schuster G, Becker C. Dihydroorotat-ubiquinone oxidoreductase links mitochondria in the biosynthesis of pyrimidine nucleotides. *Mol Cell Biochem* 1997;**174**(1–2):125–9.
136. Olgun A. Converting NADH to NAD+ by nicotinamide nucleotide transhydrogenase as a novel strategy against mitochondrial pathologies during aging. *Biogerontology* 2009;**10**(4):531–4.
137. Bentinger M, Tekle M, Dallner G. Coenzyme Q–biosynthesis and functions. *Biochem Biophys Res Commun* 2010;**396**(1):74–9.
138. Villalba JM, Parrado C, Santos-Gonzalez M, Alcain FJ. Therapeutic use of coenzyme Q 10 and coenzyme Q 10 -related compounds and formulations. *Expert Opin Investig Drugs* 2010;**19**(4):535–54.
139. Kwong LK, Kamzalov S, Rebrin I, Bayne A-CV, Jana CK, Morris P, et al. Effects of coenzyme Q10 administration on its tissue concentrations, mitochondrial oxidant generation, and oxidative stress in the rat. *Free Radic Biol Med* 2002;**33**(5):627–38.
140. Parrado-Fernández C, López-Lluch G, Rodríguez-Bies E, Santa-Cruz S, Navas P, Ramsey JJ, et al. Calorie restriction modifies ubiquinone and COQ transcript levels in mouse tissues. *Free Radic Biol Med* 2011;**50**(12):1728–36.
141. Bentinger M, Dallner G, Chojnacki T, Swiezewska E. Distribution and breakdown of labeled coenzyme Q10 in rat. *Free Radic Biol Med* 2003;**34**(5):563–75.
142. Sohal RS, Kamzalov S, Sumien N, Ferguson M, Rebrin I, Heinrich KR, et al. Effect of coenzyme Q10 intake on endogenous coenzyme Q content, mitochondrial electron transport chain, antioxidative defenses, and life span of mice. *Free Radic Biol Med* 2006;**40**(3):480–7.
143. Kamzalov S, Sumien N, Forster MJ, Sohal RS. Coenzyme Q intake elevates the mitochondrial and tissue levels of Coenzyme Q and alpha-Tocopherol in young mice. *J Nutr* 2003;**133**:3175–80.

144. Kelso GF, Porteous CM, Coulter CV, Hughes G, Porteous WK, Ledgerwood EC, et al. Selective targeting of a redox-active ubiquinone to mitochondria within cells: antioxidant and antiapoptotic properties. *J Biol Chem* 2001;**276**(7):4588–96.
145. Rossman MJ, Santos-Parker JR, Steward CAC, Bispham NZ, Cuevas LM, Rosenberg HL, et al. Chronic supplementation with a mitochondrial antioxidant (MitoQ) improves vascular function in healthy older adults. *Hypertension* 2018;**72**(1).
146. Forster MJ, Morris P, Sohal RS. Genotype and age influence the effect of caloric intake on mortality in mice. *FASEB J* 2003;**17**(6):690–2.
147. Kamzalov S, Sohal RS. Effect of age and caloric restriction on coenzyme Q and alpha-tocopherol levels in the rat. *Exp Gerontol* 2004;**39**(8):1199–205.
148. Lass A, Kwong L, Sohal RS. Mitochondrial coenzyme Q content and aging. *Biofactors* 1999;**9**(2–4):199–205.
149. Navas P, Villalba JM, de Cabo R. The importance of plasma membrane coenzyme Q in aging and stress responses. *Mitochondrion* 2007;**7**:S34–40.
150. Hyun DH, Emerson SS, Jo DG, Mattson MP, de Cabo R. Calorie restriction up-regulates the plasma membrane redox system in brain cells and suppresses oxidative stress during aging. *Proc Natl Acad Sci U S A* 2006;**103**(52):19908–12.
151. Jiménez-Hidalgo M, Santos-Ocaña C, Padilla S, Villalba JM, López-Lluch G, Martín-Montalvo A, et al. NQR1 controls lifespan by regulating the promotion of respiratory metabolism in yeast. *Aging Cell* 2009;**8**(2):140–51.
152. Ochoa JJ, Quiles JL, Ibáñez S, Martínez E, López-Frías M, Huertas JR, et al. Aging-related oxidative stress depends on dietary lipid source in rat postmitotic rissues. *J Bioenerg Biomembr* 2003;**35**(3):267–75.
153. Ochoa JJ, Quiles JL, Lopez-Frias M, Huertas JR, Mataix J. Effect of lifelong coenzyme Q10 supplementation on age-related oxidative stress and mitochondrial function in liver and skeletal muscle of rats fed on a polyunsaturated fatty acid (PUFA)-rich diet. *J Gerontol A Biol Sci Med Sci* 2007;**62**(11):1211–8.
154. Persky a M, G a B. Clinical pharmacology of the dietary supplement creatine monohydrate. *Pharmacol Rev* 2001;**53**(2):161–76.
155. Wyss M, Kaddurah-Daouk R. Creatine and creatinine metabolism. *Physiol Rev* 2000;**80**(3):1107–213.
156. Tarnopolsky MA. The mitochondrial cocktail: rationale for combined nutraceutical therapy in mitochondrial cytopathies. *Adv Drug Deliv Rev* 2008;**60**(13–14):1561–7.
157. Tarnopolsky MA, Beal MF. Potential for creatine and other therapies targeting cellular energy dysfunction in neurological disorders. *Ann Neurol* 2001;**49**(5):561–74.
158. Harris RC, Söderlund K, Hultman E. Elevation of creatine in resting and exercised muscle of normal subjects by creatine supplementation. *Clin Sci* 1992;**83**(3):367–74.
159. Tarnopolsky MA, Roy BD, MacDonald JR. A randomized, controlled trial of creatine monohydrate in patients with mitochondrial cytopathies. *Muscle Nerve* 1997;**20**(12):1502–9.
160. Rodriguez MC, MacDonald JR, Mahoney DJ, Parise G, Beal MF, Tarnopolsky MA. Beneficial effects of creatine, CoQ10, and lipoic acid in mitochondrial disorders. *Muscle Nerve* 2007;**35**(2):235–42.
161. Rodan LH, Wells GD, Banks L, Thompson S, Schneiderman JE, Tein I. L-Arginine affects aerobic capacity and muscle metabolism in MELAS (mitochondrial encephalomyopathy, lactic acidosis and stroke-like episodes) syndrome. *PLoS ONE* 2015;**10**(5).
162. Bogan KL, Brenner C. Nicotinic acid, nicotinamide, and nicotinamide riboside: a molecular evaluation of $NAD^+$ precursor vitamins in human nutrition. *Annu Rev Nutr* 2008;**28**(1):115–30.
163. Ratajczak J, Joffraud M, Trammell SAJ, Ras R, Canela N, Boutant M, et al. NRK1 controls nicotinamide mononucleotide and nicotinamide riboside metabolism in mammalian cells. *Nat Commun* 2016;**7**:13103.
164. Khan NA, Auranen M, Paetau I, Pirinen E, Euro L, Forsström S, et al. Effective treatment of mitochondrial myopathy by nicotinamide riboside, a vitamin B3. *EMBO Mol Med* 2014;**6**(6):721–31.
165. Belenky P, Racette FG, Bogan KL, McClure JM, Smith JS, Brenner C. Nicotinamide riboside promotes Sir2 silencing and extends lifespan via Nrk and Urh1/Pnp1/Meu1 pathways to NAD+. *Cell* 2007;**129**(3):473–84.
166. Bieganowski P, Brenner C. Discoveries of nicotinamide riboside as a nutrient and conserved NRK genes establish a preiss-handler independent route to NAD+ in fungi and humans. *Cell* 2004;**117**(4):495–502.
167. Zhang H, Ryu D, Wu Y, Gariani K, Wang X, Luan P, et al. NAD+ repletion improves mitochondrial and stem cell function and enhances life span in mice. *Science* 2016;**352**(6292):1436–43.
168. Cantó C, Menzies KJ, Auwerx J. NAD+ metabolism and the control of energy homeostasis: a balancing act between mitochondria and the nucleus. *Cell Metab* 2015;**22**(1):31–53.

169. Birjmohun RS, Hutten BA, Kastelein JJP, Stroes ESG. Efficacy and safety of high-density lipoprotein cholesterol-increasing compoundsA meta-analysis of randomized controlled trials. *J Am Coll Cardiol* 2005;**45**(2):185–97.
170. Belenky P, Bogan KL, Brenner C. NAD+ metabolism in health and disease. *Trends Biochem Sci* 2007;**32**(1):12–9.
171. Cerutti R, Pirinen E, Lamperti C, Marchet S, Sauve AA, Li W, et al. NAD+-dependent activation of Sirt1 corrects the phenotype in a mouse model of mitochondrial disease. *Cell Metab* 2014;**19**(6):1042–9.
172. Gariani K, Menzies KJ, Ryu D, Wegner CJ, Wang X, Ropelle ER, et al. Eliciting the mitochondrial unfolded protein response by nicotinamide adenine dinucleotide repletion reverses fatty liver disease in mice. *Hepatology* 2016;**63**(4):1190–204.
173. Fang EF, Kassahun H, Croteau DL, Scheibye-Knudsen M, Marosi K, Lu H, et al. $NAD^+$ replenishment improves lifespan and healthspan in ataxia telangiectasia models via mitophagy and DNA repair. *Cell Metab* 2016;**24**(4):566–81.
174. Mitchell SJ, Bernier M, Aon MA, Cortassa S, Kim EY, Fang EF, et al. Nicotinamide improves aspects of healthspan, but not lifespan, in mice. *Cell Metab* 2018;**27**(3):667–676.e4.
175. Tarragó MG, Chini CCS, Kanamori KS, Warner GM, Caride A, de Oliveira GC, et al. A potent and specific CD38 inhibitor ameliorates age-related metabolic dysfunction by reversing tissue NAD+ decline. *Cell Metab* 2018;**27**(5):1081–1095.e10.
176. Majamaa K, Rusanen H, Remes AM, Pyhtinen J, Hassinen IE. Increase of blood NAD+ and attenuation of lactacidemia during nicotinamide treatment of a patient with the MELAS syndrome. *Life Sci* 1996;**58**(8):691–9.
177. Penn AMW, Lee JWK, Thuillier P, Wagner M, Maclure KM, Menard MR, et al. MELAS syndrome with mitochondrial tRNA Leu[UUR] mutation: correlation of clinical state, nerve conduction, and muscle 31P magnetic resonance spectroscopy during treatment with nicotinamide and riboflavin. *Neurology* 1992;**42**(11):2147–52.
178. Zhang Z, Tsukikawa M, Peng M, Polyak E, Nakamaru-Ogiso E, Ostrovsky J, et al. Primary respiratory chain disease causes tissue-specific dysregulation of the global transcriptome and nutrient-sensing signaling network. *PLoS ONE* 2013;**8**(7):e69282.
179. Felici R, Lapucci A, Cavone L, Pratesi S, Berlinguer-Palmini R, Chiarugi A. Pharmacological NAD-boosting strategies improve mitochondrial homeostasis in human complex I-mutant Fibroblasts. *Mol Pharmacol* 2015;**87**(6):965–71.
180. Karamanlidis G, Lee CF, Garcia-Menendez L, Kolwicz SC, Suthammarak W, Gong G, et al. Mitochondrial complex I deficiency increases protein acetylation and accelerates heart failure. *Cell Metab* 2013;**18**(2):239–50.
181. Liu J. The effects and mechanisms of mitochondrial nutrient α-lipoic acid on improving age-associated mitochondrial and cognitive dysfunction: an overview. *Neurochem Res* 2008;**33**(1):194–203.
182. Quinn JF, Bussiere JR, Hammond RS, Montine TJ, Henson E, Jones RE, et al. Chronic dietary α-lipoic acid reduces deficits in hippocampal memory of aged Tg2576 mice. *Neurobiol Aging* 2007;**28**(2):213–25.
183. Barbiroli B, Medori R, Tritschler HJ, Klopstock T, Seibel P, Reichmann H, et al. Lipoic (thioctic) acid increases brain energy availability and skeletal muscle performance as shown by in vivo 31P-MRS in a patient with mitochondrial cytopathy. *J Neurol* 1995;**242**(7):472–7.
184. Smith AR, Shenvi SV, Widlansky M, Suh JH, Hagen TM. Lipoic acid as a potential therapy for chronic diseases associated with oxidative stress. *Curr Med Chem* 2004;**11**(9):1135–46.
185. Moncada S, Erusalimsky JD. Does nitric oxide modulate mitochondrial energy generation and apoptosis? *Nat Rev Mol Cell Biol* 2002;**3**(3):214–20.
186. Larsen FJ, Weitzberg E, Lundberg JO, Ekblom B. Effects of dietary nitrate on oxygen cost during exercise. *Acta Physiol* 2007;**191**(1):59–66.
187. Lundberg JO, Gladwin MT, Ahluwalia A, Benjamin N, Bryan NS, Butler A, et al. Nitrate and nitrite in biology, nutrition and therapeutics. *Nat Chem Biol* 2009;**5**(12):865–9.
188. Larsen FJ, Schiffer TA, Borniquel S, Sahlin K, Ekblom B, Lundberg JO, et al. Dietary inorganic nitrate improves mitochondrial efficiency in humans. *Cell Metab* 2011;**13**(2):149–59.
189. Wallace DC. A mitochondrial paradigm of metabolic and degenerative diseases, aging, and cancer: a dawn for evolutionary medicine. *Annu Rev Genet* 2005;**39**:359–407.
190. Barja G. Aging in vertebrates, and the effect of caloric restriction: a mitochondrial free radical production-DNA damage mechanism? *Biol Rev Camb Philos Soc* 2004;**79**(2):235–51.
191. Guarente L. Mitochondria-a nexus for aging, calorie restriction, and sirtuins? *Cell* 2008;**132**(2):171–6.
192. Nisoli E, Tonello C, Cardile A, Cozzi V, Bracale R, Tedesco L, et al. Calorie restriction promotes mitochondrial biogenesis by inducing the expression of eNOS. *Science* 2005;**310**(5746):314–7.

193. Wang J, Jiang JC, Jazwinski SM. Gene regulatory changes in yeast during life extension by nutrient limitation. *Exp Gerontol* 2010;**45**(7–8):621–31.
194. Lin SJ, Ford E, Haigis M, Liszt G, Guarente L. Calorie restriction extends yeast life span by lowering the level of NADH. *Genes Dev* 2004;**18**(1):12–6.
195. Schulz TJ, Zarse K, Voigt A, Urban N, Birringer M, Ristow M. Glucose restriction extends *Caenorhabditis elegans* life span by inducing mitochondrial respiration and increasing oxidative stress. *Cell Metab* 2007;**6**(4):280–93.
196. Civitarese AE, Carling S, Heilbronn LK, Hulver MH, Ukropcova B, Deutsch WA, et al. Calorie restriction increases muscle mitochondrial biogenesis in healthy humans. *PLoS Med* 2007;**4**(3):485–94.
197. Harrison DE, Strong R, Sharp ZD, Nelson JF, Astle CM, Flurkey K, et al. Rapamycin fed late in life extends lifespan in genetically heterogeneous mice. *Nature* 2009;**460**:392–5.
198. Dahl HHM, Thorburn DR. Clinical spectrum and diagnosis of mitochondrial disorders. *Am J Med Genet—Semin Med Genet* 2001;**106**(1):4–17.
199. Kerr DS. Review of clinical trials for mitochondrial disorders: 1997–2012. *Neurotherapeutics* 2013;**10**(2):307–19.
200. Libri V, Yandim C, Athanasopoulos S, Loyse N, Natisvili T, Law PP, et al. Epigenetic and neurological effects and safety of high-dose nicotinamide in patients with Friedreich's ataxia: an exploratory, open-label, dose-escalation study. *Lancet* 2014;**384**(9942):504–13.
201. Lynch DR, Willi SM, Wilson RB, Cotticelli MG, Brigatti KW, Deutsch EC, et al. A0001 in Friedreich ataxia: biochemical characterization and effects in a clinical trial. *Mov Disord* 2012;**27**(8):1026–33.
202. Kaufmann P, Thompson JLP, Levy G, Buchsbaum R, Shefner J, Krivickas LS, et al. Phase II trial of CoQ10 for ALS finds insufficient evidence to justify phase III. *Ann Neurol* 2009;**66**(2):235–44.
203. Zamani P, Rawat D, Shiva-Kumar P, Geraci S, Bhuva R, Konda P, et al. Effect of inorganic nitrate on exercise capacity in heart failure with preserved ejection fraction. *Circulation* 2015;**131**(4):371–80.
204. Hernández-Ojeda J, Cardona-Muñoz EG, Román-Pintos LM, Troyo-Sanromán R, Ortiz-Lazareno PC, Cárdenas-Meza MA, et al. The effect of ubiquinone in diabetic polyneuropathy: a randomized double-blind placebo-controlled study. *J Diabetes Complicat* 2012;**26**(4):352–8.
205. Applegarth DA, Toone JR, Lowry RB. Incidence of inborn errors of metabolism in British Columbia, 1969–1996. *Pediatrics* 2000;**105**(1):e10.

CHAPTER

# 5

# Mitochondrial DNA: Structure, Genetics, Replication and Defects

*Jan-Willem Taanman, Albert M. Kroon*

**Department of Clinical and Movement Neurosciences, Institute of Neurology, University College London, London, United Kingdom**

## 1 INTRODUCTION

Mitochondria are essential, double membrane-bound organelles found in the cytosol of eukaryotic cells. The outer membrane separates the mitochondrion from the cytosol, with specific parts being associated with the endoplasmic reticulum. The inner membrane is elaborately folded inward into the mitochondrial matrix to form a series of parallel cristae membranes. Depending on the cell type, mitochondria have different shapes and sizes, but often form a dynamic reticulate structure regulated by frequent fission and fusion events. This enables the organelles to share their content and be transported to different areas of the cell. When impaired or redundant, mitochondria can be removed through a dedicated macroautophagy process, known as mitophagy.

Mitochondria are the energy-transducing organelles of the cell in which the fuels that drive cellular metabolism are broken down sequentially through the pyruvate dehydrogenase complex or the fatty acid β-oxidation system, the tricarboxylic acid (TCA) cycle, and the oxidative phosphorylation pathway to generate adenosine triphosphate (ATP). The five enzyme complexes of the oxidative phosphorylation pathway are embedded in the cristae membranes. The first four complexes (I–IV) form the mitochondrial respiratory chain. They are coupled electronically by two smaller components, coenzyme $Q_{10}$ and cytochrome-c. Reducing equivalents formed in the TCA cycle and during β-oxidation are oxidized by complexes I and II. Electrons from the oxidation are transferred sequentially to complex III and complex IV. The latter complex, otherwise known as cytochrome-c oxidase, donates the electrons to oxygen, reducing it to water. The free energy of the redox reactions is used by complexes I, III, and IV to maintain a proton electrochemical gradient across the inner membrane, which is used by complex V to drive ATP synthesis.[1] In addition to the energy-transducing pathways, mitochondria control numerous other metabolic processes, including the heme

*Mitochondria in Obesity and Type 2 Diabetes*
https://doi.org/10.1016/B978-0-12-811752-1.00005-5

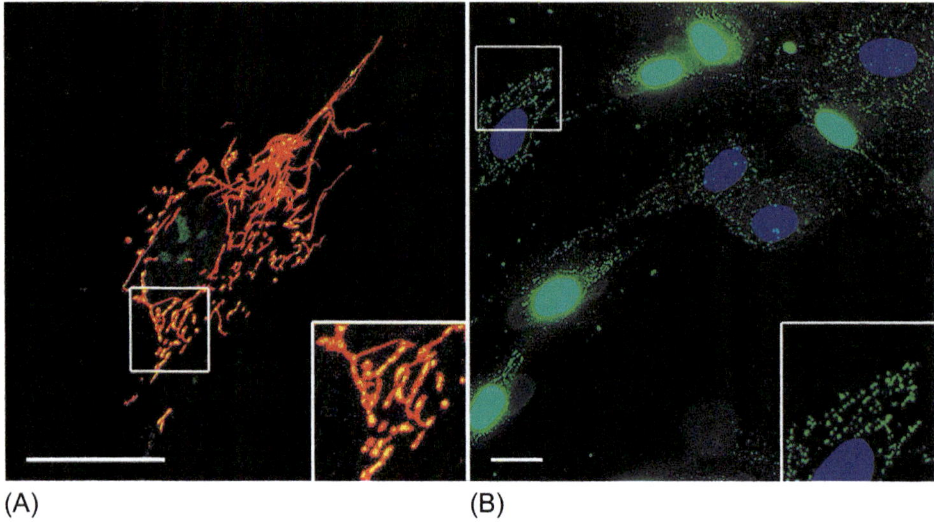


FIG. 1 MtDNA is dispersed throughout the mitochondrial network and replicates independently of the cell cycle. (A) MtDNA in a cultured human fibroblast was stained fluorescent green with PicoGreen, followed by fluorescent red staining of mitochondria with tetramethylrhodamine methyl ester, resulting in a yellow staining of mtDNA co-localizing with the organelles. (B) Human fibroblasts were cultured for 16h with 5-bromo-2′-deoxyuridine to label replicating DNA, followed by detection of incorporated bromodeoxyuridine with a fluorescent green-labeled antibody and nuclear DNA counterstaining with fluorescent blue 4′,6-diamidino-2-phenylindole. Note that many cells passed through S-phase during the labeling, as indicated by incorporation of bromodeoxyuridine in their nuclear DNA, whereas other cells were in $G_1$ or $G_2$. MtDNA that replicated during the labeling is visualized as small green specks in the cytoplasm. Scale bars: 10 μm.

and steroid biosynthesis. Furthermore, the organelles play key roles in thermogenesis, calcium ion homeostasis, and the intrinsic apoptosis signaling pathway.

Mitochondria contain their own genome, consisting of multiple copies of mitochondrial DNA (mtDNA) distributed evenly throughout the mitochondrial network (Fig. 1A). In this chapter, we review the structure, function, replication, and transmission of human mtDNA. Additionally, we discuss defects of mtDNA and their association with disease. Lastly, we consider new technologies that eliminate mutant mtDNA and methodologies to reduce transmission of mtDNA-inherited diseases.

## 2 GENERAL FEATURES OF THE MITOCHONDRIAL GENOME

### 2.1 Evolutionary Origin, Structure, Organization, and Copy Number

Overwhelming phylogenomic evidence indicates that mitochondria originate from a single endosymbiotic event between the ancestral eukaryotic organism and an aerobic α-proteobacterium ~$2 \times 10^9$ years ago.[2] Whether this event took place early or late during eukaryogenesis is still a matter of debate.[3, 4] The precise group of α-proteobacteria also is still unclear, although recent phylogenomic analysis suggest that mitochondria diverged from α-proteobacteria before the diversification of all currently known α-proteobacterial lineages.[5]

During its evolution into the present-day energy converters of the eukaryotic cell, the mitochondrial progenitor transferred many of its genes to the nucleus of its host. Nevertheless, some DNA has remained within mitochondria. The presence of this DNA was first suggested in September of 1963, by experiments that demonstrated DNA-dependent protein synthesis in isolated mitochondria[6] and was confirmed by electron microscopy later that year.[7,8]

Since its discovery, it has become clear that mtDNA has very different properties compared to nuclear DNA. Mammalian mtDNA is a covalently closed, circular DNA duplex and is generally compacted into a supercoiled structure.[9–11] One of the strands of the mtDNA duplex is named the heavy- or H-strand; the other strand is named the light- or L-strand. This nomenclature dates back to the 1960s, when it was discovered that the two strands have different buoyant densities on equilibrium cesium chloride (CsCl) gradients as a result of their different base content.[12] CsCl gradient centrifugation combined with electron microscopy further revealed that, in metabolically active cells, a large proportion of mtDNA molecules contain a short three-stranded structure, termed the displacement- or D-loop. In the D-loop, a short nucleic acid strand, complementary to the L-strand, displaces the H-strand.[13,14] This short nucleic acid strand traditionally is called the 7S DNA and is ~650 nucleotides long.[15] The D-loop region has evolved as a major control site for mtDNA replication and transcription (Fig. 2).

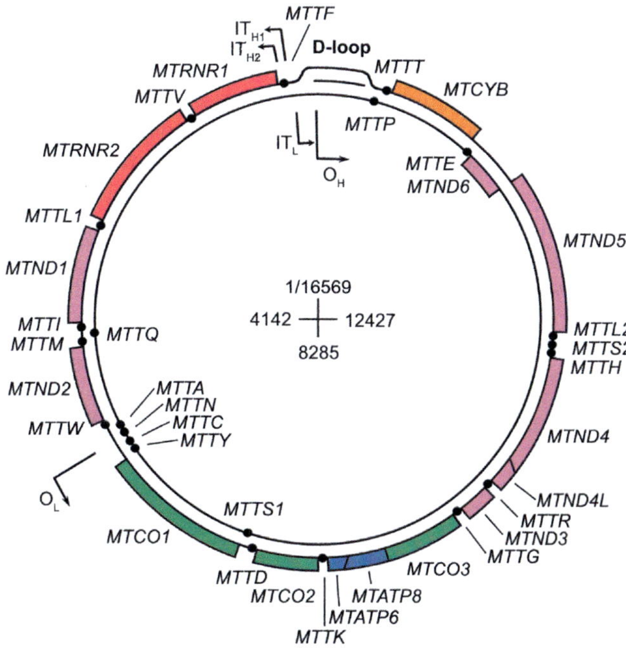


FIG. 2 Genetic map of the human mitochondrial genome. The outer circle represents the H-strand; the inner circle represents the L-strand. The D-loop is depicted as a three-stranded structure. Numbering of the nucleotides is shown in the center. The origins of H-strand ($O_H$) and L-strand ($O_L$) replication and the direction of DNA synthesis are indicated by long bent arrows; the initiation of transcription sites ($IT_{H1}$, $IT_{H2}$, $IT_L$) and the direction of RNA synthesis are indicated by short bent arrows. Genes coding for the 2 rRNA species and the 13 proteins are shown as shaded boxes; the 22 tRNA genes are shown as *black dots*. Gene names are given in italics. *Adapted from Taanman J-W. The mitochondrial genome: structure, transcription, translation and replication. Biochim Biophys Acta 1999;1410:103–23.*

While the human diploid nuclear genome is $6.5 \times 10^9$ base pairs (bp) in size and divided over 23 pairs of chromosomes, human mtDNA is only 16,569 bp but is present in numerous copies in somatic cells. The mtDNA molecule is arranged within nucleoprotein complexes, known as mitochondrial nucleoids, with a mean size of $80 \times 80 \times 100$ nm.[16] Nucleoids are located in the mitochondrial matrix and are associated tightly with the inner membrane. Each organelle appears to contain at least one nucleoid.[17] The average number of mtDNA molecules per nucleoid is 1.4, suggesting that the majority of nucleoids contain just a single copy. Unlike nuclear DNA, mtDNA is not packaged by histones but by mitochondrial transcription factor A or TFAM. Packaging is essential for mtDNA maintenance in vivo. Proteomic analyses have revealed that, in addition to TFAM, a number of other mitochondrial proteins are enriched in nucleoid preparations, including proteins of the mtDNA transcription and replication machinery, RNA processing enzymes, and quality control proteases.[18] The protein composition of nucleoids is likely to be highly dynamic and dependent on functional necessity, such as transcription or replication.

First attempts to estimate the amount of mtDNA per cell date back to the mid-1960s. The relative proportion of total cellular DNA found in mitochondria was estimated to be ~1% in chicken liver.[19] Based on this percentage and the known sizes of chicken nuclear and mtDNA, the cellular mtDNA copy number in chicken liver is ~1500. Some recent estimates of the mtDNA copy number in different human tissues are given in Table 1.[20–24] Tissues with a high level of aerobic metabolism (e.g., heart) have a high mtDNA copy number, whereas tissues with a lower level of aerobic metabolism (e.g., liver) have a lower copy number.[32] Compared to solid tissues, the mtDNA copy number in peripheral blood mononuclear cells is quite low and has been estimated to be ~400 copies.[24] This number, however, is probably even lower because of the presence of contaminating platelets that contain mtDNA but no nuclear DNA.[33] In contrast to anuclear platelets, anuclear erythrocytes contain no mtDNA.[25]

The mtDNA copy number is severely downregulated during spermatogenesis.[34] A mature spermatozoon contains <10 mtDNA molecules (Table 1).[29–31] During oogenesis, however, the mtDNA copy number increases steadily, >1000-fold, following growth activation

TABLE 1 Copy Number of mtDNA in Human Tissues

| Tissue | Cellular mtDNA Copy Number (Mean ± SD or Range) | Reference |
|---|---|---|
| Cardiac muscle | 6970 ± 920 | [20] |
| Skeletal muscle | 1420–5400 | [20–23] |
| Liver | 738–4100 | [22, 23] |
| Subcutaneous fat | 2042 ± 391 | [24] |
| Erythrocytes | 0 | [25] |
| Platelets | ~4 | [25] |
| Peripheral blood mononuclear cells | 409 ± 148 | [24] |
| Metaphase II oocytes | 143,000–679,176 | [26–28] |
| Spermatozoa | 0–7 | [29–31] |

of primordial germ cells to mature metaphase II oocytes.[35, 36] Estimates of the mtDNA copy number of a metaphase II oocyte vary from an average of $1.43 \times 10^5$ to $6.79 \times 10^5$ (Table 1).[26–28] Following fertilization, the total mtDNA copy number of the developing embryo remains stable until implantation of the blastocyst.[37, 38] Therefore, the mtDNA copy number per cell decreases rapidly with every cell division until gastrulation. During further fetal development, there is a continuous increase of the cellular mtDNA copy number in the different tissues.[39] This might reflect a response to increased bioenergetic demands of the developing tissues. The cellular mtDNA copy number increases markedly over the perinatal period,[40, 41] but does not change further with increasing age.[20, 22, 42]

## 2.2 Gene Content and Gene Function

The sequence of the human mtDNA first was published in 1981.[43] Most genetic information is encoded on the H-strand, with genes for 2 ribosomal RNAs (rRNAs), 14 transfer RNAs (tRNAs), and 12 proteins, while the L-strand codes for 8 tRNAs and 1 protein (Fig. 2).[44] All mitochondrial gene names start with the letters *MT*, followed by the rest of the gene name in italicized capitals; mtDNA nucleotide positions start with m. followed by the position number. The mitochondrial genome shows extraordinary economy of organization because genes are tightly packed and lack introns. Two pairs of protein genes (*MTATP6* and *MTATP8*, and *MTND4* and *MTND4L*) overlap and intergene sequences between the other genes are absent or limited to a few bases except for the 1123-bp noncoding D-loop region. In many protein genes, the last one or two bases of the stop codon are not encoded but generated posttranscriptionally by polyadenylation of the mRNA.[45]

Shortly after the first mtDNA sequences became available, comparisons with mitochondrial protein sequences exposed deviations from the universal genetic code. In mammalian mtDNA, AUA codes for methionine rather than isoleucine and UGA specifies a tryptophan instead of a stop codon.[46] It originally was thought that AGR (R=A, G) was used as a stop instead of an arginine codon, but more recent investigations revealed that these codons promote a -1frameshift of human mitochondrial ribosomes and, as a consequence, use the standard UAG stop codon.[47] The 2 mitochondrially encoded rRNAs and 22 mitochondrially encoded tRNAs are required for translation of the 13 mitochondrial protein genes. Mitochondria use a simplified decoding mechanism, in which a modified tRNA wobble base interaction with mRNA codons allows the translation of all codons with the 22 tRNA species instead of the 32 tRNA species required according to Crick's wobble hypothesis.[48] All 13 mitochondrially encoded proteins are crucial subunits of the oxidative phosphorylation enzyme complexes.[48] These subunits are synthesized on ribosomes in the mitochondrial matrix. All other mitochondrial proteins are encoded by nuclear DNA. The mitochondrial proteome contains ~1200 different proteins.[49] Thus, the vast majority of mitochondrial proteins is synthesized on cytosolic ribosomes and actively imported into the organelle. These proteins include additional subunits of the oxidative phosphorylation enzymes, constituents of the mitochondrial transcription, translation and replication machinery, and other mitochondrial enzyme and signaling pathway proteins.

Sedimentation studies of the early 1970s indicated that the small (28S) and large (39S) mitochondrial ribosomal subunits respectively contain the 12S and 16S rRNA species coded for by the *MTRNR1* and *MTRNR2* genes; however, a 5S rRNA species was not detected in mitochondrial ribosomes,[50–52] despite its presence in all bacterial, chloroplast, and cytosolic ribosomes. This conundrum finally was solved >40 years later, when high-resolution structures of the large mitochondrial ribosomal subunit revealed that, in humans, mitochondrial $tRNA^{Val}$ and, in pig, $tRNA^{Phe}$ usurps the position of 5S rRNA.[53, 54] Remarkably, human mitochondrial ribosomes have been shown to exhibit adaptive plasticity when mitochondrial $tRNA^{Val}$ levels are reduced by switching to the integration of mitochondrial $tRNA^{Phe}$ to generate translationally competent ribosomes.[55] The genes for mitochondrial $tRNA^{Phe}$ (*MTTF*), 12S rRNA, $tRNA^{Val}$ (*MTTV*), and 16S rRNA are positioned sequentially on the mtDNA (Fig. 2). They are transcribed as a polycistronic transcript at a much higher rate than the other genes on the H-strand by using a dedicated initiation of transcription site ($IT_{H1}$).[48] Thus, the gene order allows coordinated expression of all RNA components of mitochondrial ribosomes.

In addition to the 37 mitochondrial genes identified in the early 1980s, more recent in silico inspection of the human mitochondrial genome revealed several short protein coding open reading frames nestled within the ribosomal RNA genes like Russian dolls. These new genes express 16–38 amino acids long circulating polypeptides.[56] The first member of this class of mitochondrially derived small polypeptides was called humanin, and its gene resides in *MTRNR2*. Humanin has been demonstrated to have a number of metabolic effects, including an increase in glucose-stimulated insulin secretion from pancreatic β-cells, and a decrease in weight gain and visceral fat.[57, 58] Further scrutiny of *MTRNR2* uncovered short reading frames of six additional polypeptides, which have been named small humanin-like peptides (SHLP) 1–6. SHLP2 and SHLP3 have been shown to have similar protective effects as humanin.[59] In addition, the mitochondrial-derived polypeptide MOTS-c (mitochondrial open reading frame of the 12S rRNA-c) has been found to prevent age-dependent and high fat-diet-induced insulin resistance, as well as diet-induced obesity in mice by functioning as an exercise mimetic and activator of AMP-activated protein kinase (AMPK).[60] The m.1382A > C polymorphism, which results in a Lys14Gln substitution in MOTS-c and is unique to the northeast Asian population, has been found to correlate with longevity through its assumed endocrine action,[61] though additional work is required to clarify the actual mechanism.

## 3 REPLICATION OF THE MITOCHONDRIAL GENOME

In contrast to nuclear DNA replication, which is cell cycle dependent, mtDNA replication occurs at a constant rate throughout the cell cycle and continues in fully differentiated, post-mitotic cells, such as neurons and muscle fibers.[62, 63] Replication of mtDNA even persists for several hours in enucleated cells.[63] This relaxed replication of mtDNA is illustrated in Fig. 1B, which shows fibroblasts that have been cultured in the presence of the thymidine analog 5-bromo-2′-deoxyuridine. Immunocytochemical detection of incorporated bromodeoxyuridine reveals massive incorporation in nuclear DNA of cells that passed through S-phase during the labeling, when nuclear DNA is replicated. Other cells show no incorporation of bromodeoxyuridine in their nuclear DNA and were in $G_1$ or $G_2$ of the cell cycle. All cells, however, show bromodeoxyuridine incorporation in granules in their cytoplasm, representing replicated mtDNA. Despite the relaxed nature of mtDNA replication, the mtDNA copy number is maintained at a constant level in proliferating and quiescent cells,[64] indicating that mtDNA turnover is strictly regulated. TFAM levels vary concurrently with mtDNA levels,[65] suggesting that TFAM levels control mtDNA levels. Lon, the major protease of the mitochondrial matrix, might indirectly regulate the mtDNA copy number by selective degradation of TFAM.[66]

Early studies suggested that the D-loop region is attached to the inner mitochondrial membrane by a protein structure.[67] This membrane association was confirmed in later experiments, which demonstrated that mtDNA is replicated in association with a discrete proteinaceous structure that spans the outer and inner mitochondrial membranes at sites rich in cholesterol.[68, 69] Mitochondrial nucleoids actively engaged in mtDNA replication are spatially and temporally linked to a subset of endoplasmic reticulum-mitochondria contacts destined for mitochondrial division.[70] Thus, these contact sites couple mtDNA replication with mitochondrial division to distribute newly synthesized nucleoids to daughter organelles.

The mtDNA replication mechanism has been studied since the 1970s. The prevailing view is that mtDNA molecules are replicated unidirectionally from two spatially and temporally distinct, strand-specific origins.[12, 48, 71] Replication starts in the D-loop at the origin of H-strand replication ($O_H$) with the synthesis of a daughter H-strand and consequent displacement of the parental H-strand. The origin of L-strand replication ($O_L$) is located two-thirds of the genome downstream of $O_H$ (Fig. 2). Synthesis of the daughter L-strand is initiated when the replication fork of the leading (daughter H) strand passes $O_L$ and, as a result, exposes $O_L$ on the displaced (parental) H-strand in single-stranded form.

The molecular machinery responsible for mtDNA replication is complex.[71] RNA primer synthesis for replication initiation is performed by the mitochondrial transcription apparatus composed of mitochondrial RNA polymerase (POLRMT), TFAM (which serves as a transcription factor in addition to its packaging role), mitochondrial transcription factor B2 (TFB2M), and mitochondrial transcription elongation factor (TEFM). Synthesis of the mtDNA strands is catalyzed by DNA polymerase γ. The enzyme is an $\alpha_1\beta_2$ heterotrimer composed of one POLG subunit and two POLG2 subunits. POLG is the catalytic subunit, containing the $5' \rightarrow 3'$ DNA polymerase activity as well as a proofreading $3' \rightarrow 5'$ exonuclease activity to safeguard faithful replication. Further enzymes needed during synthesis are the helicase Twinkle (TWNK) to unwind the DNA duplex; mitochondrial topoisomerase 1 (TOP1MT) to relax DNA supercoils; and mitochondrial single-stranded binding protein (SSBP1) to stabilize single-stranded regions of mtDNA replicative intermediates. After DNA polymerase γ comes full circle, it encounters the 5′-end of the nascent mtDNA strand. RNA primers are removed by RNase H1 (RNASEH1). Further processing of the 5′-end is performed by mitochondrial genome maintenance exonuclease 1 (MGME1). When synthesis of the two nascent strands is near completion, two different daughter molecules are formed: a duplex circle with a nascent H-strand and a gapped duplex circle with a partial nascent L-strand. After the gaps are filled in and ligated by DNA ligase III (LIG3), the daughter duplexes are topologically linked at the $O_H$ region as hemicatenanes, that is, two circular DNA duplexes bound together via a single-stranded linkage. The decatenation of the daughter mtDNA molecules is catalyzed by the mitochondrial isoform of topoisomerase 3α (TOP3A). This step allows the daughter mtDNA molecules to separate and adopt their tertiary structure through the introduction of superhelical turns.

# 4 TRANSMISSION OF THE MITOCHONDRIAL GENOME

## 4.1 Maternal Transmission

In nearly all mammals, the entire spermatozoon, including the midpiece packed with mitochondria, enters the ovum at fertilization.[72,73] Nonetheless, it has been known for decades that only maternal mtDNA is passed on to progeny in intraspecies crossings.[74–77] As the mtDNA copy number of an oocyte is more than five magnitudes higher than that of a spermatozoon (Table 1), the maternal mtDNA inheritance observed in early studies could simply have been the result of dilution of paternal mtDNA beyond the detection limits of the restriction fragment length polymorphism analysis on which these studies relied. Recent massively parallel resequencing of mtDNA, however, revealed no evidence of paternal mtDNA transmission to offspring in humans.[78] In fact, it is now known that, additionally to passive dilution, three distinct molecular mechanisms actively prevent paternal mtDNA transmission. First, it has been shown in several animal phyla that paternal mtDNA is degraded shortly before or just after fertilization and that mitochondrial endonuclease G plays a key role in this process.[79–82] Second, studies in flies and mammals have revealed that sperm-derived mitochondria are selectively eliminated from the early embryo, mediated by the ubiquitin-proteasome system and the mitophagy pathway.[83–85] Third, studies in flies have suggested that DNA polymerase $\gamma$ promotes the abrupt elimination of mitochondrial genomes through an unknown mechanism during spermatogenesis.[86] The reason for exclusion of paternal mtDNA might be that it is compromised because of oxidative injury during spermatogenesis and fertilization.[87] There is evidence that spermatozoa harbor more mutated mtDNA than oocytes.[88]

The dogma of strict maternal inheritance of human mtDNA has been put in doubt by reports of paternal contributions to the offspring in some families.[88a,88b] Even though these findings demonstrate that partial paternal transmission of mtDNA is in principle possible, the families appear rare exceptions and these observations should not affect genetic counseling.

## 4.2 Muller's Ratchet

According to Muller's ratchet, detrimental mutations accumulate in an irreversible manner, in asexual populations that do not undergo other forms of recombination.[89] The predicted mutational meltdown will decrease fitness and, eventually, result in extinction of the population. As a consequence of its maternal inheritance, the mitochondrial genome is asexual. Recombination of mtDNA is vanishingly rare or absent,[90,91] therefore mtDNA is expected to be subject to Muller's ratchet. A comparative study of mitochondrial and nuclear tRNA genes demonstrated that the former accumulate nucleotide substitutions much more rapidly than the latter.[92] The elevated mutation rate of mtDNA is thought to be the result of a combination of factors, including limited mtDNA repair mechanisms, absence of protective histones, and proximity to the reactive oxygen species-generating oxidative phosphorylation system.[93,94] Because eukaryotes have thrived since the endosymbiotic event that established mitochondria, there must be mitigating circumstances that offset Muller's ratchet. These might include the occurrence of compensatory back mutations; the polyploidy of mtDNA, which ensures that not all mtDNA molecules are affected by a mutation; a purifying mechanism that prevents transmission of deleterious mutations; and the diminutive size of mtDNA.[95] In fact, the threat from extinction caused by Muller's ratchet might have been a driving force for the transfer of the majority of α-proteobacterial genes to the nucleus during mitochondrial evolution.

### 4.3 The Mother's Curse

Because the mitochondrial genome is transmitted maternally, it can make a direct and adaptive response to selection only through females. Therefore, mutations in mtDNA that will affect only males will not respond to natural (female) selection, enabling deleterious mutations to accumulate if they exert male-specific effects. This effect has been dubbed the mother's curse.[96] The existence of the mother's curse has been supported by nuclear gene expression effects observed within male (but not female) reproductive tissues of *Drosophila melanogaster* strains that differed only in the origin of their mtDNA.[97] In addition to putative effects on male fertility, it has been shown that mtDNA harbors variation that accelerates male (but not female) aging in *D. melanogaster*.[98] Thus, the mother's curse might contribute to the shorter lifespans of adult males compared to adult females, observed across the animal kingdom. Coadaptation between nuclear and mitochondrial genomes, with nuclear gene mutations compensating for male-harming mtDNA mutations, might counterbalance mother's curse.[99] Even so, males still are expected to suffer a transient fitness cost during the mitonuclear coevolution.

## 5 DEFECTS OF THE MITOCHONDRIAL GENOME

### 5.1 Mitochondrial Diseases

Inherited mitochondrial diseases are caused by mutations in mitochondrial and nuclear genes that impair mitochondrial function.[100] Facilitated by next generation DNA sequencing technologies, the list of genes involved in mitochondrial diseases is expanding rapidly. The sequencing results have revealed that every aspect of mitochondrial biology might be affected by a mutation, including mtDNA maintenance, transcription, translation, cofactor biosynthesis, metabolism, protein import, morphology, and quality control.[44] Because mitochondria generate the lion's share of ATP, organs with a high energy demand (e.g., heart, muscle, and brain) are predominantly affected. Collectively, mitochondrial diseases are among the most common of all genetic disorders, affecting 1 in ~4300 individuals.[101] They are responsible for devastating illnesses associated with severe disability and shortened lifespan in children and adults. Because of their extreme clinical, biochemical, and genetic heterogeneity, the diagnostic odyssey is long, burdensome, and complex. Unfortunately, no effective treatments currently exist to halt or reverse disease progression. Investigation into the underlying disease mechanisms and the search for potential treatments represent formidable challenges for both basic and clinical scientists. We will discuss quantitative and qualitative defects of mtDNA.

### 5.2 Mitochondrial DNA Depletion

Depletion of mtDNA first was documented in 1991, in infants with a fatal mitochondrial disease.[102] Children with mtDNA depletion syndrome often have <5% of normal mtDNA levels in liver and skeletal muscle (Fig. 3A).[42, 102, 103] Since the first reports, it has become clear that this

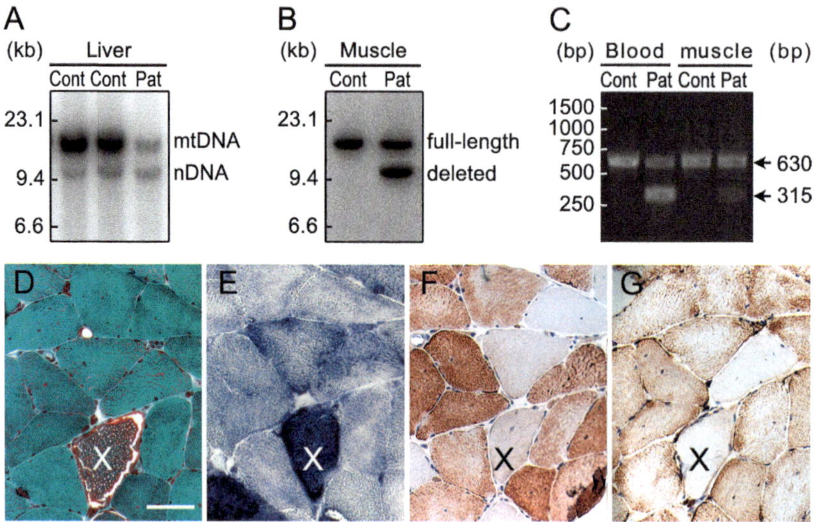


FIG. 3 Defects of mtDNA. (A) Southern blot analysis of *Pvu*II-linearized, liver mtDNA from two control subjects (Cont) and an Alpers disease patient with mtDNA depletion (Pat). In addition to a probe for mtDNA, the blot was cohybridized with a probe for the nuclear 18S rRNA gene (nDNA) to demonstrate equal loading. Migration of DNA size markers is indicated in kilobases (kb). (B) Southern blot analysis of *Pvu*II-linearized, skeletal muscle mtDNA from a control subject and a patient with PEO, revealing faster migrating, deleted mtDNA in addition to full-length mtDNA in the patient. (C) Agarose gel electrophoretic analysis of the PCR amplicon of skeletal muscle and blood mtDNA from a control subject and a MELAS patient. A 630-base pair (bp) mtDNA region encompassing the mutation site was amplified, followed by digestion with the restriction enzyme *Apa*I. The mutation produces an *Apa*I restriction site in the middle of the amplicon, resulting in two 315-bp fragments after digestion. Note that the patient harbors both wild-type and mutant mtDNA at different proportions in muscle and blood. Migration of a DNA ladder is indicated in bp. (D)–(G) Traverse serial sections of a skeletal muscle biopsy from a MERRF patient. Histological staining with Gömöri trichrome to reveal ragged-red fibers (D). Histological staining for succinate dehydrogenase (E) and cytochrome-c oxidase (F) activity. Immunohistological detection of the mtDNA-encoded cytochrome-c oxidase subunit MTCO1 (G). Note that the fiber marked X shows accumulation of mitochondria and increased succinate dehydrogenase activity but lacks cytochrome-c oxidase activity and MTCO1. Scale bar: 50 μm.

condition is caused by autosomal recessive mutations in genes coding for proteins of the mitochondrial replication machinery (*POLG, TWNK, TFAM*), enzymes engaged in the maintenance of mitochondrial deoxyribonucleotides (*DGUOK, TK2, SUCLA2, SUCLG1, RRM2B, TYMP, MPV17*) and proteins that play a role in mitochondrial morphology (*OPA1* and *FBXL4*).[104] Mutations in these genes are very rare, except for missense mutations in *POLG*, which have been identified in numerous neonates and infants.[105] The two phenotypes most commonly reported with *POLG* mutations in the pediatric population are Alpers syndrome and myocerebrohepatopathy spectrum (MCHS).[106] Alpers syndrome is characterized by refractory epilepsy, progressive encephalopathy with psychomotor regression, and liver disease that present during infancy after an initial period of normal development. Hepatic involvement often progresses rapidly to end-stage liver failure. MCHS is also severe and fatal, and is characterized by a triad of hypotonia or myopathy, encephalopathy or developmental delay, and liver dysfunction.

Mitochondria need a steady supply of deoxyribonucleotides for mtDNA synthesis because mtDNA is constantly turned over. Because there is no de novo synthesis of these DNA building blocks in mitochondria, the organelles have their own deoxyribonucleoside salvage pathway to synthesize deoxyribonucleotides.[107] Defects of salvage pathway enzymes, such as deoxyguanosine kinase (DGUOK) and thymidine kinase 2 (TK2), result in severe mtDNA depletion.[104] Mutations in the *DGUOK* gene cause either a multiorgan disease with neonatal onset or an isolated hepatic disease with infantile/early childhood-onset. In both forms, liver failure is the most common cause of death. Children with mutations in *TK2* present with isolated myopathy. The most common cause of death in these patients is pulmonary infection. Mutations in the *SUCLA2* and *SUCLG1* genes, which code for subunits of the TCA cycle enzyme succinyl-CoA synthase, also might lead to mtDNA depletion. This is because succinyl-CoA synthase provides the ATP for the kinase reaction of the mitochondrial deoxyribonucleoside salvage pathway enzyme, nucleotide diphosphate kinase.

The mitochondrial salvage pathway provides insufficient DNA building blocks to maintain mtDNA copy number, necessitating further import from the cytosol.[108] During S-phase, the cytosolic deoxyribonucleotide pool size is large because cytosolic de novo synthesis is upregulated. In postmitotic cells, mitochondria depend on deoxyribonucleoside diphosphates synthesized by ribonucleotide reductase in the cytosol. Mutations in the *RRM2B* gene, coding for a subunit of ribonucleotide reductase, have been identified in pediatric patients with severe mtDNA depletion associated with fatal encephalomyopathy.[104]

## 5.3 Mitochondrial DNA Deletions

Many of the genes involved in mtDNA depletion in children also are involved in adult-onset diseases associated with multiple mtDNA deletions.[104] As with mtDNA depletion, gene mutations leading to multiple mtDNA deletion are very rare, except for missense mutations in *POLG*.[105] The estimated birth prevalence of a *POLG* disease is 1 in ~50,000. Most adult patients with *POLG* mutations present with autosomal dominant or autosomal recessive progressive external ophthalmoplegia (PEO) with or without parkinsonism. Autosomal dominant mutations also might cause premature menopause.[109] Remarkably, *POLG* mutations associated with mtDNA depletion in children overlap with *POLG* mutations associated with multiple mtDNA deletions in adults.[105]

In addition to multiple mtDNA deletions caused by nuclear gene defects, postmitotic cells might acquire somatic large-scale single mtDNA deletions. These deletions first were described in 1988, in patients with mitochondrial myopathies.[110] Since this discovery, >130 different single mtDNA rearrangements (deletions and duplications) have been associated with disease.[44] Most deletions are flanked by short direct repeats and virtually all map to the long arc of the mtDNA, between $O_H$ and $O_L$. The most frequently occurring deletion is a 4977-bp deletion, known as the common deletion. This deletion is flanked by two 13-bp repeats at positions m.8470 and m.13477, and affects five tRNA and seven protein-coding genes.[111] Deletions are assumed to be formed through slipped mispairing during H-strand replication or, alternatively, during L-strand replication.[112]

Deleted mtDNA always co-exists with full-length mtDNA at varying proportions in a cell (Fig. 3B). This is known as heteroplasmy, whereas homoplasmy describes a condition where all mtDNA copies are identical in sequence. The mutation load (proportion of mutated mtDNA) might vary between different tissues, with some harboring no deleted mtDNA, while others contain high levels of deleted mtDNA. Only when the mutation load exceeds a critical threshold (50%–60% of the total mtDNA) will the deletion result in a clinical phenotype. Because of the relaxed nature of mtDNA replication, deletions accumulate with age through clonal expansion and random intracellular drift in postmitotic cells.[113] Thus, the proportion of cells in a tissue harboring deleterious levels of mutated mtDNA will increase with age and might cause a disease later in life.

Biochemically, mtDNA deletions cause oxidative phosphorylation deficiency resulting in an ATP deficit. Clinically, deletions have been associated mainly with PEO, Kearns-Sayre syndrome (KSS), and Pearson marrow-pancreas syndrome (PMPS).[114] Patients with PEO show a progressive inability to move their eyes and eye lids because of paralysis of the eye muscles. Patients present with ptosis or a drooping of the upper eye lids. KSS is a more severe syndromic variant of PEO. Additionally, these patients develop heart disease and retinitis pigmentosa causing blindness. PMPS is a severe and often fatal disorder of infancy/early childhood characterized by sideroblastic anemia (bone marrow forms ringed sideroblasts rather than healthy red blood cells) together with exocrine pancreatic dysfunction.

## 5.4 Mitochondrial DNA Point Mutations

In addition to deletions, mtDNA might contain point mutations. Since the first report of a mtDNA point mutation in 1988,[115] >650 different point mutations have been documented to cause disease.[44] Most mtDNA point mutations are heteroplasmic. As with mtDNA deletions, the clinical phenotype associated with mtDNA point mutations depends on the mutation load and tissue distribution. Whereas the majority of patients carrying single mtDNA deletions present as sporadic cases, however, mtDNA point mutations are maternally inherited in most patients. Ultradeep sequencing of the mitochondrial genome has revealed that <1% heteroplasmy is universal.[116] Therefore, ostensibly de novo somatic mtDNA mutations are most likely due to clonal expansion of low-level inherited variants.[117] About 1 in 200 individuals in the general population bear a pathogenic mtDNA point mutation at >1% heteroplasmy, whereas in 1 in ~5000 individuals, the mutation load exceeds the phenotypic threshold level of 70%–90% and will cause disease.[101] These diseases are progressive and severe.[118]

The most prevalent mtDNA mutation is the m.3243A>G mutation in the *MTTL1* gene coding for $tRNA^{Leu(UUR)}$.[101] This mutation first was described in patients presenting with mitochondrial encephalomyopathy, lactic acidosis, and stroke-like episodes (MELAS),[119] but the spectrum of disease phenotypes caused by this mutation is extremely broad, extending from hypertrophic cardiomyopathy and retinitis pigmentosa to diabetes with or without sensorineural deafness. In fact, the m.3243A>G mutation is associated with ~1.5% of the diabetic population.[120] Some patients with maternally inherited diabetes and deafness (MIDD) associated with the m.3243A>G mutation have additional clinical features common in other mitochondrial disorders, such as ptosis, renal deficiency, retinitis pigmentosa, and cardiomyopathy. The mutation causes a UUG codon-specific translational deficit, resulting in mitochondrial protein synthesis failure and subsequent oxidative phosphorylation impairment, with dire consequences for the cell's physiology.[121] Molecular genetic diagnosis does not require mtDNA sequencing. The *Apa*I restriction enzyme site generated by the m.3243A>G transition enables detection by restriction length polymorphism analysis of the PCR-amplified region surrounding the m.3243 position (Fig 3C).

Another relatively widespread point mutation is the m.8344A>G transition in the *MTTK* gene coding for $tRNA^{Lys}$. This mutation originally was identified in patients with myoclonic epilepsy associated with ragged-red fibers (MERRF)[122] and remains the most frequent cause of this syndrome. Skeletal muscle biopsies from MERRF patients show so-called ragged-red fibers when histologically stained with Gömöri trichrome dye (Fig. 3D), which is indicative of a buildup of defective mitochondria in the subsarcolemmal region. The mitochondrial accumulation in these fibers is accompanied by intense activity staining of the TCA and oxidative phosphorylation enzyme succinate dehydrogenase, which subunits are solely nuclear-encoded (Fig. 3E). In contrast, the ragged-red and numerous other fibers do not stain for cytochrome-c oxidase activity and do not contain the mtDNA-encoded cytochrome-c oxidase subunit MTCO1 (Fig. 3F, G). This mosaic expression of cytochrome-c oxidase in muscle is characteristic of patients carrying mtDNA mutations affecting the expression of mitochondrial cytochrome-c oxidase genes. Different fibers have different mtDNA mutation loads. A high load induces the biogenesis of mitochondria as compensatory response, resulting in ragged-red fibers with high succinate dehydrogenase activity, but cytochrome-c oxidase is no longer expressed.

Patients carrying the m.8993T>C/G missense mutations in their *MTATP6* gene coding for a subunit of oxidative phosphorylation complex V (ATP synthase) present with neuropathy, ataxia, and retinitis pigmentosa (NARP) in childhood or early adulthood.[123] Intriguingly, a high burden of this mutation (>90%) has been associated with maternally inherited Leigh syndrome (MILS).[124] Leigh syndrome or subacute necrotizing encephalomyopathy typically presents in infancy, and is characterized by lesions of the brain stem, basal ganglia, and cerebellum. In ~80% of Leigh syndrome patients, the disease is caused by mutations in nuclear genes involved in the biosynthesis of pyruvate dehydrogenase or oxidative phosphorylation enzymes. The remaining ~20% of patients harbor missense mutations in *MTATP6* or other mitochondrial genes coding for oxidative phosphorylation complex I subunits.

The maternally inherited disease Leber hereditary optic neuropathy (LHON) commonly is associated with the mtDNA missense mutations m.3460A>G (*MTND1*), m.11778A>G (*MTND4*) and m.14484T>C (*MTND6*).[115, 125] In contrast to the aforementioned mtDNA mutations, which are almost always heteroplasmic, mutations causing LHON tend to be homoplasmic in maternally related members of affected families. Patients present with a subacute loss of central vision in early adulthood, because of selective degeneration of retinal ganglion cells. This cell type specificity might be related to the skewed distribution of axonal mitochondria in retinal ganglion cells, making the segment devoid of mitochondria vulnerable to metabolic insults.[118] The mutations, which all affect subunits of complex I, are thought to impair enzymatic function, resulting in excessive reactive oxygen species production and neuronal cell death through oxidative stress.[126] The penetrance of the mutations is incomplete and is higher in males (~50%) than in females (~10%). The incomplete penetrance and sex bias have been proposed to result from a compensatory response based on increased mitochondrial biosynthesis,[127] which is further enhanced by estrogens.[128] Identification of a nuclear genetic modifier responsible for the boosted biosynthesis has so far been unpersuasive.

Aminoglycoside antibiotics, which are used worldwide to treat Gram-negative sepsis, will induce rapid, permanent, and profound sensorineural hearing loss in individuals who carry the homoplasmic m.1555A>G mutation in their *MTRNR1* gene coding for 12S rRNA.[129] Aminoglycosides exert their bactericidal effects by interacting with bacterial 16S rRNA, causing translational infidelity. The m.1555A>G transition is predicted to create a new C-G base pair in mitochondrial 12S rRNA, making its secondary structure more closely resemble that of bacterial 16S rRNA.[130] As a result, aminoglycosides bind with high affinity to mutant 12S rRNA, while aminoglycosides show low affinity to wild-type 12S rRNA. Binding of the antibiotic produces a mitochondrial translation defect, which compromises the biosynthesis of the oxidative phosphorylation complexes. The selective ototoxicity of aminoglycosides is explained by their selective accumulation in the cochlea. In fact, vestibulocochlear damage is common and independent of the m.1555A > G mutation when high doses of the drug are administered for prolonged periods. The prevalence of the m.1555A > G mutation is 1 in ~500 in individuals of European descent.[131, 132] This high incidence warrants general screening of neonates because the genetic test is inexpensive, but the potential burden to a carrier and costs to society are huge.

In some families, the same homoplasmic m.1555A > G mutation is associated with spontaneous, nonsyndromic sensorineural deafness, although exposure to aminoglycosides will accelerate the development of the deficiency.[129, 130] In these families, some maternally related members might exhibit severe congenital deafness, while others show late-onset moderate hearing impairment or are unaffected. This indicates the presence of additional genetic or environmental factors that determine the penetrance of the mutation. Several nuclear modifier genes have been proposed,[118, 130] suggesting that more than one gene might determine whether the mutation causes deafness when not triggered by aminoglycosides.

## 5.5 Transmission of Heteroplasmic mtDNA Mutations

The transmission of heteroplasmic mtDNA mutations is an intensely studied subject because clarification of the mechanism will help to predict the recurrence risk of mtDNA mutations relevant to genetic counseling. In 1982, Hauswirth and Laipis[133] reported large heteroplasmic shifts within a few generations of a single maternal lineage of Holstein cows. The authors suggested that this change could be because of a massive reduction in mtDNA content during oogenesis, leading to segregation of mtDNA variants by random genetic drift. Since this first observation, rapid shifts in heteroplasmy levels have been found in many animal models and humans.[134] These findings led to the germline genetic bottleneck hypothesis of mtDNA inheritance, which proposes that during female germ line development a small number of mtDNA molecules are sampled from a larger population for amplification and transmission, resulting in a mtDNA mutation load that is highly variable between offspring. Several studies of animal models have shown that there is a marked reduction in mtDNA content during female germ line development. Single-cell deep mtDNA sequencing of early primordial germ cells (mtDNA copy number: ~1400) from early-gestation human female embryos has revealed rare variants that reached higher heteroplasmy levels in late primordial germ cells, consistent with the genetic bottleneck theory.[135] In addition, there appeared to be selection against severely deleterious mutations, concomitant with a progressive upregulation of mtDNA transcription and replication, and linked to a change from glycolytic to oxidative metabolism. It is thought that the metabolic transition during germ line development exposes severely deleterious mutations to a filtering mechanism, averting the relentless buildup of mtDNA mutations predicted by Muller's ratchet. Mutations evading this process, however, will show rapid shifts in heteroplasmy levels, explaining the dramatic phenotypic variation seen in families with inherited mtDNA disorders (Fig. 4).

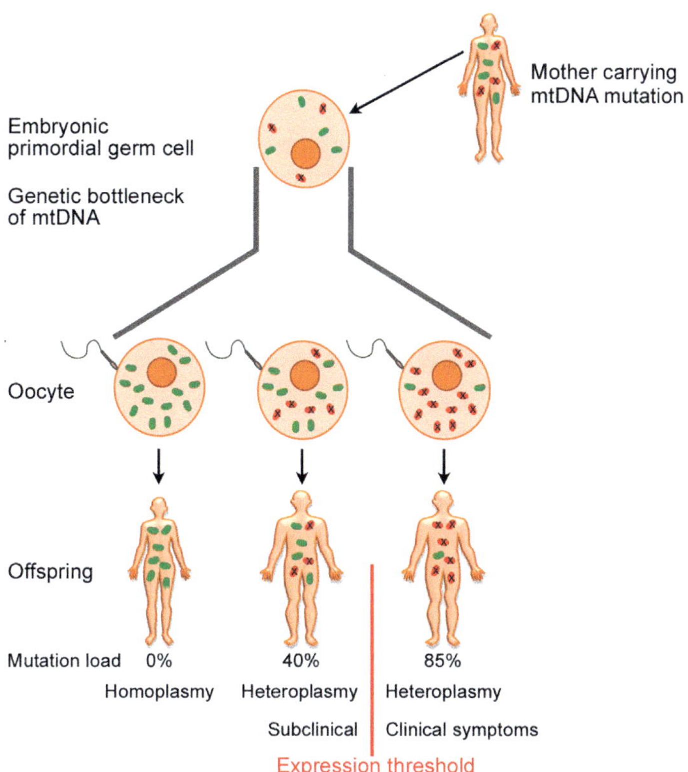


FIG. 4 Inheritance of mtDNA point mutations. MtDNA in a cell or tissue can be either homoplasmic (all copies are identical in sequence) or heteroplasmic (a mixture of wild-type *(green)* and mutated *(red/X)* mtDNA). A mtDNA mutation will result in a clinical phenotype only when the mutation load exceeds a critical threshold. A reduction in mtDNA content during female germ line development, known as the mitochondrial genetic bottleneck, and random genetic drift leads to a wide variation of heteroplasmy levels in oocytes. Therefore, the mtDNA mutation load passed on from mother to child will vary, resulting in extensive phenotypic heterogeneity among siblings.

Several studies of oocytes and embryonic tissues from females carrying mtDNA mutations have indicated that segregation of the m.3243A > G and m.8993T > G point mutations are governed by random genetic drift.[136–138] This suggests that these mutations are not subjected to purifying selection.[139] Analyses of human pedigrees transmitting a number of common mtDNA point mutations (m.3243A > G, m.3460G > A, m. 8344A > G, m. 8993T > C/G, m.11778G > A) confirmed the absence of selection during transmission but suggested that the segregation rate varies between different mutations.[140] The predicted range of heteroplasmy levels for these mutations in offspring is wide and, consequently, genetic counseling is difficult. In contrast, other studies found evidence of purifying selection acting on deleterious mtDNA variants during germline development.[135, 141, 142] Notably, there appears to be a strong selection against mtDNA deletions because the recurrence risk of a mtDNA deletion disorder is small (~4%).[143] The nature of the filtering mechanism remains unresolved.

# 6 MANAGEMENT AND PREVENTION OF mtDNA DISEASES

## 6.1 Treatments and Clinical Trials

For patients with a mtDNA disease, only symptomatic and supportive treatments are available. More than 1300 studies comparing nonpharmacological and pharmacological treatments for mitochondrial diseases have been conducted; however, the comprehensive 2012 Cochrane review of mitochondrial therapies[144] assessed only 12. Most studies were excluded because of methodological biases, or lack of randomization or blinding. Sadly, dramatic improvements were not observed in any of the included studies. A combination of vitamins and cofactors often is used as part of the treatment, but the Cochrane review found little evidence supporting their efficacy.[144] Coenzyme $Q_{10}$ or its water-soluble analog idebenone is prescribed commonly as antioxidant to defuse excessive reactive oxygen species production resulting from respiratory chain dysfunction but its efficacy remains controversial. Nevertheless, consensus recommendations from the Mitochondrial Medicine Society[145] suggest the use of coenzyme $Q_{10}$. Idebenone has been approved for treatment of LHON in Europe.

Aerobic endurance training stimulates mitochondrial biogenesis and increases mitochondrial enzyme activities in muscle.[146] Endurance training has been proven safe and beneficial in patients harboring mtDNA mutations, and is recommended by the Mitochondrial Medicine Society.[145]

Vasodysregulation is thought to contribute to the etiology of stroke-like episodes in MELAS patients. This notion is supported by the low L-arginine and L-citrulline levels during acute stoke episodes.[147] Low levels of these amino acids might result in vasoconstriction because L-arginine is a precursor of the signaling molecule nitric oxide (NO), which induces vasodilation. L-Citrulline also raises NO production as it stimulates de novo L-arginine biosynthesis. In fact, preliminary investigations have suggested that L-citrulline might have a more potent effect than L-arginine in MELAS patients.[148] Therefore, although placebo-controlled randomized clinical trials thus far have not been performed, administration of L-citrulline or L-arginine should be considered in MELAS patients.[145]

A number of clinical trials are either ongoing or have been completed, but their outcomes have yet to be reported.[149] Current therapeutic strategies tested in clinical trials include non-tailored treatments that can be applied across multiple mtDNA diseases, such as reduction of reactive oxygen species or induction of mitochondrial biogenesis or mitophagy to purge defective mitochondria. Tailored therapeutic approaches are tested to treat specific mtDNA diseases with gene therapy. These investigations include allotopic expression of genes of mitochondrial origin in the nuclear genome and modulation of mtDNA heteroplasmy levels with engineered nuclease technologies.[150, 151] In the latter approach, mtDNA mutations are eliminated selectively by mitochondrially targeted zinc finger nucleases (mtZFNs) or transcription activator-like effector nucleases (mitoTALENs) that specifically cleave mutated mtDNA. Destruction of mutant mtDNA has been shown to lead to a repopulation of wild-type genomes and recovery of mitochondrial function.[152, 153] Barriers for therapeutic application of mitochondrial-targeted nucleases are their large size and dimeric architecture, making it challenging to package their coding sequences in viral vectors for in vivo delivery. To overcome this problem, smaller monomeric forms have been developed (mitoTev-TALE), which are able to robustly shift mtDNA heteroplasmy levels.[154]

A promising therapy, tailored for mtDNA depletion syndromes caused by deoxyribonucleotide synthesis gene mutations, is based on the original observation that the depletion can be prevented by deoxyribonucleotide (dAMP+dGMP) supplementation in DGUOK-deficient patient fibroblast cultures.[64] TK2-deficient mice treated with deoxyribonucleosides (dC + dT) showed delayed disease onset, prolonged life span, and restored mtDNA content, as well as respiratory chain function.[155] This approach is known as nucleoside therapy and a clinical trial with TK2-deficient patients is ongoing (NCT03639701).

Mitochondrial neurogastrointestinal encephalomyopathy (MNGIE) is an uncommon, progressive, autosomal recessive disease caused by mutations in the thymidine phosphorylase gene (*TYMP*) that produce loss of thymidine phosphorylase activity.[156] The pathological accumulation of thymidine resulting from the enzyme deficiency induces a mitochondrial deoxyribonucleotide imbalance, which in turn causes mtDNA depletion/deletion abnormalities. Therapeutic options for patients include enzyme replacement therapy and dialysis, but these approaches only transiently reverse the deoxyribonucleotide imbalance. Allogeneic hematopoietic stem cell transplantation has been shown to restore thymidine phosphorylase activity and alleviate symptoms in patients; however, transplant-related complications result in a high mortality rate. Adeno-associated viral vector and hematopoietic stem cell gene therapies are able to stop disease progression in murine disease models but, unfortunately, do not result in a recovery of already established tissue degenerations.

## 6.2 Methodologies to Prevent mtDNA Disease Transmission

Because there is no cure for mtDNA disorders, prevention of germline transmission is a priority. Several reproductive options are now available to women carrying pathogenic mtDNA mutations.[157] Egg donation is a safe choice to eliminate transmission but has the disadvantage that the child is genetically related only to the father. Prenatal diagnosis (PND), in which the mitochondrial genome is tested through amniocentesis or chorion villus sampling, is another option but interpretation of the results is complex as the mtDNA mutation load in extra-fetal tissue might not accurately represent that of the fetal tissues. When intermediate levels of heteroplasmic mtDNA are found, it will be difficult for the parents to decide whether to continue or terminate the pregnancy. PND could be offered to mothers with a low mutation load to provide reassurances about fetal health. In addition, it might be offered when the mutation shows rapid shifts in heteroplasmy levels, resulting in a strongly skewed distribution among fetuses, as has been found for the m.8993T > C/G mutations.[137,140]

Preimplantation genetic diagnosis (PGD) could be offered to select embryos created by in vitro fertilization (IVF). Typically, two blastomeres are taken from an eight-cell embryo and screened for the mtDNA mutation. The highest cellular mutation load is used to predict the mutation load in the future child. Although a single case study has been reported in which the child's mutant load exceeded that of the trophectoderm biopsy,[158] currently available data suggest that the mutation load in a blastomere from the eight-cell stage reflects that of the future child.[157] PGD appears a reliable procedure to reduce transmission of heteroplasmic mtDNA mutations; however, it can be ethically challenging to determine which embryos should be implanted.

PND and PGD cannot be used to prevent germline transmission of homoplasmic mtDNA mutations, such as those associated with LHON. Moreover, women might not produce oocytes, and hence embryos, with heteroplasmy levels that are sufficiently low to be suitable for implantation. Emerging mitochondrial replacement therapies (MRTs), however, are able to avert transmission of mutations that are homoplasmic or present at a high heteroplasmic load.[159] MRTs, aka three-parent baby therapies, are a modification of conventional IVF. There are two main procedures (Fig. 5). In the pronuclear transfer procedure, the mother's oocyte with mutated mtDNA is fertilized with the father's sperm by intracytoplasmic sperm injection (ICSI), and a donated oocyte with wild-type mtDNA also is fertilized with the father's sperm by ICSI. Then, the parental pronuclei are removed from the zygote and transferred into the enucleated zygote with wild-type mtDNA. Finally, the embryo with wild-type mtDNA is transferred to the uterus. The maternal spindle transfer technique starts at an earlier stage of oocyte development. The metaphase-II spindle and associated chromosomes are transferred from the mother's oocyte into an enucleated donor oocyte. Then, the reconstructed oocyte is fertilized by ICSI and the embryo with wild-type mtDNA is implanted. Maternal spindle transfer technically is more challenging then pronuclear transfer because the spindle is more difficult to visualize and remove. Furthermore, the spindle is less resistant to physical damage than the pronuclei. A zygote, however, is discarded with the pronuclear transfer technique, which could raise ethical concerns, whereas no zygote is destroyed with the spindle transfer technique.

MRT involves replacement of the entire mitochondrial genome. Considering that the haploid nuclear genome is $3.24 \times 10^9$ bp and an oocyte contains $\sim 5 \times 10^5$ copies of 16,569-bp mtDNA (Table 1), this involves replacement of ~70% of the total oocyte DNA. Therefore, it is not surprising that some mtDNA carryover has been detected (<2%) in proof-of-concept studies.[160–162] In spite of this low level of mtDNA carryover, a reversion from donor mtDNA to patient mtDNA has been observed in some embryonic stem cell lines.[162, 163] It is unclear whether these findings are clinically relevant, but it is obvious that minimizing mtDNA carryover is crucial. While a large body of research in animals and in vitro with human oocytes and zygotes suggests that MRTs are safe and ready for clinical evaluation,[159] risks associated with this approach still exist. Notwithstanding these uncertainties, the first reported case of MRT was carried out in Mexico in 2016. The child was healthy at 7 months with an m.8993T > G mutation load of 2.4%–9.2% in his tested tissue.[164] Long-term clinical followup will be vital because mtDNA mitotic segregation might occur later in life.

MRT reduces the risk but cannot currently guarantee prevention of transmission of mtDNA diseases. Moreover, the availability of sufficient donor oocytes is a persistent problem. MitoTALENs have been used to selectively reduce mutated mtDNA in the germline.[165, 166] Although this strategy offers a potential alternative to MRT, currently the remaining mutation load is still too high and variable. With technological advance, however, nuclease technologies might be able to fully eliminate mutant mitochondrial genomes from oocytes or zygotes and prevent transmission of mtDNA disorders.

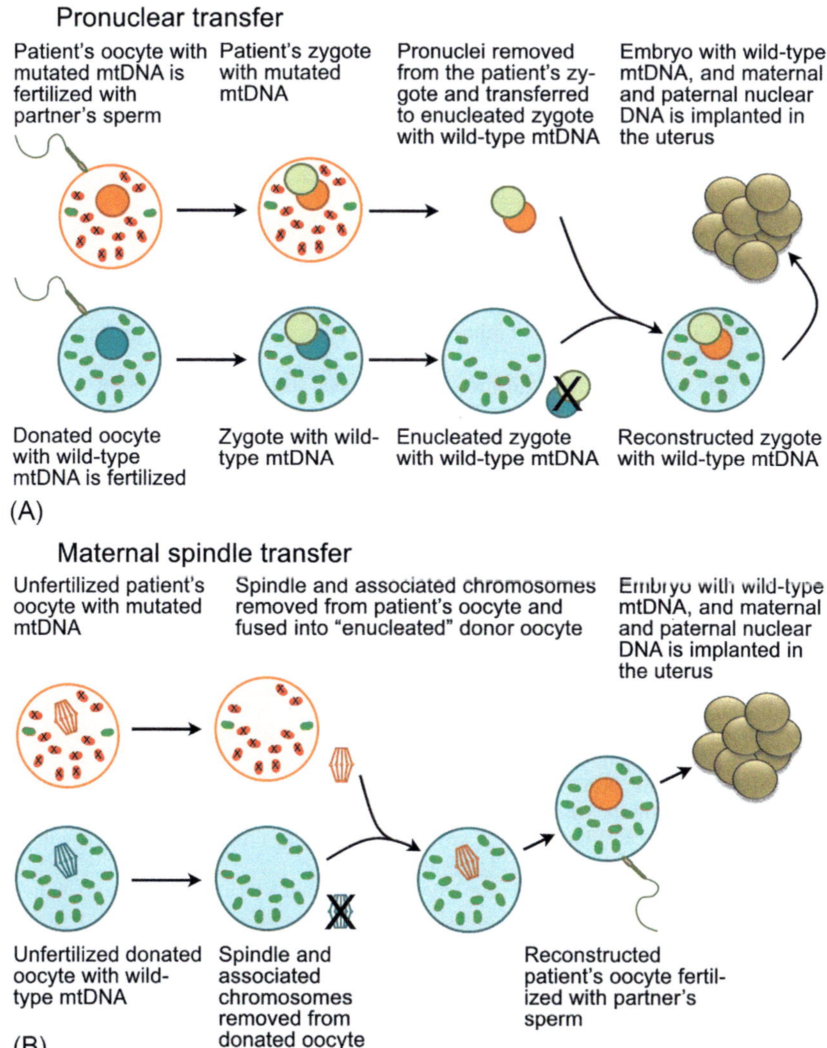


FIG. 5 Mitochondrial replacement therapies. (A) Pronuclear transfer. Transplantation of the nuclear genome is performed after fertilization. The patient's oocyte with mutated *(red/X)* and wild-type *(green)* mtDNA is fertilized with the partner's sperm and a donated egg with wild-type mtDNA also is fertilized with the partner's sperm. The pronuclei are removed from the patient's zygote and transferred to the enucleated zygote with wild-type mtDNA. The embryo with wild-type mtDNA is implanted. (B) Maternal spindle transfer. Transplantation of the nuclear genome is performed before fertilization. Starting at an earlier stage of oocyte development than pronuclear transfer, the nuclear spindle with associated chromosomes are removed from the patient's oocyte with mutated and wild-type mtDNA, and fused into an enucleated donor oocyte with wild-type mtDNA. The reconstructed patient's oocyte is fertilized and the embryo with wild-type mtDNA is implanted.

## Acknowledgments

We thank Dr. Philip Campbell and John Harvey for careful review of the manuscript. This work was supported by Royal Free Charity Fund 42.


## References


1. Taanman J-W, Williams SL. Structure and function of the mitochondrial oxidative phosphorylation system. In: Schapira AHV, DiMauro S, editors. *Mitochondrial disorders in neurology*. Boston: Butterworth Heinemann; 2002. p. 1–34.
2. Gray MW, Burger G, Lang BF. Mitochondrial evolution. *Science* 1999;**283**:1476–81.
3. Pittis AA, Gabaldón T. Late acquisition of mitochondria by a host with chimaeric prokaryotic ancestry. *Nature* 2016;**531**:101–4.
4. Degli Esposti M. Late mitochondrial acquisition, really? *Genome Biol Evol* 2016;**8**:2031–5.
5. Martijn J, Vosseberg J, Guy L, Offre P, Ettema TJG. Deep mitochondrial origin outside the sampled alphaproteobacteria. *Nature* 2018;**557**:101–5.
6. Kroon AM. Inhibitors of mitochondrial protein synthesis. *Biochim Biophys Acta* 1963;**76**:165–6.
7. Nass MMK, Nass S. Intramitochondrial fibers with DNA characteristics. I. Fixation and electron staining reactions. *J Cell Biol* 1963;**19**:593–611.
8. Nass S, Nass MMK. Intramitochondrial fibers with DNA characteristics. II. Enzymatic and other hydrolytic treatments. *J Cell Biol* 1963;**19**:613–29.
9. van Bruggen EF, Borst P, Ruttenberg GJ, Gruber M, Kroon AM. Circular mitochondrial DNA. *Biochim Biophys Acta* 1966;**119**:437–9.
10. Kroon AM, Borst P, van Bruggen EF, Ruttenberg GJ. Mitochondrial DNA from sheep heart. *Proc Natl Acad Sci U S A* 1966;**56**:1836–43.
11. Radloff R, Bauer W, Vinograd J. A dye-buoyant-density method for the detection and isolation of closed circular duplex DNA: the closed circular DNA in HeLa cells. *Proc Natl Acad Sci U S A* 1967;**57**:1514–21.
12. Kasamatsu H, Vinograd J. Replication of circular DNA in eukaryotic cells. *Annu Rev Biochem* 1974;**43**:695–719.
13. Arnberg A, van Bruggen EF, Borst P. The presence of DNA molecules with a displacement loop in standard mitochondrial DNA preparations. *Biochim Biophys Acta* 1971;**246**:353–7.
14. Kasamatsu H, Robberson DL, Vinograd J. A novel closed-circular mitochondrial DNA with properties of a replicating intermediate. *Proc Natl Acad Sci U S A* 1971;**68**:2252–7.
15. Walberg MW, Clayton DA. Sequence and properties of the human KB cell and mouse L cell D-loop regions of mitochondrial DNA. *Nucleic Acids Res* 1981;**9**:5411–21.
16. Bonekamp NA, Larsson N-G. SnapShot: mitochondrial nucleoid. *Cell* 2018;**172**:388.
17. Margineantu DH, Gregory CW, Sundell L, Sherwood SW, Beechem JM, Capaldi RA. Cell cycle dependent morphology changes and associated mitochondrial DNA redistribution in mitochondria of human cell lines. *Mitochondrion* 2002;**1**:425–35.
18. Hensen F, Cansiz S, Gerhold JM, Spelbrink JN. To be or not to be a nucleoid protein: a comparison of mass-spectrometry based approaches in the identification of potential mtDNA-nucleoid associated proteins. *Biochimie* 2014;**100**:219–26.
19. Borst P, Kroon AM. Mitochondrial DNA: physicochemical properties, replication, and genetic function. *Int Rev Cytol* 1969;**26**:107–90.
20. Miller FJ, Rosenfeldt FL, Zhang C, Linnane AW, Nagley P. Precise determination of mitochondrial DNA copy number in human skeletal and cardiac muscle by a PCR-based assay: lack of change of copy number with age. *Nucleic Acids Res* 2003;**31**:e61.
21. Barthélémy C, Ogier de Baulny H, Diaz J, Cheval MA, Frachon P, Romero N, et al. Late-onset mitochondrial DNA depletion: DNA copy number, multiple deletions, and compensation. *Ann Neurol* 2001;**49**:607–17.
22. Dimmock D, Tang LY, Schmitt ES, Wong LJ. Quantitative evaluation of the mitochondrial DNA depletion syndrome. *Clin Chem* 2010;**56**:1119–27.
23. Chabi B, Mousson de Camaret B, Duborjal H, Issartel JP, Stepien G. Quantification of mitochondrial DNA deletion, depletion, and overreplication: application to diagnosis. *Clin Chem* 2003;**49**:1309–17.

24. Gahan ME, Miller F, Lewin SR, Cherry CL, Hoy JF, Mijch A, et al. Quantification of mitochondrial DNA in peripheral blood mononuclear cells and subcutaneous fat using real-time polymerase chain reaction. *J Clin Virol* 2001;**22**:241–7.
25. Shuster RC, Rubenstein AJ, Wallace DC. Mitochondrial DNA in anucleate human blood cells. *Biochem Biophys Res Commun* 1988;**155**:1360–5.
26. Duran HE, Simsek-Duran F, Oehninger SC, Jones HW, Castora FJ. The association of reproductive senescence with mitochondrial quantity, function, and DNA integrity in human oocytes at different stages of maturation. *Fertil Steril* 2011;**96**:384–8.
27. Monnot S, Samuels DC, Hesters L, Frydman N, Gigarel N, Burlet P, et al. Mutation dependance of the mitochondrial DNA copy number in the first stages of human embryogenesis. *Hum Mol Genet* 2013;**22**:1867–72.
28. Murakoshi Y, Sueoka K, Takahashi K, Sato S, Sakurai T, Tajima H, et al. Embryo developmental capability and pregnancy outcome are related to the mitochondrial DNA copy number and ooplasmic volume. *J Assist Reprod Genet* 2013;**30**:1367–75.
29. May-Panloup P, Chrétien MF, Savagner F, Vasseur C, Jean M, Malthièry Y, et al. Increased sperm mitochondrial DNA content in male infertility. *Hum Reprod* 2003;**18**:550–6.
30. Amaral A, Ramalho-Santos J, St John JC. The expression of polymerase gamma and mitochondrial transcription factor A and the regulation of mitochondrial DNA content in mature human sperm. *Hum Reprod* 2007;**22**:1585–96.
31. Pavili L, Daudin M, Moinard N, Walschaerts M, Cuzin L, Massip P, et al. Decrease of mitochondrial DNA level in sperm from patients infected with human immunodeficiency virus-1 linked to nucleoside analogue reverse transcriptase inhibitors. *Fertil Steril* 2010;**94**:2151–6.
32. D'Erchia AM, Atlante A, Gadaleta G, Pavesi G, Chiara M, De Virgilio C, et al. Tissue-specific mtDNA abundance from exome data and its correlation with mitochondrial transcription, mass and respiratory activity. *Mitochondrion* 2015;**20**:13–21.
33. Banas B, Kost BP, Goebel FD. Platelets, a typical source of error in real-time PCR quantification of mitochondrial DNA content in human peripheral blood cells. *Eur J Med Res* 2004;**9**:371–7.
34. Rantanen A, Larsson N-G. Regulation of mitochondrial DNA copy number during spermatogenesis. *Hum Reprod* 2000;**15**(Suppl 2):86–91.
35. Mahrous E, Yang Q, Clarke HJ. Regulation of mitochondrial DNA accumulation during oocyte growth and meiotic maturation in the mouse. *Reproduction* 2012;**144**:177–85.
36. Cotterill M, Harris SE, Collado Fernandez E, Lu J, Huntriss JD, Campbell BK, et al. The activity and copy number of mitochondrial DNA in ovine oocytes throughout oogenesis *in vivo* and during oocyte maturation *in vitro*. *Mol Hum Reprod* 2013;**19**:444–50.
37. Pikó L, Taylor KD. Amounts of mitochondrial DNA and abundance of some mitochondrial gene transcripts in early mouse embryos. *Dev Biol* 1987;**123**:364–74.
38. Thundathil J, Filion F, Smith LC. Molecular control of mitochondrial function in preimplantation mouse embryos. *Mol Reprod Dev* 2005;**71**:405–13.
39. Heerdt BG, Augenlicht LH. Changes in the number of mitochondrial genomes during human development. *Exp Cell Res* 1990;**186**:54–9.
40. Poulton J, Sewry C, Potter CG, Bougeron T, Chretien D, Wijburg FA, et al. Variation in mitochondrial DNA levels in muscle from normal controls. Is depletion of mtDNA in patients with mitochondrial myopathy a distinct clinical syndrome. *J Inherit Metab Dis* 1995;**18**:4–20.
41. Morten KJ, Ashley N, Wijburg F, Hadzic N, Parr J, Jayawant S, et al. Liver mtDNA content increases during development: a comparison of methods and the importance of age- and tissue-specific controls for the diagnosis of mtDNA depletion. *Mitochondrion* 2007;**7**:386–95.
42. Macmillan CJ, Shoubridge EA. Mitochondrial DNA depletion: prevalence in a pediatric population referred for neurologic evaluation. *Pediatr Neurol* 1996;**14**:203–10.
43. Anderson S, Bankier AT, Barrell BG, de Bruijn MHL, Coulson AR, Drouin J, et al. Sequence and organization of the human mitochondrial genome. *Nature* 1981;**290**:457–65.
44. *MITOMAP: a human mitochondrial genome database*. http://www.mitomap.org; 2018 [accessed 07.09.18].
45. Ojala D, Montoya J, Attardi G. tRNA punctuation model of RNA processing in human mitochondria. *Nature* 1981;**290**:470–4.
46. *The genetic codes*. http://www.ncbi.nlm.nih.gov/Taxonomy/Utils/wprintgc.cgi; 2018 [accessed 07.09.18].

47. Temperley R, Richter R, Dennerlein S, Lightowlers RN, Chrzanowska-Lightowlers ZM. Hungry codons promote frameshifting in human mitochondrial ribosomes. *Science* 2010;**327**:301.
48. Taanman J-W. The mitochondrial genome: structure, transcription, translation and replication. *Biochim Biophys Acta* 1999;**1410**:103–23.
49. *Human MitoCarta2.0: 1158 mitochondrial genes*. https://www.broadinstitute.org/files/shared/metabolism/mitocarta/human.mitocarta2.0.html; 2018 [accessed 09.09.18].
50. Attardi G, Ojala D. Mitochondrial ribosome in HeLa cells. *Nat New Biol* 1971;**229**:133–6.
51. Brega A, Vesco C. Ribonucleoprotein particles involved in HeLa mitochondrial protein synthesis. *Nat New Biol* 1971;**229**:136–9.
52. Lizardi PM, Luck DJ. Absence of a 5S RNA complement in the mitochondrial ribosomes of *Neurospora crassa*. *Nat New Biol* 1971;**229**:140–2.
53. Brown A, Amunts A, Bai XC, Sugimoto Y, Edwards PC, Murshudov G, et al. Structure of the large ribosomal subunit from human mitochondria. *Science* 2014;**346**:718–22.
54. Greber BJ, Boehringer D, Leibundgut M, Bieri P, Leitner A, Schmitz N, et al. The complete structure of the large subunit of the mammalian mitochondrial ribosome. *Nature* 2014;**515**:283–6.
55. Rorbach J, Gao F, Powell CA, D'Souza A, Lightowlers RN, Minczuk M, et al. Human mitochondrial ribosomes can switch their structural RNA composition. *Proc Natl Acad Sci U S A* 2016;**113**:12198–201.
56. Kim S-J, Xiao J, Wan J, Cohen P, Yen K. Mitochondrially derived peptides as novel regulators of metabolism. *J Physiol* 2017;**595**:6613–21.
57. Kuliawat R, Klein L, Gong Z, Nicoletta-Gentile M, Nemkal A, Cui L, et al. Potent humanin analog increases glucose-stimulated insulin secretion through enhanced metabolism in the β cell. *FASEB J* 2013;**27**:4890–8.
58. Gong Z, Su K, Cui L, Tas E, Zhang T, Dong HH, et al. Central effects of humanin on hepatic triglyceride secretion. *Am J Physiol Endocrinol Metab* 2015;**309**:E283–92.
59. Cobb LJ, Lee C, Xiao J, Yen K, Wong RG, Nakamura HK, et al. Naturally occurring mitochondrial-derived peptides are age-dependent regulators of apoptosis, insulin sensitivity, and inflammatory markers. *Aging (Albany NY)* 2016;**8**:796–809.
60. Lee C, Zeng J, Drew BG, Sallam T, Martin-Montalvo A, Wan J, et al. The mitochondrial-derived peptide MOTS-c promotes metabolic homeostasis and reduces obesity and insulin resistance. *Cell Metab* 2015;**21**:443–54.
61. Fuku N, Pareja-Galeano H, Zempo H, Alis R, Arai Y, Lucia A, et al. The mitochondrial-derived peptide MOTS-c: a player in exceptional longevity? *Aging Cell* 2015;**14**:921–3.
62. Bogenhagen D, Clayton DA. Mouse L cell mitochondrial DNA molecules are selected randomly for replication throughout the cell cycle. *Cell* 1977;**11**:719–27.
63. Magnusson J, Orth M, Lestienne P, Taanman J-W. Replication of mitochondrial DNA occurs throughout the mitochondria of cultured human cells. *Exp Cell Res* 2003;**289**:133–42.
64. Taanman J-W, Muddle JR, Muntau AC. Mitochondrial DNA depletion can be prevented by dGMP and dAMP supplementation in a resting culture of deoxyguanosine kinase-deficient fibroblasts. *Hum Mol Genet* 2003;**12**:1839–45.
65. Campbell CT, Kolesar JE, Kaufman BA. Mitochondrial transcription factor A regulates mitochondrial transcription initiation, DNA packaging, and genome copy number. *Biochim Biophys Acta* 2012;**1819**:921–9.
66. Matsushima Y, Goto Y, Kaguni LS. Mitochondrial Lon protease regulates mitochondrial DNA copy number and transcription by selective degradation of mitochondrial transcription factor A (TFAM). *Proc Natl Acad Sci U S A* 2010;**107**:18410–5.
67. Albring M, Griffith J, Attardi G. Association of a protein structure of probable membrane derivation with HeLa cell mitochondrial DNA near its origin of replication. *Proc Natl Acad Sci U S A* 1977;**74**:1348–52.
68. Meeusen S, Nunnari J. Evidence for a two membrane-spanning autonomous mitochondrial DNA replisome. *J Cell Biol* 2003;**163**:503–10.
69. Gerhold JM, Cansiz-Arda S, Lõhmus M, Engberg O, Reyes A, van Rennes H, et al. Human mitochondrial DNA-protein complexes attach to a cholesterol-rich membrane structure. *Sci Rep* 2015;**5**:15292.
70. Lewis SC, Uchiyama LF, Nunnari J. ER-mitochondria contacts couple mtDNA synthesis with mitochondrial division in human cells. *Science* 2016;**353**:aaf5549.
71. Falkenberg M. Mitochondrial DNA replication in mammalian cells: overview of the pathway. *Essays Biochem* 2018;**63**:287–96.
72. Szollosi D. The fate of sperm middle-piece mitochondria in the rat egg. *J Exp Zool* 1965;**159**:367–77.
73. Sathananthan AH, Ng SC, Edirisinghe R, Ratnam SS, Wong PC. Human sperm-egg interaction in vitro. *Gamete Res* 1986;**15**:317–26.

74. Buzzo K, Fouts DL, Wolstenholme DR. *Eco*RI cleavage site variants of mitochondrial DNA molecules from rats. *Proc Natl Acad Sci U S A* 1978;**75**:909–13.
75. Kroon AM, de Vos WM, Bakker H. The heterogeneity of rat-liver mitochondrial DNA. *Biochim Biophys Acta* 1978;**519**:269–73.
76. Hayashi J-I, Yonekawa H, Gotoh O, Watanabe J, Tagashira Y. Strictly maternal inheritance of rat mitochondrial DNA. *Biochem Biophys Res Commun* 1978;**83**:1032–8.
77. Giles RE, Blanc H, Cann HM, Wallace DC. Maternal inheritance of human mitochondrial DNA. *Proc Natl Acad Sci U S A* 1980;**77**:6715–9.
78. Pyle A, Hudson G, Wilson IJ, Coxhead J, Smertenko T, Herbert M, et al. Extreme-depth re-sequencing of mitochondrial DNA finds no evidence of paternal transmission in humans. *PLoS Genet* 2015;**11**:e1005040.
79. Nishimura Y, Yoshinari T, Naruse K, Yamada T, Sumi K, Mitani H, et al. Active digestion of sperm mitochondrial DNA in single living sperm revealed by optical tweezers. *Proc Natl Acad Sci U S A* 2006;**103**:1382–7.
80. DeLuca SZ, O'Farrell PH. Barriers to male transmission of mitochondrial DNA in sperm development. *Dev Cell* 2012;**22**:660–8.
81. Luo S-M, Ge Z-J, Wang Z-W, Jiang Z-Z, Wang Z-B, Ouyang Y-C, et al. Unique insights into maternal mitochondrial inheritance in mice. *Proc Natl Acad Sci U S A* 2013;**110**:13038–43.
82. Zhou Q, Li H, Li H, Nakagawa A, Lin JLJ, Lee E-S, et al. Mitochondrial endonuclease G mediates breakdown of paternal mitochondria upon fertilization. *Science* 2016;**353**:394–9.
83. Politi Y, Gal L, Kalifa Y, Ravid L, Elazar Z, Arama E. Paternal mitochondrial destruction after fertilization is mediated by a common endocytic and autophagic pathway in *Drosophila*. *Dev Cell* 2014;**29**:305–20.
84. Song W-H, Yi Y-J, Sutovsky M, Meyers S, Sutovsky P. Autophagy and ubiquitin-proteasome system contribute to sperm mitophagy after mammalian fertilization. *Proc Natl Acad Sci U S A* 2016;**113**:E5261–70.
85. Rojansky R, Cha M-Y, Chan DC. Elimination of paternal mitochondria in mouse embryos occurs through autophagic degradation dependent on PARKIN and MUL1. *eLife* 2016;**5**:e17896.
86. Yu Z, O'Farrell PH, Yakubovich N, DeLuca SZ. The mitochondrial DNA polymerase promotes elimination of paternal mitochondrial genomes. *Curr Biol* 2017;**27**:1033–9.
87. Aitken RJ. Free radicals, lipid peroxidation and sperm function. *Reprod Fertil Dev* 1995;**7**:659–68.
88. Reynier P, Chrétien MF, Savagner F, Larcher G, Rohmer V, Barrière P, et al. Long PCR analysis of human gamete mtDNA suggests defective mitochondrial maintenance in spermatozoa and supports the bottleneck theory for oocytes. *Biochem Biophys Res Commun* 1998;**252**:373–7.

88a. Schwartz M, Vissing J. Paternal inheritance of mitochondrial DNA. *N Engl J Med* 2002;**347**:576–80.

88b. Luo S, Valencia CA, Zhang J, Lee N-C, Stone V, Gui B, et al. Biparental inheritance of mitochondrial DNA in humans. *Proc Natl Acad Sci U S A* 2018;**155**:13039–44.

89. Muller HJ. The relation of recombination to mutational advance. *Mutat Res* 1964;**1**:2–9.
90. Elson JL, Andrews RM, Chinnery PF, Lightowlers RN, Turnbull DM, Howell N. Analysis of European mtDNAs for recombination. *Am J Hum Genet* 2001;**68**:145–53.
91. Hagström E, Freyer C, Battersby BJ, Stewart JB, Larsson N-G. No recombination of mtDNA after heteroplasmy for 50 generations in the mouse maternal germline. *Nucleic Acids Res* 2014;**42**:1111–6.
92. Lynch M. Mutation accumulation in transfer RNAs: molecular evidence for Muller's ratchet in mitochondrial genomes. *Mol Biol Evol* 1996;**13**:209–20.
93. Shigenaga MK, Hagen TM, Ames BN. Oxidative damage and mitochondrial decay in aging. *Proc Natl Acad Sci U S A* 1994;**91**:10771–8.
94. Larsen NB, Rasmussen M, Rasmussen LJ. Nuclear and mitochondrial DNA repair: similar pathways? *Mitochondrion* 2005;**5**:89–108.
95. Loewe L. Quantifying the genomic decay paradox due to Muller's ratchet in human mitochondrial DNA. *Genet Res* 2006;**87**:133–59.
96. Gemmell NJ, Metcalf VJ, Allendorf FW. Mother's curse: the effect of mtDNA on individual fitness and population viability. *Trends Ecol Evol* 2004;**19**:238–44.
97. Innocenti P, Morrow EH, Dowling DK. Experimental evidence supports a sex-specific selective sieve in mitochondrial genome evolution. *Science* 2011;**332**:845–8.
98. Camus MF, Clancy DJ, Dowling DK. Mitochondria, maternal inheritance, and male aging. *Curr Biol* 2012;**22**:1717–21.
99. Patel MR, Miriyala GK, Littleton AJ, Yang H, Trinh K, Young JM, et al. A mitochondrial DNA hypomorph of cytochrome oxidase specifically impairs male fertility in *Drosophila melanogaster*. *eLife* 2016;**5**:e16923.

100. Gorman GS, Chinnery PF, DiMauro S, Hirano M, Koga Y, McFarland R, et al. Mitochondrial diseases. *Nat Rev Dis Primers* 2016;**2**:16080.
101. Gorman GS, Schaefer AM, Ng Y, Gomez N, Blakely EL, Alston CL, et al. Prevalence of nuclear and mitochondrial DNA mutations related to adult mitochondrial disease. *Ann Neurol* 2015;**77**:753–9.
102. Moraes CT, Shanske S, Tritschler HJ, Aprille JR, Andreetta F, Bonilla E, et al. mtDNA depletion with variable tissue expression: a novel genetic abnormality in mitochondrial diseases. *Am J Hum Genet* 1991;**48**:492–501.
103. Morris AAM, Taanman J-W, Blake J, Cooper JM, Lake BD, Malone M, et al. Liver failure associated with mitochondrial DNA depletion. *J Hepatol* 1998;**28**:556–63.
104. El-Hattab AW, Craigen WJ, Scaglia F. Mitochondrial DNA maintenance defects. *Biochim Biophys Acta* 2017;**1863**:1539–55.
105. *Human DNA polymerase gamma mutation database*. https://tools.niehs.nih.gov/polg/; 2018 [accessed 07.09.18].
106. Hikmat O, Tzoulis C, Chong WK, Chentouf L, Klingenberg C, Fratter C, et al. The clinical spectrum and natural history of early-onset diseases due to DNA polymerase gamma mutations. *Genet Med* 2017;**19**:1217–25.
107. Arnér ESJ, Eriksson S. Mammalian deoxyribonucleoside kinases. *Pharmacol Ther* 1995;**67**:155–86.
108. Gandhi VV, Samuels DC. Enzyme kinetics of the mitochondrial deoxyribonucleoside salvage pathway are not sufficient to support rapid mtDNA replication. *PLoS Comput Biol* 2011;**7**.
109. Pagnamenta AT, Taanman J-W, Wilson CJ, Anderson NE, Marotta R, Duncan AJ, et al. Dominant inheritance of premature ovarian failure associated with mutant mitochondrial DNA polymerase gamma. *Hum Reprod* 2006;**21**:2467–73.
110. Holt IJ, Harding AE, Morgan-Hughes JA. Deletions of muscle mitochondrial DNA in patients with mitochondrial myopathies. *Nature* 1988;**331**:717–9.
111. Schon EA, Rizzuto R, Moraes CT, Nakase H, Zeviani M, DiMauro S. A direct repeat is a hotspot for large-scale deletion of human mitochondrial DNA. *Science* 1989;**244**:346–9.
112. Shoffner JM, Lott MT, Voljavec AS, Soueidan SA, Costigan DA, Wallace DC. Spontaneous Kearns-Sayre/chronic external ophthalmoplegia plus syndrome associated with a mitochondrial DNA deletion: a slip-replication model and metabolic therapy. *Proc Natl Acad Sci U S A* 1989;**86**:7952–6.
113. Elson JL, Samuels DC, Turnbull DM, Chinnery PF. Random intracellular drift explains the clonal expansion of mitochondrial DNA mutations with age. *Am J Hum Genet* 2001;**68**:802–6.
114. Pitceathly RDS, Rahman S, Hanna MG. Single deletions in mitochondrial DNA—molecular mechanisms and disease phenotypes in clinical practice. *Neuromuscul Disord* 2012;**22**:577–86.
115. Wallace DC, Singh G, Lott MT, Hodge JA, Schurr TG, Lezza AM, et al. Mitochondrial DNA mutation associated with Leber's hereditary optic neuropathy. *Science* 1988;**242**:1427–30.
116. Payne BAI, Wilson IJ, Yu-Wai-Man P, Coxhead J, Deehan D, Horvath R, et al. Universal heteroplasmy of human mitochondrial DNA. *Hum Mol Genet* 2013;**22**:384–90.
117. Guo Y, Li C-I, Sheng Q, Winther JF, Cai Q, Boice JD, et al. Very low-level heteroplasmy mtDNA variations are inherited in humans. *J Genet Genomics* 2013;**40**:607–15.
118. Carelli V, La MC. Clinical syndromes associated with mtDNA mutations: where we stand after 30 years. *Essays Biochem* 2018;**62**:235–54.
119. Goto Y, Nonaka I, Horai S. A mutation in the $tRNA^{(Leu)(UUR)}$ gene associated with the MELAS subgroup of mitochondrial encephalomyopathies. *Nature* 1990;**348**:651–3.
120. Gerbitz K-D, van den Ouweland JMW, Maassen JA, Jaksch M. Mitochondrial diabetes mellitus: a review. *Biochim Biophys Acta* 1995;**1271**:253–60.
121. Yasukawa T, Kirino Y, Ishii N, Holt IJ, Jacobs HT, Makifuchi T, Fukuhara N, Ohta S, Suzuki T, Watanabe K. Wobble modification deficiency in mutant tRNAs in patients with mitochondrial diseases. *FEBS Lett* 2005;**579**:2948–52.
122. Shoffner JM, Lott MT, Lezza AM, Seibel P, Ballinger SW, Wallace DC. Myoclonic epilepsy and ragged-red fiber disease (MERRF) is associated with a mitochondrial DNA $tRNA^{(Lys)}$ mutation. *Cell* 1990;**61**:931–7.
123. Holt IJ, Harding AE, Petty RK, Morgan-Hughes JA. A new mitochondrial disease associated with mitochondrial DNA heteroplasmy. *Am J Hum Genet* 1990;**46**:428–33.
124. Tatuch Y, Christodoulou J, Feigenbaum A, Clarke JTR, Wherret J, Smith C, et al. Heteroplasmic mtDNA mutation (T→G) at 8993 can cause Leigh disease when the percentage of abnormal mtDNA is high. *Am J Hum Genet* 1992;**50**:852–8.
125. Yen M-Y, Wang A-G, Wei Y-H. Leber's hereditary optic neuropathy: a multifactorial disease. *Prog Retin Eye Res* 2006;**25**:381–96.

126. Lin CS, Sharpley MS, Fan W, Waymire KG, Sadun AA, Carelli V, et al. Mouse mtDNA mutant model of Leber hereditary optic neuropathy. *Proc Natl Acad Sci U S A* 2012;**109**:20065–70.
127. Giordano C, Iommarini L, Giordano L, Maresca A, Pisano A, Valentino ML, et al. Efficient mitochondrial biogenesis drives incomplete penetrance in Leber's hereditary optic neuropathy. *Brain* 2014;**137**:335–53.
128. Giordano C, Montopoli M, Perli E, Orlandi M, Fantin M, Ross-Cisneros FN, et al. Oestrogens ameliorate mitochondrial dysfunction in Leber's hereditary optic neuropathy. *Brain* 2011;**134**:220–34.
129. Prezant TR, Agapian JV, Bohlman MC, Bu X, Oztas S, Qiu WQ, et al. Mitochondrial ribosomal RNA mutation associated with both antibiotic-induced and non-syndromic deafness. *Nat Genet* 1993;**4**:289–94.
130. Taanman J-W. A nuclear modifier for a mitochondrial DNA disorder. *Trends Genet* 2001;**17**:609–11.
131. Bitner-Glindzicz M, Pembrey M, Duncan A, Heron J, Ring SM, Hall A, et al. Prevalence of mitochondrial 1555A→G mutation in European children. *N Engl J Med* 2009;**360**:640–2.
132. Vandebona H, Mitchell P, Manwaring N, Griffiths K, Gopinath B, Wang JJ, et al. Prevalence of mitochondrial 1555A→G mutation in adults of European descent. *N Engl J Med* 2009;**360**:642–4.
133. Hauswirth WW, Laipis PJ. Mitochondrial DNA polymorphism in a maternal lineage of Holstein cows. *Proc Natl Acad Sci U S A* 1982;**79**:4686–90.
134. Zhang H, Burr SP, Chinnery PF. The mitochondrial DNA genetic bottleneck: inheritance and beyond. *Essays Biochem* 2018;**62**:225–34.
135. Floros VI, Pyle A, Dietmann S, Wei W, Tang WCW, Irie N, et al. Segregation of mitochondrial DNA heteroplasmy through a developmental genetic bottleneck in human embryos. *Nat Cell Biol* 2018;**20**:144–51.
136. Brown DT, Samuels DC, Michael EM, Turnbull DM, Chinnery PF. Random genetic drift determines the level of mutant mtDNA in human primary oocytes. *Am J Hum Genet* 2001;**68**:533–6.
137. Steffann J, Gigarel N, Corcos J, Bonniere M, Encha-Razavi F, Sinico M, et al. Stability of the m.8993T→G mtDNA mutation load during human embryofetal development has implications for the feasibility of prenatal diagnosis in NARP syndrome. *J Med Genet* 2007;**44**:664–9.
138. Monnot S, Gigarel N, Samuels DC, Burlet P, Hesters L, Frydman N, et al. Segregation of mtDNA throughout human embryofetal development: m.3243A>G as a model system. *Hum Mutat* 2011;**32**:116–25.
139. Burr SP, Pezet M, Chinnery PF. Mitochondrial DNA heteroplasmy and purifying selection in the mammalian female germ line. *Develop Growth Differ* 2018;**60**:21–32.
140. Wilson IJ, Carling PJ, Alston CL, Floros VI, Pyle A, Hudson G, Sallevelt SCEH, et al. Mitochondrial DNA sequence characteristics modulate the size of the genetic bottleneck. *Hum Mol Genet* 2016;**25**:1031–41.
141. Li M, Rothwell R, Vermaat M, Wachsmuth M, Schröder R, Laros JFJ, et al. Transmission of human mtDNA heteroplasmy in the genome of the Netherlands families: support for a variable-size bottleneck. *Genome Res* 2016;**26**:417–26.
142. De Fanti S, Vicario S, Lang M, Simone D, Magli C, Luiselli D, et al. Intra-individual purifying selection on mitochondrial DNA variants during human oogenesis. *Hum Reprod* 2017;**32**:1100–7.
143. Chinnery PF, DiMauro S, Shanske S, Schon EA, Zeviani M, Mariotti C, et al. Risk of developing a mitochondrial DNA deletion disorder. *Lancet* 2004;**364**:592–6.
144. Pfeffer G, Majamaa K, Turnbull DM, Thorburn D, Chinnery PF. Treatment for mitochondrial disorders. *Cochrane Database Syst Rev* 2012;CD004426.
145. Parikh S, Goldstein A, Koenig MK, Scaglia F, Enns GM, Saneto R, et al. Diagnosis and management of mitochondrial disease: a consensus statement from the Mitochondrial Medicine Society. *Genet Med* 2015;**17**:689–701.
146. Taivassalo T, Shoubridge EA, Chen J, Kennaway NG, DiMauro S, Arnold DL, et al. Aerobic conditioning in patients with mitochondrial myopathies: physiological, biochemical, and genetic effects. *Ann Neurol* 2001;**50**:133–41.
147. Koga Y, Akita Y, Nishioka J, Yatsuga S, Povalko N, Tanabe Y, et al. L-arginine improves the symptoms of strokelike episodes in MELAS. *Neurology* 2005;**64**:710–2.
148. El-Hattab AW, Emrick LT, Hsu JW, Chanprasert S, Almannai M, Craigen WJ, et al. Impaired nitric oxide production in children with MELAS syndrome and the effect of arginine and citrulline supplementation. *Mol Genet Metab* 2016;**117**:407–12.
149. Hirano M, Emmanuele V, Quinzii CM. Emerging therapies for mitochondrial diseases. *Essays Biochem* 2018;**62**:467–81.
150. Wan X, Pei H, Zhao M-J, Yang S, Hu W-K, He H, et al. Efficacy and safety of rAAV2-ND4 treatment for Leber's hereditary optic neuropathy. *Sci Rep* 2016;**6**.

151. Pereira CV, Moraes CT. Current strategies towards therapeutic manipulation of mtDNA heteroplasmy. *Front Biosci (Landmark Ed)* 2017;**22**:991–1010.
152. Bacman SR, Williams SL, Pinto M, Peralta S, Moraes CT. Specific elimination of mutant mitochondrial genomes in patient-derived cells by mitoTALENs. *Nat Med* 2013;**19**:1111–3.
153. Gammage PA, Rorbach J, Vincent AI, Rebar EJ, Minczuk M. Mitochondrially targeted ZFNs for selective degradation of pathogenic mitochondrial genomes bearing large-scale deletions or point mutations. *EMBO Mol Med* 2014;**6**:458–66.
154. Pereira CV, Bacman SR, Arguello T, Zekonyte U, Williams SL, Edgell DR, et al. mitoTev-TALE: a monomeric DNA editing enzyme to reduce mutant mitochondrial DNA levels. *EMBO Mol Med* 2018;**10**:e8084.
155. Lopez-Gomez C, Levy RJ, Sanchez-Quintero MJ, Juanola-Falgarona M, Barca E, Garcia-Diaz B, et al. Deoxycytidine and deoxythymidine treatment for thymidine kinase 2 deficiency. *Ann Neurol* 2017;**81**:641–52.
156. Yadak R, Sillevis Smitt P, van Gisbergen MW, van Til NP, de Coo IFM. Mitochondrial neurogastrointestinal encephalomyopathy caused by thymidine phosphorylase enzyme deficiency: from pathogenesis to emerging therapeutic options. *Front Cell Neurosci* 2017;**11**:31.
157. Smeets HJM, Sallevelt SCEH, Dreesen JCFM, de Die-Smulders CEM, de Coo IFM. Preventing the transmission of mitochondrial DNA disorders using prenatal or preimplantation genetic diagnosis. *Ann N Y Acad Sci* 2015;**1350**:29–36.
158. Mitalipov S, Amato P, Parry S, Falk MJ. Limitations of preimplantation genetic diagnosis for mitochondrial DNA diseases. *Cell Rep* 2014;**7**:935–7.
159. Greenfield A, Braude P, Flinter F, Lovell-Badge R, Ogilvie C, Perry ACF. Assisted reproductive technologies to prevent human mitochondrial disease transmission. *Nat Biotechnol* 2017;**35**:1059–68.
160. Craven L, Tuppen HA, Greggains GD, Harbottle SJ, Murphy JL, Cree LM, et al. Pronuclear transfer in human embryos to prevent transmission of mitochondrial DNA disease. *Nature* 2010;**465**:82–5.
161. Hyslop LA, Blakeley P, Craven L, Richardson J, Fogarty NME, Fragouli E, et al. Towards clinical application of pronuclear transfer to prevent mitochondrial DNA disease. *Nature* 2016;**534**:383–6.
162. Kang E, Wu J, Gutierrez NM, Koski A, Tippner-Hedges R, Agaronyan K, et al. Mitochondrial replacement in human oocytes carrying pathogenic mitochondrial DNA mutations. *Nature* 2016;**540**:270–5.
163. Yamada M, Emmanuele V, Sanchez-Quintero MJ, Sun B, Lallos G, Paull D, et al. Genetic drift can compromise mitochondrial replacement by nuclear transfer in human oocytes. *Cell Stem Cell* 2016;**18**:749–54.
164. Zhang J, Lui H, Luo S, Lu Z, Chávez-Badiola A, Lui Z, et al. Live birth derived from oocyte spindle transfer to prevent mitochondrial disease. *Reprod BioMed Online* 2017;**34**:361–8.
165. Reddy P, Ocampo A, Suzuki K, Luo J, Bacman SR, Williams SL, et al. Selective elimination of mitochondrial mutations in the germline by genome editing. *Cell* 2015;**161**:459–69.
166. Yang Y, Wu H, Kang X, Liang Y, Lan T, Li T, et al. Targeted elimination of mutant mitochondrial DNA in MELAS-iPSCs by mitoTALENs. *Protein Cell* 2018;**9**:283–97.

PART II

# TISSUES INVOLVED IN THE PROGRESSION OF THE PATHOLOGIES

CHAPTER

# 6

# Role of Mitochondria in the Skeletal Muscle Metabolism in Obesity and Type 2 Diabetes☆

*Paula M. Miotto, Graham P. Holloway*

**Department of Human Health and Nutritional Sciences, University of Guelph, Guelph, ON, Canada**

Mitochondria represent a key organelle in almost all cells within the human body and have a well-defined role for maintaining metabolic homeostasis. During the past decade, the physiological relevance of mitochondria has slowly been redefined, and it now appears that mitochondria are dynamic organelles, both with respect to structure and function, influencing several processes within tissues that extend beyond ATP homeostasis, including apoptosis and redox balance. As a result, mitochondria represent a key organelle in overall tissue health, and a better understanding of the regulation of this dynamic structure could provide novel insights into the development and/or treatment of a number of clinical pathologies, including insulin resistance and type 2 diabetes.

The incidence of diabetes is increasing at an alarming rate, however, the exact etiology of this pathology remains unknown, limiting success of various intervention strategies. Skeletal muscle, by virtue of its mass and high rate of insulin stimulated glucose transport, represents an important tissue in the development of insulin resistance, and skeletal muscle insulin resistance represents a strong risk factor for the development of type 2 diabetes. This chapter aims to provide an overview of the relationship between altered mitochondrial bioenergetics and skeletal muscle insulin resistance, with a particular emphasis on mechanisms associated with high-fat diet-induced obesity. We also will discuss the benefits of exercise in preventing insulin resistance, because a major focal point of training literature involves mitochondrial adaptations.

☆ This chapter will focus on the involvement of mitochondria in the regulation of muscle metabolic flexibility and insulin resistance.

*Mitochondria in Obesity and Type 2 Diabetes*
https://doi.org/10.1016/B978-0-12-811752-1.00006-7

# 1 REGULATION OF INSULIN SENSITIVITY IN SKELETAL MUSCLE

The maintenance of blood glucose homeostasis depends on the balance between liver glucose production and peripheral glucose uptake by a variety of tissues. Skeletal muscle, by virtue of its mass, plays a major role in regulating whole body glucose homeostasis. In this regard, resting skeletal muscle facilitates glucose uptake in response to insulin, a process that depends on muscle insulin signaling and the subsequent translocation of a glucose transporter (GLUT4) to the plasma membrane.[1, 2] Specifically, insulin binds to the insulin receptor on the sarcolemma, resulting in autophosphorylation of the receptor. This results in recruitment of insulin receptor substrate 1 (IRS1). Phosphorylation of IRS1 results in activation of phosphatidylinositol 3-kinase (PI3K), leading to the production of a peptide known as phosphatidylinositol 3,4,5-trisphosphate (PIP3),[3] and ultimately activation of protein kinase B (PKB: also known as AKT)[4, 5] which phosphorylates and inhibits AS160 (aka TBC1D4) and/or TBC1D1, resulting in translocation of GLUT4 to the sarcolemma[6] (summarized in Fig. 1).

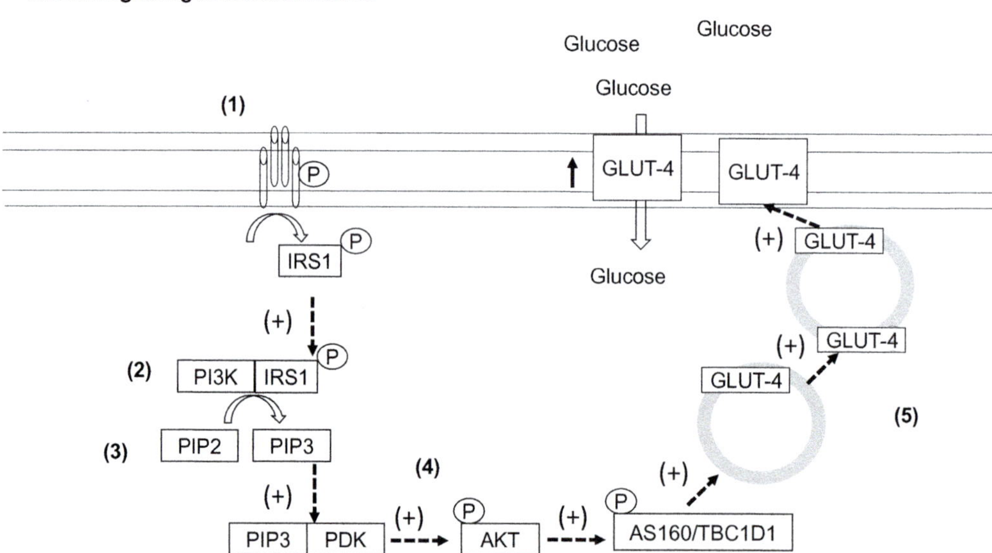


FIG. 1 **Insulin signaling in skeletal muscle.** Summary of insulin signaling in skeletal muscle. (1) Insulin binds to the insulin receptor on the sarcolemma, resulting in autophosphorylation of the receptor. This results in recruitment of insulin receptor substrate 1 (IRS1). (2) Phosphorylation of IRS1 results in activation of phosphatidylinositol 3-kinase (PI3K). (3) The activated PI3K/IRS1 complex produces a peptide known as phosphatidylinositol 3,4,5-trisphosphate (PIP3), resulting in PDK activation. (4) The activated PIP3/PDK complex phosphorylates and activates protein kinase B (AKT), and AKT phosphorylates and inactivates AS160 or TBC1D1. (5) phosphorylation of AS160/TBC1D1 removes the suppression of GLUT-4, and GLUT-4 translocates to the sarcolemma to promote glucose uptake.

During the development of type 2 diabetes, insulin sensitivity of skeletal muscle (and other tissues) is attenuated, resulting in less GLUT4 on the plasma membrane, directly contributing to compromised whole body glucose homeostasis. While the explanation for impaired insulin signaling within muscle is incompletely understood, an accumulation of reactive lipids and excessive reactive oxygen species (ROS) production both have been implicated as key metabolic events in the induction of skeletal muscle insulin resistance. Alterations in mitochondrial function have been linked to lipid accumulation and increased mitochondrial ROS production, suggesting alterations within this organelle substantially contribute to the induction of insulin resistance.

## 2 SKELETAL MUSCLE LIPID ACCUMULATION AND INSULIN RESISTANCE

Historically, the most widely studied/implicated mechanism in the development of insulin resistance is the accumulation of lipids within muscle. It has been known for decades that a high-fat environment results in impaired glucose homeostasis,[7,8] because the consumption of a high-fat diet induces white adipose/liver insulin resistance, which results in excessive free fatty acid delivery to muscle[9] (as reviewed in Ref. 10). In combination with the increased redistribution of fatty acid transporters to the plasma membrane in muscle during obesity,[11] this results in a dramatic increase in the transport of lipids into skeletal muscle. Although lipids are important for a variety of metabolic functions, excessive lipids are known to attenuate insulin signaling. In support of this, acute intralipid infusions,[12] or intramuscular incubations in the presence of a high fatty acid concentration,[13] dramatically attenuate insulin responsiveness. Moreover, insulin-resistant skeletal muscles in humans have high lipid accumulation.[14,15] Although it originally was thought that total lipid accumulation in skeletal muscle would negatively affect insulin signaling, lipid composition might be an important consideration. It has been shown that increased triacylglycerol (TAG) accumulation as a result of overexpressing DGAT1 (involved in lipid synthesis)[16] or PLIN2 (involved in the regulation of lipolysis)[17] in skeletal muscle is not detrimental for insulin signaling. Elevated intramuscular TAG stores as a result of endurance training actually coincides with improved insulin sensitivity.[18] In contrast, incomplete lipolysis[19] or synthesis of TAG can result in elevated diacylglycerol (DAG) accumulation, a reactive lipid, which activates protein kinase C (PKC)[20] and impairs insulin signaling through inducing serine phosphorylation of IRS1.[21,22] This observation is supported in mice lacking PKC, which are protected from high-fat diet induced insulin resistance.[23] Ceramides also have been shown to activate PKC and prevent Akt phosphorylation,[24] while simultaneously activating protein phosphatase-mediated dephosphorylation of Akt, which combined dramatically attenuates a terminal component in the insulin signaling cascade (see Fig. 2). A direct role for ceramides is supported by the pharmacological inhibition of serine palmitoyltransferase, an enzyme involved in the biosynthesis of ceramides, that protected against diet-induced insulin resistance.[25] Collectively, it appears that the accumulation of reactive lipids and not necessarily TAG are associated with insulin resistance.

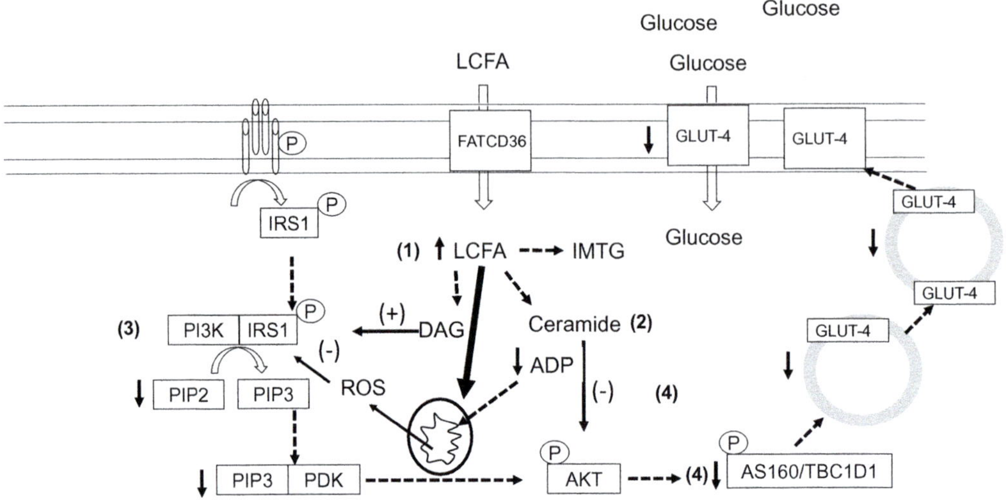


FIG. 2 **Influence of reactive lipids on skeletal muscle insulin signaling.** (1) High lipid uptake into skeletal muscle can be stored as intramuscular lipids or oxidized by the mitochondria. (2) Incomplete synthesis of triacylglycerol (TAG) can result in diacylglycerol (DAG) and ceramide accumulation. (3) DAG can cause serine phosphorylation of IRS1, resulting in inhibition of insulin signaling. In contrast, (4) ceramides can prevent AKT phosphorylation (activation) as well as promote AKT dephosphorylation, resulting in sequestered vesicles containing GLUT4 and less GLUT4 on the sarcolemma.

# 3 SKELETAL MUSCLE ROS EMISSION AND INSULIN RESISTANCE

In addition to reactive lipids, a key observation suggesting mitochondria have a direct role in the development of insulin resistance extends from the finding that pharmacological inhibition of carnitine palmitoyltransferase-I (CPT-I), and therefore lipid transport into mitochondria, protected cells from palmitate-induced insulin resistance.[26] This effect of etomoxir was further supported in vivo in diet-induced obese mice,[27] and potentially in humans.[28] Together, these data implicate the movement of lipids into mitochondria as a contributing factor to the development of insulin resistance. As a result, an increased production of mitochondrial-derived ROS has been proposed to contribute to the induction of insulin resistance. With the consumption of a high-fat diet, it is believed that increased mitochondrial ROS production occurs because of an imbalance between substrate delivery[26] and electron entry into the electron transport chain relative to ADP availability within the matrix.[29] A high-fat diet is particularly worrisome for the generation of ROS because all enzymes involved in fatty acid catabolism within the mitochondria are near-equilibrium, and therefore are influenced directly by substrate concentration. In addition, the electron transferring flavoprotein, a component of beta-oxidation, dramatically increases ROS production rates as a result of minor increases in lipid availability.[30, 31] For these reasons, it is likely that mitochondrial $H_2O_2$ emission is increased dramatically in obese insulin-resistant individuals.[29, 32] Although the exact manner with which mitochondrial ROS causes insulin resistance have not been delineated, similar to DAG, it has been suggested that ROS-mediated activation of the NF-κB/IκB/IKKβ pathway results in serine phosphorylation of IRS1, attenuating insulin signaling at a proximal step.[33, 34]

Several further lines of evidence suggest that mitochondrial bioenergetics and ROS have a primary role in the etiology of insulin resistance. First, mitochondrial ROS generation increases in response to fatty acid exposure and high-fat feeding,[35] either in cell culture,[36, 37] rodents,[35] or humans,[32] and this can lead to insulin resistance.[26, 38] Second, attenuating mitochondrial ROS emission in vivo using either an exogenous mitochondrial targeted antioxidant (SS31)[29] or overexpression of mitochondrial antioxidant enzymes,[29, 39, 40] prevents diet-induced insulin resistance. In humans, infusing reduced glutathione also has been shown to transiently increase glucose infusion rates,[41] further implicating redox stress as a key process influencing insulin sensitivity. Nevertheless, controversy surrounds the notion that mitochondrial ROS directly contributes to the development of insulin resistance, as in contrast to SS31, consumption of an alternative putative mitochondrial targeted antioxidant, the Skulachev ion (SkQ), does not protect against diet-induced insulin resistance.[42] The relationship between mitochondrial ROS and insulin resistance, however, primarily extends from studies examining the relatively acute effects of attenuating mitochondrial ROS with respect to insulin resistance, while the efficacy of SkQ was tested with a 16-week high-fat diet.[42] The lack of a beneficial effect of attenuating ROS following this prolonged intervention likely coincides with reactive lipid accumulation, which can induce insulin resistance independently. This same interpretation might help rectify the discrepancy that, while aging is associated with an increased prevalence of insulin resistance,[43, 44] human muscle does not display an increased propensity to produce ROS.[45–47]

## 4 MITOCHONDRIAL CONTENT IN SKELETAL MUSCLE: INFLUENCE ON REACTIVE LIPID ACCUMULATION, ROS EMISSION, AND INSULIN RESISTANCE

Mitochondria are key organelles found in a variety of tissues that are required for aerobic energy production and regulate lipid metabolism and redox balance within skeletal muscle. For many years, scientists have speculated that a dysfunction in mitochondrial fatty acid oxidation, because both a reduction in the number of mitochondria and their intrinsic activity, contributes to the development of insulin resistance. In support of this, mitochondrial DNA content, electron transport chain abundance, and mitochondrial volume (TEM analysis) were shown to be lower in diabetic individuals.[48] Although these data clearly suggest less mitochondria in metabolically compromised individuals, these authors also used the ratio of succinate dehydrogenase and mtDNA as an index to suggest compromised intrinsic mitochondrial respiratory function.[48] Direct assessments of mitochondrial function, however, provide little evidence to support the view that an intrinsic impairment in the ability of mitochondria to oxidize fatty acids contributes to insulin resistance. For example, although rates of oxygen consumption are lower in diabetic individuals when assessed in permeabilized muscle fibers,[49] this measurement is influenced by mitochondrial content, and when respiration is normalized to mtDNA, the differences with diabetes are lost.[49] In isolated mitochondria, moreover, rates of oxygen consumption, P/O ratios, and palmitate oxidation are not compromised in obese/insulin resistant individuals.[50, 51] Nevertheless, mitochondrial content is reduced in most reports examining the skeletal muscle of insulin resistant/obese individuals.[51–55] The potential for decreased mitochondrial content to participate and/or exacerbate insulin resistance has been supported by recent reports using microarray approaches, which found components of the electron transport chain reduced in the skeletal muscle of insulin resistant/diabetic individuals.[56, 57] This is matched by reductions in mitochondrial respiration[50, 58, 59] and ATP synthesis.[60, 61] In addition, peroxisome proliferator activated receptor $\gamma$ co-activator-1 $\alpha$ (PGC-1$\alpha$), a protein thought to play a critical role in coordinating nuclear and mitochondrial genomes, is reduced in the skeletal muscle of type 2 diabetic individuals, a finding that coincides with reduced expression of mitochondrial proteins.[56, 57] In muscle cell lines, increasing the expression of PGC-1$\alpha$ improves insulin-sensitivity, an effect attributed to the resulting induction of mitochondrial biogenesis, as well as genes involved in fatty acid oxidation, oxidative phosphorylation, and glucose transport.[62, 63] Further, PGC-1$\alpha$[64] overexpression reduces skeletal muscle DAG and ceramide content while improving insulin-signaling. Combined, these studies provide compelling evidence that mitochondrial content influences skeletal muscle insulin sensitivity.

Although these data suggest a relationship between mitochondrial content and insulin sensitivity, this is not a uniform finding. This is best exemplified by the fact that consumption of a high-fat diet, while inducing insulin resistance, actually increases mitochondrial content.[35] Therefore, reductions in mitochondrial content are not required for the development of insulin resistance. It is important to consider that mitochondrial content could indirectly influence reactive lipid accumulation or ROS emission, and therefore, consumption of a high-fat diet still can result in insulin resistance in the absence of reductions in mitochondrial content. Ablation of proteins involved in plasma membrane fatty acid uptake,[65] or the overexpression of proteins involved in lipid synthesis,[16, 17] attenuate reactive lipid accumulation and the induction of insulin resistance in the absence of changes in mitochondrial content. The finding that genetic approaches that increase mitochondrial content protect against the detrimental effects of a high-fat diet[66] strongly suggests that the increase in mitochondrial content observed with a high-fat diet is likely a beneficial compensatory response.

It remains unclear why obese/insulin-resistant individuals display a decrease in mitochondrial content, however, several possibilities exist; the easiest explanation is their lack of physical activity. Although exercise training increases mitochondrial content,[67] detraining has been shown to rapidly result in a loss of mitochondrial volume.[68] Alternatively, a growing body of literature has suggested that mitophagy and the formation of an extensive mitochondrial reticulum network are important for the maintenance of healthy mitochondria throughout the lifespan. In support of this, mitofusin-2 has been shown to be lower in insulin-resistant humans and correlate with rates of glucose oxidation.[69] Moreover, in cell cultures, the pharmacological inhibition of mitochondrial fission prevented palmitic acid-induced insulin resistance.[70] Altogether, obesity and insulin resistance are associated with reductions in mitochondrial content that alone could influence insulin resistance through greater lipid accumulation and excessive ROS emissions. Given that reductions in mitochondrial content are not required for insulin resistance to occur, it is also possible that altered mitochondrial function and the regulation of substrate sensitivity could play a role.

# 5 MITOCHONDRIAL FUNCTION: INFLUENCE ON REACTIVE LIPIDS AND ROS PRODUCTION

The biological processes influenced by alterations in mitochondrial content have not been identified, however, in 1984, Dr. John Holloszy postulated that an increase in mitochondrial content would alter the sensitivity of oxidative metabolism to ADP.[71] This basic working model has been supported in several ways to explain exercise-training responses,[72–74] and direct assessments of mitochondrial respiration using permeabilized muscle fibers have shown an improvement in respiration at a submaximal ADP concentration following training,[67,75] suggesting mitochondrial content directly contributes to the improvement in ADP sensitivity. Exercise-induced mitochondrial biogenesis has become synonymous with training adaptations, because it has repeatedly been shown that exercise training increases mitochondrial content, improves mitochondrial ADP sensitivity, and attenuates cytosolic ADP concentrations,[73,76,77] resulting in less activation of rate-limiting enzymes involved in carbohydrate breakdown and ultimately a greater reliance on fatty acid metabolism. Within this framework, it would be rational to hypothesize that a reduction in mitochondrial content, as seen in obese/insulin-resistant individuals,[51–55] would compromise fatty acid oxidation and impair ADP sensitivity. Although it might appear contradictory that direct in vitro measurements of skeletal muscle fatty acid oxidation indicate an uncompromised ability to metabolize lipids with obesity, these measurements determine the oxidation of exogenous $^{14}$C-palmitate,[78,79] and therefore can be influenced by plasma membrane transport and mitochondrial metabolism. Because rates of plasma membrane fatty acid transport are increased with obesity,[11] in association with greater sarcolemmal FAT/CD36 protein content,[11] it appears that a higher palmitate delivery to the mitochondria is required to maintain fat oxidation, supporting the notion that obese individuals display a compromised ability to metabolize fatty acids. In turn, this might contribute to an increase in reactive lipid accumulation and insulin resistance. The notion that ADP sensitivity is altered with insulin resistance as a mechanism contributing to this observation, however, has not been addressed adequately, because previous studies have primarily examined mitochondrial respiratory function in insulin-resistant muscle in the presence of saturating ADP concentrations, an approach that likely prevented the detection of changes in the sensitivity of rate-limiting steps in oxidative phosphorylation such as ADP transport rates. Some evidence exists that this working model is accurate, however, as respiration in the presence of nonsaturating ADP concentrations is impaired in diabetic rats and following a high-fat diet, even in the absence of reductions in mitochondrial content.[13,79a] Although the notion of impaired mitochondrial lipid or ADP sensitivity in human muscle has not been addressed, considering the consistent finding of reduced mitochondrial content in individuals with compromised insulin signaling,[51–55] there is a high likelihood that substrate sensitivity is compromised in these individuals. In turn, this could contribute to increased lipid accumulation, elevated ROS emission, and impairments in insulin signaling (Fig. 3).

This interpretation also might help rectify the discrepancy with aging, because the absence of increased mitochondrial ROS in older individuals has been assessed exclusively in the absence of ADP, and therefore has examined the capacity for mitochondrial ROS production, which likely does not reflect the in vivo situation. From a physiological standpoint, it is more appropriate to

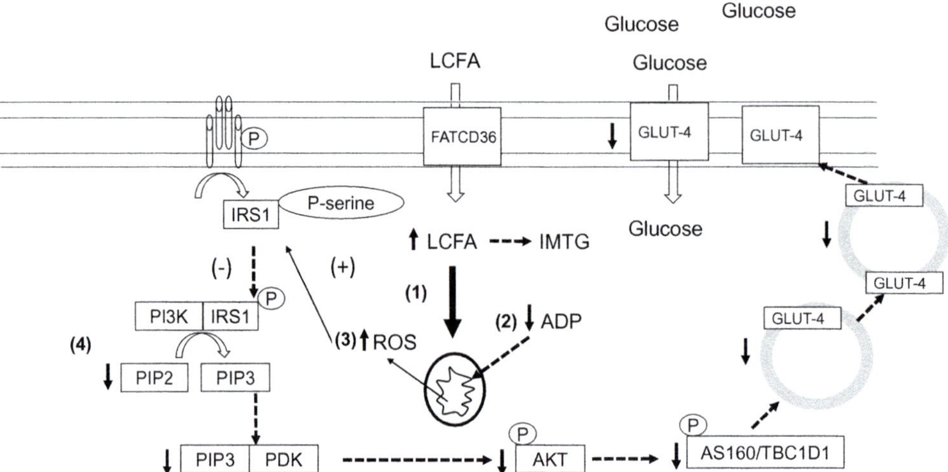


FIG. 3 **Influence of mitochondrial ROS on skeletal muscle insulin signaling.** Following lipid uptake into skeletal muscle, especially after high fat consumption where lipid delivery is elevated, (1) a greater lipid influx into the mitochondria combined with (2) a reduction in ADP transport during obesity can (3) promote mitochondrial ROS emission. In turn, (4) ROS can suppress insulin signaling by causing serine phosphorylation of IRS1.

examine mitochondrial bioenergetics (respiration and ROS emission) in the presence of submaximal concentrations of ADP, because ADP binding to $F_1F_0$ ATP synthase decreases membrane potential and the overall rate of superoxide production.[80–82] An increase in mitochondrial content has been suggested to improve mitochondrial ADP sensitivity, while a reduction in mitochondrial content impairs ADP sensitivity (Fig. 4). In this manner, alterations in mitochondrial content might influence ADP sensitivity to indirectly influence reactive lipid accumulation and/or ROS production, and therefore provide a mechanism for the protective effects of increasing mitochondrial content described in the previous section. Very little evidence, however, has been generated with respect to mitochondrial ADP sensitivity and insulin resistance.

# 6 REGULATION OF SUBSTRATE TRANSPORT: POTENTIAL CONTRIBUTION TO REACTIVE LIPID ACCUMULATION, ROS PRODUCTION, AND INSULIN RESISTANCE

## 6.1 Lipid Transport

In addition to mitochondrial content indirectly influencing rates of reactive lipid accumulation, considerable effort has been focused on understanding the rate-limiting steps in mitochondrial fatty acid oxidation, and potential alterations with insulin resistance. Mitochondrial lipid transport is regulated by CPT-I, an enzyme located on the outer mitochondrial membrane that is responsible for transesterifying acyl-CoA and L-carnitine, producing a more neutral lipid, acylcarnitine, that is able to traverse the inner mitochondrial membrane. CPT-I is

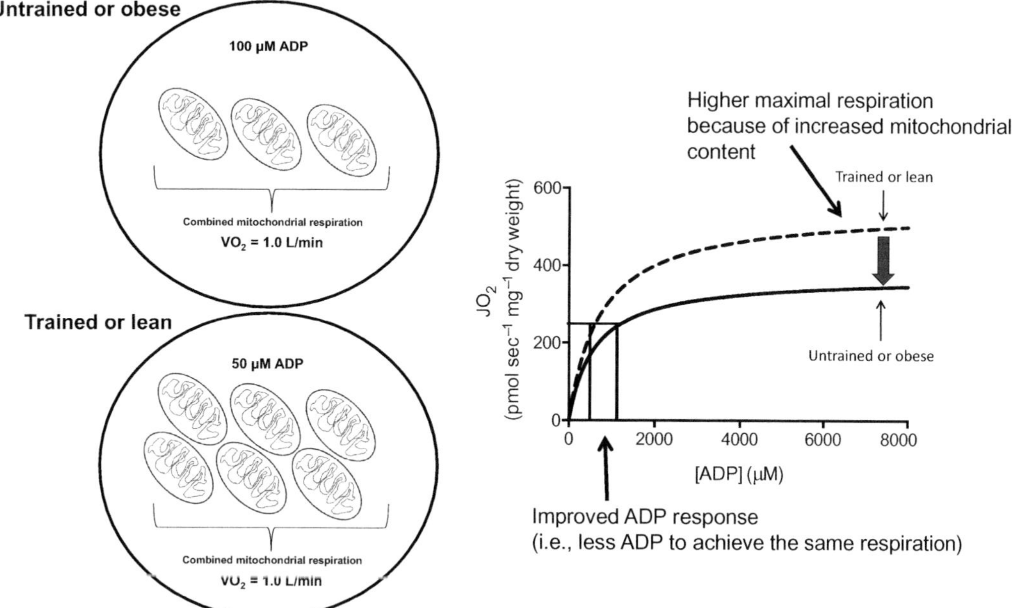


FIG. 4 **Effect of training and obesity on mitochondrial content and ADP stimulated respiration.** Relative to an untrained state, exercise training increases mitochondrial content and coincides with greater ADP-stimulated respiration, whereby a specific $VO_2$ is achieved at a lower ADP concentration. Similarly, obese individuals tend to have lower mitochondrial content than lean, and in this can contribute to less ADP sensitivity than lean.

considered the rate-limiting step in fatty acid oxidation, because the maximal activity of this enzyme is orders of magnitude lower than matrix localized enzymes (e.g., β-HAD) and CPT-I is inhibited by a naturally occurring compound, malonyl-CoA (M-CoA).[83] The binding of M-CoA to CPT-I occurs at a distal location from the catalytic site, and is known to decrease the catalytic flux of this enzyme through allosteric regulation. In the context of insulin resistance, it appears that M-CoA content increases with obesity,[84] which in theory would be expected to attenuate the movement of lipids into the mitochondrial matrix. In stark contrast, however, insulin resistance is associated with a marked increase in P-CoA content[85, 86] and the production of acylcarnitine, which would be indicative of increased lipid flux through CPT-I. Although these observations appear difficult to rectify, the concentration of P-CoA has been shown to influence M-CoA sensitivity. It now appears that increases in P-CoA can override any potential inhibition exerted by M-CoA.[87] In this manner, the increase in acyl-CoA moieties that occurs with high-fat diets/obesity would result in a dramatic movement of lipids into the mitochondrial matrix.[85] It therefore appears that an inability to transport lipids into the mitochondria does not contribute to the development of insulin resistance. Instead, an increase in lipid transport into the matrix appears to contribute directly to diet-induced insulin resistance, independent of reactive lipid accumulation, because C2C12 cells exposed to palmitate and L-carnitine develop impaired insulin action,[88, 89] while palmitate exposure in the absence of CPT-I flux (absence of L-carnitine or etomoxir-mediated inhibition of CPT-I) was not associated with impaired sensitivity.[26] Combined, these data suggest that the movement of fatty acids into the mitochondria is increased with high-fat environments, and contributes to the development of insulin resistance independent of reactive lipids.

## 6.2 ADP Transport

Although a detrimental effect of increased mitochondrial lipid transport appears at odds with the notion of impaired mitochondrial fatty acid oxidation contributing to insulin resistance, in resting muscle mitochondrial respiration/substrate oxidation is limited primarily by the presence of matrix localized ADP. In this respect, increased lipid transport cannot independently increase fatty acid oxidation without a concomitant increase in ADP availability. As previously stated, however, insulin resistance is likely associated with a reduction in mitochondrial ADP sensitivity, which would prevent a dramatic increase in fatty acid oxidation. The movement of excessive lipids into the mitochondrial matrix also would be expected to increase the abundance of reducing equivalents for the electron transport chain (e.g., NADH), because the enzymes involved in matrix localized lipid metabolism are all near equilibrium. As described in detail in Chapter 1, electrons are passed through a series of reactions within the electron transport chain, and free energy is released and used to transport protons against their concentration gradient from the matrix into the intermembrane space, generating a proton motive force. As protons are transported down their concentration gradient through $F_1F_0$ ATP synthase, the kinetic energy is used to drive ATP production. ADP binding to $F_1F_0$ ATP synthase decreases proton motive force, which also directly reduces the overall rate of superoxide production.[80–82] The increased lipid transport, in combination with impaired ADP sensitivity, likely contributes to redox stress with obesity. Although the reasons for reduced ADP sensitivity are not fully understood, and might be influenced by mitochondrial content, altered regulation of proteins involved in ADP transport likely also play a role.

The phosphate shuttling mechanism for energy transfer between matrix and cytosolic compartments includes three major protein complexes: voltage-dependent anion channel (VDAC) on the outer mitochondrial membrane; mitochondrial creatine kinase (miCK) in the inter membrane space (IMS); and adenine nucleotide translocase (ANT) on the inner mitochondrial membrane. A leading model describing energy transfer from the matrix to cytosolic compartments proposes that ATP/ADP freely diffuse across concentration gradients through VDAC and ANT, and it is estimated that as much as 80% of the energy transfer from the matrix to cytoplasm occurs through miCK-dependent phosphate shuttling in cardiac[90] and oxidative skeletal muscles. Direct in vivo evidence implicating a prominent role for miCK has not been generated in skeletal muscle, because muscle lacking the miCK enzyme do not display compromised force production, altered fiber type, or metabolite concentrations,[91] suggesting a limited role of miCK in regulating basal metabolism. Alternatively, ANT is required for the transport of ADP/ATP across the inner mitochondrial membrane, and historically ADP flux through ANT was attributed solely to substrate concentrations. Although the simplicity of this proposed working model is appealing, external regulation of ANT likely exists. ANT has several potential regulatory mechanisms, including acetylation of lysine 23,[92] tyrosine 194 phosphorylation,[93] and glutathionylation/carbonylation.[94] Although the functional roles for these regulatory points remain unknown, an acute bout of high-intensity interval exercise has been shown to decrease lysine 23 acetylation, a response predicted to improve the sensitivity to ADP binding.[92] In support of this, in vitro assessments of mitochondrial ADP sensitivity indicate an improvement in ADP respiratory sensitivity in the absence of Cr after high-intensity exercise.[95] Combined, these data suggest that the regulation of ANT is highly complex.

It has been shown in diabetic rats that reductions in ANT2, independent of changes in mitochondrial content, aligned with reduced ADP sensitivity and an impaired suppression of ROS production in parallel with insulin resistance.[13] In contrast, resveratrol supplementation increased ANT2 and attenuated ROS production and insulin resistance.[13] These data suggest that the ADP transport axis could be important for modulating ROS production and insulin resistance. Moreover, the regulation of ANT itself has been speculated to play a role in regulating ROS. In this regard, P-CoA has been shown to inhibit ANT[96–98] and reduce ADP sensitivity in permeabilized muscle fibers.[67, 99] Given that ADP provision to the mitochondrial matrix reduces membrane potential and attenuates ROS production,[80–82] it is possible that the P-CoA that accumulates during obesity or high-fat consumption[85, 86] inhibits ANT and impairs the ability of the system to attenuate ROS production. This, combined with an increase in mitochondrial P-CoA accumulation, could exacerbate mitochondrial ROS production and insulin resistance. In addition, while aging is associated with oxidative stress, and the absence of a change in maximal mitochondrial ROS emission rates, ANT appears to be inhibited in rats and houseflies,[99a] which might suggest ROS is increased with aging because of a reduction in ADP sensitivity. Collectively, it appears that both an increase in reactive lipids and mitochondrial derived ROS production influence insulin resistance. Given that lipid accumulation could inhibit ANT and reduce ADP sensitivity, and reductions in ADP sensitivity can promote mitochondrial-derived ROS emission, it is possible that ANT serves as a nexus merging these two mechanisms and combined can contribute to insulin resistance.

# 7 EXERCISE TRAINING AS A MEANS TO PREVENT METABOLIC DYSFUNCTION AND INSULIN RESISTANCE

Exercise training is a well-known strategy to improve insulin sensitivity, and therefore represents an ideal intervention/model to further study mechanistic relationships in muscle. Exercise has been proposed to improve skeletal muscle glucose uptake through several mechanisms, including the induction of GLUT4 trafficking to the plasma membrane in an insulin-independent manner.[100] Regardless of the presence of reactive lipids and redox stress, exercise can transiently improve glucose homeostasis in diabetic individuals.[101] Chronically, however, exercise training improves insulin responsiveness,[67] potentially by decreasing reactive lipids and redox stress. Key to these responses is likely the induction of mitochondrial biogenesis and the expansion of the mitochondrial network. The pioneering work of Dr. John Holloszy almost 60 years ago clearly delineated the ability of exercise training to increase mitochondrial content.[102] Since the observation that exercise increased mitochondrial proteins, our understanding of the molecular mechanisms involved in the induction of mitochondrial biogenesis has grown considerably. PGC-1α is thought to represent a key protein coordinating the nuclear and mitochondrial genomes,[103–106] and is transiently, but continuously, increased following bouts of exercise.[105] It is likely for this reason that overexpressing PGC-1α improves skeletal muscle insulin sensitivity in obese rats, a response partially mediated through mitochondrial biogenesis and a reduction in reactive lipids.[64] In the context of insulin resistance/diabetes, several reports have shown that exercise or exercise training improves glucose homeostasis,[107–109] and training increases PGC-1α and mitochondrial proteins in humans.[105, 110] In addition, exercise training in obese individuals results in a reduction in DAG and ceramide content within muscle.[111, 112] Although the increase in fat oxidation and mitochondria that occurs during exercise training likely contributes to this response, exercise training also increases the abundance of enzymes involved in fatty acid synthesis/storage, promoting intramuscular TAG accumulation.[113] Indeed, the abundance of neutral lipid droplets increases with training,[113] resulting in a similar phenotype to overexpressing DGAT,[16] namely improved insulin sensitivity, reduced reactive lipids, and increased TAG.[113] Although mitochondrial biogenesis is a fundamental response to exercise training, several mechanisms induced by exercise likely contribute to the improvement in glucose homeostasis. Nevertheless, mitochondrial adaptations to exercise are thought to represent a key event, in large part because of an improvement in ADP sensitivity (see Fig. 4). Direct assessments of mitochondrial respiratory function in permeabilized muscle fibers has revealed that at submaximal concentrations, trained individuals display greater respiration in the presence of nonsaturating ADP concentrations[67] (Fig. 4).

Given that insulin resistance, at least in rodents, is associated with impairments in submaximal ADP-supported respiration,[13, 79a] it is possible that exercise training could restore ADP sensitivity and improve redox balance at the level of the mitochondria because of a greater ability for ADP to suppress ROS rates. While this possibility remains to be determined in humans, exercise training in obese rats resulted in a greater ability for ADP to suppress ROS rates.[114] Another important aspect of ADP sensitivity/redox balance pertains to the biological effects of acyl-CoA content. These lipid species also are increased in the skeletal muscle of insulin-resistant individuals[85] and are known to have diverse actions, including promoting CPT-I flux and attenuating ADP sensitivity. Exercise training has been shown to decrease the detrimental effects of P-CoA on mitochondrial ADP sensitivity in healthy individuals.[67] Therefore, exercise appears to induce mitochondrial biogenesis, improve mitochondrial ADP sensitivity, reduce reactive lipids, and improve mitochondrial redox balance. These changes would be improved further as a result of exercise inducing the expression of genes involved in lipid storage, as well as antioxidants.

## 8 SUMMARY/CONCLUSIONS

Overall, the induction of skeletal muscle insulin resistance is associated with reactive lipid accumulation and increased mitochondrial ROS production. It appears that both of these scenarios can be influenced by a reduction in mitochondrial content and/or altered regulation of substrate sensitivity. Given that ADP transport can directly influence ROS emission and indirectly affect mitochondrial lipid oxidation, it is tempting to speculate that the regulation of ANT serves as a nexus connecting lipid accumulation and mitochondrial ROS emission. The ability for P-CoA to inhibit ANT, and therefore regulate the ability for ADP to suppress ROS rates, supports a potential link between these two mechanisms underlying insulin resistance. Although the classic notion of mitochondrial dysfunction has not been supported in the literature, compelling evidence has been generated to suggest an impairment in mitochondrial bioenergetics exists in the skeletal muscle of diabetic/insulin resistant individuals. Nevertheless, individuals with compromised glucose homeostasis still respond positively to exercise training, which potentially is a result of the induction of mitochondrial biogenesis.

## References


1. Cushman SW, Wardzala LJ. Potential mechanism of insulin action on glucose transport in the isolated rat adipose cell. Apparent translocation of intracellular transport systems to the plasma membrane. *J Biol Chem* 1980;**255**:4758–62.
2. Suzuki K, Kono T. Evidence that insulin causes translocation of glucose transport activity to the plasma membrane from an intracellular storage site. *Proc Natl Acad Sci U S A* 1980;**77**:2542–5.
3. Tengholm A, Meyer T. A PI3-kinase signaling code for insulin-triggered insertion of glucose transporters into the plasma membrane. *Curr Biol* 2002;**12**:1871–6.
4. Alessi DR, Deak M, Casamayor A, Caudwell FB, Morrice N, Norman DG, Gaffney P, Reese CB, MacDougall CN, Harbison D, Ashworth A, Bownes M. 3-Phosphoinositide-dependent protein kinase-1 (PDK1): structural and functional homology with the Drosophila DSTPK61 kinase. *Curr Biol* 1997;**7**:776–89.
5. Stokoe D, Stephens LR, Copeland T, Gaffney PR, Reese CB, Painter GF, Holmes AB, McCormick F, Hawkins PT. Dual role of phosphatidylinositol-3,4,5-trisphosphate in the activation of protein kinase B, *Science* 1997;**277**:567–70.
6. Ramm G, Larance M, Guilhaus M, James DE. A role for 14-3-3 in insulin-stimulated GLUT4 translocation through its interaction with the RabGAP AS160. *J Biol Chem* 2006;**281**:29174–80.
7. Randle PJ, Garland PB, Hales CN, Newsholme EA. The glucose fatty-acid cycle. Its role in insulin sensitivity and the metabolic disturbances of diabetes mellitus. *Lancet* 1963;**1**:785–9.
8. Boden G, Chen X, Ruiz J, White JV, Rossetti L. Mechanisms of fatty acid-induced inhibition of glucose uptake. *J Clin Invest* 1994;**93**:2438–46.
9. Turner N, Kowalski GM, Leslie SJ, Risis S, Yang C, Lee-Young RS, Babb JR, Meikle PJ, Lancaster GI, Henstridge DC, White PJ, Kraegen EW, Marette A, Cooney GJ, Febbraio MA, Bruce CR. Distinct patterns of tissue-specific lipid accumulation during the induction of insulin resistance in mice by high-fat feeding. *Diabetologia* 2013;**56**:1638–48.
10. Cusi K. The role of adipose tissue and lipotoxicity in the pathogenesis of type 2 diabetes. *Curr Diab Rep* 2010;**10**:306–15.
11. Bonen A, Parolin ML, Steinberg GR, Calles-Escandon J, Tandon NN, Glatz JF, Luiken JJ, Heigenhauser GJ, Dyck DJ. Triacylglycerol accumulation in human obesity and type 2 diabetes is associated with increased rates of skeletal muscle fatty acid transport and increased sarcolemmal FAT/CD36. *FASEB J* 2004;**18**:1144–6.
12. Dresner A, Laurent D, Marcucci M, Griffin ME, Dufour S, Cline GW, Slezak LA, Andersen DK, Hundal RS, Rothman DL, Petersen KF, Shulman GI. Effects of free fatty acids on glucose transport and IRS-1-associated phosphatidylinositol 3-kinase activity. *J Clin Invest* 1999;**103**:253–9.
13. Smith BK, Perry CG, Herbst EA, Ritchie IR, Beaudoin MS, Smith JC, Neufer PD, Wright DC, Holloway GP. Submaximal ADP-stimulated respiration is impaired in ZDF rats and recovered by resveratrol. *J Physiol* 2013;**591**:6089–101.
14. Petersen KF, Dufour S, Befroy D, Garcia R, Shulman GI. Impaired mitochondrial activity in the insulin-resistant offspring of patients with type 2 diabetes. *N Engl J Med* 2004;**350**:664–71.
15. Petersen KF, Dufour S, Shulman GI. Decreased insulin-stimulated ATP synthesis and phosphate transport in muscle of insulin-resistant offspring of type 2 diabetic parents. *PLoS Med* 2005;**2**.
16. Liu L, Zhang Y, Chen N, Shi X, Tsang B, Yu YH. Upregulation of myocellular DGAT1 augments triglyceride synthesis in skeletal muscle and protects against fat-induced insulin resistance. *J Clin Invest* 2007;**117**:1679–89.
17. Bosma M, Hesselink MK, Sparks LM, Timmers S, Ferraz MJ, Mattijssen F, van Beurden D, Schaart G, de Baets MH, Verheyen FK, Kersten S, Schrauwen P. Perilipin 2 improves insulin sensitivity in skeletal muscle despite elevated intramuscular lipid levels. *Diabetes* 2012;**61**:2679–90.
18. Goodpaster BH, He J, Watkins S, Kelley DE. Skeletal muscle lipid content and insulin resistance: evidence for a paradox in endurance-trained athletes. *J Clin Endocrinol Metab* 2001;**86**:5755–61.

19. Moro C, Bajpeyi S, Smith SR. Determinants of intramyocellular triglyceride turnover: implications for insulin sensitivity. *Am J Physiol Endocrinol Metab* 2008;**294**:E203–13.
20. Griffin ME, Marcucci MJ, Cline GW, Bell K, Barucci N, Lee D, Goodyear LJ, Kraegen EW, White MF, Shulman GI. Free fatty acid-induced insulin resistance is associated with activation of protein kinase C theta and alterations in the insulin signaling cascade. *Diabetes* 1999;**48**:1270–4.
21. Itani SI, Ruderman NB, Schmieder F, Boden G. Lipid-induced insulin resistance in human muscle is associated with changes in diacylglycerol, protein kinase C, and IkappaB-alpha. *Diabetes* 2002;**51**:2005–11.
22. Yu C, Chen Y, Cline GW, Zhang D, Zong H, Wang Y, Bergeron R, Kim JK, Cushman SW, Cooney GJ, Atcheson B, White MF, Kraegen EW, Shulman GI. Mechanism by which fatty acids inhibit insulin activation of insulin receptor substrate-1 (IRS-1)-associated phosphatidylinositol 3-kinase activity in muscle. *J Biol Chem* 2002;**277**:50230–6.
23. Kim JK, Fillmore JJ, Sunshine MJ, Albrecht B, Higashimori T, Kim DW, Liu ZX, Soos TJ, Cline GW, O'Brien WR, Littman DR, Shulman GI. PKC-theta knockout mice are protected from fat-induced insulin resistance. *J Clin Invest* 2004;**114**:823–7.
24. Summers SA. Ceramides in insulin resistance and lipotoxicity. *Prog Lipid Res* 2006;**45**:42–72.
25. Kurek K, Miklosz A, Lukaszuk B, Chabowski A, Gorski J, Zendzian-Piotrowska M. Inhibition of ceramide de novo synthesis ameliorates diet induced skeletal muscles insulin resistance. *J Diabetes Res* 2015;**2015**:154762.
26. Koves TR, Ussher JR, Noland RC, Slentz D, Mosedale M, Ilkayeva O, Bain J, Stevens R, Dyck JR, Newgard CB, Lopaschuk GD, Muoio DM. Mitochondrial overload and incomplete fatty acid oxidation contribute to skeletal muscle insulin resistance. *Cell Metab* 2008;**7**:45–56.
27. Keung W, Ussher JR, Jaswal JS, Raubenheimer M, Lam VH, Wagg CS, Lopaschuk GD. Inhibition of carnitine palmitoyltransferase-1 activity alleviates insulin resistance in diet-induced obese mice. *Diabetes* 2013;**62**:711–20.
28. Hubinger A, Weikert G, Wolf HP, Gries FA. The effect of etomoxir on insulin sensitivity in type 2 diabetic patients. *Horm Metab Res* 1992;**24**:115–8.
29. Anderson EJ, Lustig ME, Boyle KE, Woodlief TL, Kane DA, Lin CT, Price JW, 3rd LK, Rabinovitch PS, Szeto HH, Houmard JA, Cortright RN, Wasserman DH, Neufer PD. Mitochondrial $H_2O_2$ emission and cellular redox state link excess fat intake to insulin resistance in both rodents and humans. *J Clin Invest* 2009;**119**:573–81.
30. Rodrigues JV, Gomes CM. Mechanism of superoxide and hydrogen peroxide generation by human electron-transfer flavoprotein and pathological variants. *Free Radic Biol Med* 2012;**53**:12–9.
31. Seifert EL, Estey C, Xuan JY, Harper ME. Electron transport chain-dependent and -independent mechanisms of mitochondrial $H_2O_2$ emission during long-chain fatty acid oxidation. *J Biol Chem* 2010;**285**:5748–58.
32. Hey-Mogensen M, Hojlund K, Vind BF, Wang L, Dela F, Beck-Nielsen H, Fernstrom M, Sahlin K. Effect of physical training on mitochondrial respiration and reactive oxygen species release in skeletal muscle in patients with obesity and type 2 diabetes. *Diabetologia* 2010;**53**:1976–85.
33. Sinha S, Perdomo G, Brown NF, O'Doherty RM. Fatty acid-induced insulin resistance in L6 myotubes is prevented by inhibition of activation and nuclear localization of nuclear factor kappa B. *J Biol Chem* 2004;**279**:41294–301.
34. Yuan M, Konstantopoulos N, Lee J, Hansen L, Li ZW, Karin M, Shoelson SE. Reversal of obesity- and diet-induced insulin resistance with salicylates or targeted disruption of Ikkbeta. *Science* 2001;**293**:1673–7.
35. Jain SS, Paglialunga S, Vigna C, Ludzki A, Herbst EA, Lally JS, Schrauwen P, Hoeks J, Tupling AR, Bonen A, Holloway GP. High-fat diet-induced mitochondrial biogenesis is regulated by mitochondrial-derived reactive oxygen species activation of CaMKII. *Diabetes* 2014;**63**:1907–13.
36. Maddux BA, See W, Lawrence Jr. JC, Goldfine AL, Goldfine ID, Evans JL. Protection against oxidative stress-induced insulin resistance in rat L6 muscle cells by mircomolar concentrations of alpha-lipoic acid. *Diabetes* 2001;**50**:404–10.
37. Evans JL, Goldfine ID, Maddux BA, Grodsky GM. Oxidative stress and stress-activated signaling pathways: a unifying hypothesis of type 2 diabetes. *Endocr Rev* 2002;**23**:599–622.
38. Bonnard C, Durand A, Peyrol S, Chanseaume E, Chauvin MA, Morio B, Vidal H, Rieusset J. Mitochondrial dysfunction results from oxidative stress in the skeletal muscle of diet-induced insulin-resistant mice. *J Clin Invest* 2008;**118**:789–800.
39. Lee HY, Lee JS, Alves T, Ladiges W, Rabinovitch PS, Jurczak MJ, Choi CS, Shulman GI, Samuel VT. Mitochondrial targeted catalase protects against high-fat diet-induced muscle insulin resistance by decreasing intramuscular lipid accumulation. *Diabetes* 2017.

40. Paglialunga S, Ludzki A, Root-McCaig J, Holloway GP. In adipose tissue, increased mitochondrial emission of reactive oxygen species is important for short-term high-fat diet-induced insulin resistance in mice. *Diabetologia* 2015;**58**:1071–80.
41. Paolisso G, Giugliano D, Pizza G, Gambardella A, Tesauro P, Varricchio M, D'Onofrio F. Glutathione infusion potentiates glucose-induced insulin secretion in aged patients with impaired glucose tolerance. *Diabetes Care* 1992;**15**:1–7.
42. Paglialunga S, van Bree B, Bosma M, Valdecantos MP, Amengual-Cladera E, Jorgensen JA, van Beurden D, den Hartog GJM, Ouwens DM, Briede JJ, Schrauwen P, Hoeks J. Targeting of mitochondrial reactive oxygen species production does not avert lipid-induced insulin resistance in muscle tissue from mice. *Diabetologia* 2012;**55**:2759–68.
43. Fink RI, Kolterman OG, Griffin J, Olefsky JM. Mechanisms of insulin resistance in aging. *J Clin Invest* 1983;**71**:1523–35.
44. Nair KS. Aging muscle. *Am J Clin Nutr* 2005;**81**:953–63.
45. Capel F, Rimbert V, Lioger D, Diot A, Rousset P, Mirand PP, Boirie Y, Morio B, Mosoni L. Due to reverse electron transfer, mitochondrial $H_2O_2$ release increases with age in human vastus lateralis muscle although oxidative capacity is preserved. *Mech Ageing Dev* 2005;**126**:505–11.
46. Gouspillou G, Sgarioto N, Kapchinsky S, Purves-Smith F, Norris B, Pion CH, Barbat-Artigas S, Lemieux F, Taivassalo T, Morais JA, Aubertin-Leheudre M, Hepple RT. Increased sensitivity to mitochondrial permeability transition and myonuclear translocation of endonuclease G in atrophied muscle of physically active older humans. *FASEB J* 2014;**28**:1621–33.
47. Gram M, Vigelso A, Yokota T, Helge JW, Dela F, Hey-Mogensen M. Skeletal muscle mitochondrial $H_2O_2$ emission increases with immobilization and decreases after aerobic training in young and older men. *J Physiol* 2015;**593**:4011–27.
48. Ritov VB, Menshikova EV, He J, Ferrell RE, Goodpaster BH, Kelley DE. Deficiency of subsarcolemmal mitochondria in obesity and type 2 diabetes. *Diabetes* 2005;**54**:8–14.
49. Boushel R, Gnaiger E, Schjerling P, Skovbro M, Kraunsoe R, Dela F. Patients with type 2 diabetes have normal mitochondrial function in skeletal muscle. *Diabetologia* 2007;**50**:790–6.
50. Mogensen M, Sahlin K, Fernstrom M, Glintborg D, Vind BF, Beck-Nielsen H, Hojlund K. Mitochondrial respiration is decreased in skeletal muscle of patients with type 2 diabetes. *Diabetes* 2007;**56**:1592–9.
51. Holloway GP, Thrush AB, Heigenhauser GJ, Tandon NN, Dyck DJ, Bonen A, Spriet LL. Skeletal muscle mitochondrial FAT/CD36 content and palmitate oxidation are not decreased in obese women. *Am J Physiol Endocrinol Metab* 2007;**292**:E1782–9.
52. Kelley DE, He J, Menshikova EV, Ritov VB. Dysfunction of mitochondria in human skeletal muscle in type 2 diabetes. *Diabetes* 2002;**51**:2944–50.
53. Kim JY, Hickner RC, Cortright RL, Dohm GL, Houmard JA. Lipid oxidation is reduced in obese human skeletal muscle. *Am J Physiol Endocrinol Metab* 2000;**279**:E1039–44.
54. Morino K, Petersen KF, Dufour S, Befroy D, Frattini J, Shatzkes N, Neschen S, White MF, Bilz S, Sono S, Pypaert M, Shulman GI. Reduced mitochondrial density and increased IRS-1 serine phosphorylation in muscle of insulin-resistant offspring of type 2 diabetic parents. *J Clin Invest* 2005;**115**:3587–93.
55. Pongratz RL, Kibbey RG, Kirkpatrick CL, Zhao X, Pontoglio M, Yaniv M, Wollheim CB, Shulman GI, Cline GW. Mitochondrial dysfunction contributes to impaired insulin secretion in INS-1 cells with dominant-negative mutations of HNF-1alpha and in HNF-1alpha-deficient islets. *J Biol Chem* 2009;**284**:16808–21.
56. Patti ME, Butte AJ, Crunkhorn S, Cusi K, Berria R, Kashyap S, Miyazaki Y, Kohane I, Costello M, Saccone R, Landaker EJ, Goldfine AB, Mun E, DeFronzo R, Finlayson J, Kahn CR, Mandarino LJ. Coordinated reduction of genes of oxidative metabolism in humans with insulin resistance and diabetes: potential role of PGC1 and NRF1. *Proc Natl Acad Sci U S A* 2003;**100**:8466–71.
57. Mootha VK, Lindgren CM, Eriksson KF, Subramanian A, Sihag S, Lehar J, Puigserver P, Carlsson E, Ridderstrale M, Laurila E, Houstis N, Daly MJ, Patterson N, Mesirov JP, Golub TR, Tamayo P, Spiegelman B, Lander ES, Hirschhorn JN, Altshuler D, Groop LC. PGC-1alpha-responsive genes involved in oxidative phosphorylation are coordinately downregulated in human diabetes. *Nat Genet* 2003;**34**:267–73.
58. Phielix E, Jelenik T, Nowotny P, Szendroedi J, Roden M. Reduction of non-esterified fatty acids improves insulin sensitivity and lowers oxidative stress, but fails to restore oxidative capacity in type 2 diabetes: a randomised clinical trial. *Diabetologia* 2014;**57**:572–81.
59. Rabol R, Larsen S, Hojberg PM, Almdal T, Boushel R, Haugaard SB, Andersen JL, Madsbad S, Dela F. Regional anatomic differences in skeletal muscle mitochondrial respiration in type 2 diabetes and obesity. *J Clin Endocrinol Metab* 2010;**95**:857–63.

60. Fink BD, Bai F, Yu L, Sivitz WI. Impaired utilization of membrane potential by complex II-energized mitochondria of obese, diabetic mice assessed using ADP recycling methodology. *Am J Phys Regul Integr Comp Phys* 2016;**311**:R756–63.
61. Szendroedi J, Schmid AI, Meyerspeer M, Cervin C, Kacerovsky M, Smekal G, Graser-Lang S, Groop L, Roden M. Impaired mitochondrial function and insulin resistance of skeletal muscle in mitochondrial diabetes. *Diabetes Care* 2009;**32**:677–9.
62. Michael LF, Wu Z, Cheatham RB, Puigserver P, Adelmant G, Lehman JJ, Kelly DP, Spiegelman BM. Restoration of insulin-sensitive glucose transporter (GLUT4) gene expression in muscle cells by the transcriptional coactivator PGC-1. *Proc Natl Acad Sci U S A* 2001;**98**:3820–5.
63. Koves TR, Li P, An J, Akimoto T, Slentz D, Ilkayeva O, Dohm GL, Yan Z, Newgard CB, Muoio DM. Peroxisome proliferator-activated receptor-gamma co-activator 1alpha-mediated metabolic remodeling of skeletal myocytes mimics exercise training and reverses lipid-induced mitochondrial inefficiency. *J Biol Chem* 2005;**280**:33588–98.
64. Benton CR, Holloway GP, Han XX, Yoshida Y, Snook LA, Lally J, Glatz JF, Luiken JJ, Chabowski A, Bonen A. Increased levels of peroxisome proliferator-activated receptor gamma, coactivator 1 alpha (PGC-1alpha) improve lipid utilisation, insulin signalling and glucose transport in skeletal muscle of lean and insulin-resistant obese Zucker rats. *Diabetologia* 2010;**53**:2008–19.
65. Kim JK, Gimeno RE, Higashimori T, Kim HJ, Choi H, Punreddy S, Mozell RL, Tan G, Stricker-Krongrad A, Hirsch DJ, Fillmore JJ, Liu ZX, Dong J, Cline G, Stahl A, Lodish HF, Shulman GI. Inactivation of fatty acid transport protein 1 prevents fat-induced insulin resistance in skeletal muscle. *J Clin Invest* 2004;**113**:756–63.
66. Lagouge M, Argmann C, Gerhart-Hines Z, Meziane H, Lerin C, Daussin F, Messadeq N, Milne J, Lambert P, Elliott P, Geny B, Laakso M, Puigserver P, Auwerx J. Resveratrol improves mitochondrial function and protects against metabolic disease by activating SIRT1 and PGC-1alpha. *Cell* 2006;**127**:1109–22.
67. Ludzki A, Paglialunga S, Smith BK, Herbst EA, Allison MK, Heigenhauser GJ, Neufer PD, Holloway GP. Rapid repression of ADP transport by palmitoyl-CoA is attenuated by exercise training in humans: a potential mechanism to decrease oxidative stress and improve skeletal muscle insulin signaling. *Diabetes* 2015;**64**:2769–79.
68. Coyle EF, Martin 3rd WH, Bloomfield SA, Lowry OH, Holloszy JO. Effects of detraining on responses to submaximal exercise. *J Appl Physiol* 1985;**59**(1985):853–9.
69. Zorzano A, Liesa M, Palacin M. Role of mitochondrial dynamics proteins in the pathophysiology of obesity and type 2 diabetes. *Int J Biochem Cell Biol* 2009;**41**:1846–54.
70. Jheng HF, Tsai PJ, Guo SM, Kuo LH, Chang CS, Su IJ, Chang CR, Tsai YS. Mitochondrial fission contributes to mitochondrial dysfunction and insulin resistance in skeletal muscle. *Mol Cell Biol* 2012;**32**:309–19.
71. Holloszy JO, Coyle EF. Adaptations of skeletal muscle to endurance exercise and their metabolic consequences. *J Appl Physiol Respir Environ Exerc Physiol* 1984;**56**:831–8.
72. Dudley GA, Tullson PC, Terjung RL. Influence of mitochondrial content on the sensitivity of respiratory control. *J Biol Chem* 1987;**262**:9109–14.
73. Phillips SM, Green HJ, Tarnopolsky MA, Heigenhauser GJ, Grant SM. Progressive effect of endurance training on metabolic adaptations in working skeletal muscle. *Am J Phys* 1996;**270**:E265–72.
74. Green HJ, Jones S, Ball-Burnett M, Farrance B, Ranney D. Adaptations in muscle metabolism to prolonged voluntary exercise and training. *J Appl Physiol* 1985;**78**(1995):138–45.
75. Zoll J, Sanchez H, N'Guessan B, Ribera F, Lampert E, Bigard X, Serrurier B, Fortin D, Geny B, Veksler V, Ventura-Clapier R, Mettauer B. Physical activity changes the regulation of mitochondrial respiration in human skeletal muscle. *J Physiol* 2002;**543**:191–200.
76. Chesley A, Heigenhauser GJ, Spriet LL. Regulation of muscle glycogen phosphorylase activity following short-term endurance training. *Am J Phys* 1996;**270**:E328–35.
77. Phillips SM, Green HJ, Tarnopolsky MA, Heigenhauser GF, Hill RE, Grant SM. Effects of training duration on substrate turnover and oxidation during exercise. *J Appl Physiol (1985)* 1996;**81**:2182–91.
78. Holloway GP, Lally J, Nickerson JG, Alkhateeb H, Snook LA, Heigenhauser GJ, Calles-Escandon J, Glatz JF, Luiken JJ, Spriet LL, Bonen A. Fatty acid binding protein facilitates sarcolemmal fatty acid transport but not mitochondrial oxidation in rat and human skeletal muscle. *J Physiol* 2007;**582**:393–405.
79. Dyck DJ, Bonen A. Muscle contraction increases palmitate esterification and oxidation and triacylglycerol oxidation. *Am J Phys* 1998;**275**:E888–96.

79a. Miotto PM, LeBlanc PJ, Holloway GP. High-fat diet causes mitochondrial dysfunction as a result of impaired ADP sensitivity. *Diabetes* 2018;**67**(11):2199–205.
80. Anderson EJ, Neufer PD. Type II skeletal myofibers possess unique properties that potentiate mitochondrial H(2)O(2) generation. *Am J Phys Cell Physiol* 2006;**290**:C844–51.
81. Anderson EJ, Yamazaki H, Neufer PD. Induction of endogenous uncoupling protein 3 suppresses mitochondrial oxidant emission during fatty acid-supported respiration. *J Biol Chem* 2007;**282**:31257–66.
82. Picard M, Ritchie D, Wright KJ, Romestaing C, Thomas MM, Rowan SL, Taivassalo T, Hepple RT. Mitochondrial functional impairment with aging is exaggerated in isolated mitochondria compared to permeabilized myofibers. *Aging Cell* 2010;**9**:1032–46.
83. McGarry JD, Brown NF. The mitochondrial carnitine palmitoyltransferase system. From concept to molecular analysis. *Eur J Biochem* 1997;**244**:1–14.
84. Bandyopadhyay GK, Yu JG, Ofrecio J, Olefsky JM. Increased malonyl-CoA levels in muscle from obese and type 2 diabetic subjects lead to decreased fatty acid oxidation and increased lipogenesis; thiazolidinedione treatment reverses these defects. *Diabetes* 2006;**55**:2277–85.
85. Ellis BA, Poynten A, Lowy AJ, Furler SM, Chisholm DJ, Kraegen EW, Cooney GJ. Long-chain acyl-CoA esters as indicators of lipid metabolism and insulin sensitivity in rat and human muscle. *Am J Physiol Endocrinol Metab* 2000;**279**:E554–60.
86. Kim JK, Kim HJ, Park SY, Cederberg A, Westergren R, Nilsson D, Higashimori T, Cho YR, Liu ZX, Dong J, Cline GW, Enerback S, Shulman GI. Adipocyte-specific overexpression of FOXC2 prevents diet-induced increases in intramuscular fatty acyl CoA and insulin resistance. *Diabetes* 2005;**54**:1657–63.
87. Smith BK, Perry CG, Koves TR, Wright DC, Smith JC, Neufer PD, Muoio DM, Holloway GP. Identification of a novel malonyl-CoA IC(50) for CPT-I: implications for predicting in vivo fatty acid oxidation rates. *Biochem J* 2012;**448**:13–20.
88. Crunkhorn S, Dearie F, Mantzoros C, Gami H, da Silva WS, Espinoza D, Faucette R, Barry K, Bianco AC, Patti ME. Peroxisome proliferator activator receptor gamma coactivator-1 expression is reduced in obesity: potential pathogenic role of saturated fatty acids and p38 mitogen-activated protein kinase activation. *J Biol Chem* 2007;**282**:15439–50.
89. Hommelberg PP, Plat J, Sparks LM, Schols AM, van Essen AL, Kelders MC, van Beurden D, Mensink RP, Langen RC. Palmitate-induced skeletal muscle insulin resistance does not require NF-kappaB activation. *Cell Mol Life Sci* 2011;**68**:1215–25.
90. Aliev M, Guzun R, Karu-Varikmaa M, Kaambre T, Wallimann T, Saks V. Molecular system bioenergics of the heart: experimental studies of metabolic compartmentation and energy fluxes versus computer modeling. *Int J Mol Sci* 2011;**12**:9296–331.
91. Steeghs K, Oerlemans F, de Haan A, Heerschap A, Verdoodt L, de Bie M, Ruitenbeek W, Benders A, Jost C, van Deursen J, Tullson P, Terjung R, Jap P, Jacob W, Pette D, Wieringa B. Cytoarchitectural and metabolic adaptations in muscles with mitochondrial and cytosolic creatine kinase deficiencies. *Mol Cell Biochem* 1998;**184**:183–94.
92. Mielke C, Lefort N, McLean CG, Cordova JM, Langlais PR, Bordner AJ, Te JA, Ozkan SB, Willis WT, Mandarino LJ. Adenine nucleotide translocase is acetylated in vivo in human muscle: modeling predicts a decreased ADP affinity and altered control of oxidative phosphorylation. *Biochemistry* 2014;**53**:3817–29.
93. Feng J, Zhu M, Schaub MC, Gehrig P, Roschitzki B, Lucchinetti E, Zaugg M. Phosphoproteome analysis of isoflurane-protected heart mitochondria: phosphorylation of adenine nucleotide translocator-1 on Tyr194 regulates mitochondrial function. *Cardiovasc Res* 2008;**80**:20–9.
94. Yan LJ, Sohal RS. Mitochondrial adenine nucleotide translocase is modified oxidatively during aging. *Proc Natl Acad Sci U S A* 1998;**95**:12896–901.
95. Ydfors M, Hughes MC, Laham R, Schlattner U, Norrbom J, Perry CG. Modelling in vivo creatine/phosphocreatine in vitro reveals divergent adaptations in human muscle mitochondrial respiratory control by ADP after acute and chronic exercise. *J Physiol* 2016;**594**:3127–40.
96. Ho CH, Pande SV. On the specificity of the inhibition of adenine nucleotide translocase by long chain acyl-coenzyme A esters. *Biochim Biophys Acta* 1974;**369**:86–94.
97. Morel F, Lauquin G, Lunardi J, Duszynski J, Vignais PV. An appraisal of the functional significance of the inhibitory effect of long chain acyl-CoAs on mitochondrial transports. *FEBS Lett* 1974;**39**:133–8.
98. Shrago E, Shug A, Elson C, Spennetta T, Crosby C. Regulation of metabolic transport in rat and Guinea pig liver mitochondria by long chain fatty acyl coenzyme A esters. *J Biol Chem* 1974;**249**:5269–74.

99. Miotto PM, Holloway GP. In the absence of phosphate shuttling, exercise reveals the in vivo importance of creatine-independent mitochondrial ADP transport. *Biochem J* 2016;**473**:2831–43.

99a. Yan LJ, Sohal RS. Mitochondrial adenine nucleotide translocase is modified oxidatively during aging. *Proc Natl Acad Sci USA* 1998;**95**(22):12896–901.

100. Lund S, Holman GD, Schmitz O, Pedersen O. Contraction stimulates translocation of glucose transporter GLUT4 in skeletal muscle through a mechanism distinct from that of insulin. *Proc Natl Acad Sci U S A* 1995;**92**:5817–21.

101. Martin IK, Katz A, Wahren J. Splanchnic and muscle metabolism during exercise in NIDDM patients. *Am J Phys* 1995;**269**:E583–90.

102. Holloszy JO. Biochemical adaptations in muscle. Effects of exercise on mitochondrial oxygen uptake and respiratory enzyme activity in skeletal muscle. *J Biol Chem* 1967;**242**:2278–82.

103. Wright DC, Han DH, Garcia-Roves PM, Geiger PC, Jones TE, Holloszy JO. Exercise-induced mitochondrial biogenesis begins before the increase in muscle PGC-1alpha expression. *J Biol Chem* 2007;**282**:194–9.

104. Safdar A, Little JP, Stokl AJ, Hettinga BP, Akhtar M, Tarnopolsky MA. Exercise increases mitochondrial PGC-1alpha content and promotes nuclear-mitochondrial cross-talk to coordinate mitochondrial biogenesis. *J Biol Chem* 2011;**286**:10605–17.

105. Perry CG, Lally J, Holloway GP, Heigenhauser GJ, Bonen A, Spriet LL. Repeated transient mRNA bursts precede increases in transcriptional and mitochondrial proteins during training in human skeletal muscle. *J Physiol* 2010;**588**:4795–810.

106. Little JP, Safdar A, Bishop D, Tarnopolsky MA, Gibala MJ. An acute bout of high-intensity interval training increases the nuclear abundance of PGC-1alpha and activates mitochondrial biogenesis in human skeletal muscle. *Am J Phys Regul Integr Comp Phys* 2011;**300**:R1303–10.

107. Bonen A, Tan MH, Watson-Wright WM. Effects of exercise on insulin binding and glucose metabolism in muscle. *Can J Physiol Pharmacol* 1984;**62**:1500–4.

108. Heath GW, Gavin 3rd JR, Hinderliter JM, Hagberg JM, Bloomfield SA, Holloszy JO. Effects of exercise and lack of exercise on glucose tolerance and insulin sensitivity. *J Appl Physiol Respir Environ Exerc Physiol* 1983;**55**:512–7.

109. Reitman JS, Vasquez B, Klimes I, Nagulesparan M. Improvement of glucose homeostasis after exercise training in non-insulin-dependent diabetes. *Diabetes Care* 1984;**7**:434–41.

110. Russell AP, Feilchenfeldt J, Schreiber S, Praz M, Crettenand A, Gobelet C, Meier CA, Bell DR, Kralli A, Giacobino JP, Deriaz O. Endurance training in humans leads to fiber type-specific increases in levels of peroxisome proliferator-activated receptor-gamma coactivator-1 and peroxisome proliferator-activated receptor-alpha in skeletal muscle. *Diabetes* 2003;**52**:2874–81.

111. Shepherd SO, Cocks M, Meikle PJ, Mellett NA, Ranasinghe AM, Barker TA, Wagenmakers AJM, Shaw CS. Lipid droplet remodelling and reduced muscle ceramides following sprint interval and moderate-intensity continuous exercise training in obese males. *Int J Obes* 2017;**41**:1745–54.

112. Dube JJ, Amati F, Stefanovic-Racic M, Toledo FG, Sauers SE, Goodpaster BH. Exercise-induced alterations in intramyocellular lipids and insulin resistance: the athlete's paradox revisited. *Am J Physiol Endocrinol Metab* 2008;**294**:E882–8.

113. Devries MC, Lowther SA, Glover AW, Hamadeh MJ, Tarnopolsky MA. IMCL area density, but not IMCL utilization, is higher in women during moderate-intensity endurance exercise, compared with men. *Am J Phys Regul Integr Comp Phys* 2007;**293**:R2336–42.

114. Monaco CMF, Proudfoot R, Miotto PM, Herbst EAF, MacPherson REK, Holloway GP. Alpha-linolenic acid supplementation prevents exercise-induced improvements in white adipose tissue mitochondrial bioenergetics and whole-body glucose homeostasis in obese Zucker rats. *Diabetologia* 2018;**61**:433–44.

CHAPTER

# 7

# Role of Mitochondria in Adipose Tissues Metabolism and Plasticity

*Audrey Carrière, Louis Casteilla*

**STROMALab, Université de Toulouse, EFS, ENVT, Inserm U1031, ERL CNRS 5311, CHU Rangueil, Toulouse, France**

## 1 INTRODUCTION

In mammals, one classically distinguishes two main classes of adipose tissues—white adipose tissues (WAT) and brown adipose tissues (BAT)—that together control whole body energetic homeostasis, through metabolic and endocrine activities. While WAT represents the main energy store of the organism through the capacity of white adipocytes to store or release lipids, BAT is specialized in energy dissipation and heat production thanks to the expression of the mitochondrial uncoupling protein-1 (UCP1) in brown adipocytes.[1] The recent rediscovery of BAT in human adults relaunched investigations on its putative benefic role for whole body metabolic homeostasis.[2–5] One key feature of adipose tissues is their extraordinary plasticity and their capacity to adapt their size, phenotype, and function to changing environmental, metabolic and stresses conditions. This tissue plasticity is sustained by great plasticity of adipose cells including adipose progenitors (present in the stroma vascular fraction together with immune and endothelial cells) and mature adipocytes. Cellular processes including proliferation, differentiation, or transdifferentiation events support such adipose tissue plasticity.[6, 7] For example, apparent hyperplasia (increase adipocyte number) and hypertrophia (increase size of existing adipocytes) together enable WAT to store the excess of energy and to prevent lipotoxicity and related damages to other tissues and organs, until WAT reaches its limits of expandability.[8] In addition, the conversion of some WAT deposits into brown-like or beige adipose tissues that display similar functions to classical BAT participate to energy expenditure and ameliorate metabolic profile in mice.[2, 3] Therefore, ameliorating WAT expandability and its functional metabolic flexibility and/or activating BAT and WAT browning to take in charge nutrient overload constitute therapeutic strategies to counteract obesity and its associated lipotoxic effects and metabolic alterations. The adequate responses of both WAT and BAT to


https://doi.org/10.1016/B978-0-12-811752-1.00007-9

nutrient overload are closely dependent on mitochondria functioning. Mitochondria number, morphology, activity, cellular distribution, and turnover have been shown to be determinant for cell metabolism, physiology, and fate in different tissues and organs, including adipose tissues.[9–11] Mitochondria are highly dynamic and can be restructured to meet the metabolic demands. Therefore, mitochondrial responses to stress and metabolic signals determine the proper adaptation of adipose cells to changing metabolic conditions. Because both WAT and BAT act in concert to ensure optimal metabolic status of the organism, any perturbation of mitochondria functioning will be associated with adipose tissue dysfunctions that directly contribute to obesity and associated metabolic diseases, including type 2 diabetes.[9–11] Therefore, understanding the intimate functions and dysfunctions of mitochondria within adipose cells and tissues is crucial. After a brief description of the different adipose tissues and their specific role in metabolic homeostasis, we will describe how mitochondria support their physiology.

# 2 WHITE, BROWN AND BEIGE ADIPOSE TISSUES

## 2.1 White Adipose Tissue

WAT is considered to be the main energy store of the organism and a high-endocrine organ, secreting numerous adipokines including leptin and adiponectin,[12] and therefore affects the function of several tissues in the body. Such activities are supported mainly by mature white adipocytes that exhibit a large unilocular lipid droplet, nucleus, and mitochondria being pushed near the plasma membrane.[3] WAT is distributed into different fat pads that can be classified either as subcutaneous or visceral depots and that display specific biological functions. Subcutaneous WAT generally is viewed as a protective tissue, whereas visceral WAT is associated with an adverse metabolic profile, probably because it drains free fatty acids directly into the liver through the portal vein and because higher release of adipokines such as the interleukin-8 or the plasminogen activating factor.[13, 14] Whether the two depots have different mitochondrial content is not clear.

Although the subcutaneous inguinal fat pad of rats has been shown to contain less mitochondria than the visceral one,[15] decreased mitochondrial content associated with high oxidative stress has been shown in visceral fat pad of C57Bl6 mice and could account for mechanisms mediating the elevated metabolic risk associated with this fat depot.[16] WAT ensures healthy storage of excess nutrients during positive energetic balance and rapid energy mobilization through fatty acids release to supply other organs in case of metabolic needs. WAT has a great ability to store surplus energy in the form of lipids and to expand because of apparent adipocyte hyperplasia, which corresponds to an increase in the number of adipocytes because of the proliferation and differentiation of adipose progenitors, and hypertrophia that corresponds to enlargement of existing adipocytes (by de novo lipogenesis or fatty acids uptake from the circulation).[17] In the case of sustained energy overload and obesity, however, WAT undergoes fibro-inflammation and becomes dysfunctional, reaches its expandability limits, leading to insulin resistance, ectopic lipids accumulation, and dysfunction of organs, including liver and muscle.[18] This WAT alteration contributes to increase risks of cardiovascular diseases and type 2diabetes.[8] Mitochondrial dysfunctions within adipose cells might be involved in such adipose tissue dysfunction, as developed in paragraphs 4.2.

## 2.2 Brown Adipose Tissue

In contrast to WAT that acts as an energy reservoir, BAT burns it. BAT is specialized in heat production through nonshivering thermogenesis, and its main role is to prevent hypothermia in small mammals, including human neonates.[19–21] Heat production occurs through cold-induced activation of the UCP1 protein, present in the inner membrane of the numerous mitochondria of brown adipocytes. In addition to the enrichment in mitochondria, brown adipocytes exhibit a multilocular phenotype that easily distinguishes them from white adipocytes.[3] UCP1 dissipates the $H^+$ gradient across the inner mitochondrial membrane, resulting in the uncoupling of the electron transport chain from the ATP synthase activity (Fig. 1). This accelerates respiration and converts the energy of substrates oxidation into heat, instead of ATP.[22, 23] The sympathetic nervous system, activated during cold exposure, is the main activator of BAT.[24] In addition to the transcriptional upregulation of the UCP1 gene,[25] noradrenaline liberated by the sympathetic nerves surrounding brown adipocytes induces lipolysis through activation of the adipose tissue triglyceride lipase (ATGL), through a β3-adrenoreceptor/cAMP/Protein Kinase A pathway inducing phosphorylation of hormone-sensitive lipase (LHS) and perilipin. Free long-chain fatty acids then will release UCP1 from the inhibitory action of purine nucleotides (ADP, ATP, or GDP) and activate its $H^+$ carrier activity.[26] The fact that intracellular fatty acids generated within brown adipocytes are activators of UCP1 has been challenged recently by two independent studies.[27, 28] These works showed that invalidation of ATGL or ATGL-activating protein CGI58 within brown adipocytes did not compromise the β3-adrenergic thermogenic response and cold-induced nonshivering thermogenesis. These findings suggest that alternative models of UCP1 activation might exist and that circulating energy substrates in particular fatty acids generated by WAT lipolysis are sufficient to fuel BAT and to activate UCP1.[27, 28] The liver has been shown to release plasma acylcarnitines as a fuel source for BAT.[29] These findings, however, also open the exciting possibility that other activators of the UCP1 activity than fatty acids might exist,[30] as demonstrated for retinoic acid.[31] In addition to noradrenaline, other molecules control brown adipocytes activity, including thyroid hormone, fibroblast growth factor-21, and an increasing number of others reviewed in.[4] A recent discovery in the field of adipose tissues is the demonstration that brown adipocytes derive from a common progenitor with muscle cells, expressing the myogenic transcription factor myf5.[32] This proximity of lineages might mirror the important thermogenic functions assumed by both tissues. From a thermodynamic point of view, the futile cycle assumed by muscle contractions (synthesized ATP is degraded immediately by the contraction mechanism) might be as effective as uncoupling to produce heat. One could postulate that during evolution, cells with muscular potential gradually have lost their contractile equipment in favor of uncoupling mechanisms, the most specialized of which is the one based on UCP1. This cellular link between skeletal muscle and BAT also is underlined by the presence of brown adipocyte precursors in human muscle.[33]

In addition to maintaining body core temperature, an additional role of BAT might be to counterbalance the energy overload through nutrients dissipation by the diet-induced thermogenesis process. Although this role of BAT is difficult to reconcile with evolutionary biology (premodern humans spent a high amount of energy to survive and obesity was essentially nonexistent[34]), several arguments provided by experiences in mice and recent data from humans support a role of BAT in diet-induced thermogenesis.[35, 36] UCP1 knockout mice

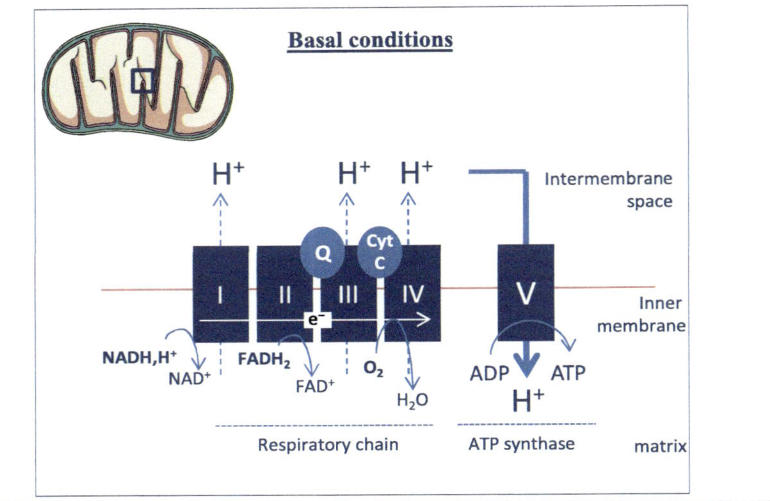


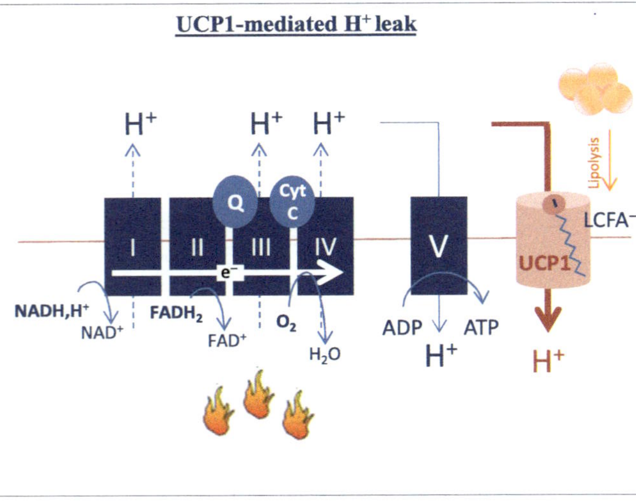


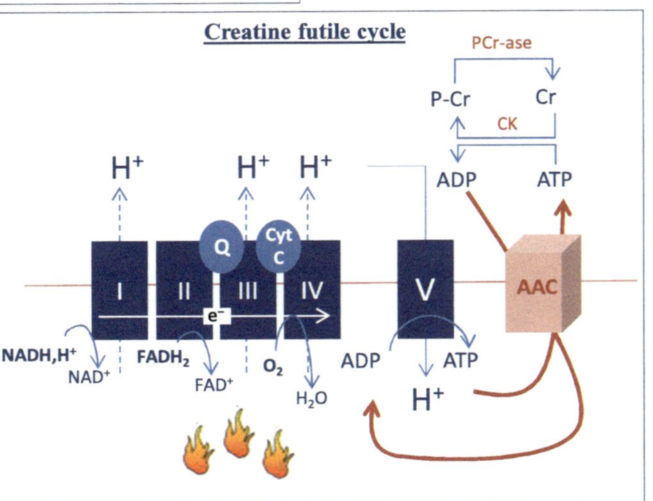


FIG. 1 Mitochondria support thermogenesis. Two mechanisms of thermogenesis are described: proton leak mediated by UCP1 whose activation depends on long chain fatty acids and the creatine phosphorylation-dephosphorylation futile cycle. Both processes increase electron flux into the respiratory chain and upstream catabolic reactions, leading to heat production. *LCFA*, long chain fatty acids; *UCP1*, uncoupling protein-1; *AAC*, ATP/ADP carrier; *PCr-ase*, phosphocreatine phosphatase; *CK*, creatine kinase.

rising to thermoneutrality develop obesity[37]; BAT transplantation ameliorates the metabolic profile of mice submitted to a high-fat diet.[38] In addition to its high oxidative properties and high energetic substrate consumption, including glucose and fatty acids,[39] BAT might have beneficial effects through the release of molecules named batokines, which can have a positive impact on the metabolic status of the organism.[40] Although BAT has been thought for many years to be present in humans only in the neonates or in rare exceptions such as workers exposed to cold temperatures or patients suffering from pheochromocytoma,[41, 42] 18-fluorodeoxyglucose-PET-SCAN imaging associated with immunohistochemistry experiments targeting UCP1 showed that BAT is present in human adults in paravertebral, supraclavicular, and perirenal regions.[43–46] The lower content/activity of BAT in obese patients in association with hyperglycemic profile, aging, and visceral fat increase suggest that defective BAT might be involved, contributes, or exacerbates the development of obesity and metabolic complications.[2–5] Obesity-associated BAT impairment has been observed in both mice and humans[47, 48] and is characterized by impaired β-adrenergic signaling, reduced UCP1 protein, and enlargement and whitening of BAT, which might be caused by the rarefaction of the vasculature.[49] These discoveries renewed the old idea that activating BAT might be a strategy to fight against metabolic diseases associated with obesity. Although improvement in glucose metabolism has been demonstrated in type 2 diabetes subjects exposed to cold,[50] no compelling evidence so far exists for any benefic role of BAT on weight loss in humans. More research is needed to definitively demonstrate its relevance as a target to treat metabolic diseases. In any case, the balance of benefits/risks (cardiovascular side effects are induced with treatments such as β3-adrenoreceptor agonists, as shown by the clinical trial using mirabegron[51]) has to be evaluated carefully.

## 2.3 Beige Adipose Tissue

The discovery of BAT in human adults relaunched investigations about the browning of WAT, first discovered in the 1990s.[52–54] WAT browning corresponds to the transformation of certain white fat pads into brown-like adipose tissues during cold exposure or β3-adrenergic agonist treatments. In mice, the subcutaneous fat pad, which displays a remarkable structural heterogeneity, is one of the tissues most responsive to browning, and UCP1-expressing adipocytes arise in specific regions of the tissue.[55] These cells have been named inducible brown adipocytes, brite adipocytes, or beige adipocytes. Although their transcriptional signature differs from the one of classical brown adipocytes,[56] their morphology (multilocular) and mitochondria content are similar to brown adipocytes, and they are functionally thermogenic.[57] Recruitment of beige progenitors[7, 58, 59] and/or transformation of some mature adipocytes that exhibit a white phenotype and that do not express the UCP1 protein at room temperature or thermoneutrality but that will turn mitochondrial-enriched and UCP1 positive after cold exposure[60, 61] represent the main cellular processes at the origin of their appearance. Whether the conversion of UCP1-negative into UCP1-positive adipocytes corresponds to a true transdifferentiation event or activation of dormant or hypoactive beige adipocytes that have the capacity to quickly initiate the thermogenic program in response to cold still are debated. This conversion is reversible.[61–63] Using a single-cell monitoring system enabling the tracking of morphological changes of individual adipocytes ex vivo, it was observed that, contrary to brown adipocytes isolated from interscapular BAT, beige adipocytes isolated from the inguinal fat rapidly lose their multilocular phenotype and become unilocular, bypassing an intermediate precursor stage.[62] One unresolved point is the developmental origin of beige adipocytes; some studies highlighted a myf-5 independent lineage for beige (and white) adipocytes, while others found pools of white and beige adipocytes arising from myf-5 positive progenitors.[32, 64, 65] The existence of distinct subpopulations of white, beige, and brown adipocytes, exhibiting different developmental origins, distributed differentially among the distinct adipose fat pads, and maybe associated with different functions, might reasonably be postulated. In support of this, metabolic heterogeneity among beige adipocytes has been demonstrated.[66, 67] This heterogeneity is evident for white adipocytes, because not all of them are able to become beige.[55, 61] Although they are thought to play a marginal role in the regulation of whole-body energy metabolism given their under-representation compared to classical brown adipocytes gathered in BAT,[57, 68] several studies demonstrate that they play a significant role. Mice exhibiting specific activation or inactivation of beige adipocytes (with no or very weak effect on brown adipocytes activity) through modulation of Prdm16 activity are resistant or develop obesity and insulin resistance when exposed to high-fat diet, respectively.[69, 70] In human adults, UCP1-positive adipocytes would be both classical brown adipocytes and beige adipocytes (based on expression profile comparison with murine cells), depending on the anatomical situation.[71] In addition to cold, other metabolic stresses, including physical exercise,[72, 73] massive burns,[74, 75] or cancer-associated cachexia[76, 77] induced WAT browning, highlighting the possibility that beige adipocytes assume functions in addition to thermogenesis, among which metabolites and redox pressure dissipation have been proposed.[2, 78]

This short overview illustrates how the last decade has been associated with groundbreaking discoveries about adipose tissues and their plasticity, heterogeneity, and functional roles in whole body homeostasis. The complexity of their biology, however, still needs to be unraveled to better understand their dysfunction.

## 3 ROLE OF MITOCHONDRIA IN METABOLIC ACTIVITIES OF ADIPOCYTES

Before describing how mitochondria control adipose cell physiology and fate in WAT or in BAT/beige adipose tissues, we will briefly explain how mitochondria support the metabolic activities of white and brown/beige adipocytes.

### 3.1 Lipogenesis

Mitochondria play a critical role in de novo lipogenesis,[79] that is, the conversion of excess carbohydrates into fatty acids, a process occurring in adipose tissues and liver (the main site in humans[80]). Mitochondria enable the anaplerotic generation of metabolic intermediates feeding lipogenesis, in particular through citrate production and export into the cytoplasm. After feeding and in a high-energy state, high ATP/ADP and NADH,$H^+$/$NAD^+$ ratio reduces the activity of the isocitrate dehydrogenase that belongs to the tricarboxylic acid cycle, thereby increasing citrate concentration within the mitochondrial matrix, and its diffusion in the cytosol. Cytosolic citrate is split by the ATP citrate lyase into oxaloacetate and acetylCoA, which is the precursor for fatty acid synthesis after its processing through the activity of acetyl-CoA carboxylase (increased by citrate) and fatty acid synthase complex that generates fatty acids. Three fatty acid molecules then are esterified with a molecule of glycerol 3-phosphate to produce triacylglycerol. In the cytosol, citrate inhibits the activity of the glycolytic phosphofructokinase-1, thereby redirecting glucose 6 phosphate into the pentose pathway, which generates the NADPH required for fatty acid synthesis. Any mitochondria defects in ATP generation will have an impact on lipogenesis as demonstrated by the effect of the uncoupler 2,4-dinitrophenol that decreases lipogenesis, probably as a result of inhibition of pyruvate carboxylase activity because of ATP reduction.[81]

## 3.2 Glyceroneogenesis

Because of low glycerol kinase activity—even if its expression is induced by rosiglitazone in murine and human white adipocytes[82, 83]—glyceroneogenesis is the main source of glycerol in white adipocytes.[79, 84] In the fed state, glycerol 3-phosphate is produced from the reduction of dihydroxyacetone phosphate (from glycolysis or the pentose phosphate pathway) by the glycerol-3-phosphate dehydrogenase activity. During fasting, precursors other than glucose can be metabolized to form glycerol-3-phosphate such as pyruvate, lactate, and some amino acids. One of the rate-limiting steps in glyceroneogenesis is the synthesis of phosphoenolpyruvate from oxaloacetate, which is catalyzed by the phosphoenolpyruvate carboxykinase (PEPCK), present both in cytosol and mitochondria. PEPCK protein content is much greater in BAT than in WAT, maybe because the primary function of BAT is thermogenesis, which is fueled by beta-oxidation of fatty acids supplied by the triacylglycerol/fatty acid futile cycle (an ATP-consuming process involved in thermogenesis). In addition, BAT exhibits important glycerol kinase activity that enables direct recycling of glycerol produced by hydrolysis of stored triacylglycerol into glycerol-3-phosphate.[85] In addition to providing ATP, mitochondria support glyceroneogenesis by providing tricarboxylic acid cycle intermediates such as pyruvate and malate that feed cytosolic oxaloacetate by malate transport into cytosol.[79]

## 3.3 Lipolysis and β-Oxidation

During exercise, periods of stress, or fasting, triacylglycerol reserves in adipose tissues are mobilized through activation of lipolysis that releases fatty acids and glycerol in the blood stream to serve as respiratory fuel for other organs, including heart, muscle, and liver. Although during fasting lipolysis is stimulated by catecholamine that activate LHS and ATGL through cyclic AMP-protein kinase A pathway,[86] during exercise, the heart muscle releases atrial natriuretic factor that stimulates hormone sensitive lipase by cyclic GMP increase.[87] In a fed state, however, lipolysis is inhibited by insulin through the PI3K/AKT pathway that contributes to the decrease in cAMP levels through activation of the phosphodiesterase 3B.[88] Lipolysis is critical for BAT because it provides fatty acids that are not only energetic substrates for uncoupled respiration following mitochondrial β-oxidation but that also act as direct activators of the activity of UCP1. After fatty acids conjugation with a CoA group in the cytosol and conversion to long-chain acyl carnitines, these latter are transported into mitochondria and converted back to long chain acyl-CoA which enter the β-oxidation pathway that provides acetyl-CoA, NADH,$H^+$, and FADH2 and two carbons shorter acyl-CoA.

Conversion of human white fat cells into beige adipocytes is associated with high increase of fatty acid esterification into triglycerides together with activation of lipolysis and fatty acid oxidation metabolic pathways, indicating the existence of a futile cycle.[89]

From a biochemistry point of view, mitochondria are essential for each type of adipose tissues, because they support metabolic activities typical to adipocytes, whatever their nature. This is only the tip of the iceberg, however, as detailed in the following paragraphs.

# 4 MITOCHONDRIA AND WHITE ADIPOSE TISSUE PHYSIOLOGY

Although the role of mitochondria in BAT has been studied intensively, their involvement in the development and biology of WAT has been largely neglected. The oxidative potential of mitochondria isolated from white adipocytes is far from being negligible,[90] however, and the following paragraphs illustrate how mitochondria deeply affect white adipose cell biology. Although the role of mitochondria in calcium homeostasis can affect the biology of adipocytes and insulin sensitivity, we will not develop this topic in this review but will refer to excellent reviews.[91, 92]

## 4.1 Mitochondria Participates to the Adipogenic Program

### *4.1.1 Mitochondria Biogenesis and Network Remodeling*

Mitochondria bioenergetics and metabolic profiles determine the control of stem cell fate, that is, quiescence, proliferation, or differentiation,[93] this latter process often accompanied by an increase in mitochondria mass and a metabolic shift toward mitochondrial respiration.[94] This is true during adipogenesis as shown by electron microscopy experiments revealing an increase in the number of mitochondria.[95] Proteomic approaches confirmed an enrichment of mitochondrial proteins in differentiated adipocytes of more than 20-fold compared to adipose progenitors, in the 3T3-L1 murine cell line.[95, 96] Adipogenic differentiation also has been shown to be associated with higher oxygen consumption and increased mitochondrial activity, in human and murine cells.[95, 97, 98] Mitochondrial content is regulated by the balance between mitochondrial biogenesis and degradation. In addition to sirtuin-1 (SIRT1), AMP-activated protein kinase (AMPK), and the mammalian target of rapamycin (mTORC1) that represent key molecular mechanisms controlling mitochondrial biogenesis, turnover, and dynamics,[99] the peroxisome proliferative-activated receptor ψ (PPARψ) and its co-activator (peroxisome proliferator- activated receptor gamma coactivator 1-alpha, PGC1α), as well as nuclear respiratory factor-1 (NRF1) induce mitochondrial biogenesis during adipocyte differentiation.[95, 100] This mitochondria biogenesis might occur to meet the energy need associated with the differentiation process by providing ATP and by feeding de novo lipogenesis through the production of citrate from acetylCoa. Citrate export out of mitochondria is essential during commitment of adipose progenitors into adipogenesis.[101] Although the number of mitochondria increases during adipogenesis, it is reduced during terminal differentiation consistent with the functional phenotype of mature white adipocytes more adapted to lipid storage than to use.[9] This is concomitant with the emergence of the antilipolytic system.[102]

Mitochondria are dynamic organelles, organized in networks, that go throughout repetitive cycles of fusion and fission mechanisms to maintain a proper activity according to metabolic conditions and energetic needs of the cell.[103] Mitochondrial organization within cells is affected by the adipogenic program. Although mitochondria are organized as a continuous reticulum in preadipocytes, they appear fragmented and redistributed around the lipid droplets during adipocyte differentiation.[95] The upregulation of the expression of dynamin-related protein-1 and mitofusin-2 during differentiation of murine preadipocyte cell line is in accordance with mitochondrial remodeling.[104] Targeting mitochondrial fusion and fission mechanisms has an impact on lipid accumulation[105] suggesting that the specific mitochondrial remodeling during differentiation could have a permissive role on cellular triacylglycerol accumulation, maybe to attend the energy needs for lipid droplets formation. Prohibitins, localized in the inner mitochondrial membrane, are also involved in adipogenesis,[106] although the mechanisms involved required further investigations.

### 4.1.2 Mitochondria Regulate Adipogenic Gene Expression Through Redox Mechanisms and as Support of Nuclear Epigenetic Modifications

Mitochondria play central roles in modulating redox-dependent cellular processes through mitochondrial ROS production and the control of the reduced/oxidized form of preponderant redox couples such as NADH,$H^+$/$NAD^+$. We will develop how mitochondria can affect adipose cell fate through these mechanisms (Fig. 2).

According to their concentration, their subcellular site of production as well as their nature, ROS can have distinct effects, ranging from toxic molecules to true second messengers.[107, 108] After the discovery of the role of H2O2 on insulin signaling, the role of ROS produced from different subcellular sources on adipose biology has been investigated intensively, as reviewed.[109] The intimate functioning of the mitochondrial electron transport chain is associated with electron leak and subsequent mono-electronic reduction of oxygen, leading to generation of anion superoxide ($O2^-$), which is subsequently diverted into $H_2O_2$ (hydrogen peroxide). Any electron slowing and accumulation within the respiratory chain associated with high proton-motive force (e.g., high NADH/$NAD^+$ ratio in conditions of low ATP demand or low oxygen availability) will increase mitochondrial ROS production. An increase in their concentration has been shown during the first steps of adipogenic differentiation of human bone marrow-derived mesenchymal progenitors.[98] Treatment of cells with mitochondrial-targeted antioxidants lead to significant reduction in lipid accumulation, suggesting that mitochondrial ROS, physiologically produced during commitment of cells into differentiation, are not simply a consequence of increased mitochondrial metabolism but also causal in the differentiation program.[98, 110, 111] Using pharmacological inhibitors targeting the different complexes of the electron transport chain, however, mitochondrial ROS produced in excess have been shown to inhibit murine preadipocytes proliferation[112] and differentiation through upregulation of CHOP10, a dominant negative factor of the CEBP family.[113] This mitochondrial ROS-CHOP10 signaling cascade also partly mediated hypoxia-dependent inhibition of adipocyte differentiation[113] as well as adiponectin expression.[114] The uncoupling protein-2 (UCP2), that downregulate ROS production[115–117] probably through mechanisms independent on its uncoupling activity,[116] increased adiponectin expression through this mechanism, what might influence systemic insulin sensitivity.[114]

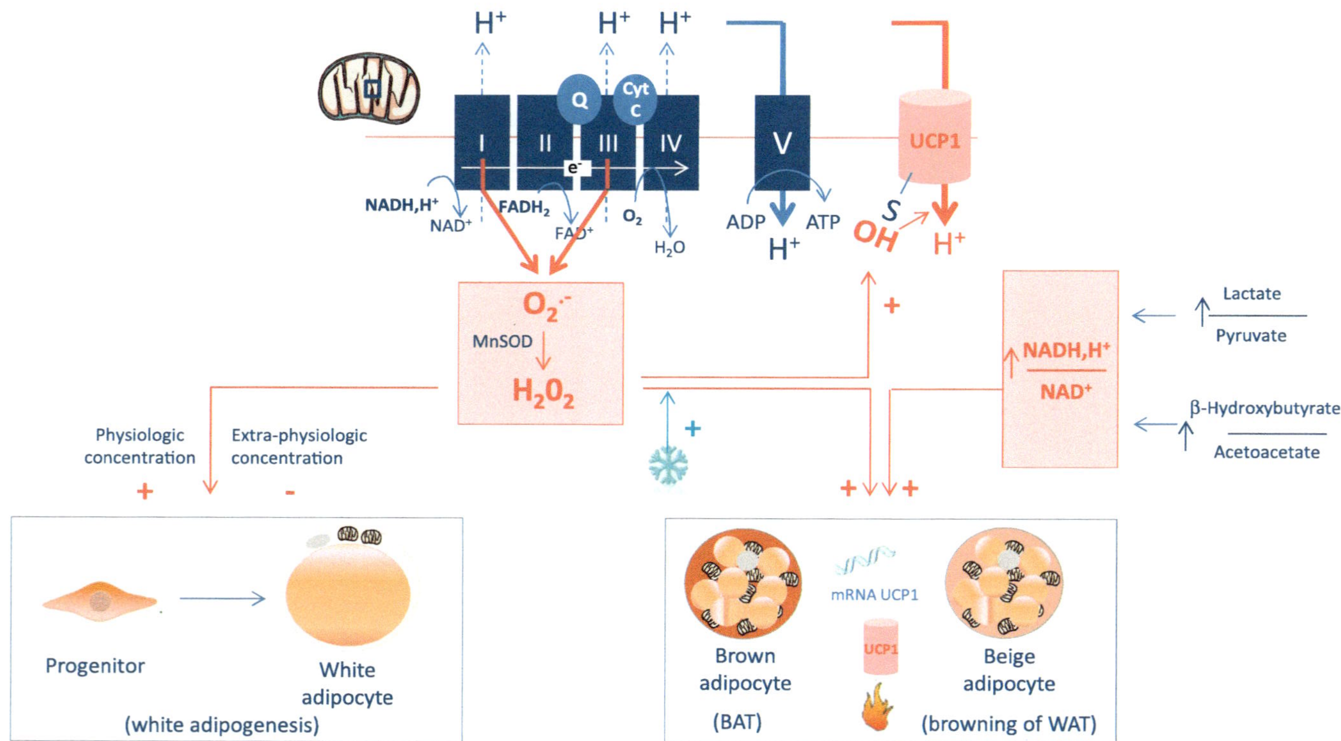


FIG. 2 Mitochondrial ROS and redox state control adipose cells biology. When physiologically produced, mitochondrial ROS promote white adipose progenitors commitment into adipogenesis. At high doses, however, they inhibit proliferation and differentiation. Mitochondrial ROS content increase upon cold exposure and activate the thermogenic program in both brown and beige adipocytes, through increase of UCP1 expression and cysteine residues oxidation (–S–OH). The redox state (NADH,$H^+$/$NAD^+$) is also an intimate regulator of UCP1 expression.

Many mitochondrial enzymes rely on the NADH,$H^+$/$NAD^+$ redox couple which is tightly controlled by mitochondria functioning. Because mitochondrial membrane is impermeable to NADH,$H^+$/$NAD^+$, the malate-aspartate and glycerol-3 phosphate shuttles connect mitochondrial to cytoplasmic NADH,$H^+$/$NAD^+$ pools. This preponderant redox couple links metabolism to gene expression as enzymes such as sirtuins and poly(ADP-ribose) polymerases (Parps) are strictly dependent on $NAD^+$ availability. These proteins have opposite effects on metabolism and their role in adipogenesis, lipid homeostasis, inflammation, adipokine secretion, and plasticity of both WAT and BAT has been well reviewed[118] and won't be further detailed here.

Mitochondria also regulate gene expression by providing intermediate metabolites necessary to generate and modify epigenetic marks in the nucleus.[119] AcetylCoA serves as a substrate for histone acetyltransferases, and α-ketoglutarate enables histone methylation processes. Any perturbation in mitochondria activity and acetylCoA production can influence cell behavior. The role of mitochondria functioning in epigenetic modifications driving adipose cell fate has been shown clearly by the knockdown of ATP citrate lyase, which induced a decrease in cytosolic acetylCoA, caused a reduction in histone acetylation, and prevented the upregulation of genes encoding glucose transporters, therefore impairing adipogenesis.[120]

## 4.2 Mitochondrial Impairments Within WAT During Obesity/Type 2 Diabetes

Several studies reported a decrease in mitochondrial mass, mitochondrial DNA, and mitochondrial oxidative capacities in WAT of distinct models of obese and diabetic mice as well as in humans.[121–128] This is consistent with the fact that PGC1α, whose overexpression converts human white adipocytes into UCP1-expressing adipocytes,[129] is decreased in subcutaneous fat in obese mice and humans.[122, 128, 130] Obesity also has been associated with increased oxidative stress in WAT in mice and humans.[131] A proteomic approach performed on isolated mitochondria from white adipocytes derived from diabetic patients showed, in addition to a decrease content of specific components of the respiratory complexes, oxidative modifications of mitochondrial proteins.[132] As a response to chronic oxidative stress, an adaptive increase of the antioxidant response has been shown within WAT fat pads from nondiabetic obese rats.[133] The deterioration of NADH,$H^+$/$NAD^+$ metabolism in adipose tissues of obese mice and humans also is consistent with mitochondrial impairments and might necessarily affect sirtuins/Parps activities.[118] Data about mitochondrial turnover are less clear as opposite regulation of proteins involved in mitochondrial turnover has been described during obesity.[130, 134]

Although the link between mitochondrial abnormalities in WAT and obesity appears strong, whether these mitochondrial defects are causal in the development of metabolic alterations associated with obesity including insulin resistance or simply a consequence still are debated. Because the invalidation of the mitochondrial transcription factor A (TFAM, a major controller of mitochondrial mass and function) specifically in adipose tissue induced lipoatrophy and whole-body insulin resistance, however, this is in favor of a causative role of mitochondria dysfunction in metabolic alterations.[135, 136] This putative causative role also is highlighted by the lipodystrophy syndrome occurring during treatment of human immunodeficiency virus patients with antiretroviral drugs that generate mitochondrial toxicity in adipose tissue, a process well reviewed in Ref. 137. Nucleoside analogues has been shown to induce mitochondrial dysfunction and increased oxidative stress together with fat inflammation and fat loss, as shown by in vitro and in vivo experiments.[138, 139] Site-specific differential mitochondrial effects of AZT also have been published and proposed to be related to the pharmacological-induced decrease of mitochondrial DNA content.[140]

All these findings together illustrate that mitochondria within WAT are far from being inert organelles and that their number, form, and activities are strictly controlled by dedicated molecular and cellular processes orchestrated during the adipogenic program. Because their activities tightly control gene expression through multiple mechanisms, it is reasonable to claim, therefore, that any perturbation of mitochondria homeostasis can participate in WAT dysfunction and associated metabolic disorders.

# 5 SPECIFIC ROLES OF MITOCHONDRIA IN BROWN/BEIGE ADIPOCYTES

In contrast to the underestimation of their role in WAT for many years, mitochondria function in BAT activity and browning of WAT always has been considered, because they host UCP1 and support thermogenesis. We will detail additional aspects that render them so important for brown and beige adipocytes.

## 5.1 Mitochondrial Remodeling and Adipocyte Plasticity

Mitochondrial network of brown adipocytes is dense and tightly associated with the lipid droplets, as revealed by electron microscopy images.[141] This mitochondrial network is dynamic, and its remodeling directly contributes to adrenergically induced changes in energy expenditure. Noradrenaline-induced mitochondrial fragmentation in murine brown adipocytes associated with opa1 cleavage enhance the susceptibility to fatty acid-induced uncoupling and thereby heat production and energy expenditure.[142] Mitochondrial fission also has been shown to occur during human white to beige adipocyte conversion.[143] Mitochondrial content, regulated by the balance between mitochondrial biogenesis and degradation also, is tightly controlled during activation/inactivation of brown adipocytes.[144] Although mitochondria biogenesis is stimulated by PGC1α during cold exposure,[145] mitochondrial removal by autophagy is repressed during cold-induced BAT activation; decreasing autophagy in brown adipocytes in culture increased UCP1 expression,[146] suggesting that repression of autophagy participates to adaptive mechanisms to activate thermogenesis in brown adipocytes. Mice with genetic invalidation of genes involved in autophagy, however, gave discordant phenotypes and conclusions, as reviewed,[144] maybe because the different models target adipose cells at different stages (undifferentiated or differentiated), maybe because they are not specific to adipose cells, or maybe because some proteins involved in autophagy act on additional signalization pathways beyond autophagy.[147] It has been clearly demonstrated, however, that removal of mitochondria by autophagy participates to the reversal of beige adipocytes toward a white phenotype. This cellular plasticity is associated with loss of mitochondria underlined by the whitening of fat pad visualized by optical tissue clearing.[62] Autophagy-mediated mitochondrial clearance is needed for beige to white adipocyte reversal as genetic and pharmacological inhibition of autophagy retain beige adipocytes for a prolonged period of time even after withdrawal of β3 agonist treatments or cold stimulus and suppressed diet-induced obesity and insulin resistance.[62] These findings plus the fact that autophagy has been found to be modified in obese patients (increased or decreased autophagy flux has been reported[148, 149]) suggest that exploring autophagy regulation during BAT activation and WAT browning might provide novel tools for management of energy balance.[144]

## 5.2 Mitochondrial ROS and Redox State Control UCP1 Expression and Activity

Although the inhibitory effect of mild uncoupling on ROS production is well known, the role of UCP1 in the control of ROS is still being debated.[150–152] Considering ROS generation for the same electron flux and with the same oxygen concentration might help to resolve this debate. First, according to the law of mass action, the production of superoxide anion is correlated strictly to the oxygen concentration. Second, for a same electron flux, any uncoupling activity results in a decrease in ROS production. As uncoupling increases respiratory rate, catabolism, and then in electrons inlet and flux into the respiratory chain, however, this can generate an increase in ROS production. Despite this, increasing evidence suggests that mitochondrial ROS control UCP1 expression and activity (Fig. 2). It is known that brown adipocytes mitochondria have higher level of ROS (see previous comments) and a more oxidized status when compared to other cell types, and their activation by cold exposure further increased such oxidative state.[153] This might be explained by the high increase of electrons flux through the respiratory chain together with the decrease of reduced glutathione content (a major thioldisulphide redox system) upon cold exposure.[154] Treating mice with a glutathione depleting agent (buthionine sulfoximine) increased UCP1 expression in WAT and BAT.[154] Consistently, mice deficient for the nuclear factor erythroid 2 related factor (NRF2), which regulates the expression of several antioxidant enzymes, are resistant to diet-induced obesity through increased energy expenditure and higher UCP1 expression within WAT.[155] Suppression of ROS by Sestrins, a family of highly conserved stress-responsive proteins that reduce oxidative stress through the regulation of peroxiredoxins and oxidoreductases activity, however, decreased UCP1 expression by downregulating the p38-MAPK pathway.[156] Although the previously mentioned studies did not specifically target ROS produced by mitochondria, some other studies demonstrate that mitochondrial oxidative stress is an intimate regulator of the thermogenic program. Mice with adipocyte-specific invalidation of the mitochondrial manganese-dependent superoxide dismutase are resistant to high-fat diet-induced weight gain and insulin resistance through elevation of thermogenic energy expenditure.[157] Consistent with this, mice deficient for the mitochondrial glutathione peroxydase (an enzyme that decomposes $H_2O_2$ into $H_2O$) or for the mitochondrial $NADP^+$-dependent isocitrate dehydrogenase display the same phenotype.[158, 159] Although these studies converge toward a positive role of ROS on UCP1 expression, strong genetic evidence also demonstrates that mitochondrial ROS are key for posttranslational modifications of UCP1 that promote its uncoupling activity in vivo[160, 161] (Fig. 2). This is consistent with the fact that acute activation of BAT by cold exposure or β3 adrenergic agonist treatment induced an increase in mitochondrial ROS content in BAT, as shown by the decrease in aconitase activity, a mitochondrial enzyme inactivated by the superoxide anion, confirming previous studies.[162] Treatment of mice before cold exposure with a mitochondrial targeted antioxidant (MitoQ) induced hypothermia upon cold exposure, without affecting muscle shivering activity. Using a redox gel-shift method, the authors showed an increase in cysteines oxidation in the UCP1 protein upon cold challenge and using mutagenesis, they identified Cys253 as a target of mitochondrial ROS. Redox modification of this cysteine residue sensitizes UCP1-dependent proton leak in the mitochondria of brown adipocytes.[160, 161]

The impact of redox state on UCP1 regulation also has been demonstrated by increasing $NADH/NAD^+$ ratio following exposure of adipocytes to an excess of redox metabolites such as lactate or the ketone body β-hydroxybutyrate,[163] two major intermediate metabolites exhibiting significant roles in mediating intertissue and intercellular metabolic coupling.[164, 165] The addition of pyruvate prevented lactate-induced UCP1 while acetoacetate blunted β-hydroxybutyrate effect, suggesting that an increase in $NADH,H^+/NAD^+$ ratio triggers UCP1 upregulation, probably through redox sensitive signalization pathways (Fig. 2). As the uncoupling activity of UCP1 accelerates electron transport chain activity and therefore decrease $NADH,H^+/NAD^+$ ratio, browning, therefore, might constitute an adaptive mechanism to alleviate high redox pressure within tissue and cells.[78] The role of beige adipocytes in ROS/redox balance is supported by recent findings that demonstrate negative correlation between thermogenic markers and oxidative stress-related genes in the human epicardial adipose tissue.[166] Lactate-induced UCP1 regulation seem to be a general mechanism, because it has been recently shown to participate in programmed cell death 4-induced differentiation of progenitors into beige adipocytes,[167] to occur in muscle cells[168] where brown progenitors are present,[33] but also during intermittent fasting, with gut microbia-derived lactate (and acetate) driving WAT browning.[169]

## 5.3 UCP1-Independent Thermogenesis

Several studies support the existence of UCP1-independent and alternative thermogenesis mechanisms (Fig. 1). UCP1-deficient mice can adapt to cold after a progressive decline of ambient temperatures from 28°C to 4°C.[170–172] In addition to the existence of nonshivering thermogenic mechanisms in skeletal muscle mediated by sarcolipin,[173] an adaptive remodeling is observed in the inguinal fat pad of $UCP1^{-/-}$ mice after cold adaptation, as shown by increased mitochondrial content, sarco(endo)plasmic reticulum $Ca^{2+}$-ATPase, and mitochondrial glycerol 3-phosphate dehydrogenase that might participate in alternative thermogenic programs.[5, 174] One study also reported that cold exposure increased the expression and activity of mitochondrial creatine kinases (CKMT1/CKMT2) in the inguinal fat pad and demonstrated that creatine phosphorylation and dephosphorylation cycle represent a futile cycle that dissipates energy in the form of heat through stimulation of mitochondrial ATP turnover.[175] The fact that mice deficient for the rate-limiting enzyme of creatine biosynthesis (glycine amidinotransferase), specifically in adipose tissue, are prone to diet-induced obesity provide the demonstration about the physiological role of creatine metabolism in energy expenditure.[176] This UCP1-independent thermogenesis also exists in humans, because exclusive expression of the mitochondrial creatine kinases CK-MT1A/B and CK- MT2 has been shown in human adult BAT compared to adjacent WAT.[177] Furthermore, the patch clamp technique enabling the measure of protons fluxes through the inner mitochondrial membrane revealed that the vast majority of epidydimal adipocytes do not express the UCP1 protein but exhibit alternative thermogenesis through the creatine futile cycle.[67] This led the authors to conclude that two types of heat-producing beige adipocytes co-exist, some that do express the UCP1 protein and some that do not.[67, 178] The creatine kinase futile cycle suggests the common lineage shared by some brown adipocytes and muscle cells, further reinforcing the link.

These findings highlight the diversity of the mechanisms by which mitochondria support the biology of brown and beiges adipocytes, UCP1 among them.

## 6 CONCLUSION AND PERSPECTIVES

Although numerous reports describe the link between mitochondria dysfunction in adipocytes and metabolic diseases, the complete demonstration of a causative link between the elements is missing. The lack of a clear conclusion also is demonstrated by contradictory reports about the adiposity and primary mitochondrial diseases including mitochondrial DNA diseases.[179] The recent rediscovery of UCP1-positive cells in human adults and the increasing literature about the importance of such cells in energy homeostasis, however, is consistent with a determinant and causative role of mitochondria in metabolic diseases. To reach the step where some metabolic disorders could be analyzed as adipocyte mitochondriopathies, however, investigations are lacking; further efforts are required, among which are a better understanding of the biology of mitochondrial transporters and shuttles, as well as the genetic specific to this organelle. We could speculate that a putative heteroplasmy linked to differences in organelle genome, turnover, and dynamics could explain the heterogeneity of adipocytes and their respective turnover such as largely described for muscle in mitochondrial myopathies.[180] This is consistent with a case report describing a heteroplasmic mitochondrial DNA mutation in a patient with juvenile-onset metabolic syndrome,[181] although the definitive demonstration is still lacking. This clearly shows that the field is largely open and could bring unexpected discoveries.

## Acknowledgments

We apologize to authors whose work could not be cited because of space limitations. We would like to thank Professors M. Rigoulet and X. Leverve for helpful discussions as well as all the members of STROMALab. We also thank our financial supports (European Union FP7 project DIABAT (HEALTH-F2-2011-278373) and "La Société Française de Nutrition"). This work is dedicated to Professor X. Leverve.

## References

1. Rosen ED, Spiegelman BM. What we talk about when we talk about fat. *Cell* 2014;**156**(1-2):20–44.
2. Kajimura S, Spiegelman BM, Seale P. Brown and beige fat: physiological roles beyond heat generation. *Cell Metab* 2015;**22**(4):546–59.
3. Bartelt A, Heeren J. Adipose tissue browning and metabolic health. *Nat Rev Endocrinol* 2013.
4. Bhatt PS, Dhillo WS, Salem V. Human brown adipose tissue-function and therapeutic potential in metabolic disease. *Curr Opin Pharmacol* 2017;**37**:1–9.
5. Betz MJ, Enerback S. Targeting thermogenesis in brown fat and muscle to treat obesity and metabolic disease. *Nat Rev Endocrinol* 2018;**14**(2):77–87.
6. Giordano A, Smorlesi A, Frontini A, Barbatelli G, Cinti S. White, brown and pink adipocytes: the extraordinary plasticity of the adipose organ. *Eur J Endocrinol* 2014;**170**(5):R159–71.
7. Berry DC, Jiang Y, Graff JM. Emerging roles of adipose progenitor cells in tissue development, homeostasis, expansion and thermogenesis. *Trends Endocrinol Metab* 2016;**27**(8):574–85.

8. Carobbio S, Pellegrinelli V, Vidal-Puig A. Adipose tissue function and expandability as determinants of lipotoxicity and the metabolic syndrome. *Adv Exp Med Biol* 2017;**960**:161–96.
9. Boudina S, Graham TE. Mitochondrial function/dysfunction in white adipose tissue. *Exp Physiol* 2014;**99**(9):1168–78.
10. Cedikova M, Kripnerova M, Dvorakova J, Pitule P, Grundmanova M, Babuska V, et al. Mitochondria in white, brown, and beige adipocytes. *Stem Cells Int* 2016;**2016**:6067349.
11. De Pauw A, Tejerina S, Raes M, Keijer J, Arnould T. Mitochondrial (dys)function in adipocyte (de)differentiation and systemic metabolic alterations. *Am J Pathol* 2009;**175**(3):927–39.
12. Kershaw EE, Flier JS. Adipose tissue as an endocrine organ. *J Clin Endocrinol Metab* 2004;**89**(6):2548–56.
13. Pellegrinelli V, Carobbio S, Vidal-Puig A. Adipose tissue plasticity: how fat depots respond differently to pathophysiological cues. *Diabetologia* 2016;**59**(6):1075–88.
14. Yang X, Smith U. Adipose tissue distribution and risk of metabolic disease: does thiazolidinedione-induced adipose tissue redistribution provide a clue to the answer? *Diabetologia* 2007;**50**(6):1127–39.
15. Deveaud C, Beauvoit B, Salin B, Schaeffer J, Rigoulet M. Regional differences in oxidative capacity of rat white adipose tissue are linked to the mitochondrial content of mature adipocytes. *Mol Cell Biochem* 2004;**267**(1-2):157–66.
16. Schottl T, Kappler L, Braun K, Fromme T, Klingenspor M. Limited mitochondrial capacity of visceral versus subcutaneous white adipocytes in male C57BL/6N mice. *Endocrinology* 2015;**156**(3):923–33.
17. Rutkowski JM, Stern JH, Scherer PE. The cell biology of fat expansion. *J Cell Biol* 2015;**208**(5):501–12.
18. Sun K, Tordjman J, Clement K, Scherer PE. Fibrosis and adipose tissue dysfunction. *Cell Metab* 2013;**18**(4):470–7.
19. Casteilla L, Champigny O, Bouillaud F, Robelin J, Ricquier D. Sequential changes in the expression of mitochondrial protein mRNA during the development of brown adipose tissue in bovine and ovine species. Sudden occurrence of uncoupling protein mRNA during embryogenesis and its disappearance after birth. *Biochem J* 1989;**257**(3):665–71.
20. Casteilla L, Forest C, Robelin J, Ricquier D, Lombet A, Ailhaud G. Characterization of mitochondrial-uncoupling protein in bovine fetus and newborn calf. *Am J Phys* 1987;**252**(5 Pt 1):E627–36.
21. Ricquier D. Neonatal brown adipose tissue, UCP1 and the novel uncoupling proteins. *Biochem Soc Trans* 1998;**26**(2):120–3.
22. Nicholls DG, Locke RM. Thermogenic mechanisms in brown fat. *Physiol Rev* 1984;**64**(1):1–64.
23. Ricquier D. Uncoupling protein 1 of brown adipocytes, the only uncoupler: a historical perspective. *Front Endocrinol (Lausanne)* 2011;**2**:85.
24. Cannon B, Nedergaard J. Brown adipose tissue: function and physiological significance. *Physiol Rev* 2004;**84**(1):277–359.
25. Inagaki T, Sakai J, Kajimura S. Transcriptional and epigenetic control of brown and beige adipose cell fate and function. *Nat Rev Mol Cell Biol* 2017;**18**(8):527.
26. Fedorenko A, Lishko PV, Kirichok Y. Mechanism of fatty-acid-dependent UCP1 uncoupling in brown fat mitochondria. *Cell* 2012;**151**(2):400–13.
27. Schreiber R, Diwoky C, Schoiswohl G, Feiler U, Wongsiriroj N, Abdellatif M, et al. Cold-induced thermogenesis depends on ATGL-mediated lipolysis in cardiac muscle, but not brown adipose tissue. *Cell Metab* 2017;**26**(5):753–63. e7.
28. Shin H, Ma Y, Chanturiya T, Cao Q, Wang Y, Kadegowda AKG, et al. Lipolysis in brown adipocytes is not essential for cold-induced thermogenesis in mice. *Cell Metab* 2017;**26**(5):764–77. e5.
29. Simcox J, Geoghegan G, Maschek JA, Bensard CL, Pasquali M, Miao R, et al. Global analysis of plasma lipids identifies liver-derived acylcarnitines as a fuel source for brown fat thermogenesis. *Cell Metab* 2017;**26**(3):509–22. e6.
30. Cannon B, Nedergaard J. What ignites UCP1? *Cell Metab* 2017;**26**(5):697–8.
31. Rial E, Gonzalez-Barroso M, Fleury C, Iturrizaga S, Sanchis D, Jimenez-Jimenez J, et al. Retinoids activate proton transport by the uncoupling proteins UCP1 and UCP2. *EMBO J* 1999;**18**(21):5827–33.
32. Seale P, Bjork B, Yang W, Kajimura S, Chin S, Kuang S, et al. PRDM16 controls a brown fat/skeletal muscle switch. *Nature* 2008;**454**(7207):961–7.
33. Crisan M, Casteilla L, Lehr L, Carmona M, Paoloni-Giacobino A, Yap S, et al. A reservoir of brown adipocyte progenitors in human skeletal muscle. *Stem Cells* 2008;**26**(9):2425–33.
34. Kozak LP. Brown fat and the myth of diet-induced thermogenesis. *Cell Metab* 2010;**11**(4):263–7.
35. Trayhurn P. Origins and early development of the concept that brown adipose tissue thermogenesis is linked to energy balance and obesity. *Biochimie* 2017;**134**:62–70.

36. Himms-Hagen J. Brown adipose tissue thermogenesis: interdisciplinary studies. *FASEB J* 1990;**4**(11):2890–8.
37. Feldmann HM, Golozoubova V, Cannon B, Nedergaard J. UCP1 ablation induces obesity and abolishes diet-induced thermogenesis in mice exempt from thermal stress by living at thermoneutrality. *Cell Metab* 2009;**9**(2):203–9.
38. Liu X, Wang S, You Y, Meng M, Zheng Z, Dong M, et al. Brown adipose tissue transplantation reverses obesity in Ob/Ob mice. *Endocrinology* 2015;**156**(7):2461–9.
39. Bartelt A, Bruns OT, Reimer R, Hohenberg H, Ittrich H, Peldschus K, et al. Brown adipose tissue activity controls triglyceride clearance. *Nat Med* 2011;**17**(2):200–5.
40. Villarroya F, Gavalda-Navarro A, Peyrou M, Villarroya J, Giralt M. The lives and times of brown adipokines. *Trends Endocrinol Metab* 2017;**28**(12):855–67.
41. Ricquier D, Nechad M, Mory G. Ultrastructural and biochemical characterization of human brown adipose tissue in pheochromocytoma. *J Clin Endocrinol Metab* 1982;**54**(4):803–7.
42. Bouillaud F, Villarroya F, Hentz E, Raimbault S, Cassard AM, Ricquier D. Detection of brown adipose tissue uncoupling protein mRNA in adult patients by a human genomic probe. *Clin Sci (Lond)* 1988;**75**(1):21–7.
43. Cypess AM, Lehman S, Williams G, Tal I, Rodman D, Goldfine AB, et al. Identification and importance of brown adipose tissue in adult humans. *N Engl J Med* 2009;**360**(15):1509–17.
44. Saito M, Okamatsu-Ogura Y, Matsushita M, Watanabe K, Yoneshiro T, Nio-Kobayashi J, et al. High incidence of metabolically active brown adipose tissue in healthy adult humans: effects of cold exposure and adiposity. *Diabetes* 2009;**58**(7):1526–31.
45. van Marken Lichtenbelt WD, Vanhommerig JW, Smulders NM, Drossaerts JM, Kemerink GJ, Bouvy ND, et al. Cold-activated brown adipose tissue in healthy men. *N Engl J Med* 2009;**360**(15):1500–8.
46. Virtanen KA, Lidell ME, Orava J, Heglind M, Westergren R, Niemi T, et al. Functional brown adipose tissue in healthy adults. *N Engl J Med* 2009;**360**(15):1518–25.
47. Himms-Hagen J, Desautels M. A mitochondrial defect in brown adipose tissue of the obese (ob/ob) mouse: reduced binding of purine nucleotides and a failure to respond to cold by an increase in binding. *Biochem Biophys Res Commun* 1978;**83**(2):628–34.
48. Vijgen GH, Bouvy ND, Teule GJ, Brans B, Schrauwen P, van Marken Lichtenbelt WD. Brown adipose tissue in morbidly obese subjects. *PLoS ONE* 2011;**6**(2):e17247.
49. Shimizu I, Aprahamian T, Kikuchi R, Shimizu A, Papanicolaou KN, MacLauchlan S, et al. Vascular rarefaction mediates whitening of brown fat in obesity. *J Clin Invest* 2014;**124**(5):2099–112.
50. Hanssen MJ, Hoeks J, Brans B, van der Lans AA, Schaart G, van den Driessche JJ, et al. Short-term cold acclimation improves insulin sensitivity in patients with type 2 diabetes mellitus. *Nat Med* 2015;**21**(8):863–5.
51. Cypess AM, Weiner LS, Roberts-Toler C, Franquet Elia E, Kessler SH, Kahn PA, et al. Activation of human brown adipose tissue by a beta3-adrenergic receptor agonist. *Cell Metab* 2015;**21**(1):33–8.
52. Loncar D. Convertible adipose tissue in mice. *Cell Tissue Res* 1991;**266**(1):149–61.
53. Young P, Arch JR, Ashwell M. Brown adipose tissue in the parametrial fat pad of the mouse. *FEBS Lett* 1984;**167**(1):10–4.
54. Cousin B, Cinti S, Morroni M, Raimbault S, Ricquier D, Penicaud L, et al. Occurrence of brown adipocytes in rat white adipose tissue: molecular and morphological characterization. *J Cell Sci* 1992;**103**:931–42.
55. Barreau C, Labit E, Guissard C, Rouquette J, Boizeau ML, Gani Koumassi S, et al. Regionalization of browning revealed by whole subcutaneous adipose tissue imaging. *Obesity (Silver Spring)* 2016;**24**(5):1081–9.
56. Walden TB, Hansen IR, Timmons JA, Cannon B, Nedergaard J. Recruited vs. nonrecruited molecular signatures of brown, "brite," and white adipose tissues. *Am J Physiol Endocrinol Metab* 2012;**302**(1):E19–31.
57. Shabalina IG, Petrovic N, de Jong JM, Kalinovich AV, Cannon B, Nedergaard J. UCP1 in brite/beige adipose tissue mitochondria is functionally thermogenic. *Cell Rep* 2013;**5**(5):1196–203.
58. Lee YH, Petkova AP, Mottillo EP, Granneman JG. In vivo identification of bipotential adipocyte progenitors recruited by beta3-adrenoceptor activation and high-fat feeding. *Cell Metab* 2012;**15**(4):480–91.
59. Wang QA, Tao C, Gupta RK, Scherer PE. Tracking adipogenesis during white adipose tissue development, expansion and regeneration. *Nat Med* 2013;**19**(10):1338–44.
60. Barbatelli G, Murano I, Madsen L, Hao Q, Jimenez M, Kristiansen K, et al. The emergence of cold-induced brown adipocytes in mouse white fat depots is determined predominantly by white to brown adipocyte transdifferentiation. *Am J Physiol Endocrinol Metab* 2010;**298**(6):E1244–53.
61. Rosenwald M, Perdikari A, Rulicke T, Wolfrum C. Bi-directional interconversion of brite and white adipocytes. *Nat Cell Biol* 2013;**15**(6):659–67.

62. Altshuler-Keylin S, Shinoda K, Hasegawa Y, Ikeda K, Hong H, Kang Q, et al. Beige adipocyte maintenance is regulated by autophagy-induced mitochondrial clearance. *Cell Metab* 2016;**24**(3):402–19.
63. Gospodarska E, Nowialis P, Kozak LP. Mitochondrial turnover: a phenotype distinguishing brown adipocytes from interscapular brown adipose tissue and white adipose tissue. *J Biol Chem* 2015;**290**(13):8243–55.
64. Sanchez-Gurmaches J, Guertin DA. Adipocytes arise from multiple lineages that are heterogeneously and dynamically distributed. *Nat Commun* 2014;**5**:4099.
65. Sanchez-Gurmaches J, Guertin DA. Adipocyte lineages: tracing back the origins of fat. *Biochim Biophys Acta* 2014;**1842**(3):340–51.
66. Lee YH, Kim SN, Kwon HJ, Granneman JG. Metabolic heterogeneity of activated beige/brite adipocytes in inguinal adipose tissue. *Sci Rep* 2017;**7**:39794.
67. Bertholet AM, Kazak L, Chouchani ET, Bogaczynska MG, Paranjpe I, Wainwright GL, et al. Mitochondrial patch clamp of beige adipocytes reveals UCP1-positive and UCP1-negative cells both exhibiting futile creatine cycling. *Cell Metab* 2017;**25**(4):811–22. e4.
68. Nedergaard J, Cannon B. UCP1 mRNA does not produce heat. *Biochim Biophys Acta* 2013;**1831**(5):943–9.
69. Cohen P, Levy JD, Zhang Y, Frontini A, Kolodin DP, Svensson KJ, et al. Ablation of PRDM16 and beige adipose causes metabolic dysfunction and a subcutaneous to visceral fat switch. *Cell* 2014;**156**(1-2):304–16.
70. Seale P, Conroe HM, Estall J, Kajimura S, Frontini A, Ishibashi J, et al. Prdm16 determines the thermogenic program of subcutaneous white adipose tissue in mice. *J Clin Invest* 2011;**121**(1):96–105.
71. Nedergaard J, Cannon B. How brown is brown fat? It depends where you look. *Nat Med* 2013;**19**(5):540–1.
72. Bostrom P, Wu J, Jedrychowski MP, Korde A, Ye L, Lo JC, et al. A PGC1-alpha-dependent myokine that drives brown-fat-like development of white fat and thermogenesis. *Nature* 2012;**481**(7382):463–8.
73. Aldiss P, Betts J, Sale C, Pope M, Symonds ME. Exercise-induced 'browning' of adipose tissues. *Metabolism* 2018;**81**:63–70.
74. Porter C, Herndon DN, Bhattarai N, Ogunbileje JO, Szczesny B, Szabo C, et al. Severe burn injury induces thermogenically functional mitochondria in murine white adipose tissue. *Shock* 2015;**44**(3):258–64.
75. Sidossis LS, Porter C, Saraf MK, Borsheim E, Radhakrishnan RS, Chao T, et al. Browning of subcutaneous white adipose tissue in humans after severe adrenergic stress. *Cell Metab* 2015;**22**(2):219–27.
76. Petruzzelli M, Schweiger M, Schreiber R, Campos-Olivas R, Tsoli M, Allen J, et al. A switch from white to brown fat increases energy expenditure in cancer-associated cachexia. *Cell Metab* 2014;**20**(3):433–47.
77. Kir S, White JP, Kleiner S, Kazak L, Cohen P, Baracos VE, et al. Tumour-derived PTH-related protein triggers adipose tissue browning and cancer cachexia. *Nature* 2014;**513**(7516):100–4.
78. Jeanson Y, Carriere A, Casteilla L. A new role for browning as a redox and stress adaptive mechanism? *Front Endocrinol (Lausanne)* 2015;**6**:158.
79. Salway JG. *Metabolism at a glance*. 4th ed. Wiley Blackwell; 2017.
80. Diraison F, Yankah V, Letexier D, Dusserre E, Jones P, Beylot M. Differences in the regulation of adipose tissue and liver lipogenesis by carbohydrates in humans. *J Lipid Res* 2003;**44**(4):846–53.
81. Kopecky J, Rossmeisl M, Flachs P, Bardova K, Brauner P. Mitochondrial uncoupling and lipid metabolism in adipocytes. *Biochem Soc Trans* 2001;**29**(Pt 6):791–7.
82. Guan HP, Li Y, Jensen MV, Newgard CB, Steppan CM, Lazar MA. A futile metabolic cycle activated in adipocytes by antidiabetic agents. *Nat Med* 2002;**8**(10):1122–8.
83. Tan GD, Debard C, Tiraby C, Humphreys SM, Frayn KN, Langin D, et al. A "futile cycle" induced by thiazolidinediones in human adipose tissue? *Nat Med* 2003;**9**(7):811–2. author reply 2.
84. Nye C, Kim J, Kalhan SC, Hanson RW. Reassessing triglyceride synthesis in adipose tissue. *Trends Endocrinol Metab* 2008;**19**(10):356–61.
85. Bertin R, Andriamihaja M, Portet R. Glycerokinase activity in brown and white adipose tissues of cold-adapted obese Zucker rats. *Biochimie* 1984;**66**(7-8):569–72.
86. Collins S, Cao W, Robidoux J. Learning new tricks from old dogs: beta-adrenergic receptors teach new lessons on firing up adipose tissue metabolism. *Mol Endocrinol* 2004;**18**(9):2123–31.
87. Moro C, Lafontan M. Natriuretic peptides and cGMP signaling control of energy homeostasis. *Am J Physiol Heart Circ Physiol* 2013;**304**(3):H358–68.
88. Langin D. Adipose tissue lipolysis as a metabolic pathway to define pharmacological strategies against obesity and the metabolic syndrome. *Pharmacol Res* 2006;**53**(6):482–91.
89. Barquissau V, Beuzelin D, Pisani DF, Beranger GE, Mairal A, Montagner A, et al. White-to-brite conversion in human adipocytes promotes metabolic reprogramming towards fatty acid anabolic and catabolic pathways. *Mol Metab* 2016;**5**(5):352–65.

90. Prunet-Marcassus B, Moulin K, Carmona MC, Villarroya F, Penicaud L, Casteilla L. Inverse distribution of uncoupling proteins expression and oxidative capacity in mature adipocytes and stromal-vascular fractions of rat white and brown adipose tissues. *FEBS Lett* 1999;**464**(3):184–8.
91. Wang CH, Tsai TF, Wei YH. Role of mitochondrial dysfunction and dysregulation of Ca(2+) homeostasis in insulin insensitivity of mammalian cells. *Ann N Y Acad Sci* 2015;**1350**:66–76.
92. Arruda AP, Hotamisligil GS. Calcium homeostasis and organelle function in the pathogenesis of obesity and diabetes. *Cell Metab* 2015;**22**(3):381–97.
93. Shyh-Chang N, Daley GQ, Cantley LC. Stem cell metabolism in tissue development and aging. *Development* 2013;**140**(12):2535–47.
94. Parker GC, Acsadi G, Brenner CA. Mitochondria: determinants of stem cell fate? *Stem Cells Dev* 2009;**18**(6):803–6.
95. Wilson-Fritch L, Burkart A, Bell G, Mendelson K, Leszyk J, Nicoloro S, et al. Mitochondrial biogenesis and remodeling during adipogenesis and in response to the insulin sensitizer rosiglitazone. *Mol Cell Biol* 2003;**23**(3):1085–94.
96. Newton BW, Cologna SM, Moya C, Russell DH, Russell WK, Jayaraman A. Proteomic analysis of 3T3-L1 adipocyte mitochondria during differentiation and enlargement. *J Proteome Res* 2011;**10**(10):4692–702.
97. Drehmer DL, de Aguiar AM, Brandt AP, Petiz L, Cadena SM, Rebelatto CK, et al. Metabolic switches during the first steps of adipogenic stem cells differentiation. *Stem Cell Res* 2016;**17**(2):413–21.
98. Tormos KV, Anso E, Hamanaka RB, Eisenbart J, Joseph J, Kalyanaraman B, et al. Mitochondrial complex III ROS regulate adipocyte differentiation. *Cell Metab* 2011;**14**(4):537–44.
99. Lopez-Lluch G. Mitochondrial activity and dynamics changes regarding metabolism in ageing and obesity. *Mech Ageing Dev* 2017;**162**:108–21.
100. Rong JX, Klein JL, Qiu Y, Xie M, Johnson JH, Waters KM, et al. Rosiglitazone induces mitochondrial biogenesis in differentiated murine 3T3-L1 and C3H/10T1/2 adipocytes. *PPAR Res* 2011;**2011**:179454.
101. Kajimoto K, Terada H, Baba Y, Shinohara Y. Essential role of citrate export from mitochondria at early differentiation stage of 3T3-L1 cells for their effective differentiation into fat cells, as revealed by studies using specific inhibitors of mitochondrial di- and tricarboxylate carriers. *Mol Genet Metab* 2005;**85**(1):46–53.
102. Saulnier-Blache JS, Dauzats M, Daviaud D, Gaillard D, Ailhaud G, Negrel R, et al. Late expression of alpha 2-adrenergic-mediated antilipolysis during differentiation of hamster preadipocytes. *J Lipid Res* 1991;**32**(9):1489–99.
103. Friedman JR, Nunnari J. Mitochondrial form and function. *Nature* 2014;**505**(7483):335–43.
104. Ducluzeau PH, Priou M, Weitheimer M, Flamment M, Duluc L, Iacobazi F, et al. Dynamic regulation of mitochondrial network and oxidative functions during 3T3-L1 fat cell differentiation. *J Physiol Biochem* 2011;**67**(3):285–96.
105. Kita T, Nishida H, Shibata H, Niimi S, Higuti T, Arakaki N. Possible role of mitochondrial remodelling on cellular triacylglycerol accumulation. *J Biochem* 2009;**146**(6):787–96.
106. Liu D, Lin Y, Kang T, Huang B, Xu W, Garcia-Barrio M, et al. Mitochondrial dysfunction and adipogenic reduction by prohibitin silencing in 3T3-L1 cells. *PLoS ONE* 2012;**7**(3).
107. Droge W. Free radicals in the physiological control of cell function. *Physiol Rev* 2002;**82**(1):47–95.
108. Holmstrom KM, Finkel T. Cellular mechanisms and physiological consequences of redox-dependent signalling. *Nat Rev Mol Cell Biol* 2014;**15**(6):411–21.
109. Jankovic A, Korac A, Buzadzic B, Otasevic V, Stancic A, Daiber A, et al. Redox implications in adipose tissue (dys)function—a new look at old acquaintances. *Redox Biol* 2015;**6**:19–32.
110. Wang W, Zhang Y, Lu W, Liu K. Mitochondrial reactive oxygen species regulate adipocyte differentiation of mesenchymal stem cells in hematopoietic stress induced by arabinosylcytosine. *PLoS ONE* 2015;**10**(3).
111. Kim JH, Kim SH, Song SY, Kim WS, Song SU, Yi T, et al. Hypoxia induces adipocyte differentiation of adipose-derived stem cells by triggering reactive oxygen species generation. *Cell Biol Int* 2014;**38**(1):32–40.
112. Carriere A, Fernandez Y, Rigoulet M, Penicaud L, Casteilla L. Inhibition of preadipocyte proliferation by mitochondrial reactive oxygen species. *FEBS Lett* 2003;**550**(1-3):163–7.
113. Carriere A, Carmona MC, Fernandez Y, Rigoulet M, Wenger RH, Penicaud L, et al. Mitochondrial reactive oxygen species control the transcription factor CHOP-10/GADD153 and adipocyte differentiation: a mechanism for hypoxia-dependent effect. *J Biol Chem* 2004;**279**(39):40462–9.
114. Chevillotte E, Giralt M, Miroux B, Ricquier D, Villarroya F. Uncoupling protein-2 controls adiponectin gene expression in adipose tissue through the modulation of reactive oxygen species production. *Diabetes* 2007;**56**(4):1042–50.

115. Arsenijevic D, Onuma H, Pecqueur C, Raimbault S, Manning BS, Miroux B, et al. Disruption of the uncoupling protein-2 gene in mice reveals a role in immunity and reactive oxygen species production. *Nat Genet* 2000;**26**(4):435–9.
116. Bouillaud F. UCP2, not a physiologically relevant uncoupler but a glucose sparing switch impacting ROS production and glucose sensing. *Biochim Biophys Acta* 2009;**1787**(5):377–83.
117. Duval C, Negre-Salvayre A, Dogilo A, Salvayre R, Penicaud L, Casteilla L. Increased reactive oxygen species production with antisense oligonucleotides directed against uncoupling protein 2 in murine endothelial cells. *Biochem Cell Biol* 2002;**80**(6):757–64.
118. Jokinen R, Pirnes-Karhu S, Pietilainen KH, Pirinen E. Adipose tissue NAD(+)-homeostasis, sirtuins and poly(ADP-ribose) polymerases-important players in mitochondrial metabolism and metabolic health. *Redox Biol* 2017;**12**:246–63.
119. Matilainen O, Quiros PM, Auwerx J. Mitochondria and epigenetics—crosstalk in homeostasis and stress. *Trends Cell Biol* 2017;**27**(6):453–63.
120. Wellen KE, Hatzivassiliou G, Sachdeva UM, Bui TV, Cross JR, Thompson CB. ATP-citrate lyase links cellular metabolism to histone acetylation. *Science* 2009;**324**(5930):1076–80.
121. Wilson-Fritch L, Nicoloro S, Chouinard M, Lazar MA, Chui PC, Leszyk J, et al. Mitochondrial remodeling in adipose tissue associated with obesity and treatment with rosiglitazone. *J Clin Invest* 2004;**114**(9):1281–9.
122. Rong JX, Qiu Y, Hansen MK, Zhu L, Zhang V, Xie M, et al. Adipose mitochondrial biogenesis is suppressed in db/db and high-fat diet-fed mice and improved by rosiglitazone. *Diabetes* 2007;**56**(7):1751–60.
123. Dahlman I, Forsgren M, Sjogren A, Nordstrom EA, Kaaman M, Naslund E, et al. Downregulation of electron transport chain genes in visceral adipose tissue in type 2 diabetes independent of obesity and possibly involving tumor necrosis factor-alpha. *Diabetes* 2006;**55**(6):1792–9.
124. Yin X, Lanza IR, Swain JM, Sarr MG, Nair KS, Jensen MD. Adipocyte mitochondrial function is reduced in human obesity independent of fat cell size. *J Clin Endocrinol Metab* 2014;**99**(2):E209–16.
125. Chattopadhyay M, Guhathakurta I, Behera P, Ranjan KR, Khanna M, Mukhopadhyay S, et al. Mitochondrial bioenergetics is not impaired in nonobese subjects with type 2 diabetes mellitus. *Metabolism* 2011;**60**(12):1702–10.
126. Choo HJ, Kim JH, Kwon OB, Lee CS, Mun JY, Han SS, et al. Mitochondria are impaired in the adipocytes of type 2 diabetic mice. *Diabetologia* 2006;**49**(4):784–91.
127. Lindinger PW, Christe M, Eberle AN, Kern B, Peterli R, Peters T, et al. Important mitochondrial proteins in human omental adipose tissue show reduced expression in obesity. *Data Brief* 2015;**4**:40–3.
128. Heinonen S, Muniandy M, Buzkova J, Mardinoglu A, Rodriguez A, Fruhbeck G, et al. Mitochondria-related transcriptional signature is downregulated in adipocytes in obesity: a study of young healthy MZ twins. *Diabetologia* 2017;**60**(1):169–81.
129. Tiraby C, Tavernier G, Lefort C, Larrouy D, Bouillaud F, Ricquier D, et al. Acquirement of brown fat cell features by human white adipocytes. *J Biol Chem* 2003;**278**(35):33370–6.
130. Cummins TD, Holden CR, Sansbury BE, Gibb AA, Shah J, Zafar N, et al. Metabolic remodeling of white adipose tissue in obesity. *Am J Physiol Endocrinol Metab* 2014;**307**(3):E262–77.
131. Furukawa S, Fujita T, Shimabukuro M, Iwaki M, Yamada Y, Nakajima Y, et al. Increased oxidative stress in obesity and its impact on metabolic syndrome. *J Clin Invest* 2004;**114**(12):1752–61.
132. Gomez-Serrano M, Camafeita E, Lopez JA, Rubio MA, Breton I, Garcia-Consuegra I, et al. Differential proteomic and oxidative profiles unveil dysfunctional protein import to adipocyte mitochondria in obesity-associated aging and diabetes. *Redox Biol* 2017;**11**:415–28.
133. Galinier A, Carriere A, Fernandez Y, Carpene C, Andre M, Caspar-Bauguil S, et al. Adipose tissue proadipogenic redox changes in obesity. *J Biol Chem* 2006;**281**(18):12682–7.
134. Tol MJ, Ottenhoff R, van Eijk M, Zelcer N, Aten J, Houten SM, et al. A PPARgamma-Bnip3 axis couples adipose mitochondrial fusion-fission balance to systemic insulin sensitivity. *Diabetes* 2016;**65**(9):2591–605.
135. Vernochet C, Damilano F, Mourier A, Bezy O, Mori MA, Smyth G, et al. Adipose tissue mitochondrial dysfunction triggers a lipodystrophic syndrome with insulin resistance, hepatosteatosis, and cardiovascular complications. *FASEB J* 2014;**28**(10):4408–19.
136. Wang CH, Wang CC, Huang HC, Wei YH. Mitochondrial dysfunction leads to impairment of insulin sensitivity and adiponectin secretion in adipocytes. *FEBS J* 2013;**280**(4):1039–50.
137. Giralt M, Domingo P, Villarroya F. Adipose tissue biology and HIV-infection. *Best Pract Res Clin Endocrinol Metab* 2011;**25**(3):487–99.

138. Caron-Debarle M, Boccara F, Lagathu C, Antoine B, Cervera P, Bastard JP, et al. Adipose tissue as a target of HIV-1 antiretroviral drugs. Potential consequences on metabolic regulations. *Curr Pharm Des* 2010;**16**(30):3352–60.
139. Caron M, Auclairt M, Vissian A, Vigouroux C, Capeau J. Contribution of mitochondrial dysfunction and oxidative stress to cellular premature senescence induced by antiretroviral thymidine analogues. *Antivir Ther* 2008;**13**(1):27–38.
140. Deveaud C, Beauvoit B, Hagry S, Galinier A, Carriere A, Salin B, et al. Site specific alterations of adipose tissue mitochondria in 3′-azido-3′-deoxythymidine (AZT)-treated rats: an early stage in lipodystrophy? *Biochem Pharmacol* 2005;**70**(1):90–101.
141. Cinti S. *The Adipose Organ*. Milan, Italy: Kurtis; 1999.
142. Wikstrom JD, Mahdaviani K, Liesa M, Sereda SB, Si Y, Las G, et al. Hormone-induced mitochondrial fission is utilized by brown adipocytes as an amplification pathway for energy expenditure. *EMBO J* 2014;**33**(5):418–36.
143. Pisani DF, Barquissau V, Chambard JC, Beuzelin D, Ghandour RA, Giroud M, et al. Mitochondrial fission is associated with UCP1 activity in human brite/beige adipocytes. *Mol Metab* 2018;**7**:35–44.
144. Altshuler-Keylin S, Kajimura S. Mitochondrial homeostasis in adipose tissue remodeling. *Sci Signal* 2017;**10**(468).
145. Puigserver P, Wu Z, Park CW, Graves R, Wright M, Spiegelman BM. A cold-inducible coactivator of nuclear receptors linked to adaptive thermogenesis. *Cell* 1998;**92**(6):829–39.
146. Cairo M, Villarroya J, Cereijo R, Campderros L, Giralt M, Villarroya F. Thermogenic activation represses autophagy in brown adipose tissue. *Int J Obes* 2016;**40**(10):1591–9.
147. Subramani S, Malhotra V. Non-autophagic roles of autophagy-related proteins. *EMBO Rep* 2013;**14**(2):143–51.
148. Kovsan J, Bluher M, Tarnovscki T, Kloting N, Kirshtein B, Madar L, et al. Altered autophagy in human adipose tissues in obesity. *J Clin Endocrinol Metab* 2011;**96**(2):E268–77.
149. Soussi H, Clement K, Dugail I. Adipose tissue autophagy status in obesity: expression and flux–two faces of the picture. *Autophagy* 2016;**12**(3):588–9.
150. Oelkrug R, Kutschke M, Meyer CW, Heldmaier G, Jastroch M. Uncoupling protein 1 decreases superoxide production in brown adipose tissue mitochondria. *J Biol Chem* 2010;**285**(29):21961–8.
151. Shabalina IG, Vrbacky M, Pecinova A, Kalinovich AV, Drahota Z, Houstek J, et al. ROS production in brown adipose tissue mitochondria: the question of UCP1-dependence. *Biochim Biophys Acta* 2014;**1837**(12):2017–30.
152. Bouillaud F, Alves-Guerra MC, Ricquier D. UCPs, at the interface between bioenergetics and metabolism. *Biochim Biophys Acta* 2016;**1863**(10):2443–56.
153. Mailloux RJ, Adjeitey CN, Xuan JY, Harper ME. Crucial yet divergent roles of mitochondrial redox state in skeletal muscle vs. brown adipose tissue energetics. *FASEB J* 2012;**26**(1):363–75.
154. Lettieri Barbato D, Tatulli G, Maria Cannata S, Bernardini S, Aquilano K, Ciriolo MR. Glutathione decrement drives thermogenic program in adipose cells. *Sci Rep* 2015;**5**:13091.
155. Schneider K, Valdez J, Nguyen J, Vawter M, Galke B, Kurtz TW, et al. Increased energy expenditure, Ucp1 expression, and resistance to diet-induced obesity in mice lacking nuclear factor-erythroid-2-related transcription factor-2 (Nrf2). *J Biol Chem* 2016;**291**(14):7754–66.
156. Ro SH, Nam M, Jang I, Park HW, Park H, Semple IA, et al. Sestrin2 inhibits uncoupling protein 1 expression through suppressing reactive oxygen species. *Proc Natl Acad Sci U S A* 2014;**111**(21):7849–54.
157. Han YH, Buffolo M, Pires KM, Pei S, Scherer PE, Boudina S. Adipocyte-specific deletion of manganese superoxide dismutase protects from diet-induced obesity through increased mitochondrial uncoupling and biogenesis. *Diabetes* 2016;**65**(9):2639–51.
158. Loh K, Deng H, Fukushima A, Cai X, Boivin B, Galic S, et al. Reactive oxygen species enhance insulin sensitivity. *Cell Metab* 2009;**10**(4):260–72.
159. Lee SJ, Kim SH, Park KM, Lee JH, Park JW. Increased obesity resistance and insulin sensitivity in mice lacking the isocitrate dehydrogenase 2 gene. *Free Radic Biol Med* 2016;**99**:179–88.
160. Chouchani ET, Kazak L, Jedrychowski MP, Lu GZ, Erickson BK, Szpyt J, et al. Mitochondrial ROS regulate thermogenic energy expenditure and sulfenylation of UCP1. *Nature* 2016;**532**(7597):112–6.
161. Chouchani ET, Kazak L, Spiegelman BM. Mitochondrial reactive oxygen species and adipose tissue thermogenesis: bridging physiology and mechanisms. *J Biol Chem* 2017;**292**(41):16810–6.
162. Barja de Quiroga G, Lopez-Torres M, Perez-Campo R, Abelenda M, Paz Nava M, Puerta ML. Effect of cold acclimation on GSH, antioxidant enzymes and lipid peroxidation in brown adipose tissue. *Biochem J* 1991;**277** (Pt 1):289–92.

163. Carriere A, Jeanson Y, Berger-Muller S, Andre M, Chenouard V, Arnaud E, et al. Browning of white adipose cells by intermediate metabolites: an adaptive mechanism to alleviate redox pressure. *Diabetes* 2014;**63**(10):3253–65.
164. Leverve XM, Mustafa I. Lactate: a key metabolite in the intercellular metabolic interplay. *Crit Care* 2002;**6**(4):284–5.
165. Romero-Garcia S, Moreno-Altamirano MM, Prado-Garcia H, Sanchez-Garcia FJ. Lactate contribution to the tumor microenvironment: mechanisms, effects on immune cells and therapeutic relevance. *Front Immunol* 2016;**7**:52.
166. Chechi K, Voisine P, Mathieu P, Laplante M, Bonnet S, Picard F, et al. Functional characterization of the Ucp1-associated oxidative phenotype of human epicardial adipose tissue. *Sci Rep* 2017;**7**(1):15566.
167. Bai Y, Shang Q, Zhao H, Pan Z, Guo C, Zhang L, et al. Pdcd4 restrains the self-renewal and white-to-beige transdifferentiation of adipose-derived stem cells. *Cell Death Dis* 2016;**7**:e2169.
168. Kim N, Nam M, Kang MS, Lee JO, Lee YW, Hwang GS, et al. Piperine regulates UCP1 through the AMPK pathway by generating intracellular lactate production in muscle cells. *Sci Rep* 2017;**7**:41066.
169. Li G, Xie C, Lu S, Nichols RG, Tian Y, Li L, et al. Intermittent fasting promotes white adipose browning and decreases obesity by shaping the gut microbiota. *Cell Metab* 2017;**26**(4):672–85. e4.
170. Ukropec J, Anunciado RP, Ravussin Y, Hulver MW, Kozak LP. UCP1-independent thermogenesis in white adipose tissue of cold-acclimated Ucp1−/− mice. *J Biol Chem* 2006;**281**(42):31894–908.
171. Meyer CW, Willershauser M, Jastroch M, Rourke BC, Fromme T, Oelkrug R, et al. Adaptive thermogenesis and thermal conductance in wild-type and UCP1-KO mice. *Am J Phys Regul Integr Comp Phys* 2010;**299**(5):R1396–406.
172. Golozoubova V, Hohtola E, Matthias A, Jacobsson A, Cannon B, Nedergaard J. Only UCP1 can mediate adaptive nonshivering thermogenesis in the cold. *FASEB J* 2001;**15**(11):2048–50.
173. Rowland LA, Bal NC, Kozak LP, Periasamy M. Uncoupling protein 1 and sarcolipin are required to maintain optimal thermogenesis, and loss of both systems compromises survival of mice under cold stress. *J Biol Chem* 2015;**290**(19):12282–9.
174. Ikeda K, Kang Q, Yoneshiro T, Camporez JP, Maki H, Homma M, et al. UCP1-independent signaling involving SERCA2b-mediated calcium cycling regulates beige fat thermogenesis and systemic glucose homeostasis. *Nat Med* 2017;**23**(12):1454–65.
175. Kazak L, Chouchani ET, Jedrychowski MP, Erickson BK, Shinoda K, Cohen P, et al. A creatine-driven substrate cycle enhances energy expenditure and thermogenesis in beige fat. *Cell* 2015;**163**(3):643–55.
176. Kazak L, Chouchani ET, Lu GZ, Jedrychowski MP, Bare CJ, Mina AI, et al. Genetic depletion of adipocyte creatine metabolism inhibits diet-induced thermogenesis and drives obesity. *Cell Metab* 2017;**26**(4):693.
177. Muller S, Balaz M, Stefanicka P, Varga L, Amri EZ, Ukropec J, et al. Proteomic analysis of human brown adipose tissue reveals utilization of coupled and uncoupled energy expenditure pathways. *Sci Rep* 2016;**6**:30030.
178. Szabo I, Zoratti M. Now UCP(rotein), now you don't: UCP1 is not mandatory for thermogenesis. *Cell Metab* 2017;**25**(4):761–2.
179. Klopstock T, Naumann M, Seibel P, Shalke B, Reiners K, Reichmann H. Mitochondrial DNA mutations in multiple symmetric lipomatosis. *Mol Cell Biochem* 1997;**174**(1-2):271–5.
180. Stewart JB, Chinnery PF. The dynamics of mitochondrial DNA heteroplasmy: implications for human health and disease. *Nat Rev Genet* 2015;**16**(9):530–42.
181. Ye W, Chen S, Jin S, Lu J. A novel heteroplasmic mitochondrial DNA mutation, A8890G, in a patient with juvenileonset metabolic syndrome: a case report. *Mol Med Rep* 2013;**8**(4):1060–6.

CHAPTER

# 8

# Role of Mitochondria in the Liver Metabolism in Obesity and Type 2 Diabetes

*Hisayuki Katsuyama*[*,†], *Juliane K. Czeczor*[*,†], *Michael Roden*[*,†,‡]

*Institute for Clinical Diabetology, German Diabetes Center, Leibniz Center for Diabetes Research at Heinrich-Heine University, Düsseldorf, Germany †German Center for Diabetes Research, Düsseldorf, Germany ‡Division of Endocrinology and Diabetology, Medical Faculty, Heinrich-Heine University, Düsseldorf, Germany

## 1 INTRODUCTION

Obesity and type 2 diabetes mellitus (T2DM) relate to ectopic triglyceride storage and impaired insulin sensitivity, that is, insulin resistance. In the liver, insulin resistance leads to reduced glycogen storage, but favors excess triglyceride accumulation, also called hepatic steatosis. Hepatic steatosis associates with whole-body and hepatic insulin resistance in non-obese[1] and obese individuals,[2] as well as patients with T2DM.[3]

Nonalcoholic fatty liver diseases (NAFLD) are characterized by hepatic steatosis (>5% of steatotic hepatocytes identified by histology or >5.6% by nuclear magnetic resonance techniques) in the absence of specific causes of hepatic triglyceride accumulation, such as relevant alcohol consumption, use of steatogenic medication, or hereditary disorders.[4–6] NAFLD includes simple steatosis, nonalcoholic steatohepatitis (NASH), fibrosis/cirrhosis, and hepatocellular carcinoma (HCC). Based on a meta-analysis of data collected between 1989 and 2015, the global prevalence of NAFLD is estimated to be 25% with 3%–16% of adults affected in Europe and the USA,[6,7] with 10%–25% of nonalcoholic steatohepatitis patients progressing to cirrhosis.[8] Overall, NASH associates with a >10-fold increased risk of liver-related death.[9] NAFLD also is a risk factor for an increased incidence of cardiovascular events,[10] which might at least partly relate to the atherogenic lipoprotein profile of NAFLD patients[4] (Fig. 1).


https://doi.org/10.1016/B978-0-12-811752-1.00008-0

The presence of T2DM also positively associates with NAFLD and, therefore, can be considered one of its comorbidities.[11,12] About 50%–87% of obese patients with T2DM also have NAFLD,[13–15] which accelerates the progression to more severe forms, such as NASH, cirrhosis, or HCC.[15–17] NAFLD also increases the long-term risk of other T2DM comorbidities, such as chronic kidney disease, by nearly 40%. Recent reports have revealed an association of lower cardiovagal tone and baroreflex sensitivity with hepatic steatosis in T2DM patients, suggesting a role of autoimmune neuropathy in NAFLD.[18]

Various factors have been implicated in the pathogenesis of ectopic lipid storage and insulin resistance. Alterations in hepatic energy metabolism recently have been linked to the development and progression of NAFLD.[11,19–21] This chapter will focus on the role of hepatic mitochondria in the regulation of insulin resistance and molecular mechanisms of steatosis in NAFLD.

We searched the PubMed online database for relevant articles published in English until December 2017, using the following terms: "hepatic mitochondrial function and insulin resistance," "hepatic mitochondrial function and obesity," "hepatic mitochondrial function and NAFLD," "hepatic mitochondrial function and diabetes," "hepatic energy metabolism and insulin resistance." We also retrieved articles from the reference lists of original articles, reviews, and our own expertise. Our literature search addressed mainly human studies, but we also included certain animal studies, which contribute to the understanding the pathophysiology and relationship between hepatic mitochondrial function and insulin resistance.

## 2 HEPATIC INSULIN RESISTANCE IN HUMANS

The liver plays a crucial role in whole-body energy homeostasis by regulating the metabolism of nutrients. During fasting, the liver is responsible for almost all endogenous glucose production (EGP) by de novo synthesis of glucose (gluconeogenesis) and by breakdown of glycogen (glycogenolysis).[11] Glycogenolysis initially accounts for ~50% of EGP; the contribution of gluconeogenesis increases with the duration of fasting.[22] This is a result of a lower portal vein insulin:glucagon ratio.[23] After meal ingestion with high insulin:glucagon ratios, the liver rapidly shifts toward glucose storage by suppression of EGP because of an insulin-mediated increase in glycogen synthesis and decreased gluconeogenesis. In the liver, insulin action leads to Akt phosphorylation, with subsequent inactivation of glycogen synthase kinase, which in turn stimulates glycogen synthase. The reduction in gluconeogenesis involves both insulin-mediated inactivation of FOXO1 in the liver and insulin-mediated inhibition of adipose tissue lipolysis.[24] Recent studies suggest that adipose tissue insulin resistance seems to be mainly responsible for the impairment of the rapid meal-dependent suppression of EGP via unrestrained fluxes of acetate and glycerol to the liver.[24,25] Patients with T2DM feature higher EGP and lower hepatic glycogen synthesis rates after meal ingestion, even during a hyperglycemic-hyperinsulinemic clamp compared with nondiabetic, but otherwise matched, individuals.[3] This study further showed a negative correlation between insulin-stimulated hepatic glycogen synthesis and hepatic triglyceride content. Together, these data suggest a close relationship between hepatic triglyceride storage and insulin resistance.

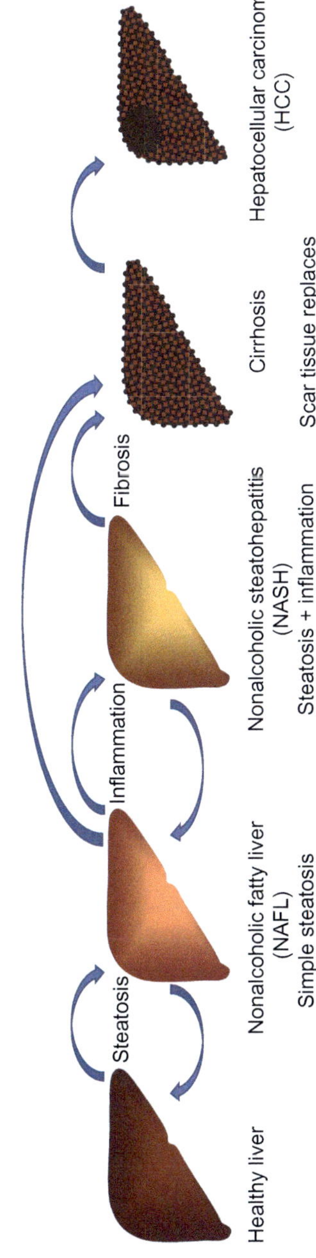


FIG. 1 The spectrum of NAFLD. NAFL can progress to NASH, cirrhosis, and hepatocellular carcinoma. Each stage is defined by pathological features.[6]

Storage of triglycerides depends on the hepatocellular fatty acid pool, which results from balanced formation from circulating free fatty acids (FFA) and from hepatic sources because of de novo lipogenesis, lipoprotein uptake, and triglyceride breakdown, as well as utilization through hepatic triglyceride synthesis and oxidation.[11] In insulin-resistant states, insulin-mediated suppression of adipose tissue lipolysis is impaired, giving rise to increased circulating FFA,[26] which contributes to the development of hepatic insulin resistance as reflected by impaired insulin-mediated suppression of EGP.[27] During de novo lipogenesis in the liver, fatty acids are synthesized from acetyl-coenzyme A (acetyl-CoA), originating from excess hepatic carbohydrates uptake. In nonobese humans, adipose tissue lipolysis accounts for ~90% of plasma FFA during fasting and ~60% after meal ingestion, leading to the conclusion that adipose lipolysis serves as the most important hepatic source of fatty acids.[28] The contribution of splanchnic lipolysis to hepatic FFA delivery ranges from 10% to 50%.[29] De novo lipogenesis accounts for 5% of hepatic triglyceride during fasting and up to 23% after a meal.[30] In obese patients with NAFLD, de novo lipogenesis is three times higher than obese persons with low liver triglyceride contents,[31] contributing to 23% of the hepatocellular fatty acid pool, whereas circulating FFA and dietary fat contribute to 59% and 15% of hepatocellular triglycerides, respectively.[28] Insulin and glucose upregulate the transcription of lipogenic enzymes such as sterol regulatory binding protein-1c (SREBP-1c) and carbohydrate response element binding protein (ChREBP), respectively.[11,32] Consequently, chronic hyperinsulinemia and hyperglycemia, as well as high-carbohydrate and high-fat diets will stimulate de novo lipogenesis in obese and insulin-resistant individuals. In addition, high-caloric meals will lead to lipid-induced insulin resistance of skeletal muscle, which in turn stimulates insulin secretion and shifts lipids toward the liver.[33] Although removal of hepatic triglycerides and secretion of very low-density lipoprotein (VLDL) is increased in obese patients with NASH,[34] these mechanisms do not normalize hepatic triglyceride content, which might be because of impaired apolipoprotein B (ApoB) synthesis[35] and/or inadequate lipid oxidation.[11]

The mechanisms underlying hepatic insulin resistance in the context of NAFLD have been studied mostly in animal models because of limitations on access to human liver tissue. In addition to abnormal mitochondrial function and oxidative stress, accumulation of excess lipid intermediates (lipotoxicity) or glucose metabolism (glucotoxicity), abnormal secretion of adipokines and inflammatory pathways likely are involved. Lipotoxic insulin resistance can result from accumulation of long-chain fatty acyl-CoA (LCFA-CoA), diacylglycerols (DAG), ceramides and other sphingolipids and/or acyl-carnitines, which link lipid metabolism to incomplete mitochondrial lipid oxidation. Similar, but not identical to skeletal muscle,[36] hepatic DAG activate novel protein kinase C (PKC) isoforms such as PKCε, which inhibits tyrosine phosphorylation of the insulin receptor, causing insulin resistance.[37] Some human studies also suggest the interaction of the DAG-PKCε pathway. Obese patients with T2DM feature higher PKC isoforms (α, ε, ζ) in the membrane fraction of liver biopsies.[38] Obese individuals present with higher hepatic DAG content, which negatively correlates to insulin-mediated suppression of hepatic EGP and activation of hepatic PKCε.[39,40] Studies about the relationship of hepatic ceramides and hepatic insulin resistance published controversial results with increased ceramides from the synthetic de novo ceramide pathway in insulin-resistant individuals with obesity and NAFLD[41] or no such relationship at all.[39] We have found that total serum ceramides, dihydroceramides, and hepatic dihydroceramides 22:0 and 24:1 negatively correlate with whole-body, but not with hepatic insulin sensitivity. Certain sphingolipids are

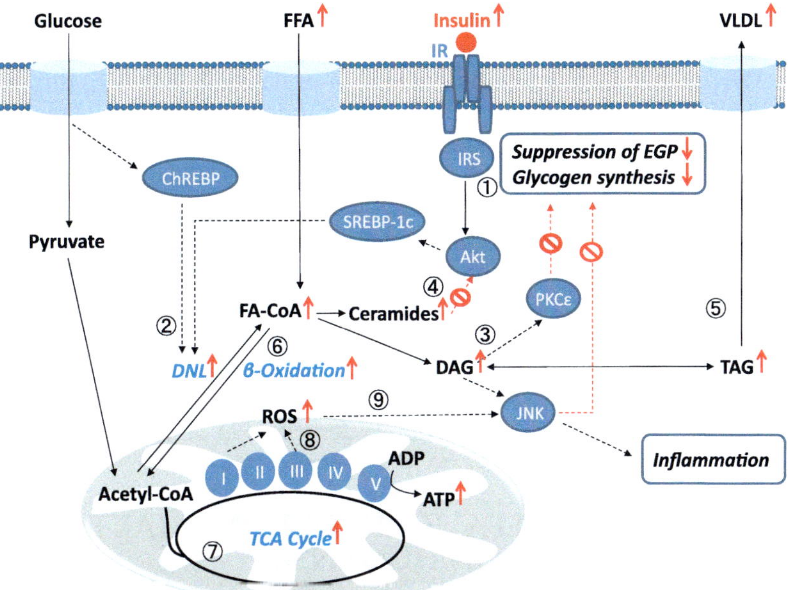


FIG. 2 Features of hepatic insulin resistance. Impairment of insulin-mediated suppression of endogenous glucose production (EGP) and glycogen synthesis characterizes hepatic insulin resistance.[1] Hyperinsulinemia and hyperglycemia increase de novo lipogenesis (DNL) by upregulation of sterol regulatory element binding-protein 1c (SREBP-1c) and carbohydrate response element binding-protein (ChREBP).[2] Higher availability of free fatty acids (FFA) gives rise to diacylglycerol (DAG) and ceramides, which promote insulin resistance through the activation of protein kinase C (PKC)ε[3] or c-Jun N-terminal kinase (JNK).[4] Secretion of very low-density lipoprotein (VLDL) is increased to remove excess hepatic triglyceride content.[5] In addition, the hepatocyte attempts to limit fatty acyl-CoA (FA-CoA) by providing acetyl-CoA for mitochondrial β-oxidation.[6] In turn, tricarboxylic acid (TCA) cycle in the mitochondria is upregulated.[7] Excessive mitochondrial respiration though the electron transport chain at the expense of coupling along with impaired antioxidative capacity can cause reactive oxygen species (ROS) generation,[8] which promotes inflammation via JNK.[9]

increased only in NASH patients and correlated with hepatic maximal respiration (lactosylceramides subspecies, sphinganine, and lactosylceramide 14:0), $H_2O_2$ emission (total ceramides and species 14:0, 20:0, and 24:0, certain dihydro- and lactosylceramides), lipid peroxides (total ceramides, sphingomyelin 22:0), and inflammatory pathways (ceramide 24:0, hexosylceramides 22:0, 24:0, 24:1). An association of hepatic acylcarnitine accumulation and insulin resistance was reported mostly in rodent models.[42] A report on obese individuals revealed no association between hepatic insulin sensitivity and intrahepatic acylcarnitine[40] (Fig. 2).

## 3 HEPATIC LIPID OXIDATION AND ATP GENERATION

Hepatic oxidation of lipids occurs mainly in mitochondria (β-oxidation), but also in peroxisomes (β-oxidation) and endoplasmic reticulum (ω-oxidation), and is stimulated by glucagon and other hormones, but inhibited by insulin.[43–45] Mitochondria play a crucial role in substrate oxidation and energy conversion into adenosine triphosphate (ATP). Mitochondrial β-oxidation generates acetyl-CoA to feed the tricarboxylic acid (TCA) cycle or to yield ketone bodies (acetoacetate and β-hydroxy-butyrate) by both β-oxidation and the TCA cycle increasing NADH and FADH2 for oxidative phosphorylation (OXPHOS) through mitochondrial electron transport chain (ETC) complexes to produce ATP. Peroxisomal β-oxidation is involved in chain shortening of very long-chain fatty acids ($>C_{20}$) for subsequent mitochondrial oxidation, but also metabolizes long-chain dicarboxylic acids generated by microsomal ω-oxidation via acyl-CoA oxidase. Lack of acyl-CoA oxidase induces severe hepatic steatosis.[46] Furthermore, increased peroxisomal β-oxidation produces $H_2O_2$ and contributes to oxidative stress.[47,48] Microsomal ω-oxidation metabolizes very long-chain fatty acids through cytochrome P4504A enzymes to dicarboxylic acids.[43,49] Excessive fat availability can give rise to these toxic dicarboxylic acids and to hepatic oxidative stress.[50] Activation of peroxisomal proliferator activated receptor (PPAR)α controls mitochondrial, peroxisomal, and microsomal oxidation, which leads to reduction of hepatic triglyceride storage and favors energy supply to peripheral tissues.[51] Mitochondrial function is described in detail in Chapter 1.

# 4 ASSESSMENT OF HEPATIC MITOCHONDRIAL FUNCTION IN HUMANS

Mitochondria comprise various features that require different methodologies for adequate assessment.[19] Human mitochondria can be examined by several in vivo and ex vivo techniques ranging from morphometry to functional analyses. Basal mitochondrial activity refers to the resting $O_2$ consumption or oxidative phosphorylation flux, which depends both on supplied substrates such as adenosine diphosphate (ADP) or $O_2$ and cellular demand for ATP synthesis (ADP:ATP ratio). Stimulation by energy depletion in vivo and exogenous substrate excess allows for assessment of sub- to maximal mitochondrial activity. Mitochondrial plasticity is defined by changes of mitochondrial activity upon altered metabolic conditions such as hyperinsulinemia.[52] The inability to adapt fuel oxidation to substrate availability is termed mitochondrial inflexibility and compromises mitochondrial plasticity.

## 4.1 Mitochondrial Morphology

Transmission electron microscopy (TEM) makes it possible to assess mitochondrial morphology and content as the percentage of whole-cell volume occupied by mitochondria in biopsy samples.[53,54] Several other markers of mitochondrial content also are in use, such as the ratio of mitochondrial DNA to nuclear DNA copy numbers,[55] cardiolipin content, complex I–V protein content, and complex I–IV activity, or enzymes involved in oxidative phosphorylation, including citrate synthase activity (CSA).[56–59]

## 4.2 Phosphorus Magnetic Resonance Spectroscopy ($^{31}$P-MRS)

Spectra obtained by $^{31}$P-MRS contain three nonequivalent phosphate groups of ATP (α, β and γ), which differ in resonance frequencies, inorganic phosphate (Pi), phosphomonoesters (PME), and phosphodiesters (PDE). $^{31}$P-MRS can provide a measure of resting mitochondrial activity by measuring either absolute hepatic concentrations of ATP[60–62] or flux rates through hepatic ATP synthase (ATP turnover) from the hepatic Pi-to-ATP exchange ratio, obtained from experimental magnetization saturation transfer.[63]

$^{31}$P-MRS allows the measurement of submaximal oxidative phosphorylation after a metabolic challenge by administration of gluconeogenic substrates such as fructose or L-alanine.[64–69] Upon entering hepatocytes, fructose is phosphorylated rapidly by fructokinase, thereby lowering hepatic ATP concentration. ATP depletion and the time course of its regeneration, defined as the ratio of final to minimum ATP and expressed as fractional ATP recovery, reflects submaximal OXPHOS capacity of the liver. Proton magnetic resonance spectroscopy ($^{1}$H-MRS) provides complementary information. Because triglycerides yield the strongest signal in the $^{1}$H-spectrum, $^{1}$H-MRS allows for the noninvasive quantification of lipid content in various organs, including the liver.[60,63,69,70]

## 4.3 Positron Emission Tomography (PET)

[$^{11}$C]palmitate PET imaging allows monitoring of fatty acid uptake, oxidation, and esterification in the liver.[71] This method, however, has some limitations resulting from high costs and time constraints because of the short-lived tracer and ionizing radiation, and it does not specifically study mitochondria.

## 4.4 Stable Isotope Tracer Techniques

Mass isotope distribution analysis (MIDA) of glucose isolated from a blood sample upon [U-$^{13}$C]propionate administration allows quantification of the flux rates through TCA cycle flux and anaplerosis.[72,73] Another method combined continuous infusion of [1-$^{13}$C]acetate with direct monitoring of $^{13}$C label incorporation into hepatic [5-$^{13}$C]glutamate and [1-$^{13}$C] glutamate using in vivo $^{13}$C-MRS (Schumann 1991), which measures flux rates through citrate synthase and pyruvate kinase.[74,75] The methods yield discrepant results raising questions about their application.[76] A new simpler method, a positional isotopomer NMR tracer analysis (PINTA), has been introduced to noninvasively assess rates of hepatic mitochondrial oxidation and anaplerotic flux from infusion of [3-$^{13}$C]lactate.[77] Finally, the [$^{13}$C]octanoate breath test has been employed to assess mitochondrial β-oxidation by measuring $^{13}CO_2$ enrichment in exhaled air.[78] This is an indirect measure of oxidation rates and can be expressed only in terms of body mass. Thus, application in humans with differing body composition is limited.

## 4.5 High-Resolution Respirometry (HRR)

HRR allows for the direct quantification of mitochondrial respiration in permeabilized liver tissue and in isolated hepatic mitochondria. This technique measures maximal oxidative capacity from monitoring $O_2$ flux rates upon sequential exposure of samples to substrates such as malate, octanoyl-carnitine, glutamate, succinate carbonyl cyanide 4-trifluoromethoxy-phenylhydrazone, leak, and respiratory control or titrating ADP concentrations. State-of-the-art oxygraphs provide standardized measurements in temperature-controlled respiratory chambers.[57] The two chambers enable measurement of oxygen concentration by polarographic oxygen sensors. Oxygen consumption is corrected for tissue wet mass or cell number.

### 4.6 Gene Expression Profiling

Microarray analyses and quantitative real-time polymerase chain reaction are used to measure expression of mRNA coding for proteins related to oxidative phosphorylation, mitochondrial biogenesis, and dynamics in liver biopsy samples.[55]

## 5 MITOCHONDRIAL FUNCTION IN LEAN AND OBESE HUMANS WITHOUT NAFLD

Only a few studies have been published about human hepatic mitochondrial function in lean and obese individuals without NAFLD. To investigate the role of hepatic energy metabolism in metabolically healthy humans, our research group examined 76 middle-age humans by combining oral glucose tolerance tests with liver $^{31}$P-MRS.[79] After adjustment for age, sex, and body mass, hepatic ATP and Pi related to postglucose challenge glycemia and Pi as well as FFA, but not with measures of insulin sensitivity. Circulating leucine and palmitoleic acid accounted for 26% and 15% of the variance in ATP and Pi, respectively. This indicates that specific circulating amino acids and FFA partly affect hepatic mitochondrial function in nondiabetic humans.

As excessive lipid availability is known to induce insulin resistance in skeletal muscle and liver,[80] we further monitored hepatic metabolism using in vivo $^{13}$C/$^{31}$P/$^{1}$H-MRS and ex vivo $^{2}$H-MRS before and during hyperinsulinemic-euglycemic clamps in lean healthy individuals, who randomly received either palm oil or vehicle as control.[81] Palm oil administration increased hepatic triglycerides and ATP rose by 35% and 16%, respectively, along with 70% higher hepatic gluconeogenesis and 20% lower net glycogenolysis. In parallel experiments, our research group performed identical clamps and hepatic transcriptome analyses in nondiabetic lean mice. Decreased hepatic insulin sensitivity, higher gluconeogenesis and lower glycogenolysis also were observed in these mice. Liver transcriptomics in mice revealed that palm oil differentially regulated inflammatory and PPAR pathways. These findings suggest that ingestion of saturated fat rapidly increased hepatic lipid storage and energy metabolism along with insulin resistance and alterations of the regulation of hepatic genes, possibly related to the development of NAFLD.

To evaluate the role of chronically increased lipid availability, we compared absolute concentrations of hepatic phosphorus compounds in nine elderly persons ranging from overweight to obese and nine young lean individuals.[60] The elderly group exhibited lower whole-body insulin sensitivity, but similar hepatic lipid volume fraction (HLVF) compared with the younger group. No significant differences were found in hepatic ATP, Pi, PDE, PME, or related ratios. The mean values of ATP tended to be higher in the elderly overweight-obese individuals ($2.50 \pm 0.61$ vs. $2.26 \pm 0.29$ mmoL/L), of whom three already had hepatic steatosis, whereas none in the young lean individuals had hepatic steatosis. Subsequently, we applied HRR to measure mitochondrial activity in liver samples from severely obese humans without NAFLD undergoing bariatric surgery compared with age-matched lean humans undergoing surgery for nonmalignant diseases such as cholecystectomy and herniotomy.[57] In spite of comparable liver triglyceride content below 5%, the obese individuals had lower whole-body insulin sensitivity than the lean individuals. In the face of similar mitochondrial content, rates of $O_2$ flux were 4.3- to 5-fold higher for both β-oxidation and TCA cycle in isolated hepatic mitochondria from obese compared with lean humans.[57] Nevertheless, hepatic respiratory control ratio, reflecting mitochondrial coupling activity, was reduced, and lipid peroxidation products were increased in the obese group. These data show that hepatic mitochondrial respiration is upregulated in obesity even in the absence of hepatic steatosis. In spite of hepatic mitochondria adapting to altered substrate availability and energy demand in obesity, there is evidence for early abnormalities in hepatic mitochondrial function in obese people (Table 1).

TABLE 1 Studies on Hepatic Mitochondrial Function in the Context of Obesity and Hepatic Steatosis

| 1st Author (Year of Publication) | Cohorts | Insulin Sensitivity | Methodology | Mitochondrial Function |
|---|---|---|---|---|
| Sanyal (2001) | 6–10 NASH<br>6 NAFL<br>6 CON | GIR (mg/kg FFM/min)<br>4.5 ± 1.3<br>7.7 ± 1.3<br>8.9 ± 1.5 | Serum β-OHB, lipid peroxidation by immunohistochemical staining for 3-NT, TEM | ↑β-Oxidation in NAFL |
| Nair (2003) | 7 obese<br>7 overweight<br>5 CON | NA | $^{31}$P-MRS before and after fructose challenge | Inverse correlation of ATP content with BMI, ≈ATP recovery |
| Chiappini (2006) | 20 NAFL<br>20 CON | NA | Global gene expression, ratio of mitochondrial DNA to nuclear DNA content | ↑Mitochondrial content and OXPHOS gene expression in NAFL |
| Misu (2007) | 14 T2DM<br>14 NGT | HOMA-IR<br>3.9 ± 3.6<br>2.2 ± 1.0 | SAGE, DNA chip analysis | ↑OXPHOS genes in T2DM, correlation with fasting glucose |
| Takamura (2008) | 21 T2DM<br>(10 obese, 11 nonobese) | NA | DNA chip analysis | ↑OXPHOS genes in obesity, correlation with HOMA-IR and QUICKI |
| Szendroedi (2009) | 9 T2DM<br>9 age, BMI-matched CON<br>9 young lean CON | M (mg/kg FFM/min)<br>4.5 ± 0.9<br>8.0 ± 1.1<br>12.5 ± 1.2 | $^{31}$P-MRS | ↓γ-ATP and Pi content in T2DM |
| Iozzo (2010) | 8 obese<br>7 CON | HOMA-IR<br>2.0 ± 0.4<br>0.9 ± 0.1 | $^{11}$C-palmitate PET | ↑FA oxidation in obesity |
| Schmid (2011) | 9 T2DM<br>8 CON | M (mg/kg/min)<br>5.3 ± 1.3<br>7.5 ± 1.9 | $^{31}$P-MRS | ↓fATP in T2DM, correlation with hepatic insulin sensitivity |
| Sunny (2011) | 8 NAFL<br>8 CON | Minimal model<br>2.1 ± 1.3<br>3.2 ± 1.6 | $^{13}$C NMR with $^{13}$C tracer | ↑Oxidative flux through TCA cycle in NAFL |

*Continued*

TABLE 1 Studies on Hepatic Mitochondrial Function in the Context of Obesity and Hepatic Steatosis—cont'd

| 1st Author (Year of Publication) | Cohorts | Insulin Sensitivity | Methodology | Mitochondrial Function |
|---|---|---|---|---|
| Abdelmalek (2012) | 25 T2DM including 16 NAFL, with high or low dietary fructose consumption | NA | $^{31}$P-MRS before and after fructose challenge | ↓ATP recovery in patients with high fructose consumption |
| Koliaki (2015) | 7 obese NASH<br>16 obese NAFL<br>18 obese no NAFL<br>12 CON | M (mg/min/kg)<br>1.4 ± 0.9<br>2.5 ± 0.3<br>3.5 ± 0.3<br>7.7 ± 0.8 | HRR in liver tissue and isolated mitochondria | ↑Maximal respiration in obese |
| Fritsch (2015) | 10 T2DM<br>10 Obese<br>10 CON | M (mg/min/kg)<br>1.9 ± 2.5<br>3.5 ± 1.4<br>8.8 ± 1.4 | $^{31}$P-MRS before and after high-caloric mixed meal | ↑hepatic ATP in obese, ≈hepatic ATP in T2DM |
| Satapati (2015) | 8 MS | NA | $^{13}$C NMR with $^{13}$C tracer | Correlation between hepatic oxidative flux and NAS |
| Petersen (2016) | 10 NAFL<br>20 CON | HOMA-IR<br>4.13 ± 0.82<br>2.47 ± 0.21 | $^{13}$C-MRS with $^{13}$C-acetate tracer | ≈Hepatic mitochondrial oxidation |
| Lund (2016) | 8 Obese T2DM<br>9 Obese<br>6 CON | Fasting insulin (pmol/l)<br>91 ± 16<br>88 ± 14<br>NA | HRR in liver tissue | ≈OXPHOS capacity |

*ATP*, adenosine triphosphate; *β-OHB*, β-hydroxybutyrate; *CON*, controls; *fATP*, flux through ATP synthesis; *FFM*, fat free mass; *HOMA-IR*, homeostasis model assessment index of insulin resistance; *IR*, insulin resistance; *MRS*, magnetic resonance spectroscopy; *MS*, metabolic syndrome; *NA*, not available; *NAFL*, nonalcoholic fatty liver; *NAS*, NAFLD activity score; *NASH*, nonalcoholic steatohepatitis; *NMR*, nuclear magnetic resonance; *3-NT*, 3-nitrotyrosine; *OXPHOS*, oxidative phosphorylation; *PET*, positron emission tomography; *Pi*, inorganic phosphate; *QUICKI*, quantitative insulin sensitivity check index; *SAGE*, serial analysis of gene expression; *TCA cycle*: tricarboxylic acid cycle; *TEM*, transmission electron microscopy; *T2DM*, type 2 diabetes mellitus.

# 6 MITOCHONDRIAL FUNCTION IN HEPATIC STEATOSIS AND T2DM

Starting in the late 1950s, studies in animal models revealed alterations of hepatic mitochondrial function in diabetes, and a relationship between impaired mitochondrial function and hepatic fat accumulation.[82–84] Later, several studies using electron microscopy reported morphological changes in liver mitochondria, by describing megamitochondria, intramitochondrial crystalline inclusions, mitochondrial matrix granules, and foamy cytoplasm in steatosis.[85]

The number of studies addressing hepatic energy metabolism using functional assays in human steatosis and T2DM is limited. These studies reported mixed results on decreased, unchanged, or increased hepatic mitochondrial function, which raised questions about methodologies and the quality of these studies. Recent evidence indicates that the concept of transient alterations of hepatic mitochondrial function might help to explain such differences by the phenotype of the study populations.[21]

There is evidence that obese nondiabetic humans with simple steatosis exhibit various features of increased hepatic mitochondrial function. The [$^{11}$C]palmitate PET method revealed doubled rates of hepatic fat oxidation in insulin-resistant obese humans, whereas fatty acid uptake and esterification rates were not different from lean volunteers.[71] Hepatic fat oxidation correlated positively to systemic insulin resistance. In line with this finding, the [U-$^{13}$C] propionate MIDA method also found doubled hepatic oxidative flux through the TCA, along with 25% higher gluconeogenic flux and 50% higher mitochondrial anaplerosis in obese patients with steatosis compared with volunteers without NAFLD. Mitochondrial oxidative and anaplerotic fluxes positively correlated to hepatic triglyceride content.[73] In our study using HRR in liver samples of bariatric surgery patients, we found that severely obese insulin-resistant humans with or without histologically proven steatosis, in absence of NASH, had similarly up to fivefold higher rates of maximal rates of mitochondrial respiration and greater lipid peroxidation products than lean humans.[57] Compared to obese humans without steatosis, patients with steatosis also featured lower expression of transcription factors regulating mitochondrial biogenesis and ETC complexes, but an increase in interleukin-1 receptor agonist (IL-1Ra). Another study found higher expression of interleukins, but also increased expression of OXPHOS genes and mitochondrial content in hepatic steatosis.[55] Nevertheless, this set of data suggests early signs of mitochondrial abnormalities in spite of elevated maximal oxidative capacity in hepatic steatosis without NASH (Fig. 3).

It is of interest to what extent chronically progressing insulin resistance in the context of steatosis would affect hepatic mitochondrial function. T2DM represents such a metabolic

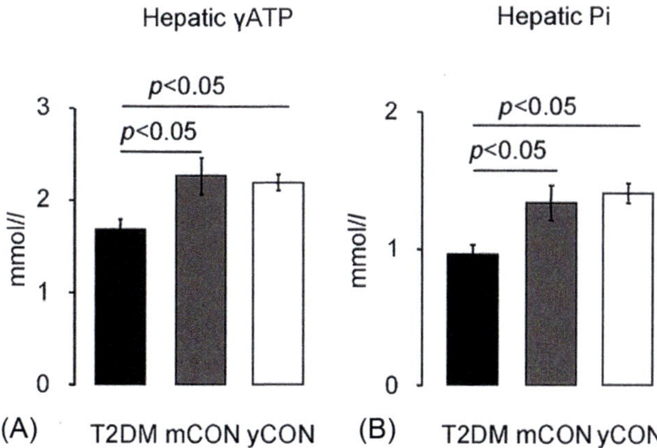


FIG. 3 Noninvasive measurements of hepatic energy metabolism in humans. Hepatic concentrations of (A) ATP and (B) inorganic phosphate (Pi) in patients with type 2 diabetes mellitus (T2DM), age- and BMI-matched nondiabetic (mCON) and young humans (yCON). Data are means ± SEM and derived from reference (60).

state combined with chronic hyperglycemia, which might further interfere with hepatic energy metabolism. We found that even T2DM patients with only moderate overweight/obesity with good glycemic control already had 26% lower ATP and 28% lower Pi than age- and body mass index (BMI, defined as body weight divided by the square of the body height)-matched nondiabetic humans.[60] Multiple linear regression analysis for the dependent variable Pi and ATP, including clinical characteristics, and HLVF identified hepatic insulin sensitivity as the single significant independent predictor explaining 57% of Pi and ATP variances.[60] In a similar group of T2DM patients, flux through hepatic ATP synthase was reduced 42% compared to age- and BMI-matched humans, despite comparable hepatic triglyceride content.[63] The hepatocellular Pi concentrations primarily determined the variability of ATP synthase flux. Flux through hepatic ATP synthase correlated positively with both hepatic and peripheral insulin sensitivity, and negatively with waist circumstance, fasting plasma glucose levels, BMI, liver triglycerides, and hemoglobin A1c.[63]

In order to examine the adaptation of mitochondrial function to acute metabolic challenges, other studies investigated the effects of insulin, fructose, or mixed meals on hepatic energy metabolism in T2DM. As cerebral insulin action might affect hepatic insulin sensitivity,[86] we monitored hepatic energy metabolism upon intranasal insulin application.[87] While intranasal insulin did not affect fasting hepatic insulin sensitivity, liver triglycerides decreased by 35% and hepatic ATP rose by 18% in nondiabetics, but not in T2DM patients. Another study examined the influence of dietary fructose consumption on hepatic ATP levels before and after an intravenous fructose challenge in obese patients with T2DM.[67] High dietary fructose consumers ($\geq$15 g/day) had slightly lower baseline hepatic ATP, greater ATP depletion, and impaired ATP regeneration after fructose challenge when compared with low dietary fructose individuals ($<$15 g/day). These data underline the role of mitochondrial function regarding dietary fructose-dependent initiation and progression of NAFLD.[6] Ingestion of a high-caloric liquid mixed meal significantly increased hepatic ATP concentrations only in young obese nondiabetics, but only marginally in obese T2DM individuals when compared to young lean persons.[68] Serial analysis of gene expression (SAGE) and DNA chip analysis in livers of obese T2DM patients revealed simultaneous upregulation of genes involved in OXPHOS and in pathways of glucose and lipid metabolism.[88,89] Although these findings are difficult to interpret, because all participants had undergone surgical treatments for malignant tumors and liver triglyceride concentrations were not reported, they point to possible time-dependent alterations of mitochondrial function in insulin-resistant states.

Animal studies revealed that hepatic insulin resistance already is present in the prediabetic state and is associated with a transient rise in mitochondrial respiration.[90] Furthermore, severe insulin deficiency and hyperglycemia activate the mitochondrial respiratory chain and stimulate coupling,[91] which might explain the transient adaptation of mitochondria in the liver of patients with diabetes. In this context, animal models of insulin resistance and/or diabetes mellitus revealed an initial upregulation followed by downregulation of mitochondrial function during progression of metabolic diseases.[92] These results suggest that prolonged hyperglycemia and hyperinsulinemia, as well as an increased hepatic lipid availability under conditions of chronic insulin resistance, progressively impair hepatic mitochondrial flexibility (Fig. 4).

This concept also might help to understand the failure of some studies to detect significant differences in hepatic mitochondrial function in certain studies. Nevertheless, methodological

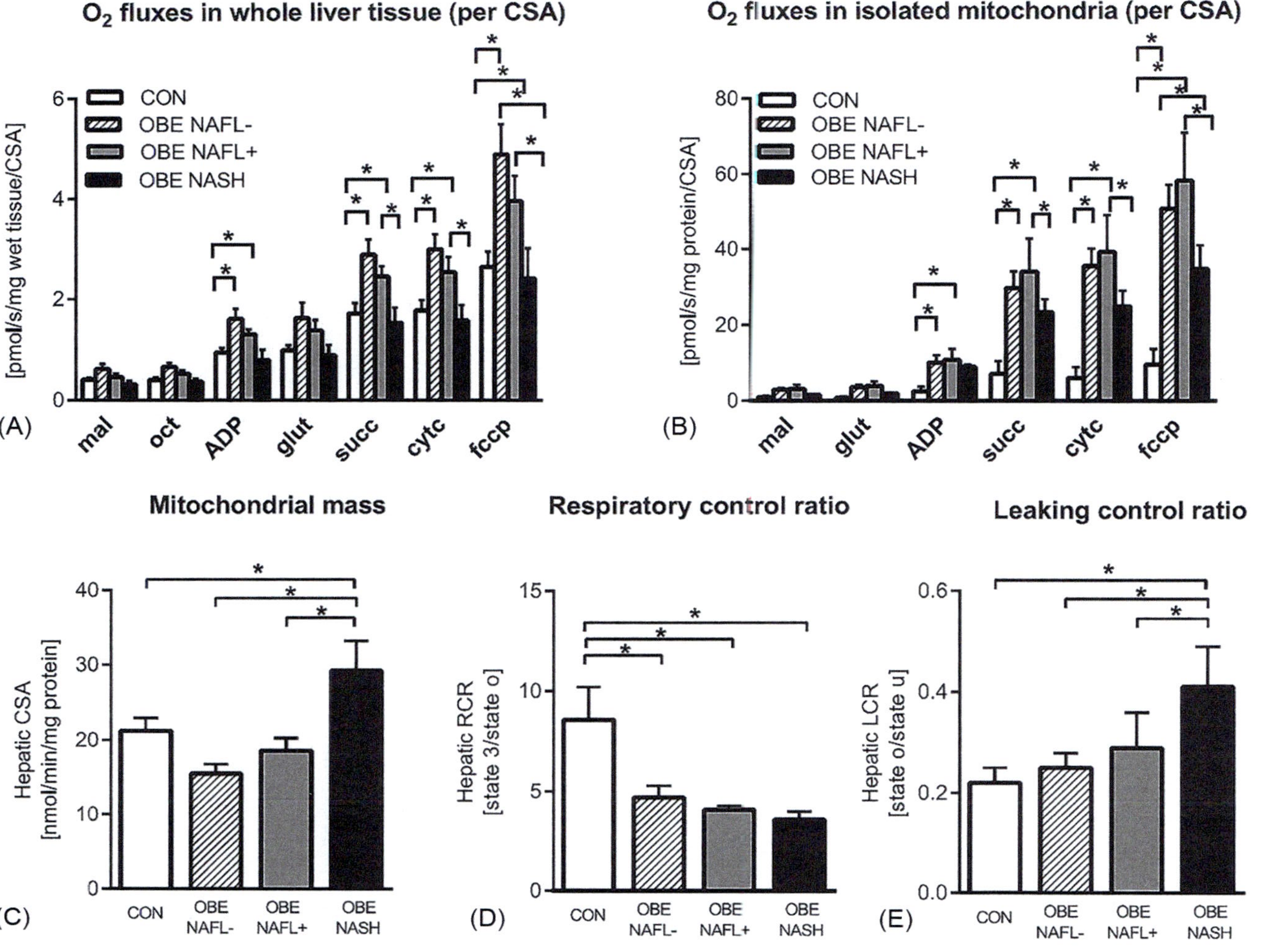


FIG. 4 Hepatic mitochondrial function and content in lean (CON) or obese (OBE) humans with or without steatosis (NAFL) or OBE with NASH. $O_2$ fluxes in (A) whole tissue and (B) isolated mitochondria upon adenosine diphosphate (ADP), cytochrome c (cyt c), substrates (*mal*, malate; *oct*, octanoyl-carnitine; *glut*, glutamate; *succ*, succinate), and carbonyl cyanide 4-trifluoromethoxy-phenylhydrazone (FCCP) as uncoupling factor. All fluxes are normalized to citrate synthase activity (CSA). (C) CSA as a measure of mitochondrial mass. (D) Respiratory control ratio (RCR) defined as state 3/state o serving as marker of mitochondrial coupling. (E) Leaking control ratio (LCR) defined as state o/state u as an index of proton leak. *Data are means ± SEM and derived from Koliaki C, Szendroedi J, Kaul K, Jelenik T, Nowotny P, Jankowiak F, et al. Adaptation of hepatic mitochondrial function in humans with non-alcoholic fatty liver is lost in steatohepatitis. Cell Metab. 2015;21(5):739–46.*

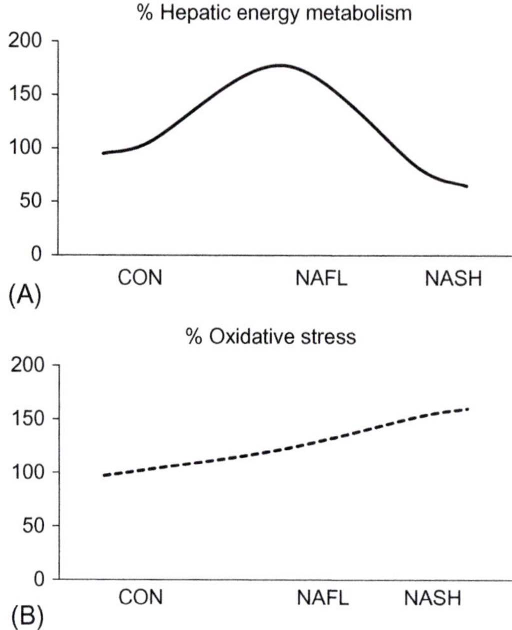


FIG. 5 Alterations in (A) hepatic energy metabolism and (B) oxidative stress in the liver ranging from lean humans with low liver fat content (CON) to obese humans with nonalcoholic fatty liver (NAFL) or nonalcoholic steatohepatitis (NASH). Different features of hepatic energy metabolism, such as oxidative phosphorylation capacity, β-oxidation, respiration rates, electron transport chain complex activities, oxidative or nonoxidative mitochondrial metabolism, were obtained from studies including healthy CON.[21] Oxidative stress was calculated from hepatic $H_2O_2$ emission.[57] The respective % changes are compared to the data of the respective CON, which were set as 100%. Data are means ± SEM.

aspects also need to be considered to explain the discrepancies among studies. Early fructose challenge experiments found comparable hepatic ATP recovery between obese/overweight and lean humans.[69] This study, however, lacked measurements of insulin sensitivity and hepatic steatosis. Another study found no differences in hepatic oxidative capacity by HRR between individuals with obesity and T2DM and nondiabetic humans.[93] The mean BMI of the nondiabetic group indicated that not all control volunteers were lean, the hepatic lipid droplet densities and areas exhibited a broad variation and were not different between groups, and insulin sensitivity was not reported in this study. Finally, the $^{13}$C tracer method combined with $^{13}$C-MRS detected no differences in mitochondrial oxidation and pyruvate cycling in livers from volunteers with or without steatosis.[74] The people with hepatic steatosis were mostly lean or only slightly overweight and thereby likely represented a specific NAFLD subcohort (Fig. 5, Table 2).

## 7 MITOCHONDRIAL FUNCTION IN NASH

Abnormal mitochondria have been particularly implicated in the progression of NAFL to NASH, fibrosis, and cirrhosis.[44,94] Morphological changes in NASH comprise swollen mitochondria, intramitochondrial crystalline inclusions, mitochondrial matrix granules,

TABLE 2 Studies on Hepatic Mitochondrial Function in NASH

| 1st Author (Year of Publication) | Cohorts | Insulin Sensitivity | Methodology | Mitochondrial Function |
|---|---|---|---|---|
| Cortez-Pinto (1999) | 8 NASH<br>7 CON | 4 of 8 NASH patients FPG > 120 mg/dL | $^{31}$P-MRS before and after fructose challenge | ↓ATP recovery in NASH |
| Sanyal (2001) | 6–10 NASH<br>6 NAFL<br>6 CON | GIR (mg/kg FFM/min)<br>4.5 ± 1.3<br>7.7 ± 1.3<br>8.9 ± 1.5 | Serum β-OHB, lipid peroxidation by immunohistochemical staining for 3-NT, TEM | ↑β-Oxidation in NASH, structural defects in NASH |
| Miele (2003) | 10 NASH<br>20 CON | NA | $^{13}$C-octanoate breath test | ↑β-Oxidation in NASH |
| Pérez-Carreras (2003) | 43 NASH<br>16 CON | HOMA-IR<br>4.5 ± 2.38<br>1.41 ± 0.71 | MRC complexes activities in liver tissue | ↓Activity of MRC complexes in NASH |
| Serviddio (2008) | 10 NASH<br>8 CON | NA | UCP-2 expression, redox balance, protein oxidation | ↑UCP-2 and oxidative stress in NASH |
| Koliaki (2015) | 7 obese NASH<br>16 obese NAFL<br>18 obese no NAFL<br>12 CON | M (mg/min/kg)<br>1.4 ± 0.9<br>2.5 ± 0.3<br>3.5 ± 0.3<br>7.7 ± 0.8 | HRR in liver tissue and isolated mitochondria | ↓Maximal respiration in NASH |
| Traussnigg (2017) | 22 NASH<br>8 NAFL | HOMA-IR<br>3.8 ± 2.9<br>1.7 ± 0.6 | $^{31}$P-MRS | ↓ATP flux in NASH |

*ATP*, adenosine triphosphate; *β-OHB*, β-hydroxybutyrate; *CON*, controls; *FPG*, fasting plasma glucose; *HOMA-IR*, homeostasis model assessment index of insulin resistance; *IR*, insulin resistance; *MRS*, magnetic resonance spectroscopy; *NA*, not available; *NAFL*, nonalcoholic fatty liver; *NASH*, nonalcoholic steatohepatitis; *3-NT*, 3-nitrotyrosine; *TEM*, transmission electron microscopy; *UCP-2*, uncoupling protein 2.

and foamy cytoplasm.[95,96] Functional changes have been reported first by monitoring hepatic ATP before and one hour after fructose infusion using $^{31}$P-MRS.[97] In spite of identical ATP levels before and after fructose infusion, patients with biopsy-proven NASH displayed impaired fructose-induced ATP depletion compared with age- and sex-matched healthy humans. Likewise, mitochondrial ETC complex activity was lower in patients with biopsy-proven NASH compared with nonobese humans in spite of comparable mitochondrial content as assessed by CSA.[56] The impairment of ETC complex activity positively correlated with BMI, whole-body insulin sensitivity calculated as HOMA-IR, and inflammation, that is, higher serum tumor necrosis factor (TNF)-α levels.[56] Other studies reported increased rates of mitochondrial β-oxidation rate along with structural mitochondrial defects in NASH patients.[78,98]

The bariatric surgery study by our group provided the possibility to directly compare hepatic oxidative capacity using HRR across obese people with different degrees of liver triglyceride content and NASH score to that of lean humans.[57] This approach identified maximal hepatic mitochondrial respiration to be 31%–40% lower in biopsy-proven NASH patients than in obese individuals with or without NAFLD, but similar compared to lean humans. The latter study underlines the importance of examining both lean and obese people to enable comparison of mitochondrial function under different metabolic conditions. Recently, a study using $^{31}$P-MRS at 7 Tesla supported these findings by reporting a lower Pi:ATP exchange rate constant and ATP turnover in biopsy-proven NASH compared with steatosis.[70] These data collectively suggest that the hepatic lipid and/or metabolic flexibility present in early NAFLD is lost in NASH. In addition, the NASH patients also featured greater hepatic insulin resistance, inflammatory response (from serum interleukin-6), mitochondrial uncoupling, and proton leakage, as well as augmented hepatic oxidative stress (from both $H_2O_2$ emission and lipid peroxides) and oxidative DNA damage (from 8-OH-deoxyguanosine).[57] These findings confirm previous studies showing increased proton leak across ETC because of upregulation of uncoupling protein-2 (UCP2) along with augmented oxidative stress[44] as well as an abnormality in mitochondrial redox homeostasis.[99] Under conditions of lower hepatic antioxidant defense capacity (from catalase activity) in NASH,[57] chronic lipid overload could lead to sustained induction of the TCA cycle, and its uncoupling from mitochondrial respiration and ATP synthesis might cause excessive ROS generation and induce inflammation as well as necrosis, apoptosis, and cell proliferation, resulting in hepatic fibrosis and carcinogenesis in hepatocytes.[100–102] This hypothesis is supported by the simultaneous presence of elevated hepatic TCA cycle flux, impaired ketogenesis, elevated hepatic DAG and ceramides in mice on a high-trans-fat, high-fructose diet.[103] Our group also reported that mice with genetically increased de novo lipogenesis exhibit liver-specific insulin resistance, which is correlated with hepatic DAG, but not ceramide levels.[104] Mice with genetically increased lipolysis in adipose tissue feature whole-body insulin resistance and accumulation of extrahepatic lipids, which are associated with portal/lobular inflammation. Both models show higher, rather than impaired, rates of hepatic mitochondrial respiration. These results clearly demonstrated that hepatic insulin resistance does not originate primarily from lower mitochondrial capacity but rather from lipotoxicity.

NASH patients presented with elevated sphingolipid species in the liver, which positively correlated with both hepatic oxidative stress and inflammation.[105] Sustained upregulation of mitochondrial oxidation is not sufficient to adapt to the chronic energy overload and, therefore, might favor accumulation of lipid intermediates, which would in turn accelerate insulin resistance, inflammation, and progression of NAFLD.

## 8 CONCLUSION

Human studies have shown that hepatic triglyceride storage positively associates with hepatic insulin resistance, which likely is mediated by lipid intermediates. Recent studies provided evidence that hepatic mitochondrial function was altered under hepatic insulin resistance. Hepatic mitochondria seem to transiently adapt to energy overload by upregulating oxidative capacity at the expense of coupling efficiency. Loss of mitochondrial flexibility likely worsens insulin resistance and triggers hepatic oxidative stress and inflammation, both of which positively associate with NAFLD progression. In spite of this evidence from human cross-sectional and prospective rodent studies, well-designed prospective clinical studies employing state-of-the-art techniques are required to elucidate the mutual relationships of mitochondrial function and dynamics with metabolic liver diseases and insulin resistance in humans.

## Acknowledgments

The work of the authors is supported in part by the Ministry of Science and Research of the State of North Rhine-Westphalia (MIWF NRW), the German Federal Ministry of Health (BMG), a grant of the Federal Ministry for Research (BMBF) to the German Center for Diabetes Research (DZD e.V.), the German Research Foundation (DFG, SFB 1116), German Diabetes Association (DDG), the Schmutzler-Stiftung, the Alexander von Humboldt-Stiftung, and the Bureau of International Health Cooperation, National Center for Global Health and Medicine, Japan.

## References

1. Sargin M, Uygur-Bayramicli O, Sargin H, Orbay E, Yayla A. Association of nonalcoholic fatty liver disease with insulin resistance: is OGTT indicated in nonalcoholic fatty liver disease? *J Clin Gastroenterol* 2003;**37**(5):399–402.
2. Seppala-Lindroos A, Vehkavaara S, Hakkinen AM, Goto T, Westerbacka J, Sovijarvi A, et al. Fat accumulation in the liver is associated with defects in insulin suppression of glucose production and serum free fatty acids independent of obesity in normal men. *J Clin Endocrinol Metab* 2002;**87**(7):3023–8.
3. Krssak M, Brehm A, Bernroider E, Anderwald C, Nowotny P, Dalla Man C, et al. Alterations in postprandial hepatic glycogen metabolism in type 2 diabetes. *Diabetes* 2004;**53**(12):3048–56.
4. Bril F, Sninsky JJ, Baca AM, Superko HR, Portillo Sanchez P, Biernacki D, et al. Hepatic steatosis and insulin resistance, but not steatohepatitis, promote atherogenic dyslipidemia in NAFLD. *J Clin Endocrinol Metab* 2016;**101**(2):644–52.
5. Tilg H, Moschen AR, Roden M. NAFLD and diabetes mellitus. *Nat Rev Gastroenterol Hepatol* 2017;**14**(1):32–42.
6. EASL-EASD-EASO. Clinical practice guidelines for the management of non-alcoholic fatty liver disease. *J Hepatol* 2016;**64**(6):1388–402.
7. Younossi ZM, Koenig AB, Abdelatif D, Fazel Y, Henry L, Wymer M. Global epidemiology of nonalcoholic fatty liver disease—meta-analytic assessment of prevalence, incidence, and outcomes. *Hepatology (Baltimore, Md)* 2016;**64**(1):73–84.
8. Wree A, Broderick L, Canbay A, Hoffman HM, Feldstein AE. From NAFLD to NASH to cirrhosis-new insights into disease mechanisms. *Nat Rev Gastroenterol Hepatol* 2013;**10**(11):627–36.
9. Anstee QM, Targher G, Day CP. Progression of NAFLD to diabetes mellitus, cardiovascular disease or cirrhosis. *Nat Rev Gastroenterol Hepatol* 2013;**10**(6):330–44.
10. Targher G, Bertolini L, Rodella S, Tessari R, Zenari L, Lippi G, et al. Nonalcoholic fatty liver disease is independently associated with an increased incidence of cardiovascular events in type 2 diabetic patients. *Diabetes Care* 2007;**30**(8):2119–21.
11. Roden M. Mechanisms of disease: hepatic steatosis in type 2 diabetes—pathogenesis and clinical relevance. *Nat Clin Pract Endocrinol Metab* 2006;**2**(6):335–48.
12. Cusi K. Treatment of patients with type 2 diabetes and non-alcoholic fatty liver disease: current approaches and future directions. *Diabetologia* 2016;**59**(6):1112–20.
13. Assy N, Kaita K, Mymin D, Levy C, Rosser B, Minuk G. Fatty infiltration of liver in hyperlipidemic patients. *Dig Dis Sci* 2000;**45**(10):1929–34.
14. Prashanth M, Ganesh HK, Vima MV, John M, Bandgar T, Joshi SR, et al. Prevalence of nonalcoholic fatty liver disease in patients with type 2 diabetes mellitus. *J Assoc Physicians India* 2009;**57**:205–10.
15. Portillo-Sanchez P, Bril F, Maximos M, Lomonaco R, Biernacki D, Orsak B, et al. High prevalence of nonalcoholic fatty liver disease in patients with type 2 diabetes mellitus and normal plasma aminotransferase levels. *J Clin Endocrinol Metab* 2015;**100**(6):2231–8.
16. Adams LA, Lymp JF, St Sauver J, Sanderson SO, Lindor KD, Feldstein A, et al. The natural history of nonalcoholic fatty liver disease: a population-based cohort study. *Gastroenterology* 2005;**129**(1):113–21.
17. Wang C, Wang X, Gong G, Ben Q, Qiu W, Chen Y, et al. Increased risk of hepatocellular carcinoma in patients with diabetes mellitus: a systematic review and meta-analysis of cohort studies. *Int J Cancer* 2012;**130**(7):1639–48.

18. Ziegler D, Strom A, Kupriyanova Y, Bierwagen A, Bonhof GJ, Bodis K, et al. Association of lower cardiovagal tone and baroreflex sensitivity with higher liver fat content early in type 2 diabetes. *J Clin Endocrinol Metab* 2017.
19. Szendroedi J, Phielix E, Roden M. The role of mitochondria in insulin resistance and type 2 diabetes mellitus. *Nat Rev Endocrinol* 2011;**8**(2):92–103.
20. Koliaki C, Roden M. Hepatic energy metabolism in human diabetes mellitus, obesity and non-alcoholic fatty liver disease. *Mol Cell Endocrinol* 2013;**379**(1–2):35–42.
21. Koliaki C, Roden M. Alterations of mitochondrial function and insulin sensitivity in human obesity and diabetes mellitus. *Annu Rev Nutr* 2016;**36**:337–67.
22. Petersen KF, Laurent D, Rothman DL, Cline GW, Shulman GI. Mechanism by which glucose and insulin inhibit net hepatic glycogenolysis in humans. *J Clin Invest* 1998;**101**(6):1203–9.
23. Roden M, Perseghin G, Petersen KF, Hwang JH, Cline GW, Gerow K, et al. The roles of insulin and glucagon in the regulation of hepatic glycogen synthesis and turnover in humans. *J Clin Invest* 1996;**97**(3):642–8.
24. Samuel VT, Shulman GI. The pathogenesis of insulin resistance: integrating signaling pathways and substrate flux. *J Clin Invest* 2016;**126**(1):12–22.
25. Rebrin K, Steil GM, Mittelman SD, Bergman RN. Causal linkage between insulin suppression of lipolysis and suppression of liver glucose output in dogs. *J Clin Invest* 1996;**98**(3):741–9.
26. Armstrong MJ, Hazlehurst JM, Hull D, Guo K, Borrows S, Yu J, et al. Abdominal subcutaneous adipose tissue insulin resistance and lipolysis in patients with non-alcoholic steatohepatitis. *Diabetes Obes Metab* 2014;**16**(7):651–60.
27. Roden M, Stingl H, Chandramouli V, Schumann WC, Hofer A, Landau BR, et al. Effects of free fatty acid elevation on postabsorptive endogenous glucose production and gluconeogenesis in humans. *Diabetes* 2000;**49**(5):701–7.
28. Donnelly KL, Smith CI, Schwarzenberg SJ, Jessurun J, Boldt MD, Parks EJ. Sources of fatty acids stored in liver and secreted via lipoproteins in patients with nonalcoholic fatty liver disease. *J Clin Invest* 2005;**115**(5):1343–51.
29. Nielsen S, Guo Z, Johnson CM, Hensrud DD, Jensen MD. Splanchnic lipolysis in human obesity. *J Clin Invest* 2004;**113**(11):1582–8.
30. Timlin MT, Parks EJ. Temporal pattern of de novo lipogenesis in the postprandial state in healthy men. *Am J Clin Nutr* 2005;**81**(1):35–42.
31. Lambert JE, Ramos-Roman MA, Browning JD, Parks EJ. Increased de novo lipogenesis is a distinct characteristic of individuals with nonalcoholic fatty liver disease. *Gastroenterology* 2014;**146**(3):726–35.
32. Tamura Y, Tanaka Y, Sato F, Choi JB, Watada H, Niwa M, et al. Effects of diet and exercise on muscle and liver intracellular lipid contents and insulin sensitivity in type 2 diabetic patients. *J Clin Endocrinol Metab* 2005;**90**(6):3191–6.
33. Petersen KF, Dufour S, Savage DB, Bilz S, Solomon G, Yonemitsu S, et al. The role of skeletal muscle insulin resistance in the pathogenesis of the metabolic syndrome. *Proc Natl Acad Sci U S A* 2007;**104**(31):12587–94.
34. Fabbrini E, Mohammed BS, Magkos F, Korenblat KM, Patterson BW, Klein S. Alterations in adipose tissue and hepatic lipid kinetics in obese men and women with nonalcoholic fatty liver disease. *Gastroenterology* 2008;**134**(2):424–31.
35. Charlton M, Sreekumar R, Rasmussen D, Lindor K, Nair KS. Apolipoprotein synthesis in nonalcoholic steatohepatitis. *Hepatology (Baltimore, Md)* 2002;**35**(4):898–904.
36. Szendroedi J, Yoshimura T, Phielix E, Koliaki C, Marcucci M, Zhang D, et al. Role of diacylglycerol activation of PKCtheta in lipid-induced muscle insulin resistance in humans. *Proc Natl Acad Sci U S A* 2014;**111**(26):9597–602.
37. Petersen MC, Shulman GI. Roles of diacylglycerols and ceramides in hepatic insulin resistance. *Trends Pharmacol Sci* 2017;**38**(7):649–65.
38. Considine RV, Nyce MR, Allen LE, Morales LM, Triester S, Serrano J, et al. Protein kinase C is increased in the liver of humans and rats with non-insulin-dependent diabetes mellitus: an alteration not due to hyperglycemia. *J Clin Invest* 1995;**95**(6):2938–44.
39. Kumashiro N, Erion DM, Zhang D, Kahn M, Beddow SA, Chu X, et al. Cellular mechanism of insulin resistance in nonalcoholic fatty liver disease. *Proc Natl Acad Sci U S A* 2011;**108**(39):16381–5.
40. Magkos F, Su X, Bradley D, Fabbrini E, Conte C, Eagon JC, et al. Intrahepatic diacylglycerol content is associated with hepatic insulin resistance in obese subjects. *Gastroenterology* 2012;**142**(7):1444–6 [e2].
41. Luukkonen PK, Zhou Y, Sadevirta S, Leivonen M, Arola J, Oresic M, et al. Hepatic ceramides dissociate steatosis and insulin resistance in patients with non-alcoholic fatty liver disease. *J Hepatol* 2016;**64**(5):1167–75.

42. Hammond LE, Neschen S, Romanelli AJ, Cline GW, Ilkayeva OR, Shulman GI, et al. Mitochondrial glycerol-3-phosphate acyltransferase-1 is essential in liver for the metabolism of excess acyl-CoAs. *J Biol Chem* 2005;**280**(27):25629–36.
43. Reddy JK, Rao MS. Lipid metabolism and liver inflammation. II. Fatty liver disease and fatty acid oxidation. *Am J Physiol Gastrointest Liver Physiol* 2006;**290**(5):G852–8.
44. Serviddio G, Bellanti F, Vendemiale G. Free radical biology for medicine: learning from nonalcoholic fatty liver disease. *Free Radic Biol Med* 2013;**65**:952–68.
45. Nassir F, Ibdah JA. Role of mitochondria in nonalcoholic fatty liver disease. *Int J Mol Sci* 2014;**15**(5):8713–42.
46. Fan CY, Pan J, Usuda N, Yeldandi AV, Rao MS, Reddy JK. Steatohepatitis, spontaneous peroxisome proliferation and liver tumors in mice lacking peroxisomal fatty acyl-CoA oxidase. Implications for peroxisome proliferator-activated receptor alpha natural ligand metabolism. *J Biol Chem* 1998;**273**(25):15639–45.
47. Mannaerts GP, Debeer LJ, Thomas J, De Schepper PJ. Mitochondrial and peroxisomal fatty acid oxidation in liver homogenates and isolated hepatocytes from control and clofibrate-treated rats. *J Biol Chem* 1979;**254**(11):4585–95.
48. De Craemer D, Pauwels M, Van den Branden C. Alterations of peroxisomes in steatosis of the human liver: a quantitative study. *Hepatology (Baltimore, Md)* 1995;**22**(3):744–52.
49. Reddy JK, Hashimoto T. Peroxisomal beta-oxidation and peroxisome proliferator-activated receptor alpha: an adaptive metabolic system. *Annu Rev Nutr* 2001;**21**:193–230.
50. Natarajan SK, Eapen CE, Pullimood AB, Balasubramanian KA. Oxidative stress in experimental liver microvesicular steatosis: role of mitochondria and peroxisomes. *J Gastroenterol Hepatol* 2006;**21**(8):1240–9.
51. Pawlak M, Lefebvre P, Staels B. Molecular mechanism of PPARalpha action and its impact on lipid metabolism, inflammation and fibrosis in non-alcoholic fatty liver disease. *J Hepatol* 2015;**62**(3):720–33.
52. Szendroedi J, Roden M. Mitochondrial fitness and insulin sensitivity in humans. *Diabetologia* 2008;**51**(12): 2155–67.
53. Toledo FG, Watkins S, Kelley DE. Changes induced by physical activity and weight loss in the morphology of intermyofibrillar mitochondria in obese men and women. *J Clin Endocrinol Metab* 2006;**91**(8):3224–7.
54. Chomentowski P, Coen PM, Radikova Z, Goodpaster BH, Toledo FG. Skeletal muscle mitochondria in insulin resistance: differences in intermyofibrillar versus subsarcolemmal subpopulations and relationship to metabolic flexibility. *J Clin Endocrinol Metab* 2011;**96**(2):494–503.
55. Chiappini F, Barrier A, Saffroy R, Domart MC, Dagues N, Azoulay D, et al. Exploration of global gene expression in human liver steatosis by high-density oligonucleotide microarray. *Lab Investig: J Techn Meth Pathol* 2006;**86**(2):154–65.
56. Perez-Carreras M, Del Hoyo P, Martin MA, Rubio JC, Martin A, Castellano G, et al. Defective hepatic mitochondrial respiratory chain in patients with nonalcoholic steatohepatitis. *Hepatology (Baltimore, Md)* 2003;**38**(4):999–1007.
57. Koliaki C, Szendroedi J, Kaul K, Jelenik T, Nowotny P, Jankowiak F, et al. Adaptation of hepatic mitochondrial function in humans with non-alcoholic fatty liver is lost in steatohepatitis. *Cell Metab* 2015;**21**(5):739–46.
58. Benard G, Faustin B, Passerieux E, Galinier A, Rocher C, Bellance N, et al. Physiological diversity of mitochondrial oxidative phosphorylation. *Am J Physiol Cell Physiol* 2006;**291**(6):C1172–82.
59. Larsen S, Nielsen J, Hansen CN, Nielsen LB, Wibrand F, Stride N, et al. Biomarkers of mitochondrial content in skeletal muscle of healthy young human subjects. *J Physiol* 2012;**590**(14):3349–60.
60. Szendroedi J, Chmelik M, Schmid AI, Nowotny P, Brehm A, Krssak M, et al. Abnormal hepatic energy homeostasis in type 2 diabetes. *Hepatology (Baltimore, Md)* 2009;**50**(4):1079–86.
61. Schmid AI, Chmelik M, Szendroedi J, Krssak M, Brehm A, Moser E, et al. Quantitative ATP synthesis in human liver measured by localized 31P spectroscopy using the magnetization transfer experiment. *NMR Biomed* 2008;**21**(5):437–43.
62. Chmelik M, Schmid AI, Gruber S, Szendroedi J, Krssak M, Trattnig S, et al. Three-dimensional high-resolution magnetic resonance spectroscopic imaging for absolute quantification of 31P metabolites in human liver. *Magn Reson Med* 2008;**60**(4):796–802.
63. Schmid AI, Szendroedi J, Chmelik M, Krssak M, Moser E, Roden M. Liver ATP synthesis is lower and relates to insulin sensitivity in patients with type 2 diabetes. *Diabetes Care* 2011;**34**(2):448–53.
64. Dagnelie PC, Menon DK, Cox IJ, Bell JD, Sargentoni J, Coutts GA, et al. Effect of L-alanine infusion on 31P nuclear magnetic resonance spectra of normal human liver: towards biochemical pathology in vivo. *Clin Sci (London, England: 1979)* 1992;**83**(2):183–90.

65. Terrier F, Vock P, Cotting J, Ladebeck R, Reichen J, Hentschel D. Effect of intravenous fructose on the P-31 MR spectrum of the liver: dose response in healthy volunteers. *Radiology* 1989;**171**(2):557–63.
66. Cortez-Pinto H, Chatham J, Chacko VP, Arnold C, Rashid A, Diehl AM. Alterations in liver ATP homeostasis in human nonalcoholic steatohepatitis: a pilot study. *JAMA* 1999;**282**(17):1659–64.
67. Abdelmalek MF, Lazo M, Horska A, Bonekamp S, Lipkin EW, Balasubramanyam A, et al. Higher dietary fructose is associated with impaired hepatic adenosine triphosphate homeostasis in obese individuals with type 2 diabetes. *Hepatology (Baltimore, Md)* 2012;**56**(3):952–60.
68. Fritsch M, Koliaki C, Livingstone R, Phielix E, Bierwagen A, Meisinger M, et al. Time course of postprandial hepatic phosphorus metabolites in lean, obese, and type 2 diabetes patients. *Am J Clin Nutr* 2015;**102**(5):1051–8.
69. Nair S, Chacko PV, Arnold C, Diehl AM. Hepatic ATP reserve and efficiency of replenishing: comparison between obese and nonobese normal individuals. *Am J Gastroenterol* 2003;**98**(2):466–70.
70. Traussnigg S, Kienbacher C, Gajdosik M, Valkovic L, Halilbasic E, Stift J, et al. Ultra-high-field magnetic resonance spectroscopy in non-alcoholic fatty liver disease: novel mechanistic and diagnostic insights of energy metabolism in non-alcoholic steatohepatitis and advanced fibrosis. *Liver Int: Off J Int Assoc Study Liver* 2017;**37**(10):1544–53.
71. Iozzo P, Bucci M, Roivainen A, Nagren K, Jarvisalo MJ, Kiss J, et al. Fatty acid metabolism in the liver, measured by positron emission tomography, is increased in obese individuals. *Gastroenterology* 2010;**139**(3):846–56. 56.e1–6.
72. Landau BR, Schumann WC, Chandramouli V, Magnusson I, Kumaran K, Wahren J. 14C-labeled propionate metabolism in vivo and estimates of hepatic gluconeogenesis relative to Krebs cycle flux. *Am J Physiol* 1993;**265**(4):E636–47. Pt 1.
73. Sunny NE, Parks EJ, Browning JD, Burgess SC. Excessive hepatic mitochondrial TCA cycle and gluconeogenesis in humans with nonalcoholic fatty liver disease. *Cell Metab* 2011;**14**(6):804–10.
74. Befroy DE, Perry RJ, Jain N, Dufour S, Cline GW, Trimmer JK, et al. Direct assessment of hepatic mitochondrial oxidative and anaplerotic fluxes in humans using dynamic 13C magnetic resonance spectroscopy. *Nat Med* 2014;**20**(1):98–102.
75. Petersen KF, Befroy DE, Dufour S, Rothman DL, Shulman GI. Assessment of hepatic mitochondrial oxidation and pyruvate cycling in NAFLD by (13)C magnetic resonance spectroscopy. *Cell Metab* 2016;**24**(1):167–71.
76. Previs SF, Kelley DE. Tracer-based assessments of hepatic anaplerotic and TCA cycle flux: practicality, stoichiometry, and hidden assumptions. *Am J Physiol Endocrinol Metab* 2015;**309**(8):E727–35.
77. Perry RJ, Peng L, Cline GW, Butrico GM, Wang Y, Zhang XM, et al. Non-invasive assessment of hepatic mitochondrial metabolism by positional isotopomer NMR tracer analysis (PINTA). *Nat Commun* 2017;**8**(1):798.
78. Miele L, Grieco A, Armuzzi A, Candelli M, Forgione A, Gasbarrini A, et al. Hepatic mitochondrial beta-oxidation in patients with nonalcoholic steatohepatitis assessed by 13C-octanoate breath test. *Am J Gastroenterol* 2003;**98**(10):2335–6.
79. Kahl S, Nowotny B, Strassburger K, Bierwagen A, Kluppelholz B, Hoffmann B, et al. Amino acid and fatty acid levels affect hepatic phosphorus metabolite content in metabolically healthy humans. *J Clin Endocrinol Metab* 2018;**103**(2):460–8.
80. Ritter O, Jelenik T, Roden M. Lipid-mediated muscle insulin resistance: different fat, different pathways? *J Mol Med (Berlin, Germany)* 2015;**93**(8):831–43.
81. Hernandez EA, Kahl S, Seelig A, Begovatz P, Irmler M, Kupriyanova Y, et al. Acute dietary fat intake initiates alterations in energy metabolism and insulin resistance. *J Clin Invest* 2017;**127**(2):695–708.
82. Vester JW, Stadie WC. Studies of oxidative phosphorylation by hepatic mitochondria from the diabetic cat. *J Biol Chem* 1957;**227**(2):669–76.
83. Hall JC, Sordahl LA, Stefko PL. The effect of insulin on oxidative phosphorylation in normal and diabetic mitochondria. *J Biol Chem* 1960;**235**:1536–9.
84. Parks Jr RE, Adler J, Copenhaver Jr JH. The efficiency of oxidative phosphorylation in mitochondria from diabetic rats. *J Biol Chem* 1955;**214**(2):693–8.
85. Ahishali E, Demir K, Ahishali B, Akyuz F, Pinarbasi B, Poturoglu S, et al. Electron microscopic findings in non-alcoholic fatty liver disease: is there a difference between hepatosteatosis and steatohepatitis. *J Gastroenterol Hepatol* 2010;**25**(3):619–26.
86. Heni M, Wagner R, Kullmann S, Gancheva S, Roden M, Peter A, et al. Hypothalamic and striatal insulin action suppresses endogenous glucose production and may stimulate glucose uptake during hyperinsulinemia in lean but not in overweight men. *Diabetes* 2017;**66**(7):1797–806.

87. Gancheva S, Koliaki C, Bierwagen A, Nowotny P, Heni M, Fritsche A, et al. Effects of intranasal insulin on hepatic fat accumulation and energy metabolism in humans. *Diabetes* 2015;**64**(6):1966–75.
88. Misu H, Takamura T, Matsuzawa N, Shimizu A, Ota T, Sakurai M, et al. Genes involved in oxidative phosphorylation are coordinately upregulated with fasting hyperglycaemia in livers of patients with type 2 diabetes. *Diabetologia* 2007;**50**(2):268–77.
89. Takamura T, Misu H, Matsuzawa-Nagata N, Sakurai M, Ota T, Shimizu A, et al. Obesity upregulates genes involved in oxidative phosphorylation in livers of diabetic patients. *Obesity (Silver Spring, Md)* 2008;**16**(12):2601–9.
90. Jelenik T, Sequaris G, Kaul K, Ouwens DM, Phielix E, Kotzka J, et al. Tissue-specific differences in the development of insulin resistance in a mouse model for type 1 diabetes. *Diabetes* 2014;**63**(11):3856–67.
91. Franko A, von Kleist-Retzow JC, Neschen S, Wu M, Schommers P, Bose M, et al. Liver adapts mitochondrial function to insulin resistant and diabetic states in mice. *J Hepatol* 2014;**60**(4):816–23.
92. Bouderba S, Sanz MN, Sanchez-Martin C, El-Mir MY, Villanueva GR, Detaille D, et al. Hepatic mitochondrial alterations and increased oxidative stress in nutritional diabetes-prone Psammomys obesus model. *Exp Diabetes Res* 2012;**2012**:430176.
93. Lund MT, Kristensen M, Hansen M, Tveskov L, Floyd AK, Stockel M, et al. Hepatic mitochondrial oxidative phosphorylation is normal in obese patients with and without type 2 diabetes. *J Physiol* 2016;**594**(15):4351–8.
94. Begriche K, Igoudjil A, Pessayre D, Fromenty B. Mitochondrial dysfunction in NASH: causes, consequences and possible means to prevent it. *Mitochondrion* 2006;**6**(1):1–28.
95. Petersen P. Abnormal mitochondria in hepatocytes in human fatty liver. *Acta Pathol Microbiol Scand Sect A: Pathol* 1977;**85**(3):413–20.
96. Caldwell SH, Swerdlow RH, Khan EM, Iezzoni JC, Hespenheide EE, Parks JK, et al. Mitochondrial abnormalities in non-alcoholic steatohepatitis. *J Hepatol* 1999;**31**(3):430–4.
97. Cortez-Pinto H, Camilo ME, Baptista A, De Oliveira AG, De Moura MC. Non-alcoholic fatty liver: another feature of the metabolic syndrome? *Clin Nutr (Edinburgh, Scotland)* 1999;**18**(6):353–8.
98. Sanyal AJ, Campbell-Sargent C, Mirshahi F, Rizzo WB, Contos MJ, Sterling RK, et al. Nonalcoholic steatohepatitis: association of insulin resistance and mitochondrial abnormalities. *Gastroenterology* 2001;**120**(5):1183–92.
99. Morris EM, Rector RS, Thyfault JP, Ibdah JA. Mitochondria and redox signaling in steatohepatitis. *Antioxid Redox Signal* 2011;**15**(2):485–504.
100. Satapati S, Kucejova B, Duarte JA, Fletcher JA, Reynolds L, Sunny NE, et al. Mitochondrial metabolism mediates oxidative stress and inflammation in fatty liver. *J Clin Invest* 2015;**125**(12):4447–62.
101. Koyama Y, Brenner DA. Liver inflammation and fibrosis. *J Clin Invest* 2017;**127**(1):55–64.
102. Satapati S, Sunny NE, Kucejova B, Fu X, He TT, Mendez-Lucas A, et al. Elevated TCA cycle function in the pathology of diet-induced hepatic insulin resistance and fatty liver. *J Lipid Res* 2012;**53**(6):1080–92.
103. Patterson RE, Kalavalapalli S, Williams CM, Nautiyal M, Mathew JT, Martinez J, et al. Lipotoxicity in steatohepatitis occurs despite an increase in tricarboxylic acid cycle activity. *Am J Physiol Endocrinol Metab* 2016;**310**(7):E484–94.
104. Jelenik T, Kaul K, Sequaris G, Flogel U, Phielix E, Kotzka J, et al. Mechanisms of insulin resistance in primary and secondary nonalcoholic fatty liver. *Diabetes* 2017;**66**(8):2241–53.
105. Apostolopoulou M, Gordillo R, Koliaki C, Gancheva S, Jelenik T, De Filippo E, et al. Specific hepatic sphingolipids relate to insulin resistance, oxidative stress, and inflammation in nonalcoholic steatohepatitis. *Diabetes Care* 2018;**41**(6):1235–43.

CHAPTER

# 9

# Contributions of Mitochondrial Dysfunction to β Cell Failure in Diabetes Mellitus

*Julia Parnis*[*,†,‡,§], *Guy A. Rutter*[*]

*Section of Cell Biology and Functional Genomics, Imperial College London, Hammersmith Hospital, London, United Kingdom †Division of Cardiovascular Medicine, Radcliffe Department of Medicine, University of Oxford, Oxford, United Kingdom ‡Wellcome Trust Centre for Human Genetics, University of Oxford, Oxford, United Kingdom §Oxford Centre for Diabetes, Endocrinology and Metabolism, University of Oxford, Churchill Hospital, Oxford, United Kingdom

## 1 INTRODUCTION

Several organs and tissues, such as liver, fat, muscle, brain, and gut, adjust body metabolism to nutrient access.[1] Pancreatic β cells, located within the pancreatic islet microorgans, are uniquely positioned to sense and regulate nutrient load by secreting proportional amount of insulin.[2] Insulin is the only hormone capable of lowering blood glucose in mammals by promoting its uptake and use by insulin-responsive organs and tissues, such as muscle and adipose tissue, and reducing glucose production by the liver. The functional significance of β cells is evident from the metabolic consequences of type 1 diabetes (T1D), an autoimmune disease, usually resulting in near-complete (70%–80% at the time of diagnosis) loss of β cells.[3]

While T1D accounts for only ~5%–10% of diabetes cases, type 2 diabetes (T2D) is more common and appears in ~90% cases of diabetes.[4] T2D affects more than 8% of the adult population worldwide,[5] a number expected to double in the next ~20 years. The disease leads to compromised glucose homeostasis and is associated with cardiovascular diseases, stroke, nephropathy, neuropathy, and retinopathy.[6] A combination of lifestyle, environmental, genetic, and epigenetic risk factors contribute to the development of T2D. Impaired insulin secretion



https://doi.org/10.1016/B978-0-12-811752-1.00009-2

because of decreased β cell mass and/or function leads to hyperglycemia, a characteristic of T2D, and its complications.

In this chapter, we discuss the pivotal role of mitochondria in the control of β cell function and mass and the possible contribution of altered mitochondrial activity in diabetes mellitus.

## 2 MITOCHONDRIA AND PANCREATIC β CELLS

Mitochondria constitute a dynamic population of organelles, existing partly as separate units and partly as an interconnected network[7] appearing under a microscope as constantly moving. Responding to multiple stimuli, mitochondria unite via fusion or divide via fission. Mitochondria move within the cell on microtubule tracks via dynein or kinesin motors to regions of high-energy demand, with $Ca^{2+}$ uptake by mitochondria regulating ATP production.[8, 9]

In spite of the well-characterized role for mitochondria in β cell stimulus-secretion coupling,[2, 10, 11] there is a relative paucity of genetic data in humans to implicate mitochondrial dysfunction in the pathogenesis of T2D. Nonetheless, it has been shown that mutations in the mitochondrial tRNA synthase $t_{RNA}^{Leu}$ lead to inherited diabetes and deafness,[11] and variants in the mitochondrial transcription factor TFB1M have been implicated by genome-wide association studies (GWAS).[11] Additionally, the loss of the mitochondrial protein frataxin in Friedreich ataxia patients impairs the activity of the electron transport chain (ETC), leading to a higher incidence of diabetes.[12]

Human β cell ultrastructure changes during the development of frank T2D.[13, 14] In T2D patients, mitochondria appear round and swollen. Both the ER and mitochondria occupy a larger volume and appear to be closer to each other, possibly reflecting ER stress and mitochondrial dysfunction. The swollen, larger mitochondria also appear in mice fed a high-fat diet for 12weeks.[15] The possibility that these changes drive impaired insulin secretion and β cell survival in T2D are discussed in the following sections.

## 3 MODELS FOR STUDYING β CELL FUNCTION

Given near-epidemic proportions of T2D in industrialized nations, it has become vital to study β cell dysfunction in models equating to the human disease. Investigations of primary human β cells within human islets obtained post mortem are thus critical.[16] Nevertheless, this is challenging for three reasons. Human islets consist of heterogeneous cell population and only ~50% of the islet are insulin-secreting β cells.[17] In contrast, β cells represent 60%–70% of the neuroendocrine cells in rodent islets.[17] Because human islets face hypoxia during isolation, their function and viability can be altered. And obtaining meaningful data with these preparations is challenging, given the limited availability of islets, interindividual differences in genetics, lifestyle, and medical history.

While rodent-derived pancreatic islets provide a more convenient model, however, important differences must be considered. Most rodent models of T2D develop diabetes only following extreme obesity, and, as such, do not fully recapitulate human T2D. In humans, the disease involves a combination of both genetic effects and increased environmental risks, including obesity.[18] The glucose set point of humans is lower than that of rodents, and significant differences exist between human and mouse islet architecture and innervation.[4]

Until quite recently, studies in the field have depended heavily upon the use of insulinoma-derived cell lines. The most commonly used include rat INS-1 cells,[19] including INS-1-derived clones,[20, 21] and mouse MIN6 cells.[22] Though having many similarities to primary human β cells, these cells nonetheless lack important features of their human relatives. The recent development of a human-derived pancreatic β cell line EndoC-β$H_1$,[23] and later variants[24, 25] resulted in the boost of β cell research. Though the cells still differ from their human prototypes because of the genetic manipulations used to obtain them,[23–26] they still, along with human embryonic stem cell-derived β-like cells,[27, 28] represent more relevant models for human β cell research.

## 4 INSULIN SECRETION: BASIC PRINCIPLES

Pancreatic β cells maintain glucose homeostasis via tightly matching insulin output to the extracellular glucose concentration. Elevated blood glucose levels trigger biphasic insulin secretion in vivo in humans[29] and in vitro in perifused islets.[30] The first phase lasts only for 3–10 min, while the slowly increasing second phase lasts for 60 min or more. The first phase is impaired preferentially in T2D.[31] The work of Coore and Randle,[32] Grodsky,[33] and Matschinsky[34] first established the necessity of glucose uptake and its metabolism (the substrate site hypothesis) as opposed to receptor-mediated insulin secretion.[2, 7] Another critical discovery by Ashcroft et al.[35] as well as Hales and Cook,[36] was that of the role of ATP-sensitive $K^+$ channels ($K_{ATP}$), which act as critical transducers of the increased metabolism prompted by glucose metabolism. These studies consequently led to the proposal of the canonical glucose-stimulated insulin secretion (GSIS) pathway (Fig. 1).[35, 36] More recent refinements have included the definition of an amplifying pathway for insulin secretion.[37]

Glucose uptake and its consequent conversion into pyruvate precedes its oxidation in mitochondria via the tricarboxylate (TCA) cycle and the electron transport chain (ETC) (Fig. 1). The TCA product NAD(P)H is used in the ETC to increase the proton-motive force (pmf) and produce energy-rich ATP, which is transferred to the cytosol. Following the rise in the cytosolic ATP/ADP ratio, $K_{ATP}$ channels close, leading to cell depolarization, and the consequent rise in intracellular free $Ca^{2+}$ concentration triggers insulin release. The fastest rates of glucose metabolism are observed in the physiological range of glucose concentration (5–10 mM).[38, 39]

Glucose is the principal β cell secretagogue. Other secretagogues include leucine and its transamination product α-ketoacid ketoisocaproate.[2] Insulin secretion and insulin's effects are regulated by gastrointestinal hormones, the autonomic nervous system, and other islet endocrine hormones, such as glucagon and somatostatin (Fig. 1).[2, 4]

Stimulatory potentiators of insulin release increase the latter only at permissive glucose concentrations (usually above 6 mM).[2] This group includes four categories: amino acids, such as arginine and glycine; incretin hormones glucagon-like peptide-1 (GLP-1), glucose-dependent insulinotropic peptide (GIP), cholecystokinin (CCK), peptide YY (PYY), and oxyntomodulin released by the gut following food ingestion; neurotransmitters, such as

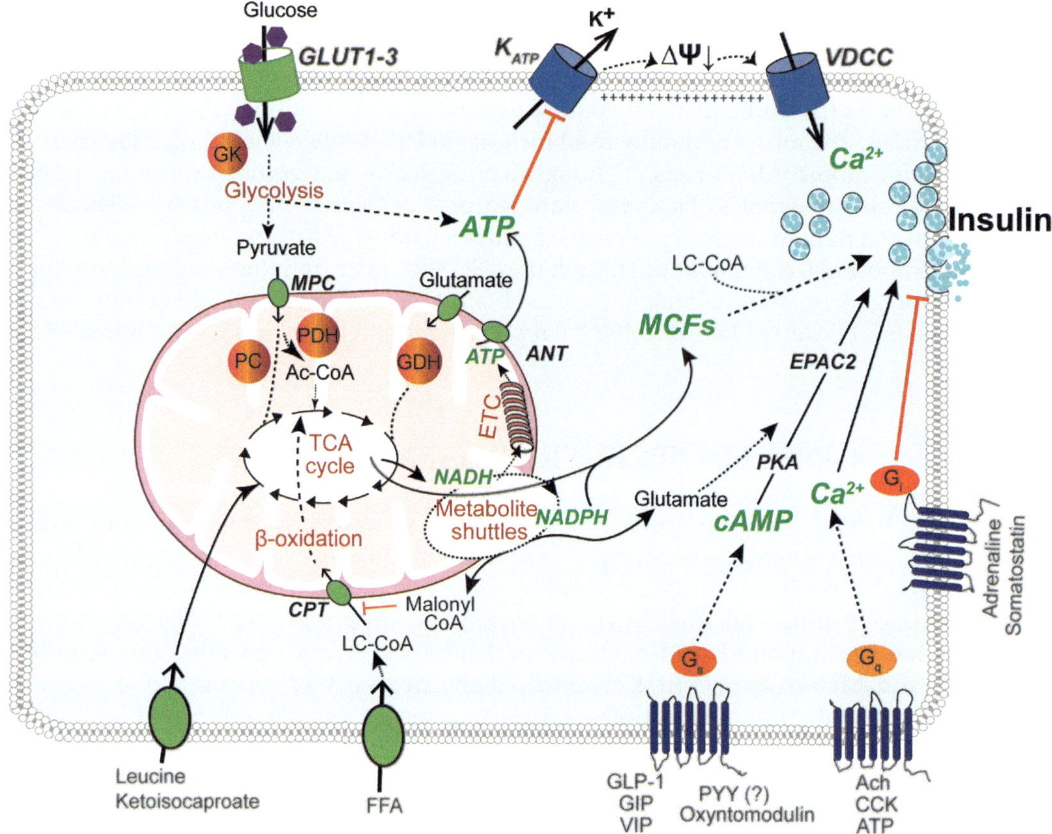


FIG. 1 Overview of stimulus-secretion coupling in pancreatic β cell. Glucose is transported into the β cell via glucose transporter GLUT 1, 2, or 3. Following glycolysis and the tricarboxylic acid cycle (TCA), resulting NAD(P)H enters electron transport chain (ETC). The following rise in ATP/ADP ratio closes plasma membrane ATP-dependent $K^+$ channels ($K_{ATP}$), leading to membrane depolarization and opening of voltage-dependent $Ca^{2+}$ channels (VDCC), triggering insulin exocytosis. In addition, metabolic coupling factors (MCFs), generated by metabolite shuttles associated with the TCA cycle, amplify insulin secretion. Stimulating effectors of insulin secretion include free fatty acids (FFAs), incretins glucagon-like peptide 1 (GLP-1), glucose-dependent insulinotropic peptide (GIP), and vasoactive intestinal polypeptide (VIP), peptide YY (PYY), oxyntomodulin. and cholecystokinin (CCK). and neurotransmitters acetylcholine (Ach) and ATP. Adrenalin and somatostatin inhibit insulin secretion. Abbreviations: *GLUT*, glucose transporter; *GK*, glucokinase; *MPC*, mitochondrial pyruvate carrier; *AcCoA*, acetyl CoA; *TCA*, tricarboxylic acid; *PC*, pyruvate carboxylase; *PDH*, pyruvate dehydrogenase; *GDH*, glutamate dehydrogenase; *ETC*, electron transport chain; *ANT*, adenine nucleotide transporter; *MCF*, metabolic coupling factors; *KATP*, ATP-dependent $K^+$ channels; *VDCC*, voltage-dependent $Ca^{2+}$ channels; *CPT*, carnitine palmitoyltransferase; *LC-CoA*, long-chain acyl-CoA; *FFA*, free fatty acids; *PKA*, protein kinase A; *EPAC2*, exchange protein directly activated by cAMP 2; *Δψ*, plasma membrane potential. Dotted lines represent processes.

acetylcholine and ATP; and fatty acids[40].[2] GLP-1 and GIP activate intracellular signaling cascades by binding to their corresponding G-protein coupled receptors (GPCRs), followed by activation of adenylyl cyclase and consequent intracellular rise in cAMP levels, resulting in stimulation of protein kinase A (PKA), exchange protein activated by cAMP 2 (EPAC2), and other pathways mediated by β-arrestin-1 and phosphatidylinositol 3′ kinase (PI3K).[40]

PKA and EPAC2 potentiate insulin granule exocytosis. The stimulatory neurotransmitters acetylcholine and ATP act to raise intracellular free $Ca^{2+}$ concentration through $G_q$-linked GPCR activation and the stimulation or phospholipase C.

Hormones inhibiting insulin release include somatostatin and the catecholamines adrenaline (epinephrine) and noradrenaline (norepinephrine).[2, 4] Somatostatin acts partially by binding to the extracellular domain of the somatostatin receptor (SSTR), which is an inhibitory GPCR that inhibits insulin secretion via an as-yet unknown mechanism. Adrenaline and noradrenaline release are modulated by exercise, and upon binding to adrenergic receptors on β cell plasma membrane, transiently repolarize the plasma membrane, therefore inhibiting insulin secretion.[41, 42]

## 5 ROLES OF MITOCHONDRIA IN GLUCOSE-STIMULATED INSULIN SECRETION

The importance of mitochondrial metabolism for GSIS is supported by evidence that cell permeant mitochondrial substrates methylpyruvate and methylsuccinate potently stimulate GSIS,[43] while ETC uncouplers and inhibitors diminish GSIS.[44] Glucose is the main insulin secretagogue, and its uptake and metabolic degradation by β cells are prerequisites of insulin secretion. Following meal ingestion and the consequent elevation in blood glucose levels, glucose is taken up by β cells via glucose transporters, GLUT2 in rodents or GLUT1 or GLUT3 in humans.[4] Glucose then is processed via glycolysis to pyruvate.

The majority (>85%) of glycolytic pyruvate enters mitochondria and is oxidized via the TCA cycle[38] and ETC. As previously mentioned, this remarkably high conversion rate is a unique property of β cells. Mitochondrial oxidative metabolism, as opposed to the aerobic production of lactate, is strongly favored in β cells because of the suppression (disallowance) of genes, including lactate dehydrogenase A, and the pyruvate/lactate (monocarboxylate) transporter MCT-1 (Slc16a1), which are expressed at high levels in almost all other tissues.[45–48] An increase in their expression parallels impaired GSIS.[2, 47] Lower levels of MCT-1 in the β cell membrane ensure that, following exercise or fasting when blood levels of pyruvate or lactate rise, insulin secretion is not triggered to cause hypoglycemia.[39]

Pyruvate enters the mitochondria via mitochondrial pyruvate transporters MPC-1 and MPC-2.[49, 50] It has been shown that genetic ablation of MPC-2 in mouse islet cells or insulin-secreting cells in *Drosophila* results in glucose intolerance and reduced plasma insulin levels, characterized by reduced islet intrinsic GSIS.[51]

In the mitochondria, pyruvate is either decarboxylated by pyruvate dehydrogenase (PDH) or carboxylated by pyruvate carboxylase (PC) to create acetyl-CoA and oxaloacetate, respectively (Figs. 1 and 2). Both reactions occur equally in β cells[38, 39] and both substrates enter TCA. β cell-specific PDH knockout resulted in glucose intolerance in mice and impaired in vitro GSIS.[52] PC has low activity at low glucose levels because matrix ATP levels, required for the reaction, and the levels of acetyl-CoA, an allosteric activator of PC, are not sufficient to activate PC.[53] Along with glutamate dehydrogenase (GDH) and aspartate transaminase, PC catalyze anaplerotic reactions, capable of replenishing TCA components. These two enzymes supply α-ketoglutarate and oxaloacetate to TCA, respectively.

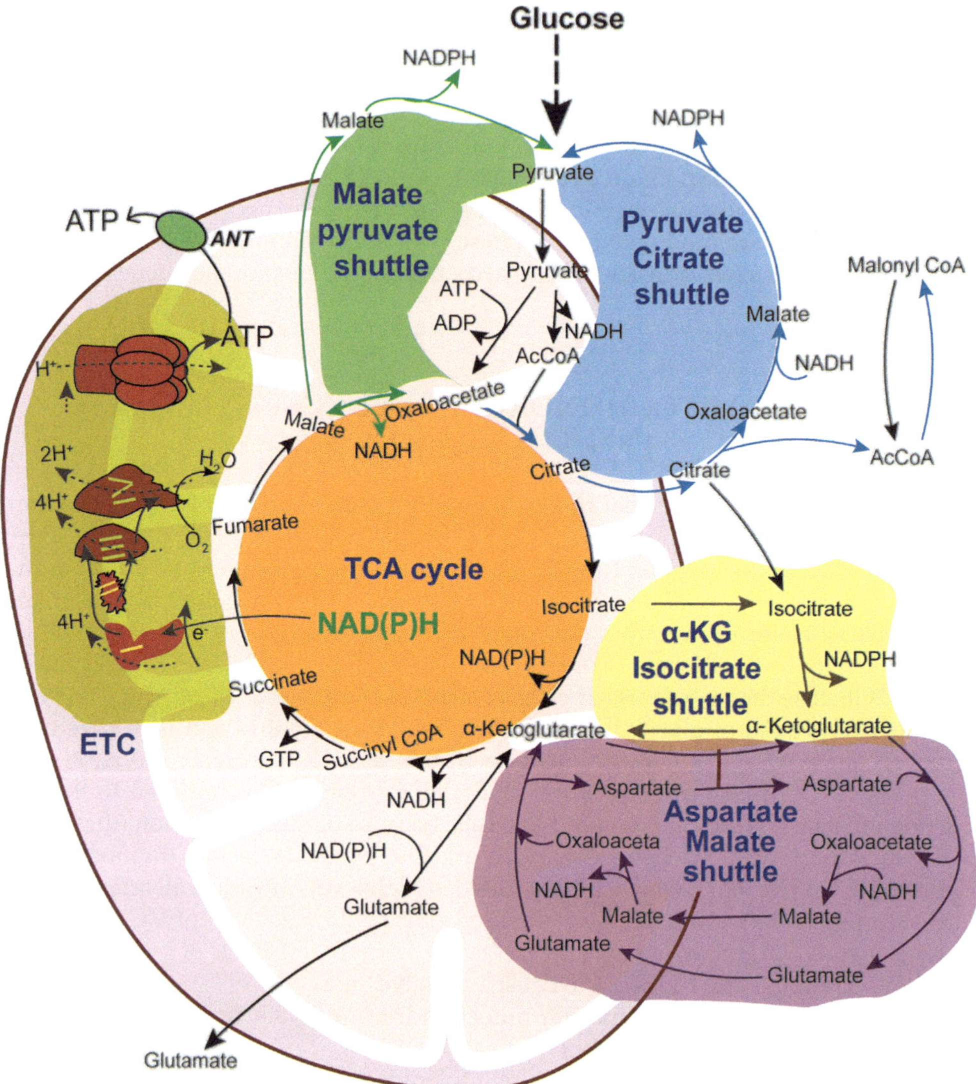


FIG. 2 Mitochondrial biochemical pathways that affect GSIS. Following glycolysis, pyruvate enters mitochondria and is decarboxylated to form acetyl CoA (AcCoA) or carboxylated to generate oxaloacetate. Both products enter tricarboxylic acid cycle (TCA) for subsequent oxidation and parallel reduction of $NAD(P)^+$ to NAD(P)H, which enters electron transport chain (ETC), consisted of complexes I-IV for the generation of electrochemical $H^+$ gradient. Use of this gradient by ATP synthase generates ATP, which is transported from mitochondria via the adenine nucleotide transporter (ANT). Alongside the ETC and TCA cycle, additional metabolites are formed via processes cycling between mitochondria and cytosol (shuttles). These metabolites are employed in nutrient-secretion coupling and anaplerotic reactions, destined to replenish TCA intermediates. The shuttles include malate pyruvate shuttle *(green)*, pyruvate citrate shuttle *(blue)*, α-ketoglutarate (α-KG) isocitrate shuttle *(yellow)*, and aspartate malate shuttle *(purple)*.

The products of the matrix dehydrogenases, NAD(P)H and $FADH_2$, donate electrons to complexes I to IV of the ETC, encoded by both nuclear and mitochondrial DNA.[11, 39] The subsequent $H^+$ extrusion through complexes I, III, and IV from the mitochondrial matrix to intermembrane space generates the $H^+$ gradient. The ATP synthase (complex V) uses this gradient, allowing phosphorylation of ADP to ATP at the expense of proton flow back into the matrix (Fig. 2). Our unpublished work indicates that ETC components, such as NDUFB8 and ATP5A1, belonging to complex I and V, respectively, are positively correlated with basal insulin levels in mice and their lack can contribute to GSIS impairment via slowing the generation of mitochondrial membrane potential and consequently ATP.

It generally has been thought that these complexes are distributed randomly in the inner mitochondrial membrane. In 2000, however, Schägger and Pfeiffer found that complexes I, III, and IV are arranged in supercomplexes or respirasomes.[54] This arrangement allows substrate channeling or direct transfer of electrons from one enzyme to another, providing kinetic advantage. Furthermore, there are suggestions that even higher levels of organization might exist (megacomplexes or respiratory strings).[55] Respirasomes stabilize complex I and decrease reactive oxygen species (ROS) production. Moreover, assembly and disassembly of the complexes are dynamic and can be modulated by mitochondrial fusion, fission, and lipid content.[55] Because of the importance of respirasomes in cellular respiration and ROS generation, their characterization might be important for better understanding of T2D pathogenesis. The existence and functional significance of respirasomes in islet β cells has yet to be reported.

ATP is transported from mitochondria via the adenine nucleotide transporter (ANT). Subsequent rise of ATP/ADP ratio is sensed by plasma membrane $K_{ATP}$ channels, which are inhibited by ATP. Decreased $K^+$ efflux depolarizes plasma membrane, opening voltage gated $Ca^{2+}$ channels. The resulting $Ca^{2+}$ influx drastically increases $Ca^{2+}$ levels in the subplasmalemmal region in the vicinity of ready-to-be-released insulin granules, evoking vesicle exocytosis and insulin release into the blood stream.

# 6 METABOLIC COUPLING FACTORS

## 6.1 Free Fatty Acids, Citrate and Malonyl-CoA

Free fatty acid (FFA) metabolism plays an important role in β cell function. FFAs can derive either from exogenous source (e.g., secreted by adipocytes) or via lipolysis of β cell triglyceride stores. At high glucose levels, enhanced metabolism inhibits mitochondrial FFA uptake, rerouting cytosolic processing of FFA to produce long chain acyl-CoA (LC-CoA). LC-CoA potentiates GSIS[56] by the following mechanism: Enhanced TCA generation of citrate and its extrusion from the mitochondria via citrate carrier leads to its cleavage by cytosolic citrate lyase to create acetyl-CoA. Next, acetyl-CoA carboxylase converts acetyl-CoA to malonyl-CoA, which, in turn, inhibits FFA mitochondrial transporter carnitine-palmitoyl transferase-1 (CPT-1),[57] the rate-limiting step for transport and oxidation of FFAs in mitochondria. Inhibition of FFA uptake follows its processing in the cytosol to LC-CoA potentiating GSIS.[56] One of the possible explanations is that monoacylglycerol (MAG), generated from LC-CoA, binds to vesicle-priming protein Munc13, promoting insulin granule exocytosis.[58]

Therefore, β oxidation correlates negatively with GSIS. For example, inactivating mutations in the gene encoding short-chain 3-hydroxyacyl-CoA dehydrogenase (SCHAD), a ubiquitously expressed enzyme involved in fatty acid oxidation cause hypoglycemia because of hyperinsulinism stemming from islet-intrinsic mechanisms.[59] Therefore, β oxidation is inhibited in β cells, and FFA-derived LC-CoA serves as a key molecule linking β oxidation and insulin release.

## 6.2 NAD(P)H

The $NAD(P)^+$/NAD(P)H pair is well known for its electron transferring properties and as an essential cofactor in numerous redox reactions. Analysis of UniProtKB/Swiss-Prot database indicates existence of 30,000 proteins able to bind $NAD(P)^+$,[60] emphasizing a key and universal role for this metabolite. Mitochondrial and cytosolic NADH pools are separated. The inner mitochondrial membrane is impermeable to $NAD^+$ and NADH, and the mitochondrial pool is preserved even during cellular stresses, which deplete cytosolic $NAD^+$ pool.[61] Moreover, although cytosolic and mitochondrial $NAD^+$ levels are similar, a much larger NADH pool is present in the mitochondria,[61] and NADPH is located mainly in the cytosol.

Glucose stimulation results in an increased intracellular NAD(P)H/$NAD(P)^+$ ratio,[39, 62] starting from the cytosol and propagating to the mitochondria.[63] A glucose-induced elevation in NAD(P)H/$NAD(P)^+$ ratio precedes an intracellular [$Ca^{2+}$] rise.[39] In addition, NAD-kinase activity, stimulated by glucose, augments the cytosolic NADPH pool.[62] Glucose also activates a cytosolic pentose-phosphate shunt and mitochondrial pyruvate-citrate and malate-aspartate shuttles, which reduce $NAD(P)^+$ to NAD(P)H[62] (Fig. 2).

NADPH plays an important role in GSIS since the NAD(P)H/$NAD(P)^+$ ratio strongly correlates with GSIS.[39] An increased NAD(P)H/$NAD(P)^+$ ratio was shown to instigate $Ca^{2+}$-dependent exocytosis of insulin granules via possible reduction of cytosolic redox sensors glutaredoxin and thioredoxin.[39] A recent study[64] demonstrated that the generation of NADPH by isocitrate dehydrogenase, and subsequent glutathione reduction, promotes insulin exocytosis via sentrin/SUMO-specific protease-1 (SENP1). NAD(P)H, therefore, is vital for β cell function and the mechanisms regulating its generation, consumption, and intracellular location are important for GSIS.

## 6.3 Glutamate

In the mitochondria, glutamate can be generated either by reversible amination reaction of α-ketoglutarate by glutamate dehydrogenase 1 (GDH1) or by conversion from glutamine by glutaminase. Glutamate also can be produced in the cytosol by aspartate aminotransferase 1 (AA1), which catalyzes transamination reaction from α-ketoglutarate. AA1 is a part of the malate-aspartate shuttle, important for anaplerosis. In addition to glutamate production, this shuttle plays an important role in replenishing $NAD(P)^+$ pools, which are being exhausted during glucose stimulation of β cells because of high activity of TCA and a limited rate of complex I. Reactions leading to glutamate production, as well as reactions of TCA, ETC, and mitochondrial shuttles, are shown in Fig. 2.

Glutamate's role in GSIS is supported by two findings: glutamate application to permeabilized cells increased GSIS[65]; and treatment of β cells with cell-permeable dimethyl glutamate augmented GSIS.[65] An elegant study by Gheni et al.[65] suggests that glutamate intracellular location is important for its function in GSIS because application of the malate-aspartate shuttle inhibitor aminooxyacetate and AA1 knockdown, but not knockdown of other malate-aspartate shuttle components, decreases cytosolic glutamate levels in pancreatic β cells and abolishes cAMP-dependent GLP-1-enhanced GSIS. Application of glutamine, however, only increased intracellular glutamate levels without affecting GSIS in both rat islets and rat β cell line INS1.[66]

The authors propose the following mechanism to explain these findings. Glutamate is taken up by insulin granules via vesicular glutamate transporters (vGLUTs) and released via vesicular excitatory amino acid transporters. Glutamate vesicular uptake is stimulated by PKA, which is activated by GLP-1-mediated cAMP increases. Glutamate entry into the granules close to the sites of the malate-aspartate shuttle, arrival of the insulin granules to exocytic site, and following glutamate efflux, therefore, ensure high glutamate concentration near membrane fusion events and the possibility to regulate insulin exocytosis.

The role of mitochondrially derived glutamate is supported by evidence that activating mutations in GDH results in neonatal hyperinsulinism.[67] Furthermore, GDH loss-of-function mutations impair insulin secretion in murine β cells, while its overexpression improves GSIS.[11] It is, however not clear whether glutamate, α-ketoglutarate, or $NAD^+$, formed by GDH, contribute to insulin secretion.[11]

## 7 BIOENERGETIC PROPERTIES OF β CELL MITOCHONDRIA

β cell mitochondria are unique in their property of translating changes in intracellular glucose levels to an altered ATP/ADP ratio and changes in the levels of related metabolites (see excellent review about this topic by Nicholls[57]). A combination of both elements proportionally increases the intrinsic secretory capacity of the β cell. Tight coupling of the glucose and mitochondrial metabolism can be achieved via three factors: (1) equalization of the intracellular and extracellular glucose concentrations, when extracellular glucose reaches a threshold; (2) glycolysis proceeds only at physiological intracellular glucose levels (which are low at fasting glucose levels)[2] and is elevated proportionally upon increases in intracellular glucose levels; and (3) limitation of the routes, removing or delivering glycolysis' substrates or mitochondrial metabolic intermediates, which are independent of glucose elevations.

Glucose transport into β cells is dictated by the kinetics of glucose transporters. Although GLUT2/SLC2A2 is the predominant glucose transporter in murine β cells,[57, 68, 69] GLUT1 (SLC2A1) and GLUT3 (SLC2A3) are expressed predominantly in human β cells, though GLUT2 expression also is detected, and might represent the most physiologically important transporter.[69] The kinetic properties of these transporters are different. GLUT2 is characterized by low affinity (Km ~11–17 mM) and high velocity; GLUT1 (Km ~6.9 mM) and GLUT3 (Km ~1.4 mM) are characterized by high affinity and low velocity uptake.[69]

These dissimilarities explain lower threshold for GSIS in human islets (Km 6.5 mM)[43] compared to rodent islets. In spite of the differences in glucose uptake between rodent and human β cells, glycolytic flux and total glucose use are similar in both species.[70] This is because of the expression of hexokinase IV (glucokinase) as a predominant enzyme phosphorylating glucose to glucose 6-phosphate, a rate-limiting step in glycolysis. Hexokinase IV, unlike other hexokinases, is characterized by low affinity, positive cooperativity with glucose, and lack of allosteric inhibition by its product, glucose 6-phosphate.[68] These data support that glucokinase is the main regulator, translating extracellular glucose elevations into increase in glycolysis.

The activity of ETC generates both a $H^+$ gradient across the inner mitochondrial membrane and negative charge of the membrane because of the extrusion of cations. Together these produce the protonmotive force:

$$\Delta \text{pmf}(\text{mV}) = \Delta\psi - 61\Delta\text{pH}$$

where pmf is protonmotive force and $\psi$ is membrane potential. Subsequently, we will use $\Delta\psi_m$ for mitochondrial membrane potential and $\Delta\psi_p$ for plasma membrane potential.

The use of fast fluorescent potentiometric probes, such as tetramethylrhodamine ethyl ester (TMRE) or tetramethylrhodamine methyl ester (TMRM), to monitor $\Delta\psi_m$ has become standard practice.[57, 71, 72] Because these cationic probes enter the cytoplasm first proportional to their $\Delta\psi_p$ and then from the cytosol to mitochondria proportional to their $\Delta\psi_m$, however, special care is required to distinguish changes in fluorescence because of alterations in $\Delta\psi_m$ from artifacts related to nonspecific staining. An additional obstacle in analysis of $\Delta\psi_m$ is the very dynamic nature of mitochondria, and confocal microscopy is best-used to explore changes in mitochondrial structure over time.

Glucose, as well as a variety of other insulin secretagogues such as permeable mitochondrial substrates and α-ketoisocaproic acid, are able to hyperpolarize mitochondrial membrane.[2, 57] An estimated $\Delta\psi_m$ in rat INS-1832/13 cells in response to a glucose rise from 3 to 16 mM is about 14 mV; in response to 9 mM pyruvate, $\Delta\psi_m$ was altered by 9 mV.[57] In dispersed murine β cells, glucose elevation from 3 to 8 mM resulted in hyperpolarization of about 6 mV.[57] In human β cells, increase of glucose levels from 3 to 16 mM leads to increase in $\Delta\psi_m$ from 124–138 mV to 152–174 mV.[57] Combining matrix ΔpH changes, observed by Wiederkehr et al.[73] in rat islets, Nichols[57] proposed that Δpmf in β cells might increase from ~140 to ~180 mV in response to high glucose.

Dispersed β cells generate very heterogeneous changes in $\Delta\psi_m$ in response to glucose; the heterogeneity decreases with an increase in glucose levels.[74] These results align well with our own unpublished data. We observed that murine β cell line MIN6 responds with higher heterogeneity in 7 mM glucose than in 15 mM glucose. This heterogeneity has been shown to be important for the coordinated response of all β cells to glucose.[71] Johnson et al. describe two populations of β cells: hubs that constitute less than 10% of β cell population and followers. Metabolically superior hub β cells responded to glucose elevation with higher $\Delta\psi_m$. Connected to the followers' population via gap junctions, these hubs dictate an orchestrated response of an islet to glucose stimulus, releasing pulses of insulin into the bloodstream.[71] Connectivity of β cells is disturbed during T2D or following prolonged exposure of human islets to palmitate.[16, 71]

## 8 MITOCHONDRIAL $CA^{2+}$

Mitochondria are intimately involved in intracellular $Ca^{2+}$ signaling, therefore, they also are involved in all $Ca^{2+}$-dependent cell functions, with a particular role in cell death. In 1961, Deluca and Engstrom measured for the first time $Ca^{2+}$ uptake by isolated mitochondria.[75] Decades of research and advances in molecular and imaging techniques allowed significant progress in the physiological characterization and in the identification of molecular machinery shaping mitochondrial $Ca^{2+}$ homeostasis. Mitochondria accumulate $Ca^{2+}$ during any cytosolic $Ca^{2+}$ increases proportionally to an amplitude of the cytosolic rise[72, 76] (Fig. 3). The shape and the speed of mitochondrial $Ca^{2+}$ accumulation depend on the mitochondrial $Ca^{2+}$ cell-specific machinery and the distance to open ER and plasma membrane $Ca^{2+}$ channels. Such channels create $Ca^{2+}$ gradients with high $Ca^{2+}$ concentration close to the open mouth of a channel, so-called $Ca^{2+}$ microdomains. Positioning of mitochondria, in turn, is controlled by fusion-fission processes and mitochondrial transport.

By targeting ATP/ADP or $Ca^{2+}$ sensors to different compartments within cells, it is possible to estimate the kinetics of metabolite and ion signaling and interplay between them. For example, Fig. 3A shows how primary murine β cells infected with adenoviruses expressing cytosolic ATP/ADP sensor Perceval and loaded with cytosolic $Ca^{2+}$-sensitive dye Fura Red react to glucose elevation. The kinetics of such a response, shown in Fig. 3B, demonstrates that ATP/ADP elevation precedes intracellular $Ca^{2+}$ rise in β cells. By targeting $Ca^{2+}$ sensor 2mt8RP into mitochondria, we successfully demonstrated that mitochondrial $Ca^{2+}$ rises follow cytosolic $Ca^{2+}$ increase (Fig. 3C and D).

Mitochondria take up $Ca^{2+}$ mainly via mitochondrial $Ca^{2+}$ uniporter (MCU) complex, composed of six different protein components.[76] Inside the mitochondria, $Ca^{2+}$ substantially is buffered possibly by acetate or phosphate, which might lead to $Ca^{2+}$ precipitates.[77] $Ca^{2+}$ is extruded from mitochondria via the $Na^{+}/Ca^{2+}$ exchanger NCLX.[76] The molecular identity of $H^{+}/Ca^{2+}$ exchanger is still debated, but might be leucine zipper-EF-hand-containing transmembrane protein1 (LETM1).[76] The mitochondrial permeability transition pore (mPTP) also contributes to $Ca^{2+}$ efflux.

Mitochondrial $Ca^{2+}$ elevations, in turn, activate mitochondrial metabolism by stimulating three mitochondrial dehydrogenases: PDH, α-ketoglutarate dehydrogenase, and isocitrate dehydrogenase.[78] In pancreatic β cells mitochondrial $Ca^{2+}$ plays an important role in GSIS by possibly supplying ATP for $K_{ATP}$ channels closure and for exocytosis. Mitochondrial $Ca^{2+}$ buffering impairs respiration in clonal β cells and decreases the second phase of GSIS in primary rat islets.[79] Similarly, MCU silencing impairs the second phase of GSIS,[72, 80] while NCLX knockdown increases insulin secretion.[81]

## 9 ER-MITOCHONDRIAL LINKS

Tight links between ER and mitochondria first were observed and characterized in 1959 by Copeland and Dalton.[82] Both organelles are coupled physically and functionally at mitochondria-associated ER membranes (MAMs). MAMs are enriched in proteins, involved in $Ca^{2+}$ homeostasis, cell signaling, lipid biosynthesis and trafficking, and mitochondrial dynamics.

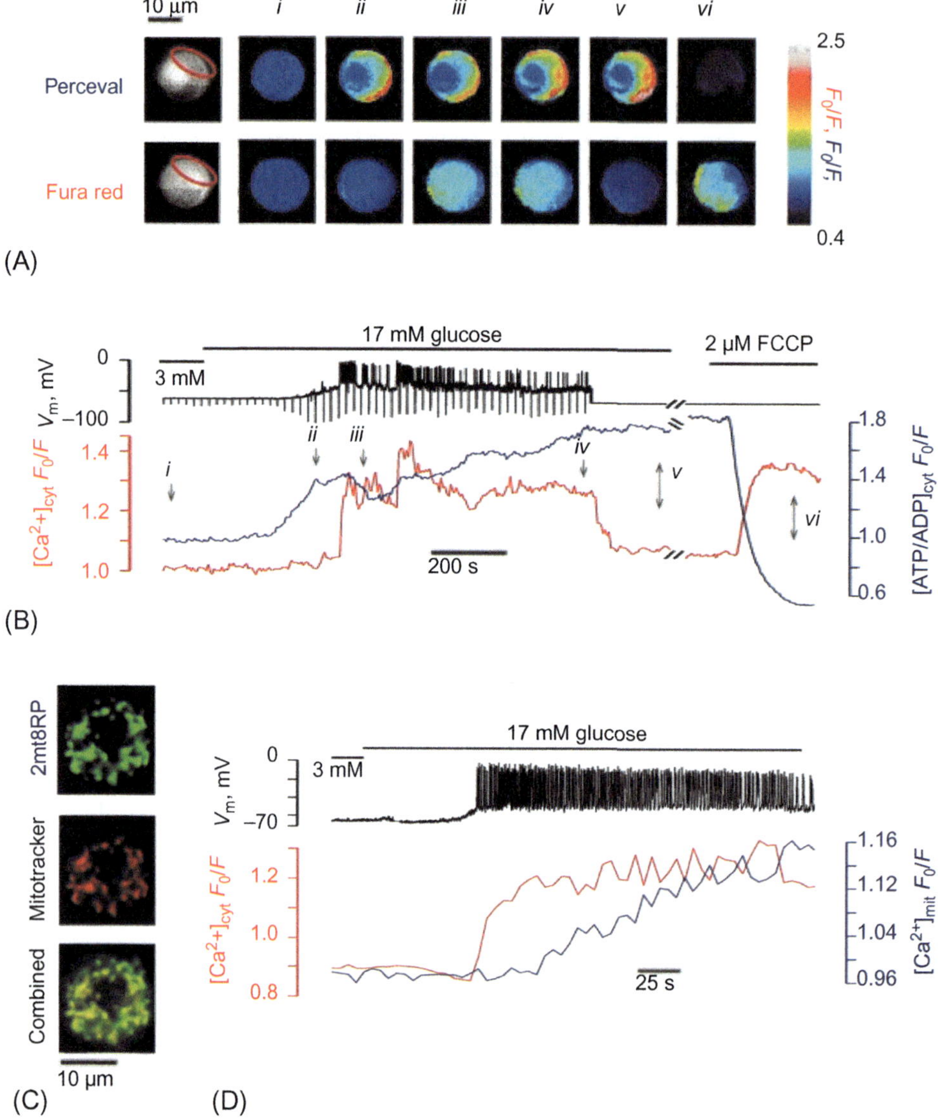


FIG. 3 Coupling between glucose-evoked ATP generation, $Ca^{2+}$ signaling and rhythmic action potential firing. (A-B) Glucose-induced changes in cytosolic [ATP/ADP] ($[ATP/ADP]_{cyt}$), $[Ca^{2+}]$ ($[Ca^{2+}]_{cyt}$) and membrane potential ($V_m$) in a single murine islet dispersed β-cell as presented by (A) pseudo-color images of the patched cell cluster at the time points indicated by arrows (i–vi) and (B) representative individual traces. Region of interest (ROI) is indicated with red oval. (C) Co-localization of mitochondrial $Ca^{2+}$ ($Ca^{2+}{}_{mit}$) sensor, ratiometric pericam 2mt8RP, and mitotracker orange in intact murine islets. (D) Glucose-evoked changes in $V_m$, $[Ca^{2+}]_{cyt}$ and $[Ca^{2+}]_{mit}$. To follow these changes, cells were infected with adenoviruses expressing pericam or 2mt8RP to measure $[ATP/ADP]_{cyt}$ or $[Ca^{2+}]_{mit}$, respectively, and loaded with Fura Red to acquire $[Ca^{2+}]_{cyt}$. The results are taken from Tarasov et al.[71]

Inositol 1,4,5-trisphosphate ($IP_3$) receptors ($IP_3Rs$) on the ER side and voltage-dependent anion channels (VDACs) on the mitochondrial site, connected via chaperones, such as GRP75 (75kDa glucose regulated protein) allow $Ca^{2+}$ transfer from the ER to mitochondria. Uncoupling $IP_3Rs$ and VDACs via GRP75 silencing impairs mitochondrial $Ca^{2+}$ uptake.[83] Moreover, $IP_3Rs$ placement on MAMs enables the formation of high [$Ca^{2+}$] microdomains, fostering $Ca^{2+}$ uptake by low-affinity MCU.[83] PDZD8[84] also has been shown to be a critical part of the tethering complex.

The MAMs also contain multiple important cell signaling proteins, such as PKR-like ER-localized kinase (PERK), an essential component of ER stress response; mTOR, a vital nutrient sensing molecule and important regulator of β cell proliferation and survival, which has been found to play an important role in β cell failure in face of the nutritional overload; AKT, an important signaling molecule mediating incretin effect on GSIS; and β cell proliferation and other proteins.[83, 85, 86] MAMs are enriched in the enzymes, such as PEMT2, DGAT2, and phosphatidylserine synthase (PSS), involved in lipid synthesis and trafficking.[83] Close interaction between ER and mitochondria is essential for facilitating lipid synthesis. Phosphatidylserine is synthesized in the ER, but it is converted into phosphatidylethanolamine in the mitochondria. Phosphatidylcholine is produced from phosphatidylethanolamine in the ER. MAMs also are involved in the control of mitochondrial dynamics by recruitment of the protein involved in mitochondrial fission, DRP-1.[87] Playing such multiple roles, MAMs integrate multiple pathways and are involved in signals mediated by hormonal actions, including insulin signaling.[86]

In spite of the importance of MAMs for cellular physiology, very little is known about the function of MAMs in β cells. The current literature indicates that, in islets from T2D donors, ER and mitochondria appear closer to each other than in healthy subjects,[14] but the appearance of ER-mitochondrial tethers, as quantified from the existence of $IP_3R2$-VDAC1 complexes, is decreased in T2D islets.[88] Moreover, there is a significant upregulation of $IP_3R2$ protein and downregulation of VDAC1 in diabetic islets.[88]

# 10 MITOCHONDRIAL INVOLVEMENT IN ISLET GLUCOTOXICITY AND LIPOTOXICITY AND THE DEVELOPMENT OF T2D

## 10.1 ER Stress and Mitochondrial Dysfunction

Impaired insulin sensitivity, progressive (if limited) loss of β cell mass, and gradual dysfunction of the remaining β cells all contribute to impaired insulin release and, consequently, to T2D pathogenesis.[2] Chronic nutrient overload is one of the key factors in the development of the disease. Initially, β cells adapt to increasing glucose loads by increasing their insulin output and by increasing in number during obesity.[2, 89] It also has been shown[90] that prolonged exposure to high glucose promotes glucose metabolism. In genetically susceptible individuals, however, a decompensation period follows by compensatory mechanisms as previously described. In the following sections, we describe the mechanisms involved in decompensation, rooted from glucose and lipid overload, with the focus on mitochondria. The summary of the events leading to β cell dysfunction and altered structure of mitochondria and the ER during T2D are illustrated in Fig. 4.

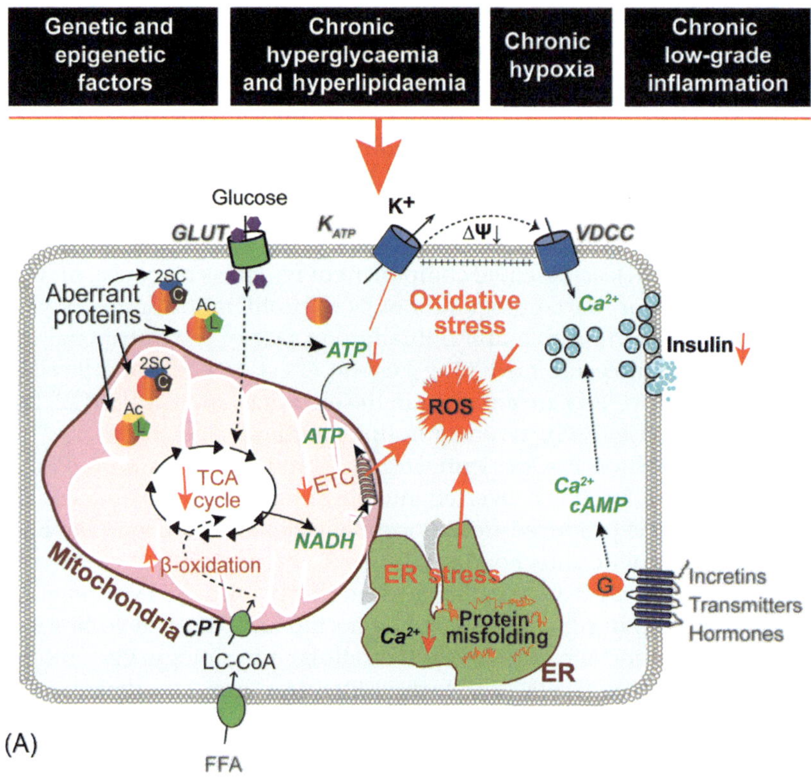


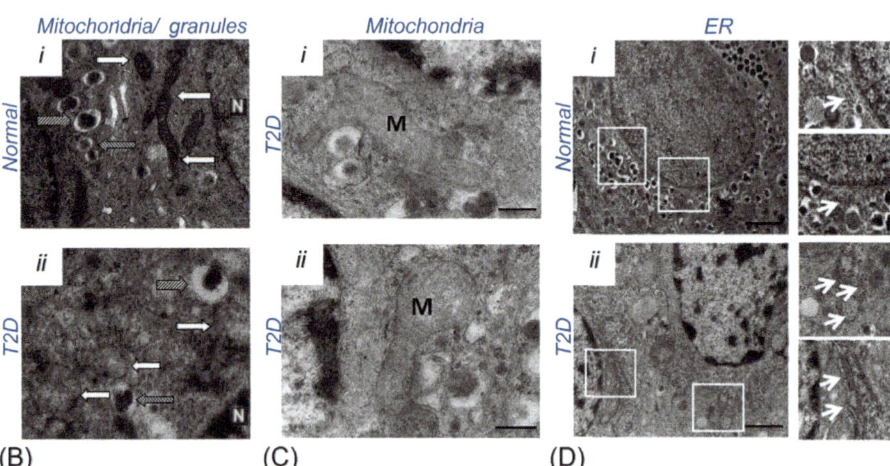


FIG. 4 Pathways leading to β cell destruction. (A) Overview of the pathways leading to β cell impairment and death. Genetic/epigenetic factors, chronic hyperglycemia, hyperlipidemia, hypoxia, and low-grade inflammation lead to ER stress, compromised mitochondrial function, and oxidative stress. ER stress is characterized by enlarged ER,

See the legend in facing page

Prolonged hyperglycemia both in vivo and in vitro has been shown to downregulate TCA enzymes and ETC components, while increasing the expression of glycolytic, gluconeogenic and glycogen-metabolizing enzymes.[91, 92] In addition, downregulation of genes involved in mitochondrial metabolism is observed in β cells of T2D patients.[93] This leads to the shift toward glycolytic metabolism and impaired mitochondrial function. For example, lower ATP levels, ATP/ADP ratio, and impaired mitochondrial hyperpolarization in response to glucose was detected in islets of T2D patients.[11] Diminished activity of the enzymes pyruvate carboxylase, glycerol phosphate dehydrogenase, and succinyl-CoA:3-ketoacid CoA transferase also was observed in T2D islets.[94] Moreover, it has been shown that the switch from oxidative metabolism leads to the glycogen accumulation in cells and islets exposed to high glucose or in T2D donors.[92] Excessive glycogen accumulation rather than hyperglycemia has been shown to contribute to β cell apoptosis.[92]

Lipid excess or lipotoxicity also plays a role in promoting β cell dysfunction. Exposure of islets to saturated FFAs, such as palmitate, has been demonstrated to deteriorate β cell function and lead to apoptosis. Lipotoxicity triggers ER stress with accompanied $Ca^{2+}$ dysregulation and increased oxidative stress. Moreover, impaired protein folding and altered posttranslational modulation contribute to reduced protein function in face of lipotoxicity.

Pancreatic β cells are highly specialized in insulin synthesis. During glucose elevations, insulin synthesis increases and represents more than 50% of total protein synthesis. In the context of hyperglycemia, proinsulin production can reach about million molecules per minute.[95] Such an increase in biosynthetic demand activates ER stress response to increase production of chaperones. ER stress above the threshold, however, causes pathological protein misfolding. Both elevated glucose levels and circulating FFAs and cytokines have been reported to decrease sarcolemmal ER $Ca^{2+}$ ATPase 2b (SERCA2b) expression, which pumps $Ca^{2+}$ into the ER, and to diminish ER $Ca^{2+}$ stores.[83, 96] ER stress results in closer association with mitochondria and increased $Ca^{2+}$ transfer from the ER to mitochondria, aggravating oxidative stress and exacerbating oxidative phosphorylation.[83]

FIG. 4, CONT'D reduced intraluminal $Ca^{2+}$, and protein misfolding. ER stress is one of the factors, contributing to diminished mitochondrial function, evident from the diminished rate of tricarboxylic acid (TCA) cycle and electron transport chain (ETC). In addition, stressed mitochondria appear swollen and with distorted cristae structure. Altered mitochondrial metabolism results in posttranslational protein modifications, specifically succination (2SC) of active cysteine residues (C) and acetylation (Ac) of lysine residues (L), leading to aggravation in protein function. Reduced inhibition of carnitine palmitoyltransferase (CPT) by malonyl CoA results in higher rate of β-oxidation, negatively affecting GSIS. Diminished levels of ATP impair inhibition of ATP-dependent $K^+$ channels ($K_{ATP}$), further aggravating GSIS. Moreover, increased ROS generation by ETC and NAD(P)H oxidases present in the ER and cytosol add to oxidative stress and cell damage. Abbreviations: *GLUT*, glucose transporter; *TCA*, tricarboxylic acid; *ETC*, electron transport chain; *KATP*, ATP-dependent $K^+$ channels; *VDCC*, voltage-dependent $Ca^{2+}$ channels; *CPT*, carnitine palmitoyltransferase; *LC-CoA*, long-chain acyl-CoA; *FFA*, free fatty acids; *Δψ*, plasma membrane potential; *ROS*, reactive oxygen species; *L*, lysine; *Ac*, acetyl; *C*, cysteine; *2SC*, succinyl. Dotted lines represent processes. (B–D) Electron micrographs of (B–C) mitochondria, insulin granules and (D) ER in human normal (i in B and D) or T2D (i in C and ii in B–D) β cell. White arrows in B show mitochondria and hashed arrows depict insulin granules. Right panels in D show magnification of the areas enwrapped by white quadrats, arrows in these panels indicate details of the ER cisternae. *M*, mitochondria; *N*, nucleus. Scale bars correspond to 0.26 μm in C and 1 μm in D. *The images in B–D are taken from Anello M, Lupi R, Spampinato D, et al. Functional and morphological alterations of mitochondria in pancreatic beta cells from type 2 diabetic patients. Diabetologia 2005; 48: 282–9. and Masini M, Martino L, Marselli L, et al. Ultrastructural alterations of pancreatic beta cells in human diabetes mellitus. Diabetes Metab Res Rev 2017; 33.*

Two recent studies make a link between posttranslational modifications mediated by accumulation of mitochondrial intermediates, instigated by overload in glucose or lipids.[97, 98] Nutrient overload can increase acetyl-CoA, which provides a source for protein acetylation.[98] First identified as playing an important role in chromatin modification and gene expression, lysine acetylation was demonstrated to modulate cellular metabolic processes.[98] Prolonged palmitate exposure of human islets results in over-acetylation of several mitochondrial proteins, including GDH1, mitochondrial superoxide dismutase, and SREBP-1.[98] These modifications might play an important role in β cell damage. Likewise, hyperglycemia was shown to increase intracellular fumarate levels in rodent and human islets.[97] Increased fumarate levels also were detected in T2D islets.[97] Fumarate, one of the TCA intermediates, promotes protein succination,[97] a chemical modification of cysteine residues to form S-(2-succino)-cysteine (2SC) that is associated with impaired protein function. In β cells, knockout of fumarate hydratase results in increased levels of fumarate, increased succination of glyceraldehyde 3-phosphate dehydrogenase (GAPDH), guanosine monophosphate reductase (GMPR), and Parkinson Protein 7 (PARK 7/DJ-1).[97] In addition, the knockout leads to the appearance of age-progressing diabetes in mice, immense impairment in mitochondrial structure and function, resulting in the complete loss of GSIS in response to 6 mM glucose and near-complete abolition of the response to 20 mM glucose.[97] These two studies show an importance of post-translational modifications in β cell impairment in T2D.

## 10.2 Reactive Oxygen Species (ROS)

The term "ROS" was developed to describe $O_2$-derived free radicals, which are able to oxidize other substrates. The major ROS include superoxide ($O_2^-$), hydroxyl radical (OH), and hydrogen peroxide ($H_2O_2$). About 1%–2% of the mitochondrial $O_2$ consumed is converted into ROS because of the activity of ETC, mainly complexes I and III, glycerol 3-phosphate dehydrogenase, PDH, α-ketoglutarate dehydrogenase, and via other mitochondrial sources.[99] Elevated NADH levels are associated with increased $H_2O_2$ production by α-ketoglutarate dehydrogenase, fostering an additional ROS production by complex I.[99] Additional cellular ROS sources include, but are not restricted to, peroxisomes, cytosolic NADPH oxidase (NOX), and ER monoxygenases.[99]

As a byproduct of numerous biological reactions, ROS exist in all the cells and serve as important signaling molecules in numerous pathways. High ROS levels, however, lead to the damage of proteins, lipids, and nucleic acids by modifications including nitration, carbonylation, peroxidation, and nitrosylation, damaging cell function and evoking cell death. Therefore, mitochondrial ROS levels are tightly regulated. First, several mitochondrial enzymes reduce ROS levels[99] and constitute antioxidant defenses. Superoxide dismutase is able to convert superoxide, the most reactive ROS, into $H_2O_2$, which can be converted to oxygen and water by catalase, glutathione peroxidase, and peroxiredoxin. Second, ROS production is controlled. The main source of mitochondrial ROS is ETC. The regulation of ROS production by ETC in vivo is still obscure, however, increased ROS formation is observed upon rise in pmf.[99] Exposure to very low glucose levels is also associated with increased ROS generation in β cell lines[100] and in dissociated primary rat β cells[101] in the low glucose range (2–10 mM for primary β cells). In addition, since ROS are short-lived molecules, close location of their synthesis and their site of function likely plays a role in ROS signaling.[102]

Pancreatic islets have much lower expression of superoxide dismutase, glutathione peroxidase, and catalase than most other tissues, and the latter is considered a disallowed gene in these cells.[2, 47, 103] β cells, however, have abundant levels of other antioxidant defenses, such as glutaredoxin and thioredoxin.[103] The expression of antioxidant defenses was shown to be lower in females than in males in both mice and humans.[104]

Using organelle-targeted redox-sensitive probes, glucose-dependent alterations in a glutathione-redox state were shown to be confined to mitochondria and not to the cytosolic/nuclear compartment.[105] Changes in the uncoupling protein-2 (UCP2) expression alter ROS levels in β cells.[106] UCPs are inner mitochondrial membrane $H^+$ transporters that dissipate the $H^+$ gradient upon opening, therefore uncoupling it from ATP synthesis. As expected, UCP2 deletion decreased, and its overexpression increased intracellular ROS levels. The effects of altering UCP2 level on β cell function, however, were not consistent in different studies,[106] requiring additional investigation into the effect of UCP2 on islet physiology.

Mitochondrial ROS are known for their involvement in a plethora of cellular functions, including adaptation to hypoxia, control of autophagy, cell differentiation, proliferation, and apoptosis.[99, 102] ROS can serve as a double-edged sword in β cell life and function. For example, microinjection of glutaredoxin stimulated the effects of NADPH effect on exocytosis, while thioredoxin reduced this effect.[103] ROS also have been shown to regulate β cell differentiation. An increase in ROS production was required for the differentiation of rat β cells in vivo and of progenitors in cultures of pancreatic islets.[107] ROS mediated their effects via ERK1/2 signaling. A recent study demonstrated that ROS are required for β cell proliferation.[108] By using genetically expressed $H_2O_2$ sensors, and pharmacological and genetic modulators of ROS, Ninov and colleagues showed that glucose increase to 18 mM resulted in elevated intracellular $H_2O_2$ levels in zebrafish and in the rat cell line INS-1. These increases corresponded to a glucose-mediated increase in proliferation.

ROS were suggested to alter protein function by generating reversible posttranslational modifications. Acting with glutathione, $H_2O_2$ can lead thiol groups (–SH) on cysteine residues to form a disulphide (–SS–) or a sulphenyl amide (–SN–) bond.[102] Phosphatases appear to be particularly susceptible to such modifications because of the existence of reactive cysteine residues in their catalytic domain.[102] Examples include PTP1b, PTEN, and MAPK phosphatases.[102]

Exposure of β cells to high glucose, high lipid content, or elevated insulin levels because of insulin resistance is known to result in oxidative stress and increased apoptosis.[109] During a hyperglycemic condition, increased oxidative metabolism and increased intracellular $Ca^{2+}$ both can increase ROS to damaging levels. $Ca^{2+}$ rise leads to the plasma membrane recruitment and activation of PKC, which in turn stimulates NOX-dependent ROS production.[109] ROS production was instigated by treatment of the rat islets or insulinoma cells with high glucose, or palmitate-rich medium.[109] This elevation was accompanied by increased protein expression of the p47phox component of the NOX.

## 10.3 Hypoxia

Pancreatic islets and β cells are heterogeneous with respect to size, differentiation status, metabolism, mitochondrial function, redox potential, and their ability to secrete insulin. For example, islets are not distributed evenly within the pancreas; islet density is higher in the tail than in the body or the head.[110] If we compare the islet population within the tail and the head, β cell loss is greater in the tail area during T1D,[111] but the opposite is true for T2D.[110] Functional differences also have been described for individual β cells. Existence of four different populations of β cells differing in their cell surface markers and transcriptome profiles has been described,[112] and our recent work shows existence of a hub population (<10%), having higher metabolic activity, but lower secretory capacity dictating the synchronicity of insulin release from highly connected to hubs population of followers, which have lower metabolic and higher secretory abilities.[71]

Oxygenation levels of whole islets and individual β cells within islets differ. A study by Olsson and Carlsson showed that, in a normal state, about 20%–25% of rat islets are low-oxygenated and have impaired metabolic phenotype.[113] This percentage decreased following pancreatomy. The authors suggest that low-oxygenated islets present a dormant reserve pool, which awakens upon metabolic challenges. High metabolic activity parallels higher oxygen demand, and, therefore, because of cell heterogeneity, there is a metabolically linked oxygenation diversity.

Prolonged hypoxia, however, is correlated with β cell impairment and death, typically by necrosis. One of the causes of hypoxia is a common sleep disorder—obstructive sleep apnea syndrome, characterized by recurrent episodes of upper airway obstructions during sleep. The syndrome appears in 3%–7% of the general population, and it results in intermittent hypoxia, sleep fragmentation, and daytime sleepiness.[114] The disorder is associated with several metabolic changes, such as dyslipidemia, insulin resistance, glucose intolerance, and T2D.[114]

Both in human patients and in mouse models, chronic intermittent hypoxia increases insulin secretion in basal/fasting state, elevates blood glucose levels, in part because of reduced insulin sensitivity.[109, 115, 116] The mechanism of this phenomenon has been investigated with a larger scrutiny in mouse models. In rodents, hypoxia correlated with β cell death and proliferation.[117] Chronic intermittent hypoxia results in mitochondrial oxidative stress, augmented basal insulin secretion, insulin resistance, flawed proinsulin processing, and reduced GSIS.[116] Pancreatic islets exposed to such hypoxia were characterized by elevated levels of mitochondrial ROS, and treatment of mice with mitochondrial ROS scavenger mito-tempol abolished hypoxia-dependent deterioration. Using rat and human islets and the INS1832/13 cell line, another study found that elevated basal insulin secretion because of hypoxia was accompanied by increased protein expression of complex I and complex II components and more efficient coupling of complex I- and complex II-linked respiration.[118] Amelioration of oxidative stress using antioxidants proved to be efficient in rodent models of diabetes, such as Zukker diabetic fatty rats and diabetic db/db mice, in improving glucose homeostasis. In humans, however, these beneficial effects of antioxidants were not preserved consistently.[109]

## 10.4 Inflammatory Responses and T2D

Chronic low-grade tissue inflammation is one of the key features of obesity and T2D. Among other tissues, signs of inflammation are observed in T2D islets and include immune cell infiltration, amyloid deposition, cell death, and fibrosis.[119, 120] In human T2D islets and animal models of diabetes, macrophages infiltrate pancreatic islets, playing a crucial role in islet inflammation.[119] Animal models of obesity and T2D show signs of macrophage infiltration before the appearance of any clinical signs.[120]

Increased amounts of pro-inflammatory cytokines, released by immune cells and with a major role for interleukin-1β (IL-1β), negatively affect islet function.[119] Human islet amyloid peptide, palmitate, and endocannabinoids stimulate IL-1β secretion by islet macrophages.[119–121] Although β cells have the necessary machinery for IL-1β synthesis, the main source for IL-1β is macrophages.[119, 120] In mouse islets, high ROS levels because of elevated concentration of glucose, upregulate the expression of thioredoxin-interacting protein TXNIP, activating NLP3 inflammasome, which is required for IL-1β synthesis.[121] IL-1R1 expression in pancreatic islets is higher than in other tissues examined.[121] All these factors make islet β cells particularly sensitive to inflammatory attack by IL-1β. Treatment of diabetic and prediabetic patients with IL-1 receptor antagonists reduced inflammation and improved glycemic control and β cell function.[119, 121] Such a beneficial effect of IL-1R antagonists was also observed in diabetic GK rats.[119] β cell mass, however, remained unaltered. Moreover, treatment of islets by liposomes filled with macrophage toxin clodronate in db/db and KKAy mice improved glucose tolerance and insulin secretion in vivo and in vitro.[119] These results provide evidence for a key role of inflammation in the T2D pathogenesis.

Constantly high levels of IL-1β result in mitochondrial function impairment and glucose-induced impairment of β cell function, leading to apoptosis.[121, 122] IL-1R stimulation releases NF-κB from the inhibition by inhibitory κB (IκB),[3] and NF-κB translocates into the nucleus, promoting expression of the inducible nitric oxide synthase (iNOS, encoded by NOS2 gene) in islet β cells,[3] and alters the expression of BCL-2 family proteins, intimately involved in apoptosis. Increased NO production by iNOS contributes to ETC impairment in cells. NO can reversibly bind to $O_2$-binding domains on cytochrome C.[123] Respiratory inhibition by NO induces the generation of reactive nitrogen and oxygen species, contributing to oxidative stress and apoptosis.[123] Reduced respiration depletes energy supply, which can result in necrosis. Reactive nitrogen species can also contribute to the opening of mPTP, furthering cell death via apoptosis and necrosis.[123] Moreover, NF-κB can induce p53 upregulated modulator of apoptosis (PUMA) expression in pancreatic β cells, resulting in the oligomerization of pro-apoptotic proteins BAX and BAK, contributing to apoptosis.[124]

## 11 MITOCHONDRIAL INVOLVEMENT IN β CELL DEATH

Both T1D and T2D are characterized by β cell loss. Although T1D results in significant eradication of insulin-secreting cells, the extent of β cell loss in T2D and its significance for development of the disease is currently under debate.[89, 125] β cell mass varies greatly among individuals—between 0.6 and 2.1 g.[125] The estimate for β cell mass reduction during T2D is between 24% and 65%.[89, 125] Although β cell loss contributes to the impaired glucose homeostasis observed during T2D, it is unlikely to be the principal driver of the disease. First, surgical removal of 50% of the pancreas leads to T2D only in obese patients.[89] Hemipancreatomy, however, impairs glucose homeostasis. Second, β cell function improvement following bariatric surgery or short-term caloric restriction often results in improved glucose homeostasis and even T2D remission.[89] Several possible explanations to reduced β cell mass observed in T2D include β cell death or dedifferentiation, inadequate capacity of proliferation in face of increased demand, or that individuals with genetically or environmentally predetermined low β cell mass are susceptible to diabetes.[126]

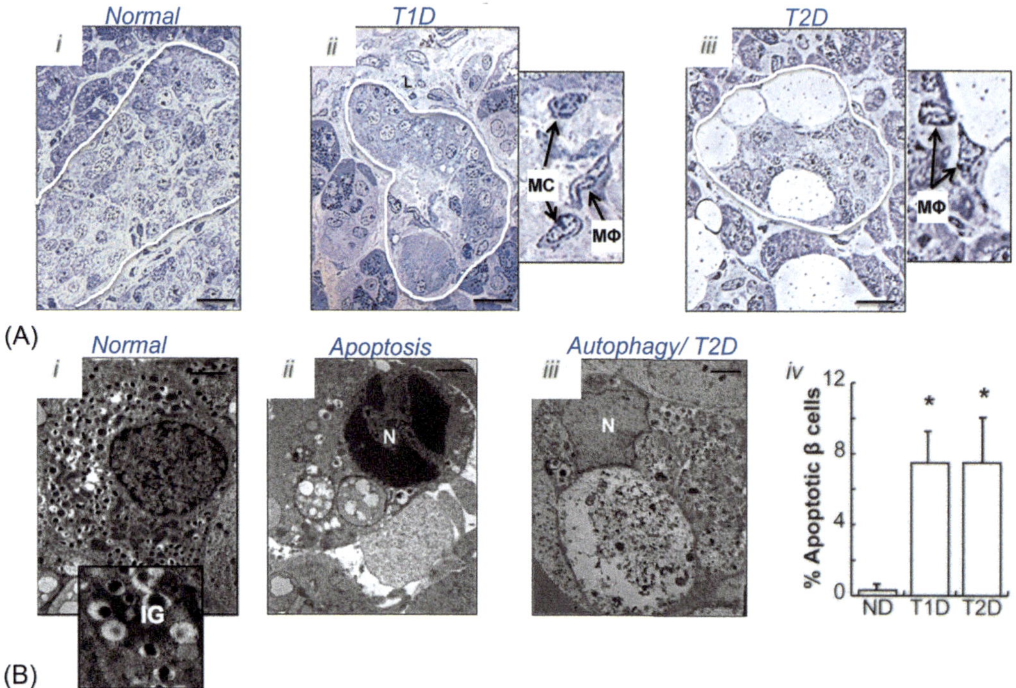


FIG. 5 Pathology of β cell death in diabetes. (A) Pancreatic tissue slices from (i) nondiabetic (ND), (ii) type 1 (T1D), and (iii) type 2 (T2D) diabetic organ donors. Islets of Langerhans are encircled in white. Right panel insets show infiltrating immune cells. *MC*, mast cell; *MΦ*, macrophage. (B) Electron micrographs of (i) normal and (ii) apoptotic β cells, and (iii) β cell of T2D donor showing signs of altered autophagy. (iv) Percentage of apoptotic β cells from ND, T1D and T2D organ donors. Scale bars correspond to 10 μm in A and 1 μm in B. *The images are taken from Masini M, Martino L, Marselli L, et al. Ultrastructural alterations of pancreatic beta cells in human diabetes mellitus. Diabetes Metab Res Rev 2017; 33.*

β cell death, however, represents an important feature of T2D and, as with any tissue damage, associated increased inflammatory response can aggravate β cell deterioration further. As discussed previously, factors contributing to β cell death are glucotoxicity, lipotoxicity, and inflammatory cytokines. The death pathways of β cells include necrosis, apoptosis, and autophagy. Fig. 5A shows images of islets from normal, T1D, and T2D human donors, revealing loss of β cell mass and signs of inflammation. Fig. 5B illustrates morphological features of apoptotic and autophagic β cell death.

Programmed cell death by apoptosis is characterized by the involvement of BCL-2 family, which contains at least 12 anti-apoptotic and pro-apoptotic proteins, located at the mitochondria.[127] The balance between these two types of proteins determines whether the cell will live or die. Two of the key BCL-2 proteins that induce apoptosis are BAX and BAK.[127] A third protein, sharing homology to BAX and BAK and linking ER stress to apoptosis, is BOK.[127] Activation of these proteins via upstream pathways stimulates their oligomerization at the outer mitochondrial membrane, resulting in mitochondrial outer membrane permeabilization (MOMP),[127] which is considered a committed step of apoptosis. MOMP results in cytochrome C release to cytosol, required for the apoptotic protease activating factor 1 (APAF-1) oligomerization and formation of a wheel-shaped apoptosome.[128] The apoptosome serves as a signaling platform for dATP or ATP-dependent activation of caspase proteases, followed by cleavage of genomic DNA, exposure of phosphatidylserine and other signals on the cell surface, and promotion of engulfment of apoptotic cells by local immune cells.[128] BCL-2 proteins directly activating BAX and BAK are BIM, BID, PUMA, and NOXA.[127] These proteins can be sequestered by anti-apoptotic BCL-2 proteins, such as BCL-XL, MCL-1 and BCL-W.[127]

Incubation of human pancreatic islets in high glucose results in the overexpression of pro-apoptotic proteins BAD and BID, specifically in β cells.[129] Overexpression of pro-apoptotic protein BIK and downregulation of anti-apoptotic protein BCL-XL also was observed, but is not restricted to β cells.[129] Modification of BIM splicing by knockdown of human Gli-similar (GLIS) 3 protein increases cytochrome C release and β cell apoptosis.[130] GLIS3 mutations have been reported to cause neonatal diabetes,[131] and GLIS3 gene locus has been identified as a susceptibility risk locus for T1D and T2D in genomewide association studies.[132, 133] The role of mitochondrial BCL-2 proteins in β cell function has been expanded recently. For example, BAD was shown to control glucokinase activity and mitochondrial respiration.[134] BCL-2 and BCL-XL expression mediated reduction in GSIS.[135] Both proteins also regulate mitochondrial fusion, therefore regulating mitochondrial shape and metabolism. Genetic deletion of BMF, which is able to bind BCL-2 and BCL-XL in mice, results in impaired GSIS and hyperglycemia.[136] Estrogen, a female sex hormone, was found to inhibit β cell apoptosis, favoring β cell pro-survival and protection from both T1D and T2D.[137, 138]

Because of its involvement in cell death, mPTP is studied extensively. Although mPTP usually flickers between open and closed states, long-lasting mPTP opening results in swelling and rupture of mitochondria, causing release of its components including pro-apoptotic cytochrome C and ATP. If ATP levels are sufficient, apoptosis will ensue, otherwise, the cell can die via necrosis. mPTP composition has been investigated for several decades. Initially, VDAC, ANT, and cyclophylin D were proposed to constitute the pore. Recent studies using genetic manipulations, however, disproved these molecules as mPTP components and put forward mitochondrial ATP synthase as the main candidate for mitochondrial pore component.

## 12 CONCLUDING REMARKS

This chapter describes the critical involvement of mitochondria in β cell metabolism, secretory function, and death. Because insulin is a vital hormone that can lower blood glucose, its secretion by pancreatic β cells is tightly regulated. Inadequate insulin secretion because of β cell dysfunction and β cell mass loss in face of nutrient overload leads to T2D.

Highly adaptive to environmental changes and challenges, mitochondria are able to integrate cellular cues into appropriate cellular responses. β cell mitochondrial metabolism is uniquely adjusted to fit specialized requirements of insulin-secreting cells. Therefore, mitochondrial metabolism in β cells is preferred over glycolytic metabolism or pentose-phosphate pathway. Mutations in the components of the machinery linking glucose metabolism to oxidative phosphorylation can lead to β cell impairment. Because of TCA's importance for insulin secretion, mitochondrial-cytosolic shuttles ensure adequate amounts of TCA intermediates. Among crucial metabolites potentiating GSIS are FFAs, malonyl-CoA, LC-CoA, NAD(P)H, and glutamate. Their importance is demonstrated by deregulations caused by mutations in the enzymes responsible for these molecules.

Ion signaling and $Ca^{2+}$ signaling, in particular, contribute to GSIS. Mitochondria react to cytosolic $Ca^{2+}$ elevations and, in turn, upregulate mitochondrial metabolism. Removal of mitochondrial $Ca^{2+}$ machinery alters β cell function. In spite of the immense importance of ER-mitochondrial contacts MAMs for $Ca^{2+}$ signaling, lipid biogenesis, lipid transfer, and signaling cascades, the role of MAMs for insulin secretion and β cell fate and survival is still unknown.

Prolonged insults because of nutritional overload result in a deterioration of mitochondrial function and, ultimately, in β cell dysfunction and death. ER stress, oxidative stress, decreased TCA activity, aggravated protein function, and increase in ROS levels because of, in part, altered ETC function, chronic hypoxia, and chronic inflammation, all contribute to β cell deterioration and could lead to T2D in genetically susceptible individuals. More studies, focusing on β cell physiology in health and disease with particular emphasis on mitochondria, might help us to understand better the pathogenesis of T2D and, we hope, lead to better therapeutics.

## Acknowledgments

Prof. G.A.R. is supported by Diabetes UK (Project BDA 11/0004210), the Wellcome Trust (Programme 081958/Z/07/Z; Senior Investigator Award WT098424AIA), and the MRC (UK; Project GO401641; Programme MR/J0003042/1). J.P., PhD is supported by Novo-Nordisk postdoctoral research fellowship.

## References

1. Thorens B. Glucose sensing and the pathogenesis of obesity and type 2 diabetes. *Int J Obes* 2008;**32**(Suppl. 6): S62–71.
2. Rutter GA, Pullen TJ, Hodson DJ, Martinez-Sanchez A. Pancreatic beta-cell identity, glucose sensing and the control of insulin secretion. *Biochem J* 2015;**466**:203–18.
3. Cnop M, Welsh N, Jonas JC, Jorns A, Lenzen S, Eizirik DL. Mechanisms of pancreatic beta-cell death in type 1 and type 2 diabetes—many differences, few similarities. *Diabetes* 2005;**54**:S97–S107.
4. Rorsman P, Braun M. Regulation of insulin secretion in human pancreatic islets. *Annu Rev Physiol* 2013;**75**:155–79.
5. Guariguata L, Whiting DR, Hambleton I, Beagley J, Linnenkamp U, Shaw JE. Global estimates of diabetes prevalence for 2013 and projections for 2035. *Diabetes Res Clin Pract* 2014;**103**:137–49.
6. Nathan DM. Long-term complications of diabetes mellitus. *N Engl J Med* 1993;**328**:1676–85.
7. Rutter GA, Rizzuto R. Regulation of mitochondrial metabolism by ER $Ca^{2+}$ release: an intimate connection. *Trends Biochem Sci* 2000;**25**:215–21.
8. Denton RM. Regulation of mitochondrial dehydrogenases by calcium ions. *BBA-Bioenergetics* 1787;**2009**:1309–16.
9. Mishra P, Chan DC. Metabolic regulation of mitochondrial dynamics. *J Cell Biol* 2016;**212**:379–87.
10. Maechler P, Wollheim CB. Mitochondrial function in normal and diabetic beta-cells. *Nature* 2001;**414**:807–12.
11. Mulder H. Transcribing beta-cell mitochondria in health and disease. *Mol Metab* 2017;**6**:1040–51.
12. Cnop M, Mulder H, Igoillo-Esteve M. Diabetes in Friedreich ataxia. *J Neurochem* 2013;**126**(Suppl. 1):94–102.
13. Anello M, Lupi R, Spampinato D, et al. Functional and morphological alterations of mitochondria in pancreatic beta cells from type 2 diabetic patients. *Diabetologia* 2005;**48**:282–9.
14. Masini M, Martino L, Marselli L, et al. Ultrastructural alterations of pancreatic beta cells in human diabetes mellitus. *Diabetes Metab Res Rev* 2017;**33**.
15. Fex M, Nitert MD, Wierup N, Sundler F, Ling C, Mulder H. Enhanced mitochondrial metabolism may account for the adaptation to insulin resistance in islets from C57BL/6J mice fed a high-fat diet. *Diabetologia* 2007;**50**:74–83.

16. Hodson DJ, Mitchell RK, Bellomo EA, et al. Lipotoxicity disrupts incretin-regulated human beta cell connectivity. *J Clin Invest* 2013;**123**:4182–94.
17. Cabrera O, Berman DM, Kenyon NS, Ricordi C, Berggren PO, Caicedo A. The unique cytoarchitecture of human pancreatic islets has implications for islet cell function. *Proc Natl Acad Sci U S A* 2006;**103**:2334–9.
18. Matveyenko AV, Butler PC. Islet amyloid polypeptide (IAPP) transgenic rodents as models for type 2 diabetes. *ILAR J* 2006;**47**:225–33.
19. Asfari M, Janjic D, Meda P, Li G, Halban PA, Wollheim CB. Establishment of 2-mercaptoethanol-dependent differentiated insulin-secreting cell lines. *Endocrinology* 1992;**130**:167–78.
20. Hohmeier HE, Mulder H, Chen G, Henkel-Rieger R, Prentki M, Newgard CB. Isolation of INS-1-derived cell lines with robust ATP-sensitive K+ channel-dependent and -independent glucose-stimulated insulin secretion. *Diabetes* 2000;**49**:424–30.
21. Merglen A, Theander S, Rubi B, Chaffard G, Wollheim CB, Maechler P. Glucose sensitivity and metabolism-secretion coupling studied during two-year continuous culture in INS-1E insulinoma cells. *Endocrinology* 2004;**145**:667–78.
22. Miyazaki J, Araki K, Yamato E, et al. Establishment of a pancreatic beta cell line that retains glucose-inducible insulin secretion: special reference to expression of glucose transporter isoforms. *Endocrinology* 1990;**127**:126–32.
23. Ravassard P, Hazhouz Y, Pechberty S, et al. A genetically engineered human pancreatic beta cell line exhibiting glucose-inducible insulin secretion. *J Clin Invest* 2011;**121**:3589–97.
24. Benazra M, Lecomte MJ, Colace C, et al. A human beta cell line with drug inducible excision of immortalizing transgenes. *Mol Metab* 2015;**4**:916–25.
25. Scharfmann R, Pechberty S, Hazhouz A, et al. Development of a conditionally immortalized human pancreatic beta cell line. *J Clin Invest* 2014;**124**:2087–98.
26. Krizhanovskii C, Kristinsson H, Elksnis A, et al. EndoC-betaH1 cells display increased sensitivity to sodium palmitate when cultured in DMEM/F12 medium. *Islets* 2017;**9**:e1296995.
27. Pagliuca FW, Millman JR, Gurtler M, et al. Generation of functional human pancreatic beta cells in vitro. *Cell* 2014;**159**:428–39.
28. Rezania A, Bruin JE, Arora P, et al. Reversal of diabetes with insulin-producing cells derived in vitro from human pluripotent stem cells. *Nat Biotechnol* 2014;**32**:1121–33.
29. Cerasi E, Luft R. Plasma-insulin response to sustained hyperglycemia induced by glucose infusion in human subjects. *Lancet* 1963;**2**:1359–61.
30. Curry DL, Bennett LL, Grodsky GM. Dynamics of insulin secretion by the perfused rat pancreas. *Endocrinology* 1968;**83**:572–84.
31. Nesher R, Cerasi E. Modeling phasic insulin release: immediate and time-dependent effects of glucose. *Diabetes* 2002;**51**(Suppl. 1):S53–9.
32. Coore HG, Randle PJ. Regulation of insulin secretion studied with pieces of rabbit pancreas incubated in vitro. *Biochem J* 1964;**93**:66–78.
33. Grodsky GM, Batts AA, Bennett LL, Vcella C, McWilliams NB, Smith DF. Effects of carbohydrates on secretion of insulin from isolated rat pancreas. *Am J Phys* 1963;**205**:638–44.
34. Matschinsky FM, Ellerman JE. Metabolism of glucose in the islets of Langerhans. *J Biol Chem* 1968;**243**:2730–6.
35. Ashcroft FM, Harrison DE, Ashcroft SJ. Glucose induces closure of single potassium channels in isolated rat pancreatic beta-cells. *Nature* 1984;**312**:446–8.
36. Cook DL, Hales CN. Intracellular ATP directly blocks K+ channels in pancreatic B-cells. *Nature* 1984;**311**:271–3.
37. Henquin JC. Triggering and amplifying pathways of regulation of insulin secretion by glucose. *Diabetes* 2000;**49**:1751–60.
38. Schuit F, De Vos A, Farfari S, et al. Metabolic fate of glucose in purified islet cells. Glucose-regulated anaplerosis in beta cells. *J Biol Chem* 1997;**272**:18572–9.
39. Wiederkehr A, Wollheim CB. Mitochondrial signals drive insulin secretion in the pancreatic beta-cell. *Mol Cell Endocrinol* 2012;**353**:128–37.
40. Campbell JE, Drucker DJ. Pharmacology, physiology, and mechanisms of incretin hormone action. *Cell Metab* 2013;**17**:819–37.
41. Rorsman P, Bokvist K, Ammala C, et al. Activation by adrenaline of a low-conductance G protein-dependent K+ channel in mouse pancreatic B cells. *Nature* 1991;**349**:77–9.
42. Straub SG, Sharp GW. Evolving insights regarding mechanisms for the inhibition of insulin release by norepinephrine and heterotrimeric G proteins. *Am J Phys Cell Physiol* 2012;**302**:C1687–98.

43. Henquin JC, Dufrane D, Nenquin M. Nutrient control of insulin secretion in isolated normal human islets. *Diabetes* 2006;**55**:3470–7.
44. Hutton JC, Sener A, Herchuelz A, et al. Similarities in the stimulus-secretion coupling mechanisms of glucose- and 2-keto acid-induced insulin release. *Endocrinology* 1980;**106**:203–19.
45. Ainscow EK, Zhao C, Rutter GA. Acute overexpression of lactate dehydrogenase-A perturbs beta-cell mitochondrial metabolism and insulin secretion. *Diabetes* 2000;**49**:1149–55.
46. Ishihara H, Wang HY, Drewes LR, Wollheim CB. Overexpression of monocarboxylate transporter and lactate dehydrogenase alters insulin secretory responses to pyruvate and lactate in beta cells. *J Clin Invest* 1999;**104**:1621–9.
47. Pullen TJ, Rutter GA. When less is more: the forbidden fruits of gene repression in the adult beta-cell. *Diabetes Obes Metab* 2013;**15**:503–12.
48. Sekine N, Cirulli V, Regazzi R, et al. Low lactate dehydrogenase and high mitochondrial glycerol phosphate dehydrogenase in pancreatic beta-cells. Potential role in nutrient sensing. *J Biol Chem* 1994;**269**:4895–902.
49. Bricker DK, Taylor EB, Schell JC, et al. A mitochondrial pyruvate carrier required for pyruvate uptake in yeast, Drosophila, and humans. *Science* 2012;**337**:96–100.
50. Herzig S, Raemy E, Montessuit S, et al. Identification and functional expression of the mitochondrial pyruvate carrier. *Science* 2012;**337**:93–6.
51. McCommis KS, Hodges WT, Bricker DK, et al. An ancestral role for the mitochondrial pyruvate carrier in glucose-stimulated insulin secretion. *Mol Metab* 2016;**5**:602–14.
52. Srinivasan M, Choi CS, Ghoshal P, et al. ss-Cell-specific pyruvate dehydrogenase deficiency impairs glucose-stimulated insulin secretion. *Am J Physiol Endocrinol Metab* 2010;**299**:E910–7.
53. Jitrapakdee S, St Maurice M, Rayment I, Cleland WW, Wallace JC, Attwood PV. Structure, mechanism and regulation of pyruvate carboxylase. *Biochem J* 2008;**413**:369–87.
54. Schagger H, Pfeiffer K. Supercomplexes in the respiratory chains of yeast and mammalian mitochondria. *EMBO J* 2000;**19**:1777–83.
55. Genova ML, Lenaz G. Functional role of mitochondrial respiratory supercomplexes. *Biochim Biophys Acta* 1837;**2014**:427–43.
56. Yaney GC, Corkey BE. Fatty acid metabolism and insulin secretion in pancreatic beta cells. *Diabetologia* 2003;**46**:1297–312.
57. Nicholls DG. The pancreatic beta-cell: a bioenergetic perspective. *Physiol Rev* 2016;**96**:1385–447.
58. Zhao S, Mugabo Y, Iglesias J, et al. alpha/beta-Hydrolase domain-6-accessible monoacylglycerol controls glucose-stimulated insulin secretion. *Cell Metab* 2014;**19**:993–1007.
59. Molven A, Hollister-Lock J, Hu J, et al. The hypoglycemic phenotype is islet cell-autonomous in short-chain hydroxyacyl-CoA dehydrogenase-deficient mice. *Diabetes* 2016;**65**:1672–8.
60. Hua YH, Wu CY, Sargsyan K, Lim C. Sequence-motif detection of NAD(P)-binding proteins: discovery of a unique antibacterial drug target. *Sci Rep* 2014;**4**:6471.
61. Anderson KA, Madsen AS, Olsen CA, Hirschey MD. Metabolic control by sirtuins and other enzymes that sense NAD(+), NADH, or their ratio. *Biochim Biophys Acta* 1858;**2017**:991–8.
62. Brun T, Maechler P. Beta-cell mitochondrial carriers and the diabetogenic stress response. *Biochim Biophys Acta* 1863;**2016**:2540–9.
63. Patterson GH, Knobel SM, Arkhammar P, Thastrup O, Piston DW. Separation of the glucose-stimulated cytoplasmic and mitochondrial NAD(P)H responses in pancreatic islet beta cells. *Proc Natl Acad Sci U S A* 2000;**97**:5203–7.
64. Ferdaoussi M, Dai X, Jensen MV, et al. Isocitrate-to-SENP1 signaling amplifies insulin secretion and rescues dysfunctional beta cells. *J Clin Invest* 2015;**125**:3847–60.
65. Gheni G, Ogura M, Iwasaki M, et al. Glutamate acts as a key signal linking glucose metabolism to incretin/cAMP action to amplify insulin secretion. *Cell Rep* 2014;**9**:661–73.
66. MacDonald MJ, Fahien LA. Glutamate is not a messenger in insulin secretion. *J Biol Chem* 2000;**275**:34025–7.
67. Stanley CA, Lieu YK, Hsu BY, et al. Hyperinsulinism and hyperammonemia in infants with regulatory mutations of the glutamate dehydrogenase gene. *N Engl J Med* 1998;**338**:1352–7.
68. Lenzen S. A fresh view of glycolysis and glucokinase regulation: history and current status. *J Biol Chem* 2014;**289**:12189–94.
69. McCulloch LJ, van de Bunt M, Braun M, Frayn KN, Clark A, Gloyn AL. GLUT2 (SLC2A2) is not the principal glucose transporter in human pancreatic beta cells: implications for understanding genetic association signals at this locus. *Mol Genet Metab* 2011;**104**:648–53.

70. De Vos A, Heimberg H, Quartier E, et al. Human and rat beta cells differ in glucose transporter but not in glucokinase gene expression. *J Clin Invest* 1995;**96**:2489–95.
71. Johnston NR, Mitchell RK, Haythorne E, et al. Beta cell hubs dictate pancreatic islet responses to glucose. *Cell Metab* 2016;**24**:389–401.
72. Tarasov AI, Semplici F, Ravier MA, et al. The mitochondrial $Ca^{2+}$ uniporter MCU is essential for glucose-induced ATP increases in pancreatic beta-cells. *PLoS ONE* 2012;**7**:e39722.
73. Wiederkehr A, Park KS, Dupont O, et al. Matrix alkalinization: a novel mitochondrial signal for sustained pancreatic beta-cell activation. *EMBO J* 2009;**28**:417–28.
74. Wikstrom JD, Katzman SM, Mohamed H, et al. beta-Cell mitochondria exhibit membrane potential heterogeneity that can be altered by stimulatory or toxic fuel levels. *Diabetes* 2007;**56**:2569–78.
75. Deluca HF, Engstrom GW. Calcium uptake by rat kidney mitochondria. *Proc Natl Acad Sci U S A* 1961;**47**:1744–50.
76. De Stefani D, Rizzuto R, Pozzan T. Enjoy the trip: calcium in mitochondria back and forth. *Annu Rev Biochem* 2016;**85**:161–92.
77. Starkov AA. The molecular identity of the mitochondrial $Ca^{2+}$ sequestration system. *FEBS J* 2010;**277**:3652–63.
78. Denton RM, McCormack JG. The role of calcium in the regulation of mitochondrial metabolism. *Biochem Soc Trans* 1980;**8**:266–8.
79. Wiederkehr A, Szanda G, Akhmedov D, et al. Mitochondrial matrix calcium is an activating signal for hormone secretion. *Cell Metab* 2011;**13**:601–11.
80. Alam MR, Groschner LN, Parichatikanond W, et al. Mitochondrial $Ca^{2+}$ uptake 1 (MICU1) and mitochondrial $Ca^{2+}$ uniporter (MCU) contribute to metabolism-secretion coupling in clonal pancreatic beta-cells. *J Biol Chem* 2012;**287**:34445–54.
81. Nita II, Hershfinkel M, Fishman D, et al. The mitochondrial $Na^{+}/Ca^{2+}$ exchanger upregulates glucose dependent $Ca^{2+}$ signalling linked to insulin secretion. *PLoS ONE* 2012;**7**:e46649.
82. Copeland DE, Dalton AJ. An association between mitochondria and the endoplasmic reticulum in cells of the pseudobranch gland of a teleost. *J Biophys Biochem Cytol* 1959;**5**:393–6.
83. Arruda AP, Hotamisligil GS. Calcium homeostasis and organelle function in the pathogenesis of obesity and diabetes. *Cell Metab* 2015;**22**:381–97.
84. Hirabayahi Y, Kwon SK, Paek H, et al. ER-mitochondria tethering by PDZD8 regulates $Ca^{2+}$ dynamics in mammalian neurons. *Science* 2017;**358**:623–9.
85. Ardestani A, Lupse B, Kido Y, Leibowitz G, Maedler K. mTORC1 signaling: a double-edged sword in diabetic beta cells. *Cell Metab* 2017;.
86. Rutter GA, Pinton P. Mitochondria-associated endoplasmic reticulum membranes in insulin signaling. *Diabetes* 2014;**63**:3163–5.
87. Friedman JR, Lackner LL, West M, DiBenedetto JR, Nunnari J, Voeltz GK. ER tubules mark sites of mitochondrial division. *Science* 2011;**334**:358–62.
88. Thivolet C, Vial G, Cassel R, Rieusset J, Madec AM. Reduction of endoplasmic reticulum-mitochondria interactions in beta cells from patients with type 2 diabetes. *PLoS ONE* 2017;**12**:e0182027.
89. Chen C, Cohrs CM, Stertmann J, Bozsak R, Speier S. Human beta cell mass and function in diabetes: recent advances in knowledge and technologies to understand disease pathogenesis. *Mol Metab* 2017;**6**:943–57.
90. Porat S, Weinberg-Corem N, Tornovsky-Babaey S, et al. Control of pancreatic beta cell regeneration by glucose metabolism. *Cell Metab* 2011;**13**:440–9.
91. Bensellam M, Van Lommel L, Overbergh L, Schuit FC, Jonas JC. Cluster analysis of rat pancreatic islet gene mRNA levels after culture in low-, intermediate- and high-glucose concentrations. *Diabetologia* 2009;**52**:463–76.
92. Brereton MF, Rohm M, Shimomura K, et al. Hyperglycaemia induces metabolic dysfunction and glycogen accumulation in pancreatic beta-cells. *Nat Commun* 2016;**7**:13496.
93. Segerstolpe A, Palasantza A, Eliasson P, et al. Single-cell transcriptome profiling of human pancreatic islets in health and type 2 diabetes. *Cell Metab* 2016;**24**:593–607.
94. MacDonald MJ, Longacre MJ, Langberg EC, et al. Decreased levels of metabolic enzymes in pancreatic islets of patients with type 2 diabetes. *Diabetologia* 2009;**52**:1087–91.
95. Schuit FC, In't Veld PA, Pipeleers DG. Glucose stimulates proinsulin biosynthesis by a dose-dependent recruitment of pancreatic beta cells. *Proc Natl Acad Sci U S A* 1988;**85**:3865–9.
96. Hara T, Mahadevan J, Kanekura K, Hara M, Lu S, Urano F. Calcium efflux from the endoplasmic reticulum leads to beta-cell death. *Endocrinology* 2014;**155**:758–68.
97. Adam J, Ramracheya R, Chibalina MV, et al. Fumarate hydratase deletion in pancreatic beta cells leads to progressive diabetes. *Cell Rep* 2017;**20**:3135–48.

98. Ciregia F, Bugliani M, Ronci M, et al. Palmitate-induced lipotoxicity alters acetylation of multiple proteins in clonal beta cells and human pancreatic islets. *Sci Rep* 2017;**7**:13445.
99. Circu ML, Aw TY. Reactive oxygen species, cellular redox systems, and apoptosis. *Free Radic Biol Med* 2010;**48**:749–62.
100. Sarre A, Gabrielli J, Vial G, Leverve XM, Assimacopoulos-Jeannet F. Reactive oxygen species are produced at low glucose and contribute to the activation of AMPK in insulin-secreting cells. *Free Radic Biol Med* 2012;**52**:142–50.
101. Roma LP, Duprez J, Takahashi HK, Gilon P, Wiederkehr A, Jonas JC. Dynamic measurements of mitochondrial hydrogen peroxide concentration and glutathione redox state in rat pancreatic beta-cells using ratiometric fluorescent proteins: confounding effects of pH with HyPer but not roGFP1. *Biochem J* 2012;**441**:971–8.
102. Sena LA, Chandel NS. Physiological roles of mitochondrial reactive oxygen species. *Mol Cell* 2012;**48**:158–67.
103. Ivarsson R, Quintens R, Dejonghe S, et al. Redox control of exocytosis: regulatory role of NADPH, thioredoxin, and glutaredoxin. *Diabetes* 2005;**54**:2132–42.
104. Tonooka N, Oseid E, Zhou H, Harmon JS, Robertson RP. Glutathione peroxidase protein expression and activity in human islets isolated for transplantation. *Clin Transpl* 2007;**21**:767–72.
105. Takahashi HK, Santos LR, Roma LP, et al. Acute nutrient regulation of the mitochondrial glutathione redox state in pancreatic beta-cells. *Biochem J* 2014;**460**:411–23.
106. Robson-Doucette CA, Sultan S, Allister EM, et al. Beta-cell uncoupling protein 2 regulates reactive oxygen species production, which influences both insulin and glucagon secretion. *Diabetes* 2011;**60**:2710–9.
107. Hoarau E, Chandra V, Rustin P, Scharfmann R, Duvillie B. Pro-oxidant/antioxidant balance controls pancreatic beta-cell differentiation through the ERK1/2 pathway. *Cell Death Dis* 2014;**5**:e1487.
108. Ahmed Alfar E, Kirova D, Konantz J, Birke S, Mansfeld J, Ninov N. Distinct levels of reactive oxygen species coordinate metabolic activity with beta-cell mass plasticity. *Sci Rep* 2017;**7**:3994.
109. Gerber PA, Rutter GA. The role of oxidative stress and hypoxia in pancreatic beta-cell dysfunction in diabetes mellitus. *Antioxid Redox Signal* 2017;**26**:501–18.
110. Wang X, Misawa R, Zielinski MC, et al. Regional differences in islet distribution in the human pancreas–preferential beta-cell loss in the head region in patients with type 2 diabetes. *PLoS ONE* 2013;**8**:e67454.
111. Poudel A, Savari O, Striegel DA, et al. Beta-cell destruction and preservation in childhood and adult onset type 1 diabetes. *Endocrine* 2015;**49**:693–702.
112. Dorrell C, Schug J, Canaday PS, et al. Human islets contain four distinct subtypes of beta cells. *Nat Commun* 2016;**7**:11756.
113. Olsson R, Carlsson PO. A low-oxygenated subpopulation of pancreatic islets constitutes a functional reserve of endocrine cells. *Diabetes* 2011;**60**:2068–75.
114. Briancon-Marjollet A, Weiszenstein M, Henri M, Thomas A, Godin-Ribuot D, Polak J. The impact of sleep disorders on glucose metabolism: endocrine and molecular mechanisms. *Diabetol Metab Syndr* 2015;**7**.
115. Pallayova M, Steele KE, Magnuson TH, et al. Sleep apnea predicts distinct alterations in glucose homeostasis and biomarkers in obese adults with normal and impaired glucose metabolism. *Cardiovasc Diabetol* 2010;**9**:83.
116. Wang N, Khan SA, Prabhakar NR, Nanduri J. Impairment of pancreatic-cell function by chronic intermittent hypoxia. *Exp Physiol* 2013;**98**:1376–85.
117. Xu J, Long YS, Gozal D, Epstein PN. Beta-cell death and proliferation after intermittent hypoxia: role of oxidative stress. *Free Radic Biol Med* 2009;**46**:783–90.
118. Hals IK, Bruerberg SG, Ma ZH, Scholz H, Bjorklund A, Grill V. Mitochondrial respiration in insulin-producing beta-cells: general characteristics and adaptive effects of hypoxia. *PLoS ONE* 2015;**10**.
119. Eguchi K, Nagai R. Islet inflammation in type 2 diabetes and physiology. *J Clin Invest* 2017;**127**:14–23.
120. Morris DL. Minireview: emerging concepts in islet macrophage biology in type 2 diabetes. *Mol Endocrinol* 2015;**29**:946–62.
121. Donath MY, Shoelson SE. Type 2 diabetes as an inflammatory disease. *Nat Rev Immunol* 2011;**11**:98–107.
122. Scarim AL, Heitmeier MR, Corbett JA. Irreversible inhibition of metabolic function and islet destruction after a 36-hour exposure to interleukin-1beta. *Endocrinology* 1997;**138**:5301–7.
123. Brown GC. Nitric oxide and mitochondria. *Front Biosci* 2007;**12**:1024–33.
124. Gurzov EN, Germano CM, Cunha DA, et al. p53 up-regulated modulator of apoptosis (PUMA) activation contributes to pancreatic beta-cell apoptosis induced by proinflammatory cytokines and endoplasmic reticulum stress. *J Biol Chem* 2010;**285**:19910–20.
125. Marchetti P, Bugliani M, De Tata V, Suleiman M, Marselli L. Pancreatic beta cell identity in humans and the role of type 2 diabetes. *Front Cell Dev Biol* 2017;**5**:55.

126. Avrahami D, Klochendler A, Dor Y, Glaser B. Beta cell heterogeneity: an evolving concept. *Diabetologia* 2017;**60**:1363–9.
127. Bhola PD, Letai A. Mitochondria-judges and executioners of cell death sentences. *Mol Cell* 2016;**61**:695–704.
128. Riedl SJ, Salvesen GS. The apoptosome: signalling platform of cell death. *Nat Rev Mol Cell Biol* 2007;**8**:405–13.
129. Federici M, Hribal M, Perego L, et al. High glucose causes apoptosis in cultured human pancreatic islets of Langerhans: a potential role for regulation of specific Bcl family genes toward an apoptotic cell death program. *Diabetes* 2001;**50**:1290–301.
130. Nogueira TC, Paula FM, Villate O, et al. GLIS3, a susceptibility gene for type 1 and type 2 diabetes, modulates pancreatic beta cell apoptosis via regulation of a splice variant of the BH3-only protein Bim. *PLoS Genet* 2013;**9**.
131. Senee V, Chelala C, Duchatelet S, et al. Mutations in GLIS3 are responsible for a rare syndrome with neonatal diabetes mellitus and congenital hypothyroidism. *Nat Genet* 2006;**38**:682–7.
132. Barrett JC, Clayton DG, Concannon P, et al. Genome-wide association study and meta-analysis find that over 40 loci affect risk of type 1 diabetes. *Nat Genet* 2009;**41**:703–7.
133. Dupuis J, Langenberg C, Prokopenko I, et al. New genetic loci implicated in fasting glucose homeostasis and their impact on type 2 diabetes risk. *Nat Genet* 2010;**42**:105–16.
134. Danial NN, Walensky LD, Zhang CY, et al. Dual role of proapoptotic BAD in insulin secretion and beta cell survival. *Nat Med* 2008;**14**:144–53.
135. Luciani DS, White SA, Widenmaier SB, et al. Bcl-2 and Bcl-xL suppress glucose signaling in pancreatic beta-cells. *Diabetes* 2013;**62**:170–82.
136. Pfeiffer S, Halang L, Dussmann H, Byrne MM, Prehn J. BH3-only protein bmf is required for the maintenance of glucose homeostasis in an in vivo model of HNF1alpha-MODY diabetes. *Cell Death Dis* 2015;**1**:15041.
137. Le May C, Chu K, Hu M, et al. Estrogens protect pancreatic beta-cells from apoptosis and prevent insulin-deficient diabetes mellitus in mice. *Proc Natl Acad Sci U S A* 2006;**103**:9232–7.
138. Tiano JP, Mauvais-Jarvis F. Importance of oestrogen receptors to preserve functional beta-cell mass in diabetes. *Nat Rev Endocrinol* 2012;**8**:342–51.

CHAPTER

# 10

# Role of Mitochondria in Brain Nutrient Sensing: Control of Energy Balance and Dysregulation in Obesity and Type 2 Diabetes

*Corinne Leloup*, Luc Pénicaud*[†]

*Center of Taste Sciences and Feeding Behaviour (CSGA), UMR CNRS 6265, INRA 1324, Bourgogne Franche-Comté University, AgroSup Dijon, Dijon, France [†]STROMALab, Université de Toulouse, EFS, ENVT, Inserm U1031, ERL CNRS 5311, CHU Rangueil, Toulouse, France

## 1 INTRODUCTION

During the last 20 years, studies have revealed the important role of the central nervous system (CNS) in the regulation of homeostatic mechanisms, including food intake, energy expenditure, and glucose, lipid and protein homeostasis. Brain regions involved in the regulation of energy balance, such as the arcuate (ARC), the ventromedial nuclei of the hypothalamus (VMN), and the hindbrain, have the ability to monitor the amount of available fuel in the body, to modulate energy intake and expenditure, and to modify peripheral tissue/organ function via alteration of neuronal firing.[1–4] For that, they must be informed about the level of nutrients. Within specific brain sites, neuronal populations have been described as able to sense all types of nutrients, enabling proper peripheral tissue adaptations to the fluctuating nutrient environment. Therefore, when glucose or fatty acids rise in the blood because of food consumption, specific brain circuits promote cessation of feeding (satiety), insulin secretion, and glucose use by peripheral tissues while inhibiting hepatic glucose production.[1–4] When the level of glucose decreases between meals, however, another set of neurons is activated to promote hunger and appropriate changes in peripheral glucose and lipid metabolisms.

The main driver of these adaptations is the arcuate nucleus of the hypothalamus. Although this nucleus contains various cell types, two main opposing types of neurons have been defined: those that express pro-opiomelanocortin and those that contain agouti-related protein


https://doi.org/10.1016/B978-0-12-811752-1.00010-9

(AgRP) and neuropeptide-Y. In the arcuate nucleus, pro-opiomelanocortin is processed to produce α-melanocyte stimulating factor (α-MSH)—an agonist for the melanocortin receptors (MC3R and MC4R)—which stimulates energy expenditure and suppresses food intake. AgRP is an inverse agonist for melanocortin receptors, opposing central melanocortin action, and neuropeptide-Y is an inhibitory neuropeptide that suppresses energy use and promotes food intake. AgRP neurons also contain and release the inhibitory neurotransmitter GABA (γ-aminobutyric acid), as do most of the other non-pro-opiomelanocortin neurons in the arcuate nucleus.[2-4] NPY/POMC are inversely sensitive to nutrients and hormones. Two types of glucose-sensitive neurons have been characterized: glucose-excited (GE) and glucose-inhibited (GI) neurons.[5, 6] They respond directly to changes in extracellular glucose levels. Whereas a large amount of data demonstrates that GI neurons are all NPY neurons, the identity of GE-neurons is still a matter of debate.[7] They have been shown to be a heterogeneous population, depending on different characteristics, among which is the glucose level to which they respond. Further studies are needed to finely define the identity of ARC-GE neurons.

Recent data have underlined mitochondria as critical regulators of cell functions in the brain cell populations involved in nutrient sensing. Mitochondria can be considered to be nutrient sensors, because, depending on the nature and amount of nutrients, it will be able to produce quantitatively various factors (ATP, ROS, etc.) that have been shown to be important signals involved in the regulation of neural cell activity. Mitochondrial mechanisms underlying nutrient responsiveness of these neurons share similarities with other peripheral nutrient sensitive cells. For a long time, pancreatic beta-cells have been identified for using the canonical ATP-sensitive $K^+$ channels ($K_{ATP}$ pathway) as a primary signal for exocytosis, a mechanism retrieved for some nutrient sensitive neurons, but in a completely different range of concentrations. During the last decade, however, nutrient-responsive signaling in neurons has been demonstrated not only dependent on ATP production, but also on mitochondrial reactive oxygen species (mROS).

## 2 HYPOTHALAMIC ATP SIGNALING IN NUTRIENT SENSING

Both glucose-sensitive and lipid-sensitive neurons have the ability to depolarize through the inactivation of the $K_{ATP}$ channels. Although the nature and concentration for lipids to trigger this depolarization is not fully documented, much more is known regarding glucose.

Brain glucose concentrations under normal conditions are ≈1–2.5 mM, but might fall to very low level (0.2 mM) under hypoglycemic conditions.[8] Hypothalamic gluco-excited neurons (and some neurons of the hindbrain, amygdala, or basal ganglia), responding electrically to a drop in brain glucose concentrations, have a constitutive closure of $K_{ATP}$ channels under normal concentrations (i.e., 1–2.5 mM). Glucose concentration allows ATP to synthesize in the mitochondria and to maintain ATP to ADP ratio high enough within the cytoplasm. This ratio keeps the $K_{ATP}$ channels closed by decreasing the hyperpolarizing outward $K^+$ flux,[2] through the binding of ATP to $K^+$ channel subunit Kir 6.2 and the sulfonylurea receptor, SUR1. As a result of the closure, plasma membrane is depolarized, with influx of extracellular $Ca^+$, and subsequent release of neuropeptide/neurotransmitter; gluco-excited neurons spike when glucose level is maintained (Fig. 1). Other less well-defined changes occur, such as the activation of the cytosolic AMPKα2 that depends on the AMP to ATP ratio.[9] Such mechanisms,

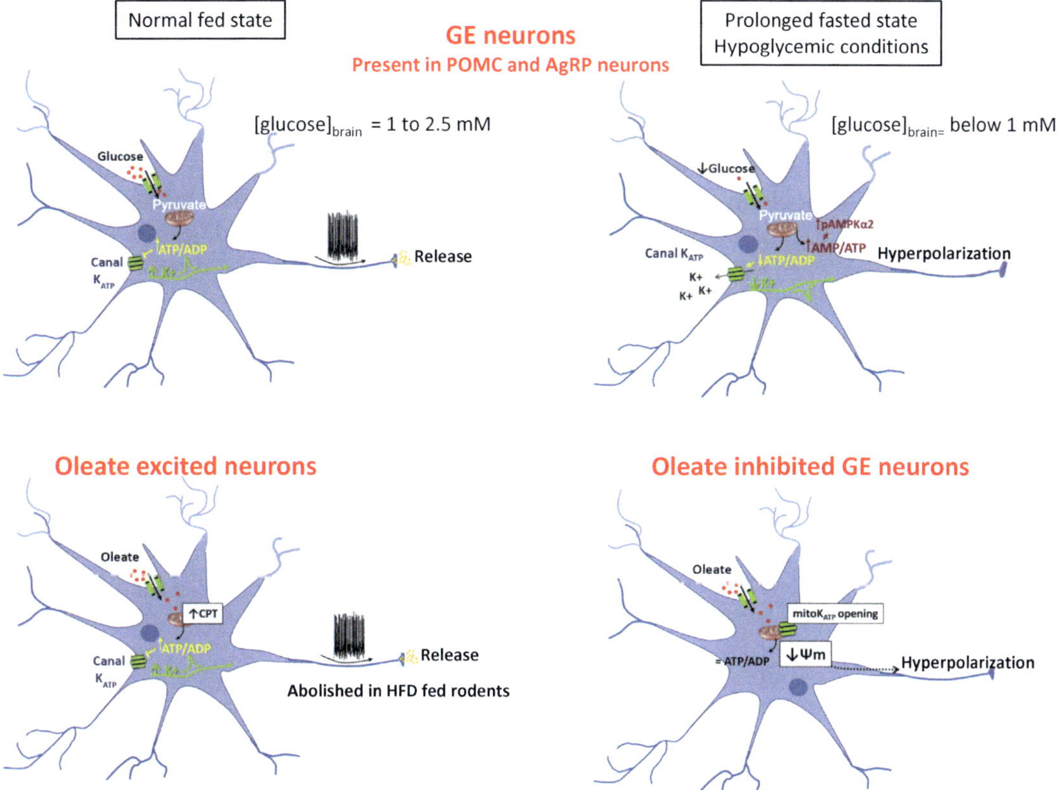


FIG. 1 ATP signaling in hypothalamic nutrient-sensing neurons. Glucose and lipid sensitive neurons have been identified, defined by their ability to depolarize to increased nutrient level. Gluco-excited (GE) neurons are active under normal blood glucose level, $K_{ATP}$ being closed by elevated ATP/ADP ratio. Excitability refers to the hypoglycemic state: $K_{ATP}$ channels are opened at low glucose level (low ATP/ADP ratio), increases intracellular potassium concentration, and hyperpolarizes GE neurons. Additionally, AMPKα2 activation through increased AMP/ATP ratio participated in hyperpolarizing these neurons. They become excited when glucose returns to normal value, closing $K_{ATP}$ channels. Some of these neurons belong to AgRP/NPY and POMC neurons. Oleate excited neurons depend on oleate β-oxidation, altering ATP/ADP ratio, thus closing $K_{ATP}$ at the membrane; in oleate inhibited GE neurons, oleate might open mito$K_{ATP}$ channels, hyperpolarizing the neurons.

as well as AgRP neurons, have been described in POMC. In these neurons, the mechanism allows the detection of a fall in blood glucose, corresponding to a hypoglycemic stimulus. In that case, ATP to ADP ratio falls, resulting in the opening of the $K_{ATP}$ and in hyperpolarization and inhibition.

Regarding lipid sensing, mounting evidence during the last decade has suggested fatty acids also are informative molecules in the brain, as is glucose. Their level and threshold for activating or inhibiting neurons is less clear than what is now known in the brain for glucose level. This is in part because fatty acids have a double origin in the blood and in situ via a brain lipase action. Fatty acid excited and inhibited neurons have been identified in specific brain areas related to energy and food behavior control (hypothalamus, hippocampus, striatum).[10, 11] Many studies have identified the long-chain fatty acid oleate as an important substrate,[12] and mitochondrial oxidation has been described as a key step in this mechanism. Inhibiting carnitine-palmitoyl transferase 1 relieves the inhibitory effect of fatty acids on food intake or hepatic production.[13] These effects of fatty acids also appear to be at least partly dependent on $K_{ATP}$ channels. Detailed studies have revealed multiple types of responses, either excited or inhibited VMN neurons, at least for oleate (OA).[6] Inhibition of carnitine palmitoyltransferase, long-chain acetyl-CoA synthetase, and ATP-sensitive K(+) channel activity accounted for only 20% of OA's excitatory effects and approximately 40% of its inhibitory effects (Fig. 1).[12] Recently, oleate was shown to decrease mitochondrial membrane potential in gluco-excited neurons with no change in total cellular ATP or ATP/ADP ratios even if hyperpolarization occurred with $K_{ATP}$ activation, thus suggesting oleate induces $K_{ATP}$-dependent hyperpolarization and inhibition of firing of a subgroup of GE hypothalamic neurons that happens without altering cellular energy charge (Fig. 1).[14]

In addition to carbohydrates and lipids, brain amino acids sensing has been found to require the ATP-sensitive $K^+$ channels. Their concentration in the brain tissue reflects their circulating levels,[15] and most works have focused on the branched-chain ketogenic leucine, known to activate the mammalian target of rapamycin (mTOR) pathway.[16] Leucine modulates food intake by activating mTOR in the hypothalamus, but the central sensing of leucine also is coupled with the control of hepatic glucose production to maintain euglycemia, a mechanism insensitive to rapamycin but one that necessitates functional $K_{ATP}$ channels.[17]

In obesity and diabetes, various models have been studied to decipher the dysfunction operating at the level of the $K_{ATP}$ channels. Diabetic Zucker fatty (ZDF) rats compared to nondiabetic Zucker fatty (ZF) rats have a decrease expression of the hypothalamic $K_{ATP}$ channel component Kir6.2, while its immunofluorescence colocalized with NPY neurons throughout the hypothalamus, suggesting that chronic changes in hypothalamic Kir6.2 expression might be associated with the development of hyperinsulinemia and hyperglycemia.[18] Both insulin and leptin, as fuel sensors, act through their hypothalamic receptors to inhibit food intake and stimulate energy expenditure. Like leptin, insulin hyperpolarizes hypothalamic GE neurons by opening $K_{ATP}$ channels.[19] In POMC neurons of mice lacking the dephosphorylative Pten pathway, animals become hyperphagic and POMC neurons exhibit a marked hyperpolarization because of increased $K_{ATP}$ channel activity. Tolbutamide, a blocker of these channels, has the ability to restore electrical activity and leptin-evoked firing of POMC neurons. Moreover, intracerebroventricular administration of tolbutamide abolished hyperphagia.[20]

Regarding hypothalamic fatty acid sensing, infusion of oleate into the mediobasal hypothalamus has been shown to activate $K_{ATP}$-channel and to suppress VLDL-triglycerides (TG) secretion in rats. In a high-fat diet (HFD), plasma TG and VLDL-TG secretion are elevated and abolish oleic acid sensing to lower VLDL-TG. Importantly, HFD-induced dysregulation is restored with direct activation of $K_{ATP}$ channels (Fig. 1).[21] Finally, a role for the brain mitochondrial $K_{ATP}$ channels has been revealed in rats fed a HFD—induced obesity. Molecular studies demonstrate that the expression of Kir6.2, a mito$K_{ATP}$ channel component, is reduced in various areas of the brain of HF-fed rats that are less protected against cerebral ischemic neuronal injury, because of an impaired brain mito$K_{ATP}$ channel.[22]

# 3 BRAIN MITOCHONDRIAL ROS SIGNALING

The mitochondrial respiratory chain represents one of the main sources of ROS. For more a decade, ROS have been known to be a major and crucial redox-signaling pathway in detecting an increase in nutrient blood level.[23–25] Increasing reduced equivalents (NADH, $H^+$, and $FADH_2$) as a result of glucose or lipid oxidation leads to increased electron transfer chain activity, finally generating elevated but physiological superoxide anions production.[24, 26, 27] Although the specificity of this signaling is largely assigned to a very low activity of the enzymes MnSOD2 and catalase in the β-cell compared to any other oxidative cells,[28] these characteristics remain to be established in the hypothalamic sensing mechanism. Whatever process favors mROS signaling in hypothalamic nutrient detection it is firmly established both in vivo and in vitro. Cerebral injection of glucose in rodents, via a physiological (carotid toward the brain) or intrathecal (hypothalamic) route, leads to short-lived mROS signaling, in form of $H_2O_2$, the predominant molecular and diffusible pathway, that might be counteracted specifically by catalase (reviewed in Ref. 29). This increase occurs both in ex-vivo hypothalamic slices and in vivo after a glucose load,[23, 30] which coincides with the timing of the physiological response, that is, an arcuate neuronal activation (a twofold increase of the firing rate) and a subsequent peak of insulin (through the activation of the vagal nerve). This mROS production is reversed by antioxidants or mitochondrial uncouplers; the effect of the latter compound indicating the respiratory chain origin of the mROS. In these conditions, both the increased firing rate in the arcuate nucleus and the subsequent insulin secretion are abolished, showing that the responses required the mROS signaling. The fact that direct hypothalamic mROS generation by respiratory chain (inhibitors antimycin or rotenone) mimics the effect of glucose, independently of NADH or ATP levels, has firmly established this ATP-independent mechanism. Thus, an increased mROS level in response to glucose, rather than just the ATP/ADP ratio, mediates the excitatory effect of glucose on hypothalamic neurons. Therefore, mROS signaling is consistent with the NADH mechanism suggested earlier for the gluco-sensing mechanism.[31, 32] This reflects the integration of both reduced NADH/ $FADH_2$ (when brain glucose or other nutrient oxidation occurs) and the electrochemical proton gradient, which reflects both the phosphate potential and the uncoupling status of the mitochondria (Fig. 2). This proposal is consistent with other studies that describe expression of UCP2 in hypothalamic areas and its relationship with mROS signaling.[33, 34] Recently, we confirmed in vitro that the mROS signaling is necessary for the response to increased glucose level in high glucose-excited neurons (HGE, neurons spiking in response to a step from 2.5 to 5–10 mM, mimicking increased blood glucose level) by showing not only that the $K_{ATP}$ channel opener diazoxide does not inhibit the magnitude of glucose response in hypothalamic HGE neurons, but also that it responds to glucose in ~95% of HGE neurons robustly blocked by antioxidants, in particular the $H_2O_2$-decomposing enzyme catalase. These results confirm that $H_2O_2$ is the specific transducer of the responses,[35] although it does not exclude upstream regulation from anion superoxides, especially at the mitochondrial level. Therefore, these data confirm that $H_2O_2$ represents the molecular signal, but also that $K_{ATP}$ channels are not necessary to detect increased glucose levels by HGE neurons. Finally, the redox-sensitive TRPC3 channel has been shown to be at least partly responsible for the responses because the TRPC3 channel inhibitor Pyr3 reduced the response to glucose in ~70% of HGE neurons.[35]

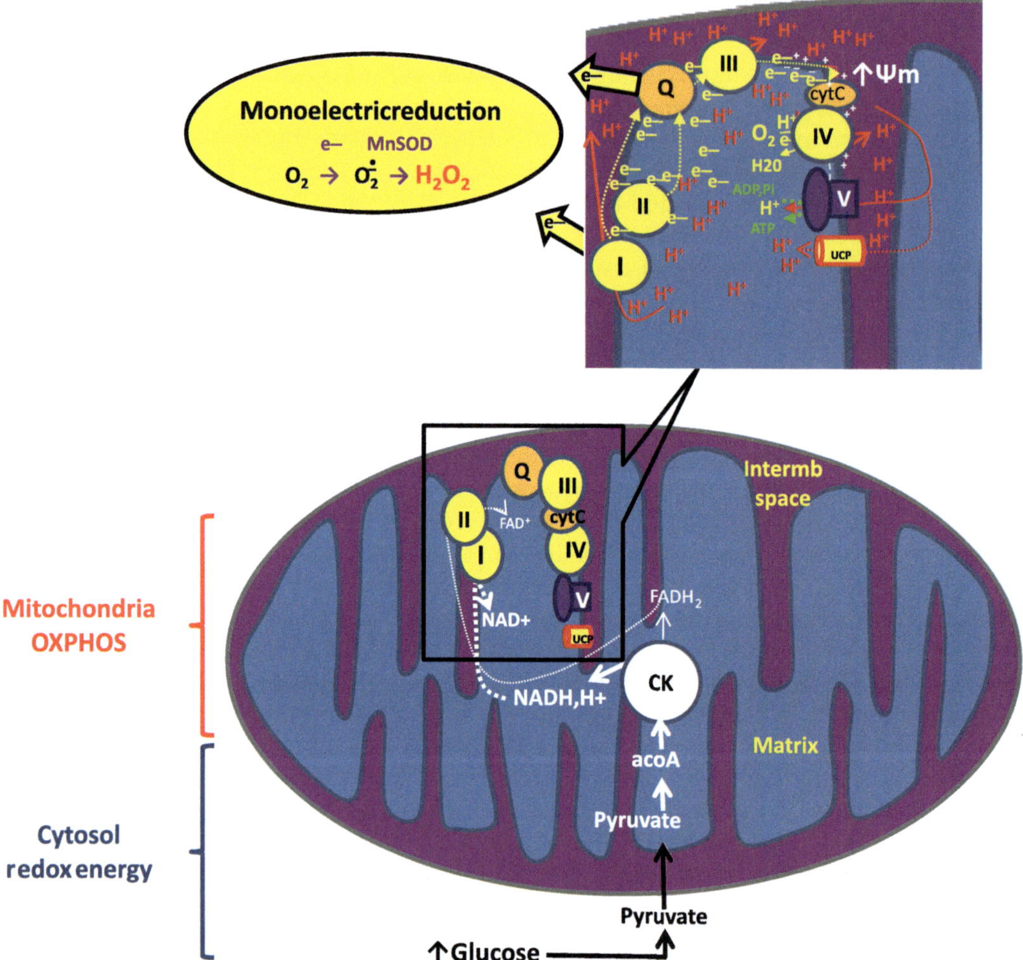


FIG. 2 Glucose induced mitochondrial ROS signaling in hypothalamic neurons. Oxidation of nutrients increases reduced equivalents (NADH,$H^+$ and $FADH_2$) leading to increased electron transfer chain activity from complex I to IV. Electron transfer, however, is restrained by the coupling with the phosphorylative activity at complex V, finally generating an elevated but physiological superoxide anions production because of an increase probability of molecular oxygen to react with electrons, mainly originating from complex I and III (a probability that might decrease after UCP activation, by uncoupling respiration from $H^+$ gradient at complex V). Superoxide anions are secondarily dismutated into hydrogen peroxide, the predominant molecular pathway (catalase that specifically degrades $H_2O_2$ abolishes cellular and physiological responses).

The implication of ROS signaling also has been described in regards to lipid sensing. In vivo, acute hypertriglyceridemia in rats leads to a rapid increase in mitochondrial respiration in the ventral hypothalamus together with a transient production of ROS. Cerebral inhibition of fatty acids-CoA mitochondrial uptake prevents hypertriglyceridemia-stimulated ROS production, indicating that ROS derive from mitochondrial metabolism.[27] Hypertriglyceridemia-stimulated ROS production is associated with change in the intracellular redox state, suggesting that ROS functions acutely as a signal. Moreover, cerebral inhibition of hypertriglyceridemia-stimulated ROS production fully abolishes the satiety related to the hypertriglyceridemia, demonstrating that hypothalamic ROS production is required to restrain food intake during hypertriglyceridemia. Last, fasting disrupts the hypertriglyceridemia-stimulated ROS production, indicating that the redox mechanism of brain lipid sensing could be modulated under physiological conditions.[27]

The nature of GE neurons (depolarizing upper 2.5 mM glucose) is still a matter of debate. However, studies support the importance of mROS signaling in conjunction with UCP2 activity in POMC neurons.[36] Present at the inner mitochondrial membrane, this protein can be assimilated to a channel that allows the passage of protons from the intermembrane space to the internal matrix (no change in mitochondrial potential might be recorded, nor ATP level), therefore reducing the constraint of the respiratory chain to transfer electrons. UCP2 indirectly accelerates the transfer of electrons slightly and reduces their probability of reducing $O_2$.[29] This effect is called mild uncoupling, because it is not accompanied by a reduction in the mitochondrial membrane potential or a decrease in the synthesis of ATP (reviewed in Ref. 37). The mechanism for obesity-induced loss of glucose sensing in arcuate POMC neurons involves UCP2, which negatively regulates glucose sensing in POMC neurons. In particular, genetic deletion of UCP2 or treatment with genipin, a UCP2 inhibitor, prevents obesity-induced loss of glucose sensing. In this case, UCP2 scavenging of mROS impairs glucose sensing in POMC neurons and has a pathogenic role in the development of T2D. This role is independent of the ATP-induced $K_{ATP}$ channel signaling, because no primary relationships between ATP to ADP ratio appear to link an energy deficit and $K_{ATP}$ channel activity with that of UCP2.[37]

In another obese and insulin-resistant model, the Zücker rat, mROS signaling was shown to be altered.[30] This model exhibits a hypothalamic hypersensitivity to glucose highlighted by an enhanced electrical activity in the arcuate nucleus and a subsequent vagal stimulation of insulin secretion at low glucose levels, inefficient in lean rats. These abnormal responses were associated with increased hypothalamic ROS levels at low glucose concentrations, a constitutive oxidized environment at both the cellular and mitochondrial levels, as measured through the glutathione ratio of the reduced form to the total form, as well as through the mitochondrial aconitase and MnSOD superoxidase activities. An overexpression of most subunits of the respiratory chain was present, together with a dysfunction in mitochondrial respiration. In this study, no difference in the number of mitochondria or in uncoupling respiration was observed between the hypothalamus of obese versus control animals. Therefore, a UCP2-independent mechanism explains excess ROS production in this model, in which intolerance to glucose and insulin-resistance is superimposed to transient glucose levels rises. The recovery of the redox status through reduced glutathione intracerebroventricular infusion fully reverses MBH hypersensitivity to glucose, demonstrating the tight control exerted via this redox signaling.[30] Hypersensitivity to glucose explains the elevated parasympathetic tone generally present in the obese and insulin-resistant state that contributes to the development of hyperinsulinemia, and finally, to the onset of type 2 diabetes.

A high level of the gut-derived hormone ghrelin (in an acylated form, mainly octanoyled ghrelin) associated with a fasting state was shown to be involved in mROS signaling. In the arcuate NPY/AgRP orexigenic neurons, ghrelin has been suggested to exert its effects through the modulation of mitochondrial respiration and mROS production.[34] In this study, ghrelin decreased the mitochondrial membrane potential in normal mice and increased neuronal activation, as assessed by c-fos expression, and the firing rate of NPY neurons, effects that are completely abolished in UCP2$^{-/-}$ mice, in which mROS is still produced.[34] Although the ability of UCP2 to decrease mitochondrial potential is a matter of debate (reviewed in 37), other cerebral UCPs (UCP4 and UCP5) also might be involved, although their potential role has not been studied yet. These isoforms are well represented at the hypothalamic level (as much as UCP2 in normal rodents) and have been demonstrated for their uncoupling activity, capability to reduce the mitochondrial potential or ATP synthesis. The interference of UCP2 deletion on these isoforms is unknown, but questions the interpretations in regards of the biochemical properties of these other UCPs.

These results show that mitochondria of orexigenic NPY/AgRP neurons need at least UCP2 to promote a normal response. Conversely, POMC neurons show a higher mROS production with saline than with ghrelin and need well-coupled mitochondria for their activation. In the fasting state, ghrelin indirectly hyperpolarizes POMC neurons by activating inhibitory NPY/AgRP (GABAergic) inputs. The hyperpolarization leaves POMC neurons less active, leading to a drop in the respiration rate and mROS production. Finally, mROS scavenging by an antioxidant treatment reverses the results described in neurons from UCP2$^{-/-}$ mice.[34] The anorexigenic POMC neurons function inversely to the NPY/AgRP neurons (Fig. 3). They can respond to a rise in glucose levels through an mROS increase only under fed conditions (while NPY neurons are inhibited), with low ghrelin and well-coupled mitochondria. Although the mechanisms underlying the different mROS production responses in NPY and POMC neurons are clear regarding ghrelin (POMC population exhibited few ghrelin receptors), a comprehensive view of the differential effects of nutrient fluxes (lipids vs. glucose) according to negative or positive energy balance on these two neuronal populations remains to be fully explored.

Together, these data demonstrate the master role of brain mROS signaling with regard to the control of both food intake and metabolism. This master component of brain energy sensing is crucial in the development of both obesity and type 2 diabetes. Growing data reveal that numerous mechanisms interplay in this signaling, and one of them has become major within the last decade: the role of mitochondrial morphology in modulating most aspects of mitochondrial bioenergetics.

## 4 BRAIN MITOCHONDRIAL DYNAMICS AND NUTRIENT SENSING

Mitochondrial dynamics (fusion and fission) have been linked to the balance between energy demand and nutrient supply.[38, 39] Many studies have highlighted how interconnected or fragmented mitochondria regulate bioenergetics adaptation: Oxidative cells exposed to a high-glucose environment harbor fragmented mitochondria, and cells under basal-glucose level exhibit elongated ones (reviewed in[40]). Mitochondria are organized into a tubular network that continuously changes its shape and motility, mediated by fission and fusion mechanisms. Numerous proteins participate in these processes, with the dynamin family proteins of large GTPases being the most important.[41] Mitochondrial fusion implies mitofusins (Mfn1 and Mfn2 in the outer membrane) and Opa1 (in the inner membrane); fission is mediated by fission proteins (mainly Fis1 and Mff, mitochondrial fission factor, in the outer membrane), and Drp1, cytosolic, which translocates and associates to Fis1 to trigger fission.[42] The study of

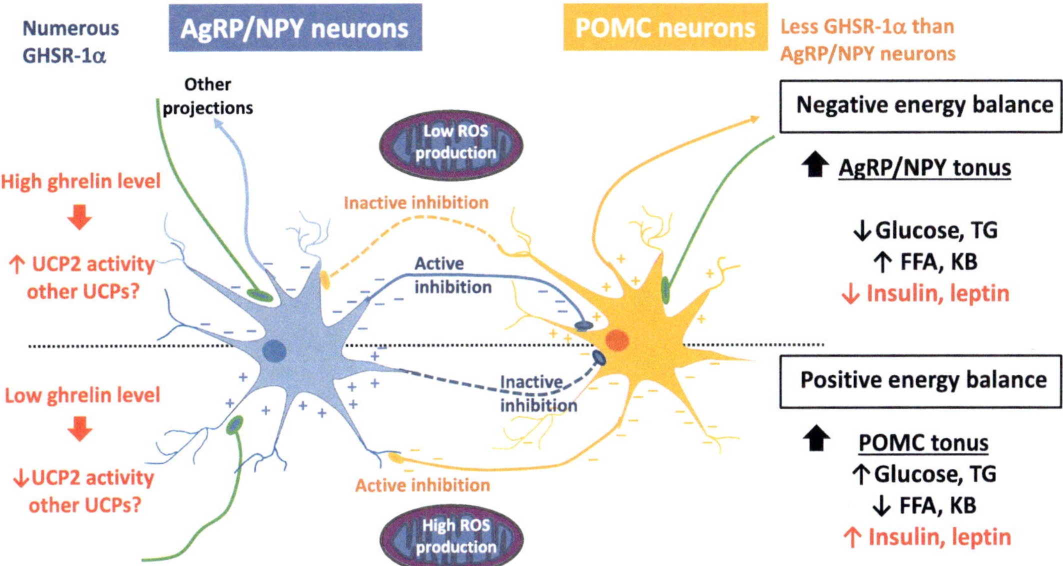


FIG. 3 Ghrelin drives mROS signaling in the melanocortinergic system. Negative energy balance is associated high level of the gut-derived hormone ghrelin that binds to the growth hormone secretagogue receptor 1α (GHSR-1α), mostly on AgRP/NPY neurons. In neurons, ghrelin exerts its effects through mitochondrial respiration by uncoupling respiration from ATP synthesis, thus decreasing mROS production. Low mROS level is a prerequisite to the firing of NPY neurons (abolished in $UCP2^{-/-}$ mice, mROS still being produced), that secondarily inactivate POMC neurons through mainly GABA release. At the same time, POMC neurons have a low mROS production (mitochondria are uncoupled), shown to be necessary for their firing and still inactivated (in addition to GABAergic inputs that hyperpolarize them). Conversely, inhibiting UCP (positive energy balance or through pharmacological treatment such as genipin) leads to activation of POMC neurons (restoring glucose sensing and α-melanocyte stimulating hormone release in these neurons). NPY neurons become silent, hyperpolarized through POMC inhibitory inputs, in addition to their own increased mROS production. In both cases, how mROS signaling silences AgRP or activates POMC neurons is unknown. The effect of additional varying levels of nutrients and hormones is less well-defined. UCP: uncoupling protein, TG: triglycerides, FFA: free fatty acids, KB: ketone bodies.

Yoon and colleagues revealed that hyperglycemic conditions induce mitochondrial fragmentation, which is causal for rapid ROS overproduction.[43] Using time-lapse imaging, they found in vitro that GFP-tagged hepatocyte cell lines and myoblasts overexpressing DLP1-K38A, a dominant-negative mutant form of DRP1, exhibited both inhibition of hyperglycemia-induced fragmentation and ROS overproduction normally shown in mitochondria of control cells. These results show that hyperglycemic conditions (from minutes to hours) trigger mitochondrial fragmentation through the fission protein DRP1, an event required to increased ROS production.

## 4.1 Mitochondrial Fission in Hypothalamic Nutrient Sensing and Dysregulation

In the mediobasal hypothalamus, where most of the hypothalamic glucose-excited neurons are housed, in vivo the importance of Drp1 translocation to the mitochondria in the short-term response to increased hypothalamic glucose level (at least 1 min) and for mROS signaling has been shown, leading to the right adaptive responses.[44] Although the exact signal between cell glucose entry and recruitment of cytosolic DRP1 proteins remains unknown, a transient intracarotid glucose injection induces the recruitment of DRP1 to MBH mitochondria. The knocked-down of DRP1 expression by intra-MBH DRP1 siRNA revealed that food intake after fasting, as well as insulin secretion induced by MBH or intracarotid glucose injection were lost, respectively. Concomitantly, mROS production in response to glucose was not present.

Finally, the evaluation of MBH mitochondrial function by oxygen consumption measurements after DRP1 knock-down revealed substrate-driven respiration was impaired in siDRP1 rats, associated with an alteration of the coupling mechanism.[44] These results show glucose-induced DRP1-dependent mitochondrial fission is an upstream regulator for mROS signaling, and consequently, a key mechanism in hypothalamic glucose sensing (Fig. 4). A recent study by Diano's team confirmed and specified more finely part of these mechanisms in VMH SF1 neurons that house both some GE and GI neurons.[45] They confirmed that systemic glucose injections decreased mitochondrial size, and thus, triggered mitochondrial fragmentation (total mitochondrial area being unchanged in the cytosol), the ratio of phosphorylated DRP1 (pDRP1) to unphosphorylated form increasing. Furthermore, electron micrographs showed that pDRP1 staining associated with mitochondria membrane in response to glucose was increased in ventro-medial hypothalamic (VMH) neurons. $SF1^{cre}$ crossed with $Ucp2^{KO}$ mice had no increased ratio of pDRP1/DRP1 in response to glucose, while re-expressing UCP2 in the VMH restored this ratio to levels similar to those observed in control mice, suggesting UCP2 activation is a prerequisite for mitochondrial fission. Moreover, glucose injection induced an elevation of ROS levels in SF1 neurons of $Ucp2^{KO}$ mice compared to controls, and VMH re-expression of UCP2 was sufficient to reduce ROS to levels measured in control mice. Although UCP2-induced reduced brain ROS levels is well documented,[46] the glucose-excited response observed in SF1 neurons contradicts this finding. Nutrient excited cells metabolize nutrients according to their systemic levels and not to their energy demand; during physiological sensing, increased glucose raises DRP1 translocation to mitochondria, and finally fission, might be dependent on a cytosolic factor (as yet unknown in these neurons), therefore, upstream the activation of UCP2, as already suggested.[43] These discrepancies could be because of the use of animals harboring constitutive deletions, which are useful for the understanding of chronic pathologies but poorly instructive for the study of transient and physiological mechanisms, particularly the sensing of nutrients. Toyama et al. recently have identified that the energy-gauging AMPK is required genetically for most cells to promote rapid mitochondrial fragmentation even in the absence of mitochondrial stress. MFF, one of the mitochondrial outer-membrane receptors for DRP1, was identified as a substrate of AMPK to mediate mitochondrial fission, independently of OPA1 activity, a major fusion protein.[47] Moreover, they showed that OPA1 (a fusion protein) cleavage is induced after a loss of the mitochondrial membrane potential (CCCP treatment, a decoupling agent), independently of AMPK, suggesting a relationship between UCP2 and OPA1 activities might exist in nutrient-sensing neurons. In their study, they demonstrated that triggering fission did not require a change of the mitochondrial membrane potential.

Regarding POMC neurons, reduced expression of phosphorylated dynamin-related protein (pDRP1) is observed in neurons of fed mice with increased mitochondrial size compared to POMC neurons of fasting animals. Inducible deletion of DRP1 of mature POMC neurons

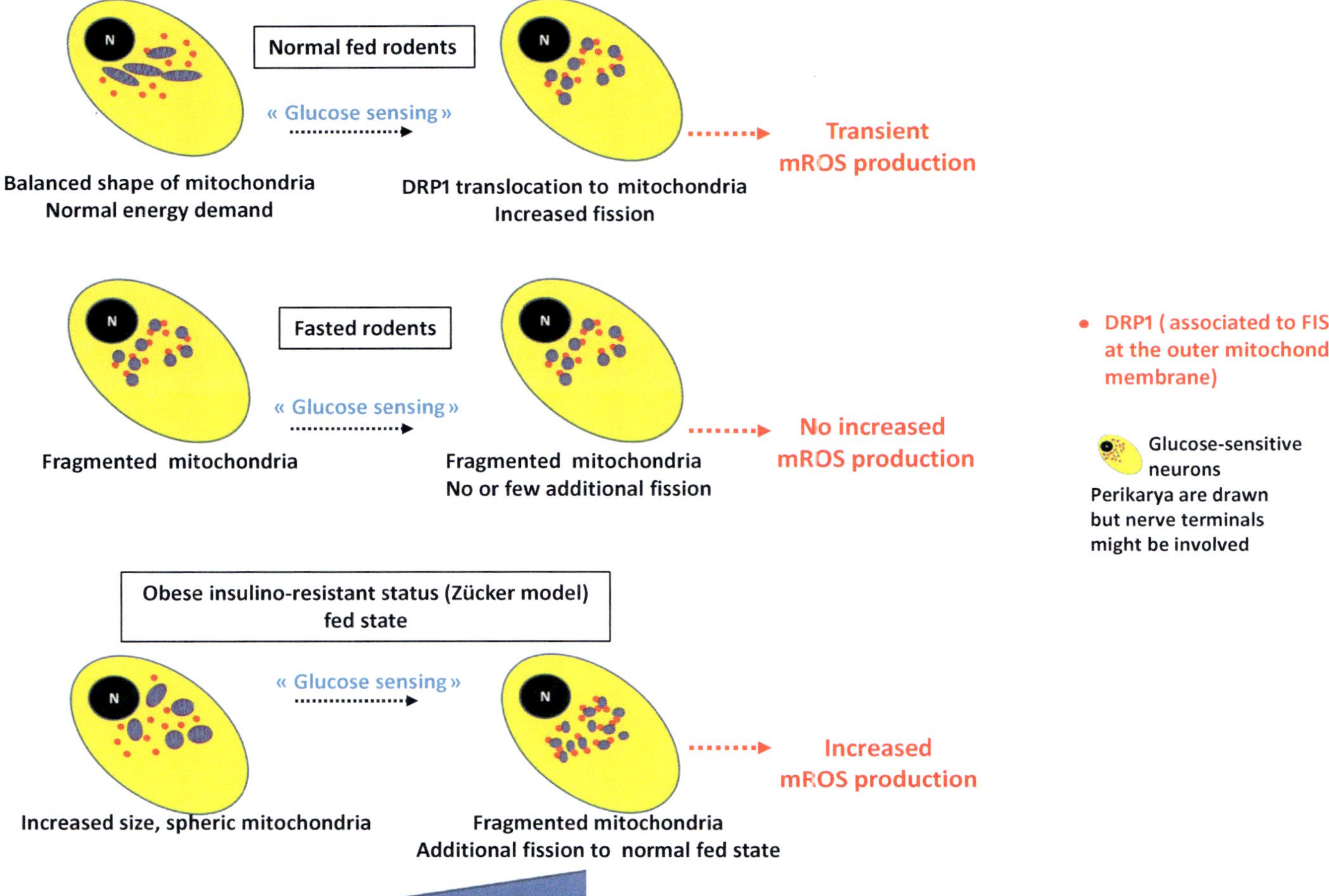


FIG. 4 The role of mitochondrial fission in hypothalamic glucose-sensing mechanism and dysfunction in insulin-resistant state. In normal fed animals, hypothalamic mitochondria exhibit a balanced shape (neither elongated, nor punctuate), although a systematic review according to cellular type and place has never been explored. An increased brain glucose level triggers the translocation of the DRP1 protein to mitochondria where it associates to FIS1, allowing the separation of the mitochondria in smaller mitochondria (fission), looking more punctuate. This mechanism is accompanied by a transient increase of mROS that could be inhibited by decreasing DRP1, thus decreasing DRP1/FIS1 association, and thus fission. In fasting animals, hypothalamic mitochondria appear fragmented, and an increased brain glucose level does not increase mROS production (personal data), suggesting that no more or not enough additional fission might occur. In the obese insulino-resistant Zücker rat, mitochondria appear bigger and more spheric, the hypothalamus is hypersensitive to glucose and exhibits a higher mROS production, suggesting additional fission has occurred.

resulted in increased leptin sensitivity and glucose responsiveness, with increased mitochondrial size, ROS production, and neuronal activation. Furthermore, deletion of DRP1 enhanced the gluco-privic stimulus in these neurons, causing their stronger inhibition and a greater activation of counter-regulatory responses to hypoglycemia that were PPAR dependent, thus revealing a role for mitochondrial fission in leptin sensitivity and glucose sensing of POMC neurons.[48]

### 4.2 Mitochondrial Fusion in Hypothalamic Nutrient Sensing and Dysregulation

Other studies about hypothalamic POMC and NPY/AgRP populations, although not focused on the transient nutrient sensing, revealed the importance of the fusion proteins MFN1 and MFN2.[49, 50] The work of Claret's team established a critical role for MFN2 in both mitochondrial fusion and mitochondria-endoplasmic reticulum (ER) interactions. They showed that mitochondria-ER contacts in POMC neurons were decreased in diet-induced obesity, and POMC-specific deletion of Mfn2 resulted in loss of mitochondria-ER contacts, ER stress-induced leptin resistance, hyperphagia, decreased energy expenditure, and finally, obesity. ER stress and the activation of the UPR signaling network directly block leptin signaling by reducing pStat3 levels.[51] Consistently, mice with a POMC-specific deletion of Mfn2 displayed reduced pStat3 staining in POMC neurons. Furthermore, the ER is involved in the synthesis, folding, and transport of secretory proteins suggesting dysfunctional ER might interfere with proper synthesis and release of key neuropeptides that mediate the anorectic effects of leptin, thereby altering whole-body energy balance.[52] ER stress in POMC neurons results in defective POMC processing and/or α-MSH production, with mutant mice harboring reduced total α-MSH content and processing rate. This study highlighted an expanding role of mitochondrial dynamics in membranes other than those of the mitochondria itself. More recently, the same team showed that mice lacking MFN1 in POMC neurons do not sense glucose, affecting glucose homeostasis.[53] These mice exhibit mitochondria with normal ultrastructural conformations, although smaller than in wild type. The transition from fed to fasting conditions, however, is not associated with an increased elongation of mitochondria. A transcriptome analysis of hypothalamic transcripts during this transition revealed a pleiotropic effect of Mfn1 deletion, with no adaptation ranging from plasma membrane, intracellular signaling cascade, cell proliferation, or immune response of related-genes. Regarding glucose sensing, POMC-Mfn1KO mice exhibit enhanced satiety in response to an intracerebroventricular glucose load that is inefficient in control mice, suggesting hypersensitivity to glucose. Conversely, glucopenia does not increase food intake. These results demonstrate MFN1-mediated mitochondrial dynamics is required for responses to bidirectional variations in glucose level.

## 5 PERSPECTIVES

### 5.1 Relationships Between Mitochondria and Synaptic Vesicles

Along the same lines, at the synaptic vesicles level, a recent work revealed that the recycling of neurotransmitters (a balance between vesicles retrieval from plasma membranes and recruitment from reserve pools within the synapse) is dependent on the Bcl-2 family protein

Bcl-$_{XL}$ by regulating endocytotic vesicle trafficking. The study showed that, during synaptic stimulation, Bcl-2 forms a complex of proteins with DRP1, Mff and clathrin. Depletion of DRP1 induced misformed endocytotic vesicles,[54] suggesting this mechanism might be of great interest for synaptic plasticity, especially in nutrient sensitive neurons.

### 5.2 Relationships Between Astroglial Mitochondria and Sensing Neurons

Mitochondria are important for matching local energy demand but also for $Ca^2$+ buffering in ad equation with neuronal demand and signaling. Although less studied than in neurons, understanding the role of mitochondria in astrocytes might open new mechanisms regarding the complex bidirectional relationship between astrocytes and neurons. Astrocytes have been suspected to play an important role in nutrient sensing.[55–57] Their cellular location influence intracellular $Ca^{2+}$ levels, affecting gliotransmission, and thus, the signaling of nearby neurons.[58] Astroglial mitochondria also provide ATP for important astrocytic functions such as the glutamate-glutamine shuttle (a major player in glutamatergic and GABAergic transmission) and ion homoeostasis at the tripartite synapse.[59] Altering mitochondrial $Ca^{2+}$ buffering with FCCP (a mitochondrial uncoupler) results in increased glutamate release by astrocytes, suggesting a direct correlation between $Ca^{2+}$ levels and glutamate release might be present.[60]

Under normal conditions in vitro, astrocytic mitochondria exhibit continuous cycles of fission and fusion with an equal frequency, while the intracellular environment has the ability to vary this balance. For example, $Ca^{2+}$ elevations in cultured astrocytes results in rounding mitochondria.[58] Longer mitochondria might increase $Ca^{2+}$ buffering and ATP dispersal over a larger cytoplasmic area. The balance between mitochondrial fusion and fission events is altered in astrocytes after cellular injuries. Treatment with pro-inflammatory stimuli in acute slices triggers an increase in mitochondrial fission, along with an increase in the phosphorylated/activated form of Drp1.[58]

At the level of the hypothalamus, postnatal ablation of insulin receptors (IRs) in glial cells affects hypothalamic astrocyte morphology, but also mitochondrial function through the neuronal network. Indeed, glial IR ablation reduces glucose-induced activation of POMC neurons.[61] Extensive studies of mitochondrial functions and dysfunctions, however, have not been yet investigated as far as in hypothalamic neurons.

## 6 CONCLUDING REMARKS

During the past decade, a growing number of studies have been published about the role of mitochondria in brain nutrient sensing mechanisms, and the critical role of this organelle is recognized by most of the authors in this field. There are still concerns, however, and avenues of research to be explored. First, the necessity of delineating the exact contribution of ATP and ROS production in the control of electrical activity of the various classes of nutrient sensitive neurons involved. Second, to understand whether mitochondria in pathological conditions in which brain nutrient sensing might play a role is cause or consequence of disease. Third, do deficiencies observed in the absence of fusion and fission components arise from disturbed mitochondrial dynamics or from impaired mitochondrial functions? Studying systems in which fusion/fission can be induced physiologically without respiratory dysfunctions is necessary to explore the mechanisms involved. Fourth, mitochondrial dynamics, being functionally interdependent with the endoplasmic reticulum or peroxisomes, makes it difficult to clearly define the sequence of events in vivo for mitochondrial function per se.

## Acknowledgments

The authors received continuous support from the CNRS, the University Paul Sabatier, Toulouse and the University Bourgogne Franche-Comté, Dijon, the ANR and various grants from societies and foundations (Société Francophone du Diabète, NRJ prize-Institut de France). The works reported in this review will not have been possible without the implication of other members of the former team (A Carrière, L Casteilla, A Benani, Y Fernandez, X Fioramonti, A Galinier, A Lorsignol) as well as technicians, PhD students, and post-doctoral fellows. This work is dedicated to the memory of professor Xavier Leverve, without whom the concepts and ideas developed here would never have existed.

## References

1. Pénicaud L, Leloup C, Fioramonti X, Lorsignol A, Benani A. Brain glucose sensing: a subtle mechanism. *Curr Opin Clin Nutr Metab Care* 2006;**9**(4):458–62.
2. Levin BE, Magnan C, Dunn-Meynell A, Le Foll C. Metabolic sensing and the brain: who, what, where, and how? *Endocrinology* 2011 Jul;**152**(7):2552–7.
3. Schneeberger M, Gomis R, Claret M. Hypothalamic and brainstem neuronal circuits controlling homeostatic energy balance. *J Endocrinol* 2014;**220**(2):T25–46.
4. Seoane-Collazo P, Fernø J, Gonzalez F, Diéguez C, Leis R, Nogueiras R, et al. Hypothalamic-autonomic control of energy homeostasis. *Endocrine* 2015;**50**(2):276–91.
5. Song Z, Levin BE, McArdle JJ, Bakhos N, Routh VH. Convergence of pre- and postsynaptic influences on glucosensing neurons in the ventromedial hypothalamic nucleus. *Diabetes* 2001;**50**:2673–81.
6. Fioramonti X, Song Z, Vazirani RP, Beuve A, Routh VH. Hypothalamic nitric oxide in hypoglycemia detection and counterregulation: a two-edged sword. *Antioxid Redox Signal* 2011;**14**(3):505–17.
7. Fioramonti X, Chrétien C, Leloup C, Pénicaud L. Recent advances in the cellular and molecular mechanisms of hypothalamic neuronal glucose detection. *Front Physiol* 2017;**8**:875.
8. McCrimmon RJ, Sherwin RS. Hypoglycemia in type 1 diabetes. *Diabetes* 2000;**59**:2333–9.
9. Claret M, Smith MA, Batterham RL, Selman C, Choudhury AI, Fryer LG, et al. AMPK is essential for energy homeostasis regulation and glucose sensing by POMC and AgRP neurons. *J Clin Invest* 2007;**117**:2325–36.
10. Lam TK, Pocai A, Gutierrez-Juarez R, Obici S, Bryan J, Aguilar-Bryan L, et al. Hypothalamic sensing of circulating fatty acids is required for glucose homeostasis. *Nat Med* 2005;**11**:320–7.
11. Lam TK, Schwartz GJ, Rossetti L. Hypothalamic sensing of fatty acids. *Nat Neurosci* 2005;(5):579–84.
12. Le Foll C, Irani BG, Magnan C, Dunn-Meynell AA, Levin BE. Characteristics and mechanisms of hypothalamic neuronal fatty acid sensing. *Am J Phys Regul Integr Comp Phys* 2009;(3):R655–64.
13. Obici S, Feng Z, Arduini A, Conti R, Rossetti L. Inhibition of hypothalamic carnitine palmitoyltransferase-1 decreases food intake and glucose production. *Nat Med* 2003;(6):756–61.
14. Dadak S, Beall C, Vlachaki Walker JM, Soutar MPM, McCrimmon RJ, et al. Oleate induces KATP channel-dependent hyperpolarization in mouse hypothalamic glucose-excited neurons without altering cellular energy charge. *Neuroscience* 2017;**346**:29–42.
15. Choi YH, Fletcher PJ, Anderson GH. Extracellular amino acid profiles in the paraventricular nucleus of the rat hypothalamus are influenced by diet composition. *Brain Res* 2001;**892**:320–8.
16. Cota D, Proulx K, Smith KA, Kozma SC, Thomas G, Woods SC, et al. Hypothalamic mTOR signaling regulates food intake. *Science* 2006;**312**:927–30.
17. Su Y, Lam TK, He W, Pocai A, Bryan J, Aguilar-Bryan L, et al. Hypothalamic leucine metabolism regulates liver glucose production. *Diabetes* 2012;**61**:85–93.
18. Gyte A, Pritchard LE, Jones HB, Brennand JC, White A. Reduced expression of the KATP channel subunit, Kir6.2, is associated with decreased expression of neuropeptide Y and agouti-related protein in the hypothalami of Zucker diabetic fatty rats. *J Neuroendocrinol* 2007;**19**(12):941–51.

19. Spanswick D, Smith MA, Mirshamsi S, Routh VH, Ashford ML. Insulin activates ATP-sensitive K+ channels in hypothalamic neurons of lean, but not obese rats. *Nat Neurosci* 2000;**3**:757–8.
20. Plum L, Ma X, Hampel B, Balthasar N, Coppari R, et al. Enhanced PIP3 signaling in POMC neurons causes KATP channel activation and leads to diet-sensitive obesity. *J Clin Invest* 2006;**116**:1886–901.
21. Yue JT, Abraham MA, LaPierre MP, Mighiu PI, Light PE, Filippi BM, Lam TK. A fatty acid-dependent hypothalamic-DVC neurocircuitry that regulates hepatic secretion of triglyceride-rich lipoproteins. *Nat Commun* 2015;**6**:5970.
22. Yang Z, Chen Y, Zhang Y, Jiang Y, Fang X, Xu J. Sevoflurane postconditioning against cerebral ischemic neuronal injury is abolished in diet-induced obesity: role of brain mitochondrial KATP channels. *Mol Med Rep* 2014;(3):843–50.
23. Leloup C, Magnan C, Benani A, Bonnet E, Alquier T, Offer G, et al. Mitochondrial reactive oxygen species are required for hypothalamic glucose sensing. *Diabetes* 2006;**55**:2084–90.
24. Pi J, Bai Y, Zhang Q, Wong V, Floering LM, Daniel K, et al. Reactive oxygen species as a signal in glucose-stimulated insulin secretion. *Diabetes* 2007;**56**:1783–91.
25. Leloup C, Tourrel-Cuzin C, Magnan C, Karaca M, Castel J, Carneiro L, et al. Mitochondrial reactive oxygen species are obligatory signals for glucose-induced insulin secretion. *Diabetes* 2009;**58**:673–81.
26. Bindokas VP, Kuznetsov A, Sreenan S, Polonsky KS, Roe MW, Philipson LH. Visualizing superoxide production in normal and diabetic rat islets of Langerhans. *J Biol Chem* 2003;**278**:9796–801.
27. Benani A, Troy S, Carmona MC, Fioramonti X, Lorsignol A, Leloup C, et al. Role for mitochondrial reactive oxygen species in brain lipid sensing: redox regulation of food intake. *Diabetes* 2007;**56**:152–60.
28. Tiedge M, Lortz S, Drinkgern J, Lenzen S. Relation between antioxidant enzyme gene expression and antioxidative defense status of insulin-producing cells. *Diabetes* 1997;**46**:1733–42.
29. Leloup C, Casteilla L, Carrière A, Galinier A, Benani A, Carneiro L, et al. Balancing mitochondrial redox signaling: a key point in metabolic regulation. *Antioxid Redox Signal* 2011;**14**:519–30.
30. Colombani A-L, Carneiro L, Benani A, Galinier A, Jaillard T, Duparc T, et al. Enhanced hypothalamic glucose sensing in obesity: alteration of redox signaling. *Diabetes* 2009;**58**(10):2189–97.
31. Yang XJ, Kow LM, Funabashi T, Mobbs CV. Hypothalamic glucose sensor: similarities to and differences from pancreatic beta-cell mechanisms. *Diabetes* 1999;**48**:1763–72.
32. Ainscow EK, Mirshamsi S, Tang T, Ashford ML, Rutter GA. Dynamic imaging of free cytosolic ATP concentration during fuel sensing by rat hypothalamic neurones: evidence for ATP-independent control of ATP-sensitive K(+) channels. *J Physiol* 2002;**544**:429–45.
33. Coppola A, Liu ZW, Andrews ZB, Paradis E, Roy MC, Friedman JM, et al. A central thermogenic-like mechanism in feeding regulation: an interplay between arcuate nucleus T3 and UCP2. *Cell Metab* 2007;**5**:21–33.
34. Andrews ZB, Liu ZW, Walllingford N, Erion DM, Borok E, Friedman JM, et al. UCP2 mediates ghrelin's action on NPY/AgRP neurons by lowering free radicals. *Nature* 2008;**454**:846–51.
35. Chrétien C, Fenech C, Liénard F, Grall S, Chevalier C, Chaudy S, et al. Transient receptor potential canonical 3 (TRPC3) channels are required for hypothalamic glucose detection and energy homeostasis. *Diabetes* 2017;**66**:314–24.
36. Parton LE, Ye CP, Coppari R, Enriori PJ, Choi B, Zhang CY, et al. Glucose sensing by POMC neurons regulates glucose homeostasis and is impaired in obesity. *Nature* 2007;**449**:228–32.
37. Ramsden DB, Ho PW, Ho JW, Liu HF, So DH, Tse HM, et al. Human neuronal uncoupling proteins 4 and 5 (UCP4 and UCP5): structural properties, regulation, and physiological role in protection against oxidative stress and mitochondrial dysfunction. *Brain Behav* 2012;**2**:468–78.
38. Sebastián D, Zorzano A. When MFN2 (mitofusin 2) met autophagy: a new age for old muscles. *Autophagy* 2016;**12**(11):2250–1.
39. Schrepfer E, Scorrano L. Mitofusins, from mitochondria to metabolism. *Mol Cell* 2016;**61**(5):683–94.
40. Liesa M, Shirihai OS. Mitochondrial dynamics in the regulation of nutrient utilization and energy expenditure. *Cell Metab* 2013;**17**(4):491–506.
41. Frezza C, Cipolat S, Martins de Brito O, Micaroni M, Beznoussenko GV, et al. OPA1 controls apoptotic cristae remodeling independently from mitochondrial fusion. *Cell* 2006;**126**(1):177–89.
42. Losón OC, Song Z, Chen H, Chan DC. Fis1, Mff, MiD49, and MiD51 mediate Drp1 recruitment in mitochondrial fission. *Mol Biol Cell* 2013;**24**(5):659–67.
43. Yu T, Robotham JL, Yoon Y. Increased production of reactive oxygen species in hyperglycemic conditions requires dynamic change of mitochondrial morphology. *Proc Natl Acad Sci U S A* 2006;**103**(8):2653–8.

44. Carneiro L, Allard C, Guissard C, Fioramonti X, Tourrel-Cuzin C, et al. Importance of mitochondrial dynamin-related protein 1 in hypothalamic glucose sensitivity in rats. *Antioxid Redox Signal* 2012;**17**(3):433–44.
45. Toda C, Kim JD, Impellizzeri D, Cuzzocrea S, Liu ZW, et al. UCP2 regulates mitochondrial fission and ventromedial nucleus control of glucose responsiveness. *Cell* 2016;**164**(5):872–83.
46. Horvath TL, Andrews ZB, Diano S. Fuel utilization by hypothalamic neurons: roles for ROS. *Trends Endocrinol Metab* 2009;**20**(2):78–87.
47. Toyama EQ, Herzig S, Courchet J, Lewis Jr. TL, Losón OC, et al. Metabolism. AMP-activated protein kinase mediates mitochondrial fission in response to energy stress. *Science* 2016;**351**(6270):275–81.
48. Santoro A, Campolo M, Liu C, Sesaki H, Meli R, et al. DRP1 suppresses leptin and glucose sensing of POMC neurons. *Cell Metab* 2017;**25**(3):647–60.
49. Dietrich MO, Liu ZW, Horvath TL. Mitochondrial dynamics controlled by mitofusins regulate Agrp neuronal activity and diet-induced obesity. *Cell* 2013;**155**(1):188–99.
50. Schneeberger M, Dietrich MO, Sebastián D, Imbernón M, Castaño C, et al. Mitofusin 2 in POMC neurons connects ER stress with leptin resistance and energy imbalance. *Cell* 2013;**155**(1):172–87.
51. Ozcan L, Ergin AS, Lu A, Chung J, Sarkar S, et al. Endoplasmic reticulum stress plays a central role in development of leptin resistance. *Cell Metab* 2009;**9**(1):35–51.
52. Rowland AA, Voeltz GK. Endoplasmic reticulum-mitochondria contacts: function of the junction. *Nat Rev Mol Cell Biol* 2012;**13**(10):607–25.
53. Ramírez S, Gómez-Valadés AG, Schneeberger M, Varela L, Haddad-Tóvolli R, et al. Mitochondrial dynamics mediated by mitofusin 1 is required for POMC neuron glucose-sensing and insulin release control. *Cell Metab* 2017;**25**(6):1390–9.
54. Li H, Alavian KN, Lazrove E, Mehta N, Jones A, et al. A Bcl-xL-Drp1 complex regulates synaptic vesicle membrane dynamics during endocytosis. *Nat Cell Biol* 2013;**15**(7):773–85.
55. Guillod-Maximin E, Lorsignol A, Alquier T, Pénicaud L. Acute intracarotid glucose injection towards the brain induces specific c-fos activation in hypothalamic nuclei: involvement of astrocytes in cerebral glucose-sensing in rats. *J Neuroendocrinol* 2004;**16**(5):464–71.
56. Marty N, Dallaporta M, Foretz M, Emery M, Tarussio D, et al. Regulation of glucagon secretion by glucose transporter type 2 (glut2) and astrocyte-dependent glucose sensors. *J Clin Invest* 2005;**115**(12):3545–53.
57. Chowen JA, Argente-Arizón P, Freire-Regatillo A, Frago LM, Horvath TL, et al. The role of astrocytes in the hypothalamic response and adaptation to metabolic signals. *Prog Neurobiol* 2016;**144**:68–87.
58. Terri-Leigh S, Gupta-Agarwal S, Kittler JT. Mitochondrial dynamics in astrocytes. *Biochem Soc Trans* 2014;**42**(5):1302–10.
59. Dienel GA. Astrocytic energetics during excitatory neurotransmission: what are contributions of glutamate oxidation and glycolysis? *Neurochem Int* 2013;**63**:244–58.
60. Reyes RC, Parpura V. Mitochondria modulate $Ca^{2+}$-dependent glutamate release from rat cortical astrocytes. *J Neurosci* 2008;**28**:9682–91.
61. García-Cáceres C, Quarta C, Varela L, Gao Y, Gruber T, et al. Astrocytic insulin signaling couples brain glucose uptake with nutrient availability. *Cell* 2016;**166**(4):867–80.

PART III

# TISSUES SUFFERING CONSEQUENCES FROM THE PATHOLOGIES

CHAPTER

# 11

# Role of Mitochondria in Cardiovascular Comorbidities Associated with Obesity and Type 2 Diabetes

*Sihem Boudina*

**Department of Nutrition and Integrative Physiology, College of Health, University of Utah, Salt Lake City, UT, United States**

## 1 INTRODUCTION

The importance of the heart has been recognized by many civilizations, without even knowing its physiological importance to the human body. Ancient Egyptians recognized the value of the heart by making it the only organ left inside the body during mummification. They believed that the heart is the center of intelligence and feeling, and the mummified person would need it in the afterlife. With the advancement of science, we have come to realize that without a heart, life cannot exist. The heart beats every second of our lives-two and a half million times during an average lifespan. It does not rest, pumping ~280 L of blood every hour in adults. As a result of this continued pumping of blood, the heart produces and consumes about 30 kg of ATP every single day.

Under normoxic conditions, ~95% of ATP supply in the heart derives from oxidative phosphorylation in the mitochondria; the remaining 5% comes from glycolysis.[1, 2] The mitochondrial supply of ATP becomes limited in pathological conditions, however, leading to energy starvation and the development of cardiac pathologies. This chapter describes the role of mitochondria in normal cardiovascular function in humans and mice and then the main disturbances in mitochondrial metabolism in the heart and the vasculature during diabetes.

*Mitochondria in Obesity and Type 2 Diabetes*
https://doi.org/10.1016/B978-0-12-811752-1.00011-0

# 2 MITOCHONDRIA AND HEART FUNCTION

The heart contains the greatest amount of mitochondria in the body. The human heart is composed of 23% mitochondria, whereas mouse heart contains 32%.[3] This high content of mitochondria is regulated to meet the energy demand in the myocardium. Mitochondria play an important role in the maintenance of cardiac function not only by supplying ATP for contractions, but also by serving as a sensor for the well-being of the cell. In order to preserve mitochondrial function, several processes are in place in the heart, including the maintenance of mitochondrial biogenesis and quality control, the preservation of mitochondrial oxidative capacity, and the prevention of oxidative stress and calcium overload (Fig. 1).

## 2.1 Mitochondrial Biogenesis

Mitochondrial content in the heart is not constant; the heart possesses a finely tuned regulatory network that orchestrates the biogenesis, maintenance, and turnover of mitochondria. The mitochondrion contains its own self-replicating genome. The mitochondrial DNA (mtDNA) encodes 13 essential components of the electron transport chain (ETC) and all the rRNAs and tRNAs necessary for translation of the mtDNA-encoded proteins.[4] More than 99% of mitochondrial proteins, however, are encoded by nuclear DNA.[5] The mitochondrial signal that triggers nuclear transcription of mitochondrial protein and mitochondrial biogenesis is not completely understood. Key regulators of mitochondrial biogenesis, however, have been identified recently and involved including peroxisome proliferator-activated receptor $\gamma$ coactivators 1$\alpha$ (PGC-1$\alpha$) and 1$\beta$ (PGC-1$\beta$). Genetic manipulation of PGC-1$\alpha$ and $\beta$ has provided important insights into the regulation of mitochondrial biogenesis and function in the heart. Cardiac-specific transgenic overexpression of PGC-1$\alpha$ in mice results in uncontrolled mitochondrial biogenesis, which led to the development of dilated cardiomyopathy and death at 6 weeks of age.[6] In contrast, germline deletion of both PGC-1$\alpha$ and PGC-1$\beta$ evokes perinatal lethal heart failure caused by a complete lack of cardiac mitochondrial biogenesis.[7] These studies suggest that mitochondrial content in the heart must be regulated adequately to maintain cardiac function and that both PGC-1$\alpha$ and PGC-1$\beta$ are necessary. When both PGC-1$\alpha$ and PGC-1$\beta$ were deleted in the adult heart, however, no quantitative changes were detected in mitochondrial volume density, but qualitative abnormalities in mitochondrial phospholipid biosynthesis were observed.[8] Altogether, these studies highlight the importance of mitochondrial biogenesis in the heart, especially during the perinatal period, coincident with the increase in substrate oxidation.

The transcriptional machinery involved in mitochondrial biogenesis has been studied extensively, especially the one mediated by PGC-1$\alpha$, and is known to involve nuclear factors such as nuclear respiratory factor-1 and 2 (NRF-1 and NRF2). NRF-1 and NRF-2 regulate expression of every ETC complex.[9, 10] In addition, PGC-1/NRF-1 interaction activates downstream factors involved in mtDNA replication. Specifically, NRF-1 activates transcription of genes encoding factors that mediate replication and transcription of the mitochondrial genome, including mitochondrial transcription factors A (TFAM) and mitochondrial transcription factors B1 and B2 (TFB1M and TFB2M).[11, 12] Deletion of TFAM in cardiac and skeletal muscle of mice resulted in oxidative phosphorylation deficiency and premature death caused by dilated cardiomyopathy.[13, 14] Similarly, heart and skeletal muscle-specific ablation of TFB1M in mice caused mitochondrial cardiomyopathy characterized by reduced ETC function and

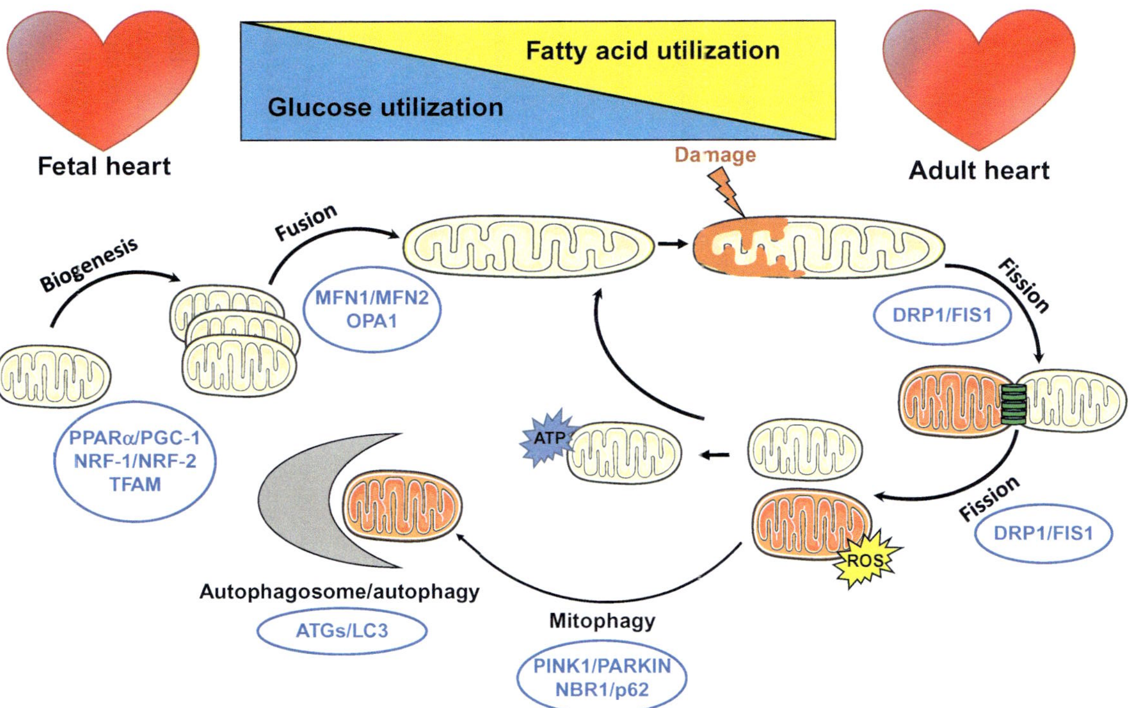


FIG. 1 Schematic representation of mitochondrial biogenesis and quality control mechanisms in the heart. During the perinatal period, mitochondrial biogenesis increases in response to increased fatty acid availability. In the adult heart, the mitochondrial dynamic represented by fission and fusion is in place to eliminate damaged mitochondria through mitophagy/autophagy. Key regulators of mitochondrial biogenesis and quality control are highlighted in blue. *PPARα*, peroxisome proliferator receptor α; *PGC-1*, peroxisome proliferator-activated receptor γ coactivator 1; *MFN1*, mitofusin 1; *MFN2*, mitofusin 2; *DRP1*, dynamin related protein 1; *FIS1*, mitochondrial fission protein 1; *PINK1*, PTEN-induced putative kinase 1; *p62*, sequestosome 1; *ATGs*, autophagy genes; *LC3*, microtubule associated protein; *ATP*, adenosine triphosphate; *ROS*, reactive oxygen species.

a compensatory increase in mitochondrial biogenesis.[15] These studies clearly demonstrated that the maintenance of a proper mitochondrial number is required for heart function. In addition to PGC-1, nitric oxide (NO) is a potent inducer of mitochondrial biogenesis in many tissues and cell lines including vessels.[16] Shear stress induces PGC-1α and mitochondrial biogenesis via activation of AMP-activated protein kinase (AMPK), sirtuin 1 (SIRT1), and endothelial nitric oxide synthase (eNOS).[17]

## 2.2 Mitochondrial Dynamics and Quality Control

After the mitochondrial number is set in the heart during the perinatal period, adequate homeostatic mechanisms are needed to preserve mitochondrial quality in postmitotic cardiomyocytes. Mitochondrial dynamics and mitophagy orchestrate with mitochondrial biogenesis to replace old mitochondria with new ones in the adult heart, thus maintaining mitochondrial quality control. Mitochondrial dynamics is executed by two opposing processes: fission and fusion. In the heart, fission is regulated by dynamin-related protein (DRP)-1, which physically segregates dysfunctional mitochondrial components into a depolarized daughter organelle targeted to mitophagy. Mitochondrial fusion, however, is mediated by mitofusins 1 and 2 (MFN1 and MFN2) for the outer mitochondrial membrane (OMM) and mitochondrial dynamin-like GTPase OPA1 for the inner mitochondrial membrane (IMM). Mitofusins and OPA1 usually work in concert to coordinately fuse both mitochondrial membranes.[18] In spite of a low frequency of these processes in the adult heart, their interruption has adverse consequences on mitochondrial stress, biogenesis, and cardiomyocyte death. The role of mitochondrial dynamics in the heart was characterized recently using genetic loss of function studies. Tamoxifen-inducible deletion of MFN1 and MFN2 in the heart caused heart failure precipitated by mitochondrial fragmentation and impaired mitochondrial respiratory capacity.[19] Similarly, deficiency of DRP1 in the adult heart caused dilated cardiomyopathy and death precipitated by mitochondrial elongation and dysfunction.[20, 21] The mechanisms for cardiac dysfunction in fusion and fission deficient hearts are distinct, with DRP1-deficient hearts exhibiting cardiomyocyte death, enhanced fibrosis, and altered autophagy, whereas MFN1/MFN2 double knockout hearts have increased mitochondrial unfolded protein response. Both mouse models, however, develop oxidative phosphorylation defects, implying that mitochondrial dynamics is required for the maintenance of mitochondrial quality and cardiac function.

At the center of mitochondrial quality control is mitophagy (also referred to as mitochondrial autophagy), which is a cellular mechanism for identifying and selectively eliminating dysfunctional mitochondria that could otherwise accumulate and become a source for cytotoxic reactive oxygen species (ROS). Mitophagy shares common steps with the autophagy pathway (i.e., autophagosomal engulfment of mitochondria and their transfer to lysosomes for degradation and component recycling), however, the proximal events that detect and select dysfunctional organelles for targeted elimination are highly specific for mitophagy. Two main mediators of detection/selection process are the cytosolic E3 ubiquitin ligase PARKIN and the mitochondrial kinase PINK1.[18] Manipulations of mitophagy through loss or gain of function of PINK1 or PARKIN in the heart had no effect on cardiac function unless the heart was subjected to metabolic stress such as ischemia-reperfusion (IR), transverse aortic banding (TAC), or aging.[22–24] These results suggest that the level of mitophagy might be too low or that other players can compensate for the loss of PINK and PARKIN in the adult heart. Other adaptor proteins (p62/SQSTM1 and NBR1) and pro-apoptotic mitochondrial proteins (NIX and BNIP3) were shown to mediate mitophagy in the heart. The implication of these proteins in homeostatic mitophagy in the heart, however, was either not investigated or was observed only when the heart was subjected to stress such as TAC, IR, or cardiac proteinopathy[25, 26]

Studies examining the role of mitochondrial dynamics in the vasculature are just starting to emerge. Therefore, deletion of MFN in endothelial cells (ECs) reduced vascular endothelial growth factor (VEGF)-induced AKT-eNOS signaling, whereas knockdown of MFN2 resulted in a decrease in ROS generation and diminished the expression of components of the electron transport chain.[27] Furthermore, an induction of fusion and a reduction in fission was observed in aortas from rats after 8 days of exercise, a response that was blunted after treatment with the NOS inhibitor L-NAME.[28] Finally, our group has shown recently that deletion of the autophagy proteins ATG3 or ATG5 induced EC dysfunction and reduced eNOS activation after shear stress through impaired glycolysis-purinergic signaling.[29] These studies highlight the importance of mitochondrial quality control and dynamics in the heart and the vessels under physiologic conditions.

## 2.3 Mitochondrial Energetics

Mitochondrial density is matched with energy demand in the heart as evidenced by the parallel increase in mitochondrial content with heart rate and oxygen consumption.[3] In addition to being the site for most of the energy production, mitochondria also are responsible for substrate oxidation, energy propagation, and metabolic regulation. The adult heart predominantly uses fatty acids (FAs) relative to other substrates such as glucose, lactate, ketone bodies, or amino acids.[30] In spite of being inefficient substrates, FAs and mitochondrial β-oxidation generate most of the reducing equivalents for oxidative phosphorylation and the subsequent ATP needed for heart function. This substrate preference, however, is altered with physiological and pathological conditions, including increased workload, hypertrophy, diabetes, ischemia/reperfusion (I/R), and heart failure, and is subject to transcriptional and posttranslational regulation. The master regulator of cardiac FAs oxidation is the peroxisome proliferator-activated receptor α (PPARα), which regulates the transcription of an array of genes involved in cellular FAs utilization pathways including transport, esterification, and oxidation. Paradoxically, whole body deletion of PPARα had no effect on left ventricular function at baseline but protected mice from diabetes or diet-induced cardiac dysfunction and lipotoxicity.[31, 32] In contrast, cardiac-restricted overexpression of PPARα enhanced FAs oxidation and reduced glucose utilization, thus mimicking diabetic cardiomyopathy.[33] These results highlight the role of PPARα in the regulation of substrate preference in the heart, especially in conditions associated with systemic elevation of FAs such as in diabetes.

Glucose utilization in the heart is regulated predominantly by growth hormones signaling, beta-adrenergic signaling, and by hypoxia. This regulation starts at the level of glucose transport, which is regulated predominantly by insulin and insulin-like growth factor 1 (IGF-1) in the myocardium. The heart expresses glucose transporters 1 and 4 (GLUT1 and GLUT4) and GLUT4-mediated glucose uptake represents the major mechanisms by which the heart increases its transport of glucose in response to insulin.[34, 35] In contrast, GLUT1 mediates basal glucose uptake in the heart but can compensate for the absence of GLUT4.[36, 37]

Indeed, cardiac-specific deletion of GLUT4 resulted in a compensatory increase in GLUT1 expression and basal glucose uptake but blunted insulin-mediated transport and caused cardiac hypertrophy with preserved function in mice.[38] In contrast, restricted deletion of GLUT1 in the heart had reduced basal glucose uptake, glycolysis, and glucose oxidation and enhanced FAs oxidation without an evident increase in GLUT4 expression.[39] Moreover, neither deletion nor overexpression of GLUT1 in the heart protected from TAC-induced cardiac dysfunction.[39, 40] Glucose transport and oxidation of both glucose and FAs in the heart are regulated by insulin. Therefore, lack of insulin receptors (IRs) in cardiac cells resulted in impaired glucose and FAs oxidation with a compensatory increase in glycolysis.[41] These results suggest that insulin via its signaling pathway might regulate genes involved in glucose and FAs metabolism and affect mitochondrial oxidative capacity. A coordinated reduction in the tricarboxylic acid (TCA) cycle and FAs enzymes was observed in mice lacking IRs specifically in cardiac cells,[42] suggesting that insulin is a key regulator of mitochondrial metabolism in the heart.

After ATP is generated in the mitochondria, it is incorporated in an energy reservoir known as creatine phosphate PCr, a reaction catalyzed by the enzyme creatine kinase.[43] This system allows the efficient transfer of energy from mitochondria to the site of utilization especially when energy demand is high. This is because PCr is more diffusible than ATP, allowing rapid energy transfer.[44] Furthermore, PCr acts as an energy buffer in the heart, providing compartmentalized energy reserve available for mobilization when ATP demand outstrips supply. Human mutations or genetic manipulations of components of this system, however, yielded conflicting results, with some showing altered cardiac function and some showing no effect,[45–47] suggesting the existence of alternative mechanisms for the maintenance of energy homeostasis in the myocardium.

## 2.4 Mitochondrial Signaling

In addition to energy generation, mitochondria now are considered a hub for cellular signaling, such as retrograde signaling for the induction of mitochondrial biogenesis, redox and antioxidative signaling, calcium and cell death signaling, posttranslational modifications, and mitochondrial-specific unfolded protein response ($UPR^{mt}$). Respiratory deficiency and a decrease in ATP are considered to be the primary activators of most of these responses, but the mediators are not fully understood. Although these signalings are complex and continue to be updated, we will focus this section on mitochondrial signaling that can directly affect cardiac function.

Heart mitochondria produce variable amounts of superoxide at the level of complexes I and III.[48] Mitochondrial ROS also can be generated through enzymatic reactions involving NOX4, p66Shc, monoamine oxidase, glycerol-3-phosphate dehydrogenase, proline dehydrogenase, and dihydroorotate dehydrogenase, as well as via reverse electron flow to complex I from complex II.[49–51] The contribution of these sources to homeostatic levels of ROS under physiological conditions, however, require further investigations. To maintain a physiologic level of ROS, the mitochondria is equipped with efficient antioxidant systems that can maintain a low level required for signaling and not sufficient to trigger damage to mitochondrial components. A recent study of heart mitochondria from mouse found that glutathione (GSH) and thioredoxin (TRX) scavenging systems are essential for keeping minimal levels of hydrogen peroxide ($H_2O_2$) emission, especially during maximal state 3 respiration.[52] Furthermore, mitochondria cannot leak superoxide because it is rapidly dismutated into $H_2O_2$ by manganese superoxide dismutase (MnSOD). The importance of this dismutation in the heart is confirmed by the development of dilated cardiomyopathy in mice lacking MnSOD in skeletal muscle and heart.[53] Physiological levels of mitochondrial ROS are involved in signaling processes that are important for an optimal response to physiological and pathological stimuli. This ROS-mediated signaling occurs in part through specific mechanisms, including modulation of redox couples ($NAD^+$/NADH, $NADP^+$/NADPH, GSSG/GSH and oxidized TRX/reduced TRX) (extensively reviewed in.[54, 55] Changes in redox potential within mitochondria regulate mitochondrial sirtuin 3 (SIRT3), which results in the deacetylation of complex I subunits, β-oxidation enzymes, and mitochondrial permeability transition pore (mPTP).[56–58] Consistent with these observations, inactivation of complex I in mouse heart reduced $NAD^+$/NADH and inhibited SIRT3, leading to an increase in protein acetylation, sensitization of mPTP, and the development of heart failure after pressure overload.[59] Consistent with these results, SIRT3 was shown to oppose cardiac hypertrophic response through suppression of ROS and enhanced expression of MnSOD and catalase.[60]

Mitochondrial stress responses sometimes occur independently of an increase in ROS[61] and can be tissue-specific. An induction of $UPR^{mt}$ associated with enhanced mitochondrial biogenesis was observed in the heart but not in skeletal muscle of mice lacking mitochondrial aspartyl-tRNA synthase (DARS2) in these organs, despite similar ETC deficiency.[62] It is currently not clear whether these stress responses are protective or detrimental to cardiac function because muscle-specific DARS2-deficient mice develop cardiomyopathy and die at 7 weeks of age.[62] This study and others[63] suggest that stress responses, including enhanced mitochondrial biogenesis, are not sufficient to compensate for ETC deficiency and the resulting cardiac dysfunction.

Another important way by which the mitochondria communicate with the rest of the cell is through the buffering of calcium.[64, 65] With the recent discovery of the mitochondrial calcium uniporter (MCU),[66, 67] our understanding of mitochondrial calcium signaling and its role in cardiac function has improved. Therefore, inactivation of myocardial MCU reduced physiological heart rate acceleration through impairment of oxidative phosphorylation, which is necessary to accelerate reloading of the cytosolic calcium reserves before each heartbeat.[68] Consistent with these studies, global deletion of MCU or conditional deletion of this transporter in the heart had no overt impact on resting cardiac function, but reduced cardiac performance in response to acute workload.[69–72] These results suggest that mitochondrial calcium serves as a sensor of cytosolic calcium to match oxidative capacity with myocardial contraction in homeostatic conditions. In disease conditions, however, when cytosolic calcium concentrations increase, leading to enhanced mitochondrial calcium uptake, this can trigger mPTP opening and can lead to myocardial cell death.[73]

Cardiomyocytes are postmitotic and their regeneration in the adult heart is minimal, so a loss of these cells is detrimental to tissue integrity and contractile function. Two main forms of cell death occur in the heart: necrosis and apoptosis, which are regulated by signaling cascades involving extracellular and intracellular signals. Mitochondrial signals play a key role in the onset of cell death, which is believed to be initiated by the permeabilization of the outer mitochondrial membrane (OMM) in the case of apoptosis. The mitochondrial event involved in the initiation of necrosis, however, is the opening of mPTP.[74, 75] OMM permeabilization allows the release of cytochrome c into the cytoplasm and the activation of the pro-apoptotic proteins B-cell lymphoma 2 (BCL2), BCL2-associated x protein (BAX), and BCL2 homologous antagonist/killer (BAK). It is important to consider that the rates of apoptosis and necrosis are very low in the adult heart, and that these pathways become prominent only in a diseased heart. This is supported by data showing that mice lacking these pro-apoptotic molecules do not have any cardiac phenotype at baseline but show protection against cardiac injury.[76–78] Finally, other forms of cell death that involve mitochondria, such as autosis (autophagy-initiated cell death) and ferroptosis, have been discovered but their molecular mechanisms are still not well understood.[79, 80]

# 3 MITOCHONDRIA AND ENDOTHELIAL FUNCTION

The heart's microvascular bed provides the endothelial surface area to facilitate the delivery of oxygen, nutrients, and hormones and the removal of metabolic waste from the myocardium. Changes in the cardiac microvascular blood volume and flow could have a profound effect on myocardial metabolism, function, and health. In fact, the microvascular compartment, which is located within the myocardium, constitutes ~90% of the myocardial blood volume.[81] Located at the interface between blood and tissue, ECs can sense changes in hemodynamic forces, ambient $PO_2$, and local blood-borne signals and respond with appropriate changes in function to maintain homeostasis. These responses include the paracrine release of diffusible mediators such as NO, prostacyclin, endothelin-1 (ET-1), and growth factors; the activity of surface enzymes such as angiotensin-converting enzyme (ACE), which regulates local levels of bioactive angiotensin II and bradykinin; and the expression of surface proteins such as adhesion molecules that interact with other cell types.[82] Consistent with a signaling rather than a metabolic role for ECs, mitochondrial content in these cells is low (2%–6% of cytoplasm volume).[83, 84] Therefore, ECs derive 80% of their ATP from glycolysis, which depends on glucose uptake through the predominantly expressed GLUT1.[85] Furthermore, it appears that vascular smooth muscle cells are responsible for the majority of vascular respirations as evidenced by similar state 3 respiration rates in intact and denuded (without ECs) vessels.[86] At rest, mitochondrial content and respiratory capacity are low in vessels, but they can increase in response to increased blood flow. Therefore, shear stress was shown to enhance mitochondrial content and oxidative capacity in vessels, effects mediated in part by NO.[87] ECs mitochondria, however, play an important role in cell signaling, as it is one of the main sources of ROS even under physiologic conditions.[82, 88] Shear stress-induced $H_2O_2$ formation and vasodilation are mediated by superoxide generated in the mitochondria.[89] Similarly, scavenging ROS (including mitochondrial ROS) ablated exercise-induced vasodilation of human brachial arteries.[90] Another node of signaling involving mitochondrial ROS in ECs is the formation of peroxynitrite (ONOO−), which leads to less NO bioavailability and a decrease in the formation of S-nitrosated proteins. This modification was shown to modulate several mitochondrial proteins in ECs, including glyceraldehyde 3-phosphate dehydrogenase (GAPDH), leading to alterations in its activity and thus glycolysis.[91] The mediators of flow-mediated increases in mitochondrial ROS generation in ECs are not well characterized but might involve changes in the cytoskeleton and/or calcium signaling. These studies highlight the importance of mitochondrial signaling in mediating aspects of the endothelial function under physiological conditions, but mitochondrial dysfunction and excessive ROS generation can become the trigger for endothelial dysfunction in pathological conditions.

# 4 DIABETES AND CARDIOVASCULAR DISEASES (CVD)

The prevalence of diabetes mellitus is growing rapidly. It is estimated that globally the number of adults affected with diabetes will increase from 135 million in 1995 to 300 million by 2025.[92] The incidence of CVD is higher in patients with diabetes compared to patients without diabetes, and CVDs are the leading cause of morbidity and mortality in this patient population.[93] Diabetes causes coronary artery disease (CAD), such as atherosclerosis, which increases the risk of myocardial infarction, stroke, and limb loss. Furthermore, microangiopathy associated with diabetes contributes to retinopathy and nephropathy and might cause cardiac pathology as well.[94] Diabetes also can cause a form of cardiomyopathy named diabetic cardiomyopathy, which is not associated with changes in blood pressure or CAD.[95, 96] Because diabetes, especially type 2 diabetes mellitus (T2DM), often is associated with obesity, these combined conditions often precipitate the development of left ventricular hypertrophy that can progress to congestive heart failure (CHF).[97–99] The mechanisms involved in the higher incidence of CVD in diabetes are multifactorial and involve structural, transcriptional, posttranslational, and metabolic alterations (Fig. 2). This chapter will narrow our focus on the mechanisms involving mitochondria.

# 5 MITOCHONDRIAL DYSFUNCTION IN THE DIABETIC HEART

## 5.1 Impaired Mitochondrial Biogenesis

Mitochondrial biogenesis measured by surface area, number, and mtDNA content was reduced in the hearts of type 1 diabetes (T1D) OVE26 mouse model.[100] Similarly, mitochondrial number and mtDNA were elevated in the hearts of obese diabetes (db/db) mice.[101]

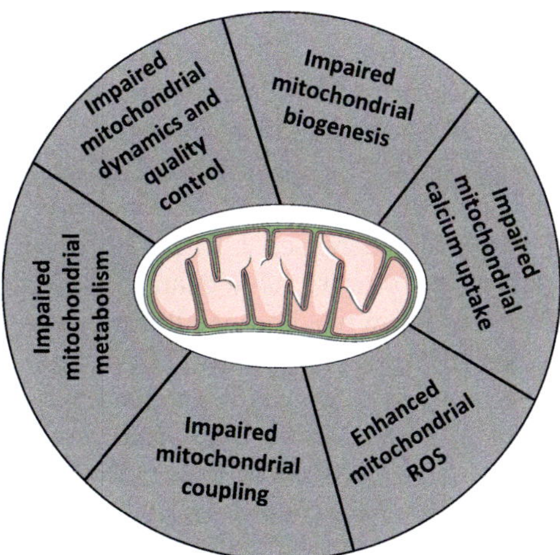


FIG. 2 Mitochondria is at the center of cardiovascular complications. Alterations in mitochondrial biogenesis, dynamics, quality control, substrate oxidation, coupling, and calcium buffering capacity contribute to cardiac and endothelial dysfunction in diabetes.

This compensatory increase in mitochondrial biogenesis can occur in the insulin-resistant heart even when diabetes is not present, as is the case for transgenic mice expressing UCP1-driven diphtheria toxin A (UCPDTA mice).[102] Mechanistically, it is suggested that the elevation of free FAs or triglycerides and the induction of PPARα are required for the increase in mitochondrial biogenesis as lack of PPARα abolished this response in UCPDTA mice in spite of no change in systemic metabolism.[102] In contrast, insulin resistance per se without changes in systemic metabolism did not increase mitochondrial biogenesis in the hearts as is the case for mice lacking IRs specifically in cardiomyocytes.[42] Furthermore, forced expression of PPARα in the heart activates mitochondrial biogenesis in the absence of insulin resistance, suggesting that activation of this pathway by FAs influx in the heart might be the trigger.[103] In contrast to these studies, others have reported either unchanged mitochondrial number or decreased PGC-1α mRNA expression in the hearts of T1D mice.[104, 105] The reasons for this discrepancy are currently unknown, but might involve the duration of diabetes and/or changes in circulating lipids. In contrast to the heart, few studies have examined mitochondrial biogenesis in vessels of diabetic animals in vivo and in ECs in vitro. Although a reduction in mtDNA content and PGC-1α and TFAM expression were reported in aortas from diabetic (db/db) mice,[106] no change in these parameters was observed in human umbilical cord ECs incubated with high glucose in vitro.[107] Both in heart and vessels, the compensatory increase in mitochondrial content in the diabetic heart is not sufficient to supply the necessary ATP, suggesting additional defects in mitochondrial substrate utilization, uncoupling, and superoxide-mediated damage.

## 5.2 Altered Mitochondrial Dynamics and Quality Control

Changes in mitochondrial dynamics characterize the diabetic myocardium. Initial studies on H9c2 cells (a myoblast cell line, derived from embryonic rat heart) showed that exposure to hyperglycemia induced mitochondrial fragmentation, leading to ROS accumulation and cell death, and these effects were prevented by over-expression of a fission-inactive mutant of DRP1.[108] Similarly, exposure of H9c2 cells to palmitate (to mimic high-fat feeding conditions) induced DRP1 expression and caused mitochondrial fragmentation partially through ROS generation.[109] Using neonatal cardiomyocytes, Makino et al.[110] showed that high glucose decreased OPA1 protein expression and induced mitochondrial fragmentation, and these effects were reversed by overexpression of OPA1. Similar findings were reported for ECs exposed to high glucose with reduced OPA1, increased DRP1, and mitochondrial fragmentation.[111] In spite of this in vitro evidence, few studies have examined mitochondrial dynamics in the diabetic heart in vivo and it is unknown whether alterations in this process contribute to diabetes-associated cardiac dysfunction. A recent study that examined mitochondrial dynamics in atrial tissue of obese patients with and without diabetes showed that the presence of diabetes enhanced mitochondrial fragmentation and reduced MFN1 protein expression.[112] Because mitochondrial dynamics are tritely linked to mitochondrial clearance and autophagy, diabetes has been shown to affect these processes. An increase in autophagy was reported in the hearts of streptozotocin-induced T1D mice[113, 114] but a further increase in autophagy by overexpression of beclin 1 led to further deterioration of cardiac function in these hearts. These results suggest that downregulation of autophagy might be an adaptive response to limit diabetic cardiac injury in T1D mice. These in vivo studies, however, are not supported by in vitro observation of reduced autophagy in cardiac cells incubated with high glucose.[115] In contrast to T1D, the role of autophagy in the pathogenesis of diabetic cardiomyopathy in T2D is controversial. High-fat feeding was shown to reduce autophagy in the heart.[116–118] In contrast, autophagy was reported to be increased in atrial tissue from diabetic patients and in hearts of mice fed high sucrose diet.[119, 120] Consistent with the high-fat studies, we and others have reported reduced autophagy level in the hearts of leptin and leptin receptor-deficient mice and in diabetic OVE26 mice.[114, 121, 122] The reasons for the discrepant results about the level of cardiac autophagy in the hearts of T2D mice are not fully understood but might involve the state of activation of main autophagy modulators including AMPK and the mammalian target of rapamycin (mTOR) as reviewed recently.[123] It is safe to conclude, however, that impairment of autophagy might exacerbate mitochondrial dysfunction because the clearance of damaged mitochondria requires intact autophagy machinery.

## 5.3 Impaired Fuel Utilization and Oxidative Capacity

The healthy heart relies on a balance between the oxidation of carbohydrates, free FAs, and ketones as fuel sources, and it has the flexibility to switch substrates depending on their availability. The flexibility of substrate utilization, however, is lost in the diabetic heart, as shown by increased reliance on FAs and decreased oxidation of glucose.[124–126] This fuel switch is imposed in part by the enhanced availability of exogenous and endogenous FAs and triglycerides (TAG) and the reduced uptake of glucose, which are caused by the development of peripheral and cardiac insulin resistance. In addition to altered substrate availability in diabetes, enhanced FAs utilization can be activated transcriptionally through FAs-mediated activation of PPARα. As such, cardiac-specific overexpression of PPARα alone leads to cardiomyopathy similar to the one seen in diabetes and resulted in enhanced expression of genes involved in FAs oxidation and genes, such as pyruvate dehydrogenase kinase 4, involved in the inhibition of glucose oxidation.[33]

Initial studies performed in the 1980s by Kuo and colleagues[127, 128] showed that isolated mitochondria from obese diabetic (db/db) mice exhibited a defect in ADP-stimulated (state 3) respiration with both palmitoyl carnitine and pyruvate as substrates. Interestingly, the decrease in oxidative capacity of isolated mitochondria from hearts of diabetic mice was associated with enhanced accumulation of long-chain beta-hydroxy fatty acids and reduced NAD+: NADH ratios. Using permeabilized cardiac fibers, we showed that state 3 respiration and ATP synthesis rates were reduced both in T1D and T2D mouse models.[42, 129, 130] Not all mitochondria within the myocardium are affected equally by diabetes. A study that examined mitochondrial respiration in subsarcolemmal (SSM) and interfibrillar (IFM) showed that T1D reduced state 4 respiration and complex III activity in IFM but not in SSM.[131] Using $^{31}$P-magnetic resonance spectroscopy, several studies reported reduced PCr/ATP ratios in the hearts of T1D and T2D patients even when cardiac dysfunction is not visible.[132, 133] Moreover, Anderson and colleagues[134] showed decreased glutamate and FAs-supported mitochondrial respiration in the arterial tissue of type 2 diabetic individuals. The mechanisms underlying diabetes-induced reduction in oxidative capacity of heart mitochondria are not completely understood but might involve the increase in mitochondrial ROS, which, in turn, can affect ETC components. It is not clear if diabetes per se versus other metabolic abnormalities, such as obesity and hyperlipidemia, also can cause these mitochondrial defects. A recent study demonstrated that diabetes reduced mitochondrial oxidative capacity and enhanced ROS production in atrial tissue of humans, whereas obesity without diabetes had no effect.[112] In addition to quantitative impairment in mitochondrial oxidative capacity, diabetes also was shown to alter the mitochondrial proteome. Alterations in the expression of TCA cycle, FAs oxidation, and oxidative phosphorylation (OXPHOS) proteins were reported in the hearts of T1D mouse models.[100, 130, 135, 136]

In ECs, studies that examined the effect of diabetes on mitochondrial respiration are sparse, but extensive data is available to suggest that hyperglycemia can affect TCA function directly.[137] Thus, high glucose significantly reduced GAPDH activity and increased flux through the hexosamine pathway in bovine aortic ECs.[137] The inhibition of GAPDH is triggered by ROS and the consequent DNA damage that activates poly(ADP-ribose) polymerase and enhanced poly(ADP-ribosyl)ation of GAPDH.[138]

Assessment of mitochondrial function in hearts or vessels from diabetic subjects or animals is sometimes confounded by two factors: It is not clear if the defect in mitochondrial respiration is causal or simply a consequence of the disease, and systemic alterations such as hyperglycemia, hyperlipidemia, and inflammation might have an additional impact on mitochondria. It is necessary to study mitochondrial function early in disease progression to see if mitochondrial dysfunction can precipitate cardiac and vascular dysfunction. Such studies have been performed in mice that exhibited insulin resistance, a condition that precedes the development of diabetes. Deletion of IRs in cardiac cells caused cardiac insulin resistance and resulted in decreased TCA cycle activity and FAs oxidative capacity even when systemic metabolism is normal.[42] These early defects in mitochondrial metabolism were associated with a progressive decline in cardiac dysfunction, suggesting a causal relationship. In contrast to the heart, absent or impaired insulin receptor signaling in vessels had no effect on endothelial function, but palmitate alone impaired eNOS phosphorylation and NO production in cultured ECs.[139] The mechanisms by which palmitate caused EC dysfunction is believed to be mediated in part by ceramides, as inhibition of their synthesis restored eNOS phosphorylation and vascular function.[140] Enhanced FAs oxidation and ROS generation also were observed in ECs under an insulin-resistant state.[141] Together, these studies suggest that lack or reduced insulin action can have a direct impact on mitochondrial energetics in the heart but not in the vasculature.

## 5.4 Reduced Metabolic Efficiency and Enhanced Mitochondrial Uncoupling

More than 100 years ago, Starling and Evans[142] reported a decrease in the respiratory quotient (RQ) in perfused hearts from diabetic dogs, to which they attributed a defect in carbohydrate utilization. Although they did not find any difference in oxygen use between normal and diabetic hearts, the resulting work per oxygen consumed was not measured in this study. It took about 90 years to realize that cardiac work in diabetic hearts comes at a much higher oxygen cost (metabolic or cardiac inefficiency).[102, 129, 143–149] The majority of studies reporting an increase in myocardial oxygen consumption ($MVO_2$) and reduced both cardiac efficiency (CE) and ATP/O ratios were performed on hearts from T2D humans and animals. Studies using T1D subjects or animals failed to see these changes.[100, 105, 150] The reasons for preserved CE and ATP/O ratios in hearts of T1D models are not fully understood but might involve ROS generation and the production of lipid peroxides in the hearts of T2D that is absent in the hearts of T1D.[105, 129] ROS are known to activate mitochondrial uncoupling in cardiac mitochondria in the presence of FAs.[151, 152] Furthermore, unchanged ATP/O ratios in the hearts of T1D animals suggest that these mitochondria retain normal coupling in spite of depressed respiratory capacity, as shown by reduced state 3 respiration and respiratory control ratios.[100, 153, 154]

The mechanisms responsible for reduced CE and enhanced $MVO_2$ in the hearts of T2D humans and animals is not completely understood, however, it is believed to occur mainly through enhanced mitochondrial uncoupling and proton leak, and increased oxygen wasting associated with basal metabolism and excitation-contraction (EC) coupling.[155, 156] The heart expresses uncoupling proteins 2 and 3 (UCP2 and UCP3), which, similar to UCP1 in brown adipose tissue, can dissipate energy as heat at the expense of ATP production by channeling protons from the intermembrane space to the matrix of mitochondria.[157] The induction of UCP3 and, to a lesser extent, UCP2 mRNA and protein expression was observed in the hearts of T1D and T2D animals and is believed to be under the transcriptional control of PPARα.[105, 158–161] Supporting evidence for PPARα-induced expression of UCPs in the heart comes from studies using PPARα$^{-/-}$ hearts that were resistant to diabetes-induced UCP3 expression.[162] The increase in UCP3 expression occurs in response to an acute elevation of circulating free FAs and PPARα activation such as in the hearts of mice fed high-fat diet, in which genetic ablation of UCP3 restored CE and mitochondrial coupling.[163] Cardiac UCP3 protein expression, however, does not always correlate with reduced CE or mitochondrial uncoupling. The hearts of the T1D AKITA mice had more than a twofold increase in UCP3 protein content but no alteration in CE or ATP/O ratios.[105] This could be because of the absence of ROS generation in the hearts of AKITA mice as ROS or ROS products are known to activate UCPs. Similarly, genetic deletion of UCP3 in the hearts of leptin-deficient (ob/ob) mice failed to restore CE or mitochondrial uncoupling,[163] suggesting that other proteins, such as the adenine nucleotide translocator (ANT) or mitochondrial thioesterase 1 (MTE-1), might be involved in mediating mitochondrial uncoupling.[101, 147, 164] In addition to mitochondrial uncoupling, the diabetic heart uses more oxygen for nonmechanical processes, such as basal metabolism and calcium transport and recycling.[165, 166]

## 5.5 Enhanced Mitochondrial ROS Generation

One common defect caused by diabetes in both cardiac and endothelial cells is enhanced mitochondrial ROS generation. Although the source of ROS in heart and the vasculature is diverse, we will focus this section on the role of mitochondrial ROS (mitoROS) in the pathogenesis of diabetic cardiomyopathy and endothelial dysfunction. Early studies demonstrated an increase in indices of oxidative stress in the blood of patients with uncomplicated T1D.[167] Similar results also were reported for patients and animals with T2D.[168, 169] The first studies, showing an involvement of mitoROS in diabetes-mediated endothelial dysfunction, were pioneered by the Brownlee laboratory. They demonstrated that hyperglycemia increases ETC flux and enhanced superoxide generation in ECs, an effect reversed by overexpressing MnSOD or UCP1.[170] Similarly, work by the Epstein laboratory, revealed that mitochondria are the main source for ROS in cardiac cells of T1D and T2D mouse models because inhibitors of complex I, complex II, or MnSOD eliminated ROS generation in these cells.[171, 172] Assessing mitochondrial ROS in vivo has been challenging and most studies relied on the measurement of mitoROS using specific dyes. Our group showed that mitoROS, as assessed by measurement of hydrogen peroxide ($H_2O_2$) in isolated heart mitochondria, are decreased in T1D AKITA mice but increased in (db/db) and UCPDTA mice.[101, 105, 173] In spite of similar systemic metabolism in both AKITA, (db/db) and UCPDTA mice, only T2D mice develop severe cardiac insulin resistance, which could have been the cause of enhanced mitoROS. Mice with cardiomyocytes restricted insulin resistance (because of lack of IRs) have reduced cardiac mitochondrial aconitase activity (an indicator of mitoROS generation).[42] In human atrial tissue from T2D patients, $H_2O_2$ emission was elevated, glutathione was depleted and lipid peroxides were enhanced.[134, 174] From these studies and others, it was assumed that the use of a general antioxidant will be sufficient to reverse the cardiovascular complications. The use of antioxidants in clinical trials has been disappointing, however, because they failed to confer protection against CVD,[175–177] which might be caused in part by the fact that these general antioxidants are not targeted to specific sites of ROS generation such as mitochondria. This conclusion is supported by recent interventions that used mitochondria-targeted antioxidants to reverse or reduced diabetes-mediated cardiac and endothelial dysfunction. Therefore, the use of SOD mimetics with enhanced capacity to accumulate in mitochondria such as mito-TEMPO and MnTBAP improved cardiac electrical activity, reduced structural abnormalities, and preserved cardiac energetic capacity in both T1D and T2D mouse models.[173, 178, 179] These studies suggest that clinical trials that use antioxidant strategies targeted to mitochondria might have therapeutic benefit in diabetic complications.

## 5.6 Altered Mitochondrial Calcium Signaling

Numerous studies have demonstrated an altered calcium flux in the heart of diabetic animals, but few have examined the effect of diabetes on mitochondrial calcium transport in the heart. Early studies showed that mitochondrial calcium uptake is reduced in the hearts of both T1D and T2D animals.[180, 181] A direct evidence for the effect of diabetes on mitochondrial calcium handling came from in vitro studies showing that hyperglycemia caused a reduction in mitochondrial calcium uptake through repression of MCU protein expression.[182] Similarly, MCU content and mitochondrial calcium transport were reduced in the hearts of T1D mice, which was associated with altered mitochondrial bioenergetics and cardiac dysfunction.[183] These studies implicate MCU-mediated calcium transport in the mitochondria in the pathogenesis of diabetes-induced cardiac dysfunction. Restoration of MCU expression in cardiomyocytes exposed to high glucose or in diabetic hearts is sufficient to prevent diabetic-mediated alterations in mitochondrial energetics, antioxidant responses, and cardiac function.[182–184] Because mitochondrial calcium stimulates several dehydrogenases in the mitochondria,[185] it was assumed that lack of MUC in cardiac cells would lead to reduced oxidative phosphorylation and glucose oxidation. Mice expressing a dominant negative form of MCU specifically in cardiac cells, however, exhibited an increase in mitochondrial oxygen consumption rate (OCR). This increase in OCR was mediated by the elevation of cytosolic calcium as incubation of cells in low calcium abolished this increase. The increase in cytosolic calcium in mice expressing dominant negative form of MCU in cardiomyocytes is because of the lack of mitochondrial MCU.[70] Mitochondria can serve as a sink for calcium to prevent cytosolic calcium overload.[186] It will be of interest for future studies to examine how a lack of cardiac MCU contributes to the development of diabetic cardiomyopathy.

# 6 SUMMARY AND CONCLUSIONS

The thorough investigations clearly have positioned the mitochondria at the center for diabetes-mediated cardiovascular complications. These organelles not only provide ATP to maintain contraction and cellular respiration but is also a hub for cellular signaling. Given its role in the pathogenesis of diabetes complications, treatment strategies that target these organelles might be efficacious in preventing or limiting cardiac and endothelial dysfunction in diabetes. It is clear that early alterations in mitochondrial metabolism, quality control, calcium handling, and ROS generation might be causal for the development of diabetic cardiomyopathy and diabetes-induced endothelial dysfunction. The triggers of these mitochondrial alterations in the heart and the vasculature in diabetes are not fully understood but might include changes in systemic metabolism (hyperglycemia, hyperinsulinemia, and hyperlipidemia) as well as alterations in insulin sensitivity (insulin resistance). Although drugs and lifestyle interventions that target mitochondria in diabetes are just emerging, there is a great promise on using mitochondria-targeted antioxidants, exercise, and dietary components (polyphenols) to minimize mitochondrial stress and to ameliorate cardiac and endothelial function. Given that CVD is the principal cause of morbidity and mortality in patients with diabetes, effective treatments would be of tremendous benefit to this growing epidemic.

## References


1. Ingwall JS. *ATP and the Heart*. Norwell, MA: Kluwer Academic Publishers; 2002.
2. Opie LH. *From Cell to Circulation*. Philadelphia, PA: Lippincott Williams & Wilkins; 2004.
3. Schaper J, Meiser E, Stammler G. Ultrastructural morphometric analysis of myocardium from dogs, rats, hamsters, mice, and from human hearts. *Circ Res* 1985;**56**(3):377–91. Epub 1985/03/01.
4. Scarpulla RC, Vega RB, Kelly DP. Transcriptional integration of mitochondrial biogenesis. *Trends Endocrinol Metab* 2012;**23**(9):459–66. Epub 2012/07/24.
5. Pagliarini DJ, Calvo SE, Chang B, Sheth SA, Vafai SB, Ong SE, et al. A mitochondrial protein compendium elucidates complex I disease biology. *Cell* 2008;**134**(1):112–23. Epub 2008/07/11.
6. Lehman JJ, Barger PM, Kovacs A, Saffitz JE, Medeiros DM, Kelly DP. Peroxisome proliferator-activated receptor gamma coactivator-1 promotes cardiac mitochondrial biogenesis. *J Clin Invest* 2000;**106**(7):847–56. Epub 2000/10/06.
7. Lai L, Leone TC, Zechner C, Schaeffer PJ, Kelly SM, Flanagan DP, et al. Transcriptional coactivators PGC-1alpha and PGC-lbeta control overlapping programs required for perinatal maturation of the heart. *Genes Dev* 2008;**22**(14):1948–61. Epub 2008/07/17.
8. Lai L, Wang M, Martin OJ, Leone TC, Vega RB, Han X, et al. A role for peroxisome proliferator-activated receptor gamma coactivator 1 (PGC-1) in the regulation of cardiac mitochondrial phospholipid biosynthesis. *J Biol Chem* 2014;**289**(4):2250–9. Epub 2013/12/18.
9. Scarpulla RC. Nuclear control of respiratory chain expression by nuclear respiratory factors and PGC-1-related coactivator. *Ann N Y Acad Sci* 2008;**1147**:321–34. Epub 2008/12/17.
10. Satoh J, Kawana N, Yamamoto Y. Pathway analysis of ChIP-Seq-based NRF1 target genes suggests a logical hypothesis of their involvement in the pathogenesis of neurodegenerative diseases. *Gene Regul Syst Biol* 2013;**7**:139–52. Epub 2013/11/20.
11. Gleyzer N, Vercauteren K, Scarpulla RC. Control of mitochondrial transcription specificity factors (TFB1M and TFB2M) by nuclear respiratory factors (NRF-1 and NRF-2) and PGC-1 family coactivators. *Mol Cell Biol* 2005;**25**(4):1354–66. Epub 2005/02/03.
12. Falkenberg M, Larsson NG, Gustafsson CM. DNA replication and transcription in mammalian mitochondria. *Annu Rev Biochem* 2007;**76**:679–99. Epub 2007/04/06.
13. Li H, Wang J, Wilhelmsson H, Hansson A, Thoren P, Duffy J, et al. Genetic modification of survival in tissue-specific knockout mice with mitochondrial cardiomyopathy. *Proc Natl Acad Sci U S A* 2000;**97**(7):3467–72. Epub 2000/03/29.

14. Wang J, Wilhelmsson H, Graff C, Li H, Oldfors A, Rustin P, et al. Dilated cardiomyopathy and atrioventricular conduction blocks induced by heart-specific inactivation of mitochondrial DNA gene expression. *Nat Genet* 1999;**21**(1):133–7. Epub 1999/01/23.
15. Metodiev MD, Lesko N, Park CB, Camara Y, Shi Y, Wibom R, et al. Methylation of 12S rRNA is necessary for in vivo stability of the small subunit of the mammalian mitochondrial ribosome. *Cell Metab* 2009;**9**(4):386–97. Epub 2009/04/10.
16. Nisoli E, Clementi E, Paolucci C, Cozzi V, Tonello C, Sciorati C, et al. Mitochondrial biogenesis in mammals: the role of endogenous nitric oxide. *Science* 2003;**299**(5608):896–9. Epub 2003/02/08.
17. Chen Z, Peng IC, Cui X, Li YS, Chien S, Shyy JY. Shear stress, SIRT1, and vascular homeostasis. *Proc Natl Acad Sci U S A* 2010;**107**(22):10268–73. Epub 2010/05/19.
18. Dorn 2nd GW. Mitochondrial dynamics in heart disease. *Biochim Biophys Acta* 2013;**1833**(1):233–41. Epub 2012/03/28.
19. Chen Y, Liu Y, Dorn 2nd GW. Mitochondrial fusion is essential for organelle function and cardiac homeostasis. *Circ Res* 2011;**109**(12):1327–31. Epub 2011/11/05.
20. Ikeda Y, Shirakabe A, Maejima Y, Zhai P, Sciarretta S, Toli J, et al. Endogenous Drp1 mediates mitochondrial autophagy and protects the heart against energy stress. *Circ Res* 2015;**116**(2):264–78. Epub 2014/10/22.
21. Song M, Mihara K, Chen Y, Scorrano L, Dorn 2nd GW. Mitochondrial fission and fusion factors reciprocally orchestrate mitophagic culling in mouse hearts and cultured fibroblasts. *Cell Metab* 2015;**21**(2):273–86. Epub 2015/01/21.
22. Billia F, Hauck L, Konecny F, Rao V, Shen J, Mak TW. PTEN-inducible kinase 1 (PINK1)/Park6 is indispensable for normal heart function. *Proc Natl Acad Sci U S A* 2011;**108**(23):9572–7. Epub 2011/05/25.
23. Kubli DA, Zhang X, Lee Y, Hanna RA, Quinsay MN, Nguyen CK, et al. Parkin protein deficiency exacerbates cardiac injury and reduces survival following myocardial infarction. *J Biol Chem* 2013;**288**(2):915–26. Epub 2012/11/16.
24. Song M, Gong G, Burelle Y, Gustafsson AB, Kitsis RN, Matkovich SJ, et al. Interdependence of Parkin-mediated mitophagy and mitochondrial fission in adult mouse hearts. *Circ Res* 2015;**117**(4):346–51. Epub 2015/06/04.
25. Diwan A, Krenz M, Syed FM, Wansapura J, Ren X, Koesters AG, et al. Inhibition of ischemic cardiomyocyte apoptosis through targeted ablation of Bnip3 restrains postinfarction remodeling in mice. *J Clin Invest* 2007;**117**(10):2825–33. Epub 2007/10/03.
26. Chaanine AH, Gordon RE, Kohlbrenner E, Benard L, Jeong D, Hajjar RJ. Potential role of BNIP3 in cardiac remodeling, myocardial stiffness, and endoplasmic reticulum: mitochondrial calcium homeostasis in diastolic and systolic heart failure. *Circ Heart Fail* 2013;**6**(3):572–83. Epub 2013/03/20.
27. Lugus JJ, Ngoh GA, Bachschmid MM, Walsh K. Mitofusins are required for angiogenic function and modulate different signaling pathways in cultured endothelial cells. *J Mol Cell Cardiol* 2011;**51**(6):885–93. Epub 2011/08/16.
28. Miller MW, Knaub LA, Olivera-Fragoso LF, Keller AC, Balasubramaniam V, Watson PA, et al. Nitric oxide regulates vascular adaptive mitochondrial dynamics. *Am J Physiol Heart Circ Physiol* 2013;**304**(12):H1624–33. Epub 2013/04/16.
29. Bharath LP, Cho JM, Park SK, Ruan T, Li Y, Mueller R, et al. Endothelial cell autophagy maintains shear stress-induced nitric oxide generation via glycolysis-dependent purinergic signaling to endothelial nitric oxide synthase. *Arterioscler Thromb Vasc Biol* 2017;**37**(9):1646–56. Epub 2017/07/08.
30. Taegtmeyer H. Energy metabolism of the heart: from basic concepts to clinical applications. *Curr Probl Cardiol* 1994;**19**(2):59–113. Epub 1994/02/01.
31. Lee SS, Pineau T, Drago J, Lee EJ, Owens JW, Kroetz DL, et al. Targeted disruption of the alpha isoform of the peroxisome proliferator-activated receptor gene in mice results in abolishment of the pleiotropic effects of peroxisome proliferators. *Mol Cell Biol* 1995;**15**(6):3012–22. Epub 1995/06/01.
32. Finck BN, Han X, Courtois M, Aimond F, Nerbonne JM, Kovacs A, et al. A critical role for PPARalpha-mediated lipotoxicity in the pathogenesis of diabetic cardiomyopathy: modulation by dietary fat content. *Proc Natl Acad Sci U S A* 2003;**100**(3):1226–31. Epub 2003/01/29.
33. Finck BN, Lehman JJ, Leone TC, Welch MJ, Bennett MJ, Kovacs A, et al. The cardiac phenotype induced by PPARalpha overexpression mimics that caused by diabetes mellitus. *J Clin Invest* 2002;**109**(1):121–30. Epub 2002/01/10.
34. Garrido MC, Riveiro-Falkenbach E, Rodriguez-Peralto JL. Primary cutaneous follicle center lymphoma with follicular mucinosis. *JAMA Dermatol* 2014;**150**(8):906–7. Epub 2014/06/20.

35. Egert S, Nguyen N, Brosius 3rd FC, Schwaiger M. Effects of wortmannin on insulin- and ischemia-induced stimulation of GLUT4 translocation and FDG uptake in perfused rat hearts. *Cardiovasc Res* 1997;**35**(2):283–93. Epub 1997/08/01.
36. Kraegen EW, Sowden JA, Halstead MB, Clark PW, Rodnick KJ, Chisholm DJ, et al. Glucose transporters and in vivo glucose uptake in skeletal and cardiac muscle: fasting, insulin stimulation and immunoisolation studies of GLUT1 and GLUT4. *Biochem J* 1993;**295**(Pt 1):287–93. Epub 1993/10/01.
37. Laybutt DR, Thompson AL, Cooney GJ, Kraegen EW. Selective chronic regulation of GLUT1 and GLUT4 content by insulin, glucose, and lipid in rat cardiac muscle in vivo. *Am J Phys* 1997;**273**(3 Pt 2):H1309–16. Epub 1997/10/10.
38. Abel ED, Kaulbach HC, Tian R, Hopkins JC, Duffy J, Doetschman T, et al. Cardiac hypertrophy with preserved contractile function after selective deletion of GLUT4 from the heart. *J Clin Invest* 1999;**104**(12):1703–14. Epub 1999/12/22.
39. Pereira RO, Wende AR, Olsen C, Soto J, Rawlings T, Zhu Y, et al. GLUT1 deficiency in cardiomyocytes does not accelerate the transition from compensated hypertrophy to heart failure. *J Mol Cell Cardiol* 2014;**72**:95–103. Epub 2014/03/04.
40. Pereira RO, Wende AR, Olsen C, Soto J, Rawlings T, Zhu Y, et al. Inducible overexpression of GLUT1 prevents mitochondrial dysfunction and attenuates structural remodeling in pressure overload but does not prevent left ventricular dysfunction. *J Am Heart Assoc* 2013;**2**(5):e000301. Epub 2013/09/21.
41. Belke DD, Betuing S, Tuttle MJ, Graveleau C, Young ME, Pham M, et al. Insulin signaling coordinately regulates cardiac size, metabolism, and contractile protein isoform expression. *J Clin Invest* 2002;**109**(5):629–39. Epub 2002/03/06.
42. Boudina S, Bugger H, Sena S, O'Neill BT, Zaha VG, Ilkun O, et al. Contribution of impaired myocardial insulin signaling to mitochondrial dysfunction and oxidative stress in the heart. *Circulation* 2009;**119**(9):1272–83. Epub 2009/02/25.
43. Ingwall JS, Kramer MF, Fifer MA, Lorell BH, Shemin R, Grossman W, et al. The creatine kinase system in normal and diseased human myocardium. *N Engl J Med* 1985;**313**(17):1050–4. Epub 1985/10/24.
44. Neubauer S, Horn M, Naumann A, Tian R, Hu K, Laser M, et al. Impairment of energy metabolism in intact residual myocardium of rat hearts with chronic myocardial infarction. *J Clin Invest* 1995;**95**(3):1092–100. Epub 1995/03/01.
45. Lygate CA, Aksentijevic D, Dawson D, ten Hove M, Phillips D, de Bono JP, et al. Living without creatine: unchanged exercise capacity and response to chronic myocardial infarction in creatine-deficient mice. *Circ Res* 2013;**112**(6):945–55. Epub 2013/01/18.
46. Weiss RG, Gerstenblith G, Bottomley PA. ATP flux through creatine kinase in the normal, stressed, and failing human heart. *Proc Natl Acad Sci U S A* 2005;**102**(3):808–13. Epub 2005/01/14.
47. Bottomley PA, Panjrath GS, Lai S, Hirsch GA, Wu K, Najjar SS, et al. Metabolic rates of ATP transfer through creatine kinase (CK Flux) predict clinical heart failure events and death. *Sci Transl Med* 2013;**5**(215):215re3. Epub 2013/12/18.
48. St-Pierre J, Buckingham JA, Roebuck SJ, Brand MD. Topology of superoxide production from different sites in the mitochondrial electron transport chain. *J Biol Chem* 2002;**277**(47):44784–90. Epub 2002/09/19.
49. Brand MD. The sites and topology of mitochondrial superoxide production. *Exp Gerontol* 2010;**45**(7–8):466–72. Epub 2010/01/13.
50. Chouchani ET, Pell VR, Gaude E, Aksentijevic D, Sundier SY, Robb EL, et al. Ischaemic accumulation of succinate controls reperfusion injury through mitochondrial ROS. *Nature* 2014;**515**(7527):431–5. Epub 2014/11/11.
51. Carpi A, Menabo R, Kaludercic N, Pelicci P, Di Lisa F, Giorgio M. The cardioprotective effects elicited by p66(Shc) ablation demonstrate the crucial role of mitochondrial ROS formation in ischemia/reperfusion injury. *Biochim Biophys Acta* 2009;**1787**(7):774–80. Epub 2009/04/14.
52. Aon MA, Stanley BA, Sivakumaran V, Kembro JM, O'Rourke B, Paolocci N, et al. Glutathione/thioredoxin systems modulate mitochondrial H2O2 emission: an experimental-computational study. *J Gen Physiol* 2012;**139**(6):479–91. Epub 2012/05/16.
53. Nojiri H, Shimizu T, Funakoshi M, Yamaguchi O, Zhou H, Kawakami S, et al. Oxidative stress causes heart failure with impaired mitochondrial respiration. *J Biol Chem* 2006;**281**(44):33789–801. Epub 2006/09/09.
54. Murphy E, Ardehali H, Balaban RS, DiLisa F, Dorn 2nd GW, Kitsis RN, et al. Mitochondrial function, biology, and role in disease: a scientific statement from the American Heart Association. *Circ Res* 2016;**118**(12):1960–91. Epub 2016/04/30.

55. Berthiaume JM, Kurdys JG, Muntean DM, Rosca MG. Mitochondrial NAD(+)/NADH redox state and diabetic cardiomyopathy. *Antioxid Redox Signal* 2017. Epub 2017/10/28.
56. Hirschey MD, Shimazu T, Goetzman E, Jing E, Schwer B, Lombard DB, et al. SIRT3 regulates mitochondrial fatty-acid oxidation by reversible enzyme deacetylation. *Nature* 2010;**464**(7285):121–5. Epub 2010/03/06.
57. Hafner AV, Dai J, Gomes AP, Xiao CY, Palmeira CM, Rosenzweig A, et al. Regulation of the mPTP by SIRT3-mediated deacetylation of CypD at lysine 166 suppresses age-related cardiac hypertrophy. *Aging* 2010;**2**(12):914–23. Epub 2011/01/08.
58. Ahn BH, Kim HS, Song S, Lee IH, Liu J, Vassilopoulos A, et al. A role for the mitochondrial deacetylase Sirt3 in regulating energy homeostasis. *Proc Natl Acad Sci U S A* 2008;**105**(38):14447–52. Epub 2008/09/17.
59. Karamanlidis G, Lee CF, Garcia-Menendez L, Kolwicz Jr. SC, Suthammarak W, Gong G, et al. Mitochondrial complex I deficiency increases protein acetylation and accelerates heart failure. *Cell Metab* 2013;**18**(2):239–50. Epub 2013/08/13.
60. Sundaresan NR, Gupta M, Kim G, Rajamohan SB, Isbatan A, Gupta MP. Sirt3 blocks the cardiac hypertrophic response by augmenting Foxo3a-dependent antioxidant defense mechanisms in mice. *J Clin Invest* 2009;**119**(9):2758–71. Epub 2009/08/05.
61. Trifunovic A, Hansson A, Wredenberg A, Rovio AT, Dufour E, Khvorostov I, et al. Somatic mtDNA mutations cause aging phenotypes without affecting reactive oxygen species production. *Proc Natl Acad Sci U S A* 2005;**102**(50):17993–8. Epub 2005/12/08.
62. Dogan SA, Pujol C, Maiti P, Kukat A, Wang S, Hermans S, et al. Tissue-specific loss of DARS2 activates stress responses independently of respiratory chain deficiency in the heart. *Cell Metab* 2014;**19**(3):458–69. Epub 2014/03/13.
63. Hansson A, Hance N, Dufour E, Rantanen A, Hultenby K, Clayton DA, et al. A switch in metabolism precedes increased mitochondrial biogenesis in respiratory chain-deficient mouse hearts. *Proc Natl Acad Sci U S A* 2004;**101**(9):3136–41. Epub 2004/02/24.
64. Drago I, De Stefani D, Rizzuto R, Pozzan T. Mitochondrial $Ca^{2+}$ uptake contributes to buffering cytoplasmic $Ca^{2+}$ peaks in cardiomyocytes. *Proc Natl Acad Sci U S A* 2012;**109**(32):12986–91. Epub 2012/07/24.
65. Maack C, Cortassa S, Aon MA, Ganesan AN, Liu T, O'Rourke B. Elevated cytosolic $Na^+$ decreases mitochondrial $Ca^{2+}$ uptake during excitation-contraction coupling and impairs energetic adaptation in cardiac myocytes. *Circ Res* 2006;**99**(2):172–82. Epub 2006/06/17.
66. Baughman JM, Perocchi F, Girgis HS, Plovanich M, Belcher-Timme CA, Sancak Y, et al. Integrative genomics identifies MCU as an essential component of the mitochondrial calcium uniporter. *Nature* 2011;**476**(7360):341–5. Epub 2011/06/21.
67. De Stefani D, Raffaello A, Teardo E, Szabo I, Rizzuto R. A forty-kilodalton protein of the inner membrane is the mitochondrial calcium uniporter. *Nature* 2011;**476**(7360):336–40. Epub 2011/06/21.
68. Wu Y, Rasmussen TP, Koval OM, Joiner ML, Hall DD, Chen B, et al. The mitochondrial uniporter controls fight or flight heart rate increases. *Nat Commun* 2015;**6**:6081. Epub 2015/01/21.
69. Pan X, Liu J, Nguyen T, Liu C, Sun J, Teng Y, et al. The physiological role of mitochondrial calcium revealed by mice lacking the mitochondrial calcium uniporter. *Nat Cell Biol* 2013;**15**(12):1464–72. Epub 2013/11/12.
70. Rasmussen TP, Wu Y, Joiner ML, Koval OM, Wilson NR, Luczak ED, et al. Inhibition of MCU forces extramitochondrial adaptations governing physiological and pathological stress responses in heart. *Proc Natl Acad Sci U S A* 2015;**112**(29):9129–34. Epub 2015/07/15.
71. Kwong JQ, Lu X, Correll RN, Schwanekamp JA, Vagnozzi RJ, Sargent MA, et al. The mitochondrial calcium uniporter selectively matches metabolic output to acute contractile stress in the heart. *Cell Rep* 2015;**12**(1):15–22. Epub 2015/06/30.
72. Luongo TS, Lambert JP, Yuan A, Zhang X, Gross P, Song J, et al. The mitochondrial calcium uniporter matches energetic supply with cardiac workload during stress and modulates permeability transition. *Cell Rep* 2015;**12**(1):23–34. Epub 2015/06/30.
73. Baines CP, Kaiser RA, Purcell NH, Blair NS, Osinska H, Hambleton MA, et al. Loss of cyclophilin D reveals a critical role for mitochondrial permeability transition in cell death. *Nature* 2005;**434**(7033):658–62. Epub 2005/04/01.
74. Kung G, Konstantinidis K, Kitsis RN. Programmed necrosis, not apoptosis, in the heart. *Circ Res* 2011;**108**(8):1017–36. Epub 2011/04/16.
75. Nakagawa T, Shimizu S, Watanabe T, Yamaguchi O, Otsu K, Yamagata H, et al. Cyclophilin D-dependent mitochondrial permeability transition regulates some necrotic but not apoptotic cell death. *Nature* 2005;**434**(7033):652–8. Epub 2005/04/01.

76. Hochhauser E, Kivity S, Offen D, Maulik N, Otani H, Barhum Y, et al. Bax ablation protects against myocardial ischemia-reperfusion injury in transgenic mice. *Am J Physiol Heart Circ Physiol* 2003;**284**(6):H2351–9. Epub 2003/05/14.
77. Brocheriou V, Hagege AA, Oubenaissa A, Lambert M, Mallet VO, Duriez M, et al. Cardiac functional improvement by a human Bcl-2 transgene in a mouse model of ischemia/reperfusion injury. *J Gene Med* 2000;**2**(5):326–33. Epub 2000/10/25.
78. Lee P, Sata M, Lefer DJ, Factor SM, Walsh K, Kitsis RN. Fas pathway is a critical mediator of cardiac myocyte death and MI during ischemia-reperfusion in vivo. *Am J Physiol Heart Circ Physiol* 2003;**284**(2):H456–63. Epub 2002/11/05.
79. Liu Y, Levine B. Autosis and autophagic cell death: the dark side of autophagy. *Cell Death Differ* 2015;**22**(3):367–76. Epub 2014/09/27.
80. Xie Y, Hou W, Song X, Yu Y, Huang J, Sun X, et al. Ferroptosis: process and function. *Cell Death Differ* 2016;**23**(3):369–79. Epub 2016/01/23.
81. Kassab GS, Lin DH, Fung YC. Morphometry of pig coronary venous system. *Am J Phys* 1994;**267**(6 Pt 2):H2100–13. Epub 1994/12/01.
82. Li JM, Shah AM. Endothelial cell superoxide generation: regulation and relevance for cardiovascular pathophysiology. *Am J Physiol Regul Integr Comp Physiol* 2004;**287**(5):R1014–30. Epub 2004/10/12.
83. Oldendorf WH, Cornford ME, Brown WJ. The large apparent work capability of the blood-brain barrier: a study of the mitochondrial content of capillary endothelial cells in brain and other tissues of the rat. *Ann Neurol* 1977;**1**(5):409–17. Epub 1977/05/01.
84. Culic O, Gruwel ML, Schrader J. Energy turnover of vascular endothelial cells. *Am J Phys* 1997;**273**(1 Pt 1):C205–13. Epub 1997/07/01.
85. Olson AL, Pessin JE. Structure, function, and regulation of the mammalian facilitative glucose transporter gene family. *Annu Rev Nutr* 1996;**16**:235–56. Epub 1996/01/01.
86. Park SY, Gifford JR, Andtbacka RH, Trinity JD, Hyngstrom JR, Garten RS, et al. Cardiac, skeletal, and smooth muscle mitochondrial respiration: are all mitochondria created equal? *Am J Physiol Heart Circ Physiol* 2014;**307**(3):H346–52. Epub 2014/06/08.
87. Park SY, Rossman MJ, Gifford JR, Bharath LP, Bauersachs J, Richardson RS, et al. Exercise training improves vascular mitochondrial function. *Am J Physiol Heart Circ Physiol* 2016;**310**(7):H821–9. Epub 2016/01/31.
88. Zhang DX, Gutterman DD. Mitochondrial reactive oxygen species-mediated signaling in endothelial cells. *Am J Physiol Heart Circ Physiol* 2007;**292**(5):H2023–31. Epub 2007/01/24.
89. Liu Y, Zhao H, Li H, Kalyanaraman B, Nicolosi AC, Gutterman DD. Mitochondrial sources of $H_2O_2$ generation play a key role in flow-mediated dilation in human coronary resistance arteries. *Circ Res* 2003;**93**(6):573–80. Epub 2003/08/16.
90. Richardson RS, Donato AJ, Uberoi A, Wray DW, Lawrenson L, Nishiyama S, et al. Exercise-induced brachial artery vasodilation: role of free radicals. *Am J Physiol Heart Circ Physiol* 2007;**292**(3):H1516–22. Epub 2006/11/23.
91. Yang Y, Loscalzo J. S-nitrosoprotein formation and localization in endothelial cells. *Proc Natl Acad Sci U S A* 2005;**102**(1):117–22. Epub 2004/12/25.
92. King H, Aubert RE, Herman WH. Global burden of diabetes, 1995–2025: prevalence, numerical estimates, and projections. *Diabetes Care* 1998;**21**(9):1414–31. Epub 1998/09/04.
93. Garcia MJ, McNamara PM, Gordon T, Kannel WB. Morbidity and mortality in diabetics in the Framingham population. Sixteen year follow-up study. *Diabetes* 1974;**23**(2):105–11. Epub 1974/02/01.
94. Iltis I, Kober F, Dalmasso C, Cozzone PJ, Bernard M. Noninvasive characterization of myocardial blood flow in diabetic, hypertensive, and diabetic-hypertensive rats using spin-labeling MRI. *Microcirculation* 2005;**12**(8):607–14. Epub 2005/11/15.
95. Fein FS. Diabetic cardiomyopathy. *Diabetes Care* 1990;**13**(11):1169–79. Epub 1990/11/01.
96. Rubler S, Dlugash J, Yuceoglu YZ, Kumral T, Branwood AW, Grishman A. New type of cardiomyopathy associated with diabetic glomerulosclerosis. *Am J Cardiol* 1972;**30**(6):595–602. Epub 1972/11/08.
97. Kannel WB, McGee DL. Diabetes and cardiovascular disease. The Framingham study. *JAMA* 1979;**241**(19):2035–8. Epub 1979/05/11.
98. Nichols GA, Hillier TA, Erbey JR, Brown JB. Congestive heart failure in type 2 diabetes: prevalence, incidence, and risk factors. *Diabetes Care* 2001;**24**(9):1614–9. Epub 2001/08/28.
99. Struthers AD, Morris AD. Screening for and treating left-ventricular abnormalities in diabetes mellitus: a new way of reducing cardiac deaths. *Lancet* 2002;**359**(9315):1430–2. Epub 2002/04/30.

100. Shen X, Zheng S, Thongboonkerd V, Xu M, Pierce Jr. WM, Klein JB, et al. Cardiac mitochondrial damage and biogenesis in a chronic model of type 1 diabetes. *Am J Physiol Endocrinol Metab* 2004;**287**(5):E896–905. Epub 2004/07/29.
101. Boudina S, Sena S, Theobald H, Sheng X, Wright JJ, Hu XX, et al. Mitochondrial energetics in the heart in obesity-related diabetes: direct evidence for increased uncoupled respiration and activation of uncoupling proteins. *Diabetes* 2007;**56**(10):2457–66. Epub 2007/07/12.
102. Duncan JG, Fong JL, Medeiros DM, Finck BN, Kelly DP. Insulin-resistant heart exhibits a mitochondrial biogenic response driven by the peroxisome proliferator-activated receptor-alpha/PGC-1alpha gene regulatory pathway. *Circulation* 2007;**115**(7):909–17. Epub 2007/01/31.
103. Finck BN, Bernal-Mizrachi C, Han DH, Coleman T, Sambandam N, LL LR, et al. A potential link between muscle peroxisome proliferator- activated receptor-alpha signaling and obesity-related diabetes. *Cell Metab* 2005;**1**(2):133–44. Epub 2005/08/02.
104. Chang LT, Sun CK, Wang CY, Youssef AA, Wu CJ, Chua S, et al. Downregulation of peroxisme proliferator activated receptor gamma co-activator 1alpha in diabetic rats. *Int Heart J* 2006;**47**(6):901–10. Epub 2007/02/03.
105. Bugger H, Boudina S, Hu XX, Tuinei J, Zaha VG, Theobald HA, et al. Type 1 diabetic akita mouse hearts are insulin sensitive but manifest structurally abnormal mitochondria that remain coupled despite increased uncoupling protein 3. *Diabetes* 2008;**57**(11):2924–32. Epub 2008/08/06.
106. Csiszar A, Labinskyy N, Pinto JT, Ballabh P, Zhang H, Losonczy G, et al. Resveratrol induces mitochondrial biogenesis in endothelial cells. *Am J Physiol Heart Circ Physiol* 2009;**297**(1):H13–20. Epub 2009/05/12.
107. Kukidome D, Nishikawa T, Sonoda K, Imoto K, Fujisawa K, Yano M, et al. Activation of AMP-activated protein kinase reduces hyperglycemia-induced mitochondrial reactive oxygen species production and promotes mitochondrial biogenesis in human umbilical vein endothelial cells. *Diabetes* 2006;**55**(1):120–7. Epub 2005/12/29.
108. Yu T, Robotham JL, Yoon Y. Increased production of reactive oxygen species in hyperglycemic conditions requires dynamic change of mitochondrial morphology. *Proc Natl Acad Sci U S A* 2006;**103**(8):2653–8. Epub 2006/02/16.
109. Watanabe T, Saotome M, Nobuhara M, Sakamoto A, Urushida T, Katoh H, et al. Roles of mitochondrial fragmentation and reactive oxygen species in mitochondrial dysfunction and myocardial insulin resistance. *Exp Cell Res* 2014;**323**(2):314–25. Epub 2014/03/19.
110. Makino A, Suarez J, Gawlowski T, Han W, Wang H, Scott BT, et al. Regulation of mitochondrial morphology and function by O-GlcNAcylation in neonatal cardiac myocytes. *Am J Physiol Regul Integr Comp Physiol* 2011;**300**(6):R1296–302. Epub 2011/02/25.
111. Makino A, Scott BT, Dillmann WH. Mitochondrial fragmentation and superoxide anion production in coronary endothelial cells from a mouse model of type 1 diabetes. *Diabetologia* 2010;**53**(8):1783–94. Epub 2010/05/13.
112. Montaigne D, Marechal X, Coisne A, Debry N, Modine T, Fayad G, et al. Myocardial contractile dysfunction is associated with impaired mitochondrial function and dynamics in type 2 diabetic but not in obese patients. *Circulation* 2014;**130**(7):554–64. Epub 2014/06/15.
113. Xu X, Kobayashi S, Chen K, Timm D, Volden P, Huang Y, et al. Diminished autophagy limits cardiac injury in mouse models of type 1 diabetes. *J Biol Chem* 2013;**288**(25):18077–92. Epub 2013/05/10.
114. Kanamori H, Takemura G, Goto K, Tsujimoto A, Mikami A, Ogino A, et al. Autophagic adaptations in diabetic cardiomyopathy differ between type 1 and type 2 diabetes. *Autophagy* 2015;**11**(7):1146–60. Epub 2015/06/05.
115. Kobayashi S, Xu X, Chen K, Liang Q. Suppression of autophagy is protective in high glucose-induced cardiomyocyte injury. *Autophagy* 2012;**8**(4):577–92. Epub 2012/04/14.
116. Sciarretta S, Zhai P, Shao D, Maejima Y, Robbins J, Volpe M, et al. Rheb is a critical regulator of autophagy during myocardial ischemia: pathophysiological implications in obesity and metabolic syndrome. *Circulation* 2012;**125**(9):1134–46. Epub 2012/02/02.
117. Jaishy B, Zhang Q, Chung HS, Riehle C, Soto J, Jenkins S, et al. Lipid-induced NOX2 activation inhibits autophagic flux by impairing lysosomal enzyme activity. *J Lipid Res* 2015;**56**(3):546–61. Epub 2014/12/23.
118. Hsu HC, Chen CY, Lee BC, Chen MF. High-fat diet induces cardiomyocyte apoptosis via the inhibition of autophagy. *Eur J Nutr* 2016;**55**(7):2245–54. Epub 2015/09/12.
119. Munasinghe PE, Riu F, Dixit P, Edamatsu M, Saxena P, Hamer NS, et al. Type-2 diabetes increases autophagy in the human heart through promotion of Beclin-1 mediated pathway. *Int J Cardiol* 2016;**202**:13–20. Epub 2015/09/20.
120. Mellor KM, Bell JR, Young MJ, Ritchie RH, Delbridge LM. Myocardial autophagy activation and suppressed survival signaling is associated with insulin resistance in fructose-fed mice. *J Mol Cell Cardiol* 2011;**50**(6):1035–43. Epub 2011/03/10.

121. Pires KM, Buffolo M, Schaaf C, David Symons J, Cox J, Abel ED, et al. Activation of IGF-1 receptors and Akt signaling by systemic hyperinsulinemia contributes to cardiac hypertrophy but does not regulate cardiac autophagy in obese diabetic mice. *J Mol Cell Cardiol* 2017;**113**:39–50. Epub 2017/10/11.
122. Xie Z, Lau K, Eby B, Lozano P, He C, Pennington B, et al. Improvement of cardiac functions by chronic metformin treatment is associated with enhanced cardiac autophagy in diabetic OVE26 mice. *Diabetes* 2011;**60**(6):1770–8. Epub 2011/05/13.
123. Kobayashi S, Liang Q. Autophagy and mitophagy in diabetic cardiomyopathy. *Biochim Biophys Acta* 2015;**1852**(2):252–61. Epub 2014/06/03.
124. Lopaschuk GD, Ussher JR, Folmes CD, Jaswal JS, Stanley WC. Myocardial fatty acid metabolism in health and disease. *Physiol Rev* 2010;**90**(1):207–58. Epub 2010/01/21.
125. Lopaschuk GD. Abnormal mechanical function in diabetes: relationship to altered myocardial carbohydrate/lipid metabolism. *Coron Artery Dis* 1996;**7**(2):116–23. Epub 1996/02/01.
126. Stanley WC, Lopaschuk GD, McCormack JG. Regulation of energy substrate metabolism in the diabetic heart. *Cardiovasc Res* 1997;**34**(1):25–33. Epub 1997/04/01.
127. Kuo TH, Moore KH, Giacomelli F, Wiener J. Defective oxidative metabolism of heart mitochondria from genetically diabetic mice. *Diabetes* 1983;**32**(9):781–7. Epub 1983/09/01.
128. Kuo TH, Giacomelli F, Wiener J. Oxidative metabolism of Polytron versus Nagarse mitochondria in hearts of genetically diabetic mice. *Biochim Biophys Acta* 1985;**806**(1):9–15. Epub 1985/01/23.
129. Boudina S, Sena S, O'Neill BT, Tathireddy P, Young ME, Abel ED. Reduced mitochondrial oxidative capacity and increased mitochondrial uncoupling impair myocardial energetics in obesity. *Circulation* 2005;**112**(17):2686–95. Epub 2005/10/26.
130. Bugger H, Chen D, Riehle C, Soto J, Theobald HA, Hu XX, et al. Tissue-specific remodeling of the mitochondrial proteome in type 1 diabetic akita mice. *Diabetes* 2009;**58**(9):1986–97. Epub 2009/06/23.
131. Dabkowski ER, Williamson CL, Bukowski VC, Chapman RS, Leonard SS, Peer CJ, et al. Diabetic cardiomyopathy-associated dysfunction in spatially distinct mitochondrial subpopulations. *Am J Physiol Heart Circ Physiol* 2009;**296**(2):H359–69. Epub 2008/12/09.
132. Metzler B, Schocke MF, Steinboeck P, Wolf C, Judmaier W, Lechleitner M, et al. Decreased high-energy phosphate ratios in the myocardium of men with diabetes mellitus type I. *J Cardiovasc Magn Reson: Off J Soc Cardiovasc Magn Reson* 2002;**4**(4):493–502. Epub 2003/01/29.
133. Perseghin G, Ntali G, De Cobelli F, Lattuada G, Esposito A, Belloni E, et al. Abnormal left ventricular energy metabolism in obese men with preserved systolic and diastolic functions is associated with insulin resistance. *Diabetes Care* 2007;**30**(6):1520–6. Epub 2007/03/27.
134. Anderson EJ, Kypson AP, Rodriguez E, Anderson CA, Lehr EJ, Neufer PD. Substrate-specific derangements in mitochondrial metabolism and redox balance in the atrium of the type 2 diabetic human heart. *J Am Coll Cardiol* 2009;**54**(20):1891–8. Epub 2009/11/07.
135. Baseler WA, Dabkowski ER, Williamson CL, Croston TL, Thapa D, Powell MJ, et al. Proteomic alterations of distinct mitochondrial subpopulations in the type 1 diabetic heart: contribution of protein import dysfunction. *Am J Physiol Regul Integr Comp Physiol* 2011;**300**(2):R186–200. Epub 2010/11/05.
136. Turko IV, Murad F. Quantitative protein profiling in heart mitochondria from diabetic rats. *J Biol Chem* 2003;**278**(37):35844–9. Epub 2003/07/10.
137. Giacco F, Brownlee M. Oxidative stress and diabetic complications. *Circ Res* 2010;**107**(9):1058–70. Epub 2010/10/30.
138. Du X, Matsumura T, Edelstein D, Rossetti L, Zsengeller Z, Szabo C, et al. Inhibition of GAPDH activity by poly(ADP-ribose) polymerase activates three major pathways of hyperglycemic damage in endothelial cells. *J Clin Invest* 2003;**112**(7):1049–57. Epub 2003/10/03.
139. Symons JD, SL MM, Riehle C, Tanner J, Palionyte M, Hillas E, et al. Contribution of insulin and Akt1 signaling to endothelial nitric oxide synthase in the regulation of endothelial function and blood pressure. *Circ Res* 2009;**104**(9):1085–94. Epub 2009/04/04.
140. Zhang QJ, Holland WL, Wilson L, Tanner JM, Kearns D, Cahoon JM, et al. Ceramide mediates vascular dysfunction in diet-induced obesity by PP2A-mediated dephosphorylation of the eNOS-Akt complex. *Diabetes* 2012;**61**(7):1848–59. Epub 2012/05/16.
141. Du X, Edelstein D, Obici S, Higham N, Zou MH, Brownlee M. Insulin resistance reduces arterial prostacyclin synthase and eNOS activities by increasing endothelial fatty acid oxidation. *J Clin Invest* 2006;**116**(4):1071–80. Epub 2006/03/11.

142. Starling EH, Evans CL. The respiratory exchanges of the heart in the diabetic animal. *J Physiol* 1914;**49**(1-2): 67–88. Epub 1914/12/22.
143. Scheuermann-Freestone M, Madsen PL, Manners D, Blamire AM, Buckingham RE, Styles P, et al. Abnormal cardiac and skeletal muscle energy metabolism in patients with type 2 diabetes. *Circulation* 2003;**107**(24):3040–6. Epub 2003/06/18.
144. Mazumder PK, O'Neill BT, Roberts MW, Buchanan J, Yun UJ, Cooksey RC, et al. Impaired cardiac efficiency and increased fatty acid oxidation in insulin-resistant ob/ob mouse hearts. *Diabetes* 2004;**53**(9):2366–74. Epub 2004/08/28.
145. Peterson LR, Herrero P, Schechtman KB, Racette SB, Waggoner AD, Kisrieva-Ware Z, et al. Effect of obesity and insulin resistance on myocardial substrate metabolism and efficiency in young women. *Circulation* 2004;**109**(18):2191–6. Epub 2004/05/05.
146. How OJ, Aasum E, Severson DL, Chan WY, Essop MF, Larsen TS. Increased myocardial oxygen consumption reduces cardiac efficiency in diabetic mice. *Diabetes* 2006;**55**(2):466–73. Epub 2006/01/31.
147. Cole MA, Murray AJ, Cochlin LE, Heather LC, McAleese S, Knight NS, et al. A high fat diet increases mitochondrial fatty acid oxidation and uncoupling to decrease efficiency in rat heart. *Basic Res Cardiol* 2011;**106**(3):447–57. Epub 2011/02/15.
148. Buchanan J, Mazumder PK, Hu P, Chakrabarti G, Roberts MW, Yun UJ, et al. Reduced cardiac efficiency and altered substrate metabolism precedes the onset of hyperglycemia and contractile dysfunction in two mouse models of insulin resistance and obesity. *Endocrinology* 2005;**146**(12):5341–9. Epub 2005/09/06.
149. Diamant M, Lamb HJ, Groeneveld Y, Endert EL, Smit JW, Bax JJ, et al. Diastolic dysfunction is associated with altered myocardial metabolism in asymptomatic normotensive patients with well-controlled type 2 diabetes mellitus. *J Am Coll Cardiol* 2003;**42**(2):328–35. Epub 2003/07/24.
150. Herrero P, Peterson LR, JB MG, Matthew S, Lesniak D, Dence C, et al. Increased myocardial fatty acid metabolism in patients with type 1 diabetes mellitus. *J Am Coll Cardiol* 2006;**47**(3):598–604. Epub 2006/02/07.
151. Echtay KS, Roussel D, St-Pierre J, Jekabsons MB, Cadenas S, Stuart JA, et al. Superoxide activates mitochondrial uncoupling proteins. *Nature* 2002;**415**(6867):96–9. Epub 2002/01/10.
152. Echtay KS, Esteves TC, Pakay JL, Jekabsons MB, Lambert AJ, Portero-Otin M, et al. A signalling role for 4-hydroxy-2-nonenal in regulation of mitochondrial uncoupling. *EMBO J* 2003;**22**(16):4103–10. Epub 2003/08/13.
153. Tomita M, Mukae S, Geshi E, Umetsu K, Nakatani M, Katagiri T. Mitochondrial respiratory impairment in streptozotocin-induced diabetic rat heart. *Jpn Circ J* 1996;**60**(9):673–82. Epub 1996/09/01.
154. Pierce GN, Dhalla NS. Heart mitochondrial function in chronic experimental diabetes in rats. *Can J Cardiol* 1985;**1**(1):48–54. Epub 1985/01/01.
155. Hafstad AD, Boardman N, Aasum E. How exercise may amend metabolic disturbances in diabetic cardiomyopathy. *Antioxid Redox Signal* 2015;**22**(17):1587–605. Epub 2015/03/05.
156. Sung MM, Hamza SM, Dyck JR. Myocardial metabolism in diabetic cardiomyopathy: potential therapeutic targets. *Antioxid Redox Signal* 2015;**22**(17):1606–30. Epub 2015/03/27.
157. Murray AJ, Anderson RE, Watson GC, Radda GK, Clarke K. Uncoupling proteins in human heart. *Lancet* 2004;**364**(9447):1786–8. Epub 2004/11/16.
158. Hidaka S, Kakuma T, Yoshimatsu H, Sakino H, Fukuchi S, Sakata T. Streptozotocin treatment upregulates uncoupling protein 3 expression in the rat heart. *Diabetes* 1999;**48**(2):430–5. Epub 1999/05/20.
159 Van Der Lee KA, Willemsen PH, Van Der Vusse GJ, Van Bilsen M. Effects of fatty acids on uncoupling protein-2 expression in the rat heart. FASEB J: Off Publ Fed Am Soc Exp Biol 2000;14(3):495-502. Epub 2000/03/04.
160. Young ME, Patil S, Ying J, Depre C, Ahuja HS, Shipley GL, et al. Uncoupling protein 3 transcription is regulated by peroxisome proliferator-activated receptor (alpha) in the adult rodent heart. *FASEB J: Off Publ Fed Am Soc Exp Biol* 2001;**15**(3):833–45. Epub 2001/03/22.
161. Somoza B, Guzman R, Cano V, Merino B, Ramos P, Diez-Fernandez C, et al. Induction of cardiac uncoupling protein-2 expression and adenosine 5′-monophosphate-activated protein kinase phosphorylation during early states of diet-induced obesity in mice. *Endocrinology* 2007;**148**(3):924–31. Epub 2006/11/04.
162. Murray AJ, Panagia M, Hauton D, Gibbons GF, Clarke K. Plasma free fatty acids and peroxisome proliferator-activated receptor alpha in the control of myocardial uncoupling protein levels. *Diabetes* 2005;**54**(12):3496–502. Epub 2005/11/25.
163. Boudina S, Han YH, Pei S, Tidwell TJ, Henrie B, Tuinei J, et al. UCP3 regulates cardiac efficiency and mitochondrial coupling in high fat-fed mice but not in leptin-deficient mice. *Diabetes* 2012;**61**(12):3260–9. Epub 2012/08/23.

164. King KL, Young ME, Kerner J, Huang H, O'Shea KM, Alexson SE, et al. Diabetes or peroxisome proliferator-activated receptor alpha agonist increases mitochondrial thioesterase I activity in heart. *J Lipid Res* 2007;**48**(7):1511–7. Epub 2007/04/18.
165. Boardman N, Hafstad AD, Larsen TS, Severson DL, Aasum E. Increased O2 cost of basal metabolism and excitation-contraction coupling in hearts from type 2 diabetic mice. *Am J Physiol Heart Circ Physiol* 2009;**296**(5):H1373–9. Epub 2009/03/17.
166. Boardman NT, Larsen TS, Severson DL, Essop MF, Aasum E. Chronic and acute exposure of mouse hearts to fatty acids increases oxygen cost of excitation-contraction coupling. *Am J Physiol Heart Circ Physiol* 2011;**300**(5):H1631–6. Epub 2011/02/22.
167. Marra G, Cotroneo P, Pitocco D, Manto A, Di Leo MA, Ruotolo V, et al. Early increase of oxidative stress and reduced antioxidant defenses in patients with uncomplicated type 1 diabetes: a case for gender difference. *Diabetes Care* 2002;**25**(2):370–5. Epub 2002/01/30.
168. Savu O, Ionescu-Tirgoviste C, Atanasiu V, Gaman L, Papacocea R, Stoian I. Increase in total antioxidant capacity of plasma despite high levels of oxidative stress in uncomplicated type 2 diabetes mellitus. *J Int Med Res* 2012;**40**(2):709–16. Epub 2012/05/23.
169. Furukawa S, Fujita T, Shimabukuro M, Iwaki M, Yamada Y, Nakajima Y, et al. Increased oxidative stress in obesity and its impact on metabolic syndrome. *J Clin Invest* 2004;**114**(12):1752–61. Epub 2004/12/16.
170. Nishikawa T, Edelstein D, Du XL, Yamagishi S, Matsumura T, Kaneda Y, et al. Normalizing mitochondrial superoxide production blocks three pathways of hyperglycaemic damage. *Nature* 2000;**404**(6779):787–90. Epub 2000/04/28.
171. Ye G, Metreveli NS, Donthi RV, Xia S, Xu M, Carlson EC, et al. Catalase protects cardiomyocyte function in models of type 1 and type 2 diabetes. *Diabetes* 2004;**53**(5):1336–43. Epub 2004/04/28.
172. Shen X, Zheng S, Metreveli NS, Epstein PN. Protection of cardiac mitochondria by overexpression of MnSOD reduces diabetic cardiomyopathy. *Diabetes* 2006;**55**(3):798–805. Epub 2006/03/01.
173. Ilkun O, Wilde N, Tuinei J, Pires KM, Zhu Y, Bugger H, et al. Antioxidant treatment normalizes mitochondrial energetics and myocardial insulin sensitivity independently of changes in systemic metabolic homeostasis in a mouse model of the metabolic syndrome. *J Mol Cell Cardiol* 2015;**85**:104–16. Epub 2015/05/26.
174. Katunga LA, Gudimella P, Efird JT, Abernathy S, Mattox TA, Beatty C, et al. Obesity in a model of gpx4 haploinsufficiency uncovers a causal role for lipid-derived aldehydes in human metabolic disease and cardiomyopathy. *Mol Metab* 2015;**4**(6):493–506. Epub 2015/06/05.
175. Sesso HD, Buring JE, Christen WG, Kurth T, Belanger C, MacFadyen J, et al. Vitamins E and C in the prevention of cardiovascular disease in men: the Physicians' Health Study II randomized controlled trial. *JAMA* 2008;**300**(18):2123–33. Epub 2008/11/11.
176. Ye Y, Li J, Yuan Z. Effect of antioxidant vitamin supplementation on cardiovascular outcomes: a meta-analysis of randomized controlled trials. *PLoS ONE* 2013;**8**(2):e56803. Epub 2013/02/26.
177. Stepaniak U, Micek A, Grosso G, Stefler D, Topor-Madry R, Kubinova R, et al. Antioxidant vitamin intake and mortality in three Central and Eastern European urban populations: the HAPIEE study. *Eur J Nutr* 2016;**55**(2):547–60. Epub 2015/03/13.
178. Luo M, Guan X, Luczak ED, Lang D, Kutschke W, Gao Z, et al. Diabetes increases mortality after myocardial infarction by oxidizing CaMKII. *J Clin Invest* 2013;**123**(3):1262–74. Epub 2013/02/22.
179. Ni R, Cao T, Xiong S, Ma J, Fan GC, Lacefield JC, et al. Therapeutic inhibition of mitochondrial reactive oxygen species with mito-TEMPO reduces diabetic cardiomyopathy. *Free Radic Biol Med* 2016;**90**:12–23. Epub 2015/11/19.
180. Flarsheim CE, Grupp IL, Matlib MA. Mitochondrial dysfunction accompanies diastolic dysfunction in diabetic rat heart. *Am J Phys* 1996;**271**(1 Pt 2):H192–202. Epub 1996/07/01.
181. Fauconnier J, Lanner JT, Zhang SJ, Tavi P, Bruton JD, Katz A, et al. Insulin and inositol 1,4,5-trisphosphate trigger abnormal cytosolic $Ca^{2+}$ transients and reveal mitochondrial $Ca^{2+}$ handling defects in cardiomyocytes of ob/ob mice. *Diabetes* 2005;**54**(8):2375–81. Epub 2005/07/28.
182. Diaz-Juarez J, Suarez J, Cividini F, Scott BT, Diemer T, Dai A, et al. Expression of the mitochondrial calcium uniporter in cardiac myocytes improves impaired mitochondrial calcium handling and metabolism in simulated hyperglycemia. *Am J Physiol Cell Physiol* 2016;**311**(6):C1005–13. Epub 2016/09/30.
183. Suarez J, Cividini F, Scott BT, Lehmann K, Diaz-Juarez J, Diemer T, et al. Restoring mitochondrial calcium uniporter expression in diabetic mouse heart improves mitochondrial calcium handling and cardiac function. *J Biol Chem* 2018. Epub 2018/04/08.

184. Ji L, Liu F, Jing Z, Huang Q, Zhao Y, Cao H, et al. MICU1 alleviates diabetic cardiomyopathy through mitochondrial Ca(2+)-dependent antioxidant response. *Diabetes* 2017;**66**(6):1586–600. Epub 2017/03/16.
185. Denton RM. Regulation of mitochondrial dehydrogenases by calcium ions. *Biochim Biophys Acta* 2009;**1787**(11):1309–16. Epub 2009/05/06.
186. Dhalla NS, Rangi S, Zieroth S, Xu YJ. Alterations in sarcoplasmic reticulum and mitochondrial functions in diabetic cardiomyopathy. *Exp Clin Cardiol* 2012;**17**(3):115–20. Epub 2013/04/27.

CHAPTER

# 12

# Role of Mitochondria in the Regulation of Kidney Function and Metabolism in Type 2 Diabetes

*Xianlin Han*[*,†,‡], *Yuguang Shi*[*,§], *Maggie Diamond-Stanic*[†], *Kumar Sharma*[†,¶]

[*]Barshop Institute for Longevity and Aging Studies, University of Texas Health San Antonio, San Antonio, TX, United States [†]Center for Renal Precision Medicine, Division of Nephrology, Department of Medicine, University of Texas Health San Antonio, San Antonio, TX, United States [‡]Department of Biochemistry, University of Texas Health San Antonio, San Antonio, TX, United States [§]Department of Pharmacology, University of Texas Health San Antonio, San Antonio, TX, United States [¶]Audie L. Murphy Memorial VA Hospital, South Texas Veterans Health Care System (STVHCS), San Antonio, TX, United States

## 1 INTRODUCTION

The kidney is the organ that regulates salt and water metabolism and finely tunes levels of electrolytes and minerals in the blood. This role of the kidney has allowed organisms to face many environmental challenges, such as reduced availability of sodium, potassium, and water. In addition to the regulation of inorganic molecules and water balance, the kidney is a major regulator of organic substances. These include organic nutrient sources such as sugars, fats, and proteins and likely many other organic acids that are byproducts of the metabolism of the mammalian host and also the microbiome. With this larger recognition of the role of the kidney, it is not surprising that the kidney is a susceptible target in nutrient stress conditions, with the prime example being type 2 diabetes.

The pathology of diabetic kidney disease (DKD) because of type 2 diabetes is similar to that seen in type 1 diabetes, with features such as mesangial matrix accumulation, podocyte dropout, glomerular basement membrane thickening, and arteriolar hyalinosis. Most therapies focus on glomerular and vascular changes, and specifically on protecting the glomerulus through use of inhibitors of the renin-angiotensin system to reduce intracapillary hypertension. Although these


https://doi.org/10.1016/B978-0-12-811752-1.00012-2

approaches consistently have been shown to be beneficial in reducing the rate of decline in renal function in both type 1 and type 2 diabetes, there remains a large number of patients in whom an accelerated decline of renal function and consequent morbidity and mortality persists. Because the onset of diabetes is occurring at younger ages in individuals who likely will live longer with a lower incidence of cardiovascular disease, a further increase in the number of patients with DKD driven by longstanding type 2 diabetes is inevitable. As such, there is a continued need to pursue greater understanding of the pathogenesis of DKD in type 2 diabetes and to develop new therapies to arrest, and hopefully reverse, renal disease because of type 2 diabetes.

Although transcriptomic and proteomic studies clearly have demonstrated a signature of inflammation that characterizes DKD, the question as to why the diabetic kidney is inflamed remains enigmatic. Clues to unravel this critical question have been provided recently by a new field of study, metabolomics. These studies suggest that mitochondrial dysfunction is a cardinal feature of DKD with type 2 diabetes.[1] Even more exciting is that potential approaches to improve mitochondrial function with direct and indirect approaches might prove to be remarkably beneficial for this large patient population.[2, 3]

It is now accepted widely that mitochondrial dysfunction is at the crossroads of many age-related diseases, including obesity, type 2 diabetes, cardiovascular diseases, neurodegeneration, and cancer. Neither the primary cause nor the major consequences of mitochondrial dysfunction in age-related metabolic disease are well understood,[4] however, and the role of mitochondrial dysfunction in the kidneys of type 2 diabetic patients is only now being elucidated.

This chapter will focus on the role of mitochondria in DKD, with a brief overview of mitochondria in nondiabetic kidney diseases, followed by discussions of novel findings from metabolomics studies, mitochondrial superoxide production, mitochondrial content, and cardiolipin, a key component of the inner mitochondrial membrane, which has been implicated in regulation of mitochondrial dysfunction in various diseases, including diabetes (Fig. 1).

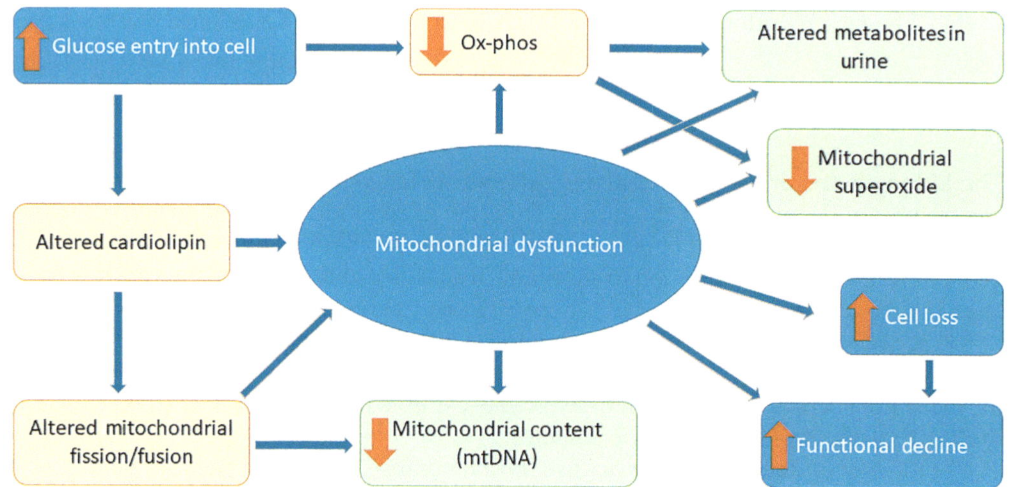


FIG. 1 Schematic showing the many pathways and mechanisms by which diabetes and obesity result in mitochondrial dysfunction and kidney failure. *Blue boxes* represent the manifestations of the disease process; *yellow boxes* represent pathways by which high glucose leads to mitochondrial dysfunction; *green boxes* are outcomes that have been measured recently through novel technologies and that have provided new insight into the disease process.

# 2 MITOCHONDRIA IN REGULATION OF KIDNEY FUNCTION-ROLE IN VARIOUS DISEASES AND MITOCHONDRIAL PHENOTYPE

The proximal tubular cells and podocytes within the kidney are exquisitely sensitive to defects in mitochondrial function.[5, 6] This might be partly a result of their high metabolic demand for constant, high-efficiency function, and also might reflect some degree of sharing of common progenitors. The role of mitochondria in many nondiabetic kidney diseases has been well characterized. Several genetic disorders linked to mitochondria have renal manifestations.[7] Although these genetic disorders often manifest in childhood, several genetic alterations affecting the mitochondrial genome have been recognized in adulthood. Along with genetic diseases, many nephrotoxins, such as the anti-cancer drug cisplatin and the antiretroviral drug tenofovir, clearly affect mitochondrial function and are major complications of these medications.[8, 9]

The phenotype of most primary disorders of mitochondrial function is fairly consistent, in spite of the variant etiologies of mitochondrial dysfunction because of genetic mutations and toxins. Most of these diseases manifest with Fanconi syndrome, with evidence of proximal tubular transport dysfunction. With reduced proximal tubular transport activity, there is often glucosuria (because of reduced function of SGLT2/1 activity), phosphaturia, alkali loss, and amino aciduria. In addition to proximal tubular dysfunction, many mitochondrial diseases also manifest as tubulointerstitial and cystic diseases, and glomerulosclerosis with podocyte dysfunction syndromes.

It is likely that any disease affecting mitochondria in the kidney will affect proximal tubular function, leading to a partial or complete Fanconi-type pattern in urinary metabolites, possibly with tubulointerstitial and glomerular disease as well. Although the histopathological lesions associated with DKD because of type 2 diabetes, including tubulointerstitial fibrosis and glomerulosclerosis, are well established, only recently have we begun to examine urinary and cellular metabolites using metabolomics to interrogate mitochondrial function in the diabetic kidney.

## 2.1 Mitochondrial Function in Diabetic Kidney Disease

The suspicion that mitochondria might play a key role in human diabetic nephropathy was first identified in 1991.[10] Subsequently, Brownlee developed a theory in which excess superoxide production from mitochondria could serve to connect multiple pathways of cell dysfunction.[11] This view suggested that excess glucose entering a cell is metabolized, leading to enhanced pyruvate entry into mitochondria, which stimulates excess activity of the tricarboxylic acid (TCA) cycle. The TCA cycle provides electron donors to the electron transport chain (ETC), and enhanced flux via the ETC will generate excess superoxide anions from complex I and complex III. These superoxide molecules, which originate from the mitochondria, can affect both intramitochondrial and extramitochondrial pathways. Therefore, the Brownlee hypothesis suggests that excessive glucose entry into cells will stimulate excess mitochondrial superoxide production, which manifests the diabetic phenotype as a result.

This unifying theory held sway for more than 15 years and still is considered to be a major pathway to explain diabetic complications. Increased cellular oxidative stress is a feature of many diseases, including diabetic kidney disease. Until recently, however, the source of the reactive oxygen species has been assumed to primarily originate from the mitochondria. Recent data from independent groups have suggested that mitochondria might have a more complex role in response to excess glucose. In living cells, excess glucose exposure results in an almost immediate inhibition of electron transport chain activity. This response, known as the Crabtree effect, is characterized by increased phosphorylation of pyruvate dehydrogenase (PDH), which shunts cellular energy production away from oxidative phosphorylation in the mitochondria and into glycolysis in the cytosol, permitting the movement of carbon skeletons out of the mitochondria into the cytosol, where they are used as building blocks for cell growth. In this scenario, one would expect increased phosphorylation of PDH and an increase in organ mass. Increased kidney weight is a hallmark of experimental models of diabetic kidney disease, and phosphorylation of PDH has been demonstrated in the diabetic kidney.[12] The implications for disease development from the activation of the Crabtree effect within the context of diabetes and sustained glucose excess has not yet been clarified.

The Crabtree effect predicts a reduction in mitochondrial superoxide production in response to excess glucose entry into the cell, which stands in direct contrast to the Brownlee hypothesis. The Crabtree effect is not observed in isolated mitochondria, suggesting that a required cytoarchitecture and/or extra-mitochondrial pathway is regulating the mitochondrial ETC. Therefore, studies to evaluate mitochondrial function in the in vivo state are required to fully understand the effect of high glucose on mitochondrial function.

Our studies were some of the first to address the question of whether mitochondrial superoxide production is elevated in response to high glucose in vivo. Using dihydroethidium (DHE) as an in vivo probe of superoxide production, we found a consistent reduction of superoxide in the diabetic kidney with type 1 diabetes.[13] These data suggest that rather than functioning normally or at an increased rate (as indicated by the Brownlee hypothesis), mitochondria in the type 1 diabetic kidney are dysfunctional.

Whether mitochondrial dysfunction occurs in patients with type 2 diabetes has been a more difficult question to address. The usual approaches to evaluate mitochondria are via electron microscopy structural analysis. Most studies have suggested that mitochondria appear either normal or more fragmented in diabetic kidneys.[14] There is not a consistent conclusion, however, and the data might be dependent largely on the stage of disease and the technique used to evaluate mitochondrial structure.

A breakthrough in the question of mitochondrial dysfunction in type 2 DKD came with the application of a targeted metabolomics platform to patients with established DKD. Using 24-hour urine collections, a robust metabolomics signature composed of a consistent reduction of a panel of 13 metabolites was found in patients with longstanding diabetes and reduced eGFR (estimated glomerular filtration rate <75) as compared with healthy controls.[12] These data were generated using a training and validation set and were subjected to stringent Bonferroni correction for multiple testing, making this study one of the first high quality, quantitative data sets to address the question of urinary metabolites in type 2 DKD. Many of the metabolites identified were not detectable in the plasma of the subjects, however, the urine signature was quite robust. Using a bioinformatics and systems biology approach, the metabolites were connected to their protein enzymes, and a large interconnected network emerged. Almost all of the proteins were found to reside or be regulated within mitochondria, and, together with the metabolites, provide evidence of alterations in the Krebs cycle (citrate, aconitate), pyrimidine metabolism (uridine), leucine catabolism (3-hydroxyisovaleric acid, 3-methylcrotonylglycine, tyrosine metabolism (vanillylmandelic acid), valine catabolism (3-hydroxyisobutyric acid), the isoleucine catabolism L-pathway (tiglylglycine, 2-methylacetoacetic acid, R-pathway (2-ethyl-3-hydroxypropionate), propionate metabolism (3-hydroxypropionate), branched-chain fatty acid metabolism (3-methyladipic acid), and oxalate metabolism (glycolic acid).[12] This strongly suggested that the metabolite pattern was intrinsically associated with altered mitochondrial function. Because the metabolites all were reduced in the urine of the subjects with type 2 DKD, it was concluded that this metabolomics signature indicated a reduction of mitochondrial function in patients with DKD. Importantly, 12/13 metabolites remained significantly reduced in diabetic patients with DKD when compared with patients with type 2 diabetes with intact renal function (eGFR>75), indicating that the signature was driven by the DKD phenotype and not just type 2 diabetes.[12] This was the first recognition based on quantitative data from a training and validation set that patients with type 2 DKD indeed have mitochondrial dysfunction.

## 3 MITOCHONDRIAL CONTENT IN DIABETIC KIDNEY DISEASE

Our studies using urinary metabolomics demonstrated reduced mitochondrial function indirectly. To more directly address the question, mitochondrial content also was measured. To date, this question has not been sufficiently answered; however, several studies suggest that patients with kidney disease do have reduced mitochondrial content. In our studies with urinary exosomes, patients with DKD vs. healthy controls showed a marked reduction of urine exosomal mitochondrial DNA (mtDNA) content (12). mtDNA also was assessed in the circulation among diabetic individuals and healthy controls in an independent study. Again, mtDNA content was significantly lower in patients with diabetes and nephropathy as compared to those with diabetes without nephropathy.[15] A potential causative role of decreased mtDNA content for DKD was suggested in longitudinal studies, in which circulating mtDNA was found to identify a quartile of subjects that would develop chronic kidney disease up to 20 years later.[16] Because this was in a community clinic cohort of >9000 individuals, there is strong evidence connecting mitochondrial content to risk of progressive renal disease.

## 4 MITOCHONDRIAL OXIDATIVE PHOSPHORYLATION DYSFUNCTION AND CARDIOLIPIN IN DIABETES

During the oxidative phosphorylation process within the mitochondria, a small number of electrons escapes from the ETC and directly leaks to the acceptor (i.e., oxygen), giving rise to free superoxide radicals (see Chapter 1 for a full review of oxidative phosphorylation). Free superoxide radicals might play important roles in signal transduction and indicate normal mitochondrial ETC activity.[17] Overproduction of free radicals, however, has been associated with the pathogenesis of many diseases and aging[18] and also might be associated with mitochondrial dysfunction.

Cardiolipin plays a major role in normal function of the ETC and also might play a role in regulating superoxide production from the ETC. Cardiolipin comprises a class of mitochondrial signature phospholipids consisting of four fatty acyl chains formed by the condensation of one phosphatidylglycerol with one phosphatidic acid. In mammalian cells, cardiolipin initially is synthesized as a premature form from the endogenously available phosphatidylglycerols in a process involving four sequential reactions catalyzed by: cytidine diphosphatediacylglycerol synthetase, phosphatidylglycerolphosphate synthase, phosphatidylglycerolphosphate phosphatases, and cardiolipin synthase.[19] Cardiolipin becomes fully functional only when it undergoes extensive remodeling catalyzed by phospholipase $A_2$, tafazzin, monolysocardiolipin acyltransferase, and/or acyl-CoA:lysocardiolipin acyltransferase.[19] Cardiolipin synthesis and acyl chain remodeling occur exclusively in mitochondria. The major cardiolipin molecular species present in the majority of mammalian mitochondria contains four linoleates. It is thought that this species is more functionally efficient than other cardiolipin species as it possesses a symmetric configuration and a balanced antioxidative property.[20, 21]

Cardiolipin is present predominantly in the mitochondrial inner membrane, and it plays essential roles in optimal mitochondrial function.[22] This essential role has been well evidenced in Barth syndrome, a rare and often fatal X-linked genetic disorder caused by tafazzin mutation, which results in reduced cardiolipin content and increased levels of lysocardiolipins and is characterized by amino aciduria, neutropenia, dilated cardiomyopathy, and myocardial noncompaction.[23] It also has been demonstrated that cardiolipin serves as a type of glue to keep the ETC together as a supercomplex, which is crucial for efficient transferring of electrons from donors to acceptors.[24] Changed content or composition of cardiolipin leads to mitochondrial dysfunction[21] and has been found in numerous disease states, such as diabetic cardiomyopathy[21] and cancer.[25]

The regulation of cardiolipin has been evaluated in several models of diabetes. For example, lipidomics analysis revealed that cardiolipin species were altered severely in the myocardium of ob/ob mice compared with wild-type littermate controls (Fig. 2).[26] Specifically, the spectral ion peaks from ob/ob mouse heart drastically shift from the lower mass region (e.g., at $m/z$ 723.5 corresponding to tetra-18:2 cardiolipin species) to the higher mass regions at the clusters of $m/z$ 760.5 and 772.5 that correspond to cardiolipin species containing two to four fatty acyl chains of arachidonate (20:4, AA) or docosahexaenoate (22:6, DHA) or their combinations. This indicates that the majority of the cardiolipin species were remodeled from the species containing the fatty acyl chains with 18 carbon atoms in the wild-type mouse heart to those of longer fatty acyl chains containing AA and DHA in the ob/ob mouse hearts. Cardiolipin species containing AA and DHA are virtually undetectable in wild-type mouse heart (Fig. 2A), and these species are more susceptible to being oxidized compared to those containing 18:1 or 18:2 fatty acyl chains. This pattern of changes in cardiolipin species also was present in the myocardium from several mouse models of diabetes, including streptozotocin-induced (type 1), high fat diet-induced, db/db, and protein kinase AKT2 knockout mice.[21, 26]

The shift of cardiolipin species with diabetes might be because of the increased remodeling required to eliminate the oxidized fatty acyl chains after increases in reactive oxygen species under diabetic conditions,[26] however, the underlying mechanism remains unclear. The consequence of such a shift in cardiolipin structure might be summarized in two aspects. First, according to the symmetrical hypothesis, the reduction of symmetrical tetra-18:2 cardiolipin species will reduce the capability of cardiolipin to glue the ETC super-complex together, leading to reduced ETC efficiency.[20] Second, increased content of cardiolipin species containing polyunsaturated fatty acyl chains such as AA and DHA could lead to the loss of

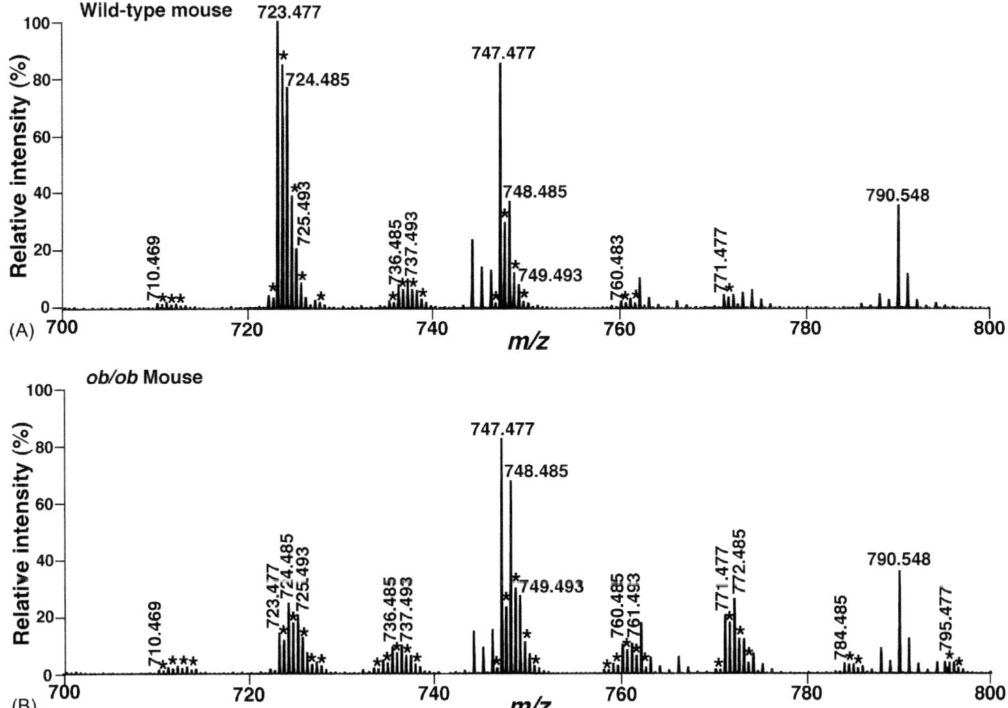


FIG. 2 Expanded negative-ion electrospray ionization mass spectra of myocardial lipid extracts from ob/ob and control mice. Myocardial lipid extracts of wild-type (panel A) and ob/ob (panel B) mice at 4 months of age were prepared by a modified Bligh and Dyer procedure.[26] Negative-ion electrospray ionization mass spectra were acquired by using a QqTOF mass spectrometer as previously described.[27] Both spectra were displayed after normalization to the cardiolipin internal standard (which is not included in the spectra). The asterisks indicate the majority of the doubly-charged cardiolipin plus-one isotopologues whose ion peak intensities can be used to quantify individual cardiolipin molecular species as previously described.[27] Other unlabeled ion peaks correspond to deprotonated molecular species of other anionic phospholipids and ethanolamine glycerophospholipids. *Reprinted from Han, 2007 with permission from the American Chemical Society, Copyright 2007.*

mitochondrial inner membrane integrity and subsequently result in electron escape and proton ion leakage, both of which represent the losses of ETC efficiency and mitochondrial dysfunction that has been well documented under diabetic conditions.[28–30] The role of cardiolipin structure and its regulation, therefore, is of major interest to understanding mitochondrial dysfunction. While the role of cardiolipin in DKD has not been studied directly, its investigation clearly is warranted.

# 5 CARDIOLIPIN AND DEFECTIVE MITOCHONDRIAL DYNAMICS IN DIABETES

Although the underlying causes of mitochondrial dysfunction in type 2 diabetes remain elusive, disruption of mitochondrial networks has been implicated in the pathogenesis of diabetes.[31–33] Mitochondria are highly dynamic networks that constantly change shape and structure because of cycles of fusion and fission, and these cycles are required for mitochondrial quality control[34] (see full discussion in Chapter 1). Disruption of mitochondrial dynamics leading to mitochondrial fragmentation has been implicated in the etiology of aging and age-related metabolic diseases, including diabetes and obesity-related complications.[31–33, 35–38] Accordingly, the onset of type 2 diabetes is associated with mitochondrial fragmentation in islet β-cells,[35] endothelial cells,[37] neurons,[31–33] and kidney cells,[14] possibly because of a mitochondrial fusion defect.[38] Likewise, mitochondrial fragmentation also has been implicated in insulin resistance in the skeletal muscle of mice with genetic and diet-induced obesity.[39] This fragmentation is accompanied by mitochondrial depolarization, loss of ATP production, and insulin resistance.

Altered expression of several of the GTPases that control mitochondrial trafficking, fusion and fission, and so are crucial for mitochondrial quality control and mitophagy have been identified in obesity and type 2 diabetic patients. These include dynamin-related protein 1 (DRP1), mitofusion 1 (MFN1), mitofusion 2 (MFN2), and optic atrophy 1 (OPA1). For example, deficiency of MFN2, which along with MFN1 is required for the fusion of the outer mitochondrial membrane,[40–42] has been identified as a common defect in obese and type 2 diabetic patients.[42, 43] In support of these findings, physical exercise, which is known to reduce obesity and related insulin resistance, significantly increases MFN2 expression and mitochondrial biogenesis.[44] MFN2 deficiency also causes oxidative stress and mitochondrial fragmentation,[45–47] whereas inhibition of mitochondrial fission prevents ROS production and insulin resistance.[38, 39] MFN2 also plays a major role in regulating mitochondrial biogenesis. Genetic mutations of MFN2 in humans, which cause Charcot-Marie-Tooth disease type 2A and autosomal dominant optic atrophy, are associated with mtDNA depletion and accumulation of stress-induced mtDNA damage.[48, 49] Consistent with these findings, targeted deletion of MFN2 in skeletal muscle of mice leads to oxidative stress, mtDNA instability and depletion, insulin resistance, and muscle atrophy.[50–52]

As the signature phospholipid of mitochondria, cardiolipin also plays a key role in regulating mitochondrial dynamics and mitophagy. Cardiolipin directly binds to several mitochondrial GTPases, which is required for activation of these GTPases.[53, 54] For example, DRP1 specifically associates with cardiolipin through a lysine-rich variable domain, which promotes DRP1 self-assembly into multimeric complexes and enhances its GTPase activity.[53] OPA1 binds tightly to cardiolipin, and the binding stimulates its dimerization.[54] Cardiolipin acyl composition plays a key role in regulating mitochondrial dynamics through modulation of MFN2 expression. Accordingly, overexpression of ALCAT1, an acyltransferase that catalyzes pathological remodeling of cardiolipin in several aging-related metabolic diseases,[47, 55, 56] significantly reduced MFN2 protein and mRNA expression.[47] Conversely, targeted deletion of ALCAT1 in mice significantly increased MFN2 protein and mRNA expression.[47] Phenotypically, ALCAT1 overexpression resulted in significant mitochondrial fragmentation, which could be reversed by transient overexpression of MFN1 and MFN2.[47] This loss was mediated by oxidative stress, because $H_2O_2$ treatment in cell culture further depleted MFN1 and MFN2 protein levels, while preincubation with diphenyleneiodonium, an NADPH oxidase inhibitor, blocked $H_2O_2$-induced decreases in MFN1 and MFN2.[47] Restoring mitochondrial dynamics by MFN2 overexpression also rescued mitochondrial respiratory defects caused by depletion of cardiolipin.[47]

Cardiolipin is highly sensitive to oxidative damage to its polyunsaturated fatty acyl chains by ROS. Oxidized cardiolipin in the form of lipid peroxides also functions as a major source of ROS, triggering a self-destructive process known as cardiolipin peroxidation. Cardiolipin is the only phospholipid in mitochondria that undergoes early oxidation during apoptosis.[57] Oxidation of cardiolipin reduces its binding to cytochrome c, leading to its detachment from cytochrome c, a key step in the onset of apoptosis.[58, 59] In direct support for a key role for cardiolipin peroxidation in the pathogenesis of diabetic complications, rodents treated with the mitochondria-targeted antioxidant SKQ1 had significantly improved aging-related metabolic diseases and diabetic complications, including kidney ischemia reperfusion injury, through attenuated oxidative stress and cardiolipin peroxidation.[60–66]

Cardiolipin also plays a key role in the initiation of mitophagy, a mitochondrial quality control process that is defective in type 2 diabetes and diabetic complications. Accordingly, cardiolipin externalization onto the outer mitochondrial membrane signals damaged mitochondria for mitophagy in response to stress.[67] As such, cardiolipin deficiency or prevention of cardiolipin externalization significantly reduced mitophagy without noticeably affecting autophagy.[67, 68] Cardiolipin remodeling also significantly affected the ability for LC3, an initiator protein for autophagosome biogenesis, to recognize the phospholipid. Specifically, LC3 binds most avidly to tetralinoleoyl cardiolipin (TLCL).[69] TLCL deficiency in diabetes and diabetic complications likely will impair mitophagy by reducing LC3 affinity for externalized cardiolipin. In support of this speculation, mitophagy defects have been demonstrated in mouse fibroblasts (MEFs) isolated from tafazzin (TAZ) knockout mice, a mouse model of defective cardiolipin remodeling, TLCL deficiency, and Barth syndrome, and in cell lines with ALCAT1 overexpression.[47, 70] Both of these models have reduced total cardiolipin amounts and a reduced proportion of TLCL present in mitochondria.[71, 72] TAZ depletion significantly reduced mitochondria colocalization with GFP-LC3 and lysosomes, but did not reduce autophagosome biogenesis.[70] Furthermore, restoration of cardiolipin maturation by overexpressing TAZ was sufficient to rescue the mitophagy defect. Therefore, consistent with the work done by Chu et al.,[67, 68] TLCL depletion impaired autophagosome recognition of mitochondria.[67]

Recent developments in the field have identified contact sites between the endoplasmic reticulum (ER) and mitochondria as a key regulatory point for mitochondrial dynamics and mitophagy.[73, 74] ER-mitochondria contacts mark mitochondrial division sites that couple mtDNA synthesis with mitochondrial division.[75–77] The first tethering structures between the ER and outer mitochondrial membrane, termed the ERMES (ER-mitochondria encounter structure), which recently were identified in yeast, play a key role in cardiolipin synthesis and remodeling, because depletion of ERMES causes deficiency of mitochondrial phospholipids, including cardiolipin.[78] In further support for a role of cardiolipin remodeling in mitochondrial dynamics, ALCAT1 protein is localized at the ER-mitochondrial contact site, where it regulates mitochondrial dynamics.[72] Identification of the mammalian counterpart of the ERMES will provide key insights into mitochondrial dysfunction in type 2 diabetes and other aging-related metabolic diseases, including DKD.

## 6 CARDIOLIPIN IN DKD

Although the role of cardiolipin in obesity-related and diabetes-related chronic kidney disease remains to be elucidated, there are suggestions that cardiolipin is related to disease progression. SS-31, the mitochondria-targeted peptide that protects cardiolipin from oxidative damage, was demonstrated to prevent apoptosis of renal tubular epithelial cells induced by ischemia/reperfusion injury.[79] Treatment of high-fat diet-fed obese mice with the same peptide, SS-31, prevented obesity-induced glomerulosclerosis and podocyte injury and loss, reduced obesity-induced renal inflammation, and, importantly, normalized mitochondrial structure as assessed by electron microscopy in proximal tubule cells, podocytes, and endothelial cells independent of an effect on weight or blood glucose.[6] These exciting data support a role for altered cardiolipin in type 2 DKD, and further validate the focus on mitochondrial dysfunction as a driving force at the heart of the pathogenesis of DKD and as a target for the development of new therapeutics.

## 7 CONCLUSIONS

There has been rapid progress in understanding mitochondrial function in the development and progression of diabetic complications. Kidney disease with diabetes and obesity has been clearly linked with mitochondrial dysfunction, although many of the mechanisms and stages of disease linked to discrete changes in mitochondrial structure and function are only emerging. With improved imaging and -omics tools, further characterization of mitochondrial structure, cardiolipin alterations, and mitochondrial dynamics likely will provide new targets for therapies that might profoundly change the course of this devastating disease.

## Disclaimer

The contents do not represent the views of the U.S. Department of Veterans Affairs or the United States Government.

## Acknowledgments

This work was supported by the National Institutes of Health/National Institute of Diabetes and Digestive and Kidney Diseases (NIDDK) grants (5R24DK082841-08 and 1UG3DK114920-01) and the Juvenile Diabetes Research Foundation and Merit Review Award #1I01 BX00323 from the United States (U.S.) Department of Veterans Affairs Biomedical Laboratory Research and Development Service.

## References

1. Sharma K. Mitochondrial dysfunction in the diabetic kidney. *Adv Exp Med Biol* 2017;**982**:553–62.
2. Szeto HH. Pharmacologic approaches to improve mitochondrial function in AKI and CKD. *J Am Soc Nephrol* 2017;**28**(10):2856–65.
3. Poyan Mehr A, Parikh SM. PPARgamma-coactivator-1alpha, nicotinamide adenine dinucleotide and renal stress resistance. *Nephron* 2017.
4. Schriner SE, Linford NJ, Martin GM, Treuting P, Ogburn CE, Emond M, Coskun PE, Ladiges W, Wolf N, Van Remmen H, et al. Extension of murine life span by overexpression of catalase targeted to mitochondria. *Science* 2005;**308**(5730):1909–11.
5. Kawakami T, Gomez IG, Ren S, Hudkins K, Roach A, Alpers CE, Shankland SJ, D'Agati VD, Duffield JS. Deficient autophagy results in mitochondrial dysfunction and FSGS. *J Am Soc Nephrol* 2015;**26**(5):1040–52.
6. Szeto HH, Liu S, Soong Y, Alam N, Prusky GT, Seshan SV. Protection of mitochondria prevents high-fat diet-induced glomerulopathy and proximal tubular injury. *Kidney Int* 2016;**90**(5):997–1011.

7. Emma F, Salviati L. Mitochondrial cytopathies and the kidney. *Nephrol Therapeut* 2017;**13**(Suppl. 1):S23–8.
8. Yang Y, Liu H, Liu F, Dong Z. Mitochondrial dysregulation and protection in cisplatin nephrotoxicity. *Arch Toxicol* 2014;**88**(6):1249–56.
9. Herlitz LC, Mohan S, Stokes MB, Radhakrishnan J, D'Agati VD, Markowitz GS. Tenofovir nephrotoxicity: acute tubular necrosis with distinctive clinical, pathological, and mitochondrial abnormalities. *Kidney Int* 2010;**78**(11):1171–7.
10. Takebayashi S, Kaneda K. Mitochondrial derangement: possible initiator of microalbuminuria in NIDDM. *J Diabet Complications* 1991;**5**(2-3):104–6.
11. Nishikawa T, Edelstein D, Du XL, Yamagishi S, Matsumura T, Kaneda Y, Yorek MA, Beebe D, Oates PJ, Hammes HP, et al. Normalizing mitochondrial superoxide production blocks three pathways of hyperglycaemic damage. *Nature* 2000;**404**(6779):787–90.
12. Sharma K, Karl B, Mathew AV, Gangoiti JA, Wassel CL, Saito R, Pu M, Sharma S, You YH, Wang L, et al. Metabolomics reveals signature of mitochondrial dysfunction in diabetic kidney disease. *J Am Soc Nephrol* 2013;**24**(11):1901–12.
13. Dugan LL, You YH, Ali SS, Diamond-Stanic M, Miyamoto S, Decleves AE, Andreyev A, Quach T, Ly S, Shekhtman G, et al. AMPK dysregulation promotes diabetes-related reduction of superoxide and mitochondrial function. *J Clin Invest* 2013;**123**(11):4888–99.
14. Wang W, Wang Y, Long J, Wang J, Haudek SB, Overbeek P, Chang BH, Schumacker PT, Danesh FR. Mitochondrial fission triggered by hyperglycemia is mediated by ROCK1 activation in podocytes and endothelial cells. *Cell Metab* 2012;**15**(2):186–200.
15. Malik AN, Parsade CK, Ajaz S, Crosby-Nwaobi R, Gnudi L, Czajka A, Sivaprasad S. Altered circulating mitochondrial DNA and increased inflammation in patients with diabetic retinopathy. *Diabetes Res Clin Pract* 2015;**110**(3):257–65.
16. Tin A, Grams ME, Ashar FN, Lane JA, Rosenberg AZ, Grove ML, Boerwinkle E, Selvin E, Coresh J, Pankratz N, et al. Association between mitochondrial DNA copy number in peripheral blood and incident CKD in the atherosclerosis risk in communities study. *J Am Soc Nephrol* 2016;**27**(8):2467–73.
17. Banerjee R. Introduction to the thematic minireview series: host-microbiome metabolic interplay. *J Biol Chem* 2017;**292**(21):8544–5.
18. Kauppila TES, Kauppila JHK, Larsson NG. mammalian mitochondria and aging: an update. *Cell Metab* 2017;**25**(1):57–71.
19. Mejia EM, Nguyen H, Hatch GM. Mammalian cardiolipin biosynthesis. *Chem Phys Lipids* 2014;**179**:11–6.
20. Schlame M, Ren M, Xu Y, Greenberg ML, Haller I. Molecular symmetry in mitochondrial cardiolipins. *Chem Phys Lipids* 2005;**138**(1–2):38–49.
21. He Q, Han X. Cardiolipin remodeling in diabetic heart. *Chem Phys Lipids* 2014;**179**:75–81.
22. Chicco AJ, Sparagna GC. Role of cardiolipin alterations in mitochondrial dysfunction and disease. *Am J Phys Cell Physiol* 2007;**292**(1):C33–44.
23. Barth PG, Valianpour F, Bowen VM, Lam J, Duran M, Vaz FM, Wanders RJ. X-linked cardioskeletal myopathy and neutropenia (Barth syndrome): an update. *Am J Med Genet A* 2004;**126a**(4):349–54.
24. Zhang M, Mileykovskaya E, Dowhan W. Gluing the respiratory chain together. Cardiolipin is required for supercomplex formation in the inner mitochondrial membrane. *J Biol Chem* 2002;**277**(46):43553–6.
25. Kiebish MA, Han X, Cheng H, Chuang JH, Seyfried TN. Cardiolipin and electron transport chain abnormalities in mouse brain tumor mitochondria: lipidomic evidence supporting the Warburg theory of cancer. *J Lipid Res* 2008;**49**(12):2545–56.
26. Han X, Yang J, Yang K, Zhao Z, Abendschein DR, Gross RW. Alterations in myocardial cardiolipin content and composition occur at the very earliest stages of diabetes: a shotgun lipidomics study. *Biochemistry* 2007;**46**(21):6417–28.
27. Han X, Yang K, Yang J, Cheng H, Gross RW. Shotgun lipidomics of cardiolipin molecular species in lipid extracts of biological samples. *J Lipid Res* 2006;**47**(4):864–79.
28. Mazumder PK, O'Neill BT, Roberts MW, Buchanan J, Yun UJ, Cooksey RC, Boudina S, Abel ED. Impaired cardiac efficiency and increased fatty acid oxidation in insulin-resistant ob/ob mouse hearts. *Diabetes* 2004;**53**(9):2366–74.
29. Blake R, Trounce IA. Mitochondrial dysfunction and complications associated with diabetes. *Biochim Biophys Acta* 2014;**1840**(4):1404–12.

30. Verma SK, Garikipati VNS, Kishore R. Mitochondrial dysfunction and its impact on diabetic heart. *Biochim Biophys Acta* 2017;**1863**(5):1098–105.
31. Edwards JL, Quattrini A, Lentz SI, Figueroa-Romero C, Cerri F, Backus C, Hong Y, Feldman EL. Diabetes regulates mitochondrial biogenesis and fission in mouse neurons. *Diabetologia* 2010;**53**(1):160–9.
32. Makino A, Scott BT, Dillmann WH. Mitochondrial fragmentation and superoxide anion production in coronary endothelial cells from a mouse model of type 1 diabetes. *Diabetologia* 2010;**53**(8):1783–94.
33. Molina AJ, Wikstrom JD, Stiles L, Las G, Mohamed H, Elorza A, Walzer G, Twig G, Katz S, Corkey BE, et al. Mitochondrial networking protects beta-cells from nutrient-induced apoptosis. *Diabetes* 2009;**58**(10):2303–15.
34. Twig G, Elorza A, Molina AJ, Mohamed H, Wikstrom JD, Walzer G, Stiles L, Haigh SE, Katz S, Las G, et al. Fission and selective fusion govern mitochondrial segregation and elimination by autophagy. *EMBO J* 2008;**27**(2):433–46.
35. Bindokas VP, Kuznetsov A, Sreenan S, Polonsky KS, Roe MW, Philipson LH. Visualizing superoxide production in normal and diabetic rat islets of Langerhans. *J Biol Chem* 2003;**278**(11):9796–801.
36. Knott AB, Perkins G, Schwarzenbacher R, Bossy-Wetzel E. Mitochondrial fragmentation in neurodegeneration. *Nat Rev Neurosci* 2008;**9**(7):505–18.
37. Shenouda SM, Widlansky ME, Chen K, Xu G, Holbrook M, Tabit CE, Hamburg NM, Frame AA, Caiano TL, Kluge MA, et al. Altered mitochondrial dynamics contributes to endothelial dysfunction in diabetes mellitus. *Circulation* 2011;**124**(4):444–53.
38. Yu T, Robotham JL, Yoon Y. Increased production of reactive oxygen species in hyperglycemic conditions requires dynamic change of mitochondrial morphology. *Proc Natl Acad Sci U S A* 2006;**103**(8):2653–8.
39. Jheng HF, Tsai PJ, Guo SM, Kuo LH, Chang CS, Su IJ, Chang CR, Tsai YS. Mitochondrial fission contributes to mitochondrial dysfunction and insulin resistance in skeletal muscle. *Mol Cell Biol* 2012;**32**(2):309–19.
40. Santel A, Fuller MT. Control of mitochondrial morphology by a human mitofusin. *J Cell Sci* 2001;**114**(Pt 5): 867–74.
41. Santel A, Frank S, Gaume B, Herrler M, Youle RJ, Fuller MT. Mitofusin-1 protein is a generally expressed mediator of mitochondrial fusion in mammalian cells. *J Cell Sci* 2003;**116**(Pt 13):2763–74.
42. Bach D, Pich S, Soriano FX, Vega N, Baumgartner B, Oriola J, Daugaard JR, Lloberas J, Camps M, Zierath JR, et al. Mitofusin-2 determines mitochondrial network architecture and mitochondrial metabolism. A novel regulatory mechanism altered in obesity. *J Biol Chem* 2003;**278**(19):17190–7.
43. Bach D, Naon D, Pich S, Soriano FX, Vega N, Rieusset J, Laville M, Guillet C, Boirie Y, Wallberg-Henriksson H, et al. Expression of Mfn2, the Charcot-Marie-Tooth neuropathy type 2A gene, in human skeletal muscle: effects of type 2 diabetes, obesity, weight loss, and the regulatory role of tumor necrosis factor alpha and interleukin-6. *Diabetes* 2005;**54**(9):2685–93.
44. Cartoni R, Leger B, Hock MB, Praz M, Crettenand A, Pich S, Ziltener JL, Luthi F, Deriaz O, Zorzano A, et al. Mitofusins 1/2 and ERRalpha expression are increased in human skeletal muscle after physical exercise. *J Physiol* 2005;**567**(Pt 1):349–58.
45. Chen H, Chomyn A, Chan DC. Disruption of fusion results in mitochondrial heterogeneity and dysfunction. *J Biol Chem* 2005;**280**(28):26185–92.
46. Chen H, Detmer SA, Ewald AJ, Griffin EE, Fraser SE, Chan DC. Mitofusins Mfn1 and Mfn2 coordinately regulate mitochondrial fusion and are essential for embryonic development. *J Cell Biol* 2003;**160**(2):189–200.
47. Li J, Liu X, Wang H, Zhang W, Chan DC, Shi Y. Lysocardiolipin acyltransferase 1 (ALCAT1) controls mitochondrial DNA fidelity and biogenesis through modulation of MFN2 expression. *Proc Natl Acad Sci U S A* 2012;**109**(18):6975–80.
48. Rouzier C, Bannwarth S, Chaussenot A, Chevrollier A, Verschueren A, Bonello-Palot N, Fragaki K, Cano A, Pouget J, Pellissier JF, et al. The MFN2 gene is responsible for mitochondrial DNA instability and optic atrophy 'plus' phenotype. *Brain J Neurol* 2012;**135**(Pt 1):23–34.
49. Renaldo F, Amati-Bonneau P, Slama A, Romana C, Forin V, Doummar D, Barnerias C, Bursztyn J, Mayer M, Khouri N, et al. MFN2, a new gene responsible for mitochondrial DNA depletion. *Brain J Neurol* 2012;**135**(Pt 8): e223. 1–4; author reply e4, 1–3.
50. Chen H, Vermulst M, Wang YE, Chomyn A, Prolla TA, McCaffery JM, Chan DC. Mitochondrial fusion is required for mtDNA stability in skeletal muscle and tolerance of mtDNA mutations. *Cell* 2010;**141**(2):280–9.
51. Sebastian D, Hernandez-Alvarez MI, Segales J, Sorianello E, Munoz JP, Sala D, Waget A, Liesa M, Paz JC, Gopalacharyulu P, et al. Mitofusin 2 (Mfn2) links mitochondrial and endoplasmic reticulum function with insulin signaling and is essential for normal glucose homeostasis. *Proc Natl Acad Sci U S A* 2012;**109**(14):5523–8.

52. Zhao T, Huang X, Han L, Wang X, Cheng H, Zhao Y, Chen Q, Chen J, Cheng H, Xiao R, et al. Central role of mitofusin 2 in autophagosome-lysosome fusion in cardiomyocytes. *J Biol Chem* 2012;**287**(28):23615–25.
53. Bustillo-Zabalbeitia I, Montessuit S, Raemy E, Basanez G, Terrones O, Martinou JC. Specific interaction with cardiolipin triggers functional activation of dynamin-related protein 1. *PLoS ONE* 2014;**9**(7):e102738.
54. DeVay RM, Dominguez-Ramirez L, Lackner LL, Hoppins S, Stahlberg H, Nunnari J. Coassembly of Mgm1 isoforms requires cardiolipin and mediates mitochondrial inner membrane fusion. *J Cell Biol* 2009;**186**(6):793–803.
55. Liu X, Ye B, Miller S, Yuan H, Zhang H, Tian L, Nie J, Imae R, Arai H, Li Y, et al. Ablation of ALCAT1 mitigates hypertrophic cardiomyopathy through effects on oxidative stress and mitophagy. *Mol Cell Biol* 2012;**32**(21):4493–504.
56. Wang L, Liu X, Nie J, Zhang J, Kimball SR, Zhang H, Zhang WJ, Jefferson LS, Cheng Z, Ji Q, et al. ALCAT1 controls mitochondrial etiology of fatty liver diseases, linking defective mitophagy to steatosis. *Hepatology (Baltimore, MD)* 2015;**61**(2):486–96.
57. Kagan VE, Tyurin VA, Jiang J, Tyurina YY, Ritov VB, Amoscato AA, Osipov AN, Belikova NA, Kapralov AA, Kini V, et al. Cytochrome c acts as a cardiolipin oxygenase required for release of proapoptotic factors. *Nat Chem Biol* 2005;**1**(4):223–32.
58. Shidoji Y, Hayashi K, Komura S, Ohishi N, Yagi K. Loss of molecular interaction between cytochrome c and cardiolipin due to lipid peroxidation. *Biochem Biophys Res Commun* 1999;**264**(2):343–7.
59. Lutter M, Fang M, Luo X, Nishijima M, Xie X, Wang X. Cardiolipin provides specificity for targeting of tBid to mitochondria. *Nat Cell Biol* 2000;**2**(10):754–61.
60. Bakeeva LE. Age-related changes in ultrastructure of mitochondria. Effect of SkQ1. *Biochemistry Biokhimiia* 2015;**80**(12):1582–8.
61. Demyanenko IA, Zakharova VV, Ilyinskaya OP, Vasilieva TV, Fedorov AV, Manskikh VN, Zinovkin RA, Pletjushkina OY, Chernyak BV, Skulachev VP, et al. Mitochondria-targeted antioxidant SkQ1 improves dermal wound healing in genetically diabetic mice. *Oxidative Med Cell Longev* 2017;**2017**:6408278.
62. Kezic A, Spasojevic I, Lezaic V, Bajcetic M. Mitochondria-targeted antioxidants: future perspectives in kidney ischemia reperfusion injury. *Oxidative Med Cell Longev* 2016;**2016**:2950503.
63. Shabalina IG, Vyssokikh MY, Gibanova N, Csikasz RI, Edgar D, Hallden-Waldemarson A, Rozhdestvenskaya Z, Bakeeva LE, Vays VB, Pustovidko AV, et al. Improved health-span and lifespan in mtDNA mutator mice treated with the mitochondrially targeted antioxidant SkQ1. *Aging* 2017;**9**(2):315–39.
64. Stefanova NA, Muraleva NA, Maksimova KY, Rudnitskaya EA, Kiseleva E, Telegina DV, Kolosova NG. An antioxidant specifically targeting mitochondria delays progression of Alzheimer's disease-like pathology. *Aging* 2016;**8**(11):2713–33.
65. Voronkova YG, Popova TN, Agarkov AA, Zinovkin RA. Effect of SkQ1 on activity of the glutathione system and NADPH-generating enzymes in an experimental model of hyperglycemia. *Biochemistry Biokhimiia* 2015;**80**(12):1614–21.
66. Zernii EY, Gancharova OS, Baksheeva VE, Golovastova MO, Kabanova EI, Savchenko MS, Tiulina VV, Sotnikova LF, Zamyatnin Jr. AA, Philippov PP, et al. Mitochondria-targeted antioxidant SkQ1 prevents anesthesia-induced dry eye syndrome. *Oxidative Med Cell Longev* 2017;**2017**:9281519.
67. Chu CT, Ji J, Dagda RK, Jiang JF, Tyurina YY, Kapralov AA, Tyurin VA, Yanamala N, Shrivastava IH, Mohammadyani D, et al. Cardiolipin externalization to the outer mitochondrial membrane acts as an elimination signal for mitophagy in neuronal cells. *Nat Cell Biol* 2013;**15**(10):1197–205.
68. Kagan VE, Jiang J, Huang Z, Tyurina YY, Desbourdes C, Cottet-Rousselle C, Dar HH, Verma M, Tyurin VA, Kapralov AA, et al. NDPK-D (NM23-H4)-mediated externalization of cardiolipin enables elimination of depolarized mitochondria by mitophagy. *Cell Death Differ* 2016;**23**(7):1140–51.
69. Cho YY, Kim DJ, Lee HS, Jeong CH, Cho EJ, Kim MO, Byun S, Lee KY, Yao K, Carper A, et al. Autophagy and cellular senescence mediated by Sox2 suppress malignancy of cancer cells. *PLoS ONE* 2013;**8**(2):e57172.
70. Hsu P, Liu X, Zhang J, Wang HG, Ye JM, Shi Y. Cardiolipin remodeling by TAZ/tafazzin is selectively required for the initiation of mitophagy. *Autophagy* 2015;**11**(4):643–52.
71. Soustek MS, Falk DJ, Mah CS, Toth MJ, Schlame M, Lewin AS, Byrne BJ. Characterization of a transgenic short hairpin RNA-induced murine model of Tafazzin deficiency. *Hum Gene Ther* 2011;**22**(7):865–71.
72. Li J, Romestaing C, Han X, Li Y, Hao X, Wu Y, Sun C, Liu X, Jefferson LS, Xiong J, et al. Cardiolipin remodeling by ALCAT1 links oxidative stress and mitochondrial dysfunction to obesity. *Cell Metab* 2010;**12**(2):154–65.
73. Belgareh-Touze N, Cavellini L, Cohen MM. Ubiquitination of ERMES components by the E3 ligase Rsp5 is involved in mitophagy. *Autophagy* 2017;**13**(1):114–32.

74. Park JS, Thorsness MK, Policastro R, McGoldrick LL, Hollingsworth NM, Thorsness PE, Neiman AM. Yeast Vps13 promotes mitochondrial function and is localized at membrane contact sites. *Mol Biol Cell* 2016;**27**(15):2435–49.
75. Friedman JR, Lackner LL, West M, DiBenedetto JR, Nunnari J, Voeltz GK. ER tubules mark sites of mitochondrial division. *Science* 2011;**334**(6054):358–62.
76. Lewis SC, Uchiyama LF, Nunnari J. ER-mitochondria contacts couple mtDNA synthesis with mitochondrial division in human cells. *Science* 2016;**353**(6296):aaf5549.
77. Murley A, Lackner LL, Osman C, West M, Voeltz GK, Walter P, Nunnari J. ER-associated mitochondrial division links the distribution of mitochondria and mitochondrial DNA in yeast. *eLife* 2013;**2**:e00422.
78. Kornmann B, Currie E, Collins SR, Schuldiner M, Nunnari J, Weissman JS, Walter P. An ER-mitochondria tethering complex revealed by a synthetic biology screen. *Science* 2009;**325**(5939):477–81.
79. Zhao WY, Han S, Zhang L, Zhu YH, Wang LM, Zeng L. Mitochondria-targeted antioxidant peptide SS31 prevents hypoxia/reoxygenation-induced apoptosis by down-regulating p66Shc in renal tubular epithelial cells. *Cell Physiol Biochem* 2013;**32**(3):591–600.

CHAPTER

# 13

# Role of Mitochondria in Neurodegeneration in Obesity and Type 2 Diabetes

*Susana Cardoso*[*,†], *Raquel M. Seiça*[‡,§], *Paula I. Moreira*[*,‡]

[*]CNC-Center for Neuroscience and Cell Biology, University of Coimbra, Coimbra, Portugal [†]Institute for Interdisciplinary Research, University of Coimbra, Coimbra, Portugal [‡]Laboratory of Physiology—Faculty of Medicine, University of Coimbra, Coimbra, Portugal [§]iCBR - Coimbra Institute for Clinical and Biomedical Research, Faculty of Medicine, University of Coimbra, Coimbra, Portugal

## 1 INTRODUCTION

More than 382 million people suffer from type 2 diabetes (T2D), and this appalling figure might further rise to 438 million by the year of 2030, supporting the notion that T2D is a pandemic disease of the 21st century.[1] Even though individual genetic backgrounds can alter the susceptibility and risk to develop certain diseases, the causation always will be strongly linked to exposure to transmittable, toxic, or environmental agents.[2] Recent estimates indicate that obesity currently affects more than 600 million people worldwide, and evidence predicts that about 80%–90% of overweight patients with abdominal fat deposition will develop diabetes. Approximately 90% of the diabetes upsurge is attributable to excess weight, largely because of poor lifestyle and nutritional habits.[3,4] In this respect, given such a sturdily interrelation, Sims et al.[5] suggested the term "diabesity" to further emphasize the intricate relationship between both metabolic conditions that can occur in the same individual. The twin epidemics of obesity and diabetes represent troublesome clinical conditions that carry with them severe outcomes, such as worse physical performance, decreased quality of life, accelerated biological aging, excess morbidity and mortality, and higher health care costs.[6]

Apart from the economic and public health burdens associated with obesity and T2D, a growing body of literature highlights the impacts that diabetes and obesity have on the development of secondary diseases affecting numerous tissues, including the peripheral (PNS) and central nervous system (CNS).[6,7] Even though there is a global awareness about the



Mitochondria in Obesity and Type 2 Diabetes
https://doi.org/10.1016/B978-0-12-811752-1.00013-4

importance of promoting healthy lifestyle habits, there is still a lack of effective, long-term treatment options to counteract the fast spread of diabesity epidemics. Therefore, there is a particular urgency to understand the underlying pathophysiologic mechanisms driving obesity and T2D development and its complications. In this scenario, although the exact molecular mechanisms of the disorders' evolution and complications in several tissues still are blurred, it is known that their cornerstone relies in a compromised mitochondria population and associated disturbances in energy metabolism and insulin resistance.[8, 9]

Mitochondria are known as the powerhouses of the cell because of their ability to produce adenosine triphosphate (ATP) via the shared efforts of the tricarboxylic acid cycle (TCA) and the respiratory chain/oxidative phosphorylation system (OxPhos). Mitochondria also are recognized as dynamic and multifunctional organelles essential for the life and death of the cells.[10, 11] From amino acid biosynthesis, fatty acid oxidation, steroid metabolism, calcium ($Ca^{2+}$) homeostasis, reactive oxygen species (ROS) production and scavenging, to regulation of apoptotic signaling pathways, mitochondria are embedded in the signaling cascades and programs that modulate cells homeostasis.[12] It is expectable, therefore, that because of the multiplicity of functions of those organelles, there is a variability in pathophysiology, severity, and age of onset of a growing number of diseases recognized to arise from primary or secondary alterations in mitochondrial-mediated pathways.[13]

This review will start with a brief introduction about how obesity and T2D-mediated perturbations are capable of disrupting neuronal homeostasis, leading to the decline in brain capacities and the occurrence of brain-related pathologies. An emphasis will be put on mitochondrial defects as major players in the pathogenesis of obesity and T2D-associated complications, namely in peripheral diabetic neuropathy (PDN) and in two major neurodegenerative disorders, namely Alzheimer's disease (AD) and Parkinson's disease (PD).

## 2 ASSOCIATION OF OBESITY AND TYPE 2 DIABETES WITH BRAIN HEALTH/COGNITION

Apart from genetic and environmental influence, obesity is a medical condition closely linked with an imbalance between energy intake and expenditure that culminates in an excessive body fat and adverse effects on health, leading to reduced life expectancy and/or increased health problems.[14] Irrespective of gender, ethnicity, and age, obesity and/or overweight affects 1 in 10 adults worldwide; its diagnosis usually is made by determining body mass index (BMI $\geq$25) and waist circumference (men with values >102 and women >88 cm).[15] Notably, both anthropometric indexes are related strongly with patients′ cognitive performance,[16] along with an increased likelihood of developing dementia when present at midlife.[17, 18] As reported, higher BMI is associated with lower brain volumes in cognitively normal overweight and obese elderly subjects[19] and with a lower neuronal viability in brain regions including frontal, parietal, and temporal lobes.[20] A direct association of waist-hip ratio with decreased hippocampal volume also has been demonstrated to occur.[21] Data from rodent models of obesity and obese humans also show that obesity is strongly related with reduced memory performance,[22] such as delayed recall and recognition[23, 24] along with visual what-where-when episodic memory tasks.[25] A recent population-based study highlighted that weight exerts a significant role in the susceptibility for mild cognitive impairment (MCI) subjects to suffer from neuropsychiatric symptoms.[26] MCI is considered a transitional state between normal aging and dementia characterized by modest impairments in memory, language, or other cognitive functions, whereas the presence of neuropsychiatric changes decreases the time of progression to dementia by 2.5-fold.[27] Although MCI is believed to not significantly interfere with daily life, its progression to dementia will cause severe alterations of cognitive abilities in multiple areas, including memory, learning, orientation, language, comprehension, and judgment, that strongly compromise daily living, which bring tremendous consequences for the affected individuals and their families.[28] In MCI subjects, the combination of increased weight with T2D significantly worsens behavioral disturbances in relation to total scores, symptom cluster frequency and severity, and changes in global cognition.[28] Yoon et al.[29] recently revealed that the concurrent presence of obesity and T2D strongly compromise brain structure and cognition during early stage T2D in comparison with normal weight participants. The authors reported that observed disruptions in cortical thickness and white matter integrity and alterations in the psychomotor speed performance are more pronounced in overweight/obese T2D individuals than in those with normal weight, suggesting that weight status might play additive roles in T2D-related brain and cognitive alterations.[29] Nevertheless, the course of dementia in the elderly population can be concomitantly associated with a decrease in body weight and poor healthcare and self-care promoted by the progression of dementia. Therefore, caution must be taken when interpreting the studies data to what concerns the age and disease stage of the participating subjects[18, 30] (Fig. 1).

T2D is known to be a multifactorial disorder in which both environmental and genetic factors contribute to pathology, which is characterized by chronic hyperglycemia, in a context of reduced tissue sensitivity to insulin (insulin resistance) and inadequate insulin secretion.[6, 8] T2D currently is considered a common age-related disorder and a major health concern, particularly in developed countries where the population is aging. In this context, converging evidence from clinical, epidemiological, and experimental studies suggests that there is a causative association between T2D and accelerated cognitive dysfunction and progression to dementia, particularly AD.[31–35] Notably, such association has been described to occur even in borderline diabetes, which is a condition of impaired glucose regulation in which blood glucose levels are higher than normal but not high enough to be diagnosed as diabetes.[36]

To better exploit the impact of diabetes on brain structure and function, various neuroimaging techniques have been employed. For example, by using magnetic resonance imaging (MRI) techniques and cognitive tests, it is possible to observe that middle-age individuals, in spite of well-controlled T2D, have deficits in hippocampal-based memory (recent or declarative) and selective MRI-based atrophy of the hippocampus relative to matched control subjects.[37] In this regard, such hippocampal damage and memory impairments can be viewed as possible early brain complications of T2D, whereas as disease progresses, other more resilient brain areas also are affected, leading to cognitive impairments that can spread to other cognitive domains.[37] A large cross-sectional study by Moran et al.[38] reported that subjects with T2D had lower total gray, white, and hippocampal volumes, which overall was associated with poor cognitive performance in cognitive tests. Likewise, as observed in diffusion MRI, T2D patients also were shown to present brain microstructure abnormalities and disruptions in the white matter network with a consequent slowing of information-processing speed.[39] In addition to structural changes in the brain, others have reported that T2D patients develop functional deficits in neuronal connectivity within brain areas susceptible to AD.[40, 41] These

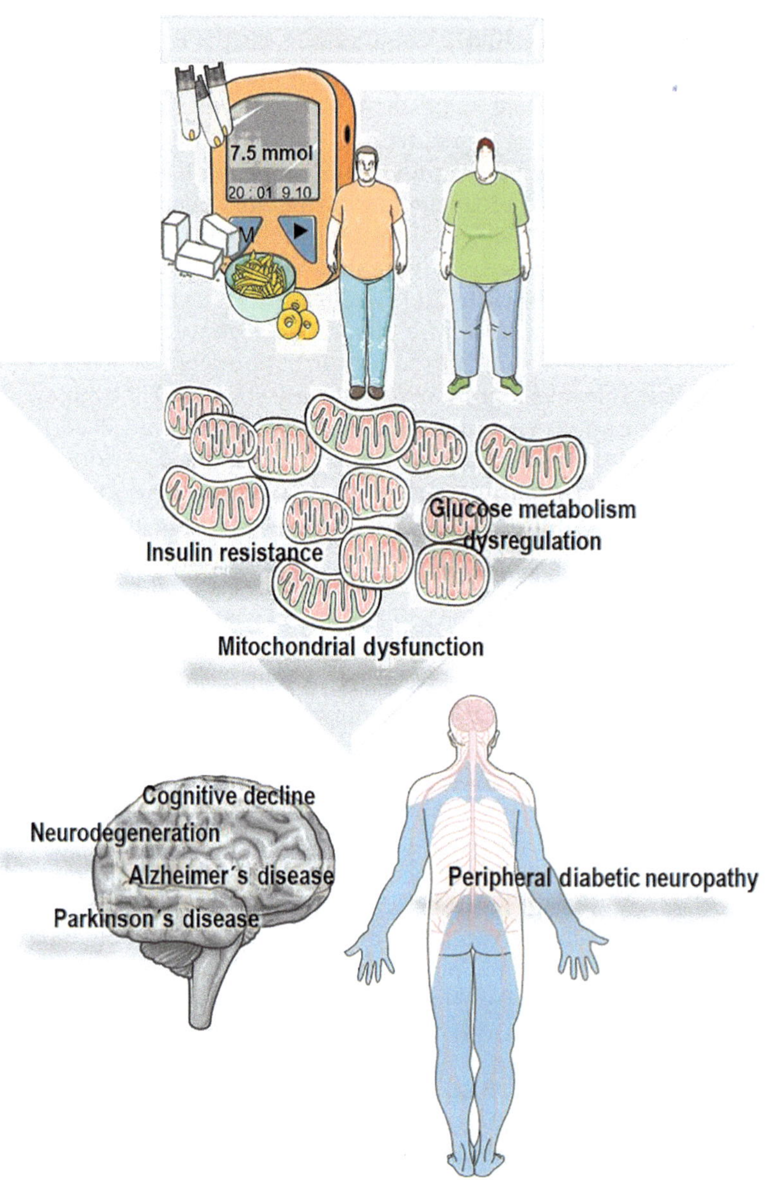


FIG. 1 Type 2 diabetes and obesity: the not so sweet relationship. Global news forecasts upcoming epidemic, as today's children are becoming overweight and grow up as obese adults. In spite of the emergent effort to raise the awareness about the importance of a healthy life and lifestyle, the obese epidemics show no signs of dwindling. Because as the saying goes "a bad thing never comes alone," obesity escalates the development of insulin resistance and type 2 diabetes (T2D). Apart from the economic and public health burdens associated with both metabolic disorders, the consequences are worrisome, including a high risk for the development of secondary diseases affecting numerous tissues including the nervous system. It is known that obesity and T2D cause several complications, including peripheral diabetic neuropathy, cognitive defects increasing the risk for dementia, namely Alzheimer's disease and Parkinson's disease, later in life. It is becoming clear that defective or insufficient mitochondrial function is inextricably implicated in the glucose dysmetabolism and altered insulin action (deficiency and/or resistance) that characterize these metabolic disorders and associated complications.

observations suggest that diabetes-related pathology might lead to neuronal loss and unusual functional connectivity within areas related to AD pathology, which might compromise pathways critical for cognition and accelerate the course of cognitive decline, resulting in an earlier manifestation of dementia.[42–44]

A recent population-based study, in which neuroimaging techniques such as single photon emission computed tomography (SPECT) and Pittsburgh compound-B positron emission tomography (PET) were applied, revealed the existence of a dementia subgroup within AD patients with characteristics predominantly associated with diabetes-related metabolic abnormalities, referred to as diabetes-related dementia.[45] Its pathophysiology seems to implicate a decrease in peripheral antioxidant levels and an increase in oxidative damage that was negatively correlated with individuals' cognitive performance.[46]

Currently, there is a strong consciousness that obesity and T2D and cognitive frailty are closely related, being displayed in a wide range of cognitive impairments, starting from cognitive decline, progressing to MCI, and ending in clinical dementia.[47, 48]

## 3 OBESITY AND TYPE 2 DIABETES-MEDIATED EFFECTS: ARE MITOCHONDRIA IN THE FRONT LINE?

As early as 1976, diabetes was considered to be an accelerated aging factor, because it increases an individual's susceptibility to degenerative diseases such as kidney disease, hypertension, coronary artery disease, stroke, atherosclerosis, and accelerated brain aging.[49, 50] With the aging of population on the rise, the scenario gets worse with the exponential increase in the number of older people diagnosed with prediabetes, a metabolic illness associated with obesity that further potentiates the risk of developing T2D.[51, 52] Aging, an unavoidable biological process, is characterized by a general decline in physiological function that leads to morbidity and mortality. In this context, as stated in Harman's free radical theory of aging (later renamed to mitochondrial theory of aging), mitochondrial dysfunction can act as a prominent and early event, promoting a chronic oxidative stress, which, in the brain, contributes to synaptic abnormalities during the aging process that might predispose to age-related disorders including neurodegenerative diseases.[53, 54] In this framework, although an age-dependent decrease of brain bioenergetics metabolism together with an impaired redox homeostasis might occur during the aging process in individuals with a history of a normal and healthy life,[55, 56] this scenario might be exacerbated in a context of metabolic challenges, pathological and environmental insults. Since the first disease associated with mitochondrial dysfunction was described in 1962,[57] a major effort has been dedicated in understanding the role that mitochondria play in health, aging, and disease.

Mitochondria are maternally inherited multifunctional organelles that work within an integrated reticulum in cells adjusted to the specific needs of the tissue.[58] In order to achieve the well-being of this vital organelle, a hierarchical system of cellular surveillance mechanisms protects mitochondria against stress, monitors mitochondrial damage, and ensures the selective removal of dysfunctional mitochondrial proteins or even total organelles.[59] In situations in which a failure of this hierarchical system occurs, however, damaged mitochondria will emerge as a central piece in cells energy deficiency and dyshomeostasis, placing mitochondria in the center of the aging process and age-related diseases.[60–62]

### 3.1 Mitochondria in Obesity and Type 2 Diabetes-Mediated Brain Defects: A Brief Insight

Diabetes and/or obesity are among the most puzzling global health problems. Even though the exact molecular mechanisms of the disorders' evolution and complications in several tissues still are blurred, it is known that the mechanisms include a compromised energy metabolism, mitochondrial dysfunction, hyperglycemia, hyperlipidemia and insulin resistance.[8] In this scenario, one of the most popular hypotheses states that mitochondrial dysfunction and oxidative stress, resulting from an increased production of mitochondrial ROS in response to hyperglycemia, are early and crucial events involved in the multiorgan damage associated to diabetes and/or obesity.[63, 64]

Under the rationale that mitochondrial dysfunction in peripheral tissues is an intervening factor for impaired insulin signaling,[65, 66] Peng et al.[67] looked into whether and how glucose levels, comparable to the extracellular levels found in diabetic brains, affect insulin signaling and mitochondrial function in cultured cortical neurons. In close agreement with observations made on peripheral organs, authors observed that hyperglycemia-induced mitochondrial function decline is the initiating step contributing to the alteration of insulin signaling in primary cortical neurons.[67] The authors also observed insulin resistance and loss of mitochondrial complexes components in cerebral cortex of db/db mice, an animal model of T2D. In the same study, it also was found that adenosine monophosphate-activated protein kinase (AMPK) inactivation is involved in high glucose-induced mitochondrial dysfunction and insulin resistance in primary cortical neurons, neuroblastoma cells, and in cerebral cortex of db/db mice. These impairments have been rescued by resveratrol, a mitochondrial activator.[67] Meanwhile, using the same animal model, Ernst et al.[68] observed CNS and peripheral molecular changes in diabetic db/db mice, including a significant reduction in mitochondria-related molecules such as mitochondrial complexes subunits, TCA cycle enzymes, and the antioxidant manganese superoxide dismutase (MnSOD). The authors concluded that these changes might contribute to the cognitive dysfunction observed in db/db mice.[68] Notwithstanding, it was found recently that the occurrence of glucose hypometabolism in brain cortical and hippocampal slices of db/db mice was associated with a significant increase in metabolism of ketone bodies and an elevated oxygen consumption and ATP synthesis rate.[69] As the authors outlined, the metabolic alterations in db/db mice might act as adaptive mechanisms in order to maintain energy metabolism in consequence of glucose hypometabolism, which suggests a possible intersection of T2D and AD pathology.[69]

Several lines of evidence have proved that mitochondria are central in the regulation of energy and metabolic homeostasis, possessing a complex quality control system to regulate mitochondrial shape, function, and mass that limits mitochondrial damage to ensure mitochondrial integrity and function.[70] An important factor driving mitochondrial biogenesis at the molecular level is the peroxisome proliferator-activated receptor gamma (PPAR$\gamma$) coactivator (PGC-1$\alpha$), which represents a coactivator of nuclear transcription factors (NRFs) 1 and 2 and mitochondrial transcription factor A (TFAM), a nuclear-encoded transcription factor crucial for replication, transcription, and maintenance of mtDNA.[71] Mitophagy, however, comprises the engulfment and digestion of small fusion-deficient mitochondria exhibiting sustained depolarization.[72] Likewise, the coordination of fusion-fission dynamic processes effectively lowers the percentage of defective mitochondria in the cell and ensure cell homeostasis. Metabolism, energy production, $Ca^{2+}$ signaling, ROS production, apoptosis, and senescence all depend on the balance of fission and fusion. Conversely, dynamic distortion (i.e., excessive fragmentation/elongation) results in cell functioning anomalies, and eventually cell death.[73] In this regard, Santos et al.[74] have reported that during the early stages of T2D, brain mitochondrial function is spared through an adaptive metabolic strategy between mitochondrial fusion-fission and biogenesis and autophagy. In detail, the authors found that, in the early stages of T2D, mitochondrial fission prevails, mitochondrial biogenesis is maintained, and autophagy decreases.[74] Mitochondrial fission could be involved in the recruitment and transport of mitochondria to critical subcellular compartments with high energy demand, such as synaptic terminals, where these organelles remain stationary, preserving synaptic and neuronal function and integrity, in part by supplying ATP.[74] It is well known that mitochondrial energy production controls cellular metabolism and changes in metabolism affect the dynamics of mitochondria.[75] In fact, data suggest that mitochondrial activity and ROS production entail ultrastructural changes of mitochondria in a way that, during excess of glucose, mitochondria acquires a more fragmented phenotype that leads to ROS overproduction via enhanced respiration and mitochondrial hyperpolarization.[76] In close agreement, a recent work demonstrated that in diabetic hippocampus, the mitochondrial shape changes in parallel with a decrease in mitochondrial respiratory enzymatic activity and ATP levels.[77] As proposed, such imbalance between mitochondrial fusion and fission processes in favor of excessive mitochondrial fission can occur through a glycogen synthase kinase 3β(GSK3β)/dynamin related protein 1 (Drp1)-dependent mechanism.[77] So, the compromise of mitochondrial function, especially by means of bioenergetics and dynamics, is a critical checkpoint for T2D-mediated brain effects, being recognized as a plausible bridge between T2D and neurodegenerative disorders (Fig. 1).

One important challenge of today's society is to increase the awareness about the importance of nutrition and health habits to promote a healthy life and aging. There is a great concern to critically appraise and understand the mechanisms by which obesity can affect brain responses, and might accelerate brain injury and age-related neurodegeneration.[78] In this context, an increase in the production of mitochondrial ROS is considered to be one of the several cellular responses to the imbalance in the energy metabolism because of excessive food intake and physical inactivity that drive obesity development.[13,79] As observed in in vivo studies, the exposure of animals to a diet-induced obesity is sufficient to promote cognitive disturbances in aging and young rodents,[80,81] along with a significant increase in ROS in brain regions, such as prefrontal cortex and hippocampus,[82] and with oxidative damage and mitochondrial dysfunction in the brain.[83] Meanwhile, others have documented that obesity induced by consumption of a HFD causes an increased brain oxidative stress and damage, detected by an increased brain corticosterone and malondialdehyde levels, and mitochondrial dysfunction, which subsequently leads to the impairment of brain insulin signaling, including insulin-induced long-term depression (LTD).[84] Along the same line, Alzoubi et al.[85] documented that a high-fat high-carbohydrate diet (HFCD) for 6 weeks induces an imbalance in the oxidative status of rat hippocampus by promoting a decrease in the levels of SOD and catalase activity and increased thiobarbituric acid reactive substances and glutathione oxidase levels. Compared to their counterparts, HFCD animals performed worse in the radial arm water maze, revealing deficits in spatial learning and memory.[85] Rats given vitamin E simultaneously with the HFCD, however, show an improvement in cognitive performance and reduced oxidative stress.[85] Described to exhibit progressive metabolic complications and insulin resistance along with several energy metabolism-associated complications, a study involving Zucker diabetic fatty (ZDF) rats demonstrated that the metabolic complications in their brains are strongly associated with increased oxidative stress, altered redox metabolism, and mitochondrial dysfunction, which are accompanied by altered cell signaling and compromised energy metabolism.[9] These studies support a role for mitochondria and oxidative damage in diet-induced cognitive defects and emphasize their involvement in obesity-mediated brain defects (Fig. 1).[82]

# 4 MITOCHONDRIAL PATHWAYS INVOLVED IN OBESITY AND TYPE 2 DIABETES-ASSOCIATED COMPLICATIONS

The current obesity and T2D epidemics across the world show no signs of dwindling, and, with an increased tendency to be diagnosed at a younger age, there is a greater exposure to the long-term adverse effects of these disorders. As research progresses, it is becoming clear that defective or insufficient mitochondrial function is inextricably linked to glucose metabolism and insulin action (deficiency and/or resistance) abnormalities that characterize those metabolic disorders and associated complications.

## 4.1 The Case of Peripheral Diabetic Neuropathy

Peripheral diabetic neuropathy (PDN) is one of the most common and distressing late complications of diabetes that results in the progressive deterioration of the sensory nervous system. Often described as the most frequent microvessel diabetic complication, PDN is present in about 10% of newly diagnosed patients and in about 50% of diabetic patients with long-standing disease.[86] Clinically, PDN often is related to incapacitating pain, sensory loss, infection, gangrene, and poor wound healing predisposing to foot ulceration and amputation, carrying with it significant health problems and morbidity when not detected correctly and in a time-dependent manner.[87] Sensory deficits usually surpass motor nerve dysfunction, appearing first in the distal portions of the extremities and progressing in a stocking-glove distribution with increasing duration or severity of diabetes.[88] It encompasses a variety of abnormalities of peripheral nerves that can arise from genetic or acquired dysfunctions of peripheral neurons and axons, dysfunctions of Schwann cells and myelin, or combinations of these defects.[89–91] Currently, PDN does not have an effective therapy and only palliative treatment is available. Because of the high obesity-mediated increase in the incidence of T2D, PDN is becoming a significant health problem.[92]

The literature shows that hyperglycemia plays a main role in the development and progression of PDN and, therefore, most of the investigation into the molecular and biochemical pathophysiology of the disease has been focused on glucose metabolic pathways and/or redox state of the cells.[88] Evidence discloses an important role for mitochondria in the progression of hyperglycemia-mediated neuronal damage of dorsal root ganglion (DRG) neurons in both cell cultures and animal models of diabetes.[93–95] In in vitro models of PDN, hyperglycemia is strongly linked with mitochondrial defects in DRG neurons, ensuing in ROS overproduction,[93] and activation of programmed cell death caspase pathway,[96] those effects being reverted by the overexpression of uncoupling protein 3 (UCP3), a member of the uncoupling proteins family with proven abilities in protecting the cells against mitochondrial ROS production.[97]

Hyperglycemia in diabetic rats and in in vitro-cultured DRG neurons promotes a strong downregulation of UCP3 expression, whereas its overexpression blocks glucose-induced programmed cell death by preventing mitochondrial hyperpolarization and formation of ROS, which suggest that the loss of UCP3 in DRG neurons might represent a significant contributing factor in high glucose-induced injury.[98] Others, however, have reported that the exposure of cultured Schwann cells to hyperglycemic conditions causes an increased expression of protein components of the respiratory chain, an effect that correlates with a decreased mitochondrial efficiency toward ATP generation and lack of glucose-induced ROS production in part because of an increase in mitochondrial MnSOD.[99] An in vivo study performed in 24-week-old db/db mice, however, revealed significant reductions in components of the glycolytic and TCA pathways and increased protein and lipid oxidation in sciatic nerves and DRG.[100] The contradictory results depend on the experimental model used, as well as the duration of hyperglycemia/diabetes.

Hyperglycemia and diabetes-induced changes in mitochondrial dynamics are major characteristics found in in vitro models and human patients with PDN.[101] In fact, evidence shows that in vitro hyperglycemia in cultured DRG neurons evokes increased mitochondrial fragmentation and reduced mitochondrial number by directly interfering with the mitochondrial fission protein Drp1.[102] Mitochondrial density is shown to increase in both myelinated and unmyelinated dorsal root axons of 24-week-old T2D mice, a situation not detected in the ventral root fibers in the same animal model or in either dorsal or ventral root fibers in young diabetic animals, prior to the development of neuropathy.[103] Although during acute hyperglycemia, (excessive) mitochondrial fission is the dominant response resulting in dysregulation of energy production, activation of caspase 3, and subsequent DRG neuron injury, during prolonged hyperglycemia, there is evidence of compensatory mitochondrial biogenesis in axons, which is insufficient to avoid the observed neurite damage.[103] Although an early mitochondrial fission promoted by the upregulation of Drp1 represents a protective or metabolic fission in order to cope with hyperglycemia, in a later stage, excessive mitochondrial fission is associated with the activation of Bim and Bax, two main pro-apoptotic proteins, culminating in apoptosis.[76] Others also have observed that DRG neurons from a genetic murine model of T2D with PDN have increased levels of mitochondrial DNA and biogenesis-related proteins and a smaller and more electron dense mitochondrial population, indicative of an increased mitochondrial fission.[104] The efficient knockdown of Drp1 in DRG neurons was able to decrease glucose-mediated neuronal injury, suggesting that, in DRG neurons, the imbalance in mitochondrial dynamics in favor of an increased mitochondrial fission plays a role in the pathogenesis of PDN.[104] It has been demonstrated recently that the genetic ablation of PGC-1α further exacerbates DPN, although the overexpression of PGC-1α in neurons prevents high glucose-induced oxidative injury.[95]

Overall, data strongly show that hyperglycemia-mediated disruption of mitochondrial function and integrity in sensory axons might underlie the development of DPN and point toward the necessity of developing therapies aimed to maintain healthy mitochondrial networks and prevent or (partially) reverse DPN.

## 4.2 Bringing Neurodegenerative Disorders Closer

It is expected that the number of people worldwide living with dementia will double nearly every 20 years, reaching 74.7 million in 2030 and 131.5 million in 2050.[105] Among the

100 neurodegenerative conditions characterized until now, there is the strong conviction that more than 20 are strongly associated with diabetes, insulin resistance, and obesity,[106] disturbed insulin sensitivity, and excessive or impaired insulin secretion, whereas estimates also show that, worldwide, approximately 2% (825,000) cases of AD, the main cause of age-related dementia, currently are attributable to diabetes.[51, 107] A systemic metabolic dysregulation overhangs the progression of life-threatening pathology of late-onset neurodegenerative diseases.[108, 109] For example, compelling evidence confirms that cerebral glucose metabolism in AD declines decades before the deterioration of cognitive functions, suggesting that energy failure is one of the earliest reversible hallmarks of AD.[110–112] In the early stages of AD, cerebral glucose use is reduced by 45%, and cerebral blood flow (CBF) by about 20%, whereas, in the later stages of the disease, metabolic and other abnormalities aggravate, resulting in 55%–65% reductions in CBF.[113] Baker et al.[114] showed that cognitively intact adults with prediabetes/T2D present a strong association between peripheral insulin resistance and disturbances in cerebral glucose metabolic rate in frontal, temporoparietal, and cingulate regions—brain areas known to be affected in AD. It also is estimated that 50%–80% of PD cases suffer from impaired glucose tolerance,[115] whereas dysregulation of glucose and energy metabolism also seems to be an early event in the pathogenesis of sporadic PD.[116] Magnetic resonance imaging and fludeoxyglucose (18F) PET studies have revealed that PD patients have a widespread cortical hypometabolism, probably occurring even at early disease stages[117, 118]; which, in turn, is closely associated to the cognitive decline experienced by PD patients.[116, 119]

The next subsections will discuss some of the most relevant findings dissecting the importance of mitochondria in the interrelation between metabolic disturbances and the most insidious neurodegenerative maladies, AD and PD (Fig. 1).

### *4.2.1 Alzheimer's Disease*

In 1906, German psychiatrist and neuropathologist Alois Alzheimer observed the postmortem brain of a woman patient who years before began complaining about deteriorating memory, even of recent events, disorientation, decreasing speech abilities, and lack of judgment. Microscopy techniques were less developed than now; however, it was possible for the young physician to detect the thinning of the cerebral cortex and the presence of two characteristic lesions, neurofibrillary tangles and extracellular deposits that later were named senile plaques.[120, 121] Since its first description, AD has gone from a rarely reported disorder to one of the most common disabling diseases among older adults, currently affecting 4%–8% of the elderly population worldwide and representing about 50%–70% of all dementia cases.[121]

Aging is the strongest known risk factor for AD, with most people diagnosed with late-onset, sporadic form, receiving the diagnosis after age 65. An early-onset familial form also can occur, but it is rarer (age of onset <65; accounting for 1%–5% of all cases). Familial AD results from mutations in one of three genes: amyloid-β protein precursor (APP), localized on chromosome 21, and presenilins 1 and 2 (PS1 and PS2), localized on chromosome 14 and 1, respectively.[122] Clinically, AD can be viewed as a continuum, beginning with memory loss of recent events (short-term memory impairment) and personality changes with a relative preservation of personal skills, which proceeds to a further decline in intellectual functioning, until it deprives the patients of their sense of self. In this latter stage, patients often experience hallucinations, delusions, and paranoia.[123] Neuropathologically, the main hallmarks found in the affected regions of AD brain are the senile plaques, which result from the accumulation of amyloid β-peptide (Aβ) and neurofibrillary tangles (NFTs), which are intracellular deposits mainly composed by hyperphosphorylated microtubule-associated protein tau (pTau), initially observed in the medial temporal lobes and eventually extending throughout the cortex.[124]

During the last decades, the research community devoted a great effort in deciphering the causes and accelerators of AD development and progression, which has contributed to the upsurge of theories about the potential triggers, risk factors, and therapies.[125] As previously stated, it has become clear that diabetes is an important risk factor for sporadic AD. Since the original Rotterdam study,[126] epidemiological/clinical observations have been gathered, showing that diabetic patients are significantly more likely to develop cognitive decline and exhibit increased susceptibility to dementia, particularly AD.[127, 128] Induction of T2D or obesity in animal models has been reported to trigger and/or accelerate AD-like pathology.[129, 130] As demonstrated by in vivo studies, it is possible to detect abnormalities in pTau levels, one of the key histopathological hallmarks of AD, in the cortex and hippocampus of T2D animal models of diabetes.[131] Accumulating evidence also suggests that low levels of circulating insulin growth factor 1 (IGF-1) and impairments of the insulin signaling pathway in the brain contribute to age-dependent cognitive decline and AD pathogenesis.[132] The chemical depletion of insulin and IGF-1 signaling mechanisms, plus oxidative injury, seems to be sufficient to cause AD-type neurodegeneration, as demonstrated in vivo using a model of intracerebroventricular (icv) administration of streptozotocin (STZ).[133] Such findings were corroborated with a microarray analysis of diabetes-related genes in the brains of postmortem AD patients and in a mouse model of AD.[134] Data revealed significant alterations in the mRNA expression profiles of genes related to insulin signaling, obesity, and diabetes in the frontal cortex, temporal cortex, and hippocampus in both AD human and mice samples.[134] The biggest differences were found to occur in the hippocampus, a key area related to memory.[134] Together, the co-existence of diabetes (or diabetes markers) and cognitive dysfunction have led some investigators to propose that AD constitutes a brain-specific form of diabetes, that is, type 3 diabetes,[135] considered a form of brain diabetes caused by a metabolic dyshomeostasis that manifests in the elderly as a result of an accumulative, lifelong impact in the brain, with molecular and biochemical features that resemble diabetes and other peripheral insulin resistance disorders.[133, 135]

It is generally accepted that insulin resistance, a hallmark of T2D and obesity, might contribute greatly to age-related cognitive deficits and AD.[136, 137] As previously outlined, insulin resistance entails disruption of metabolic homeostasis, largely because of the compromise of mitochondrial function; this further impairs cellular function at multiple levels with a broad range of consequences, from increased oxidative stress, DNA damage, to several forms of cell death.[132, 138] Data suggest that a strong correlation exists between deficits in insulin/IGF1 signaling and energy production with mitochondrial dysfunction, oxidative injury, and compensatory cytoprotective responses in AD brains in different Braak stages.[139] By using the icvSTZ administration in rats as an animal model of sporadic AD, it was possible to demonstrate that the insulin-resistant brain state that characterizes icvSTZ rats is associated with mitochondrial abnormalities and oxidative damage, both considered to be early events in AD.[140] As the authors reported, icvSTZ rats are characterized by a decline in brain mitochondrial bioenergetics function, a decrease in the activity of the three mitochondrial enzymes, pyruvate dehydrogenase (PDH) and α-ketoglutarate dehydrogenase (α-KGDH), two enzymes in the rate-limiting step of the tricarboxylic acid (TCA) cycle, and cytochrome oxidase (COX), the terminal enzyme in the mitochondrial respiratory chain that is responsible for reducing molecular oxygen, and by an increase in mitochondrial oxidative stress and damage.[140] Those alterations were associated with cognitive decline and an increase in the levels of the neurotoxic proteins Aβ and pTau.[140] In this regard, increasing data foster the idea that mitochondrial failure might represent a functional link between metabolic disorders and AD.[141] For example, previous reports indicate that brain mitochondrial deficits often encountered in T2D animals are potentiated in the presence of an additional stress, that is, the presence of Aβ peptide.[142] In this context, an earlier work by Carvalho et al.[143] demonstrated that the metabolic alterations associated with diabetes or prediabetes are linked to the occurrence of brain mitochondrial abnormalities, oxidative imbalance, and with an increase in Aβ protein levels; those alterations presenting a similar profile to the alterations found in 3xTgAD mice brains.[143] Such findings prompted the authors to suggest that the metabolic alterations associated with diabetes underpin the development of AD-like pathologic features and that mitochondria lingers in this scenario as a functional link between both pathologies.[143] The same authors observed that brain mitochondrial biogenesis and autophagy processes are significantly compromised in T2D and AD animal models, which revealed to be related to the loss of synaptic integrity and cognitive deficits found in both pathologies.[144] It has been previously reported that hippocampal PGC-1α expression is decreased as a function of AD clinical dementia and that its protein content correlates with progression of Aβ neuropathology in AD.[145] Further, increasing concentrations of $A\beta_{1-42}$ and $A\beta_{1-40}$ peptides in embryonic cortico-hippocampal neurons derived from Tg2576 AD mice cultured under hyperglycemic conditions was shown to cause a dose-dependent attenuation in PGC-1α expression.[145] The therapeutic preservation of neuronal PGC-1α expression, however, is able to promote the nonamyloidogenic processing of APP precluding the generation of Aβ peptides.[145]

As outlined earlier, obesity and consumption of an HFD are known to increase the risk of AD[22] and disease neuropathology and/or cognitive deficits in AD mouse models.[146, 147] As reported, the administration of an HFD to an animal model of AD, the Tg2576 mice, was shown to evoke a state of obesity and insulin resistance, as well as a burst in Aβ formation in their brains.[148] In the meantime, concomitant studies have demonstrated that the metabolic alterations induced by HFD have direct effects on brain insulin regulation and mitochondrial function contributing to AD pathology, with mitochondria acting as a key culprit of cognitive decline in both the HFD-treated and AD-like rodents at a relatively young age.[149] The brains of HFD-fed mice show not only markers of insulin resistance but also elevated levels of APP and $A\beta_{1-40}/A\beta_{1-42}$ together with BACE, GSK3β and tau proteins involved in APP processing and Aβ accumulation. The authors found that those alterations were accompanied by markers of increased oxidative stress and mitochondrial dysfunction and dynamics.[150] Others have demonstrated that obesity and excess energy intake shift the balance of mitochondrial dynamics, contributing to mitochondrial dysfunction and metabolic deterioration, all of which lead to insulin resistance.[151] More recently, Martins et al.[152] documented that an HFD promotes similar memory decline and impairments in mitochondrial morphology and synapse number in non-Tg mice akin to those observed in 3xTgAD mice. HFD causes a decrease in synaptic density in non-Tg mice, together with a decrease in mitochondrial number in hippocampal neuropil, effects similar to those detected in control-fed 3xTgAD mice. These observations suggest that the detrimental effects of an HFD on memory might be because of changes in mitochondrial morphology leading to the compromise of mitochondrial function and a reduction in synaptic number.[152]

### 4.2.2 *Parkinson's Disease*

PD is the second most common neurodegenerative disease, affecting 1% of individuals older than 60 years, with the risk increasing with age.[153] Clinically, the primary symptoms of PD include progressive muscle rigidity, bradykinesia, and resting tremor that result from the pathological degeneration and death of the pigmented neurons in the substantia nigra and dopamine deficiency, and by the presence of intraneuronal proteinaceous cytoplasmic inclusions that immunostain for α-synuclein and ubiquitin, designated Lewy bodies.[154] About 90% of PD cases are considered sporadic, whereas a small proportion (~10%) seems to have a monogenic form of the disease, with Mendelian inheritance, mainly because of mutations in the genes encoding for PTEN-induced kinase 1 (PINK1), Parkin, and DJ-1 proteins,[155] which cause forms of familial PD that are clinically and pathologically indistinguishable from sporadic PD. An increasing number of studies suggests that the onset of PD can begin up to 20 years before the appearance of the first clinical motor symptoms, although imaging and pathological studies suggest that the nigrostriatal degeneration can be detected 5–10 years before clinical manifestations.[154–157]

Since the first studies with 1-methyl 4-phenyl-1,2,3,6-tetrahydropyridine (MPTP), the synthetic opiate that causes Parkinsonism in drug-addicted individuals by inhibiting the complex I of the mitochondrial respiratory chain,[158] until today with the recent suggestion that the systemic administration of mitochondria can be a promising strategy for PD treatment,[159] substantial evidence reveals that mitochondrial dysfunction is a shared feature between sporadic and monogenic PD.[154, 160–162] Because of their high level of metabolic activity and complex structure, mitochondria, particularly as they age,[163] become vulnerable to a variety of external insults, such as the metabolic stress imposed by obesity and T2D.[164] Among the several disparate exogenous factors that can cause PD, some evidence points to a relationship between PD and metabolic factors, such as obesity and T2D. Recent evidence from a cross-sectional population-based analysis in a Spanish population has shown a significant positive association between PD and patients with long-duration diabetes.[165] Likewise, a Finnish prospective cohort study revealed that over time T2D nearly doubles PD risk in men and women, independently of known modifying risk factors, including body mass index, systolic blood pressure, total cholesterol, smoking, and alcohol and coffee consumption.[166] An earlier study with the same population demonstrated that excess weight is associated with an elevated risk of PD.[167] It has been documented that extreme body weight causes a deterioration in the mobility of PD patients compared with patients with a normal body weight,[168] a process that seems to involve the loss of midbrain dopaminergic neuronal function.[169] Additionally, data from animal and in vitro studies have documented that insulin dysregulation/resistance is associated with the pathophysiology, accelerated disease progression, and increased risk of PD.[153, 166, 170] Previous findings stated that the expression of DJ-1, an antioxidant protein with reduced expression in the CNS of PD patients, is significantly reduced in pancreatic islets of T2D individuals and seems to be required to maintain physiological ROS levels in an age- and diet-dependent manner.[171] It also is known that in neurons, DJ-1 is upregulated upon oxidative stress and protects neurons from mitochondrial anomalies,[172] whereas its loss leads to the early onset of PD.[173] Previous studies have shown that PINK1, a mitochondrial targeted serine/threonine kinase that protects cells against stress-induced apoptosis,[174] is one of the underlying causes of recessive familial form of PD[175] and is closely related with pancreatic β-cell function and insulin levels regulation.[176] The loss of PD-associated protein PINK1 significantly impairs the ability of mouse pancreatic β-cells and primary intact islets to take up glucose.[176] Similarly, Parkin was shown to be essential for pancreatic β-cells glucose oxidation capacity and subsequent insulin secretion signaling, leading to better glucose tolerance.[177] Given the similarities found between neurons and pancreatic islets,[178] a possible link between T2D/metabolic dysregulation and PD has been suggested. Khang et al.[179] applied a proteomic approach to investigate altered proteomes in the substantia nigra of db/db mice and HFD mice in order to identify a shared molecular pathway linking T2D to PD pathogenesis. Data revealed significant alterations in the levels of mitochondrial-related proteins in the substantia nigra of both animal models, whereas an unchanged alteration of the levels of PINK1 and DJ-1 and a significant loss of Parkin was detected.[179] Under basal conditions, PINK1 is imported into the inner mitochondrial membrane to be processed by mitochondrial proteases.[180] Upon mitochondrial membrane depolarization or damage, however, PINK1 accumulates in the outer membrane of mitochondria, where it recruits Parkin to trigger a mitophagy.[181] The impairment of PINK/Parkin-mediated mitophagy in mice lacking TP53INP1, for example, causes the release of a key stress protein with an antioxidant-associated tumor suppressive function, an accumulation of defective mitochondria, and a state of oxidative stress.[182] Under these conditions, mice become more prone to redox-driven obesity and insulin resistance.[182] Parkin loss also is linked to the accumulation of Parkin-interacting substrate (PARIS) and the reduction of PGC-1α.[179] Metformin, an antidiabetic compound, is able to restore the levels of Parkin and PGC-1α, confirming the intricate role that metabolic dysfunction plays in PD pathogenesis and the importance of mitochondria in this setting.[179]

## 5 CONCLUSIONS

Common sense tells us that the world population is aging, life expectancy is increasing, and, consequently, the social burden of health care and pharmacoeconomic systems also are increasing. As the population ages, however, there is also an exponential risk to live into the ages where the brain-related hitches and neurodegenerative diseases develop. Although most people experience a stable cognitive health and performance during their lifetime, with only a gradual decline in short-term memory and processing speed, others, however, face a severe path in which the decay in their cognitive abilities and behavior progresses to a more serious state of cognitive impairment and, ultimately, dementia.[121] Research in the last few years has been marked by the increased awareness of the importance in promoting active and healthy aging to counteract the evolving epidemics of obesity and T2D, both of which are anticipated to become more prevalent as the population ages, and in conveying an increased likelihood of suffering cognitive deficits and increased probability of developing dementia (18). In particular, it is believed that the compromise of the several mitochondrial tiers necessary for normal cell functioning, such as reduced oxidative capacity and antioxidant defense, enhanced generation of ROS, reduced oxidative phosphorylation capacity and ATP production and defects in mitochondrial transport and integrity, are in the baseline of age-related metabolic and neurodegenerative disorders (Fig. 1).[8] Targeting impaired pathways involved in mitochondrial homeostasis might represent a powerful therapeutic strategy to counteract the evolving obesity and T2D epidemics and their multiorgan-associated complications. The likelihood that metabolic disorders, such as obesity and T2D, share dysregulated mitochondrial pathways with secondary complications, such as neurodegeneration, opens up the possibility that therapies effective for one disorder also can be effective for others.

## Acknowledgments

The authors' work is supported by FEDER funds through the Operational Programme Competitiveness Factors—COMPETE and Healthy Aging 2020 (CENTRO-01-0145-FEDER-000012) and by national funds by FCT—Foundation for Science and Technology under the project (PEst-C/SAU/LA0001/2013-2014) and strategic project UID/NEU/04539/2013. Susana Cardoso is recipient of a fellowship from the FCT (SFRH/BPD/95770/2013).

## References

1. Hu FB. Globalization of diabetes: the role of diet, lifestyle, and genes. *Diabetes Care* 2011;**34**(6):1249–57.
2. Matthews DR, Matthews PC. Banting memorial lecture 2010. Type 2 diabetes as an 'infectious' disease: is this the black death of the 21st century? *Diabet Med* 2011;**28**(1):2–9.
3. Webber L, Divajeva D, Marsh T, McPherson K, Brown M, Galea G, et al. The future burden of obesity-related diseases in the 53 WHO European-region countries and the impact of effective interventions: a modelling study. *BMJ Open* 2014;**4**(7):e004787.
4. Hossain P, Kawar B, El Nahas M. Obesity and diabetes in the developing world—a growing challenge. *N Engl J Med* 2007;**356**(3):213–5.
5. Sims EA, Danforth Jr. E, Horton ES, Bray GA, Glennon JA, Salans LB. Endocrine and metabolic effects of experimental obesity in man. *Recent Prog Horm Res* 1973;**29**:457–96.
6. Everson-Rose SA, Ryan JP. Diabetes, obesity, and the brain: new developments in biobehavioral medicine. *Psychosom Med* 2015;**77**(6):612–5.
7. Calcutt NA, Cooper ME, Kern TS, Schmidt AM. Therapies for hyperglycaemia-induced diabetic complications: from animal models to clinical trials. *Nat Rev Drug Discov* 2009;**8**(5):417–29.
8. Bhatti JS, Bhatti GK, Reddy PH. Mitochondrial dysfunction and oxidative stress in metabolic disorders—a step towards mitochondria based therapeutic strategies. *Biochim Biophys Acta* 2017;**1863**(5):1066–77.
9. Raza H, John A, Howarth FC. Increased oxidative stress and mitochondrial dysfunction in zucker diabetic rat liver and brain. *Cell Physiol Biochem* 2015;**35**(3):1241–51.
10. Martin LJ. Biology of mitochondria in neurodegenerative diseases. *Prog Mol Biol Transl Sci* 2012;**107**:355–415.
11. Reddy PH, Reddy TP. Mitochondria as a therapeutic target for aging and neurodegenerative diseases. *Curr Alzheimer Res* 2011;**8**(4):393–409.
12. McBride HM, Neuspiel M, Wasiak S. Mitochondria: more than just a powerhouse. *Curr Biol* 2006;**16**(14):R551–60.
13. Patti ME, Corvera S. The role of mitochondria in the pathogenesis of type 2 diabetes. *Endocr Rev* 2010;**31**(3):364–95.
14. Haslam DW, James WP. Obesity. *Lancet* 2005;**366**(9492):1197–209.
15. Valsamakis G, Konstantakou P, Mastorakos G. New targets for drug treatment of obesity. *Annu Rev Pharmacol Toxicol* 2017;**57**:585–605.
16. Elias MF, Goodell AL, Waldstein SR. Obesity, cognitive functioning and dementia: back to the future. *J Alzheimers Dis* 2012;**30**(Suppl 2):S113–25.
17. Albanese E, Launer LJ, Egger M, Prince MJ, Giannakopoulos P, Wolters FJ, et al. Body mass index in midlife and dementia: systematic review and meta-regression analysis of 589,649 men and women followed in longitudinal studies. *Alzheimers Dement (Amst)* 2017;**8**:165–78.
18. Gustafson D, Rothenberg E, Blennow K, Steen B, Skoog I. An 18-year follow-up of overweight and risk of Alzheimer disease. *Arch Intern Med* 2003;**163**(13):1524–8.
19. Raji CA, Ho AJ, Parikshak NN, Becker JT, Lopez OL, Kuller LH, et al. Brain structure and obesity. *Hum Brain Mapp* 2010;**31**(3):353–64.
20. Gazdzinski S, Millin R, Kaiser LG, Durazzo TC, Mueller SG, Weiner MW, et al. BMI and neuronal integrity in healthy, cognitively normal elderly: a proton magnetic resonance spectroscopy study. *Obesity* 2010;**18**(4):743–8.

21. Jagust W, Harvey D, Mungas D, Haan M. Central obesity and the aging brain. *Arch Neurol* 2005;**62**(10):1545–8.
22. Jurdak N, Lichtenstein AH, Kanarek RB. Diet-induced obesity and spatial cognition in young male rats. *Nutr Neurosci* 2008;**11**(2):48–54.
23. Cournot M, Marquie JC, Ansiau D, Martinaud C, Fonds H, Ferrieres J, et al. Relation between body mass index and cognitive function in healthy middle-aged men and women. *Neurology* 2006;**67**(7):1208–14.
24. Gunstad J, Paul RH, Cohen RA, Tate DF, Gordon E. Obesity is associated with memory deficits in young and middle-aged adults. *Eat Weight Disord* 2006;**11**(1):e15–9.
25. Cheke LG, Simons JS, Clayton NS. Higher body mass index is associated with episodic memory deficits in young adults. *Q J Exp Psychol (Hove)* 2016;**69**(11):2305–16.
26. Sanderlin AH, Todem D, Bozoki AC. Obesity and co-morbid conditions are associated with specific neuropsychiatric symptoms in mild cognitive impairment. *Front Aging Neurosci* 2017;**9**:164.
27. Palmer K, Berger AK, Monastero R, Winblad B, Backman L, Fratiglioni L. Predictors of progression from mild cognitive impairment to Alzheimer disease. *Neurology* 2007;**68**(19):1596–602.
28. Hugo J, Ganguli M. Dementia and cognitive impairment: epidemiology, diagnosis, and treatment. *Clin Geriatr Med* 2014;**30**(3):421–42.
29. Yoon S, Cho H, Kim J, Lee DW, Kim GH, Hong YS, et al. Brain changes in overweight/obese and normal-weight adults with type 2 diabetes mellitus. *Diabetologia* 2017;**60**(7):1207–17.
30. Ho AJ, Raji CA, Saharan P, DeGiorgio A, Madsen SK, Hibar DP, et al. Hippocampal volume is related to body mass index in Alzheimer's disease. *NeuroReport* 2011;**22**(1):10–4.
31. Feinkohl I, Price JF, Strachan MW, Frier BM. The impact of diabetes on cognitive decline: potential vascular, metabolic, and psychosocial risk factors. *Alzheimers Res Ther* 2015;**7**(1):46.
32. Vagelatos NT, Eslick GD. Type 2 diabetes as a risk factor for Alzheimer's disease: the confounders, interactions, and neuropathology associated with this relationship. *Epidemiol Rev* 2013;**35**:152–60.
33. Li H, Wu J, Zhu L, Sha L, Yang S, Wei J, et al. Insulin degrading enzyme contributes to the pathology in a mixed model of Type 2 diabetes and Alzheimer's disease: possible mechanisms of IDE in T2D and AD. *Biosci Rep* 2018;**38**(1).
34. Scheen AJ. Central nervous system: a conductor orchestrating metabolic regulations harmed by both hyperglycaemia and hypoglycaemia. *Diabetes Metab* 2010;**36**(Suppl. 3):S31–8.
35. Alafuzoff I, Aho L, Helisalmi S, Mannermaa A, Soininen H. Beta-amyloid deposition in brains of subjects with diabetes. *Neuropathol Appl Neurobiol* 2009;**35**(1):60–8.
36. Xu W, Qiu C, Winblad B, Fratiglioni L. The effect of borderline diabetes on the risk of dementia and Alzheimer's disease. *Diabetes* 2007;**56**(1):211–6.
37. Gold SM, Dziobek I, Sweat V, Tirsi A, Rogers K, Bruehl H, et al. Hippocampal damage and memory impairments as possible early brain complications of type 2 diabetes. *Diabetologia* 2007;**50**(4):711–9.
38. Moran C, Phan TG, Chen J, Blizzard L, Beare R, Venn A, et al. Brain atrophy in type 2 diabetes: regional distribution and influence on cognition. *Diabetes Care* 2013;**36**(12):4036–42.
39. Reijmer YD, Brundel M, de Bresser J, Kappelle LJ, Leemans A, Biessels GJ, et al. Microstructural white matter abnormalities and cognitive functioning in type 2 diabetes: a diffusion tensor imaging study. *Diabetes Care* 2013;**36**(1):137–44.
40. Chen YC, Jiao Y, Cui Y, Shang SA, Ding J, Feng Y, et al. Aberrant brain functional connectivity related to insulin resistance in type 2 diabetes: a resting-state fMRI study. *Diabetes Care* 2014;**37**(6):1689–96.
41. Hoogenboom WS, Marder TJ, Flores VL, Huisman S, Eaton HP, Schneiderman JS, et al. Cerebral white matter integrity and resting-state functional connectivity in middle-aged patients with type 2 diabetes. *Diabetes* 2014;**63**(2):728–38.
42. Secnik J, Cermakova P, Fereshtehnejad SM, Dannberg P, Johnell K, Fastbom J, et al. Diabetes in a large dementia cohort: clinical characteristics and treatment from the Swedish dementia registry. *Diabetes Care* 2017;**40**(9):1159–66.
43. Baglietto-Vargas D, Shi J, Yaeger DM, Ager R, LaFerla FM. Diabetes and Alzheimer's disease crosstalk. *Neurosci Biobehav Rev* 2016;**64**:272–87.
44. Biessels GJ, Strachan MW, Visseren FL, Kappelle LJ, Whitmer RA. Dementia and cognitive decline in type 2 diabetes and prediabetic stages: towards targeted interventions. *Lancet Diabetes Endocrinol* 2014;**2**(3):246–55.
45. Fukasawa R, Hanyu H, Shimizu S, Kanetaka H, Sakurai H, Ishii K. Identification of diabetes-related dementia: longitudinal perfusion SPECT and amyloid PET studies. *J Neurol Sci* 2015;**349**(1–2):45–51.
46. Hatanaka H, Hanyu H, Fukasawa R, Sato T, Shimizu S, Sakurai H. Peripheral oxidative stress markers in diabetes-related dementia. *Geriatr Gerontol Int* 2016;**16**(12):1312–8.

47. Kravitz E, Schmeidler J, Schnaider Beeri M. Type 2 diabetes and cognitive compromise: potential roles of diabetes-related therapies. *Endocrinol Metab Clin N Am* 2013;**42**(3):489–501.
48. Whitmer RA, Gunderson EP, Barrett-Connor E, Quesenberry Jr CP, Yaffe K. Obesity in middle age and future risk of dementia: a 27 year longitudinal population based study. *BMJ* 2005;**330**(7504):1360.
49. Whitmer RA. Type 2 diabetes and risk of cognitive impairment and dementia. *Curr Neurol Neurosci Rep* 2007;**7**(5):373–80.
50. Mijajlovic MD, Aleksic VM, Sternic NM, Mirkovic MM, Bornstein NM. Role of prediabetes in stroke. *Neuropsychiatr Dis Treat* 2017;**13**:259–67.
51. Maestre GE. Reduction of cognitive decline in patients with or at high risk for diabetes. *Curr Geriatr Rep* 2017;**6**(3):188–95.
52. Portero McLellan KC, Wyne K, Villagomez ET, Hsueh WA. Therapeutic interventions to reduce the risk of progression from prediabetes to type 2 diabetes mellitus. *Ther Clin Risk Manag* 2014;**10**:173–88.
53. Harman D. Aging: a theory based on free radical and radiation chemistry. *J Gerontol* 1956;**11**(3):298–300.
54. Harman D. The biologic clock: the mitochondria? *J Am Geriatr Soc* 1972;**20**(4):145–7.
55. Grimm A, Friedland K, Eckert A. Mitochondrial dysfunction: the missing link between aging and sporadic Alzheimer's disease. *Biogerontology* 2016;**17**(2):281–96.
56. Lee HC, Wei YH. Role of mitochondria in human aging. *J Biomed Sci* 1997;**4**(6):319–26.
57. Luft R, Ikkos D, Palmieri G, Ernster L, Afzelius B. A case of severe hypermetabolism of nonthyroid origin with a defect in the maintenance of mitochondrial respiratory control: a correlated clinical, biochemical, and morphological study. *J Clin Invest* 1962;**41**:1776–804.
58. Herst PM, Rowe MR, Carson GM, Berridge MV. Functional mitochondria in health and disease. *Front Endocrinol* 2017;**8**:296.
59. Rugarli EI, Langer T. Mitochondrial quality control: a matter of life and death for neurons. *EMBO J* 2012;**31**(6):1336–49.
60. Bhatti JS, Kumar S, Vijayan M, Bhatti GK, Reddy PH. Therapeutic strategies for mitochondrial dysfunction and oxidative stress in age-related metabolic disorders. *Prog Mol Biol Transl Sci* 2017;**146**:13–46.
61. Burte F, Carelli V, Chinnery PF, Yu-Wai-Man P. Disturbed mitochondrial dynamics and neurodegenerative disorders. *Nat Rev Neurol* 2015;**11**(1):11–24.
62. Cui H, Kong Y, Zhang H. Oxidative stress, mitochondrial dysfunction, and aging. *J Signal Transduct* 2012;**2012**:646354.
63. Ceriello A. New insights on oxidative stress and diabetic complications may lead to a "causal" antioxidant therapy. *Diabetes Care* 2003;**26**(5):1589–96.
64. Brownlee M. Biochemistry and molecular cell biology of diabetic complications. *Nature* 2001;**414**(6865):813–20.
65. Hao CN, Geng YJ, Li F, Yang T, Su DF, Duan JL, et al. Insulin-like growth factor-1 receptor activation prevents hydrogen peroxide-induced oxidative stress, mitochondrial dysfunction and apoptosis. *Apoptosis* 2011;**16**(11):1118–27.
66. Franko A, Kunze A, Bose M, von Kleist-Retzow JC, Paulsson M, Hartmann U, et al. Impaired insulin signaling is associated with hepatic mitochondrial dysfunction in IR(+/−)-IRS-1(+/−) double heterozygous (IR-IRS1dh) mice. *Int J Mol Sci* 2017;**18**(6).
67. Peng Y, Liu J, Shi L, Tang Y, Gao D, Long J, et al. Mitochondrial dysfunction precedes depression of AMPK/AKT signaling in insulin resistance induced by high glucose in primary cortical neurons. *J Neurochem* 2016;**137**(5):701–13.
68. Ernst A, Sharma AN, Elased KM, Guest PC, Rahmoune H, Bahn S. Diabetic db/db mice exhibit central nervous system and peripheral molecular alterations as seen in neurological disorders. *Transl Psychiatry* 2013;**3**:e263.
69. Andersen JV, Christensen SK, Nissen JD, Waagepetersen HS. Improved cerebral energetics and ketone body metabolism in db/db mice. *J Cereb Blood Flow Metab* 2017;**37**(3):1137–47.
70. Schon EA, Przedborski S. Mitochondria: the next (neurode)generation. *Neuron* 2011;**70**(6):1033–53.
71. Scarpulla RC. Nuclear activators and coactivators in mammalian mitochondrial biogenesis. *Biochim Biophys Acta* 2002;**1576**(1–2):1–14.
72. Wild P, Dikic I. Mitochondria get a Parkin' ticket. *Nat Cell Biol* 2010;**12**(2):104–6.
73. Bonda DJ, Wang X, Perry G, Smith MA, Zhu X. Mitochondrial dynamics in Alzheimer's disease: opportunities for future treatment strategies. *Drugs Aging* 2010;**27**(3):181–92.
74. Santos RX, Correia SC, Alves MG, Oliveira PF, Cardoso S, Carvalho C, et al. Mitochondrial quality control systems sustain brain mitochondrial bioenergetics in early stages of type 2 diabetes. *Mol Cell Biochem* 2014;**394**(1–2):13–22.

75. Roy M, Reddy PH, Iijima M, Sesaki H. Mitochondrial division and fusion in metabolism. *Curr Opin Cell Biol* 2015;**33**:111–8.
76. Yu T, Robotham JL, Yoon Y. Increased production of reactive oxygen species in hyperglycemic conditions requires dynamic change of mitochondrial morphology. *Proc Natl Acad Sci U S A* 2006;**103**(8):2653–8.
77. Huang S, Wang Y, Gan X, Fang D, Zhong C, Wu L, et al. Drp1-mediated mitochondrial abnormalities link to synaptic injury in diabetes model. *Diabetes* 2015;**64**(5):1728–42.
78. Bruce-Keller AJ, Keller JN, Morrison CD. Obesity and vulnerability of the CNS. *Biochim Biophys Acta* 2009;**1792**(5):395–400.
79. de Mello AH, Costa AB, Engel JDG, Rezin GT. Mitochondrial dysfunction in obesity. *Life Sci* 2018;**192**:26–32.
80. Zhang L, Bruce-Keller AJ, Dasuri K, Nguyen AT, Liu Y, Keller JN. Diet-induced metabolic disturbances as modulators of brain homeostasis. *Biochim Biophys Acta* 2009;**1792**(5):417–22.
81. Greenwood CE, Winocur G. High-fat diets, insulin resistance and declining cognitive function. *Neurobiol Aging* 2005;**26**(Suppl. 1):42–5.
82. Freeman LR, Zhang L, Nair A, Dasuri K, Francis J, Fernandez-Kim SO, et al. Obesity increases cerebrocortical reactive oxygen species and impairs brain function. *Free Radic Biol Med* 2013;**56**:226–33.
83. Ma W, Yuan L, Yu H, Xi Y, Xiao R. Mitochondrial dysfunction and oxidative damage in the brain of diet-induced obese rats but not in diet-resistant rats. *Life Sci* 2014;**110**(2):53–60.
84. Pratchayasakul W, Kerdphoo S, Petsophonsakul P, Pongchaidecha A, Chattipakorn N, Chattipakorn SC. Effects of high-fat diet on insulin receptor function in rat hippocampus and the level of neuronal corticosterone. *Life Sci* 2011;**88**(13–14):619–27.
85. Alzoubi KH, Khabour OF, Salah HA, Hasan Z. Vitamin E prevents high-fat high-carbohydrates diet-induced memory impairment: the role of oxidative stress. *Physiol Behav* 2013;**119**:72–8.
86. Boulton AJ, Vinik AI, Arezzo JC, Bril V, Feldman EL, Freeman R, et al. Diabetic neuropathies: a statement by the American Diabetes Association. *Diabetes Care* 2005;**28**(4):956–62.
87. Said G. Diabetic neuropathy—a review. *Nat Clin Pract Neurol* 2007;**3**(6):331–40.
88. Edwards JL, Vincent AM, Cheng HT, Feldman EL. Diabetic neuropathy: mechanisms to management. *Pharmacol Ther* 2008;**120**(1):1–34.
89. Roman-Pintos LM, Villegas-Rivera G, Rodriguez-Carrizalez AD, Miranda-Diaz AG, Cardona-Munoz EG. Diabetic polyneuropathy in type 2 diabetes mellitus: inflammation, oxidative stress, and mitochondrial function. *J Diabetes Res* 2016;**2016**:3425617.
90. Sajic M. Mitochondrial dynamics in peripheral neuropathies. *Antioxid Redox Signal* 2014;**21**(4):601–20.
91. Chowdhury SK, Dobrowsky RT, Fernyhough P. Nutrient excess and altered mitochondrial proteome and function contribute to neurodegeneration in diabetes. *Mitochondrion* 2011;**11**(6):845–54.
92. Chowdhury SK, Smith DR, Fernyhough P. The role of aberrant mitochondrial bioenergetics in diabetic neuropathy. *Neurobiol Dis* 2013;**51**:56–65.
93. Vincent AM, McLean LL, Backus C, Feldman EL. Short-term hyperglycemia produces oxidative damage and apoptosis in neurons. *FASEB J* 2005;**19**(6):638–40.
94. Fernyhough P. Mitochondrial dysfunction in diabetic neuropathy: a series of unfortunate metabolic events. *Curr Diab Rep* 2015;**15**(11):89.
95. Choi J, Chandrasekaran K, Inoue T, Muragundla A, Russell JW. PGC-1alpha regulation of mitochondrial degeneration in experimental diabetic neuropathy. *Neurobiol Dis* 2014;**64**:118–30.
96. Russell JW, Sullivan KA, Windebank AJ, Herrmann DN, Feldman EL. Neurons undergo apoptosis in animal and cell culture models of diabetes. *Neurobiol Dis* 1999;**6**(5):347–63.
97. Cardoso S, Correia S, Carvalho C, Candeias E, Placido AI, Duarte AI, et al. Perspectives on mitochondrial uncoupling proteins-mediated neuroprotection. *J Bioenerg Biomembr* 2015 Apr;**47**(1–2):119–31.
98. Vincent AM, Olzmann JA, Brownlee M, Sivitz WI, Russell JW. Uncoupling proteins prevent glucose-induced neuronal oxidative stress and programmed cell death. *Diabetes* 2004;**53**(3):726–34.
99. Zhang L, Yu C, Vasquez FE, Galeva N, Onyango I, Swerdlow RH, et al. Hyperglycemia alters the schwann cell mitochondrial proteome and decreases coupled respiration in the absence of superoxide production. *J Proteome Res* 2010;**9**(1):458–71.
100. Hinder LM, Vivekanandan-Giri A, McLean LL, Pennathur S, Feldman EL. Decreased glycolytic and tricarboxylic acid cycle intermediates coincide with peripheral nervous system oxidative stress in a murine model of type 2 diabetes. *J Endocrinol* 2013;**216**(1):1–11.

101. Hamid HS, Mervak CM, Munch AE, Robell NJ, Hayes JM, Porzio MT, et al. Hyperglycemia- and neuropathy-induced changes in mitochondria within sensory nerves. *Ann Clin Transl Neurol* 2014;**1**(10):799–812.
102. Leinninger GM, Backus C, Sastry AM, Yi YB, Wang CW, Feldman EL. Mitochondria in DRG neurons undergo hyperglycemic mediated injury through Bim, Bax and the fission protein Drp1. *Neurobiol Dis* 2006;**23**(1):11–22.
103. Vincent AM, Edwards JL, McLean LL, Hong Y, Cerri F, Lopez I, et al. Mitochondrial biogenesis and fission in axons in cell culture and animal models of diabetic neuropathy. *Acta Neuropathol* 2010;**120**(4):477–89.
104. Edwards JL, Quattrini A, Lentz SI, Figueroa-Romero C, Cerri F, Backus C, et al. Diabetes regulates mitochondrial biogenesis and fission in mouse neurons. *Diabetologia* 2010;**53**(1):160–9.
105. Prince M, Ali GC, Guerchet M, Prina AM, Albanese E, Wu YT. Recent global trends in the prevalence and incidence of dementia, and survival with dementia. *Alzheimers Res Ther* 2016;**8**(1):23.
106. Ristow M. Neurodegenerative disorders associated with diabetes mellitus. *J Mol Med* 2004;**82**(8):510–29.
107. Barnes DE, Yaffe K. The projected effect of risk factor reduction on Alzheimer's disease prevalence. *Lancet Neurol* 2011;**10**(9):819–28.
108. Sadagurski M, White MF. Integrating metabolism and longevity through insulin and IGF1 signaling. *Endocrinol Metab Clin N Am* 2013;**42**(1):127–48.
109. Mattson MP, Pedersen WA, Duan W, Culmsee C, Camandola S. Cellular and molecular mechanisms underlying perturbed energy metabolism and neuronal degeneration in Alzheimer's and Parkinson's diseases. *Ann N Y Acad Sci* 1999;**893**:154–75.
110. Andersen JV, Christensen SK, Aldana BI, Nissen JD, Tanila H, Waagepetersen HS. Alterations in cerebral cortical glucose and glutamine metabolism precedes amyloid plaques in the APPswe/PSEN1dE9 mouse model of Alzheimer's disease. *Neurochem Res* 2017;**42**(6):1589–98.
111. Zhang L, Trushin S, Christensen TA, Bachmeier BV, Gateno B, Schroeder A, et al. Altered brain energetics induces mitochondrial fission arrest in Alzheimer's disease. *Sci Rep* 2016;**6**:18725.
112. Fukuyama H, Ogawa M, Yamauchi H, Yamaguchi S, Kimura J, Yonekura Y, et al. Altered cerebral energy metabolism in Alzheimer's disease: a PET study. *J Nucl Med* 1994;**35**(1):1–6.
113. Hoyer S. Abnormalities of glucose metabolism in Alzheimer's disease. *Ann N Y Acad Sci* 1991;**640**:53–8.
114. Baker LD, Cross DJ, Minoshima S, Belongia D, Watson GS, Craft S. Insulin resistance and Alzheimer-like reductions in regional cerebral glucose metabolism for cognitively normal adults with prediabetes or early type 2 diabetes. *Arch Neurol* 2011;**68**(1):51–7.
115. Sandyk R. The relationship between diabetes mellitus and Parkinson's disease. *Int J Neurosci* 1993;**69**(1–4):125–30.
116. Dunn L, Allen GF, Mamais A, Ling H, Li A, Duberley KE, et al. Dysregulation of glucose metabolism is an early event in sporadic Parkinson's disease. *Neurobiol Aging* 2014;**35**(5):1111–5.
117. Borghammer P, Chakravarty M, Jonsdottir KY, Sato N, Matsuda H, Ito K, et al. Cortical hypometabolism and hypoperfusion in Parkinson's disease is extensive: probably even at early disease stages. *Brain Struct Funct* 2010;**214**(4):303–17.
118. Borghammer P, Hansen SB, Eggers C, Chakravarty M, Vang K, Aanerud J, et al. Glucose metabolism in small subcortical structures in Parkinson's disease. *Acta Neurol Scand* 2012;**125**(5):303–10.
119. Peppard RF, Martin WR, Carr GD, Grochowski E, Schulzer M, Guttman M, et al. Cerebral glucose metabolism in Parkinson's disease with and without dementia. *Arch Neurol* 1992;**49**(12):1262–8.
120. O'Brien C. Auguste D. and Alzheimer's disease. *Science* 1996;**273**(5271):28.
121. Daviglus ML, Bell CC, Berrettini W, Bowen PE, Connolly Jr. ES, Cox NJ, et al. NIH state-of-the-science conference statement: preventing Alzheimer's disease and cognitive decline. *NIH Consens State Sci Statements* 2010;**27**(4):1–30.
122. Hampel H, Wilcock G, Andrieu S, Aisen P, Blennow K, Broich K, et al. Biomarkers for Alzheimer's disease therapeutic trials. *Prog Neurobiol* 2011;**95**(4):579–93.
123. Sperling RA, Aisen PS, Beckett LA, Bennett DA, Craft S, Fagan AM, et al. Toward defining the preclinical stages of Alzheimer's disease: recommendations from the National Institute on Aging-Alzheimer's Association workgroups on diagnostic guidelines for Alzheimer's disease. *Alzheimers Dement* 2011;**7**(3):280–92.
124. Serrano-Pozo A, Frosch MP, Masliah E, Hyman BT. Neuropathological alterations in Alzheimer disease. *Cold Spring Harb Perspect Med* 2011;**1**(1):a006189.
125. Schachter AS, Davis KL. Alzheimer's disease. *Curr Treat Options Neurol* 2000;**2**(1):51–60.
126. Ott A, Stolk RP, van Harskamp F, Pols HA, Hofman A, Breteler MM. Diabetes mellitus and the risk of dementia: the Rotterdam study. *Neurology* 1999;**53**(9):1937–42.

127. Zhang J, Chen C, Hua S, Liao H, Wang M, Xiong Y, et al. An updated meta-analysis of cohort studies: diabetes and risk of Alzheimer's disease. *Diabetes Res Clin Pract* 2017;**124**:41–7.
128. Zhao WQ, Townsend M. Insulin resistance and amyloidogenesis as common molecular foundation for type 2 diabetes and Alzheimer's disease. *Biochim Biophys Acta* 2009;**1792**(5):482–96.
129. Ho L, Qin W, Pompl PN, Xiang Z, Wang J, Zhao Z, et al. Diet-induced insulin resistance promotes amyloidosis in a transgenic mouse model of Alzheimer's disease. *FASEB J* 2004;**18**(7):902–4.
130. Bitel CL, Kasinathan C, Kaswala RH, Klein WL, Frederikse PH. Amyloid-beta and tau pathology of Alzheimer's disease induced by diabetes in a rabbit animal model. *J Alzheimers Dis* 2012;**32**(2):291–305.
131. Kim B, Backus C, Oh S, Hayes JM, Feldman EL. Increased tau phosphorylation and cleavage in mouse models of type 1 and type 2 diabetes. *Endocrinology* 2009;**150**(12):5294–301.
132. de la Monte SM. Brain insulin resistance and deficiency as therapeutic targets in Alzheimer's disease. *Curr Alzheimer Res* 2012;**9**(1):35–66.
133. Lester-Coll N, Rivera EJ, Soscia SJ, Doiron K, Wands JR, de la Monte SM. Intracerebral streptozotocin model of type 3 diabetes: relevance to sporadic Alzheimer's disease. *J Alzheimers Dis* 2006;**9**(1):13–33.
134. Hokama M, Oka S, Leon J, Ninomiya T, Honda H, Sasaki K, et al. Altered expression of diabetes-related genes in Alzheimer's disease brains: the Hisayama study. *Cereb Cortex* 2014;**24**(9):2476–88.
135. de la Monte SM. Type 3 diabetes is sporadic Alzheimers disease: mini-review. *Eur Neuropsychopharmacol* 2014;**24**(12):1954–60.
136. Craft S, Cholerton B, Baker LD. Insulin and Alzheimer's disease: untangling the web. *J Alzheimers Dis* 2013;**33**(Suppl. 1):S263–75.
137. Liu Y, Liu F, Grundke-Iqbal I, Iqbal K, Gong CX. Deficient brain insulin signalling pathway in Alzheimer's disease and diabetes. *J Pathol* 2011;**225**(1):54–62.
138. Yin F, Boveris A, Cadenas E. Mitochondrial energy metabolism and redox signaling in brain aging and neurodegeneration. *Antioxid Redox Signal* 2014;**20**(2):353–71.
139. de la Monte SM, Wands JR. Molecular indices of oxidative stress and mitochondrial dysfunction occur early and often progress with severity of Alzheimer's disease. *J Alzheimers Dis* 2006;**9**(2):167–81.
140. Correia SC, Santos RX, Santos MS, Casadesus G, Lamanna JC, Perry G, et al. Mitochondrial abnormalities in a streptozotocin-induced rat model of sporadic Alzheimer's disease. *Curr Alzheimer Res* 2013;**10**(4):406–19.
141. Moreira PI, Santos MS, Seica R, Oliveira CR. Brain mitochondrial dysfunction as a link between Alzheimer's disease and diabetes. *J Neurol Sci* 2007;**257**(1–2):206–14.
142. Moreira PI, Santos MS, Moreno AM, Seica R, Oliveira CR. Increased vulnerability of brain mitochondria in diabetic (Goto-Kakizaki) rats with aging and amyloid-beta exposure. *Diabetes* 2003;**52**(6):1449–56.
143. Carvalho C, Cardoso S, Correia SC, Santos RX, Santos MS, Baldeiras I, et al. Metabolic alterations induced by sucrose intake and Alzheimer's disease promote similar brain mitochondrial abnormalities. *Diabetes* 2012;**61**(5):1234–42.
144. Carvalho C, Santos MS, Oliveira CR, Moreira PI. Alzheimer's disease and type 2 diabetes-related alterations in brain mitochondria, autophagy and synaptic markers. *Biochim Biophys Acta* 2015;**1852**(8):1665–75.
145. Qin W, Haroutunian V, Katsel P, Cardozo CP, Ho L, Buxbaum JD, et al. PGC-1alpha expression decreases in the Alzheimer disease brain as a function of dementia. *Arch Neurol* 2009;**66**(3):352–61.
146. Knight EM, Martins IV, Gumusgoz S, Allan SM, Lawrence CB. High-fat diet-induced memory impairment in triple-transgenic Alzheimer's disease (3xTgAD) mice is independent of changes in amyloid and tau pathology. *Neurobiol Aging* 2014;**35**(8):1821–32.
147. Barron AM, Rosario ER, Elteriefi R, Pike CJ. Sex-specific effects of high fat diet on indices of metabolic syndrome in 3xTg-AD mice: implications for Alzheimer's disease. *PLoS ONE* 2013;**8**(10):e78554.
148. Kohjima M, Sun Y, Chan L. Increased food intake leads to obesity and insulin resistance in the tg2576 Alzheimer's disease mouse model. *Endocrinology* 2010;**151**(4):1532–40.
149. Petrov D, Pedros I, Artiach G, Sureda FX, Barroso E, Pallas M, et al. High-fat diet-induced deregulation of hippocampal insulin signaling and mitochondrial homeostasis deficiences contribute to Alzheimer disease pathology in rodents. *Biochim Biophys Acta* 2015;**1852**(9):1687–99.
150. Nuzzo D, Picone P, Baldassano S, Caruana L, Messina E, Marino Gammazza A, et al. Insulin resistance as common molecular denominator linking obesity to Alzheimer's disease. *Curr Alzheimer Res* 2015;**12**(8):723–35.
151. Jheng HF, Tsai PJ, Guo SM, Kuo LH, Chang CS, Su IJ, et al. Mitochondrial fission contributes to mitochondrial dysfunction and insulin resistance in skeletal muscle. *Mol Cell Biol* 2012;**32**(2):309–19.
152. Martins IV, Rivers-Auty J, Allan SM, Lawrence CB. Mitochondrial abnormalities and synaptic loss underlie memory deficits seen in mouse models of obesity and Alzheimer's disease. *J Alzheimers Dis* 2017;**55**(3):915–32.

153. Athauda D, Foltynie T. Insulin resistance and Parkinson's disease: a new target for disease modification? *Prog Neurobiol* 2016;**145–146**:98–120.
154. Zhang J, Culp ML, Craver JG, Darley-Usmar V. Mitochondrial function and autophagy: integrating proteotoxic, redox, and metabolic stress in Parkinson's disease. *J Neurochem* 2018; [Epub ahead of print].
155. Thomas B, Beal MF. Parkinson's disease. *Hum Mol Genet* 2007;**16**:R183–94.
156. Tolosa E, Gaig C, Santamaria J, Compta Y. Diagnosis and the premotor phase of Parkinson disease. *Neurology* 2009;**72**(7 Suppl):S12–20.
157. Hilker R, Schweitzer K, Coburger S, Ghaemi M, Weisenbach S, Jacobs AH, et al. Nonlinear progression of Parkinson disease as determined by serial positron emission tomographic imaging of striatal fluorodopa F 18 activity. *Arch Neurol* 2005;**62**(3):378–82.
158. Langston JW, Ballard P, Tetrud JW, Irwin I. Chronic Parkinsonism in humans due to a product of meperidine-analog synthesis. *Science* 1983;**219**(4587):979–80.
159. Shi X, Zhao M, Fu C, Fu A. Intravenous administration of mitochondria for treating experimental Parkinson's disease. *Mitochondrion* 2017;**34**:91–100.
160. Makela J, Tselykh TV, Kukkonen JP, Eriksson O, Korhonen LT, Lindholm D. Peroxisome proliferator-activated receptor-gamma (PPARgamma) agonist is neuroprotective and stimulates PGC-1alpha expression and CREB phosphorylation in human dopaminergic neurons. *Neuropharmacology* 2016;**102**:266–75.
161. Brauer S. Parkinson's disease. *J Physiother* 2015;**61**(4):227.
162. Marongiu R, Spencer B, Crews L, Adame A, Patrick C, Trejo M, et al. Mutant Pink1 induces mitochondrial dysfunction in a neuronal cell model of Parkinson's disease by disturbing calcium flux. *J Neurochem* 2009;**108**(6):1561–74.
163. Muller-Hocker J. Mitochondria and ageing. *Brain Pathol* 1992;**2**(2):149–58.
164. Vanitallie TB. Parkinson disease: primacy of age as a risk factor for mitochondrial dysfunction. *Metabolism* 2008;**57**(Suppl. 2):S50–5.
165. De Pablo-Fernandez E, Sierra-Hidalgo F, Benito-Leon J, Bermejo-Pareja F. Association between Parkinson's disease and diabetes: data from NEDICES study. *Acta Neurol Scand* 2017;**136**(6):732–6.
166. Hu G, Jousilahti P, Bidel S, Antikainen R, Tuomilehto J. Type 2 diabetes and the risk of Parkinson's disease. *Diabetes Care* 2007;**30**(4):842–7.
167. Hu G, Jousilahti P, Nissinen A, Antikainen R, Kivipelto M, Tuomilehto J. Body mass index and the risk of Parkinson disease. *Neurology* 2006;**67**(11):1955–9.
168. Forhan M, Gill SV. Obesity, functional mobility and quality of life. *Best Pract Res Clin Endocrinol Metab* 2013;**27**(2):129–37.
169. Jang Y, Lee MJ, Han J, Kim SJ, Ryu I, Ju X, et al. A high-fat diet induces a loss of midbrain dopaminergic neuronal function that underlies motor abnormalities. *Exp Neurobiol* 2017;**26**(2):104–12.
170. Bosco D, Plastino M, Cristiano D, Colica C, Ermio C, De Bartolo M, et al. Dementia is associated with insulin resistance in patients with Parkinson's disease. *J Neurol Sci* 2012;**315**(1–2):39–43.
171. Jain D, Jain R, Eberhard D, Eglinger J, Bugliani M, Piemonti L, et al. Age- and diet-dependent requirement of DJ-1 for glucose homeostasis in mice with implications for human type 2 diabetes. *J Mol Cell Biol* 2012;**4**(4):221–30.
172. McCoy MK, Cookson MR. Mitochondrial quality control and dynamics in Parkinson's disease. *Antioxid Redox Signal* 2012;**16**(9):869–82.
173. Bonifati V, Rizzu P, Squitieri F, Krieger E, Vanacore N, van Swieten JC, et al. DJ-1( PARK7), a novel gene for autosomal recessive, early onset parkinsonism. *Neurol Sci* 2003;**24**(3):159–60.
174. Deas E, Plun-Favreau H, Wood NW. PINK1 function in health and disease. *EMBO Mol Med* 2009;**1**(3):152–65.
175. Valente EM, Abou-Sleiman PM, Caputo V, Muqit MM, Harvey K, Gispert S, et al. Hereditary early-onset Parkinson's disease caused by mutations in PINK1. *Science* 2004;**304**(5674):1158–60.
176. Deas E, Piipari K, Machhada A, Li A, Gutierrez-del-Arroyo A, Withers DJ, et al. PINK1 deficiency in β-cells increases basal insulin secretion and improves glucose tolerance in mice. *Open Biol* 2014;**4**:140051.
177. Hoshino A, Ariyoshi M, Okawa Y, Kaimoto S, Uchihashi M, Fukai K, et al. Inhibition of p53 preserves Parkin-mediated mitophagy and pancreatic β-cell function in diabetes. *Proc Natl Acad Sci U S A* 2014;**111**(8):3116–21.
178. Eberhard D. Neuron and beta-cell evolution: learning about neurons is learning about beta-cells. *BioEssays* 2013;**35**(7):584.
179. Khang R, Park C, Shin JH. Dysregulation of parkin in the substantia nigra of db/db and high-fat diet mice. *Neuroscience* 2015;**294**:182–92.

180. Deas E, Wood NW, Plun-Favreau H. Mitophagy and Parkinson's disease: the PINK1-parkin link. *Biochim Biophys Acta* 2011;**1813**(4):623–33.
181. Pickrell AM, Youle RJ. The roles of PINK1, parkin, and mitochondrial fidelity in Parkinson's disease. *Neuron* 2015;**85**(2):257–73.
182. Seillier M, Pouyet L, N'Guessan P, Nollet M, Capo F, Guillaumond F, et al. Defects in mitophagy promote redox-driven metabolic syndrome in the absence of TP53INP1. *EMBO Mol Med* 2015;**7**(6):802–18.

# PART IV

# FACTORS THAT MAY TRIGGER OR AGGRAVATE THE PATHOLOGIES

CHAPTER

# 14

# Impact of Fetal Programming on Mitochondrial Function and Susceptibility to Obesity and Type 2 Diabetes

*Amita Bansal*,†, Cetewayo Rashid*,†, Rebecca A. Simmons*,†,‡*

*Center for Research on Reproduction and Women's Health, University of Pennsylvania, Philadelphia, PA, United States †Center of Excellence in Environmental Toxicology, Perelman School of Medicine, University of Pennsylvania, Philadelphia, PA, United States ‡Division of Neonatology, The Children's Hospital of Philadelphia, Philadelphia, PA, United States

## 1 INTRODUCTION

The worldwide incidence of metabolic disorders such as type 2 diabetes and obesity continues to increase. The WHO predicts that these metabolic disorders will be a major cause of death by 2030, placing a substantial economic burden on the healthcare system globally.[1, 2] It now is accepted widely that metabolic diseases of adulthood, including type 2 diabetes and obesity, might have their origins in the womb. Almost three decades ago, the concept of fetal origins of adult diseases was first proposed by Barker and colleagues, who reported that adults born at low birth weight had greater likelihood of developing cardiovascular diseases and diabetes.[3–5] Based on these pioneering observations, a field of research now popularly known as developmental origins of health and disease (DOHaD) emerged.[6, 7] It now is clear that perturbations during early life have long-lasting effects on metabolic health. Improved understanding about the role of the early life environment on the progression of metabolic diseases has triggered efforts to design preventive strategies for these diseases at the time of their origin.[8, 9]


https://doi.org/10.1016/B978-0-12-811752-1.00014-6

The challenge in the field has been to identify common mechanisms and pathways that are involved in disparate DOHaD paradigms. Three common pathways, however, have started to emerge. These pathways are mitochondrial dysfunction, epigenetics, and inflammation.[10–13] They are not mutually exclusive, and all are likely to be involved, and interact with each other. Normal mitochondrial function is critical for normal metabolic health. Mitochondrial dysfunction is associated with diabetes and obesity. Various aspects of mitochondrial function, including mitochondrial bioenergetics, mitochondrial biogenesis, mitochondrial dynamics, and mitochondrial DNA mutations, can be impaired (Fig. 1). In this chapter, we will expand on how perturbations from conception to birth impact mitochondrial function thereby increasing the subsequent risk of obesity and type 2 diabetes.

## 2 OXIDATIVE STRESS AND FETAL GROWTH-RESTRICTION

Uteroplacental insufficiency, caused by disorders such as preeclampsia, maternal smoking, and abnormalities of uteroplacental development, is one of the most common causes of fetal growth restriction. In the face of uteroplacental insufficiency, the fetus adapts to an inadequate supply of substrates (such as glucose, amino acids, fatty acids, and oxygen) by metabolic changes, redistribution of blood flow, and changes in the production of fetal and placental hormones that control growth. The fetus' immediate metabolic response to placental insufficiency is catabolism. It consumes its own substrates to provide energy, and there are shifts from accretion to oxidative metabolism and breakdown of protein/glycogen for oxidative metabolism, or both. A more prolonged reduction in availability of substrates leads to growth restriction. Although this enhances the fetus' ability to survive by reducing the use of substrates and lowering the metabolic rate, slowing of growth in late gestation leads to disproportionate organ size, because organs and tissues that are growing rapidly at the time are most affected.

Multiple studies now have shown that intrauterine growth restriction is associated with increased oxidative stress in the human fetus.[14–22] Elevated oxidative stress also has been observed in infants who are small for their gestational age born to undernourished mothers compared to infants who are appropriate for gestational age born to healthy mothers.[23] Oxidative stress was determined by increased quantities of malondialdehyde (one of the major products of lipid peroxidation), reduced quantities of the antioxidant glutathione, and decreased activity of the antioxidants superoxide dismutase and catalase in cord blood of small versus appropriately sized infants.[23] A major consequence of limited nutrient availability is an alteration in the redox state in susceptible fetal tissues leading to oxidative stress. In particular, low levels of oxygen, evident in growth-retarded fetuses, will decrease the activity of complexes of the electron transport chain, which will generate increased levels of reactive oxygen species (ROS).[24–26] Overproduction of ROS initiates many oxidative reactions that lead to oxidative damage of both cellular and mitochondrial proteins, lipids, and nucleic acids. Increased ROS levels inactivate the iron-sulfur centers of the electron transport chain complexes, and tricarboxylic acid cycle aconitase, resulting in shutdown of mitochondrial energy production.

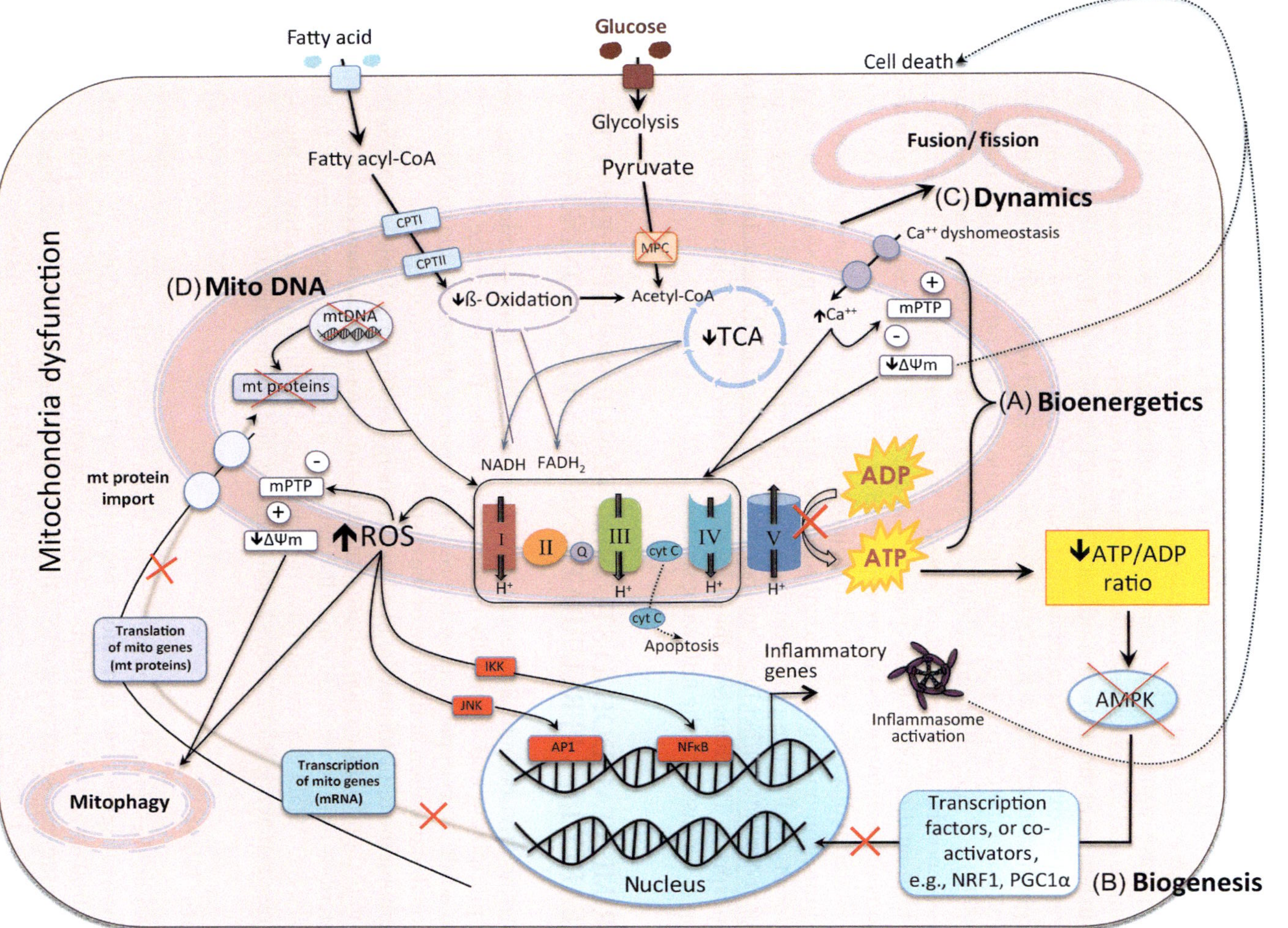


FIG. 1 A schematic of mitochondrial dysfunction. Various aspects of mitochondria that can be affected. (A) *Mitochondrial bioenergetics* involves oxidation of fatty acids via β-oxidation, and pyruvate (end product of glycolysis) to generate acetyl coA. Acetyl coA then is metabolized via the TCA cycle. NADH and $FADH_2$, the intermediary products of β-oxidation and TCA cycle, then enter oxidative phosphorylation in the mitochondrial electron transport chain, complex I–V and ATP is generated. (B) *Mitochondrial biogenesis* involves change in mitochondrial content, volume, and mass, and is stimulated by factors such as PGC1α. (C) *Mitochondrial dynamics* involves rearrangement of mitochondrial components by fusion and fission to keep the healthy components and release the dysfunctional components. (D) *Mitochondrial DNA* mutations, change in copy number, or content could consequently affect mitochondrial function. *Note: mt*, mitochondria; *MPC*, mitochondria pyruvate carrier; *CPT*, carnitine palmitoyltransferase; *ROS*, reactive oxygen species; *TCA*, tri carboxylic acid cycle; *Cyt C*, cytochrome C; *ATP*, adenosine triphosphate.

# 3 OXIDATIVE STRESS AND NUTRIENT EXCESS

Oxidative stress can be triggered not only by nutrient deficit, but also by nutrient excess. This is evident from higher levels of malondialdehyde, reduced levels of antioxidative micronutrients vitamin A and E, and reduced activity of antioxidant glutathione peroxidase and superoxide dismutase reported in pregnant diabetic (gestational diabetes or type 1 diabetes) women between 26 and 32 weeks gestation compared to controls.[27] This study did not determine changes in fetuses of the diabetic mothers. Higher leptin levels and lower concentrations of the antioxidant, paraoxonase, were reported in maternal blood of obese mothers compared to normal weight mothers.[28] These maternal changes correlated with leptin and antioxidant levels in cord blood of the offspring of obese mothers compared to normal weight mothers.[28]

Animal models have demonstrated a clear link between nutrient excess during pregnancy, oxidative stress, and the development of obesity and glucose intolerance in adulthood. Exposure to a high-fat diet in female rats prior to and during pregnancy resulted in increased levels of markers of oxidative stress in preimplantation embryos, and in fetal and newborn serum.[29] This altered redox state persisted in adipose tissue of 2-week-old and 2-month-old offspring.[29] Administration of an antioxidant supplement to the mother prevented the development of adiposity and glucose intolerance in the offspring.[29] These findings suggest that alteration in the redox state, either in the mother or embryo, mediate the later development of obesity in the offspring. The changes in the redox state can affect susceptible metabolic tissues in the offspring as discussed in the following section.

# 4 EXPERIMENTAL DOHaD STUDIES: TISSUE-SPECIFIC EFFECTS OF EARLY LIFE PERTURBATIONS ON MITOCHONDRIA

The intrauterine environment influences development of the fetus by modifying gene expression in pluripotent and terminally differentiated cells. The long-range effects on the offspring (into adulthood) depend upon the cells undergoing differentiation, proliferation, and/or functional maturation at the time of the disturbance in maternal fuel economy. A key adaptation enabling the fetus to survive in a limited energy environment might be the reprogramming of mitochondrial function. These alterations in mitochondrial function, however, can have deleterious effects, especially in cells that have a high-energy requirement. In this section, we discuss DOHaD studies that have reported mitochondrial dysfunction in four key metabolic tissues: pancreatic islets, liver, skeletal muscle, and adipose tissue.

## 4.1 Pancreatic Islets

Experimental studies in rodent models have demonstrated that uteroplacental insufficiency,[30] exposure to high-fat diet and caloric restriction[31] or low-protein diet[32–35] is associated with mitochondrial abnormalities in islets of fetal,[32, 33] newborn,[36] or adult[31, 34] offspring. The suboptimal early life exposures appear to influence various components of mitochondria in experimental studies. These include decreased mitochondrial mass and mitochondrial DNA copy number[30, 31] or content,[35] and reduction in mitochondrial transcription factor A (*Tfam*), *which* controls mitochondrial DNA replication and the overall expression of genes encoded by the mitochondria.[37] Because alterations in mitochondrial mass, DNA copy number, or expression of mitochondrial genes influence mitochondrial function, these mechanisms might underlie the reduction in oxidative phosphorylation and ATP production with consequent reduction in insulin secretion that is reported in these rodent studies.[30–34] Furthermore, as Tfam is a direct Pdx1 (a gene that previously has been shown to play a critical role in pancreatic development and function in IUGR model) target gene, some of the mitochondrial abnormalities reported in the rodent models of fetal programming might be a consequence of reduced *Pdx1* levels.[31, 38] Reduced levels of glutathione peroxidase and peroxiredoxin activity in rats exposed to a low-protein diet,[39] or increased fatty acid metabolite 3-carboxy-4-methyl-5-propyl-2-furanpropanoic acid in mice[40] have been associated with increased levels of ROS in islets. Because insulin secretion from pancreatic β-cells is dependent on active oxidative metabolism and the anti-oxidative defense systems generally are expressed at low levels in β-cells compared to other tissues,[41–43] β-cells are particularly vulnerable to increased oxidative stress. It is well accepted that oxidative stress can blunt insulin secretion.[44–47] Therefore, because of increased susceptibility of pancreatic β-cells to oxidative stress, it is believed that early and ongoing exposures to oxidative insults result in the failure of β-cells to cope with the increasing demand for insulin from peripheral tissues in fetal programming models, and contribute to development of type 2 diabetes.[48] Consistent with this, increased oxidative stress and progressive decline in mitochondrial and β-cell function have been reported in islets of adult intrauterine growth-restricted rats[30] and in islets of adult rat offspring exposed to low protein diet in utero.[34] Therefore, mitochondrial abnormalities and increased oxidative stress appear to be one of the key mechanisms underlying β-cell dysfunction and development of type 2 diabetes in growth-restricted animals. Some evidence indicates that rodent islets chronically stressed by a diabetic environment are paradoxically protected against oxidative stress.[49, 50] For example, Goto Kakizaki rat islets exhibit resistance to the cytotoxic effect of exogenous $H_2O_2$ or endogenous ROS exposures by upregulating the antioxidative response.[49, 50] This is reflected by both high glutathione content and overexpression of many genes encoding anti-oxidative proteins, including UCP2, and reduced oxidative stress-triggered β-cell death.[49, 50] A similar protection against oxidative stress, however, is not yet reported in β-cells of offspring that are exposed to an in utero suboptimal environment. Although β-cells are studied widely in a context of oxidative stress and developmental origins of type 2 diabetes, mitochondrial dysfunction is not limited to β-cells, and also has been observed in other tissues.

## 4.2 Liver

Reduced hepatic TCA substrate oxidation,[51] fatty acid oxidation,[52] and reduced hepatic mtDNA content[35, 53, 54] have been observed in IUGR offspring,[51] offspring of obese,[52, 54] or low-protein diet-fed[35, 53] rats, respectively. Hepatic mitochondrial abnormalities preceded the onset of diabetes and obesity in the offspring in these models,[51–54] suggesting that mitochondrial dysfunction might be an underlying mechanism predisposing the offspring to type 2 diabetes. Reduced activity of mitochondrial complex I, II, and III, and increased expression of lipogenic, oxidative stress and inflammatory genes also were reported in livers of 15-week-old offspring of high fat/high fat (gestation/postweaning) and high fat/control (gestation/postweaning) mice.[55] Hepatic fat content is increased in offspring of models of both under-nutrition and over-nutrition.[56, 57] Increased hepatic lipid accumulation and oxidative stress persisted in adult rat offspring born to high-fat diet-fed mothers that were switched to a balanced chow diet at weaning.[58] It is known that lipid accumulation can perturb insulin-signaling leading to insulin resistance. It is possible, therefore, that increased hepatic lipid content might be a link between mitochondrial dysfunction and insulin resistance observed in both under-nutrition and over-nutrition models.

## 4.3 Muscle

Multiple studies in animals have demonstrated that mitochondrial dysfunction in skeletal muscle is associated with increased risk of type 2 diabetes.[59–62] Similar defects are observed in skeletal muscle of IUGR rat offspring, such that reduced oxidative phosphorylation and reduced ATP production were associated with impaired glucose transporter 4 (GLUT4) recruitment, glucose uptake, and glycogen synthesis, contributing to insulin resistance and type 2 diabetes.[36] Mitochondrial abnormalities, including reduced mitochondrial DNA content, also reduced expression of mitochondrial genes. Reduced maximal respiratory capacity ($Vo_{2max}$) was observed in skeletal muscle of offspring of low-protein diet[53] or high-fat diet-fed[56] rats, or last trimester 50% food-restricted ewes.[63] In skeletal muscles of 1-year-old offspring born to rats exposed to high-fat diet from 3 weeks prior to mating until weaning, global transcriptomic analysis revealed increased expression of genes involved in cytokine signaling and inflammation, but reduced expression of genes involved in oxidative phosphorylation.[64] These transcriptomic changes were associated with reduced expression of insulin signaling and mitochondrial complex I, II, and V proteins, and increased insulin resistance in skeletal muscle in the offspring.[64] Prenatal low-protein exposure (8%), followed by postnatal high-fat diet exposure (45%) also reduced skeletal muscle mitochondrial oxidative respiration and increased the risk of type 2 diabetes in rat offspring.[65] Overall, these studies suggest that exposure to nutritional insults before or after birth can trigger skeletal muscle mitochondrial dysfunction and predispose offspring to type 2 diabetes.

## 4.4 Adipose Tissue

Compared to other tissues, relatively less research has been conducted to determine the effects of early life perturbations on adipose tissue mitochondrial function. The few studies that have looked at adipose tissue report that early life suboptimal nutrition is associated with mitochondrial abnormalities later in life. High-fat diet exposure in female mice from 8 weeks prior to mating and throughout gestation and lactation did not alter mitochondrial mass, but it did reduce mitochondrial oxygen consumption and protein expression of mitochondrial complex II and IV subunits in brown adipose tissue of exposed mice dams.[66] In brown adipose tissue of 4-week-old female offspring, however, mitochondrial mass was reduced, but activity of complex I–III and citrate synthase was increased and was associated with increased expression of multiple nuclear and mitochondrial encoded genes of the electron transport chain and increased ATP content.[66] These changes in mitochondrial function were reversed partially by switching the offspring to a normal chow diet for 6 weeks after weaning, however, the thermogenic program of the brown adipose tissue was altered in these offspring and displayed reduced responsiveness to cold.[66] Suggesting that suboptimal maternal nutrition can impair mitochondrial function of the offspring brown adipose tissue with long-term deleterious effects on regulation of metabolism and energy expenditure, which might predispose to obesity. Mitochondrial dysfunction also has been observed in subcutaneous fat of offspring born to high-calorie consuming dams,[67] or in offspring that were exposed to a low-protein diet in utero followed by a postnatal high-fat diet.[68]

Taken together, multiple studies of either fetal growth restriction or perinatal suboptimal nutrient exposure show mitochondrial dysfunction in different tissues, indicating that perhaps mitochondrial dysfunction is a common mechanism underlying the development of metabolic diseases in later life in offspring that have been exposed to an abnormal intrauterine nutrient milieu (Fig. 2).

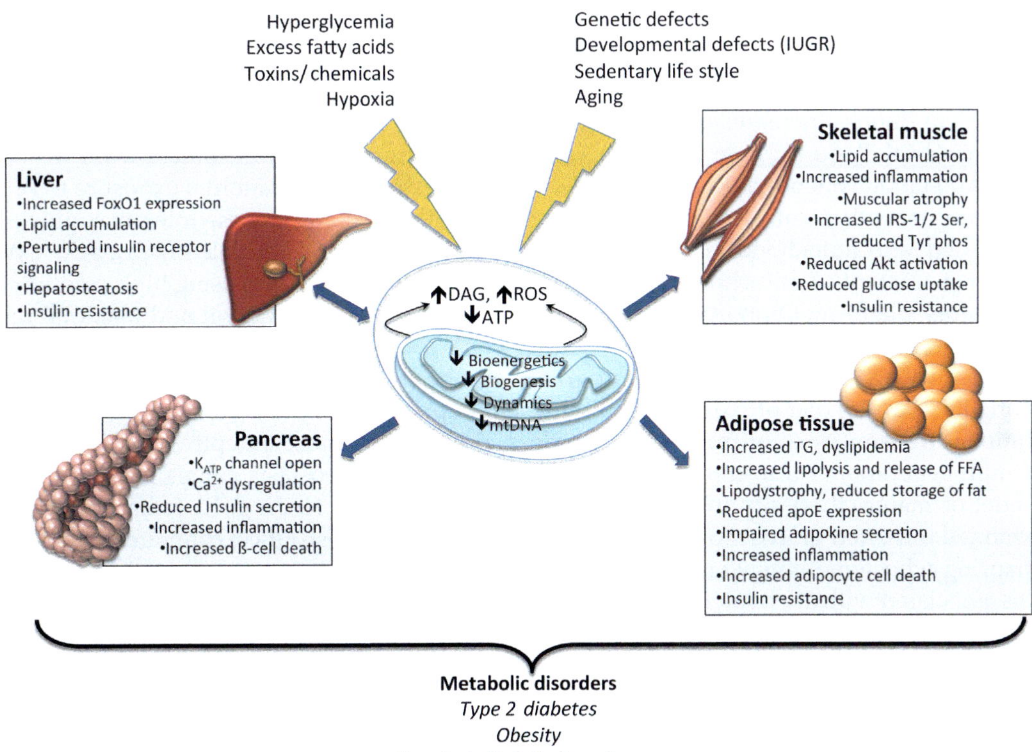


FIG. 2 Role of mitochondrial dysfunction in mediating metabolic phenotype. Exposure to a suboptimal developmental environment can alter different aspects of mitochondria (bioenergetics, biogenesis, dynamics, and mtDNA) in target metabolic tissues such as liver, skeletal muscle, adipose tissue, and pancreas, which would lead to abnormal insulin secretion and insulin resistance. Together these alterations could contribute to increased risk of metabolic disorders such as type 2 diabetes, obesity, and nonalcoholic fatty liver disease. *Note: mtDNA*, mitochondrial DNA; *TG*, triglycerides; *FFA*, free fatty acids; *IRS*, insulin receptor substrate, *Akt*, serine/threonine specific protein kinase; *FoxO1*, forkhead box protein O1; *Ser*, serine; *Tyr*, tyrosine; *phos*, phosphorylation; *ROS*, reactive oxygen species; *DAG*, diacylglycerols; *ATP*, adenosine triphosphate; *apoE*, apolipoprotein E.

## 5 TRANSMISSION OF MITOCHONDRIAL DYSFUNCTION ACROSS GENERATIONS

Among the remarkable ideas behind developmental programming is not only the ability of parental exposures to reproduce or influence phenotypes in first-generation offspring (F1) in an almost Lamarckian manner, but also influence phenotypes in subsequent generations. Indeed, studies have demonstrated parental programming of metabolic phenotypes to second (F2) and even third (F3) generations. The transmittance of a phenotype to the F2 generation upon paternal exposure and the F3 generation upon maternal exposure represents a developmental programming phenomenon called transgenerational (as opposed to intergenerational) programming.[69] Transgenerational effects are particularly intriguing because they represent effects in the generation that was not directly exposed to the initial insult/trigger. Although many studies have investigated developmental programming of obesity and diabetes in the offspring, few have looked beyond the first generation, and of those, even fewer studies explore mitochondrial contributions. Nonetheless, knowledge can be gained from looking at those few studies that explored mitochondria beyond the first generation.

Parental exposure to environmental chemicals such as bisphenol A (BPA), nutrient excess, or nutrient deficits increases obesity and diabetes risk in the offspring. Studies suggest that mitochondrial dysfunction in metabolic tissues of offspring might contribute toward the observed metabolic phenotypes. For example, we demonstrated that maternal exposure to BPA from preconception until weaning was associated with impaired glucose tolerance, glucose stimulated, and mitochondrial driven insulin secretion in F1 and F2 male offspring.[70, 71] We further characterized mitochondrial function in islets of these offspring using high-resolution respiratory platform, Oroboros, and observed significantly reduced basal and maximal respiration in islets.[71]

Other causes of developmentally programmed obesity and diabetes, such as those mimicking the western diet or parental macronutrient deprivation, have been used to study generational transmittance of metabolic disorders. Perhaps the most indepth analysis to date of multigenerational transmittance of impaired glucose homeostasis comes from a mouse model of maternal obesity induced by high-fat/high sucrose feeding. F1 female offspring exhibited impaired glucose tolerance associated with insulin resistance, compared to control offspring whose mothers were fed standard chow.[72] The insulin resistance in the offspring was associated with attenuated IRS-1 and AKT phosphorylation in skeletal muscle, indicative of peripheral insulin resistance. Skeletal muscle also had elevated MDA and lipid content, suggesting oxidative stress and reduced lipid oxidation, both of which point to mitochondrial dysfunction. Activity of β-oxidation enzyme hydroxyacyl-coenzyme A dehydrogenase type II and substrate-driven oxygen consumption were reduced.[72] Visualization of skeletal muscle mitochondria using transmission electron microscopy showed aberrant mitochondrial organization and morphology, particularly in the inner membrane.[72] Mitochondrial cristae were enlarged and some were onion-shaped, while others contained vacuoles.[72] These observations were associated with reduced DRP1 activity and OPA1 protein levels, which have roles in mitochondrial fission and inner mitochondrial membrane cristae formation, respectively.[73]

These mitochondrial defects also were observed in the oocytes of F1 offspring.[72] These oocytes also had a reduced mtDNA copy number consistent with OPA1 reduction.[74] Although metabolic parameters associated with glucose homeostasis were not investigated in subsequent generations, mitochondrial aberrations from skeletal muscle of F2 offspring closely resembled those of preceding generations.[72] There also were mitochondrial abnormalities in F2 oocytes and F3 skeletal muscle, but they were fewer in number,[72] representing a well-characterized washing out of maladaptive phenotypes at later generations.[75] It remains to be determined whether transmission of mitochondria across generations is responsible for the metabolic defects observed in the F2 and F3 generations because mitochondrial tracking was not performed. Taken together, these studies provide insights into multigenerational inheritance of impaired glucose homeostasis through the maternal lineage and suggest a role for skeletal muscle mitochondria dysfunction in offspring metabolic disorder. Although these studies allude phenotypic transmittance via oocyte mitochondrial defects, mitochondrial tracking to support this was not performed. The manifestation of metabolic dysfunction in pregnant female offspring with concomitant intrauterine perturbations could participate in multigenerational transmittance.

Although most studies show multigenerational transmission of a metabolic phenotype through the maternal lineage, few studies also suggest a role of paternal (and not maternal) lineage in this process. For example, in a murine study of maternal obesity (high-fat feeding without elevated dietary sucrose), F1 females were unaffected, but F1 males were insulin resistant.[76] Both male and female F2 offspring from the paternal lineage, but not the maternal lineage, however, were insulin resistant.[76, 77] The observation that dysregulated glucose homeostasis can occur through the paternal lineage is clear evidence that diabetic phenotypes can be transmitted independent of mitochondrial defects in the oocyte,[76,77] therefore, suggesting that paternal factors also can influence mitochondrial function in offspring. This is likely because, though mitochondria themselves are maternally derived, more than 95% of mitochondrial proteins are nuclear encoded, allowing genetic and epigenetic information carried by sperm to contribute to offspring mitochondrial dysfunction. Indeed, in vitro fertilization using sperm from obese mice resulted in 4-cell stage embryos with decreased mitochondrial membrane potential compared to embryos fertilized using sperm from normal weight sires.[78]

In summary, exposure to abnormal maternal and paternal environments is associated with metabolic defects across two to three generations, but it is unfortunate that current knowledge of mitochondrial participation in this process lacks breadth and depth. Despite a paucity of research in this area, compelling evidence from existing studies shows that a suboptimal maternal environment can alter mitochondria function in oocytes and tissues, such as pancreatic islets and skeletal muscle, that are relevant to obesity and diabetes.

## 6 STRATEGIES TO TREAT MITOCHONDRIAL DYSFUNCTION

Obesity and diabetes have adverse impacts on one's quality of life and places substantial burden on the economy, and unfortunately the prevalence of these metabolic disorders is growing at an alarming rate. In previous sections, we discussed scientific evidence showing that fetal adaptations to an aberrant intrauterine milieu compromises tissue development and hormone homeostasis, which precipitate obesity and diabetes in adulthood. In this section, we challenge the inexorability of developmentally programmed metabolic disease by examining the efficacy of behavioral, pharmacological, and nutritional interventions.

## 6.1 Behavioral

Exercise is one of the most widely studied behavioral interventions to prevent developmental programming of metabolic disorders. Different developmental timings have been targeted in this regard: exercise during pregnancy (maternal exercise) and exercise after birth (maternal exercise during lactation, or offspring exercise). We will discuss the outcomes from maternal exercise first, followed by offspring exercise

Whether the offspring is not exposed[76–79] or exposed[80] to a suboptimal environment, maternal exercise has potential benefits on offspring metabolic health and mitochondrial endpoints. Rat studies of prenatal maternal exercise continuing until weaning show beneficial effects on offspring glucose homeostasis.[81] Maternal exercise improved glucose tolerance and insulin sensitivity in both male and female offspring, but only male offspring showed improvements in body composition.[81] Female offspring exhibited decreased nonfasted hepatic glucose production (HGP) and improvements in insulin-stimulated glucose uptake predominantly in glycolytic muscles, but no difference was observed in predominantly oxidative muscles or in visceral adipose tissue.[82] Glycolytic muscles are particularly susceptible to exercise-induced adaptations compared to controls. Studies in mice showed that maternal exercise increased the size of muscle fibers and increased the proportion of slow-twitch muscle fibers by increasing the mitochondrial content in glycolytic muscles.[83] Specifically, skeletal muscles had increased mitochondria volume density without changes in mitochondrial morphology.[83] Activity of mitochondrial enzymes citrate synthase and cytochrome *c* oxidase correlated with increased basal oxidative ATP production.[83] This study also reported increased protein levels of TFAM, a mitochondrial transcription factor that contributes to mitochondrial transcription, replication and biogenesis[37, 84–86], and LON protease, a mitochondrial matrix stress response protein whose downregulation diminishes oxidative phosphorylation and is associated with aberrant morphology.[79, 80, 83] Although maternal exercise augments ATP production, hydrogen peroxide levels are diminished,[87] indicating protection against oxidative stress and potentially its metabolic sequelae. These studies provide proof of principle that maternal exercise could confer benefits to offspring that are not exposed to a postnatal suboptimal environment. Furthermore, studies also show that maternal exercise could be beneficial for offspring that are exposed to a postnatal suboptimal environment. For example, when the offspring of sedentary or exercised mothers are challenged with overnutrition in the form of high-fat feeding, we see that offspring inherit a level of metabolic protection if their mothers exercised. Specifically, when dams were exercised and their offspring consumed a high-fat diet (HFD), maternal exercise attenuated obesity, hepatic steatosis, and prevented the decrease in hepatic Tfam gene expression associated with HFD consumption in offspring from sedentary dams.[88] Regardless of the offspring nutritional status, therefore, maternal exercise seems to potentially benefit offspring metabolic and mitochondrial health.

Maternal exercise, however, has differential outcomes dependent on maternal nutritional status (i.e., it seems to be beneficial in maternal overnutritional models, but perhaps has limited efficacy in maternal undernutrition models). For example, maternal exercise prevents obesity and glucose intolerance in offspring from obese mothers.[89, 90] Also, in 12-month-old offspring from obese dams, glucose intolerance was associated with decreased skeletal muscle gene expression of Cox4 and CytC, and maternal exercise prevented this reduction in mitochondrial gene expression.[90] Because exercise diminished weight gain in dams consuming an HFD, it is possible that benefits conferred to offspring were secondary to decreasing maternal weight gain in this study. In models of maternal undernutrition, however, it is likely that maternal exercise exacerbates the offspring phenotype by further limiting offspring nutrient availability.[91] Therefore, studies have not investigated the long-term benefits of maternal exercise in the context of fetal undernourishment. Switching the exercise intervention window from in utero to postnatal might alter the level of protection conferred by exercise when looking at maternal challenges that program offspring obesity and diabetes.

In addition to understanding the impact of maternal exercise on offspring health, because maternal over/under nutrition can reduce offspring physical activity levels,[92–95] using offspring exercise as an intervention seems conceivable. Studies have shown that offspring voluntary exercise prevents obesity development in IUGR offspring.[96] Furthermore, postnatal exercise reduces weight gain but only mildly ameliorates glucose intolerance in HFD-challenged offspring from obese mothers.[97] Gene expression of liver enzymes showed that postnatal exercise decreased mitochondrial pyruvate carrier Mpc1 gene expression and increased Hadh expression.[97] HFD consumption in mothers or in offspring also decreased citrate synthase activity, but exercise was not able to normalize its activity in offspring consuming HFD.[97] The clinical relevance of these rodent studies is supported by a study in humans showing that 9 days of intense exercise improved mitochondrial citrate synthase activity and ATP production in offspring from both diabetic and nondiabetic mothers. Hyperinsulinemic-euglycemic clamp studies showed that exercise improves parameters of insulin sensitivity in offspring from nondiabetic mothers, however, this improvement in glucose homeostasis did not occur in offspring from diabetic mothers,[98] demonstrating that maternal factors can influence whether improvements in mitochondrial function translate to improvements in insulin sensitivity in offspring. This observation supports the idea of developmental programming of diabetes susceptibility, and, as such, intervention strategies must take into consideration parental risk factors as well as the timing of the intervention (during pregnancy, early postnatal period, etc.).

## 6.2 Pharmacological

In addition to behavioral interventions, pharmacological interventions in the form of glucagon-like peptide-1 (GLP-1) receptor agonists have been investigated. GLP-1 is an incretin hormone secreted by intestinal L-cells after ingestion of a meal that slows gastric emptying, suppresses appetite and potentiates GSIS. Administration of GLP-1 receptor agonist Exendin-4 (Ex-4) during the neonatal period in IUGR rats, prevents the later development of diabetes.[99] Specifically, Ex-4 prevents the decline in β-cell replication, preserves islet vascularity, and maintains GSIS both in vivo and in vitro.[99–101] Moreover, Ex-4 reduces basal HGP, abrogates hepatic insulin resistance associated with IUGR diabetes progression, and mitigates oxidative stress.[101]

Ex-4 protects male and female offspring from metabolic derangements associated with maternal and postweaning HFD in rats.[102, 103] Ex-4 intervention abrogated hyperinsulinemia and restored insulin sensitivity in 2-month-old offspring of HFD mothers.[102] Postweaning Ex-4 also reduced adiposity in offspring from HFD mothers.[102, 103] Although Ex-4 shows promise in rat models of developmentally programmed metabolic disorders, larger animals show less improvement. Neonatal Ex-4 administration to IUGR sheep actually increased fasting plasma glucose levels and did not improve glucose tolerance or insulin secretion in IUGR offspring.[104, 105] This might be because pancreas development in larger animals, such as sheep and humans, occurs earlier in gestation than in rodents, and therefore, pancreata of larger animals are more mature and less plastic during the neonatal period than in rodents.[106] As a result, the ability of neonatal Ex-4 to normalize islet physiology is compromised by temporally missing a critical window of plasticity. Although the effects of early postnatal treatment with GLP-1 or Ex-4 on mitochondria function have not been investigated in the context of developmentally programmed metabolic disease, treatment with GLP-1 or Ex-4 improves mitochondria function in INS-1 and MIN6 cell lines and adult male albino diabetic rats.[107–109]

The glucocorticoid dexamethasone decreases resting mitochondrial membrane potential and causes apoptotic cell death in the mouse INS-1 β-cell line.[110] Dexamethasone treatment decreases protein levels of phosphorylated BAD, which leads to mitochondria-assisted apoptosis.[110] The ability of the glucocorticoid to induce apoptosis is not limited to the INS-1 cell line because the apoptotic effect of dexamethasone also has been observed in isolated mouse islets.[110] Ex-4 abrogated dexamethasone-induced β-cell caspase-3 activity and apoptosis in a PKA-dependent manner.[110] In a similar study, Ex-4 prevented inflammatory cytokine-induced caspase-3 activity and oxidative stress, again in INS-1 cells.[111] Analysis of the mitochondrial proteome showed that Ex-4 counterregulated the reduction of ATP synthase subunit β and cytochrome *bc* 1 complex subunit 1, indicating modulation of electron transport and subsequent ATP synthesis by the GLP-1 receptor agonist.[111]

The Goto-kakizaki (GK) rat is a polygenic model of type 2 diabetes. Both GLP-1 and Ex-4 given only from postnatal days 2–6 increase insulin content and β-cell mass in these rats at 1 week of age.[112] At 2 months of age, both interventions modestly attenuated fasting hyperglycemia and improved glucose tolerance, but only GLP-1 statistically increased β-cell mass, in vivo and ex vivo GSIS, and improved insulin sensitivity.[112] These data show that not only can short-term GLP-1 and Ex-4 treatments during the neonatal period confer a degree of lasting protection against diabetes progression, but also that GLP-1-mediated protection might use mechanisms independent of GLP-1 receptor activation. In support of this assumption is the observation that endogenous GLP-1 cleavage nonapeptide, GLP-1(28–36)amide (not a GLP-1 receptor ligand), localizes to mitochondria in primary mouse hepatocytes and prevents oxidative stress and ROS-mediated reduction of ATP synthesis.[113] Moreover, GLP-1(28–36) attenuates stimulatory HGP.[113] GLP-1 receptor antagonism with proteolytically deactivated GLP-1 fragment (9–39) did not block stimulatory HGP nor did it prevent GLP-1(28–36)'s ability to mitigate HGP, indicating these effects are independent of GLP-1 receptor activation.[113] Taken together, these studies show that GLP-1 exerts protective effects regarding diabetes progression in rodent models of fetal programming, and confer a level of protection in tissues regulating glucose homeostasis, such as the pancreas and liver. As GLP-1's actions are pleiotropic, we cannot rule out the involvement of other tissues in curtailing the onset or severity of diabetes. Mitochondria appear to play a role in GLP-1's protection, but studies are limited as to what effect this incretin has on mitochondrial function, and more specifically, whether its effects on mitochondria actively participate in protection or are only a byproduct of other mechanisms at work.

## 6.3 Nutritional

Taurine is a nonessential sulfur-containing β-amino acid synthesized in the liver from the sulfur-containing amino acids methionine and cysteine. In mammalian tissues, taurine is highly abundant and serves diverse physiological roles including bile salt synthesis, osmoregulation, calcium regulation, and redox regulation.[114] Taurine is essential for proper fetal development, but fetal tissues do not synthesize taurine, rather it is supplied by the mother. Taurine transport in placenta is unidirectional (maternal to fetal) and mediated by the placental taurine transporter (TauT).[115] Taurine levels are decreased in SGA fetuses.[116] This reduction in fetal taurine is not the result of decreased TauT expression evidenced by comparable placental TauT protein levels in normal weight and IUGR fetuses.[117] Placental TauT activity, however, is attenuated in IUGR, which might explain the reduced levels observed in SGA fetuses.[118] Similarly, maternal obesity attenuates fetal taurine transport compared to normal weight mothers, and this decreased function is again without changes in transporter expression.[119] Because taurine is essential for proper development and levels are decreased in fetuses at risk for developmentally programmed metabolic disorders, animal models have investigated whether maternal taurine supplementation can prevent diabetes manifestation in adulthood.

Maternal low protein consumption beginning at conception reduces maternal and fetal plasma taurine concentrations and diminishes fetal insulin secretion.[120, 121] In vitro taurine supplementation of fetal islets from protein-restricted dams is unable to normalize insulin secretion.[121] If the protein-restricted maternal diet is supplemented with taurine, however, then mitochondrial ATP production and insulin secretion is restored to levels comparable to fetal islets from control dams.[121, 122] In the early postnatal period, offspring from protein-restricted dams exhibit diminished islet vascularization, GSIS, and β-cell proliferation and increased inflammatory cytokine-mediated β-cell apoptosis.[123–125] Supplementing maternal protein restriction with taurine rescues these phenotypes associated with developmentally programmed diabetes.[123–125] Taurine-supplemented maternal protein restriction also ameliorates derangements in glucose homeostasis in aged offspring.[126]

In addition to maternal protein restriction, taurine supplementation has beneficial effects regarding metabolic outcomes in offspring whose mothers were either diabetic or obese.[127–130] Taurine levels are not altered in every IUGR model, however, and fetuses rendered growth-restricted by bilateral uterine artery ligation have normal taurine levels.[131] It has been reported that taurine supplementation in pregnancies with normal TauT activity actually might increase adiposity and insulin resistance.[132]

As previously mentioned, taurine is physiologically polyvalent and deciphering which activity confers protection from an adverse intrauterine environment, to date, has been beyond the scope of scientific investigation. Among taurine's activities most relevant to glucose homeostasis is perhaps its role in calcium regulation and oxidative stress. Taurine is an antioxidant in spite of its inability to scavenge classical ROS, such as superoxide, hydroxyl radical, and hydrogen peroxide.[133, 134] Taurine accumulates in mitochondria, specifically in the mitochondrial matrix where it acts as a pH buffer.[135] Taurine depletion with β-alanine reduces mitochondrial complex I and III activities and leads to superoxide production and glutathione depletion, supporting not only the ability of taurine to abrogate oxidative stress, but also to modulate the activity of the electron transport chain.[136] Taurine also was shown to act at the level of the TCA cycle by specifically augmenting mitochondrial $Ca^{2+}$ uptake.[137, 138] Because $Ca^{2+}$ is a cofactor for several mitochondrial dehydrogenases, this increase in $Ca^{2+}$ presumably upregulates dehydrogenase activity and consequently increases the concentration of substrates for the TCA cycle.[137] Taurine also inhibits pyruvate dehydrogenase kinase (PDK), leading to pyruvate dehydrogenase (PDH) activation and pyruvate metabolism.[135] Taken together, taurine augments mitochondrial function via several possible mechanisms and also reduces mitochondria-generated ROS, both of which improve islet function and insulin action at peripheral tissues.

Maternal taurine supplementation has been shown to alter mitochondria gene expression in mouse models of maternal protein restriction. Microarray analyses of liver and skeletal muscle from newborn mice from maternal protein restriction show over-representation of genes whose proteins localize to mitochondria.[139] For both tissues, pathway analysis indicated differentially expressed genes were enriched for oxidative phosphorylation, branched-chain amino acid, and carbohydrate metabolism.[139] Differentially expressed mitochondrial genes reveal tissue-specific effects; maternal protein restriction primarily increased expression in the liver and decreased expression in skeletal muscle.[139] In both liver and skeletal muscle, addition of taurine to the protein-deficient diet normalized nearly half of the mitochondrial gene expression changes.[139] Because this study was performed in newborn mice, there is no way to tell if these changes in gene expression will persist after cessation of taurine treatment, let alone, into adulthood. Nonetheless, the study provides evidence for a significant mitochondrial effect of in-utero protein deficiency and taurine treatment in tissues vital to metabolic homeostasis.

Taken together, these studies suggest that behavioral, pharmacological, and nutritional interventions each have the potential to mitigate the deleterious consequences of a suboptimal intrauterine environment. The timing of the intervention (prenatal or postnatal) has been shown to play a critical role in success of the intervention, and should be considered carefully when implementing interventions to improve the outcome.

## 7 CONCLUSION

In conclusion, the global burden of obesity and type 2 diabetes continues to increase. Our understanding of early origins of later-life disorders provides new avenues to design novel strategies to prevent the onset of these devastating diseases. Mitochondria are essential for energy balance and redox regulation, and mitochondrial dysfunction plays a key role in phenotypes associated with fetal programming of obesity and type 2 diabetes. Mounting evidence from intervention strategies have demonstrated that the ability to normalize mitochondrial function is associated with improved metabolic health outcomes. Finally, as accumulating evidence shows transmission of abnormal metabolic phenotypes across generations, it is imperative to elucidate the mechanisms underlying this phenomenon. Mitochondrial dysfunction might be a key contributor to this process and therefore, an indepth characterization of the role of mitochondria in multigenerational transmittance of metabolic diseases likely will identify new therapeutic targets.

## Disclosure Statement

The authors declare no conflict of interest.

## References


1. Yang W, Dall TM, Halder P, Gallo P, Kowal SL, Hogan PF. Economic costs of diabetes in the U.S. in 2012. *Diabetes Care* 2013;**36**(4):1033–46. https://doi.org/10.2337/dc12-625 [Epub 013 Mar 6].
2. Mathers CD, Loncar D. Projections of global mortality and burden of disease from 2002 to 2030. *PLoS Med* 2006;**3**(11):e442.
3. Hales CN, Barker DJ, Clark PM, Cox LJ, Fall C, Osmond C, et al. Fetal and infant growth and impaired glucose tolerance at age 64. *BMJ* 1991;**303**(6809):1019–22.
4. Barker DJ, Winter PD, Osmond C, Margetts B, Simmonds SJ. Weight in infancy and death from ischaemic heart disease. *Lancet* 1989;**2**(8663):577–80.
5. Barker DJ, Osmond C. Infant mortality, childhood nutrition, and ischaemic heart disease in England and Wales. *Lancet* 1986;**1**(8489):1077–81.
6. Barker DJ. The origins of the developmental origins theory. *J Intern Med* 2007;**261**(5):412–7.
7. de Boo HA, Harding JE. The developmental origins of adult disease (Barker) hypothesis. *Aust N Z J Obstet Gynaecol* 2006;**46**(1):4–14.
8. Baird J, Jacob C, Barker M, Fall CH, Hanson M, Harvey NC, et al. Developmental origins of health and disease: a lifecourse approach to the prevention of non-communicable diseases. *Healthcare (Basel)* 2017;**5**(1): https://doi.org/10.3390/healthcare5010014. pii:E14.
9. Pinney SE, Jaeckle Santos LJ, Han Y, Stoffers DA, Simmons RA. Exendin-4 increases histone acetylase activity and reverses epigenetic modifications that silence Pdx1 in the intrauterine growth retarded rat. *Diabetologia* 2011;**54**(10):2606–14. https://doi.org/10.1007/s00125-011-2250-1 [Epub 2011 Jul 21].
10. Godfrey KM, Gluckman PD, Hanson MA. Developmental origins of metabolic disease: life course and intergenerational perspectives. *Trends Endocrinol Metab* 2010;**21**(4):199–205. https://doi.org/10.1016/j.tem.2009.12.008 [Epub 10 Jan 14].
11. Simmons RA. Developmental origins of diabetes: the role of oxidative stress. *Best Pract Res Clin Endocrinol Metab* 2012;**26**(5):701–8. https://doi.org/10.1016/j.beem.2012.03.012 [Epub May 4].
12. Warner MJ, Ozanne SE. Mechanisms involved in the developmental programming of adulthood disease. *Biochem J* 2010;**427**(3):333–47. https://doi.org/10.1042/BJ20091861.
13. Segovia SA, Vickers MH, Gray C, Reynolds CM. Maternal obesity, inflammation, and developmental programming. *Biomed Res Int* 2014;**2014**:418975. https://doi.org/10.1155/2014/418975 [Epub 2014 May 20].
14. Myatt L, Eis AL, Brockman DE, Kossenjans W, Greer IA, Lyall F. Differential localization of superoxide dismutase isoforms in placental villous tissue of normotensive, pre-eclamptic, and intrauterine growth-restricted pregnancies. *J Histochem Cytochem* 1997;**45**(10):1433–8. https://doi.org/10.177/002215549704501012.
15. Karowicz-Bilinska A, Suzin J, Sieroszewski P. Evaluation of oxidative stress indices during treatment in pregnant women with intrauterine growth retardation. *Med Sci Monit* 2002;**8**(3):CR211–6.
16. Ejima K, Nanri H, Toki N, Kashimura M, Ikeda M. Localization of thioredoxin reductase and thioredoxin in normal human placenta and their protective effect against oxidative stress. *Placenta* 1999;**20**(1):95–101. https://doi.org/10.1053/plac.998.0338.
17. Kato H, Yoneyama Y, Araki T. Fetal plasma lipid peroxide levels in pregnancies complicated by preeclampsia. *Gynecol Obstet Investig* 1997;**43**(3):158–61. https://doi.org/10.1159/000291845.
18. Bowen RS, Moodley J, Dutton MF, Theron AJ. Oxidative stress in pre-eclampsia. *Acta Obstet Gynecol Scand* 2001;**80**(8):719–25.
19. Wang Y, Walsh SW. Increased superoxide generation is associated with decreased superoxide dismutase activity and mRNA expression in placental trophoblast cells in pre-eclampsia. *Placenta* 2001;**22**(2–3):206–12. https://doi.org/10.1053/plac.2000.0608.
20. Wang Y, Walsh SW. Placental mitochondria as a source of oxidative stress in pre-eclampsia. *Placenta* 1998;**19**(8):581–6.

21. Guvendag Guven ES, Karcaaltincaba D, Kandemir O, Kiykac S, Mentese A. Cord blood oxidative stress markers correlate with umbilical artery pulsatility in fetal growth restriction. *J Matern Fetal Neonatal Med* 2013;**26**(6):576–80. https://doi.org/10.3109/14767058.2012.745497 [Epub 2012 Nov 29].
22. Kimura C, Watanabe K, Iwasaki A, Mori T, Matsushita H, Shinohara K, et al. The severity of hypoxic changes and oxidative DNA damage in the placenta of early-onset preeclamptic women and fetal growth restriction. *J Matern Fetal Neonatal Med* 2013;**26**(5):491–6. https://doi.org/10.3109/14767058.2012.733766 [Epub 2012 Nov 8].
23. Gupta P, Narang M, Banerjee BD, Basu S. Oxidative stress in term small for gestational age neonates born to undernourished mothers: a case control study. *BMC Pediatr* 2004;**4**:14.
24. Chandel NS, Budinger GR, Schumacker PT. Molecular oxygen modulates cytochrome c oxidase function. *J Biol Chem* 1996;**271**(31):18672–7.
25. Degli Esposti M, McLennan H. Mitochondria and cells produce reactive oxygen species in virtual anaerobiosis: relevance to ceramide-induced apoptosis. *FEBS Lett* 1998;**430**(3):338–42.
26. Gorgias N, Maidatsi P, Tsolaki M, Alvanou A, Kiriazis G, Kaidoglou K, et al. Hypoxic pretreatment protects against neuronal damage of the rat hippocampus induced by severe hypoxia. *Brain Res* 1996;**714**(1–2):215–25.
27. Peuchant E, Brun JL, Rigalleau V, Dubourg L, Thomas MJ, Daniel JY, et al. Oxidative and antioxidative status in pregnant women with either gestational or type 1 diabetes. *Clin Biochem* 2004;**37**(4):293–8.
28. Ferretti G, Cester AM, Bacchetti T, Raffaelli F, Vignini A, Orici F, et al. Leptin and paraoxonase activity in cord blood from obese mothers. *J Matern Fetal Neonatal Med* 2014;**27**(13):1353–6. https://doi.org/10.3109/14767058.2013.858319 [Epub 2013 Nov 26].
29. Sen S, Simmons RA. Maternal antioxidant supplementation prevents adiposity in the offspring of Western diet-fed rats. *Diabetes* 2010;**59**(12):3058–65. https://doi.org/10.2337/db10-0301 [Epub 2010 Sep 7].
30. Simmons RA, Suponitsky-Kroyter I, Selak MA. Progressive accumulation of mitochondrial DNA mutations and decline in mitochondrial function lead to beta-cell failure. *J Biol Chem* 2005;**280**(31):28785–91. https://doi.org/10.1074/jbc.M505695200 [Epub 2005 Jun 9].
31. Theys N, Ahn MT, Bouckenooghe T, Reusens B, Remacle C. Maternal malnutrition programs pancreatic islet mitochondrial dysfunction in the adult offspring. *J Nutr Biochem* 2011;**22**(10):985–94. https://doi.org/10.1016/j.jnutbio.2010.08.015 [Epub Dec 28].
32. Sparre T, Reusens B, Cherif H, Larsen MR, Roepstorff P, Fey SJ, et al. Intrauterine programming of fetal islet gene expression in rats—effects of maternal protein restriction during gestation revealed by proteome analysis. *Diabetologia* 2003;**46**(11):1497–511. https://doi.org/10.007/s00125-003-1208-3 [Epub 2003 Sep 12].
33. Reusens B, Sparre T, Kalbe L, Bouckenooghe T, Theys N, Kruhoffer M, et al. The intrauterine metabolic environment modulates the gene expression pattern in fetal rat islets: prevention by maternal taurine supplementation. *Diabetologia* 2008;**51**(5):836–45. https://doi.org/10.1007/s00125-008-0956-5 [Epub 2008 Mar 3].
34. Theys N, Bouckenooghe T, Ahn MT, Remacle C, Reusens B. Maternal low-protein diet alters pancreatic islet mitochondrial function in a sex-specific manner in the adult rat. *Am J Physiol Regul Integr Comp Physiol* 2009;**297**(5):R1516–25. https://doi.org/10.152/ajpregu.00280.2009 [Epub 2009 Sep 16].
35. Park HK, Jin CJ, Cho YM, Park DJ, Shin CS, Park KS, et al. Changes of mitochondrial DNA content in the male offspring of protein-malnourished rats. *Ann N Y Acad Sci* 2004;**1011**:205–16.
36. Selak MA, Storey BT, Peterside I, Simmons RA. Impaired oxidative phosphorylation in skeletal muscle of intrauterine growth-retarded rats. *Am J Physiol Endocrinol Metab* 2003;**285**(1):E130–7. https://doi.org/10.1152/ajpendo.00322.2002 [Epub 2003 Mar 11].
37. Gauthier BR, Wiederkehr A, Baquie M, Dai C, Powers AC, Kerr-Conte J, et al. PDX1 deficiency causes mitochondrial dysfunction and defective insulin secretion through TFAM suppression. *Cell Metab* 2009;**10**(2):110–8. https://doi.org/10.1016/j.cmet.2009.07.002.
38. Thompson RF, Fazzari MJ, Niu H, Barzilai N, Simmons RA, Greally JM. Experimental intrauterine growth restriction induces alterations in DNA methylation and gene expression in pancreatic islets of rats. *J Biol Chem* 2010;**285**(20):15111–8. https://doi.org/10.1074/jbc.M109.095133 [Epub 2010 Mar 1].
39. Theys N, Clippe A, Bouckenooghe T, Reusens B, Remacle C. Early low protein diet aggravates unbalance between antioxidant enzymes leading to islet dysfunction. *PLoS ONE* 2009;**4**(7):e6110. https://doi.org/10.1371/journal.pone.0006110.
40. Prentice KJ, Luu L, Allister EM, Liu Y, Jun LS, Sloop KW, et al. The furan fatty acid metabolite CMPF is elevated in diabetes and induces beta cell dysfunction. *Cell Metab* 2014;**19**(4):653–66. https://doi.org/10.1016/j.cmet.2014.03.008.

41. Lenzen S, Drinkgern J, Tiedge M. Low antioxidant enzyme gene expression in pancreatic islets compared with various other mouse tissues. *Free Radic Biol Med* 1996;**20**(3):463–6.
42. Tiedge M, Lortz S, Drinkgern J, Lenzen S. Relation between antioxidant enzyme gene expression and antioxidative defense status of insulin-producing cells. *Diabetes* 1997;**46**(11):1733–42.
43. Lenzen S. Oxidative stress: the vulnerable beta-cell. *Biochem Soc Trans* 2008;**36**(Pt 3):343–7. https://doi.org/10.1042/BST0360343.
44. Robertson RP. Oxidative stress and impaired insulin secretion in type 2 diabetes. *Curr Opin Pharmacol* 2006;**6**(6):615–9. https://doi.org/10.1016/j.coph.2006.09.002 [Epub Oct 10].
45. Del Guerra S, Lupi R, Marselli L, Masini M, Bugliani M, Sbrana S, et al. Functional and molecular defects of pancreatic islets in human type 2 diabetes. *Diabetes* 2005;**54**(3):727–35.
46. Maechler P, Jornot L, Wollheim CB. Hydrogen peroxide alters mitochondrial activation and insulin secretion in pancreatic beta cells. *J Biol Chem* 1999;**274**(39):27905–13.
47. Sakai K, Matsumoto K, Nishikawa T, Suefuji M, Nakamaru K, Hirashima Y, et al. Mitochondrial reactive oxygen species reduce insulin secretion by pancreatic beta-cells. *Biochem Biophys Res Commun* 2003;**300**(1):216–22.
48. Luo ZC, Fraser WD, Julien P, Deal CL, Audibert F, Smith GN, et al. Tracing the origins of "fetal origins" of adult diseases: programming by oxidative stress? *Med Hypotheses* 2006;**66**(1):38–44 [Epub 2005 Sep 27].
49. Lacraz G, Figeac F, Movassat J, Kassis N, Coulaud J, Galinier A, et al. Diabetic beta-cells can achieve self-protection against oxidative stress through an adaptive up-regulation of their antioxidant defenses. *PLoS ONE* 2009;**4**(8):e6500. https://doi.org/10.1371/journal.pone.0006500.
50. Lacraz G, Figeac F, Movassat J, Kassis N, Portha B. Diabetic GK/par rat beta-cells are spontaneously protected against $H_2O_2$-triggered apoptosis. A cAMP-dependent adaptive response. *Am J Physiol Endocrinol Metab* 2010;**298**(1):E17–27. https://doi.org/10.1152/ajpendo.90871.2008 [Epub 2009 Oct 20].
51. Peterside IE, Selak MA, Simmons RA. Impaired oxidative phosphorylation in hepatic mitochondria in growth-retarded rats. *Am J Physiol Endocrinol Metab* 2003;**285**(6):E1258–66. https://doi.org/10.152/ajpendo.00437.2002.
52. Borengasser SJ, Lau F, Kang P, Blackburn ML, Ronis MJ, Badger TM, et al. Maternal obesity during gestation impairs fatty acid oxidation and mitochondrial SIRT3 expression in rat offspring at weaning. *PLoS ONE* 2011;**6**(8):e24068. https://doi.org/10.1371/journal.pone.0024068 [Epub 2011 Aug 25].
53. Park KS, Kim SK, Kim MS, Cho EY, Lee JH, Lee KU, et al. Fetal and early postnatal protein malnutrition cause long-term changes in rat liver and muscle mitochondria. *J Nutr* 2003;**133**(10):3085–90.
54. Borengasser SJ, Faske J, Kang P, Blackburn ML, Badger TM, Shankar K. In utero exposure to prepregnancy maternal obesity and postweaning high-fat diet impair regulators of mitochondrial dynamics in rat placenta and offspring. *Physiol Genomics* 2014;**46**(23):841–50. https://doi.org/10.1152/physiolgenomics.00059.2014 [Epub 2014 Oct 21].
55. Bruce KD, Cagampang FR, Argenton M, Zhang J, Ethirajan PL, Burdge GC, et al. Maternal high-fat feeding primes steatohepatitis in adult mice offspring, involving mitochondrial dysfunction and altered lipogenesis gene expression. *Hepatology* 2009;**50**(6):1796–808. https://doi.org/10.002/hep.23205.
56. Hellgren LI, Jensen RI, Waterstradt MS, Quistorff B, Lauritzen L. Acute and perinatal programming effects of a fat-rich diet on rat muscle mitochondrial function and hepatic lipid accumulation. *Acta Obstet Gynecol Scand* 2014;**93**(11):1170–80. https://doi.org/10.1111/aogs.12458 [Epub 2014 Aug 27].
57. McCurdy CE, Bishop JM, Williams SM, Grayson BE, Smith MS, Friedman JE, et al. Maternal high-fat diet triggers lipotoxicity in the fetal livers of nonhuman primates. *J Clin Invest* 2009;**119**(2):323–35. https://doi.org/10.1172/JCI32661 [Epub 2009 Jan 19].
58. Bayol SA, Simbi BH, Fowkes RC, Stickland NC. A maternal "junk food" diet in pregnancy and lactation promotes nonalcoholic fatty liver disease in rat offspring. *Endocrinology* 2010;**151**(4):1451–61. https://doi.org/10.1210/en.2009-1192 [Epub 2010 Mar 5].
59. Kelley DE, He J, Menshikova EV, Ritov VB. Dysfunction of mitochondria in human skeletal muscle in type 2 diabetes. *Diabetes* 2002;**51**(10):2944–50.
60. Mogensen M, Sahlin K, Fernstrom M, Glintborg D, Vind BF, Beck-Nielsen H, et al. Mitochondrial respiration is decreased in skeletal muscle of patients with type 2 diabetes. *Diabetes* 2007;**56**(6):1592–9 [Epub 2007 Mar 9].
61. Yokota T, Kinugawa S, Hirabayashi K, Matsushima S, Inoue N, Ohta Y, et al. Oxidative stress in skeletal muscle impairs mitochondrial respiration and limits exercise capacity in type 2 diabetic mice. *Am J Physiol Heart Circ Physiol* 2009;**297**(3):H1069–77. https://doi.org/10.152/ajpheart.00267.2009 [Epub 2009 Jul 17].

62. Patti ME, Butte AJ, Crunkhorn S, Cusi K, Berria R, Kashyap S, et al. Coordinated reduction of genes of oxidative metabolism in humans with insulin resistance and diabetes: potential role of PGC1 and NRF1. *Proc Natl Acad Sci U S A* 2003;**100**(14):8466–71 [Epub 2003 Jun 27].
63. Jorgensen W, Gam C, Andersen JL, Schjerling P, Scheibye-Knudsen M, Mortensen OH, et al. Changed mitochondrial function by pre- and/or postpartum diet alterations in sheep. *Am J Physiol Endocrinol Metab* 2009;**297**(6):E1349–57. https://doi.org/10.152/ajpendo.00505.2009 [Epub 2009 Oct 13].
64. Latouche C, Heywood SE, Henry SL, Ziemann M, Lazarus R, El-Osta A, et al. Maternal overnutrition programs changes in the expression of skeletal muscle genes that are associated with insulin resistance and defects of oxidative phosphorylation in adult male rat offspring. *J Nutr* 2014;**144**(3):237–44. https://doi.org/10.3945/jn.113.186775 [Epub 2013 Dec 31].
65. Claycombe KJ, Roemmich JN, Johnson L, Vomhof-DeKrey EE, Johnson WT. Skeletal muscle Sirt3 expression and mitochondrial respiration are regulated by a prenatal low-protein diet. *J Nutr Biochem* 2015;**26**(2):184–9. https://doi.org/10.1016/j.jnutbio.2014.10.003 [Epub Nov 14].
66. Lettieri Barbato D, Tatulli G, Vegliante R, Cannata SM, Bernardini S, Ciriolo MR, et al. Dietary fat overload reprograms brown fat mitochondria. *Front Physiol* 2015;(6)272. https://doi.org/10.3389/fphys.2015.00272. [eCollection 2015].
67. Lettieri-Barbato D, D'Angelo F, Sciarretta F, Tatulli G, Tortolici F, Ciriolo MR, et al. Maternal high calorie diet induces mitochondrial dysfunction and senescence phenotype in subcutaneous fat of newborn mice. *Oncotarget* 2017;**8**(48):83407–18. https://doi.org/10.18632/oncotarget.9948. [eCollection 2017 Oct 13].
68. Claycombe KJ, Vomhof-DeKrey EE, Garcia R, Johnson WT, Uthus E, Roemmich JN. Decreased beige adipocyte number and mitochondrial respiration coincide with increased histone methyl transferase (G9a) and reduced FGF21 gene expression in Sprague-Dawley rats fed prenatal low protein and postnatal high-fat diets. *J Nutr Biochem* 2016;(31)113–21. https://doi.org/10.1016/j.jnutbio.2016.01.008 [Epub Feb 28].
69. Skinner MK. Role of epigenetics in developmental biology and transgenerational inheritance. *Birth Defects Res C Embryo Today* 2011;**93**(1):51–5.
70. Susiarjo M, Xin F, Bansal A, Stefaniak M, Li C, Simmons RA, et al. Bisphenol a exposure disrupts metabolic health across multiple generations in the mouse. *Endocrinology* 2015;**156**(6):2049–58.
71. Bansal A, Rashid C, Xin F, Li C, Polyak E, Duemler A, et al. Sex- and dose-specific effects of maternal bisphenol a exposure on pancreatic islets of first and second generation adult mice offspring. *Environ Health Perspect* 2017;**125**(9):0970221–09702218.
72. Saben JL, Boudoures AL, Asghar Z, Thompson A, Drury A, Zhang W, et al. Maternal metabolic syndrome programs mitochondrial dysfunction via germline changes across three generations. *Cell Rep* 2016;**16**(1):1–8.
73. Otera H, Wang C, Cleland MM, Setoguchi K, Yokota S, Youle RJ, et al. Mff is an essential factor for mitochondrial recruitment of Drp1 during mitochondrial fission in mammalian cells. *J Cell Biol* 2010;**191**(6):1141–58.
74. Cogliati S, Frezza C, Soriano ME, Varanita T, Quintana-Cabrera R, Corrado M, et al. Mitochondrial cristae shape determines respiratory chain supercomplexes assembly and respiratory efficiency. *Cell* 2013;**155**(1):160–71.
75. Drake AJ, Walker BR, Seckl JR. Intergenerational consequences of fetal programming by in utero exposure to glucocorticoids in rats. *Am J Phys Regul Integr Comp Phys* 2005;**288**(1):R34–8.
76. Dunn GA, Bale TL. Maternal high-fat diet promotes body length increases and insulin insensitivity in second-generation mice. *Endocrinology* 2009;**150**(11):4999–5009.
77. Hutchison 3rd CA, Newbold JE, Potter SS, Edgell MH. Maternal inheritance of mammalian mitochondrial DNA. *Nature* 1974;**251**(5475):536–8.
78. Binder NK, Hannan NJ, Gardner DK. Paternal diet-induced obesity retards early mouse embryo development, mitochondrial activity and pregnancy health. *PLoS ONE* 2012;**7**(12):e52304.
79. Pinti M, Gibellini L, Liu Y, Xu S, Lu B, Cossarizza A. Mitochondrial Lon protease at the crossroads of oxidative stress, ageing and cancer. *Cell Mol Life Sci* 2015;**72**(24):4807–24.
80. Gibellini L, Pinti M, Boraldi F, Giorgio V, Bernardi P, Bartolomeo R, et al. Silencing of mitochondrial Lon protease deeply impairs mitochondrial proteome and function in colon cancer cells. *FASEB J* 2014;**28**(12):5122–35.
81. Carter LG, Lewis KN, Wilkerson DC, Tobia CM, Ngo Tenlep SY, Shridas P, et al. Perinatal exercise improves glucose homeostasis in adult offspring. *Am J Physiol Endocrinol Metab* 2012;**303**(8):E1061–8.
82. Carter LG, Qi NR, De Cabo R, Pearson KJ. Maternal exercise improves insulin sensitivity in mature rat offspring. *Med Sci Sports Exerc* 2013;**45**(5):832–40.

83. Liu J, Lee I, Feng HZ, Galen SS, Huttemann PP, Perkins GA, et al. Aerobic exercise preconception and during pregnancy enhances oxidative capacity in the hindlimb muscles of mice offspring. *J Strength Cond Res* 2018.
84. Theilen NT, Kunkel GH, Tyagi SC. The role of exercise and TFAM in preventing skeletal muscle atrophy. *J Cell Physiol* 2017;**232**(9):2348–58.
85. Kim Y, Triolo M, Hood DA. Impact of aging and exercise on mitochondrial quality control in skeletal muscle. *Oxidative Med Cell Longev* 2017;**2017**:3165396.
86. Gordon JW, Rungi AA, Inagaki H, Hood DA. Effects of contractile activity on mitochondrial transcription factor A expression in skeletal muscle. *J Appl Physiol (1985)* 2001;**90**(1):389–96.
87. Chung E, Joiner HE, Skelton T, Looten KD, Manczak M, Reddy PH. Maternal exercise upregulates mitochondrial gene expression and increases enzyme activity of fetal mouse hearts. *Physiol Rep* 2017;**5**(5).
88. Sheldon RD, Nicole Blaize A, Fletcher JA, Pearson KJ, Donkin SS, Newcomer SC, et al. Gestational exercise protects adult male offspring from high-fat diet-induced hepatic steatosis. *J Hepatol* 2016;**64**(1):171–8.
89. Nathanielsz PW, Ford SP, Long NM, Vega CC, Reyes-Castro LA, Zambrano E. Interventions to prevent adverse fetal programming due to maternal obesity during pregnancy. *Nutr Rev* 2013;**71**(Suppl. 1):S78–87.
90. Laker RC, Lillard TS, Okutsu M, Zhang M, Hoehn KL, Connelly JJ, et al. Exercise prevents maternal high-fat diet-induced hypermethylation of the Pgc-1alpha gene and age-dependent metabolic dysfunction in the offspring. *Diabetes* 2014;**63**(5):1605–11.
91. Leury BJ, Chandler KD, Bird AR, Bell AW. Effects of maternal undernutrition and exercise on glucose kinetics in fetal sheep. *Br J Nutr* 1990;**64**(2):463–72.
92. Vickers MH, Breier BH, McCarthy D, Gluckman PD. Sedentary behavior during postnatal life is determined by the prenatal environment and exacerbated by postnatal hypercaloric nutrition. *Am J Phys Regul Integr Comp Phys* 2003;**285**(1):R271–3.
93. Bellinger L, Sculley DV, Langley-Evans SC. Exposure to undernutrition in fetal life determines fat distribution, locomotor activity and food intake in ageing rats. *Int J Obes* 2006;**30**(5):729–38.
94. Watkins AJ, Wilkins A, Cunningham C, Perry VH, Seet MJ, Osmond C, et al. Low protein diet fed exclusively during mouse oocyte maturation leads to behavioural and cardiovascular abnormalities in offspring. *J Physiol* 2008;**586**(8):2231–44.
95. Samuelsson AM, Matthews PA, Argenton M, Christie MR, McConnell JM, Jansen EH, et al. Diet-induced obesity in female mice leads to offspring hyperphagia, adiposity, hypertension, and insulin resistance: a novel murine model of developmental programming. *Hypertension* 2008;**51**(2):383–92.
96. Miles JL, Huber K, Thompson NM, Davison M, Breier BH. Moderate daily exercise activates metabolic flexibility to prevent prenatally induced obesity. *Endocrinology* 2009;**150**(1):179–86.
97. Uddin GM, Youngson NA, Doyle BM, Sinclair DA, Morris MJ. Nicotinamide mononucleotide (NMN) supplementation ameliorates the impact of maternal obesity in mice: comparison with exercise. *Sci Rep* 2017;**7**(1):15063.
98. Irving BA, Short KR, Nair KS, Stump CS. Nine days of intensive exercise training improves mitochondrial function but not insulin action in adult offspring of mothers with type 2 diabetes. *J Clin Endocrinol Metab* 2011;**96**(7):E1137–41.
99. Stoffers DA, Desai BM, DeLeon DD, Simmons RA. Neonatal exendin-4 prevents the development of diabetes in the intrauterine growth retarded rat. *Diabetes* 2003;**52**(3):734–40.
100. Ham JN, Crutchlow MF, Desai BM, Simmons RA, Stoffers DA. Exendin-4 normalizes islet vascularity in intrauterine growth restricted rats: potential role of VEGF. *Pediatr Res* 2009;**66**(1):42–6.
101. Raab EL, Vuguin PM, Stoffers DA, Simmons RA. Neonatal exendin-4 treatment reduces oxidative stress and prevents hepatic insulin resistance in intrauterine growth-retarded rats. *Am J Phys Regul Integr Comp Phys* 2009;**297**(6):R1785–94.
102. Chen H, Simar D, Pegg K, Saad S, Palmer C, Morris MJ. Exendin-4 is effective against metabolic disorders induced by intrauterine and postnatal overnutrition in rodents. *Diabetologia* 2014;**57**(3):614–22.
103. Chan YL, Saad S, Simar D, Oliver B, McGrath K, Reyk D, et al. Short term exendin-4 treatment reduces markers of metabolic disorders in female offspring of obese rat dams. *Int J Dev Neurosci* 2015;**46**:67–75.
104. Gatford KL, Sulaiman SA, Mohammad SN, De Blasio MJ, Harland ML, Simmons RA, et al. Neonatal exendin-4 reduces growth, fat deposition and glucose tolerance during treatment in the intrauterine growth-restricted lamb. *PLoS ONE* 2013;**8**(2):e56553.

105. Liu H, Schultz CG, De Blasio MJ, Peura AM, Heinemann GK, Harryanto H, et al. Effect of placental restriction and neonatal exendin-4 treatment on postnatal growth, adult body composition, and in vivo glucose metabolism in the sheep. *Am J Physiol Endocrinol Metab* 2015;**309**(6):E589–600.
106. Reddy S, Elliott RB. Ontogenic development of peptide hormones in the mammalian fetal pancreas. *Experientia* 1988;**44**(1):1–9.
107. Kang MY, Oh TJ, Cho YM. Glucagon-like peptide-1 increases mitochondrial biogenesis and function in INS-1 rat insulinoma cells. *Endocrinol Metab (Seoul)* 2015;**30**(2):216–20.
108. Li Z, Zhou Z, Huang G, Hu F, Xiang Y, He L. Exendin-4 protects mitochondria from reactive oxygen species induced apoptosis in pancreatic Beta cells. *PLoS ONE* 2013;**8**(10):e76172.
109. Wassef MAE, Tork OM, Rashed LA, Ibrahim W, Morsi H, Rabie DMM. Mitochondrial dysfunction in diabetic cardiomyopathy: effect of mesenchymal stem cell with PPAR-gamma agonist or Exendin-4. *Exp Clin Endocrinol Diabetes* 2018;**126**(1):27–38.
110. Ranta F, Avram D, Berchtold S, Dufer M, Drews G, Lang F, et al. Dexamethasone induces cell death in insulin-secreting cells, an effect reversed by exendin-4. *Diabetes* 2006;**55**(5):1380–90.
111. Tews D, Lehr S, Hartwig S, Osmers A, Paslack W, Eckel J. Anti-apoptotic action of exendin-4 in INS-1 beta cells: comparative protein pattern analysis of isolated mitochondria. *Horm Metab Res* 2009;**41**(4):294–301.
112. Tourrel C, Bailbe D, Lacorne M, Meile MJ, Kergoat M, Portha B. Persistent improvement of type 2 diabetes in the Goto-Kakizaki rat model by expansion of the beta-cell mass during the prediabetic period with glucagon-like peptide-1 or exendin-4. *Diabetes* 2002;**51**(5):1443–52.
113. Tomas E, Stanojevic V, Habener JF. GLP-1-derived nonapeptide GLP-1(28-36)amide targets to mitochondria and suppresses glucose production and oxidative stress in isolated mouse hepatocytes. *Regul Pept* 2011;**167**(2–3):177–84.
114. Huxtable RJ. Physiological actions of taurine. *Physiol Rev* 1992;**72**(1):101–63.
115. Karl PI, Fisher SE. Taurine transport by microvillous membrane vesicles and the perfused cotyledon of the human placenta. *Am J Phys* 1990;**258**(3 Pt 1):C443–51.
116. Economides DL, Nicolaides KH, Gahl WA, Bernardini I, Evans MI. Plasma amino acids in appropriate- and small-for-gestational-age fetuses. *Am J Obstet Gynecol* 1989;**161**(5):1219–27.
117. Roos S, Powell TL, Jansson T. Human placental taurine transporter in uncomplicated and IUGR pregnancies: cellular localization, protein expression, and regulation. *Am J Phys Regul Integr Comp Phys* 2004;**287**(4):R886–93.
118. Norberg S, Powell TL, Jansson T. Intrauterine growth restriction is associated with a reduced activity of placental taurine transporters. *Pediatr Res* 1998;**44**(2):233–8.
119. Ditchfield AM, Desforges M, Mills TA, Glazier JD, Wareing M, Mynett K, et al. Maternal obesity is associated with a reduction in placental taurine transporter activity. *Int J Obes* 2015;**39**(4):557–64.
120. Ryan EA, Liu D, Bell RC, Finegood DT, Crawford J. Long-term consequences in offspring of diabetes in pregnancy: studies with syngeneic islet-transplanted streptozotocin-diabetic rats. *Endocrinology* 1995;**136**(12):5587–92.
121. Cherif H, Reusens B, Ahn MT, Hoet JJ, Remacle C. Effects of taurine on the insulin secretion of rat fetal islets from dams fed a low-protein diet. *J Endocrinol* 1998;**159**(2):341–8.
122. Reusens B, Sparre T, Kalbe L, Bouckenooghe T, Theys N, Kruhoffer M, et al. The intrauterine metabolic environment modulates the gene expression pattern in fetal rat islets: prevention by maternal taurine supplementation. *Diabetologia* 2008;**51**(5):836–45.
123. Boujendar S, Reusens B, Merezak S, Ahn MT, Arany E, Hill D, et al. Taurine supplementation to a low protein diet during foetal and early postnatal life restores a normal proliferation and apoptosis of rat pancreatic islets. *Diabetologia* 2002;**45**(6):856–66.
124. Merezak S, Reusens B, Renard A, Goosse K, Kalbe L, Ahn MT, et al. Effect of maternal low-protein diet and taurine on the vulnerability of adult Wistar rat islets to cytokines. *Diabetologia* 2004;**47**(4):669–75.
125. Boujendar S, Arany E, Hill D, Remacle C, Reusens B. Taurine supplementation of a low protein diet fed to rat dams normalizes the vascularization of the fetal endocrine pancreas. *J Nutr* 2003;**133**(9):2820–5.
126. Tang C, Marchand K, Lam L, Lux-Lantos V, Thyssen SM, Guo J, et al. Maternal taurine supplementation in rats partially prevents the adverse effects of early-life protein deprivation on beta-cell function and insulin sensitivity. *Reproduction* 2013;**145**(6):609–20.
127. Thaeomor A, Teangphuck P, Chaisakul J, Seanthaweesuk S, Somparn N, Roysommuti S. Perinatal taurine supplementation prevents metabolic and cardiovascular effects of maternal diabetes in adult rat offspring. *Adv Exp Med Biol* 2017;**975**:295–305.

128. Shivananjappa MM. Muralidhara. Taurine attenuates maternal and embryonic oxidative stress in a streptozotocin-diabetic rat model. *Reprod BioMed Online* 2012;**24**(5):558–66.
129. Li M, Reynolds CM, Sloboda DM, Gray C, Vickers MH. Maternal taurine supplementation attenuates maternal fructose-induced metabolic and inflammatory dysregulation and partially reverses adverse metabolic programming in offspring. *J Nutr Biochem* 2015;**26**(3):267–76.
130. Li M, Reynolds CM, Sloboda DM, Gray C, Vickers MH. Effects of taurine supplementation on hepatic markers of inflammation and lipid metabolism in mothers and offspring in the setting of maternal obesity. *PLoS ONE* 2013;**8**(10):e76961.
131. Ogata ES, Bussey ME, Finley S. Altered gas exchange, limited glucose and branched chain amino acids, and hypoinsulinism retard fetal growth in the rat. *Metabolism* 1986;**35**(10):970–7.
132. Hultman K, Alexanderson C, Manneras L, Sandberg M, Holmang A, Jansson T. Maternal taurine supplementation in the late pregnant rat stimulates postnatal growth and induces obesity and insulin resistance in adult offspring. *J Physiol* 2007;**579**(Pt 3):823–33.
133. Aruoma OI, Halliwell B, Hoey BM, Butler J. The antioxidant action of taurine, hypotaurine and their metabolic precursors. *Biochem J* 1988;**256**(1):251–5.
134. Parvez S, Tabassum H, Banerjee BD, Raisuddin S. Taurine prevents tamoxifen-induced mitochondrial oxidative damage in mice. *Basic Clin Pharmacol Toxicol* 2008;**102**(4):382–7.
135. Hansen SH, Birkedal H, Wibrand F, Grunnet N. Taurine and regulation of mitochondrial metabolism. *Adv Exp Med Biol* 2015;**803**:397–405.
136. Jong CJ, Azuma J, Schaffer S. Mechanism underlying the antioxidant activity of taurine: prevention of mitochondrial oxidant production. *Amino Acids* 2012;**42**(6):2223–32.
137. Han J, Bae JH, Kim SY, Lee HY, Jang BC, Lee IK, et al. Taurine increases glucose sensitivity of UCP2-overexpressing beta-cells by ameliorating mitochondrial metabolism. *Am J Physiol Endocrinol Metab* 2004;**287**(5):E1008–18.
138. El Idrissi A, Trenkner E. Taurine regulates mitochondrial calcium homeostasis. *Adv Exp Med Biol* 2003;**526**:527–36.
139. Mortensen OH, Olsen HL, Frandsen L, Nielsen PE, Nielsen FC, Grunnet N, et al. A maternal low protein diet has pronounced effects on mitochondrial gene expression in offspring liver and skeletal muscle; protective effect of taurine. *J Biomed Sci* 2010;**17**(Suppl. 1):S38.

CHAPTER

# 15

# Mitochondrial Dysfunction Induced by Xenobiotics: Involvement in Steatosis and Steatohepatitis

*Karima Begriche*, Julie Massart†, Bernard Fromenty**

*INSERM, Univ Rennes, INRA, Institut NUMECAN (Nutrition Metabolisms and Cancer) UMR_A 1341, UMR_S 1241, Rennes, France †Department of Molecular Medicine and Surgery, Karolinska University Hospital, Karolinska Institutet, Stockholm, Sweden

## 1 INTRODUCTION

Steatosis, also referred to as fatty liver, is a hepatic lesion that can be frequently induced by ethanol overconsumption and many xenobiotics, such as drugs and environmental pollutants.[1–3] Notably, fat accumulation in the liver (mainly as triglycerides) is also present in a majority of obese individuals, even in those who are not drinking alcohol or are unexposed to any steatogenic xenobiotics. Although fatty liver itself is benign, it can progress in the long term to steatohepatitis, which is characterized by necroinflammation and fibrosis.[2, 4] The occurrence of steatohepatitis is a major issue for patients and physicians because the lesion can evolve toward cirrhosis and hepatocellular carcinoma (HCC). The large spectrum of liver lesions that can occur in obese individuals is referred to as nonalcoholic fatty liver disease (NAFLD). Some clinical and experimental investigations reported that some xenobiotics can worsen NAFLD.[5, 6]

It has been acknowledged that mitochondrial dysfunction plays a major role in the pathogenesis of steatosis and steatohepatitis induced by numerous xenobiotics. In NAFLD, mitochondrial dysfunction does not seem to play a central role in fat deposition but rather appears to have a significant pathogenic role in the progression of fatty liver to nonalcoholic steatohepatitis (NASH).[7, 8] More specifically, the upregulation of the mitochondrial oxidation of different substrates and the concomitant impairment of the mitochondrial respiratory chain (MRC) activity favor the overproduction of reactive oxygen species (ROS), which seems to play a pivotal role in necroinflammation and fibrosis.[7, 9] In this chapter, we examine how xenobiotics can induce steatosis and steatohepatitis secondary to mitochondrial dysfunction. We selected xenobiotics able to alter mitochondrial function by different mechanisms, such as



*Mitochondria in Obesity and Type 2 Diabetes*
https://doi.org/10.1016/B978-0-12-811752-1.00015-8

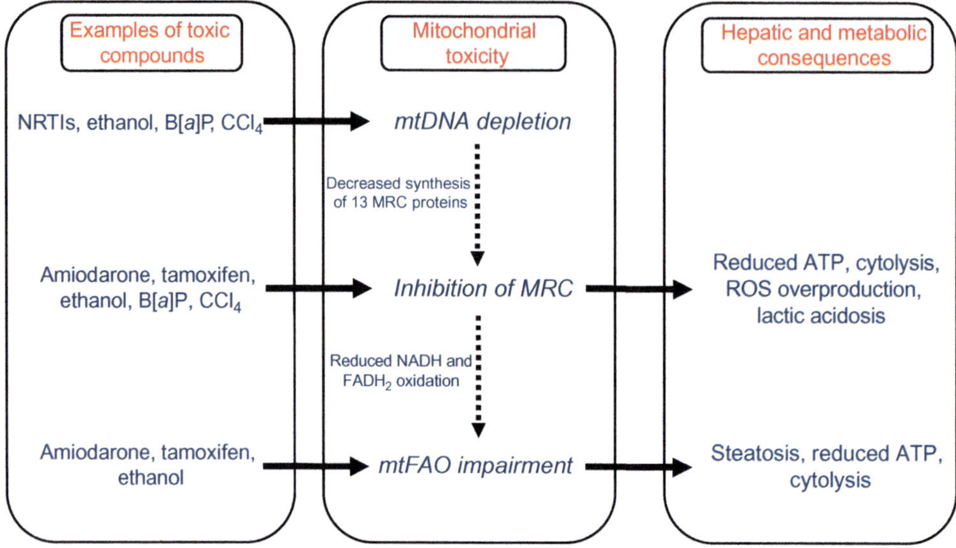


FIG. 1 Toxic compounds can impair mitochondrial function in the liver via different mechanisms, including mtDNA depletion, inhibition of MRC, and impairment of the mtFAO pathway. Ethanol overconsumption and some xenobiotics, such as amiodarone and $CCl_4$, can impair mitochondrial function by two or three different mechanisms. The main hepatic and metabolic consequences of these mitochondrial effects are indicated in the right part of the figure. Impairment of MRC and/or mtFAO can reduce liver ATP synthesis. MRC inhibition will lead to mtFAO impairment only if MRC activity is reduced severely. Depending on the severity of ATP depletion, the spectrum of hepatic cytolysis can vary from isolated increase in plasma ALT and AST to fulminant hepatitis and severe hepatic dysfunction. Impairment of mtFAO eventually leads to steatosis, corresponding to an accumulation of lipids (mainly triglycerides). Inhibition of MRC also can favor the overproduction of mitochondrial ROS, which play a major role in the progression of steatosis to steatohepatitis. Lactic acidosis, which results from tricarboxylic acid cycle impairment, is observed mainly with NRTIs such as stavudine and didanosine. Further information can be found in the text. Abbreviations: *ALT*, alanine aminotransferase; *AST*, aspartate aminotransferase; *B[a]P*, benzo[*a*]pyrene; *CCl4*, carbon tetrachloride; *MRC*, mitochondrial respiratory chain; *mtDNA*, mitochondrial DNA; *mtFAO*, mitochondrial fatty acid oxidation; *NRTIs*, nucleoside reverse transcriptase inhibitors; *ROS*, reactive oxygen species.

direct inhibition of mitochondrial fatty acid oxidation (mtFAO), impairment of mitochondrial DNA (mtDNA) replication, and oxidative stress (Fig. 1). When data are available, we indicate whether these xenobiotics were shown (or suspected) to worsen NAFLD related to obesity. Other examples of xenobiotics inducing steatosis and steatohepatitis via mitochondrial dysfunction can be found in several recent reviews.[1, 2, 10–13]

## 2 THE MAIN FEATURES OF MITOCHONDRIA

The first chapters of this book provide exhaustive information concerning mitochondrial function and structure. Therefore, the paragraphs below just give the essential information to comprehend how xenobiotics can induce mitochondrial dysfunction and the main metabolic consequences of such effect, in particular regarding fat accumulation and ROS overproduction.

## 2.1 The Oxidative Phosphorylation Process

Mitochondria provide most of the ATP required for cell function and survival. ATP synthesis is performed via the oxidative phosphorylation (OXPHOS) process, which couples the oxidation of different substrates to ATP generation.[14, 15] The constant oxidation of substrates such as fatty acids and pyruvate generates $FADH_2$ and NADH that subsequently transfer their electrons to the mitochondrial respiratory chain (MRC), thus regenerating the $NAD^+$ and FAD necessary for other cycles of fuel oxidation. The electrons provided by NADH or $FADH_2$ migrate along the respiratory chain, up to cytochrome c oxidase (COX), where they safely react with oxygen and protons to form water. Electron transfer across MRC complexes I, III, and IV is coupled with the extrusion of protons from the mitochondrial matrix into the intermembrane space of mitochondria, creating a large electrochemical potential ($\Delta\psi$) across the inner membrane. When ADP is high, protons reenter the matrix through the $F_0$ portion of ATP synthase, causing the conversion of ADP into ATP by the $F_1$ portion of this enzyme. OXPHOS can be uncoupled by different xenobiotics including drugs, as discussed below.

## 2.2 The Mitochondrial Production of ROS

MRC is deemed to be the main site of ROS production in most cells. A small fraction of electrons going through the MRC complexes I and III constantly leaks from these complexes and react with oxygen to form the superoxide anion radical ($O_2^{\cdot-}$).[7, 14] This radical, however, is mostly dismutated by manganese superoxide dismutase (MnSOD, also referred to as SOD2) into hydrogen peroxide ($H_2O_2$), which is more stable compared to the superoxide anion radical. Mitochondrial $H_2O_2$ can be subsequently detoxified into water by peroxiredoxins and glutathione peroxidases (GPXs), which necessitate reduced glutathione (GSH) as cofactor.[14] Therefore, the mitochondrial GSH store plays a major role by avoiding the excessive accumulation of hydrogen peroxide and subsequent oxidative stress.[16] Whereas mitochondrial ROS serve as key signaling molecules when produced at physiological levels,[17, 18] ROS overproduction secondary to mitochondrial dysfunction can have different deleterious effects in the liver.

## 2.3 Oxidation of Fatty Acids

Fatty acids (FAs) of different lengths can be oxidized by the mtFAO pathway, a key metabolic process mandatory for the preservation of normal energy output, especially during fasting.[19] Whereas short-chain FAs (SCFAs) and medium-chain FAs (MCFAs) freely enter mitochondria, the entry of long-chain FAs (LCFAs) into mitochondria requires carnitine palmitoyltransferase 1 (CPT1) and 2 (CPT2). During their β-oxidation, SCFAs, MCFAs, and LCFAs undergo four sequential reactions leading to the release of one acetyl-CoA molecule and a shortened FA, which can be further oxidized by other cycles of mtFAO. Two of these reactions are catalyzed by different FAD- and $NAD^+$-dependent dehydrogenases that have specific activities for SCFAs, MCFAs, or LCFAs.[19] In the liver, a significant part of the mtFAO-derived acetyl-CoA molecules generate the ketone bodies acetoacetate and β-hydroxybutyrate, especially during fasting. Importantly, some enzymes of the mitochondria β-oxidation pathway can produce a significant amount of hydrogen peroxide.[20, 21]

## 2.4 The Mitochondrial Genome

Mitochondria contain their own genome, as well as all the components (e.g., enzymes and transcription factors) required for its replication, transcription, translation, and repair. mtDNA is a 16.6kb double-strand circular genome present within each mitochondrion in several copies and encodes 13 MRC polypeptides. These polypeptides are then inserted in the inner membrane within the MRC complexes I, III, IV (cytochrome c oxidase), and V (ATP synthase), along with dozens of nuclear DNA-encoded proteins. Permanent mtDNA replication by the DNA polymerase $\gamma$ keeps levels of cellular mtDNA constant in spite of continuous degradation of the most damaged and/or dysfunctional mitochondria. mtDNA seems to be more prone to oxidative damage than nuclear DNA, which eventually can lead to the occurrence of point mutations and deletions.[22] Oxidative damage to mtDNA can also induce a strong reduction of the mtDNA copy number and subsequent OXPHOS deficiency.[22, 23]

# 3 MAIN CONSEQUENCES OF MRC AND mtFAO INHIBITION

MRC inhibition can induce significant ATP depletion, leading to cell death by necrosis.[24, 25] In the liver, ATP shortage can induce cytolytic hepatitis (also called hepatic cytolysis) that comprises a large spectrum of liver injuries of different severity (Fig. 1). Although the mildest forms of hepatic cytolysis are characterized by an isolated increase in plasma alanine aminotransferase (ALT) and aspartate aminotransferase (AST), the most severe cases can be associated with fulminant hepatitis and severe hepatic dysfunction. Moreover, hepatic cytolysis can favor inflammation through the release of endogenous damage-associated molecular patterns (DAMPS).[26] Mild hepatic cytolysis accompanied by inflammation are key features of steatohepatitis, whatever its etiology.[27–29]

Another important consequence of MRC inhibition is the secondary impairment of mtFAO and tricarboxylic acid (TCA) cycle (Fig. 1). Lower oxidation of NADH and $FADH_2$ strongly reduces FAD and $NAD^+$ levels, thus hindering the activity of the different FAD- and $NAD^+$-dependent dehydrogenases of these metabolic pathways.[2, 27] MRC inhibition, however, will secondarily lead to an impairment of mtFAO and TCA cycle only if MRC activity is reduced severely.[2, 25, 27] Inhibition of mtFAO can lead to lipid accumulation (Fig. 1) as discussed below. Impairment of the TCA cycle can induce hyperlactatemia and lactic acidosis because the conversion of unmetabolized pyruvate to lactate by lactate dehydrogenase is favored by NADH accumulation.[30, 31] Finally, a significant inhibition of MRC can also favor ROS overproduction, in particular at the level of complexes I and III.[24, 25, 32] In the liver, overproduction of mitochondrial ROS induced by xenobiotics can favor oxidative stress, inflammation, and fibrosis.[24, 30, 33] Therefore, mitochondrial ROS overproduction appears to play a key role in the progression of simple steatosis toward steatohepatitis. This key pathophysiological feature seems to be applicable for different causes of steatosis such as drugs, ethanol, different types of toxins, and obesity, or any combination of these factors (Fig. 2).[6, 7, 12]

Inhibition of mtFAO can lead to hepatic cytolysis (Fig. 1), because this metabolic pathway provides most of the ATP required for cell homeostasis and function, especially during fasting.[25, 27] Xenobiotic-induced inhibition of mtFAO also can induce steatosis (Fig. 1), or can aggravate preexisting fatty liver linked to obesity (Fig. 2).[2, 6] Although most of the stored

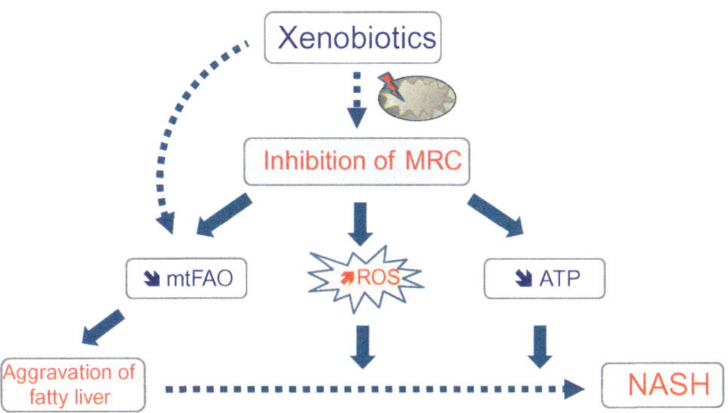


FIG. 2 Xenobiotic-induced mitochondrial dysfunction can aggravate fatty liver and promote its progression to NASH in obese patients. Some xenobiotics are able to impair the mtFAO pathway either directly or secondary to MRC inhibition, thus aggravating lipid accumulation in the fatty liver. In case of MRC inhibition, a faster progression of fatty liver to NASH can occur via a reduction of ATP synthesis and higher mitochondrial ROS production. These effects can promote hepatic cytolysis, inflammation, and fibrosis in the context of fatty liver. In addition to mitochondrial dysfunction, some xenobiotics can also aggravate fatty liver by favoring de novo lipogenesis and/or reducing VLDL secretion. Further information can be found in the text. Abbreviations: *MRC*, mitochondrial respiratory chain; *mtFAO*, mitochondrial fatty acid oxidation; *NASH*, nonalcoholic steatohepatitis; *ROS*, reactive oxygen species; *VLDL*, very-low density lipoprotein.

lipids are triglycerides, free fatty acids and dicarboxylic acids can also accumulate in a significant amount.[27, 34] This could reinforce drug-induced mitochondrial dysfunction because these lipid derivatives are able to impair OXPHOS and TCA cycle, or increase the permeability of the mitochondrial membranes.[25, 27, 35] A severe inhibition of mtFAO can also alter the hepatic gluconeogenesis pathway, leading to hypoglycemia.[27] The strong reduction in acetyl-CoA levels impairs the activity of pyruvate carboxylase, a major regulatory enzyme in the gluconeogenesis process.[25, 27]

# 4 DRUGS

## 4.1 Amiodarone

Amiodarone is a broad-spectrum antiarrhythmic drug that also presents an antianginal effect. Amiodarone-induced hepatotoxicity includes acute hepatic cytolysis, cholestasis, steatosis, steatohepatitis, and cirrhosis.[27, 36] Numerous studies have shown that mitochondrial dysfunction is a major mechanism whereby amiodarone can induce steatosis and steatohepatitis.[12, 25, 27] Investigations performed on isolated rodent mitochondria revealed that amiodarone presents a dual effect on mitochondrial respiration and OXPHOS, depending on its concentration.[37–40] At low concentrations (20–100 μM), amiodarone uncouples OXPHOS and stimulates mitochondrial respiration via a protonophoric effect. This cationic amphiphilic molecule can be protonated within the mitochondrial intermembrane space and electrophoretically transported into the mitochondrial matrix by using the membrane potential $\Delta\Psi$. At higher concentrations (>100 μM), however, the progressive intramitochondrial accumulation of amiodarone rapidly inhibits MRC activity. Notably, both OXPHOS uncoupling and MRC inhibition can reduce ATP synthesis so that amiodarone-induced energy shortage can occur even at low intracellular concentrations of this drug.[40, 41]

Amiodarone-induced MRC inhibition is most probably the main mechanism whereby this drug induces mitochondrial ROS overproduction, oxidative stress, and lipid peroxidation (Fig. 2).[39, 42, 43] Amiodarone could also favor oxidative stress via GSH depletion because of the formation of GSH conjugates with the drug and its diquinone metabolites[44] and via a reduction of GPX expression.[43] In vitro and in vivo investigations also reported that amiodarone is able to inhibit mtFAO.[38, 40, 45] Although this effect could be secondary to MRC inhibition, some experiments suggested that mtFAO could be directly inhibited by amiodarone (Fig. 2), in particular at the level of long-chain acyl-CoA dehydrogenase (LCAD) and CPT1.[38, 39, 46, 47] Lastly, perhexiline and 4,4′-diethylaminoethoxyhexestrol (DEAEH), two cationic amphiphilic drugs withdrawn from the market because of frequent steatohepatitis, were shown to alter mitochondrial function in a similar manner to amiodarone.[42, 47, 48]

## 4.2 Nucleoside Reverse Transcriptase Inhibitors

Nucleoside reverse transcriptase inhibitors (NRTIs) are the first antiretroviral drugs marketed for the treatment of human immunodeficiency virus (HIV) infection. This pharmacological class includes zidovudine (AZT), stavudine (d4T), didanosine (ddI), and zalcitabine (ddC). NRTIs are 2′,3′-dideoxynucleoside analogues in which the 3′-hydroxyl group of the sugar ring is replaced by either a hydrogen atom or another group unable to form a phosphodiester linkage. The lack of this 3′-hydroxyl group is required for the inhibition of HIV-reverse transcriptase activity.

NRTI-induced liver injury includes hepatic cytolysis, microvesicular and/or macrovacuolar steatosis, steatohepatitis, cirrhosis, and cholestasis.[2, 36] Numerous clinical and experimental investigations showed that NRTI-induced hepatotoxicity and other adverse effects are the consequence of an impairment of mtDNA replication.[22, 27, 30, 49, 50] By impairing mtDNA replication, these drugs can induce severe mtDNA depletion and OXPHOS impairment (Fig. 2). NRTIs actually act as chain terminators because their incorporation into the growing chain of mtDNA does not allow the addition of endogenous nucleotides by the DNA polymerase γ.[22, 30, 50] Therefore, the lack of a 3′-hydroxyl group is responsible not only for the antiretroviral activity of NRTIs, but also explains their mitochondrial toxicity. Zalcitabine is not used anymore because of its much higher toxicity than other NRTIs.

Some NRTIs, in particular stavudine and didanosine, are suspected to worsen NAFLD in obese patients.[6, 51] Although the mechanisms of NRTI-induced NAFLD aggravation are still unclear, it is likely that mitochondrial dysfunction could play a significant role.[6, 51] Several studies suggested that the risk of stavudine-induced lactic acidosis could be higher in overweight or obese female patients.[52, 53] Moreover, a high body mass index might increase the risk of fatal lactic acidosis.[54] It is conceivable that stavudine treatment might have significantly exacerbated NAFLD-associated mitochondrial dysfunction in obese individuals, triggering severe TCA cycle defect and lactic acidosis. Unfortunately, these investigations about stavudine-induced lactic acidosis did not determine whether this antiretroviral drug also worsened fatty liver or NASH in these patients.

## 4.3 Tamoxifen

Tamoxifen is a selective estrogen-receptor modulator used for the treatment of estrogen receptor-positive breast cancer. Tamoxifen-induced hepatotoxicity mostly includes different types of chronic liver lesions, including steatosis, steatohepatitis, fibrosis, cirrhosis, and HCC.[36, 55] Investigations performed on isolated rodent liver mitochondria have shown that tamoxifen is able to impair OXPHOS and MRC activity in a similar manner than amiodarone (Fig. 2), with OXPHOS uncoupling at relatively low tamoxifen concentrations (50–100 μM) and inhibition of MRC activity at higher concentrations.[56–58] In addition to its effect on OXPHOS, tamoxifen is also able to directly inhibit mtFAO (Fig. 2), in particular at the level of CPT1.[58] Investigations of mice showed that chronic tamoxifen treatment (i.e., 28 days) induced progressive hepatic mtDNA depletion, possibly via an impairment of mitochondrial topoisomerase activity.[58] In recent experiments performed in differentiated HepaRG cells treated for 14 days with tamoxifen, however, we failed to detect any mtDNA depletion, whereas treatment with the NRTI zalcitabine induced a strong reduction in mtDNA levels (unpublished results). Although all these mitochondrial effects most probably explain why tamoxifen is able to induce steatosis,[12, 25, 58] some investigations suggested that increased lipogenesis could also be involved in the accumulation of liver triglycerides.[59, 60]

Several studies have reported that obesity increased the risk of tamoxifen hepatotoxicity as assessed by the presence of hepatic histological alterations or increased plasma transaminases.[61–63] Two of these studies showed that obesity more specifically enhanced the risk of tamoxifen-induced steatohepatitis.[62, 63] In a study reporting three cases of tamoxifen-induced steatohepatitis, all patients were overweight, or obese.[64] The study from Saphner et al.,[63] however, suggested that tamoxifen did not worsen pre-existing NASH in a small subgroup of patients. Further investigations in larger series of obese patients undergoing serial liver biopsies would be needed to determine whether tamoxifen can actually worsen pre-existing NASH, or whether tamoxifen can induce steatohepatitis more frequently in patients with pre-existing fatty liver. Nevertheless, whatever the clinical situation, it seems likely that tamoxifen-induced mitochondrial dysfunction could be involved.

# 5 ETHANOL

Even though low levels of ethanol can be found in the blood of nonalcoholic individuals, this molecule is mentioned in this chapter because its overconsumption is a major issue for public health. The abuse of alcoholic beverages can induce a large spectrum of liver lesions, including alcoholic foamy degeneration, alcoholic hepatitis, macrovacuolar steatosis, steatohepatitis, cirrhosis, and HCC.[3, 27] A large body of evidence indicates that oxidative stress is a major mechanism whereby ethanol intoxication induces acute and chronic liver injury. Ethanol abuse is associated with ROS overproduction because of the induction of cytochrome P450 2E1 (CYP2E1) and MRC impairment.[3, 65, 66] ROS can damage different cellular components, including unsaturated fatty acids, therefore generating highly reactive aldehydes such as malondialdehyde and 4-hydroxynonenal that play a major role in alcoholic liver diseases.[3, 67] Ethanol intoxication also impairs ROS detoxification, in particular by reducing glutathione levels within the cytosol and mitochondria.[27, 68] Moreover, ethanol metabolism generates acetaldehyde and the hydroxyethyl radical, which are highly reactive metabolites able to alter many cellular components.[3, 69]

Overproduction of ROS and lipid peroxidation-derived aldehydes, as well as the generation of ethanol-derived toxic metabolites, can impair mitochondrial function and alter different mitochondrial components (Fig. 2). ROS and toxic metabolites produced after ethanol consumption could be generated directly within mitochondria because hepatic CYP2E1 is present at significant levels in these organelles.[70, 71] Ethanol-induced impairment of MRC activity and OXPHOS can be induced by different mechanisms, including direct alterations of key components of the MRC such as COX and ATP synthase, impairment of mitochondrial protein synthesis and oxidative damage to the mitochondrial genome leading to different point mutations, large deletions, and reduced mtDNA copy number (Fig. 2).[22, 23, 27, 72, 73]

Ethanol intoxication also impairs mtFAO (Fig. 2) by different mechanisms, including reduced $NAD^+$ availability secondary to ethanol metabolism by alcohol and aldehyde dehydrogenases, reduced activity of several mtFAO enzymes such as CPT1 and medium-chain acyl-CoA dehydrogenase (MCAD), and alteration of peroxisome proliferator-activated receptor-α (PPARα) signaling pathway.[27, 74–76] The latter effect seems dependent on oxidative stress and acetaldehyde generation.[74, 76] mtFAO could also be impaired secondary to reduced MRC activity. Although impairment of mtFAO undoubtedly plays a major role in ethanol-induced steatosis,[27, 73, 77] other mechanisms could also participate to fat accretion including increased de novo lipogenesis, higher esterification of fatty acids to triglycerides and reduced very-low density lipoprotein (VLDL) secretion.[3, 27, 78]

Strong evidence suggests that excess weight and obesity greatly increase the risk and the severity of alcoholic liver disease, in particular steatohepatitis.[79–81] Although several mechanisms seem to be involved, the higher basal CYP2E1 activity associated with NAFLD could play a significant role by enhancing ethanol-induced oxidative stress, lipid peroxidation, and mitochondrial dysfunction.[81–83] This, in turn, could explain stronger mitochondrial dysfunction and ATP shortage[84] and secondary necroinflammation.[83] CYP2E1-independent mechanisms, however, could also be involved because higher ethanol-induced liver injury can be observed in genetically obese fa/fa rats and ob/ob mice that do not present basal CYP2E1 induction.[82, 85]

# 6 OCCUPATIONAL AND ENVIRONMENTAL POLLUTANTS

## 6.1 Benzo[*a*]pyrene

Benzo[*a*]pyrene (B[*a*]P) is a polycyclic aromatic hydrocarbon (PAH) found in crude oil, coal tar, tobacco, and many foods, especially grilled meats. B[*a*]P is listed as a Group 1 carcinogen by the International Agency for Research on Cancer (IARC). B[*a*]P is a potent activator of the aryl hydrocarbon receptor (AhR), which is involved in different cellular effects of this PAH including induction of different CYPs (CYP1A1/2 and CYP1B1) and glycolytic reprogramming.[86, 87] Several CYP-generated oxygenated B[*a*]P metabolites are highly toxic for the cells, in particular by forming covalent DNA adducts.[86, 88]

B[*a*]P has been reported to induce lipid deposition in hepatocytes in different experimental models.[89–91] This effect seems to be linked to an increase in de novo lipogenesis,[90, 91] possibly via an activation of AhR by B[*a*]P.[90] Recent investigations in our laboratory, however, showed that a 14-day treatment with B[*a*]P reduced triglyceride content in human HepaRG cells when these cells were overloaded or not with a mixture of stearic and oleic acids.[92] These effects were associated with higher mtFAO.[92] Previous investigations in chick embryos also reported that B[*a*]P was able to increase mtFAO, possibly via the generation of an oxygenated metabolite.[93] Therefore, the exact effects of B[*a*]P on hepatic lipid metabolism require further investigations.

Our experiments also showed that B[*a*]P reduced the activity of the MRC complexes I and IV in lipid-overloaded HepaRG cells but not in nonsteatotic cells.[92] These effects in steatotic HepaRG cells were associated with ROS overproduction and a pro-inflammatory state with higher mRNA expression of interleukin-1β (IL-1β) and its receptor IL1R1, as well as increased IL-6 secretion ([92] and unpublished data). MRC impairment, ROS overproduction, and higher interleukin expression induced by the 14-day exposure of B[*a*]P were aggravated by ethanol co-exposure. Therefore, B[*a*]P might favor the pathological progression of liver steatosis to a steatohepatitis-like state, possibly by promoting MRC impairment, ROS overproduction, and inflammation.[92] We surmise that B[*a*]P-induced mitochondrial ROS overproduction might be particularly favored by the concomitant increase in the mtFAO flux and lower MRC activity.[92]

The precise mechanisms whereby B[*a*]P alters MRC activity are still unknown. Some investigations suggested the involvement of CYP-derived toxic metabolites and ROS.[94] These deleterious molecules could target different MRC complexes directly but also could damage the mitochondrial genome.[94, 95] Our recent investigations in HepaRG cells, however, suggested that B[*a*]P-induced impairment of MRC activity could be unrelated to the formation of CYP-derived metabolites, suggesting that the mere binding of B[*a*]P to AhR could alter MRC activity.[92] AhR is translocated to mitochondria and can interact with mitochondrial proteins such as ATP5α1, a subunit of ATP synthase.[96, 97] These investigations, however, did not determine whether B[*a*]P-induced MRC impairment was secondary to the interaction of this PAH with the mitochondrial AhR.

## 6.2 Carbon Tetrachloride

Carbon tetrachloride ($CCl_4$), also referred to as tetrachloromethane, is a chlorinated hydrocarbon used as a solvent, refrigerant, fire extinguisher, and dry-cleaning agent. The production and use of this compound have decreased significantly during the past decades because of the occurrence of numerous cases of multiorgan toxicity in industrial workers and the adoption of the Montreal Protocol banning chlorofluorocarbon gases. In addition to occupational exposure, $CCl_4$ can be found in the ambient air and groundwater supplies.[98]

$CCl_4$-induced hepatotoxicity in humans and animals includes acute liver injury with necrosis and steatosis as prominent pathologic features, as well as chronic liver injury including steatosis, steatohepatitis, cirrhosis, and HCC.[1, 98, 99] Repeated administration of $CCl_4$ in rodents is used as a classic experimental model of extensive fibrosis and cirrhosis. Similar to ethanol intoxication, the primary mechanisms involved in $CCl_4$-induced liver injury include CYP2E1-mediated generation of highly reactive metabolites that covalently bind to cellular macromolecules, reduce GSH levels, induce ROS overproduction, and promote lipid peroxidation.[98–100] In turn, these deleterious events induce mitochondrial dysfunction, including reduced activity of different MRC complexes, OXPHOS impairment, and mtFAO inhibition (Fig. 2).[33, 100–102] $CCl_4$ can also induce oxidative damage to the mitochondrial genome, leading to mtDNA strand breaks and reduced mtDNA copy number (Fig. 2).[33, 102] $CCl_4$-induced mitochondrial dysfunction seems to play a significant role, not only in the occurrence of steatosis and necroinflammation, but also in fibrosis progression.[33, 103, 104] In addition to mitochondrial dysfunction, $CCl_4$ could also favor hepatic triglyceride accumulation via a reduction of VLDL secretion.[1, 100]

Several experimental studies carried out in rodents showed that NAFLD and type 2 diabetes can sensitize to $CCl_4$-induced acute hepatotoxicity.[105–107] Unfortunately, these studies did not investigate the exact role of mitochondrial dysfunction.

## 6.3 Organochlorines

### 6.3.1 *Polychlorinated Biphenyls (PCBs)*

Polychlorinated biphenyls (PCBs) are organic chlorine compounds that have wide industrial use as coolants and insulating fluids, plasticizers in paints and cements, wax extenders, and flame retardants. There are more than 200 PCBs, which can be classified as dioxin-like and nondioxin-like compounds based on the similarity of their toxicity profile to 2,3,7,8-tetrachlorodibenzo-*p*-dioxin (TCDD). Examples of dioxin-like PCBs include 3,3′,4,4′-tetrachlorobiphenyl (PCB 77) and 3,3′,4,4′,5-pentachlorobiphenyl (PCB 126). PCBs were also commercially produced as mixtures, which are referred to as Aroclors. Because of PCBs' environmental toxicity and classification as persistent organic pollutants (POPs), their production was banned by the Stockholm Convention on POPs in 2001. IARC classified PCBs as Group 1 carcinogens in humans. Even though PCB levels are decreasing in the environment, PCB exposure still remains a major toxicological issue.[98] In humans, exposure to PCBs is suspected to contribute to obesity, diabetes, hypertension, and NAFLD.[98, 108, 109] The ability of some PCBs to induce steatosis and steatohepatitis has been confirmed in rodents.[1, 98, 110]

The mechanisms whereby PCBs can induce steatosis and steatohepatitis are still poorly understood and most probably differ between compounds. For example, PCBs can interact differentially with several nuclear receptors that have complex effects on lipid metabolism, such as AhR, constitutive androstane receptor (CAR), and pregnane xenobiotic receptor (PXR).[111] AhR activation can induce hepatic steatosis by different mechanisms, as discussed later for TCDD. Several in vitro studies also showed that several PCBs can directly impair mitochondrial function, in particular at the level of different MRC complexes, OXPHOS, and mtFAO.[112–115] These deleterious effects, however, were significant at rather high concentrations of PCBs (>10–20 μM), and therefore it is unclear whether mitochondrial dysfunction might play a role in PCB-induced steatosis in the context of environmental exposure.

Experimental investigations reported that some PCBs, such as PCB 77 and PCB 153, were able to worsen fatty liver in mice fed a high-fat diet (HFD).[116–118] PCB 153-induced aggravation of fatty liver was associated with a profound decrease in hepatic PPARα expression that was accompanied by a strong reduction of CPT1 and CPT2 expression.[118] The mtFAO pathway, however, was not directly investigated in this study. Previous investigations showed that another PCB, namely PCB 77, was able to potently reduce PPARα mRNA expression at very low concentrations (i.e., 1 nM).[119] Lastly, worsening of fatty liver was not observed in HFD obese mice exposed to Aroclor 1260, indicating that such hepatic effect is not observed with all PCBs.[120] Nonetheless, Aroclor 1260 caused necroinflammation in HFD obese mice, suggesting that this PCB mixture could favor the progression of fatty liver to NASH.[120]

### 6.3.2 *2,3,7,8-Tetrachlorodibenzo-*p*-Dioxin (TCDD)*

TCDD (also referred to as dioxin) is an organochloride usually formed as a byproduct during organic synthesis and waste combustion. For instance, TCDD was a contaminant in Agent Orange, an herbicide used during the Vietnam War, that contained an equal mixture of 2,4-dichlorophenoxyacetic acid (2,4-D) and 2,4,5-trichlorophenoxyacetic acid (2,4,5-T). This persistent contaminant was also massively released into the environment following different industrial accidents, including the Seveso disaster in Italy in 1976.[121] Acute intoxication by TCDD leads to different adverse effects, including severe skin lesions such as chloracne, dehydration, weight loss, peripheral neuropathy, and hepatotoxicity.[121, 122] Delayed or long-term intoxication by TCDD is also suspected to contribute to the occurrence of dyslipidemia, atherosclerosis, hypertension, diabetes, chronic liver disease, and different types of cancers.[121–123] Accordingly, TCDD is listed as a Group 1 carcinogen by IARC.

In humans, TCDD-induced hepatotoxicity includes increased plasma transaminases (reflecting hepatic cytolysis), steatosis, fibrosis, and inflammation, suggesting the occurrence of steatohepatitis in some individuals.[121, 122] Steatosis is also consistently observed in rodents intoxicated by TCDD.[124–128] Notably, necroinflammation and fibrosis were observed in some of these investigations and others,[124, 125, 128, 129] confirming that TCDD is able to induce steatohepatitis.

Mechanistic studies suggested that TCDD-induced steatosis could be induced by different mechanisms including increased hepatic fatty acid uptake, reduced VLDL secretion and decreased mtFAO.[124, 126, 128, 130] Some of these detrimental effects could be linked to AhR activation,[126, 131, 132] although other mechanisms cannot be excluded. TCDD also impairs the activity of different MRC complexes, which could participate to higher mitochondrial ROS production.[96, 133, 134] Interestingly, TCDD-induced mitochondrial ROS overproduction seems to be secondary to AhR activation.[135] Finally, a recent study reported that TCDD promotes liver fibrosis development in HFD obese mice, but the possible involvement of mitochondrial dysfunction was not addressed in this study.[136]

## 7 CONCLUDING REMARKS AND PERSPECTIVES

Strong evidence exists that many xenobiotics are able to induce steatosis and steatohepatitis. Although steatosis is a benign condition, steatohepatitis is a major issue for the affected patients because the lesion can progress to cirrhosis and eventually to HCC. Therefore, it is important to know the full spectrum of xenobiotics able to induce steatosis and steatohepatitis, as well as the populations at risk. Although prospective clinical investigations and broad epidemiologic studies could help in this task, it must be stressed that the clear identification of the involved compounds is difficult in particular in patients with polymedication, or in individuals exposed to various environmental pollutants. Experimental investigations in vitro and in vivo are important to ascertain whether a given molecule is able to induce steatosis or steatohepatitis. Regarding in vitro experiments, several investigations showed that the HepaRG cell line is a valuable model to investigate steatosis induced by drugs and environmental pollutants but also by fatty acid overload.[92, 137–141] HepaRG cells express the main xenobiotic-metabolizing enzymes (XMEs), such as CYPs and UDP-glucuronosyltransferases (UGTs), as well as the nuclear receptors that control XME expression.[141–143] Recent investigations showed that mitochondrial function in HepaRG cells is close to primary human hepatocytes, in contrast to HepG2 and Huh7 cells.[144, 145] This is a key feature because xenobiotics can induce steatosis and steatohepatitis by impairing mitochondrial function.

In contrast to obesity-associated fatty liver, the mechanisms whereby xenobiotics can induce steatosis are much less known. Although mitochondrial dysfunction is deemed to play a key role, at least for some xenobiotics, other mechanisms exist such as activation of hepatic de novo lipogenesis and impairment of VLDL secretion, as previously mentioned. In addition, some xenobiotics, including drugs, ethanol, and some environmental pollutants, could favor hepatic steatosis secondary to their deleterious effects on adipose tissue.[25, 146, 147] Therefore, although cellular models are valuable to study xenobiotic-induced steatosis, in vivo investigations in rodents could help to disclose indirect toxicity on the liver. In addition, investigations in HFD or genetically obese rodents could determine whether a given molecule can aggravate fatty liver, or can favor the progression of fatty liver to steatohepatitis.[85, 116–118, 148] Finally, epidemiological studies might also help to tackle this issue, although this approach could be difficult for populations that are exposed to many environmental pollutants.

## References


1. Kaiser JP, Lipscomb JC, Wesselkamper SC. Putative mechanisms of environmental chemical-induced steatosis. *Int J Toxicol* 2012;**31**:551–63.
2. Massart J, Begriche K, Buron N, Porceddu M, Borgne-Sanchez A, Fromenty B. Drug-induced inhibition of mitochondrial fatty acid oxidation and steatosis. *Curr Pathobiol Rep* 2013;**1**:147–57.
3. Louvet A, Mathurin P. Alcoholic liver disease: mechanisms of injury and targeted treatment. *Nat Rev Gastroenterol Hepatol* 2015;**12**:231–42.
4. Sanyal AJ, Mathurin P, Nagy LA. Commonalities and distinctions between alcoholic and nonalcoholic fatty liver disease. *Gastroenterology* 2016;**150**:1695–7.
5. Fromenty B. Drug-induced liver injury in obesity. *J Hepatol* 2013;**58**:824–6.
6. Massart J, Begriche K, Moreau C, Fromenty B. Role of nonalcoholic fatty liver disease as risk factor for drug-induced hepatotoxicity. *J Clin Transl Res* 2017;**3**:212–32.
7. Begriche K, Massart J, Robin MA, Bonnet F, Fromenty B. Mitochondrial adaptations and dysfunctions in nonalcoholic fatty liver disease. *Hepatology* 2013;**58**:1497–507.
8. Sunny NE, Bril F, Cusi K. Mitochondrial adaptation in nonalcoholic fatty liver disease: novel mechanisms and treatment strategies. *Trends Endocrinol Metab* 2017;**28**:250–60.
9. Satapati S, Kucejova B, Duarte JA, Fletcher JA, Reynolds L, Sunny NE, et al. Mitochondrial metabolism mediates oxidative stress and inflammation in fatty liver. *J Clin Invest* 2015;**125**:4447–62.
10. Satapathy SK, Kuwajima V, Nadelson J, Atiq O, Sanyal AJ. Drug-induced fatty liver disease: an overview of pathogenesis and management. *Ann Hepatol* 2015;**14**:789–806.
11. Rabinowich L, Shibolet O. Drug induced steatohepatitis: an uncommon culprit of a common disease. *Biomed Res Int* 2015;**2015**:168905.
12. Schumacher JD, Guo GL. Mechanistic review of drug-induced steatohepatitis. *Toxicol Appl Pharmacol* 2015;**289**:40–7.
13. Dash A, Figler RA, Sanyal AJ, Wamhoff BR. Drug-induced steatohepatitis. *Expert Opin Drug Metab Toxicol* 2017;**13**:193–204.
14. Wallace DC, Fan W, Procaccio V. Mitochondrial energetics and therapeutics. *Annu Rev Pathol* 2010;**5**:297–348.
15. Kühlbrandt W. Structure and function of mitochondrial membrane protein complexes. *BMC Biol* 2015;**13**:89.
16. Mari M, Morales A, Colell A, Garcia-Ruiz C, Kaplowitz N, Fernandez-Checa JC. Mitochondrial glutathione: features, regulation and role in disease. *Biochim Biophys Acta* 2013;**1830**:3317–28.

17. Imhoff BR, Hansen JM. Extracellular redox status regulates Nrf2 activation through mitochondrial reactive oxygen species. *Biochem J* 2009;**424**:491–500.
18. Diebold L, Chandel NS. Mitochondrial ROS regulation of proliferating cells. *Free Radic Biol Med* 2016;**100**:86–93.
19. Houten SM, Violante S, Ventura FV, Wanders RJ. The biochemistry and physiology of mitochondrial fatty acid β-oxidation and its genetic disorders. *Annu Rev Physiol* 2016;**78**:23–44.
20. Seifert EL, Estey C, Xuan JY, Harper ME. Electron transport chain-dependent and -independent mechanisms of mitochondrial $H_2O_2$ emission during long-chain fatty acid oxidation. *J Biol Chem* 2010;**285**:5748–58.
21. Kakimoto PA, Tamaki FK, Cardoso AR, Marana SR, Kowaltowski AJ. $H_2O_2$ release from the very long chain acyl-CoA dehydrogenase. *Redox Biol* 2015;**4**:375–80.
22. Schon E, Fromenty B. Alterations of mitochondrial DNA in liver diseases. In: Kaplowitz N, Han D, editors. *Mitochondria in liver disease*. New-York, NY: Taylor & Francis; 2016. p. 279–309.
23. Demeilliers C, Maisonneuve C, Grodet A, Mansouri A, Nguyen R, Tinel M, et al. Impaired adaptive resynthesis and prolonged depletion of hepatic mitochondrial DNA after repeated alcohol binges in mice. *Gastroenterology* 2002;**123**:1278–90.
24. Pessayre D, Mansouri A, Berson A, Fromenty B. Mitochondrial involvement in drug-induced liver injury. *Handb Exp Pharmacol* 2010;**196**:311–65.
25. Begriche K, Massart J, Robin MA, Borgne-Sanchez A, Fromenty B. Drug-induced toxicity on mitochondria and lipid metabolism: mechanistic diversity and deleterious consequences for the liver. *J Hepatol* 2011;**54**:773–94.
26. Kubes P, Mehal WZ. Sterile inflammation in the liver. *Gastroenterology* 2012;**143**:1158–72.
27. Fromenty B, Pessayre D. Inhibition of mitochondrial beta-oxidation as a mechanism of hepatotoxicity. *Pharmacol Ther* 1995;**67**:101–54.
28. Stickel F, Seitz HK. Alcoholic steatohepatitis. *Best Pract Res Clin Gastroenterol* 2010;**24**:683–93.
29. Ahmed A, Wong RJ, Harrison SA. Nonalcoholic fatty liver disease review: diagnosis, treatment, and outcomes. *Clin Gastroenterol Hepatol* 2015;**13**:2062–70.
30. Igoudjil A, Begriche K, Pessayre D, Fromenty B. Mitochondrial, metabolic and genotoxic effects of antiretroviral nucleoside reverse-transcriptase inhibitors. *Anti-Infect Agents Med Chem* 2006;**5**:273–92.
31. Margolis AM, Heverling H, Pham PA, Stolbach A. A review of the toxicity of HIV medications. *J Med Toxicol* 2014;**10**:26–39.
32. Li N, Ragheb K, Lawler G, Sturgis J, Rajwa B, Melendez JA, et al. Mitochondrial complex I inhibitor rotenone induces apoptosis through enhancing mitochondrial reactive oxygen species production. *J Biol Chem* 2003;**278**:8516–25.
33. Mitchell C, Robin MA, Mayeuf A, Mahrouf-Yorgov M, Mansouri A, Hamard M, et al. Protection against hepatocyte mitochondrial dysfunction delays fibrosis progression in mice. *Am J Pathol* 2009;**175**:1929–37.
34. Labbe G, Pessayre D, Fromenty B. Drug-induced liver injury through mitochondrial dysfunction: mechanisms and detection during preclinical safety studies. *Fundam Clin Pharmacol* 2008;**22**:335–53.
35. Dubinin MV, Adakeeva SI, Samartsev VN. Long-chain α,ω-dioic acids as inducers of cyclosporin A-insensitive nonspecific permeability of the inner membrane of liver mitochondria loaded with calcium or strontium ions. *Biochemistry* 2013;**78**:412–7.
36. Wang Y, Lin Z, Liu Z, Harris S, Kelly R, Zhang J, et al. A unifying ontology to integrate histological and clinical observations for drug-induced liver injury. *Am J Pathol* 2013;**182**:1180–7.
37. Fromenty B, Fisch C, Berson A, Letteron P, Larrey D, Pessayre D. Dual effect of amiodarone on mitochondrial respiration. Initial protonophoric uncoupling effect followed by inhibition of the respiratory chain at the levels of complex I and complex II. *J Pharmacol Exp Ther* 1990;**255**:1377–84.
38. Spaniol M, Bracher R, Ha HR, Follath F, Krähenbühl S. Toxicity of amiodarone and amiodarone analogues on isolated rat liver mitochondria. *J Hepatol* 2001;**35**:628–36.
39. Serviddio G, Bellanti F, Giudetti AM, Gnoni GV, Capitanio N, Tamborra R, et al. Mitochondrial oxidative stress and respiratory chain dysfunction account for liver toxicity during amiodarone but not dronedarone administration. *Free Radic Biol Med* 2011;**51**:2234–42.
40. Felser A, Blum K, Lindinger PW, Bouitbir J, Krähenbühl S. Mechanisms of hepatocellular toxicity associated with dronedarone: a comparison to amiodarone. *Toxicol Sci* 2013;**131**:480–90.
41. Fromenty B, Letteron P, Fisch C, Berson A, Deschamps D, Pessayre D. Evaluation of human blood lymphocytes as a model to study the effects of drugs on human mitochondria. Effects of low concentrations of amiodarone on fatty acid oxidation, ATP levels and cell survival. *Biochem Pharmacol* 1993;**46**:421–32.

42. Berson A, De Beco V, Lettéron P, Robin MA, Moreau C, El Kahwaji J, et al. Steatohepatitis-inducing drugs cause mitochondrial dysfunction and lipid peroxidation in rat hepatocytes. *Gastroenterology* 1998;**114**:764–74.
43. Yamamoto T, Kikkawa R, Yamada H, Horii I. Identification of oxidative stress-related proteins for predictive screening of hepatotoxicity using a proteomic approach. *J Toxicol Sci* 2005;**30**:213–27.
44. Parmar KR, Jhajra S, Singh S. Detection of glutathione conjugates of amiodarone and its reactive diquinone metabolites in rat bile using mass spectrometry tools. *Rapid Commun Mass Spectrom* 2016;**30**:1242–8.
45. Fromenty B, Fisch C, Labbe G, Degott C, Deschamps D, Berson A, et al. Amiodarone inhibits the mitochondrial β-oxidation of fatty acids and produces microvesicular steatosis of the liver in mice. *J Pharmacol Exp Ther* 1990;**255**:1371–6.
46. Hamdan M, Urien S, Le Louet H, Tillement JP, Morin D. Inhibition of mitochondrial carnitine palmitoyltransferase-1 by a trimetazidine derivative, S-15176. *Pharmacol Res* 2001;**44**:99–104.
47. Kennedy JA, Unger SA, Horowitz JD. Inhibition of carnitine palmitoyltransferase-1 in rat heart and liver by perhexiline and amiodarone. *Biochem Pharmacol* 1996;**52**:273–80.
48. Deschamps D, DeBeco V, Fisch C, Fromenty B, Guillouzo A, Pessayre D. Inhibition by perhexiline of oxidative phosphorylation and the beta-oxidation of fatty acids: possible role in pseudoalcoholic liver lesions. *Hepatology* 1994;**19**:948–61.
49. Walker UA, Bäuerle J, Laguno M, Murillas J, Mauss S, Schmutz G, et al. Depletion of mitochondrial DNA in liver under antiretroviral therapy with didanosine, stavudine, or zalcitabine. *Hepatology* 2004;**39**:311–7.
50. Gardner K, Hall PA, Chinnery PF, Payne BA. HIV treatment and associated mitochondrial pathology: review of 25 years of in vitro, animal, and human studies. *Toxicol Pathol* 2014;**42**:811–22.
51. Lemoine M, Serfaty L, Capeau J. From nonalcoholic fatty liver to nonalcoholic steatohepatitis and cirrhosis in HIV-infected patients: diagnosis and management. *Curr Opin Infect Dis* 2012;**25**:10–6.
52. Bolhaar MG, Karstaedt AS. A high incidence of lactic acidosis and symptomatic hyperlactatemia in women receiving highly active antiretroviral therapy in Soweto, South Africa. *Clin Infect Dis* 2007;**45**:254–60.
53. Wester CW, Eden SK, Shepherd BE, Bussmann H, Novitsky V, Samuels DC, et al. Risk factors for symptomatic hyperlactatemia and lactic acidosis among combination antiretroviral therapy-treated adults in Botswana: results from a clinical trial. *AIDS Res Hum Retrovir* 2012;**28**:759–65.
54. Coghlan ME, Sommadossi JP, Jhala NC, Many WJ, Saag MS, Johnson VA. Symptomatic lactic acidosis in hospitalized antiretroviral-treated patients with human immunodeficiency virus infection: a report of 12 cases. *Clin Infect Dis* 2001;**33**:1914–21.
55. Yang YJ, Kim KM, An JH, Lee DB, Shim JH, Lim YS, et al. Clinical significance of fatty liver disease induced by tamoxifen and toremifene in breast cancer patients. *Breast* 2016;**28**:67–72.
56. Tuquet C, Dupont J, Mesneau A, Roussaux J. Effects of tamoxifen on the electron transport chain of isolated rat liver mitochondria. *Cell Biol Toxicol* 2000;**16**:207–19.
57. Cardoso CM, Custodio JB, Almeida LM, Moreno AJ. Mechanisms of the deleterious effects of tamoxifen on mitochondrial respiration rate and phosphorylation efficiency. *Toxicol Appl Pharmacol* 2001;**176**:145–52.
58. Larosche I, Lettéron P, Fromenty B, Vadrot N, Abbey-Toby A, Feldmann G, et al. Tamoxifen inhibits topoisomerases, depletes mitochondrial DNA, and triggers steatosis in mouse liver. *J Pharmacol Exp Ther* 2007;**321**:526–35.
59. Gudbrandsen OA, Rost TH, Berge RK. Causes and prevention of tamoxifen-induced accumulation of triacylglycerol in rat liver. *J Lipid Res* 2006;**47**:2223–32.
60. Cole LK, Jacobs RL, Vance DE. Tamoxifen induces triacylglycerol accumulation in the mouse liver by activation of fatty acid synthesis. *Hepatology* 2010;**52**:1258–65.
61. Elefsiniotis IS, Pantazis KD, Ilias A, Pallis L, Mariolis A, Glynou I, et al. Tamoxifen induced hepatotoxicity in breast cancer patients with pre-existing liver steatosis: the role of glucose intolerance. *Eur J Gastroenterol Hepatol* 2004;**16**:593–8.
62. Bruno S, Maisonneuve P, Castellana P, Rotmensz N, Rossi S, Maggioni M, et al. Incidence and risk factors for non-alcoholic steatohepatitis: prospective study of 5408 women enrolled in Italian tamoxifen chemoprevention trial. *BMJ* 2005;**330**:932.
63. Saphner T, Triest-Robertson S, Li H, Holzman P. The association of nonalcoholic steatohepatitis and tamoxifen in patients with breast cancer. *Cancer* 2009;**115**:3189–95.
64. Pinto HC, Baptista A, Camilo ME, de Costa EB, Valente A, de Moura MC. Tamoxifen-associated steatohepatitis—report of three cases. *J Hepatol* 1995;**23**:95–7.
65. Cahill A, Cunningham CC, Adachi M, Ishii H, Bailey SM, Fromenty B, et al. Effects of alcohol and oxidative stress on liver pathology: the role of the mitochondrion. *Alcohol Clin Exp Res* 2002;**26**:907–15.

66. Leung TM, Nieto N. CYP2E1 and oxidant stress in alcoholic and non-alcoholic fatty liver disease. *J Hepatol* 2013;**58**:395–8.
67. Magdaleno F, Blajszczak CC, Nieto N. Key events participating in the pathogenesis of alcoholic liver disease. *Biomol Ther* 2017;**7**:9.
68. Garcia-Ruiz C, Fernandez-Checa JC. Mitochondrial glutathione: hepatocellular survival-death switch. *J Gastroenterol Hepatol* 2006;**21**(Suppl. 3):S3–6.
69. Albano E. Alcohol, oxidative stress and free radical damage. *Proc Nutr Soc* 2006;**65**:278–90.
70. Knockaert L, Fromenty B, Robin MA. Mechanisms of mitochondrial targeting of cytochrome P450 2E1: physiopathological role in liver injury and obesity. *FEBS J* 2011;**278**:4252–60.
71. Hartman JH, Miller GP, Meyer JN. Toxicological implications of mitochondrial localization of CYP2E1. *Toxicol Res* 2017;**6**:273–89.
72. Weiser B, Gonye G, Sykora P, Crumm S, Cahill A. Chronic ethanol feeding causes depression of mitochondrial elongation factor Tu in the rat liver: implications for the mitochondrial ribosome. *Am J Physiol Gastrointest Liver Physiol* 2011;**300**:G815–22.
73. Nassir F, Ibdah JA. Role of mitochondria in alcoholic liver disease. *World J Gastroenterol* 2014;**20**:2136–42.
74. Galli A, Pinaire J, Fischer M, Dorris R, Crabb DW. The transcriptional and DNA binding activity of peroxisome proliferator-activated receptor alpha is inhibited by ethanol metabolism. A novel mechanism for the development of ethanol-induced fatty liver. *J Biol Chem* 2001;**276**:68–75.
75. Li Q, Xie G, Zhang W, Zhong W, Sun X, Tan X, et al. Dietary nicotinic acid supplementation ameliorates chronic alcohol-induced fatty liver in rats. *Alcohol Clin Exp Res* 2014;**38**:1982–92.
76. Zeng T, Zhang CL, Song FY, Zhao XL, Xie KQ. CMZ reversed chronic ethanol-induced disturbance of PPAR-α possibly by suppressing oxidative stress and PGC-1α acetylation, and activating the MAPK and GSK3β pathway. *PLoS ONE* 2014;**9**.
77. Fromenty B, Pessayre D. Impaired mitochondrial function in microvesicular steatosis. Effects of drugs, ethanol, hormones and cytokines. *J Hepatol* 1997;**26**(Suppl. 2):43–53.
78. Seth D, Haber PS, Syn WK, Diehl AM, Day CP. Pathogenesis of alcohol-induced liver disease: classical concepts and recent advances. *J Gastroenterol Hepatol* 2011;**26**:1089–105.
79. Ruhl CE, Everhart JE. Joint effects of body weight and alcohol on elevated serum alanine aminotransferase in the United States population. *Clin Gastroenterol Hepatol* 2005;**3**:1260–8.
80. Hart CL, Morrison DS, Batty GD, Mitchell RJ, Davey Smith G. Effect of body mass index and alcohol consumption on liver disease: analysis of data from two prospective cohort studies. *BMJ* 2010;**340**:c1240.
81. Mahli A, Hellerbrand C. Alcohol and obesity: a dangerous association for fatty liver disease. *Dig Dis* 2016;**34**(Suppl. 1):32–9.
82. Aubert J, Begriche K, Knockaert L, Robin MA, Fromenty B. Increased expression of cytochrome P450 2E1 in nonalcoholic fatty liver disease: mechanisms and pathophysiological role. *Clin Res Hepatol Gastroenterol* 2011;**35**:630–7.
83. Minato T, Tsutsumi M, Tsuchishima M, Hayashi N, Saito T, Matsue Y, et al. Binge alcohol consumption aggravates oxidative stress and promotes pathogenesis of NASH from obesity-induced simple steatosis. *Mol Med* 2014;**20**:490–502.
84. Xu J, Lai KK, Verlinsky A, Lugea A, French SW, Cooper MP, et al. Synergistic steatohepatitis by moderate obesity and alcohol in mice despite increased adiponectin and p-AMPK. *J Hepatol* 2011;**55**:673–82.
85. Everitt H, Hu M, Ajmo JM, Rogers CQ, Liang X, Zhang R, et al. Ethanol administration exacerbates the abnormalities in hepatic lipid oxidation in genetically obese mice. *Am J Physiol Gastrointest Liver Physiol* 2013;**304**:G38–47.
86. Hardonnière K, Huc L, Sergent O, Holme JA, Lagadic-Gossmann D. Environmental carcinogenesis and pH homeostasis: not only a matter of dysregulated metabolism. *Semin Cancer Biol* 2017;**43**:49–65.
87. Shiizaki K, Kawanishi M, Yagi T. Modulation of benzo[a]pyrene-DNA adduct formation by CYP1 inducer and inhibitor. *Genes Environ* 2017;**39**:14.
88. Ewa B, Danuta MS. Polycyclic aromatic hydrocarbons and PAH-related DNA adducts. *J Appl Genet* 2017;**58**:321–30.
89. Ortiz L, Nakamura B, Li X, Blumberg B, Luderer U. In utero exposure to benzo[a]pyrene increases adiposity and causes hepatic steatosis in female mice, and glutathione deficiency is protective. *Toxicol Lett* 2013;**223**:260–7.
90. Neuschäfer-Rube F, Schraplau A, Schewe B, Lieske S, Krützfeldt JM, Ringel S, et al. Arylhydrocarbon receptor-dependent mIndy (Slc13a5) induction as possible contributor to benzo[a]pyrene-induced lipid accumulation in hepatocytes. *Toxicology* 2015;**337**:1–9.

91. Regnault C, Willison J, Veyrenc S, Airieau A, Méresse P, Fortier M, et al. Metabolic and immune impairments induced by the endocrine disruptors benzo[a]pyrene and triclosan in Xenopus tropicalis. *Chemosphere* 2016;**155**:519–27.
92. Bucher S, Le Guillou D, Allard J, Pinon G, Begriche K, Tête A, Sergent O, Lagadic-Gossmann D, Fromenty B. Possible involvement of mitochondrial dysfunction and oxidative stress in a cellular model of NAFLD progression induced by benzo[a]pyrene/ethanol coexposure. *Oxid Med Cell Longev* 2018;**2018**:4396403.
93. Westman O, Larsson M, Venizelos N, Hollert H, Engwall M. An oxygenated metabolite of benzo[a]pyrene increases hepatic β-oxidation of fatty acids in chick embryos. *Environ Sci Pollut Res Int* 2014;**21**:6243–51.
94. Bansal S, Leu AN, Gonzalez FJ, Guengerich FP, Chowdhury AR, Anandatheerthavarada HK, et al. Mitochondrial targeting of cytochrome P450 (CYP) 1B1 and its role in polycyclic aromatic hydrocarbon-induced mitochondrial dysfunction. *J Biol Chem* 2014;**289**:9936–51.
95. Backer JM, Weinstein IB. Mitochondrial DNA is a major cellular target for a dihydrodiol-epoxide derivative of benzo[a]pyrene. *Science* 1980;**209**:297–9.
96. Hwang HJ, Dornbos P, Steidemann M, Dunivin TK, Rizzo M, LaPres JJ. Mitochondrial-targeted aryl hydrocarbon receptor and the impact of 2,3,7,8-tetrachlorodibenzo-p-dioxin on cellular respiration and the mitochondrial proteome. *Toxicol Appl Pharmacol* 2016;**304**:121–32.
97. Tappenden DM, Lynn SG, Crawford RB, Lee K, Vengellur A, Kaminski NE, et al. The aryl hydrocarbon receptor interacts with ATP5α1, a subunit of the ATP synthase complex, and modulates mitochondrial function. *Toxicol Appl Pharmacol* 2011;**254**:299–310.
98. Wahlang B, Beier JI, Clair HB, Bellis-Jones HJ, Falkner KC, McClain CJ, et al. Toxicant-associated steatohepatitis. *Toxicol Pathol* 2013;**41**:343–60.
99. Brautbar N, Williams J. Industrial solvents and liver toxicity: risk assessment, risk factors and mechanisms. *Int J Hyg Environ Health* 2002;**205**:479–91.
100. Recknagel RO, Glende Jr. EA, Dolak JA, Waller RL. Mechanisms of carbon tetrachloride toxicity. *Pharmacol Ther* 1989;**43**:139–54.
101. Krähenbühl L, Ledermann M, Lang C, Krähenbühl S. Relationship between hepatic mitochondrial functions in vivo and in vitro in rats with carbon tetrachloride-induced liver cirrhosis. *J Hepatol* 2000;**33**:216–23.
102. Knockaert L, Berson A, Ribault C, Prost PE, Fautrel A, Pajaud J, et al. Carbon tetrachloride-mediated lipid peroxidation induces early mitochondrial alterations in mouse liver. *Lab Investig* 2012;**92**:396–410.
103. Demirdag K, Bahcecioglu IH, Ozercan IH, Ozden M, Yilmaz S, Kalkan A. Role of L-carnitine in the prevention of acute liver damage induced by carbon tetrachloride in rats. *J Gastroenterol Hepatol* 2004;**19**:333–8.
104. Rehman H, Liu Q, Krishnasamy Y, Shi Z, Ramshesh VK, Haque K, et al. The mitochondria-targeted antioxidant MitoQ attenuates liver fibrosis in mice. *Int J Physiol Pathophysiol Pharmacol* 2016;**8**:14–27.
105. Sawant SP, Dnyanmote AV, Shankar K, Limaye PB, Latendresse JR, Mehendale HM. Potentiation of carbon tetrachloride hepatotoxicity and lethality in type 2 diabetic rats. *J Pharmacol Exp Ther* 2004;**308**:694–704.
106. Donthamsetty S, Bhave VS, Mitra MS, Latendresse JR, Mehendale HM. Nonalcoholic fatty liver sensitizes rats to carbon tetrachloride hepatotoxicity. *Hepatology* 2007;**45**:391–403.
107. Allman M, Gaskin L, Rivera CA. $CCl_4$-induced hepatic injury in mice fed a Western diet is associated with blunted healing. *J Gastroenterol Hepatol* 2010;**25**:635–43.
108. Cave M, Appana S, Patel M, Falkner KC, McClain CJ, Brock G. Polychlorinated biphenyls, lead, and mercury are associated with liver disease in American adults: NHANES 2003–2004. *Environ Health Perspect* 2010;**118**:1735–42.
109. Donat-Vargas C, Gea A, Sayon-Orea C, Carlos S, Martinez-Gonzalez MA, Bes-Rastrollo M. Association between dietary intakes of PCBs and the risk of obesity: the SUN project. *J Epidemiol Community Health* 2014;**68**:834–41.
110. Gadupudi GS, Klaren WD, Olivier AK, Klingelhutz AJ, Robertson LW. PCB126-induced disruption in gluconeogenesis and fatty acid oxidation precedes fatty liver in male rats. *Toxicol Sci* 2016;**149**:98–110.
111. Wahlang B, Falkner KC, Clair HB, Al-Eryani L, Prough RA, States JC, et al. Human receptor activation by aroclor 1260, a polychlorinated biphenyl mixture. *Toxicol Sci* 2014;**140**:283–97.
112. Nishihara Y, Utsumi K. 2,5,2′,5′-Tetrachlorobiphenyl impairs the bioenergetic functions of isolated rat liver mitochondria. *Biochem Pharmacol* 1986;**35**:3335–9.
113. Nishihara Y, Robertson LW, Oesch F, Utsumi K. The effects of tetrachlorobiphenyls on the electron transfer reaction of isolated rat liver mitochondria. *Life Sci* 1986;**38**:627–35.
114. Mildaziene V, Nauciene Z, Baniene R, Grigiene J. Multiple effects of 2,2′,5,5′-tetrachlorobiphenyl on oxidative phosphorylation in rat liver mitochondria. *Toxicol Sci* 2002;**65**:220–7.

115. Aly HA, Domènech O. Aroclor 1254 induced cytotoxicity and mitochondrial dysfunction in isolated rat hepatocytes. *Toxicology* 2009;**262**:175–83.
116. Hennig B, Reiterer G, Toborek M, Matveev SV, Daugherty A, Smart E, et al. Dietary fat interacts with PCBs to induce changes in lipid metabolism in mice deficient in low-density lipoprotein receptor. *Environ Health Perspect* 2005;**113**:83–7.
117. Shi X, Wahlang B, Wei X, Yin X, Falkner KC, Prough RA, et al. Metabolomic analysis of the effects of polychlorinated biphenyls in nonalcoholic fatty liver disease. *J Proteome Res* 2012;**11**:3805–15.
118. Wahlang B, Falkner KC, Gregory B, Ansert D, Young D, Conklin DJ, et al. Polychlorinated biphenyl 153 is a diet-dependent obesogen that worsens nonalcoholic fatty liver disease in male C57BL6/J mice. *J Nutr Biochem* 2013;**24**:1587–95.
119. Arzuaga X, Reiterer G, Majkova Z, Kilgore MW, Toborek M, Hennig B. PPARα ligands reduce PCB-induced endothelial activation: possible interactions in inflammation and atherosclerosis. *Cardiovasc Toxicol* 2007;**7**:264–72.
120. Wahlang B, Song M, Beier JI, Cameron Falkner K, Al-Eryani L, Clair HB, et al. Evaluation of Aroclor 1260 exposure in a mouse model of diet-induced obesity and non-alcoholic fatty liver disease. *Toxicol Appl Pharmacol* 2014;**279**:380–90.
121. Pelclová D, Urban P, Preiss J, Lukás E, Fenclová Z, Navrátil T, et al. Adverse health effects in humans exposed to 2,3,7,8-tetrachlorodibenzo-p-dioxin (TCDD). *Rev Environ Health* 2006;**21**:119–38.
122. Mukerjee D. Health impact of polychlorinated dibenzo-p-dioxins: a critical review. *J Air Waste Manage Assoc* 1998;**48**:157–65.
123. Zober A, Ott MG, Messerer P. Morbidity follow up study of BASF employees exposed to 2,3,7,8-tetrachlorodibenzo-p-dioxin (TCDD) after a 1953 chemical reactor incident. *Occup Environ Med* 1994;**51**:479–86.
124. Boverhof DR, Burgoon LD, Tashiro C, Chittim B, Harkema JR, Jump DB, et al. Temporal and dose-dependent hepatic gene expression patterns in mice provide new insights into TCDD-mediated hepatotoxicity. *Toxicol Sci* 2005;**85**:1048–63.
125. Lu H, Cui W, Klaassen CD. Nrf2 protects against 2,3,7,8-tetrachlorodibenzo-p-dioxin (TCDD)-induced oxidative injury and steatohepatitis. *Toxicol Appl Pharmacol* 2011;**256**:122–35.
126. Angrish MM, Mets BD, Jones AD, Zacharewski TR. Dietary fat is a lipid source in 2,3,7,8-tetrachlorodibenzo-ρ-dioxin (TCDD)-elicited hepatic steatosis in C57BL/6 mice. *Toxicol Sci* 2012;**128**:377–86.
127. Al-Eryani L, Wahlang B, Falkner KC, Guardiola JJ, Clair HB, Prough RA, et al. Identification of environmental chemicals associated with the development of toxicant-associated fatty liver disease in rodents. *Toxicol Pathol* 2015;**43**:482–97.
128. Nault R, Fader KA, Lydic TA, Zacharewski TR. Lipidomic Evaluation of aryl hydrocarbon receptor-mediated hepatic steatosis in male and female mice elicited by 2,3,7,8-tetrachlorodibenzo-p-dioxin. *Chem Res Toxicol* 2017;**30**:1060–75.
129. Pierre S, Chevallier A, Teixeira-Clerc F, Ambolet-Camoit A, Bui LC, Bats AS, et al. Aryl hydrocarbon receptor-dependent induction of liver fibrosis by dioxin. *Toxicol Sci* 2014;**137**:114–24.
130. Lakshman MR, Ghosh P, Chirtel SJ. Mechanism of action of 2,3,7,8-tetrachlorodibenzo-p-dioxin on intermediary metabolism in the rat. *J Pharmacol Exp Ther* 1991;**258**:317–9.
131. Kawano Y, Nishiumi S, Tanaka S, Nobutani K, Miki A, Yano Y, et al. Activation of the aryl hydrocarbon receptor induces hepatic steatosis via the upregulation of fatty acid transport. *Arch Biochem Biophys* 2010;**504**:221–7.
132. Lee JH, Wada T, Febbraio M, He J, Matsubara T, Lee MJ, et al. A novel role for the dioxin receptor in fatty acid metabolism and hepatic steatosis. *Gastroenterology* 2010;**139**:653–63.
133. Senft AP, Dalton TP, Nebert DW, Genter MB, Hutchinson RJ, Shertzer HG. Dioxin increases reactive oxygen production in mouse liver mitochondria. *Toxicol Appl Pharmacol* 2002;**178**:15–21.
134. Aly HA, Domènech O. Cytotoxicity and mitochondrial dysfunction of 2,3,7,8-tetrachlorodibenzo-p-dioxin (TCDD) in isolated rat hepatocytes. *Toxicol Lett* 2009;**191**:79–87.
135. Senft AP, Dalton TP, Nebert DW, Genter MB, Puga A, Hutchinson RJ, et al. Mitochondrial reactive oxygen production is dependent on the aromatic hydrocarbon receptor. *Free Radic Biol Med* 2002;**33**:1268–78.
136. Duval C, Teixeira-Clerc F, Leblanc AF, Touch S, Emond C, Guerre-Millo M, et al. Chronic exposure to low doses of dioxin promotes liver fibrosis development in the C57BL/6J diet-induced obesity mouse model. *Environ Health Perspect* 2017;**125**:428–36.
137. Anthérieu S, Rogue A, Fromenty B, Guillouzo A, Robin MA. Induction of vesicular steatosis by amiodarone and tetracycline is associated with up-regulation of lipogenic genes in HepaRG cells. *Hepatology* 2011;**53**:1895–905.

138. Michaut A, Le Guillou D, Moreau C, Bucher S, McGill MR, Martinais S, et al. A cellular model to study drug-induced liver injury in nonalcoholic fatty liver disease: application to acetaminophen. *Toxicol Appl Pharmacol* 2016;**292**:40–55.
139. Rogue A, Anthérieu S, Vluggens A, Umbdenstock T, Claude N, de la Moureyre-Spire C, et al. PPAR agonists reduce steatosis in oleic acid-overloaded HepaRG cells. *Toxicol Appl Pharmacol* 2014;**276**:73–81.
140. Tolosa L, Gómez-Lechón MJ, Jiménez N, Hervás D, Jover R, Donato MT. Advantageous use of HepaRG cells for the screening and mechanistic study of drug-induced steatosis. *Toxicol Appl Pharmacol* 2016;**302**:1–9.
141. Bucher S, Jalili P, Le Guillou D, Begriche K, Rondel K, Martinais S, et al. Bisphenol A induces steatosis in HepaRG cells using a model of perinatal exposure. *Environ Toxicol* 2017;**32**:1024–36.
142. Aninat C, Piton A, Glaise D, Le Charpentier T, Langouët S, Morel F, et al. Expression of cytochromes P450, conjugating enzymes and nuclear receptors in human hepatoma HepaRG cells. *Drug Metab Dispos* 2006;**34**:75–83.
143. Anthérieu S, Chesné C, Li R, Guguen-Guillouzo C, Guillouzo A. Optimization of the HepaRG cell model for drug metabolism and toxicity studies. *Toxicol in Vitro* 2012;**26**:1278–85.
144. Peyta L, Jarnouen K, Pinault M, Guimaraes C, Pais de Barros JP, Chevalier S, et al. Reduced cardiolipin content decreases respiratory chain capacities and increases ATP synthesis yield in the human HepaRG cells. *Biochim Biophys Acta* 2016;**1857**:443–53.
145. Porceddu M, Buron N, Rustin P, Fromenty B, Borgne-Sanchez A. In vitro assessment of mitochondrial toxicity to predict drug-induced liver injury. In: Chen M, Will Y, editors. *Drug-induced liver toxicity*. New-York, NY: Springer; 2018. p. 283–300.
146. Nadal A, Quesada I, Tudurí E, Nogueiras R, Alonso-Magdalena P. Endocrine-disrupting chemicals and the regulation of energy balance. *Nat Rev Endocrinol* 2017;**13**:536–46.
147. Sugimoto K, Takei Y. Pathogenesis of alcoholic liver disease. *Hepatol Res* 2017;**47**:70–9.
148. Massart J, Robin MA, Noury F, Fautrel A, Lettéron P, Bado A, et al. Pentoxifylline aggravates fatty liver in obese and diabetic ob/ob mice by increasing intestinal glucose absorption and activating hepatic lipogenesis. *Br J Pharmacol* 2012;**165**:1361–74.

PART V

# PATHOLOGY PREVENTION AND CARE

CHAPTER

# 16

# Impact of Lifestyle and Clinical Interventions on Mitochondrial Function in Obesity and Type 2 Diabetes

*Brenna Osborne*[*,†], *Amanda E. Brandon*[‡], *Greg C. Smith*[*], *Nigel Turner*[*]

[*]Department of Pharmacology, School of Medical Sciences, The University of New South Wales, Sydney, Australia [†]Department of Cellular and Molecular Medicine, University of Copenhagen, Copenhagen, Denmark [‡]Sydney Medical School, Charles Perkins Centre, University of Sydney, Sydney, Australia

## 1 INTRODUCTION

Obesity and type 2 diabetes (T2D) are commonly referred to as lifestyle diseases associated with caloric excess. Lifestyle interventions can be defined as changes that patients can make to their lifestyles that improve these diseases in lieu of, or in addition to, clinical and pharmaceutical interventions, and often are prescribed as a first line of defense for patients showing diabetes and other symptoms of the metabolic syndrome. Lifestyle interventions include dietary interventions, exercise, and physical training regimes, cessation of smoking, and cessation of alcohol consumption. Dietary restrictions encompass calorie restriction, intermittent fasting, ketogenic diets, and other low calorie and very-low calorie diets that encompass a multitude of different prescriptions. There is increasing scientific interest in the role of interventions such as exercise and dietary restriction, particularly calorie restriction and intermittent fasting, on metabolic and mitochondrial health, while the effects on mitochondrial health of other lifestyle interventions, such as smoking and alcohol cessation, remain relatively understudied. Clinical interventions to treat obesity and diabetes, such as surgical intervention, include a variety of bariatric surgical techniques, including sleeve gastrectomy, Roux-en-Y gastric bypass, and adjustable gastric banding. Much scientific literature has been dedicated

*Mitochondria in Obesity and Type 2 Diabetes*
https://doi.org/10.1016/B978-0-12-811752-1.00016-X

to understanding the precise mechanisms of these surgical interventions, which since their first use as a treatment for morbid obesity in the 1950s[1] revealed that surgical interventions had beneficial effects in T2D patients that were over and above the surgical effect on calorie intake. Although the precise mechanisms of the benefit of gastric surgery still are unknown, it remains an area of intense investigation, and the effects on mitochondrial health warrant further discussion. This chapter will focus on lifestyle modifications, as well as clinical interventions such as surgery, as treatments for obesity and type 2 diabetes and their reported impacts on mitochondrial function in humans.

# 2 LIFESTYLE INTERVENTIONS

## 2.1 Exercise

Lifestyle interventions for obesity and T2D that incorporate physical training and exercise have long been shown to have positive effects on mitochondrial parameters.[2–5] Exercise often is prescribed as an initial treatment for patients with prediabetes, morbid obesity, or diabetic symptoms in the clinical setting.[6–8]

In its most recent position statement on physical activity and exercise, the American Diabetes Association (ADA) defines exercise as a structured physical activity that can be categorized as aerobic, resistance, or balance and flexibility exercise.[6] Aerobic exercise involves the repeated and continuous movement of large muscle groups and includes activities, such as walking, cycling, jogging, and swimming, that rely primarily on aerobic energy-production. Resistance or strength training involves contraction of muscles against some form of resistance and includes activities such as free-weights, weight-lifting, or body weight and resistance band exercises. Flexibility and balance exercises also have been shown to be of importance for the health of type 2-diabetes patients,[9, 10] however, these will not be covered in detail in this chapter because they have been little studied in the context of mitochondrial function. This key role of exercise in the successful treatment and prevention of obesity and T2D is highlighted by the position statements of governmental and other diabetes organizations such as the ADA[6] and its European counterpart.[8, 11, 12] Since reports emerged that type 2 diabetic patients showed differences in the mitochondrial content of their skeletal muscle compared to nondiabetic controls,[13–16] there has been a particular interest in the role of skeletal muscle mitochondria in T2D, and in interventions that might modulate this effect.

## 2.2 Aerobic Exercise

Aerobic exercise is by far the most well-studied type of exercise and long has been associated with improvements in both glycemic control and in mitochondrial function, primarily in skeletal muscle.[4, 17, 18] Although the terms aerobic and endurance exercise often are used interchangeably, aerobic exercise also might encompass high-intensity interval training (HIIT) protocols,[6] which will be discussed in a separate section. There is considerable evidence that aerobic exercise training causes changes in skeletal muscle mitochondria, although how these changes relate to insulin resistance in this tissue is more contentious,[19, 20] as markers of mitochondrial function do not always correlate with insulin sensitivity.[21] An 8-week aerobic cycling exercise study, however, was shown to increase oxidative capacity in skeletal muscle, and also was associated with improvements in glucose disposal.[22] Ten weeks of aerobic exercise training has been shown to increase intrinsic mitochondrial respiration in the skeletal muscle of previously sedentary obese patients, both with and without T2D.[18] Phielix et al. performed a 12-week exercise intervention in T2D patients and showed significant increases in both skeletal muscle mitochondrial content and in intrinsic muscle mitochondrial respiration following training.[23] Increases were reported in T2D patients following an exercise intervention of 10 weeks, in spite of no change in initial mitochondrial content between T2D and control subjects being reported.[24] The mitochondrial biogenesis observed following endurance exercise training is because of transcriptional activation of a range of genes involved in mitochondrial biogenesis and metabolism, such as peroxisome proliferator-activated receptor-gamma coactivator (PGC1α), nuclear respiratory factors (NRF-1, NRF-2), and mitochondrial transcription factor A (TFAM)[25–28] (see Fig. 1).

Endurance exercise is associated with increases in fatty acid (FA) oxidation in mitochondria.[29–31] Eight weeks of endurance training in obese subjects caused increases in FA oxidation

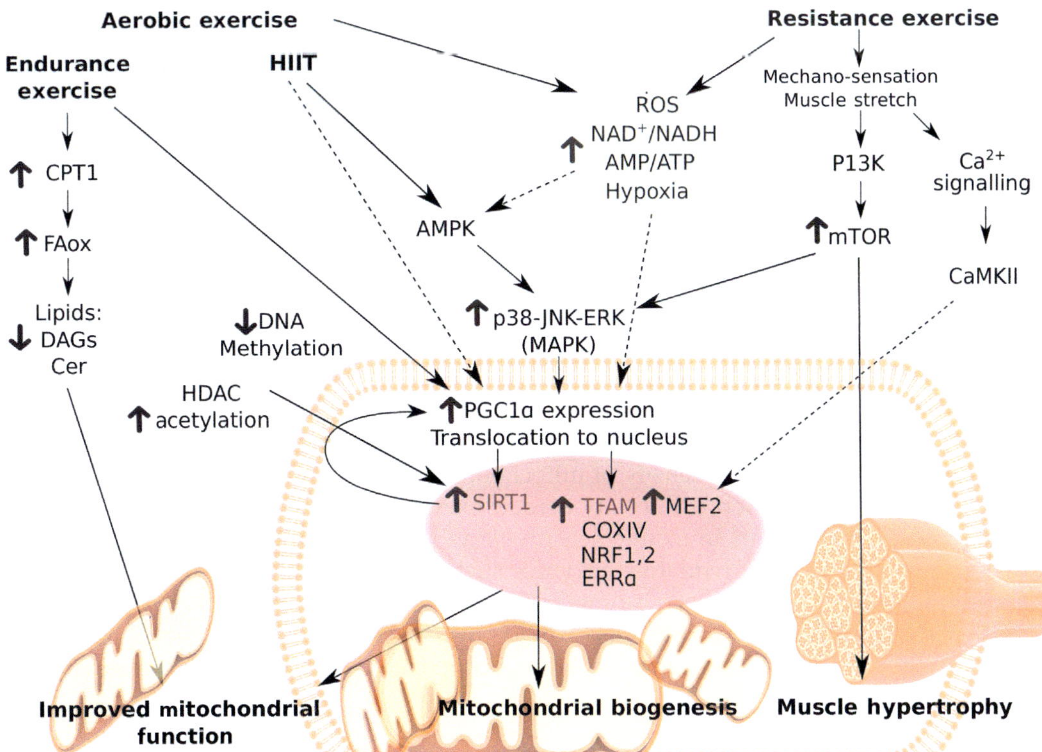


FIG. 1 Pathways of skeletal muscle adaptations to exercise: Aerobic exercise and resistance exercise involve discrete yet overlapping pathways leading to improved mitochondrial function, increased mitochondrial content, and increased muscle mass. *AMPK*, AMP kinase; *Cer*, ceramides; *CPT1*, carnitine palmitoyltransferase; *DAG*, diacylglycerol; *FAox*, fatty acid oxidation; *HDAC*, histone deacetylases; *HIIT*, high intensity interval training; *mTOR*, mammalian target of rapamycin; *ROS*, reactive oxygen species; *TFAM*, mitochondrial transcription factor A.

mediated by an increase in carnitine palmitoyltransferase (CPT1) and a decrease in bioactive lipids such as diacylglycerols (DAGs) and ceramides, with these changes in lipid storage and oxidation correlated with improvements in glucose tolerance and increases in mitochondrial enzymes citrate synthase (CS) and β-hydroxyacyl-CoA dehydrogenase (ßHAD).[30] It might be that the precise subcellular location of these intramyocellular lipids and their proximity to mitochondria is also an important adaptation to endurance exercise.[32]

Some studies, however, have failed to find differences in mitochondrial function following aerobic exercise,[33] albeit mild aerobic exercise, although other parameters such as lipid oxidation were improved.[33] Bordenave et al. also showed improvements in permeabilized muscle fiber respiration and CS activity in T2D patients using only a moderate aerobic exercise regime that was designed to target maximum lipid oxidation in individual subjects.[3] Interventions that do not require intense aerobic exercise might be more achievable for obese patients without a history of regular exercise; therefore, these types of studies are an important addition to the literature.

Endurance exercise and muscle contraction are also a potent stimulus for activation of AMP-activated protein kinase (AMPK), a key energy-sensing kinase that regulates many aspects of the adaptive response to exercise[34, 35] (see Fig. 1). Skeletal muscle-specific AMPK double knockout mice showed reduced exercise performance,[34] and studies in rodents using the AMPK-activator AICAR, have shown that AMPK can promote mitochondrial biogenesis in a PGC-1α-dependent manner.[36, 37] PGC1α also is activated during exercise via cross-talk from other signaling pathways via the action of the p38-mitogen activated protein kinase (MAPK) pathway.[38] Short et al. looked at the effects of a 16-week aerobic exercise intervention in subjects in a range of different age groups and found across all ages that there were increases in mitochondrial function, as indicated by CS and cytochrome c oxidase activity and mRNA levels of genes involved in mitochondrial metabolism in muscle including PGC1α, TFAM, and cytochrome c oxidase subunit IV (COX4).[39] Although mitochondrial function increased in all age groups, improvements in insulin sensitivity were noted only in the young age groups, and not in the middle-aged and older age groups.[39]

Longer term studies are fewer in number, however, a six-month study of aerobic exercise interventions in heart failure patients showed an increase in mitochondrial content in these patients via significant increase in the surface density of cytochrome c oxidase-positive mitochondria and in the surface density of mitochondrial cristae.[40] Toledo et al. trained type 2 diabetic subjects with moderate intensity aerobic exercise (brisk walking) for 4 months and showed both improvements in insulin sensitivity and in mitochondrial density by electron microscopy, and activity of oxidative enzymes such as CS and NADH oxidase.[4] Although there is large variation in the precise methods used to investigate mitochondrial function across different studies, as a whole, they provide overwhelming evidence that aerobic exercise has the ability to favorably remodel the mitochondrial characteristics of skeletal muscle in humans.

## 2.3 High-Intensity Interval Training (HIIT)

More recently, many studies have looked at HIIT protocols in an effort to establish the best type of activity for improving cardiometabolic health, particularly in respect to patients with insulin resistance.[41] Because of its high intensity and shorter duration, HIIT exercise might be more appealing to patients who are time poor or struggle with the prescribed volume of lower intensity but longer duration workouts,[42] although the intensity required might not be appropriate for all patients.[43] One such study showed that rapid skeletal muscle remodeling similar to that seen in endurance training (see Fig. 1) can occur rapidly in response to HIIT training in young active men following a single acute HIIT protocol (4×30s bouts of all-out cycling).[44] mRNA levels of PGC1α increased (although no change was detected in total protein), combined with changes in signaling proteins in skeletal muscle such as AMPK subunits 1 and 2 and the p38 MAPK pathway immediately following exercise.[44] Little et al. later showed that a 6-week HIIT intervention in young healthy individuals also found significant mitochondrial biogenesis and skeletal muscle remodeling, with increases in the activity of CS and COX, as well as total protein content of CS, COX subunits II and IV, and the TFAM.

The abundance of nuclear PGC1α and the protein content of its proposed activator, the sirtuin SIRT1, was increased following HIIT training.[45] Followup studies by the same group in a more clinically relevant cohort of middle-age sedentary adults found that HIIT training for 2 weeks (6 sessions) was sufficient to induce skeletal muscle remodeling and mitochondrial biogenesis reported as increases in protein content of CS, COX, and PGC1α. In this study it correlated with improvements in insulin sensitivity measured by homeostatic model assessment (HOMA), a method of assessing insulin sensitivity in human subjects.[46] Six weeks of HIIT in overweight but healthy subjects also showed increases in mitochondrial content and oxidative phosphorylation capacity in skeletal muscle fibers, but no effect in adipose tissue explants.[47] Larsen et al. also showed that, in contrast to results seen with resistance and endurance exercise,[30, 48] HIIT intervention was unable to affect the oxidation of lipid substrates in mitochondria in skeletal muscle.[47] In spite of differences in the intensity and duration of the aerobic exercise, most studies agree that aerobic exercise as a lifestyle intervention in obese and type-2 diabetic subjects has beneficial effects on mitochondrial health.

## 2.4 Resistance Exercise

The genetic and molecular mechanisms of adaptation induced by resistance and endurance training are distinct, with each mode of exercise activating and repressing specific subsets of genes and cellular signaling pathways as summarized in Fig. 1 (reviewed in[25, 49]). Pesta et al. recently reviewed the literature supporting a role for resistance or strength training as a treatment for T2D.[50] The importance of mitochondrial adaptations to resistance exercise are not as well-studied as in endurance exercise, but considerable attention has been given to how resistance exercise might complement mitochondrial changes and be beneficial for those combating T2D and obesity. One benefit of resistance exercise is that it shows benefits even when exercise is performed at a low intensity, a corollary of the fact that different pathways are activated from those in more intense aerobic or endurance-type exercise. It is well known that resistance exercise causes muscle hypertrophy because of coordinated changes in gene expression.[25, 50–52] Although as a major insulin-responsive tissue responsible for post-prandial glucose uptake, the benefits of increased muscle mass to glycemia are clear,[53, 54] resistance exercise has been shown to have other benefits to the diabetic patient above a mere increase in muscle volume, and one of these mechanisms might well be changes in mitochondrial function. A study by Sparks et al. showed that resistance training increased oxidation of both fatty-acid and glucose-derived substrates in skeletal muscle in a T2D intervention study.[55] Although this is suggestive of an increase in oxidative capacity, other studies have failed to find changes in mitochondrial parameters with resistance exercise.[56, 57] Several studies, however, have reported that when resistance training and endurance exercise are combined, increases in oxidative and mitochondrial adaptations occur in excess of that seen with either training modality alone, thus providing some support for resistance training affecting mitochondrial pathways.[55, 56, 58] One mechanism of resistance exercise ultimately affecting mitochondria is the activation of the mammalian target of rapamycin (mTOR), a kinase at the center of important metabolic signaling pathways that can regulate a variety of adaptations in skeletal muscle, including translation of mRNA, ribosomal biogenesis, and response to nutrient availability.[53] mTOR can be activated by growth factor signaling such as insulin-like growth factor (IGF-1), but also via direct mechano-sensing mechanisms induced by muscle contraction.[49] The exact downstream pathways whereby mTOR activation leads to changes in muscle fuel oxidation are still being determined, but evidence points to cross-talk between the p38-JNK-ERK signaling pathways and PGC1α expression.[38, 50] Although a direct effect of resistance exercise on mitochondrial function in skeletal muscle has not been shown conclusively, one study suggests that the response of subcutaneous adipose tissue to acute resistance exercise might be an important mediator of whole-body energy expenditure. Ormsbee et al. showed that after resistance exercise there was an increase in lipolysis from adipose tissue, and an increase in whole-body fat oxidation, although how this might relate to specific mitochondrial pathways remained untested.[59]

## 2.5 Molecular Mechanisms of Exercise Adaptation

This ability of skeletal muscle to rapidly remodel itself in response to exercise, both acutely and chronically, is remarkable and has been studied extensively. Changes in mitochondrial density/activity represent just one of these adaptations, and studies show that these effects are mediated by a range of mechanisms (reviewed in Egan et al.[60]) (see Fig. 1). Some proposed mechanisms include the mechanical effects of muscle contraction itself,[61] changes in muscle driven by depletion of ATP and other necessary cofactors, oxidative stress caused by increased load on the mitochondria, and changes triggered by hypoxia in the muscle tissue.[62, 63] Studies have shown that changes in gene expression postexercise can be rapid and transient, and include the upregulation of transcription factors and transcriptional coregulators important in coordinating muscle mitochondrial function, such as PGC1α, forkhead box (FOXO) 3a, estrogen-related receptor alpha (ERRα), TFAM, and myocyte enhancer factor 2 (MEF2).[64–68] Wright et al. showed that PGC1α is activated and translocates to the nucleus rapidly in response to exercise, much quicker than is required for transcriptional increases in its protein level.[69] Post-translational modification changes also occur acutely in the hours following intense exercise and include HDAC-mediated changes in acetylation of histones[70] and reduction in DNA-methylation at key promotor sites.[71] These transcriptional changes can occur rapidly and in response to a single acute exercise stimulus[66] and are associated with increases in mitochondrial enzyme activity that can occur within a few days.[72]

One other mechanism of mitochondrial adaptation to exercise in skeletal muscle is the role of reactive oxygen species (ROS) signaling and redox homeostasis. Superoxide and other ROS are generated as byproducts of mitochondrial respiration, and the ensuing oxidative damage has been linked strongly with pathology.[62, 73, 74] In contrast, normal physiological fluctuations in reactive species appear to be important in a signaling context, including potentially in the adaptation to an exercise stimulus.[75, 76] These signaling functions additionally have relevance to insulin action in muscle, as they inhibit phosphorylation cascades involving protein tyrosine phosphatases (PTPs), such as PTP1B and PTEN, which have a known role in PI3K-mediated downstream insulin signaling.[77] During and immediately after acute exercise, mitochondrial ROS generation is decreased.[62, 78] Trewin et al. showed that a single bout of HIIT exercise was able to elicit decreased oxidative stress as measured in permeabilized muscle fibers, which correlated with insulin sensitivity.[62] In particular, resistance exercise has achieved attention because of its ability to reduce ROS and oxidative stress in humans.[79, 80]

Most studies looking at exercise interventions and mitochondria have focused on skeletal muscle for several reasons, including the ease of sample collection from human subjects, the long-established role of skeletal muscle plasticity to exercise adaptations, and the intense interest created by the controversy about the relationship between mitochondrial dysfunction and insulin resistance in skeletal muscle.[20, 81] Other important metabolic tissues include the liver and adipose depots and, although directly assessing liver mitochondrial function in human subjects is unlikely, some studies have investigated adipose tissue in humans following exercise interventions.[47, 50] A study investigating 10 weeks of swim training in rats found an increased mitochondrial enzyme content in adipose tissue,[82] which also was reported by others in mice,[83] and possibly mediated by increased PGC1α expression following a single exercise session.[84] Studies in rats and humans suggest there also might be a role for AMPK in adipose-tissue after exercise to increase FA oxidation and promote other changes in fatty acid metabolism such as a reduction in triglyceride synthesis.[85–87] Data about mitochondrial adaptations in adipose tissues in humans following exercise interventions are few and often not in agreement, although two studies agreed that there was no difference in markers of mitochondrial content such as CS activity, but disagreed about an increase or no change in palmitate oxidation respectively following exercise intervention.[88, 89]

## 3 DIETARY RESTRICTION

In addition to exercise, weight loss is a well-established method to improve cardiometabolic health in patients with T2D, and obese patients at risk of metabolic syndrome. Dietary manipulations to induce weight loss of about 10% is seen as clinically relevant, and often used as a target in clinical studies. Caloric restriction and very-low calorie diets (VLCD) most often are used, but reports about ketogenic diet and intermittent fasting are becoming more prevalent in the literature (see Fig. 2). It is well recognized in rodent models that calorie restriction (CR; 30%–50% restriction), a dietary regime that reduces daily caloric intake without causing malnutrition, reduces body weight, and also is linked strongly with lifespan extension in yeast, flies, and rodents, as well to a lesser extent in nonhuman primates and in humans (reviewed in Ref. 90–93) Although not as well established, many laboratories have shown that this CR regime can alter mitochondrial content and/or function in metabolically active tissues such as the liver, skeletal muscle, and adipose tissue. Examples of improved function include decreases in proton leak,[94] $H_2O_2$ leak,[94–96] oxygen consumption,[96–98] and ROS production,[99] as well as enhanced activities of TCA-cycle enzymes, such as CS.[99, 100] CR also has been shown to increase mitochondrial content in aged rats[101] as well as in mice.[98, 99] Others, however, have not been able to confirm these beneficial findings.[102, 103]

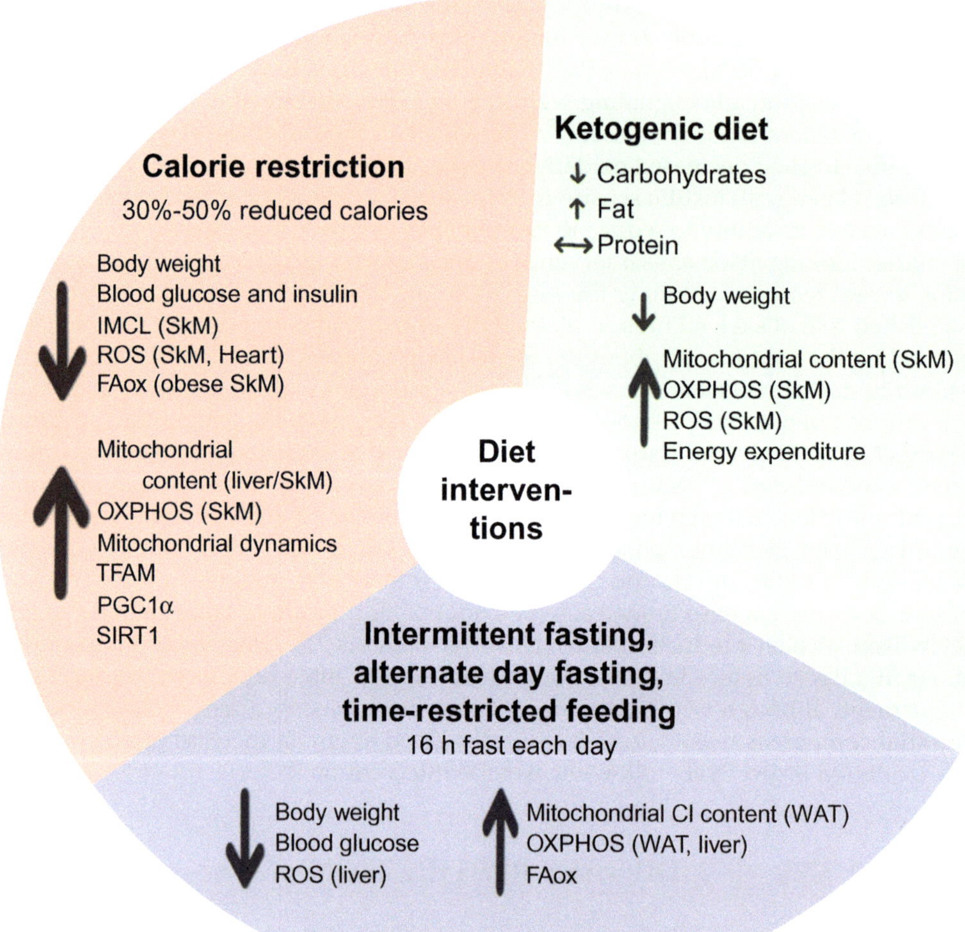


FIG. 2 Summary of diet interventions and their effects: Calorie restriction, ketogenic diets, and intermittent fasting protocols have varied effects on mitochondria and metabolic measures. Arrows indicate an increase or decrease in each measure. *FAox*, fatty acid oxidation; *IMCL*, intramyocellular lipid; *OXPHOS*, oxidative phosphorylation; *ROS*, reactive oxygen species; *SkM*, skeletal muscle; *WAT*, white adipose tissue.

These studies usually are designed to assess the effects of CR on aging and life span in normal, chow-fed animals. The effects of CR, as a tool to cause weight loss in already obese mice, is less well studied, especially when investigating mitochondria. In obese db/db mice, caloric restriction caused weight loss and decreases in fat mass and insulin levels.[104] To evaluate the effects of CR on liver mitochondrial biogenesis, the authors examined the expression of dynamin-related protein 1 (Drp1) and optic atrophy 1 (OPA1), two mitochondrial fission- and fusion-related proteins, and uncoupling protein 2 (UCP2), a mitochondrial carrier protein, which all showed a restoration in their levels after CR.[104] Unfortunately, no functional analysis was conducted. Tanajak et al. reported that CR, after obesity was established with 12 weeks of HFD, was associated with decreased ROS production, increased mitochondrial membrane potential, and reduction in mitochondrial swelling in cardiac tissue.[105]

In literature about humans, skeletal muscle mitochondria are the most studied because of the availability and ease of obtaining samples, but also because skeletal muscle is an important organ in metabolic health. Like in animal studies, employing a calorie restriction or a VLCD program in healthy overweight subjects causes weight loss, improvements in blood glucose, and plasma insulin, as well as other cardiometabolic parameters.[106, 107] Studies in the late 1990s and early 2000s, however, were controversial about whether diet alone was enough to modify mitochondrial function (measured via different markers). Simoneau et al.,[108] reported that in obese subjects who lost 10–15 kg body weight, an imbalance was seen between the capacity for the uptake and transport of fatty acids (CPT1 activity) within skeletal muscle, as was a reduced capacity of their oxidation within the mitochondria (as seen by decreases in enzyme activities). This occurred with a clear increase in whole-body insulin sensitivity as measured by the euglycemic-hyperinsulinemic clamp post weight loss.[108] The results from Larson-Meyer et al.[109] also do not support an increase in oxidative phosphorylation by diet-induced weight reduction, as measured using $^{31}$P MRS markers of mitochondrial function. Kern et al.,[110] however, reported that muscle oxidative capacity, assessed by measuring the rate of SDH activity within individual muscle fibers, increased in subjects who lost ~20% body weight, without an alteration in fiber type or size.[110] Whether the higher weight loss in this study accounts for the difference between these studies is unclear.

Civitarese et al.[111] reported that in overweight subjects subjected to 6 months of 30% calorie restriction, increases were seen in TFAM, PGC1α, and SIRT1, all involved in mitochondrial biogenesis. These increases were associated with increases in mtDNA content, but there were no changes in a range of mitochondrial enzyme activities in these subjects. This increase in mitochondrial content was not seen in a study by Toledo et al.,[112] in which overweight/obese subjects underwent caloric restriction of 25% for 16 weeks. Although this regime caused a 10% weight loss, and was associated with an improvement in insulin sensitivity and decreases in intramyocellular lipid (IMCL) content, mitochondrial content and electron transfer chain (ETC) activity (as measured by NADH-oxidase activity) were not different.[112] In a follow-up study from this group in older individuals, it again was found that diet alone was not sufficient to increase mitochondrial content and function.[113] Johnson et al.[114] also found that a 16-week CR weight loss program, which increased insulin sensitivity as assessed by clamp, did not change mitochondrial respiration rates or $H_2O_2$ emissions. A study that investigated mitochondrial respiration in permeabilized fibers showed a decrease in respiration rates after weight loss (11.5%) using a VLCD, which occurred without changes in mtDNA or CS activity.[115] With respect to caloric restriction, this variability seen in both animal studies (recently reviewed by Ingram and De Cabo[116]) and the human literature could be because of differences in diet regime, age of participants, length of intervention, and timing of studies, among other things.

On the flip side of dietary restriction, a few studies examine the effects of overfeeding in humans that also shed light on the involvement of mitochondrial pathways in changes in calorie intake. Overfeeding in healthy males with a high-fat diet for 56 days caused adaptation in skeletal muscle such as increasing lipid oxidation, mitochondrial respiration, and upregulation of mitochondrial gene expression associated with a hyperacetylation of PGC-1a,[117] along with the expected increase in body weight. A similar study for 28 days in men and women found a transient increase in mitochondrial protein content in skeletal muscle, however, no overall change in mitochondrial activity or function in response to overfeeding.[73]

The ketogenic diet (KD), a very low carbohydrate and high fat diet, is gaining traction as a weight-loss program that might be more sustainable than the classic VLCD or CR diets, because of its enhanced satiety effect[118] (see Fig. 2). KD starves the body of carbohydrates and forces the oxidation of fat as a fuel substrate, causing ketone bodies to be used as a fuel, hence the name KD. KDs have been used for decades to treat intractable seizures, and, although the mechanism of its ant-seizure activity has not been delineated clearly, it is thought that it is mediated both directly by ketones and indirectly by changes in bioenergetics and excitability of neuron activity.[119] Emerging evidence also exists for beneficial effects of KD on other diseases, such as polycystic ovary syndrome, cancer, and other neurodegenerative disorders (reviewed in Ref. 120, 121). Because of its long-time association with the brain, research into KD and mitochondrial function usually is focused on this organ. These studies also are not directed at weight management and likely cannot be extrapolated. Animal studies show that KD alone can have mild effects on skeletal muscle mitochondrial respiration (increase in complex II respiratory control ratio), however, because this was not a diet intervention study, body weight changes were not measured.[122] In another study, rats lost weight after 8 months of KD, and mitochondrial respiration was again slightly altered with a decrease in complex I-linked respiratory control ratio.[123] This was associated with a decrease in CS activity and an increase in mitochondrial ROS production in skeletal muscle, but not the liver. Hall et al. looked at metabolic parameters and energy expenditure in overweight and obese men; they found very small increases in energy expenditure and a decrease in body mass in spite of attempts to maintain isocaloric intake. This study, however, did not investigate mitochondrial parameters.[124] To the best of our knowledge, there has yet to be a study in humans that directly looks at mitochondrial function after weight loss using a ketogenic diet. With the increasing popularization of diets, such as the ketogenic diet in athletic and weight-loss circles,[125–127] it is more important than ever to understand the mechanistic value of such diets in humans, especially those who are at risk of metabolic disorders such as T2D.

Another intervention gaining popularity is intermittent fasting (IF), which includes alternate-day fasting and time-restricted feeding. IF is another dietary regime that has been shown to decrease body weight and improve glucose and lipid profiles in both rodents and humans (reviewed in Ref. 128, 129). Time-restricted feeding is when food is eaten only in a set period per day (~8h slot) with no caloric intake after this, meaning there is a 16-h fast per day (see Fig. 2). In alternate-day fasting, little to no food is consumed every second day, with no restrictions on food intake on other days, although binge eating is not recommended. In rodent studies, the predominant IF protocol used is alternate-day fasting. Boutant et al.[130] showed no change in mitochondrial function in skeletal muscle after IF, although white adipose tissue did have enhanced rates of mitochondrial respiration and an increase in complex I protein levels. Tissue-specific effects of IF also were shown by Chausse et al.,[131] with the liver having increased respiration rates and lower percentage of $H_2O_2$ release without changes in CS activity. This was not seen in skeletal muscle, brain, or heart. In a long-term study of IF in rats, although there was a decrease in body weight after 4 or 32 weeks, intra-abdominal fat increased when compared to the classic CR animals.[132] Thirty-two weeks of IF caused glucose intolerance and was associated with increases in mitochondrial $H_2O_2$ release in both skeletal muscle and adipose tissue.[132]

There is very little literature regarding IF regimes and mitochondria in humans. In one study, Heilbronn et al.[133] showed that in nonobese subjects, IF for 3 weeks caused a modest weight loss of ~2kg that did not alter insulin sensitivity. Gene expression from the muscle suggests that fatty acid oxidation pathways might be upregulated by IF, but oxidative phosphorylation genes were not affected.[133]

Although there is variability in the literature, the majority of studies suggest that weight loss alone was because of dietary restriction, even though it usually was accompanied by an increase in insulin sensitivity and decreases in IMCL. Therefore, it is unlikely to have a major impact on mitochondrial content or function.

## 3.1 Smoking Cessation

The cessation of tobacco smoking is a highly recommended lifestyle intervention for a range of diseases, including T2D.[11] Quitting smoking is one of the key lifestyle management options put forward by the ADA and EASD as a first measure for the prevention and/or management of T2DM, particularly in the context of reducing cardiovascular risk.[8] It has been shown that tobacco cessation greatly reduces the risk of disease and premature death, by reducing the risk of cancers, particularly lung cancer, reducing the risk of cardiovascular events such as heart disease and stroke, and reducing the risk of lung diseases such as chronic obstructive pulmonary disease (COPD).[134, 135]

Diabetes is associated with poor cardiometabolic health, and, on its own, diabetes is a leading cause of cardiovascular mortality.[11] Nearly two-thirds of people with diabetes will die of cardiovascular disease.[136] The Framingham study reported that the relative risk of heart failure among those with T2D was double for men, and six times higher for women.[137] There is also evidence that smoking is an independent risk factor for the development of T2D because it has been shown in several meta-analyses of patient cohorts.[138–140] Smokers are 30%–44% more likely to develop T2D than nonsmokers, and the more cigarettes consumed, the higher the risk.[135, 139]

Multiple biological mechanisms for the connection between smoking and development of T2D have been postulated[135] (see Fig. 3). These include the effects of smoking on cortisol concentrations after studies found that smokers had higher plasma cortisol than nonsmokers.[141] High cortisol also can influence the prevalence of central obesity, which is itself independently associated with diabetes in epidemiological studies.[142] This increase in central obesity, increased inflammatory markers,[143] increased oxidative stress,[144] and direct effects of smoking on blood glucose control[145] all have been suggested as causation for the association between smoking and the development of T2D.[135, 146]

Although some studies, mainly in the 1980s and 1990s, have been done about the effect of cigarette smoking on mitochondrial function, the effect of cessation of tobacco use on mitochondrial parameters remains relatively understudied compared to other lifestyle interventions covered in this chapter.

Several studies point to a disruption in normal mitochondrial respiration induced by tobacco use. In the 1970s, Gairola and Aleem showed in isolated rat liver mitochondria that mitochondrial activity was reduced after exposure to cigarette smoke preparations.[147] Gvozdják et al. showed that rabbits exposed to cigarette smoke showed a reduction in complex IV activity of their heart muscle mitochondria,[148] which also was associated with reductions in

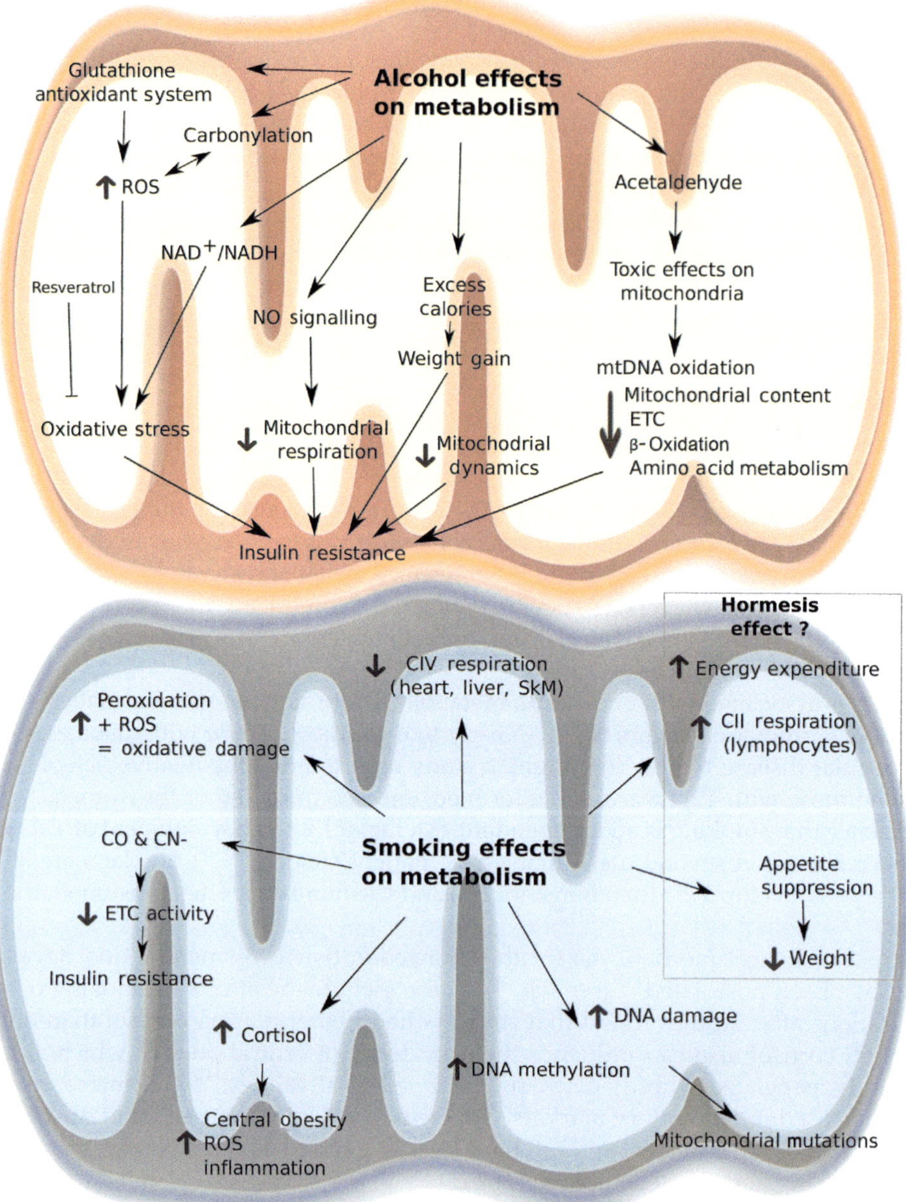


FIG. 3 Effects of alcohol and smoking consumption on mitochondrial metabolism: Graphical summary of some of the insults to mitochondria resulting from the lifestyle factors of alcohol and smoking. Metabolic effects of alcohol consumption include direct effects of alcohol metabolites such as acetaldehyde on the mitochondria, to effects of weight gain caused by excess calories, to oxidative stress caused by a variety of factors that all ultimately affect insulin resistance. Smoking has direct effects on mitochondrial respiration and oxidative stress and causes changes to DNA. In some cell types, however, smoking might elicit a hormesis effect, whereby respiration increases in response to smoking. Smoking might conversely reduce appetite and suppress weight gain. *CII*, complex II of the ETC; *CO*, carbon monoxide; *CN*, cyanide; *CIV*, complex IV of the ETC; *ETC*, electron transport chain; *ROS*, reactive oxygen species; *SkM*, skeletal muscle.

mitochondrial respiration and activity in cardiomyocytes.[149, 150] Orlander et al. compared skeletal muscle mitochondria in smokers and nonsmokers and also found decreases in complex IV activity,[151] while a followup study in twins showed that muscle fiber-type changes and reduction in complex IV activity in smokers was not because of genetic factors but appeared to be an effect of smoking.[152] Other studies have shown that complex I respiration also might be affected by smoking status, as seen in a study of platelet mitochondria.[153]

In studies in humans, smoking-induced mitochondrial dysfunction is associated with an overall increase in oxidative activity thought to be a compensatory mechanism to overcome the decreased activity of the ETC. Miró et al. looked at the effect of smoking on mitochondrial respiration in circulating lymphocytes, mainly in the context of mitochondria and their role in reactive oxygen species generation.[154] In their study of 35 smokers, they found that activity of complex IV of the mitochondrial ETC was decreased in lymphocytes. In addition, the respiration rate using succinate as a substrate was increased in smokers, which also correlated with an increase in lipid peroxidation, a key marker of oxidative damage.[154] Similarly, a possible increase in oxidative measures in smokers, including in glucose oxidation,[155] whole-body energy expenditure,[156] and oxidative stress[157] have been reported, which might constitute a hormesis effect whereby mitochondrial function is upregulated to cope with the deficit caused by tobacco's effects on the ETC, although this idea has not been thoroughly tested. These defects reported in mitochondrial respiration appear to be both rapid and reversible, with human studies showing that inhibition of complex IV of the ETC because of acute smoking occurred in blood cells within hours, but was normalized following smoking cessation.[158, 159]

One of the most common lung diseases associated with smoking is chronic obstructive pulmonary disease (COPD), which also is known to exhibit mitochondrial dysfunction.[160] COPD patients show a reduction in exercise tolerance and skeletal muscle mitochondrial function, which has been suggested to be associated with direct interference of the ETC by chemicals known to be found in cigarette smoke, including cyanide and carbon monoxide.[161, 162] COPD patients who smoke also have been shown to have increased oxidative stress occurring in muscle tissue.[163] In a study comparing primary bronchial epithelial cell lines from patients with COPD with nonsmoking controls, severe structural changes in mitochondrial morphology and function were seen, including changes in fission and fusion proteins, markers of mitochondrial biogenesis, and an increase in stress markers such as PINK1.[164] Although patients who have progressed to COPD are not ideal models for the investigation of smoking cessation, they highlight the mechanisms involved in smoking and its links to mitochondrial dysfunction.

Other cellular changes that are associated with smoking and need to be studied in the context of smoking cessation include changes in the integrity of mitochondrial DNA and epigenetic changes. Tan et al. showed that there was an increase in mutations in mitochondrial DNA in buccal cells from smokers, but the study did not specifically assess smoking cessation.[165] In addition, it appears that nicotine might have direct effects on mitochondria structure and function in beta cells as shown in a rat model that assessed fetal exposure to nicotine.[166] More recent studies have assessed epigenetic changes in methylation of DNA in humans in response to smoking status.[167] DNA methylation has been shown to increase in response to smoking, and although the changes at some loci are quite dynamic, it was shown that smoking cessation causes only partial and slow reversal of the changes (over four decades) that were induced by smoke exposure when comparing those subjects who had never smoked, were former smokers, or had continued the habit.[167]

One confounding factor that needs to be discussed is the data showing that in the short-term at least, smoking cessation actually can cause weight gain, which has been shown to exacerbate existing diabetes symptoms and to increase the risk of developing diabetes in some patients.[146, 168, 169] Although most epidemiological data support the idea that quitting smoking can cause adverse weight gain, in the long term the benefits of cessation for diabetes risk have been shown to outweigh short-term problems.[135] Several studies support the idea that smoking cessation in the long term gradually leads to a risk of developing diabetes that is similar to those who have never smoked, but agree that weight management programs need to be in place to support these patients as they quit smoking to modulate the initial increase in diabetes risk.[170–173] How this weight fluctuation might affect mitochondrial function has not been addressed, however, it is likely that adverse weight gain might have independent effects on mitochondrial function that need to be investigated.

Although the data about tobacco cessation as a beneficial lifestyle modification for health are undeniable, detailed analysis into the effect of smoking cessation on mitochondrial parameters, especially in diabetic patients, has not been adequately investigated. The literature clearly shows a link between smoking exposure and a range of mitochondrial impairments, however, the ability of these changes to be reversed requires further study. In the context of the other lifestyle and surgical modifications covered in this chapter, it is highly likely that cessation of smoking would increase the ability of patients to improve their mitochondrial health through exercise programs and through better surgical outcomes.

## 3.2 Alcohol Cessation

The adverse health effects of excess alcohol consumption are well-known,[174] and excessive alcohol consumption is a leading preventable cause of death in first-world countries such as the United States.[175] Reducing alcohol consumption is one of the interventions recommended by clinicians to reduce risk of T2D and other components of the metabolic syndrome associated with obesity.[7] Beyond the role that alcohol plays in providing unnecessary calories that can lead to weight gain, there is evidence that alcohol consumption has direct effects on cellular pathways that relate to blood sugar control and to mitochondrial function[176, 177] (see Fig. 3).

Alcohol use is a generally acceptable lifestyle factor, and many public health bodies in the western world stop short of recommending complete alcohol cessation.[174] In contrast to other lifestyle factors such as smoking, there is evidence from the US that few physicians screen their patients about alcohol use[178] in spite of evidence-based guidelines recommending such screening.[179] Few physicians seem to be aware of the precise guidelines that define low-risk or moderate drinking,[180, 181] it has been suggested that the guidelines be taught in medical school.[182] Because the data about risks of alcohol consumption for T2D vary wildly, there are few clear-cut studies in this area. Very few specific studies have been done to assess the effects of alcohol cessation on mitochondrial function in the context of human disease, and even fewer in the realm of T2D and obesity interventions.

Alcohol consumption has long been shown to have potent effects on glucose metabolism.[176, 183] How alcohol consumption affects the risk of developing T2D, however, has yielded conflicting answers in different studies, with several studies showing a reduced risk with moderate to low alcohol consumption,[184–187] and others finding an increased risk.[188–190]

Wei et al. found that the risk of diabetes is higher in men with an increased consumption of alcohol, compared to moderate drinkers,[190] however, nondrinkers conversely had a slightly higher risk for diabetes compared to moderate drinkers.[190] Similarly, another study using the Finnish twin cohort also found that low to moderate alcohol consumption actually might be protective for T2D, although the benefit was lost with higher alcohol consumption.[191] Some studies have even reported that heavy alcohol consumption is associated with a lower risk of metabolic syndrome.[192] Another study in a cohort of men found some correlations between alcohol and blood glucose levels but did not report a major finding for risk of metabolic syndrome.[193] In an attempt to reconcile these inconsistencies, Fan et al. looked at drinking patterns and compared average daily alcohol consumption in a more in depth assessment of frequency, quantity, and binge drinking episodes using data from the US National Health and Nutrition Examination Survey (NHANES) study.[194] An association between drinking in excess of the US guidelines or binge drinking and the risk of metabolic syndrome, including diabetes and other cardiovascular risk, was shown.[194] The large variation in outcomes of these studies is thought to reflect variations in definitions of alcohol intake, reporting guidelines, ethnic and gender differences in alcohol metabolism, frequency, and type of alcohol consumed, highlighting that this is an area that requires more investigation. To further cloud the issue, unreliable self-reporting is a common feature in these human studies.

Although controversy surrounds the risk of developing diabetes among drinkers, there is evidence in human patients that alcohol intake can be problematic for those who are already diabetic for control of blood glucose levels. Studies show poor glucose control in regular drinkers, which sometimes is coupled with lactic acidosis,[195] and abstinence from alcohol during initial treatment and during dose-adjustments for oral glycemic agents is recommended.[196] Alcohol abuse also has detrimental effects directly on the pancreas.[197] The term "alcohol cessation" when used in the literature also might refer to the cessation of alcohol consumption in those with a previous alcohol dependency and/or alcohol-induced liver disease, which is a very different medical setting from that of most T2D patients. Although the precise pathogenesis of alcoholic liver disease (ALD) is not entirely delineated, it has been shown that changes to mitochondria morphology and function is one of the hallmarks of the disease and its progression from stress/injury to steatosis, and on to alcoholic cirrhosis. The role of mitochondria in ALD has been reviewed elsewhere,[198] however, the importance of ROS and oxidative stress pathways, such as the mitochondrial glutathione-antioxidant pathway, to ALD pathogenesis highlight the mechanisms through which alcohol exerts its effects.

To get an insight into the role of mitochondria and alcohol consumption, we turn to animal studies where this has been studied extensively, mainly because of the difficulties in accessing the main organ of alcohol metabolism in humans, the liver. The breakdown of alcohol in the liver occurs via the enzymes alcohol dehydrogenase (ADH) in the cytoplasm and aldehyde dehydrogenase 2 (ALDH2) in the mitochondria, both of which rely on the reduction of $NAD^+$ to NADH. The intermediate of ethanol metabolism, acetaldehyde, is toxic and its further breakdown in the mitochondria is a priority to limit the buildup of this harmful byproduct. Perturbation in NAD/NADH levels in tissues following alcohol consumption has been shown in several studies and is thought to be associated with a range of oxidative deficits in mitochondria in tissues in response to ethanol.[177, 199] Studies in rats show that ethanol causes a reduction in the mitochondrial glutathione-peroxidase antioxidant system, which is associated with increased protein carbonylation.[200] Many animal studies have shown that the oxidative stress caused by alcohol leads to a multitude of cellular effects. In the liver, ethanol exposure in rats led to an increased cellular sensitivity to nitric oxide signaling, correlated with reduced mitochondrial respiration.[201] In a mouse study, ethanol caused oxidation of proteins and DNA in the mitochondria.[202] This oxidation appears to preferentially effect expression of mitochondrial components that are encoded by the mitochondrial genome, rather than the nuclear genome.[202a] Chronic alcohol consumption is linked with reduced levels of important mitochondrial proteins in heart and liver, including complexes I, III, IV, and V, and enzymes of β-oxidation, the TCA cycle, and amino acid metabolism.[203–205]

Mitochondrial pools of the antioxidant protein glutathione also has been shown in the liver in response to chronic alcohol exposure,[206] accompanied by increases in ROS production in the liver.[199] The primacy of perturbation of antioxidant systems in these effects is highlighted by the fact that resveratrol has been shown to reduce the toxic effects of alcohol in liver in a rat model via reducing ROS.[207] Studies in cell lines also have shown that ethanol causes defects in mitophagy and mitochondrial dynamics and morphology.[208]

Some early animal studies, however, reported an increase in mitochondrial respiration in response to alcohol.[209] This is thought to be an adaptation to chronic alcohol exposure because Han et al. has found similar changes in an intragastric feeding model of chronic alcohol consumption.[177] Following alcohol feeding for 4 weeks, there was an increase in state III respiration in isolated liver mitochondria.[177] Alcohol also caused dynamic changes in mitochondrial morphology and upregulation of mitochondrial biogenesis via action of PGC1α, associated with increased transcription of mitochondrial proteins including TFAM and cytochrome c.[177] In this study, an increase in the levels of both $NAD^+$ and NADH also was seen in isolated mitochondria, leading the authors to surmise that mitochondrial plasticity and increased mitochondrial respiration is an adaptive response to maintain $NAD^+$ levels and drive alcohol metabolism in the face of excess consumption.[177]

In spite of the number of studies showing the effects of alcohol on mitochondrial parameters, few studies have assessed the benefits of alcohol cessation on these pathways. Chronic alcohol consumption is known to cause problems in heart mitochondria, and the direct effects of ethanol and its toxic metabolites, such as acetaldehyde, on mitochondria are directly associated with alcoholic cardiomyopathy.[205, 210] Studies in rat myocardium showed a reduction in mitochondrial respiration following 8 weeks of chronic ethanol consumption, which was reversed upon removal of ethanol from the diet for 4 weeks,[210] highlighting that alcohol cessation in some cases might be able to treat and reverse damage caused to mitochondrial function. Similarly, ultrastructural changes in the gastrointestinal tract of rats following chronic ethanol consumption, which included mitochondrial swelling, also was found to be reversible upon cessation of the ethanol insult.[211] Many more studies are required to show how and why the cessation of alcohol consumption is a beneficial lifestyle intervention for patients with obesity and T2D.

# 4 CLINICAL INTERVENTIONS

## 4.1 Surgical Interventions

Surgical interventions that target obesity are becoming key clinical interventions for the treatment of T2D. Originally, such surgeries were viewed merely as a weight-loss treatment and the use of surgery for diabetes historically had been viewed with some skepticism in the diabetes community. It was not until relatively recently that surgical intervention was formally endorsed as a primary treatment for T2D.[212] Historically, some early clinical evidence showed that gastric surgery of various types could provide relief to the symptoms of T2D, such as a case reported in *The Lancet* in 1925 involving surgery for a peptic ulcer, a diabetic male, and a fairly unique glucose tolerance test involving "a large helping of suet pudding and plenty of syrup".[213] With the advent of more routine bariatric surgery techniques for morbid obesity from the 1950s onward, an increasing amount of evidence of the metabolic benefits of bariatric surgery began to emerge, showing that surgery had the potential to significantly reduce morbidity and mortality,[214, 215] and that the beneficial effects appeared to be independent of the weight loss itself.[216, 217] The results that can be achieved by gastric surgery in terms of diabetes symptoms and glycemic control have prompted one study by Mingrone et al. to state that surgical intervention is superior to other clinical interventions for the treatment of T2D.[218] The role of mitochondrial adaptations following bariatric surgery, and their relationship to improved metabolic health is still to be elucidated, however, this remains an area of fertile research.

Bariatric surgery has become an effective option for the therapeutic control of T2D and obesity.[219, 220] The American Society for Metabolic and Bariatric Surgery has reported that 196,000 procedures were undertaken in the United States in 2015, up from 179,000 in 2013.[221] All bariatric procedures can be undertaken laparoscopically and include Roux-en-Y gastric bypass (RYGB), sleeve gastrectomy (SG), adjustable gastric banding (AGB), and duodenal switch with biliopancretic diversion (DSBD)[222, 223] (see Table 1). The overall mortality rate for all bariatric procedures is about 0.1%,[224] which is less than gallbladder (0.7%[225]) or hip replacement surgery (0.93%[226]). The overall chance of a major complication following bariatric surgery is about 4.3%.[227] Research also has found bariatric surgery improves lifespan with a 30%–40% reduced risk of premature death and improvement in life expectancy of 89%.[228, 229]

TABLE 1 Common Bariatric Surgery Techniques and Proposed Effects

| Technique | Estimated Proportion of Surgeries[a] | Description of Procedure | Proposed Mechanisms and Effects Attributed to All Surgery Types | |
|---|---|---|---|---|
| Roux-en-Y gastric bypass (RYGB) | 18%–34% | Duodenal and proximal jejunal bypass | Gut signaling pathways: Ghrelin, GLP1, GIP, CCK etc. | • Promote satiety<br>• Enhance energy balance |
| Sleeve gastrectomy (SG) | 42%–58% | Stomach resection to ~25% of volume | | |
| Adjustable gastric banding (AGB) | 3%–14% | Silicone ring at gastroesophageal junction (reversible) | Gastric emptying Small intestine transit time | • Decreased absorption |
| Duodenal switch with biliopancreatic diversion (DSBD or BPD/DS) | 0.6%–1% | Resection at duonenal-ileum | Mitochondrial respiration in muscle increased post surgery: SIRT1/PGC1α/ PGC1β/PPARδ | • Mitochondrial biogenesis<br>• Increased energy expenditure<br>• Reduced peripheral insulin resistance |

[a]*Based on 2013–2016 estimates in US by American Society for Metabolic Bariatric Surgery, Estimate of Bariatric Surgery Numbers, 2011–2015 2016 [February 23, 2018] (available from: https://asmbs.org/resources/estimate-of-bariatric-surgery-numbers).*

RYGB is a common but technically challenging procedure that is used in approximately 40% of bariatric procedures.[221, 230] In this procedure, a proximal gastric pouch is constructed (approximately 30mls) and separated from the distal stomach. The gastric pouch is anastomosed to the distal intestinal segment and the proximal intestinal segment is anastomosed to the jejunum. Following this procedure, ingested material flows directly from the gastric pouch into the midjejunum and passes through the Roux limb, thus bypassing the duodenum and proximal jejunum.[222] These structural changes alter signaling pathways between the luminal factors and the intestinal mucosa that are thought to alter physiological process such as hunger, satiety, energy balance, and weight loss.[231–234]

SG recently has become the most common bariatric procedure, especially for weight loss, in the world because of its effectiveness and relatively standardized surgical procedure.[223] In this procedure, most of the stomach is resected, resulting in a tubular stomach with reduced curvature that is greatly resistant to stretching.[222] By removing most of the stomach, the oxyntic mucosa is reduced, resulting in alterations in signaling pathways like those found in RYGB. This procedure also accelerates gastric emptying and exposure of ingested material to the small intestine.[231–233] AGB is a less-common procedure that accounts for about 5% of bariatric procedures in the United States[223] but remains common in Australia and other countries.[235] AGB uses a locking silicone ring placed around the upper stomach just below the gastroesophageal junction. After it is in place, a subcutaneous infusion port is created, allowing saline entry or removal. This results in a 30ml upper gastric pouch that is less invasive than RYGB or SG and is reversible by removing the band.[222] More than 75% of AGB devices are removed, however, because of lack of weight loss, weight gain, and esophagogastric complications.[236] DSBD accounts for only 2% of bariatric procedures performed.[223] In this procedure, the duodenum is divided from the pylorus, which is removed, and the ileum also is divided. The ileum then is attached to the stomach and proximal ileum, delivering enzymes from the pancreas and bile from the liver.[222, 237] Like RYGB and SG, this procedure alters metabolic signaling that alters appetite, decreased nutrient ingestion, and weight loss.

Improvements in fasting glucose, insulin secretion, and insulin signaling have been reported within days and before significant weight loss for both RYGB[215, 238, 239] and SG.[240, 241] A 2004 meta-analysis of 136 studies with more than 22,094 people undergoing bariatric surgery reported a mean weight loss during an approximate 1-year period of 43 kg for RYGB, 29 kg for AGB, and 46 kg for DSBD.[242] SG is a newer technique and not included in that original study, but studies show that weight loss would fall between RYGB and AGB.[243] Long-term studies (>5 years) also have reported a 27% weight loss in RYGB and 13% for AGB compared with 1% for controls.[244–246] In most cases reported, the peak weight loss people experience is within the first year (38% RYGB and 21% AGB). More recently, two randomized trials reported 5-year outcomes comparing RYGB, SG, and DSBD.[218, 247] Interestingly, five years postsurgery RYGB had a bigger total body weight reduction compared to SG (23% vs. 18%) but similar diabetes remission rates (31% vs. 23%) and changes in glycated hemoglobin at five years (–2 vs. –2.5 from baseline).[247] The mechanisms underlying acute diabetes remission and sustained weight loss remain unknown, however, caloric deprivation, improved mitochondrial function, and reduced levels of nonesterified fatty acids all have been proposed.[248] Evidence for the importance of mitochondrial mechanisms in this improvement are studies that show that insulin resistance in skeletal muscle is linked to impairments in mitochondrial oxidation[20] that result in reduced fatty acid oxidation and the accumulation of sphingolipids, long-chain acyl-CoA, and diacylglycerol (DAG).[249, 250]

In a study looking at the acute changes in mitochondrial function following bariatric surgery, Nijhawan et al.[251] studied 20 obese people undergoing either SG or RYGB surgery for weight loss over a 12-week period. Relative to the baseline (pre-surgery measures), an average 30% weight loss was found, resulting in a change in the HOMA-IR of 3 (baseline) to 1.2 (postsurgery).[251] At the same time, they isolated peripheral blood monocytes and took a muscle sample from the vastus lateralis before and 12-weeks following surgery.[251] The bariatric surgery caused an increase in mitochondrial respiration that might form part of a mechanism to reduce weight and improve insulin sensitivity following SG or RYGB. The precise timing of these early changes in mitochondrial function and the underlying mechanisms involved remain unknown, but may help to partly explain the rapid improvements seen in glucose metabolism following bariatric surgery.

In a relatively small study, Vijgen et al.[252] looked at the long-term consequences of bariatric surgery on mitochondrial function. Ten individuals undergoing AGB surgery had body composition assessed and a muscle biopsy taken before and 1-year following surgery. An average of 30 kg of fat mass was lost 1-year following surgery, resulting in an average change of BMI of 41.3 (presurgery) to 29 (postsurgery).[252] Before surgery, mitochondrial function in muscle was decreased compared to lean controls (ADP-stimulated respiration on glutamate/succinate), but 1-year following surgery this result was comparable to lean controls, indicating prolonged improvement in mitochondrial function.[252] The improvements in mitochondrial function most likely are because of changes in signaling pathways postsurgery and not a direct consequence of rapid weight loss because there is limited evidence for mitochondrial capacity changes in skeletal muscle following weight loss through dieting, in spite of improved insulin signaling and reduced intramyocellular lipid content.[112] Therefore, acute (12 weeks[251]) or 1-year following bariatric surgery,[252] skeletal muscle mitochondrial function does improve from baseline, potentially highlighting a mechanism for weight loss and improved metabolism in these patients.

Some effort has been made to look at the long-term energy expenditure of patients following these surgical interventions. In a study conducted by Moehlecke et al.,[253] resting energy expenditure (REE) and body composition was assessed in 30 people who underwent RYGB and followed up over an 18-month period. The mean weight at the start of the study was 128 kg and more than half of this weight was composed of fat. An average weight loss of 26% was found at 6 months postsurgery increasing to 34% at 18 months. Although substantial changes in body composition were reported, mainly because of a reduced total fat mass, the REE decreased only by 405 kcal/d from a presurgery baseline of 2297 kcal/d, suggesting that the mitochondria metabolic set point remains high and might contribute to the continued weight loss.[253] In a subpopulation who had the biggest drop on REE at 6 months, only modest changes in weight loss were found. This suggests that people undergoing bariatric surgery have better weight loss outcomes if the REE set point drops only modestly from baseline; this could be used as a marker of predictive weight loss success.

Other changes in mitochondrial pathways that have been reported following bariatric surgery include changes in genes that control mitochondrial biogenesis such as PGC-1$\beta$, PGC-1$\alpha$, PGC-1$\partial$, and SIRT1, accompanied by increases in mitochondrial markers such as mitofusin 2, porin, and CS,[254, 255] although these effects were significant only in normoglycemic patients.

Exercise has been shown to contribute to improvement in muscle insulin signaling that is associated with changes in mitochondrial function with reduced intramyocellular DAG and ceramide.[39, 256] Coen et al.[257] extended this work to investigate the effect of exercise on people who had bariatric surgery. A total of 101 people underwent an RYGB procedure and were randomly assigned to an exercise or control group. Similar reductions in weight loss during the 6-month exercise intervention were found resulting in approximately 20 kg of fat mass loss between the groups.[257] Both groups showed improvements in insulin sensitivity, but the effect was greater for the exercise group. No change in mitochondrial content was found, but increased mitochondrial respiration and enzyme activity in the muscle was found after 6 months of exercise intervention.[257] The improvements in mitochondrial function in the exercise group might be linked to the more pronounced reduction in the ceramide species in the muscle of the exercise intervention group. Overall reductions in sphingolipids, however, were identified in both groups studied, indicating a common increase in mitochondrial function leading to weight loss.[257] Comparable results were found in a study by Woodlief et al.,[258] with improved insulin sensitivity, but they also found reduced abdominal deep subcutaneous fat in the exercise groups compared to the control, and improved mitochondrial function in muscle in people who underwent RYGB surgery. Likewise, Carnero et al.[259] followed 96 people who underwent RYGB surgery and were randomly assigned to an exercise intervention or control group for 6 months. They found that the most active participants in the exercise group lost more fat mass (more than 22 kg) and had reduced abdominal adipose tissue, however, mitochondrial function was not measured in this study.

All evidence points to bariatric surgery resulting in rapid improvements in metabolism and accelerated weight loss by reducing the total fat mass in obese individuals. It is clear from the limited research available that bariatric surgery alters mitochondrial function, and changes in mitochondrial respiration and activity likely are contributing to increasing energy expenditure that can be further enhanced with moderate to high-level of exercise intervention. If mitochondria have a significant role in weight loss following bariatric surgery, then one interesting avenue of future research could be the addition of agents that promote mitochondrial substrate oxidation postsurgery (e.g., mild mitochondrial uncouplers), with the potential for benefits similar to those found with surgery plus exercise intervention groups.

## 5 CONCLUSION

In the physiological setting of a human patient, it is very difficult to untangle the mechanisms that drive different benefits of a health intervention. What comes first in the etiology of a disease such as T2D? For example, is exercise-induced weight loss driving benefits in glucose metabolism, or is exercise acting independently? Interventions such as gastric surgery that were designed to reduce food intake have turned out to affect an increasingly complex network of multiorgan signaling, mitochondrial dynamics, and gut physiology. More study needs to be done to understand the mechanisms at play in these interventions, both to allow developments of pharmacological agents that might manipulate these pathways, and to understand how to design appropriate interventions that will help all patients to achieve optimal health. In conclusion, analogous to the diseases that they treat, treatments such as lifestyle and surgical interventions are multifactorial and complex in their mechanisms and research is continuing about the benefits of each of these measures in those at risk of, and currently suffering from, T2D and other metabolic disorders. The role of mitochondrial function, however, is a key factor in many of these interventions, and focusing on improving mitochondrial health during the treatment of these diseases will be enlightening.

## References


1. Kremen AJ, Linner JH, Nelson CH. An experimental evaluation of the nutritional importance of proximal and distal small intestine. *Ann Surg* 1954;**140**(3):439–48.
2. Richter EA, Turcotte L, Hespel P, Kiens B. Metabolic responses to exercise. Effects of endurance training and implications for diabetes. *Diabetes Care* 1992;**15**(11):1767–76.
3. Bordenave S, Metz L, Flavier S, Lambert K, Ghanassia E, Dupuy AM, et al. Training-induced improvement in lipid oxidation in type 2 diabetes mellitus is related to alterations in muscle mitochondrial activity. Effect of endurance training in type 2 diabetes. *Diabetes Metab* 2008;**34**(2):162–8.
4. Toledo FG, Menshikova EV, Ritov VB, Azuma K, Radikova Z, DeLany J, et al. Effects of physical activity and weight loss on skeletal muscle mitochondria and relationship with glucose control in type 2 diabetes. *Diabetes* 2007;**56**(8):2142–7.
5. van Tienen FHJ, Praet SFE, de Feyter HM, van den Broek NM, Lindsey PJ, Schoonderwoerd KGC, et al. Physical activity is the key determinant of skeletal muscle mitochondrial function in type 2 diabetes. *J Clin Endocrinol Metab* 2012;**97**(9):3261–9.
6. Colberg SR, Sigal RJ, Yardley JE, Riddell MC, Dunstan DW, Dempsey PC, et al. Physical activity/exercise and diabetes: a position statement of the American Diabetes Association. *Diabetes Care* 2016;**39**(11):2065–79.
7. Tuomilehto J, Lindström J, Eriksson JG, Valle TT, Hämäläinen H, Ilanne-Parikka P, et al. Prevention of type 2 diabetes mellitus by changes in lifestyle among subjects with impaired glucose tolerance. *New Engl J Med* 2001;**344**(18):1343–50.
8. Inzucchi SE, Bergenstal RM, Buse JB, Diamant M, Ferrannini E, Nauck M, et al. Management of hyperglycemia in type 2 diabetes: a patient-centered approach. *Diabetes Care* 2012;**35**(6):1364.
9. Morrison S, Colberg SR, Mariano M, Parson HK, Vinik AI. Balance training reduces falls risk in older individuals with type 2 diabetes. *Diabetes Care* 2010;**33**(4):748–50.
10. Herriott MT, Colberg SR, Parson HK, Nunnold T, Vinik AI. Effects of 8 weeks of flexibility and resistance training in older adults with type 2 diabetes. *Diabetes Care* 2004;**27**(12):2988–9.
11. Ryden L, Grant PJ, Anker SD, Berne C, Cosentino F, Danchin N, et al. ESC guidelines on diabetes, pre-diabetes, and cardiovascular diseases developed in collaboration with the EASD: the Task Force on diabetes, pre-diabetes, and cardiovascular diseases of the European Society of Cardiology (ESC) and developed in collaboration with the European Association for the Study of Diabetes (EASD). *Eur Heart J* 2013;**34**(39):3035–87.
12. Gunton JE, Cheung NW, Davis TM, Zoungas S, Colagiuri S. A new blood glucose management algorithm for type 2 diabetes: a position statement of the Australian Diabetes Society. *Med J Aust* 2014;**201**(11):650–3.
13. Lowell BB, Shulman GI. Mitochondrial dysfunction and type 2 diabetes. *Science* 2005;307.
14. Kelley DE, He J, Menshikova EV, Ritov VB. Dysfunction of mitochondria in human skeletal muscle in type 2 diabetes. *Diabetes* 2002;**51**.
15. Phielix E, Schrauwen-Hinderling VB, Mensink M, Lenaers E, Meex R, Hoeks J, et al. Lower intrinsic ADP-stimulated mitochondrial respiration underlies in vivo mitochondrial dysfunction in muscle of male type 2 diabetic patients. *Diabetes* 2008;**57**(11):2943–9.
16. Ritov VB, Menshikova EV, He J, Ferrell RE, Goodpaster BH, Kelley DE. Deficiency of subsarcolemmal mitochondria in obesity and type 2 diabetes. *Diabetes* 2005;**54**:1.
17. Wallberg-Henriksson H, Gunnarsson R, Henriksson J, DeFronzo R, Felig P, Ostman J, et al. Increased peripheral insulin sensitivity and muscle mitochondrial enzymes but unchanged blood glucose control in type I diabetics after physical training. *Diabetes* 1982;**31**(12):1044–50.
18. Hey-Mogensen M, Højlund K, Vind BF, Wang L, Dela F, Beck-Nielsen H, et al. Effect of physical training on mitochondrial respiration and reactive oxygen species release in skeletal muscle in patients with obesity and type 2 diabetes. *Diabetologia* 2010.

19. Holloszy JO. "Deficiency" of mitochondria in muscle does not cause insulin resistance. *Diabetes* 2013;**62**(4):1036–40.
20. Goodpaster BH. Mitochondrial deficiency is associated with insulin resistance. *Diabetes* 2013;**62**(4):1032–5.
21. Bajpeyi S, Pasarica M, Moro C, Conley K, Jubrias S, Sereda O, et al. Skeletal muscle mitochondrial capacity and insulin resistance in type 2 diabetes. *J Clin Endocrinol Metab* 2011;**96**(4):1160–8.
22. Rimbert V, Boirie Y, Bedu M, Hocquette JF, Ritz P, Morio B. Muscle fat oxidative capacity is not impaired by age but by physical inactivity: association with insulin sensitivity. *FASEB J* 2004;**18**(6):737–9.
23. Phielix E, Meex R, Moonen-Kornips E, Hesselink MKC, Schrauwen P. Exercise training increases mitochondrial content and ex vivo mitochondrial function similarly in patients with type 2 diabetes and in control individuals. *Diabetologia* 2010;**53**(8):1714–21.
24. Nielsen J, Mogensen M, Vind BF, Sahlin K, Hojlund K, Schroder HD, et al. Increased subsarcolemmal lipids in type 2 diabetes: effect of training on localization of lipids, mitochondria, and glycogen in sedentary human skeletal muscle. *Am J Physiol Endocrinol Metab* 2010;**298**(3):E706–13.
25. Hawley JA. Molecular responses to strength and endurance training: are they incompatible? *Appl Physiol Nutr Metab* 2009;34.
26. Stepto NK, Coffey VG, Carey AL, Ponnampalam AP, Canny BJ, Powell D, et al. Global gene expression in skeletal muscle from well-trained strength and endurance athletes. *Med Sci Sports Exerc* 2009;**41**(3):546–65.
27. Mahoney DJ, Parise G, Melov S, Safdar A, Tarnopolsky MA. Analysis of global mRNA expression in human skeletal muscle during recovery from endurance exercise. *FASEB J* 2005;**19**(11):1498–500.
28. Baar K, Wende AR, Jones TE, Marison M, Nolte LA, Chen MAY, et al. Adaptations of skeletal muscle to exercise: rapid increase in the transcriptional coactivator PGC-1. *FASEB J* 2002;**16**(14):1879–86.
29. Holloszy JO, Coyle EF. Adaptations of skeletal muscle to endurance exercise and their metabolic consequences. *J Appl Physiol Respir Environ Exerc Physiol* 1984;**56**(4):831–8.
30. Bruce CR, Thrush AB, Mertz VA, Bezaire V, Chabowski A, Heigenhauser GJ, et al. Endurance training in obese humans improves glucose tolerance and mitochondrial fatty acid oxidation and alters muscle lipid content. *Am J Physiol Endocrinol Metab* 2006;**291**(1):E99–e107.
31. Dumortier M, Brandou F, Perez-Martin A, Fedou C, Mercier J, Brun JF. Low intensity endurance exercise targeted for lipid oxidation improves body composition and insulin sensitivity in patients with the metabolic syndrome. *Diabetes Metab* 2003;**29**(5):509–18.
32. Devries MC, Hamadeh MJ, McCready C, Sischek S, Tarnopolsky MA. Twelve weeks of endurance training increases mitochondrial density and percent IMCL touching mitochondria and alters IMCL storage distribution. *FASEB J* 2008;**22**(1 Suppl). 753.18.
33. Trenell MI, Hollingsworth KG, Lim EL, Taylor R. Increased daily walking improves lipid oxidation without changes in mitochondrial function in type 2 diabetes. *Diabetes Care* 2008;**31**(8):1644.
34. Lantier L, Fentz J, Mounier R, Leclerc J, Treebak JT, Pehmoller C, et al. AMPK controls exercise endurance, mitochondrial oxidative capacity, and skeletal muscle integrity. *FASEB J* 2014;**28**.
35. Richter EA, Ruderman NB. AMPK and the biochemistry of exercise: implications for human health and disease. *Biochem J* 2009;**418**.
36. Winder WW, Holmes BF, Rubink DS, Jensen EB, Chen M, Holloszy JO. Activation of AMP-activated protein kinase increases mitochondrial enzymes in skeletal muscle. *J Appl Physiol (1985)* 2000;**88**(6):2219–26.
37. Leick L, Fentz J, Bienso RS, Knudsen JG, Jeppesen J, Kiens B, et al. PGC-1{alpha} is required for AICAR-induced expression of GLUT4 and mitochondrial proteins in mouse skeletal muscle. *Am J Physiol Endocrinol Metab* 2010;**299**(3):E456–65.
38. Akimoto T, Pohnert SC, Li P, Zhang M, Gumbs C, Rosenberg PB, et al. Exercise stimulates Pgc-1alpha transcription in skeletal muscle through activation of the p38 MAPK pathway. *J Biol Chem* 2005;**280**.
39. Short KR, Vittone JL, Bigelow ML, Proctor DN, Rizza RA, Coenen-Schimke JM, et al. Impact of aerobic exercise training on age-related changes in insulin sensitivity and muscle oxidative capacity. *Diabetes* 2003;**52**(8):1888–96.
40. Hambrecht R, Fiehn E, Yu J, Niebauer J, Weigl C, Hilbrich L, et al. Effects of endurance training on mitochondrial ultrastructure and fiber type distribution in skeletal muscle of patients with stable chronic heart failure. *J Am Coll Cardiol* 1997;**29**(5):1067–73.
41. Jelleyman C, Yates T, O'Donovan G, Gray LJ, King JA, Khunti K, et al. The effects of high-intensity interval training on glucose regulation and insulin resistance: a meta-analysis. *Obes Rev* 2015;**16**(11):942–61.
42. Babraj JA, Vollaard NB, Keast C, Guppy FM, Cottrell G, Timmons JA. Extremely short duration high intensity interval training substantially improves insulin action in young healthy males. *BMC Endocr Disord* 2009;**9**:3.

43. Mann S, Beedie C, Balducci S, Zanuso S, Allgrove J, Bertiato F, et al. Changes in insulin sensitivity in response to different modalities of exercise: a review of the evidence. *Diabetes Metab Res Rev* 2014;**30**.
44. Gibala MJ, McGee SL, Garnham AP, Howlett KF, Snow RJ, Hargreaves M. Brief intense interval exercise activates AMPK and p38 MAPK signaling and increases the expression of PGC-1alpha in human skeletal muscle. *J Appl Physiol (1985)* 2009;**106**(3):929–34.
45. Little JP, Safdar A, Wilkin GP, Tarnopolsky MA, Gibala MJ. A practical model of low-volume high-intensity interval training induces mitochondrial biogenesis in human skeletal muscle: potential mechanisms. *J Physiol* 2010;**588**(Pt 6):1011–22.
46. Hood MS, Little JP, Tarnopolsky MA, Myslik F, Gibala MJ. Low-volume interval training improves muscle oxidative capacity in sedentary adults. *Med Sci Sports Exerc* 2011;**43**(10):1849–56.
47. Larsen S, Danielsen JH, Sondergard SD, Sogaard D, Vigelsoe A, Dybboe R, et al. The effect of high-intensity training on mitochondrial fat oxidation in skeletal muscle and subcutaneous adipose tissue. *Scand J Med Sci Sports* 2015;**25**(1):e59–69.
48. Pesta D, Hoppel F, Macek C, Messner H, Faulhaber M, Kobel C, et al. Similar qualitative and quantitative changes of mitochondrial respiration following strength and endurance training in normoxia and hypoxia in sedentary humans. *Am J Phys Regul Integr Comp Phys* 2011;**301**.
49. Hawley JA, Hargreaves M, Joyner MJ, Zierath JR. Integrative biology of exercise. *Cell* 2014;**159**(4):738–49.
50. Pesta DH, Goncalves RLS, Madiraju AK, Strasser B, Sparks LM. Resistance training to improve type 2 diabetes: working toward a prescription for the future. *Nutr Metab* 2017;**14**(1):24.
51. Fluck M, Hoppeler H. Molecular basis of skeletal muscle plasticity—from gene to form and function. *Rev Physiol Biochem Pharmacol* 2003;**146**:159–216.
52. Ruas JL, White JP, Rao RR, Kleiner S, Brannan KT, Harrison BC, et al. A PGC-1alpha isoform induced by resistance training regulates skeletal muscle hypertrophy. *Cell* 2012;**151**(6):1319–31.
53. Goodman CA. The role of mTORC1 in regulating protein synthesis and skeletal muscle mass in response to various mechanical stimuli. *Rev Physiol Biochem Pharmacol* 2014;**166**.
54. Romero-Arenas S, Martinez-Pascual M, Alcaraz PE. Impact of resistance circuit training on neuromuscular, cardiorespiratory and body composition adaptations in the elderly. *Aging Dis* 2013;**4**.
55. Sparks LM, Johannsen NM, Church TS, Earnest CP, Moonen-Kornips E, Moro C, et al. Nine months of combined training improves ex vivo skeletal muscle metabolism in individuals with type 2 diabetes. *J Clin Endocrinol Metab* 2013;**98**(4):1694–702.
56. Irving BA, Lanza IR, Henderson GC, Rao RR, Spiegelman BM, Nair KS. Combined training enhances skeletal muscle mitochondrial oxidative capacity independent of age. *J Clin Endocrinol Metab* 2015;**100**.
57. Bateman LA, Slentz CA, Willis LH, Shields AT, Piner LW, Bales CW, et al. Comparison of aerobic versus resistance exercise training effects on metabolic syndrome (from the studies of a targeted risk reduction intervention through defined exercise—STRRIDE-AT/RT). *Am J Cardiol* 2011;**108**.
58. Wang L, Mascher H, Psilander N, Blomstrand E, Sahlin K. Resistance exercise enhances the molecular signaling of mitochondrial biogenesis induced by endurance exercise in human skeletal muscle. *J Appl Physiol (1985)* 2011;111.
59. Ormsbee MJ, Thyfault JP, Johnson EA, Kraus RM, Choi MD, Hickner RC. Fat metabolism and acute resistance exercise in trained men. *J Appl Physiol (1985)* 2007;102.
60. Egan B, Zierath JR. Exercise metabolism and the molecular regulation of skeletal muscle adaptation. *Cell Metab* 2013;**17**.
61. Hood DA. Invited review: contractile activity-induced mitochondrial biogenesis in skeletal muscle. *J Appl Physiol* 2001;**90**(3):1137–57.
62. Trewin AJ, Levinger I, Parker L, Shaw CS, Serpiello FR, Anderson MJ, et al. Acute exercise alters skeletal muscle mitochondrial respiration and $H_2O_2$ emission in response to hyperinsulinemic-euglycemic clamp in middle-aged obese men. *PLoS ONE* 2017;**12**(11). e0188421–e21.
63. Taylor CT. Mitochondria and cellular oxygen sensing in the HIF pathway. *Biochem J* 2008;**409**:19–26.
64. Pilegaard H, Saltin B, Neufer PD. Exercise induces transient transcriptional activation of the PGC-1alpha gene in human skeletal muscle. *J Physiol* 2003;**546**:851–8.
65. Little JP, Safdar A, Bishop D, Tarnopolsky MA, Gibala MJ. An acute bout of high-intensity interval training increases the nuclear abundance of PGC-1alpha and activates mitochondrial biogenesis in human skeletal muscle. *Am J Phys Regul Integr Comp Phys* 2011;**300**(6):R1303–10.
66. McGee SL, Hargreaves M. Exercise and myocyte enhancer factor 2 regulation in human skeletal muscle. *Diabetes* 2004;**53**(5):1208–14.

67. Louis E, Raue U, Yang Y, Jemiolo B, Trappe S. Time course of proteolytic, cytokine, and myostatin gene expression after acute exercise in human skeletal muscle. *J Appl Physiol (1985)* 2007;**103**(5):1744–51.
68. Hood DA. Mechanisms of exercise-induced mitochondrial biogenesis in skeletal muscle. *Appl Physiol Nutr Metab* 2009;**34**(3):465–72.
69. Wright DC, Han DH, Garcia-Roves PM, Geiger PC, Jones TE, Holloszy JO. Exercise-induced mitochondrial biogenesis begins before the increase in muscle PGC-1alpha expression. *J Biol Chem* 2007;**282**(1):194–9.
70. McGee SL, Fairlie E, Garnham AP, Hargreaves M. Exercise-induced histone modifications in human skeletal muscle. *J Physiol* 2009;**587**(Pt 24):5951–8.
71. Barres R, Yan J, Egan B, Treebak JT, Rasmussen M, Fritz T, et al. Acute exercise remodels promoter methylation in human skeletal muscle. *Cell Metab* 2012;**15**(3):405–11.
72. Spina RJ, Chi MM, Hopkins MG, Nemeth PM, Lowry OH, Holloszy JO. Mitochondrial enzymes increase in muscle in response to 7-10 days of cycle exercise. *J Appl Physiol (1985)* 1996;**80**(6):2250–4.
73. Samocha-Bonet D, Campbell LV, Mori TA, Croft KD, Greenfield JR, Turner N, et al. Overfeeding reduces insulin sensitivity and increases oxidative stress, without altering markers of mitochondrial content and function in humans. *PLoS ONE* 2012;**7**(5):e36320-e20.
74. Anderson EJ, Lustig ME, Boyle KE, Woodlief TL, Kane DA, Lin CT, et al. Mitochondrial $H_2O_2$ emission and cellular redox state link excess fat intake to insulin resistance in both rodents and humans. *J Clin Invest* 2009;**119**(3):573–81.
75. Ristow M, Zarse K, Oberbach A, Kloting N, Birringer M, Kiehntopf M, et al. Antioxidants prevent health-promoting effects of physical exercise in humans. *Proc Natl Acad Sci U S A* 2009;**106**(21):8665–70.
76. Hesselink MKC, Schrauwen-Hinderling V, Schrauwen P. Skeletal muscle mitochondria as a target to prevent or treat type 2 diabetes mellitus. *Nat Rev Endocrinol* 2016;**12**(11):633–45.
77. Loh K, Deng H, Fukushima A, Cai X, Boivin B, Galic S, et al. Reactive oxygen species enhance insulin sensitivity. *Cell Metab* 2009;**10**(4):260–72.
78. Goncalves RL, Quinlan CL, Perevoshchikova IV, Hey-Mogensen M, Brand MD. Sites of superoxide and hydrogen peroxide production by muscle mitochondria assessed ex vivo under conditions mimicking rest and exercise. *J Biol Chem* 2015;**290**:209–27.
79. Vincent HK, Bourguignon C, Vincent KR. Resistance training lowers exercise-induced oxidative stress and homocysteine levels in overweight and obese older adults. *Obesity (Silver Spring)* 2006;**14**:1921–30.
80. Vinetti G, Mozzini C, Desenzani P, Boni E, Bulla L, Lorenzetti I, et al. Supervised exercise training reduces oxidative stress and cardiometabolic risk in adults with type 2 diabetes: a randomized controlled trial. *Sci Rep* 2015;**5**:9238.
81. Holloszy JO. Skeletal muscle "mitochondrial deficiency" does not mediate insulin resistance. *Am J Clin Nutr* 2008;**89**(1):463S–6S.
82. Stallknecht B, Vinten J, Ploug T, Galbo H. Increased activities of mitochondrial enzymes in white adipose tissue in trained rats. *Am J Phys* 1991;**261**(3 Pt 1):E410–4.
83. Trevellin E, Scorzeto M, Olivieri M, Granzotto M, Valerio A, Tedesco L, et al. Exercise training induces mitochondrial biogenesis and glucose uptake in subcutaneous adipose tissue through eNOS-dependent mechanisms. *Diabetes* 2014;**63**(8):2800–11.
84. Sutherland LN, Bomhof MR, Capozzi LC, Basaraba SA, Wright DC. Exercise and adrenaline increase PGC-1 {alpha} mRNA expression in rat adipose tissue. *J Physiol* 2009;**587**(Pt 7):1607–17.
85. Park H, Kaushik VK, Constant S, Prentki M, Przybytkowski E, Ruderman NB, et al. Coordinate regulation of malonyl-CoA decarboxylase, sn-glycerol-3-phosphate acyltransferase, and acetyl-CoA carboxylase by AMP-activated protein kinase in rat tissues in response to exercise. *J Biol Chem* 2002;**277**(36):32571–7.
86. Ruderman NB, Park H, Kaushik VK, Dean D, Constant S, Prentki M, et al. AMPK as a metabolic switch in rat muscle, liver and adipose tissue after exercise. *Acta Physiol Scand* 2003;**178**(4):435–42.
87. Watt MJ, Holmes AG, Pinnamaneni SK, Garnham AP, Steinberg GR, Kemp BE, et al. Regulation of HSL serine phosphorylation in skeletal muscle and adipose tissue. *Am J Physiol Endocrinol Metab* 2006;**290**(3):E500–8.
88. Camera DM, Anderson MJ, Hawley JA, Carey AL. Short-term endurance training does not alter the oxidative capacity of human subcutaneous adipose tissue. *Eur J Appl Physiol* 2010;**109**(2):307–16.
89. Alvehus M, Boman N, Soderlund K, Svensson MB, Buren J. Metabolic adaptations in skeletal muscle, adipose tissue, and whole-body oxidative capacity in response to resistance training. *Eur J Appl Physiol* 2014;**114**(7):1463–71.
90. Ruetenik A, Barrientos A. Dietary restriction, mitochondrial function and aging: from yeast to humans. *Biochim Biophys Acta (BBA) - Bioenerg* 2015;**1847**(11):1434–47.

91. Most J, Tosti V, Redman LM, Fontana L. Calorie restriction in humans: an update. *Ageing Res Rev* 2017;**39**:36–45.
92. Redman LM, Ravussin E. Caloric restriction in humans: impact on physiological, psychological, and behavioral outcomes. *Antioxid Redox Signal* 2011;**14**(2):275–87.
93. Mattison JA, Colman RJ, Beasley TM, Allison DB, Kemnitz JW, Roth GS, et al. Caloric restriction improves health and survival of rhesus monkeys. *Nat Commun* 2017;**8**:14063.
94. Bevilacqua L, Ramsey JJ, Hagopian K, Weindruch R, Harper ME. Effects of short- and medium-term calorie restriction on muscle mitochondrial proton leak and reactive oxygen species production. *Am J Physiol Endocrinol Metab* 2004;**286**(5):E852–61.
95. Lambert AJ, Merry BJ. Effect of caloric restriction on mitochondrial reactive oxygen species production and bioenergetics: reversal by insulin. *Am J Phys Regul Integr Comp Phys* 2004;**286**(1):R71–9.
96. Lanza IR, Zabielski P, Klaus KA, Morse DM, Heppelmann CJ, Bergen 3rd HR, et al. Chronic caloric restriction preserves mitochondrial function in senescence without increasing mitochondrial biogenesis. *Cell Metab* 2012;**16**(6):777–88.
97. Hempenstall S, Page MM, Wallen KR, Selman C. Dietary restriction increases skeletal muscle mitochondrial respiration but not mitochondrial content in C57BL/6 mice. *Mech Ageing Dev* 2012;**133**(1):37–45.
98. Nisoli E, Tonello C, Cardile A, Cozzi V, Bracale R, Tedesco L, et al. Calorie restriction promotes mitochondrial biogenesis by inducing the expression of eNOS. *Science* 2005;**310**(5746):314–7.
99. López-Lluch G, Hunt N, Jones B, Zhu M, Jamieson H, Hilmer S, et al. Calorie restriction induces mitochondrial biogenesis and bioenergetic efficiency. *Proc Natl Acad Sci U S A* 2006;**103**(6):1768–73.
100. Luevano-Martinez LA, Forni MF, Peloggia J, Watanabe IS, Kowaltowski AJ. Calorie restriction promotes cardiolipin biosynthesis and distribution between mitochondrial membranes. *Mech Ageing Dev* 2017;**162**:9–17.
101. Cassano P, Sciancalepore AG, Lezza AM, Leeuwenburgh C, Cantatore P, Gadaleta MN. Tissue-specific effect of age and caloric restriction diet on mitochondrial DNA content. *Rejuvenation Res* 2006;**9**(2):211–4.
102. Drew B, Phaneuf S, Dirks A, Selman C, Gredilla R, Lezza A, et al. Effects of aging and caloric restriction on mitochondrial energy production in gastrocnemius muscle and heart. *Am J Physiol Regul Integr Comp Physiol* 2003;**284**(2):R474–80.
103. Hancock CR, Han DH, Higashida K, Kim SH, Holloszy JO. Does calorie restriction induce mitochondrial biogenesis? A reevaluation. *FASEB J* 2011;**25**(2):785–91.
104. Kim KE, Jung Y, Min S, Nam M, Heo RW, Jeon BT, et al. Caloric restriction of db/db mice reverts hepatic steatosis and body weight with divergent hepatic metabolism. *Sci Rep* 2016;**6**:30111.
105. Tanajak P, Pintana H, Siri-Angkul N, Khamseekaew J, Apaijai N, Chattipakorn SC, et al. Vildagliptin and caloric restriction for cardioprotection in pre-diabetic rats. *J Endocrinol* 2017;**232**(2):189–204.
106. Thom G, Lean M. Is There an Optimal Diet for Weight Management and Metabolic Health? *Gastroenterology* 2017;**152**(7):1739–51.
107. Holloszy JO, Fontana L. Caloric restriction in humans. *Exp Gerontol* 2007;**42**(8):709–12.
108. Simoneau JA, Veerkamp JH, Turcotte LP, Kelley DE. Markers of capacity to utilize fatty acids in human skeletal muscle: relation to insulin resistance and obesity and effects of weight loss. *FASEB J* 1999;**13**(14):2051–60.
109. Larson-Meyer DE, Newcomer BR, Hunter GR, McLean JE, Hetherington HP, Weinsier RL. Effect of weight reduction, obesity predisposition, and aerobic fitness on skeletal muscle mitochondrial function. *Am J Physiol Endocrinol Metab* 2000;**278**(1):E153–61.
110. Kern PA, Simsolo RB, Fournier M. Effect of weight loss on muscle fiber type, fiber size, capillarity, and succinate dehydrogenase activity in humans. *J Clin Endocrinol Metab* 1999;**84**(11):4185–90.
111. Civitarese AE, Carling S, Heilbronn LK, Hulver MH, Ukropcova B, Deutsch WA, et al. Calorie restriction increases muscle mitochondrial biogenesis in healthy humans. *PLoS Med* 2007;**4**(3):e76.
112. Toledo FG, Menshikova EV, Azuma K, Radikova Z, Kelley CA, Ritov VB, et al. Mitochondrial capacity in skeletal muscle is not stimulated by weight loss despite increases in insulin action and decreases in intramyocellular lipid content. *Diabetes* 2008;**57**(4):987–94.
113. Menshikova EV, Ritov VB, Dube JJ, Amati F, Stefanovic-Racic M, Toledo FGS, et al. Calorie restriction-induced weight loss and exercise have differential effects on skeletal muscle mitochondria despite similar effects on insulin sensitivity. *J Gerontol A Biol Sci Med Sci* 2017;**73**(1):81–7.
114. Johnson ML, Distelmaier K, Lanza IR, Irving BA, Robinson MM, Konopka AR, et al. Mechanism by which caloric restriction improves insulin sensitivity in sedentary obese adults. *Diabetes* 2016;**65**(1):74–84.

115. Rabol R, Svendsen PF, Skovbro M, Boushel R, Haugaard SB, Schjerling P, et al. Reduced skeletal muscle mitochondrial respiration and improved glucose metabolism in nondiabetic obese women during a very low calorie dietary intervention leading to rapid weight loss. *Metabolism* 2009;**58**(8):1145–52.
116. Ingram DK, de Cabo R. Calorie restriction in rodents: caveats to consider. *Ageing Res Rev* 2017;**39**:15–28.
117. Seyssel K, Alligier M, Meugnier E, Chanseaume E, Loizon E, Canto C, et al. Regulation of energy metabolism and mitochondrial function in skeletal muscle during lipid overfeeding in healthy men. *J Clin Endocrinol Metab* 2014;**99**(7):E1254–62.
118. Gibson AA, Seimon RV, Lee CM, Ayre J, Franklin J, Markovic TP, et al. Do ketogenic diets really suppress appetite? A systematic review and meta-analysis. *Obes Rev* 2015;**16**(1):64–76.
119. Rho JM. How does the ketogenic diet induce anti-seizure effects? *Neurosci Lett* 2017;**637**:4–10.
120. Paoli A, Rubini A, Volek JS, Grimaldi KA. Beyond weight loss: a review of the therapeutic uses of very-low-carbohydrate (ketogenic) diets. *Eur J Clin Nutr* 2013;**67**(8):789–96.
121. Vidali S, Aminzadeh S, Lambert B, Rutherford T, Sperl W, Kofler B, et al. Mitochondria: the ketogenic diet—a metabolism-based therapy. *Int J Biochem Cell Biol* 2015;**63**:55–9.
122. Hyatt HW, Kephart WC, Holland AM, Mumford P, Mobley CB, Lowery RP, et al. A ketogenic diet in rodents elicits improved mitochondrial adaptations in response to resistance exercise training compared to an isocaloric western diet. *Front Physiol* 2016;**7**:533.
123. Kephart WC, Mumford PW, Mao X, Romero MA, Hyatt HW, Zhang Y, et al. The 1-week and 8-month effects of a ketogenic diet or ketone salt supplementation on multi-organ markers of oxidative stress and mitochondrial function in rats. *Nutrients* 2017;**9**:9.
124. Hall KD, Chen KY, Guo J, Lam YY, Leibel RL, Mayer LE, et al. Energy expenditure and body composition changes after an isocaloric ketogenic diet in overweight and obese men. *Am J Clin Nutr* 2016;**104**(2):324–33.
125. McSwiney FT, Wardrop B, Hyde PN, Lafountain RA, Volek JS, Doyle L. Keto-adaptation enhances exercise performance and body composition responses to training in endurance athletes. *Metabolism* 2017;**81**:25–34.
126. Volek JS, Freidenreich DJ, Saenz C, Kunces LJ, Creighton BC, Bartley JM, et al. Metabolic characteristics of keto-adapted ultra-endurance runners. *Metabolism* 2016;**65**(3):100–10.
127. Sumithran P, Proietto J. Ketogenic diets for weight loss: a review of their principles, safety and efficacy. *Obes Res Clin Pract* 2008;**2**(1):i–ii.
128. Tinsley GM, La Bounty PM. Effects of intermittent fasting on body composition and clinical health markers in humans. *Nutr Rev* 2015;**73**(10):661–74.
129. Antoni R, Johnston KL, Collins AL, Robertson MD. Effects of intermittent fasting on glucose and lipid metabolism. *Proc Nutr Soc* 2017;**76**(3):361–8.
130. Boutant M, Kulkarni SS, Joffraud M, Raymond F, Metairon S, Descombes P, et al. SIRT1 gain of function does not mimic or enhance the adaptations to intermittent fasting. *Cell Rep* 2016;**14**(9):2068–75.
131. Chausse B, Vieira-Lara MA, Sanchez AB, Medeiros MH, Kowaltowski AJ. Intermittent fasting results in tissue-specific changes in bioenergetics and redox state. *PLoS ONE* 2015;**10**(3):e0120413.
132. Cerqueira FM, da Cunha FM, Caldeira da Silva CC, Chausse B, Romano RL, Garcia CC, et al. Long-term intermittent feeding, but not caloric restriction, leads to redox imbalance, insulin receptor nitration, and glucose intolerance. *Free Radic Biol Med* 2011;**51**(7):1454–60.
133. Heilbronn LK, Civitarese AE, Bogacka I, Smith SR, Hulver M, Ravussin E. Glucose tolerance and skeletal muscle gene expression in response to alternate day fasting. *Obes Res* 2005;**13**(3):574–81.
134. Centers for Disease C, Prevention, National Center for Chronic Disease P, Health P, Office on S, Health. Publications and Reports of the Surgeon General. *How tobacco smoke causes disease: the biology and behavioral basis for smoking-attributable disease: a report of the surgeon general.* Atlanta (GA): Centers for Disease Control and Prevention (US); 2010.
135. Centers for Disease C, Prevention, National Center for Chronic Disease P, Health P, Office on S, Health. Publications and Reports of the Surgeon General. *The health consequences of smoking—50 years of progress: a report of the surgeon general.* Atlanta (GA): Centers for Disease Control and Prevention; 2014.
136. Nathan DM, Meigs J, Singer DE. The epidemiology of cardiovascular disease in type 2 diabetes mellitus: how sweet it is ... or is it? *Lancet* 1997;**350**(Suppl. 1):Si4–9.
137. Kengne AP, Turnbull F, MacMahon S. The Framingham study, diabetes mellitus and cardiovascular disease: turning back the clock. *Prog Cardiovasc Dis* 2010;**53**(1):45–51.
138. Pan A, Wang Y, Talaei M, Hu FB, Wu T. Relation of active, passive, and quitting smoking with incident type 2 diabetes: a systematic review and meta-analysis. *Lancet Diabetes Endocrinol* 2015;**3**(12):958–67.

139. Willi C, Bodenmann P, Ghali WA, Faris PD, Cornuz J. Active smoking and the risk of type 2 diabetes: a systematic review and meta-analysis. *JAMA* 2007;**298**(22):2654–64.
140. Xie XT, Liu Q, Wu J, Wakui M. Impact of cigarette smoking in type 2 diabetes development. *Acta Pharmacol Sin* 2009;**30**(6):784–7.
141. Friedman AJ, Ravnikar VA, Barbieri RL. Serum steroid hormone profiles in postmenopausal smokers and nonsmokers. *Fertil Steril* 1987;**47**(3):398–401.
142. Canoy D, Wareham N, Luben R, Welch A, Bingham S, Day N, et al. Cigarette smoking and fat distribution in 21,828 British men and women: a population-based study. *Obes Res* 2005;**13**(8):1466–75.
143. Arnson Y, Shoenfeld Y, Amital H. Effects of tobacco smoke on immunity, inflammation and autoimmunity. *J Autoimmun* 2010;**34**(3):J258–65.
144. Neves CD, Lacerda AC, Lage VK, Lima LP, Tossige-Gomes R, Fonseca SF, et al. Oxidative stress and skeletal muscle dysfunction are present in healthy smokers. *Braz J Med Biol Res* 2016;**49**(11):e5512.
145. Madsbad S, McNair P, Christensen MS, Christiansen C, Faber OK, Binder C, et al. Influence of smoking on insulin requirement and metabolic status in diabetes mellitus. *Diabetes Care* 1980;**3**(1):41–3.
146. Bush T, Lovejoy JC, Deprey M, Carpenter KM. The effect of tobacco cessation on weight gain, obesity, and diabetes risk. *Obesity (Silver Spring)* 2016;**24**(9):1834–41.
147. Gairola C, Aleem MI. Cigarette smoke: effect of aqueous and nonaqueous fractions on mitochondrial function. *Nature* 1973;**241**(5387):287–8.
148. Gvozdjakova A, Bada V, Sany L, Kucharska J, Kruty F, Bozek P, et al. Smoke cardiomyopathy: disturbance of oxidative processes in myocardial mitochondria. *Cardiovasc Res* 1984;**18**(4):229–32.
149. Gvozdjakova A, Kucharska J, Gvozdjak J. Effect of smoking on the oxidative processes of cardiomyocytes. *Cardiology* 1992;**81**(2-3):81–4.
150. Gvozdjak J, Gvozdjakova A, Kucharska J, Bada V, Kovalikova V, Zachar A. Metabolic disorders of cardiac muscle in alcoholic and smoke cardiomyopathy. *Cor Vasa* 1989;**31**(4):312–20.
151. Orlander J, Kiessling KH, Larsson L. Skeletal muscle metabolism, morphology and function in sedentary smokers and nonsmokers. *Acta Physiol Scand* 1979;**107**(1):39–46.
152. Larsson L, Orlander J. Skeletal muscle morphology, metabolism and function in smokers and non-smokers. A study on smoking-discordant monozygous twins. *Acta Physiol Scand* 1984;**120**(3):343–52.
153. Smith PR, Cooper JM, Govan GG, Harding AE, Schapira AH. Smoking and mitochondrial function: a model for environmental toxins. *Q J Med* 1993;**86**(10):657–60.
154. Miró OS, Alonso JR, Jarreta D, Casademont J, Urbano-Márquez AL, Cardellach F. Smoking disturbs mitochondrial respiratory chain function and enhances lipid peroxidation on human circulating lymphocytes. *Carcinogenesis* 1999;**20**(7):1331–6.
155. Ludwig PW, Hoidal JR. Alterations in leukocyte oxidative metabolism in cigarette smokers. *Am Rev Respir Dis* 1982;**126**(6):977–80.
156. Hofstetter A, Schutz Y, Jequier E, Wahren J. Increased 24-hour energy expenditure in cigarette smokers. *N Engl J Med* 1986;**314**(2):79–82.
157. Reilly M, Delanty N, Lawson JA, FitzGerald GA. Modulation of oxidant stress in vivo in chronic cigarette smokers. *Circulation* 1996;**94**(1):19–25.
158. Alonso JR, Cardellach F, Casademont J, Miro O. Reversible inhibition of mitochondrial complex IV activity in PBMC following acute smoking. *Eur Respir J* 2004;**23**(2):214–8.
159. Cardellach F, Alonso JR, López S, Casademont J, Miró O. Effect of smoking cessation on mitochondrial respiratory chain function. *J Toxicol Clin Toxicol* 2003;**41**(3):223–8.
160. Meyer A, Zoll J, Charles AL, Charloux A, de Blay F, Diemunsch P, et al. Skeletal muscle mitochondrial dysfunction during chronic obstructive pulmonary disease: central actor and therapeutic target. *Exp Physiol* 2013;**98**(6):1063–78.
161. Degens H, Gayan-Ramirez G, Hees HWH. Smoking-induced skeletal muscle dysfunction. From evidence to mechanisms. *Am J Respir Crit Care Med* 2015;**191**(6):620–5.
162. Alonso J-R, Cardellach F, López S, Casademont J, Miró Ò. Carbon monoxide specifically inhibits cytochrome C oxidase of human mitochondrial respiratory Chain. *Pharmacol Toxicol* 2003;**93**(3):142–6.
163. Barreiro E, Peinado VI, Galdiz JB, Ferrer E, Marin-Corral J, Sanchez F, et al. Cigarette smoke-induced oxidative stress: a role in chronic obstructive pulmonary disease skeletal muscle dysfunction. *Am J Respir Crit Care Med* 2010;**182**(4):477–88.

164. Hoffmann RF, Zarrintan S, Brandenburg SM, Kol A, HGD B, Jafari S, et al. Prolonged cigarette smoke exposure alters mitochondrial structure and function in airway epithelial cells. *Respir Res* 2013;**14**(1):97.
165. Tan D, Goerlitz DS, Dumitrescu RG, Han D, Seillier-Moiseiwitsch F, Spernak SM, et al. Associations between cigarette smoking and mitochondrial DNA abnormalities in buccal cells. *Carcinogenesis* 2008;**29**(6):1170–7.
166. Bruin JE, Petre MA, Raha S, Morrison KM, Gerstein HC, Holloway AC. Fetal and neonatal nicotine exposure in Wistar rats causes progressive pancreatic mitochondrial damage and beta cell dysfunction. *PLoS ONE* 2008;**3**(10):e3371.
167. Wilson R, Wahl S, Pfeiffer L, Ward-Caviness CK, Kunze S, Kretschmer A, et al. The dynamics of smoking-related disturbed methylation: a two time-point study of methylation change in smokers, non-smokers and former smokers. *BMC Genomics* 2017;**18**(1):805.
168. Komiyama M, Wada H, Ura S, Yamakage H, Satoh-Asahara N, Shimatsu A, et al. Analysis of factors that determine weight gain during smoking cessation therapy. *PLoS ONE* 2013;**8**(8):e72010.
169. Tian J, Venn A, Otahal P, Gall S. The association between quitting smoking and weight gain: a systematic review and meta-analysis of prospective cohort studies. *Obes Rev* 2015;**16**(10):883–901.
170. Le Boudec J, Marques-Vidal P, Cornuz J, Clair C. Smoking cessation and the incidence of pre-diabetes and type 2 diabetes: a cohort study. *J Diabetes Complicat* 2016;**30**(1):43–8.
171. Filozof C, Fernandez Pinilla MC, Fernandez-Cruz A. Smoking cessation and weight gain. *Obes Rev* 2004;**5**(2):95–103.
172. Yeh HC, Duncan BB, Schmidt MI, Wang NY, Brancati FL. Smoking, smoking cessation, and risk for type 2 diabetes mellitus: a cohort study. *Ann Intern Med* 2010;**152**(1):10–7.
173. Harris KK, Zopey M, Friedman TC. Metabolic effects of smoking cessation. *Nat Rev Endocrinol* 2016;**12**(5):299–308.
174. Room R, Babor T, Rehm J. Alcohol and public health. *Lancet* 2005;**365**(9458):519–30.
175. Mokdad AH, Marks JS, Stroup DF, Gerberding JL. Actual causes of death in the United States, 2000. *JAMA* 2004;**291**(10):1238–45.
176. Krebs HA, Freedland RA, Hems R, Stubbs M. Inhibition of hepatic gluconeogenesis by ethanol. *Biochem J* 1969;**112**(1):117–24.
177. Han D, Ybanez MD, Johnson HS, McDonald JN, Mesropyan L, Sancheti H, et al. Dynamic adaptation of liver mitochondria to chronic alcohol feeding in mice: biogenesis, remodeling, and functional alterations. *J Biol Chem* 2012;**287**(50):42165–79.
178. Denny CH, Serdula MK, Holtzman D, Nelson DE. Physician advice about smoking and drinking: are U.S. adults being informed? *Am J Prev Med* 2003;**24**(1):71–4.
179. Babor TF, Higgins-Biddle J, Saunders JB, Monteiro M. *The alcohol use disorders identification test: guideline for use in primary care*; 2001. p. 91–105.
180. Abel EL, Kruger ML, Friedl J. How do physicians define "light," "moderate," and "heavy" drinking? *Alcohol Clin Exp Res* 1998;**22**(5):979–84.
181. Frost-Pineda K, VanSusteren T, Gold MS. Are physicians and medical students prepared to educate patients about alcohol consumption? *J Addict Dis* 2004;**23**(2):1–13.
182. Frank E, Elon L, Naimi T, Brewer R. Alcohol consumption and alcohol counselling behaviour among US medical students: cohort study. *BMJ* 2008;**337**:a2155.
183. Kaminsky YG, Kosenko EA. Blood glucose and liver glycogen in the rat. Effects of chronic ethanol consumption and its withdrawal on the diurnal rhythms. *FEBS Lett* 1986;**200**(1):217–20.
184. Djousse L, Driver JA, Gaziano JM, Buring JE, Lee IM. Association between modifiable lifestyle factors and residual lifetime risk of diabetes. *Nutr Metab Cardiovasc Dis* 2013;**23**(1):17–22.
185. Freiberg MS, Cabral HJ, Heeren TC, Vasan RS, Curtis Ellison R. Alcohol consumption and the prevalence of the metabolic syndrome in the US: a cross-sectional analysis of data from the Third National Health and Nutrition Examination Survey. *Diabetes Care* 2004;**27**(12):2954–9.
186. de Vegt F, Dekker JM, Groeneveld WJ, Nijpels G, Stehouwer CD, Bouter LM, et al. Moderate alcohol consumption is associated with lower risk for incident diabetes and mortality: the Hoorn Study. *Diabetes Res Clin Pract* 2002;**57**(1):53–60.
187. Conigrave KM, Hu BF, Camargo CA, Stampfer MJ, Willett WC, Rimm EB. A prospective study of drinking patterns in relation to risk of type 2 diabetes among men. *Diabetes* 2001;**50**(10):2390.
188. Lim J, Lee JA, Cho HJ. Association of alcohol drinking patterns with presence of impaired fasting glucose and diabetes mellitus among South Korean adults. *J Epidemiol* 2018;**28**(3):117–24.

189. Yoon YS, Oh SW, Baik HW, Park HS, Kim WY. Alcohol consumption and the metabolic syndrome in Korean adults: the 1998 Korean National Health and Nutrition Examination Survey. *Am J Clin Nutr* 2004;**80**(1):217–24.
190. Wei M, Gibbons LW, Mitchell TL, Kampert JB, Blair SN. Alcohol intake and incidence of type 2 diabetes in men. *Diabetes Care* 2000;**23**(1):18–22.
191. Carlsson S, Hammar N, Grill V, Kaprio J. Alcohol consumption and the incidence of type 2 diabetes. *Diabetes Care* 2003;**26**(10):2785.
192. Stoutenberg M, Lee DC, Sui X, Hooker S, Horigian V, Perrino T, et al. Prospective study of alcohol consumption and the incidence of the metabolic syndrome in US men. *Br J Nutr* 2013;**110**(5):901–10.
193. Shuval K, Finley CE, Chartier KG, Balasubramanian BA, Gabriel KP, Barlow CE. Cardiorespiratory fitness, alcohol intake, and metabolic syndrome incidence in men. *Med Sci Sports Exerc* 2012;**44**(11):2125–31.
194. Fan AZ, Russell M, Naimi T, Li Y, Liao Y, Jiles R, et al. Patterns of alcohol consumption and the metabolic syndrome. *J Clin Endocrinol Metab* 2008;**93**(10):3833–8.
195. Fulop M. Alcoholism, ketoacidosis, and lactic acidosis. *Diabetes Metab Rev* 1989;**5**(4):365–78.
196. Tsai CS, Oke TO, Tam CW, Olubadewo JO, Ochillo RF. The effect of regular alcohol use on the management of non-insulin diabetes mellitus. *Cell Mol Biol (Noisy-le-grand)* 2003;**49**(8):1327–32.
197. Yang AL, Vadhavkar S, Singh G, Omary MB. Epidemiology of alcohol-related liver and pancreatic disease in the United States. *Arch Intern Med* 2008;**168**(6):649–56.
198. Garcia-Ruiz C, Kaplowitz N, Fernandez-Checa JC. Role of mitochondria in alcoholic liver disease. *Curr Pathobiol Rep* 2013;**1**(3):159–68.
199. Hoek JB, Cahill A, Pastorino JG. Alcohol and mitochondria: a dysfunctional relationship. *Gastroenterology* 2002;**122**(7):2049–63.
200. Bailey SM, Patel VB, Young TA, Asayama K, Cunningham CC. Chronic ethanol consumption alters the glutathione/glutathione peroxidase-1 system and protein oxidation status in rat liver. *Alcohol Clin Exp Res* 2001;**25**(5):726–33.
201. Venkatraman A, Shiva S, Davis AJ, Bailey SM, Brookes PS, Darley-Usmar VM. Chronic alcohol consumption increases the sensitivity of rat liver mitochondrial respiration to inhibition by nitric oxide. *Hepatology* 2003;**38**(1):141–7.
202. Wieland P, Lauterburg BH. Oxidation of mitochondrial proteins and DNA following administration of ethanol. *Biochem Biophys Res Commun* 1995;**213**(3):815–9.
202a. Coleman WB, Cahill A, Ivester P, Cunningham CC. Differential effects of ethanol consumption on synthesis of cytoplasmic and mitochondrial encoded subunits of the ATP synthase. *Alcohol Clin Exp Res* 1994;**18**(4):947–50.
203. Manzo-Avalos S, Saavedra-Molina A. Cellular and mitochondrial effects of alcohol consumption. *Int J Environ Res Public Health* 2010;**7**(12):4281–304.
204. Cederbaum AI, Lieber CS, Rubin E. Effect of chronic ethanol consumption and acetaldehyde on partial reactions of oxidative phosphorylation and $CO_2$ production from citric acid cycle intermediates. *Arch Biochem Biophys* 1976;**176**(2):525–38.
205. Gvozdjak A, Bada V, Kruty F, Niederland TR, Gvozdjak J. Effect of ethanol on the metabolism of the myocardium and its relationship to development of alcoholic myocardiopathy. *Cardiology* 1973;**58**(5):290–7.
206. Hirano T, Kaplowitz N, Tsukamoto H, Kamimura S, Fernandez-Checa JC. Hepatic mitochondrial glutathione depletion and progression of experimental alcoholic liver disease in rats. *Hepatology* 1992;**16**(6):1423–7.
207. Ma Z, Zhang Y, Li Q, Xu M, Bai J, Wu S. Resveratrol improves alcoholic fatty liver disease by downregulating HIF-1α expression and mitochondrial ROS production. *PLoS ONE* 2017;**12**(8):e0183426.
208. Bonet-Ponce L, Saez-Atienzar S, da Casa C, Flores-Bellver M, Barcia JM, Sancho-Pelluz J, et al. On the mechanism underlying ethanol-induced mitochondrial dynamic disruption and autophagy response. *Biochim Biophys Acta Mol basis Dis* 2015;**1852**(7):1400–9.
209. Videla L, Bernstein J, Israel Y. Metabolic alterations produced in the liver by chronic ethanol administration. Increased oxidative capacity. *Biochem J* 1973;**134**(2):507–14.
210. Weishaar R, Sarma JS, Maruyama Y, Fischer R, Bertuglia S, Bing RJ. Reversibility of mitochondrial and contractile changes in the myocardium after cessation of prolonged ethanol intake. *Am J Cardiol* 1977;**40**(4):556–62.
211. Eversole R, Beuving L, Ulrich R. Ultrastructural alterations in ileal M cells of rats after chronic ethanol ingestion: reversal after cessation of ethanol. *J Stud Alcohol* 1992;**53**(5):519–23.
212. Rubino F, Nathan DM, Eckel RH, Schauer PR, Alberti KGMM, Zimmet PZ, et al. Metabolic surgery in the treatment algorithm for type 2 diabetes: a joint statement by International Diabetes Organizations. *Diabetes Care* 2016;**39**(6):861.

213. Leyton O. Diabetes and operation. *Lancet* 1925;**206**(5336):1162–3.
214. MacDonald Jr. KG, Long SD, Swanson MS, Brown BM, Morris P, Dohm GL, et al. The gastric bypass operation reduces the progression and mortality of non-insulin-dependent diabetes mellitus. *J Gastrointest Surg* 1997;**1**(3):213–20. discussion 20.
215. Pories WJ, Swanson MS, MacDonald KG, Long SB, Morris PG, Brown BM, et al. Who would have thought it? An operation proves to be the most effective therapy for adult-onset diabetes mellitus. *Ann Surg* 1995;**222**(3):339–50. discussion 50–2.
216. Rubino F, Gagner M. Potential of surgery for curing type 2 diabetes mellitus. *Ann Surg* 2002;**236**(5):554–9.
217. Rubino F, Marescaux J. Effect of duodenal-jejunal exclusion in a non-obese animal model of type 2 diabetes: a new perspective for an old disease. *Ann Surg* 2004;**239**(1):1–11.
218. Mingrone G, Panunzi S, De Gaetano A, Guidone C, Iaconelli A, Nanni G, et al. Bariatric-metabolic surgery versus conventional medical treatment in obese patients with type 2 diabetes: 5 year follow-up of an open-label, single-centre, randomised controlled trial. *Lancet* 2015;**386**(9997):964–73.
219. Buchwald H, Estok R, Fahrbach K, Banel D, Jensen MD, Pories WJ, et al. Weight and type 2 diabetes after bariatric surgery: systematic review and meta-analysis. *Am J Med* 2009;**122**(3):248–56. e5.
220. Kashyap SR, Bhatt DL, Schauer PR, Investigators S. Bariatric surgery vs. advanced practice medical management in the treatment of type 2 diabetes mellitus: rationale and design of the surgical therapy and medications potentially eradicate diabetes efficiently trial (STAMPEDE). *Diabetes Obes Metab* 2010;**12**(5):452–4.
221. Surgery ASfMaB. *Estimate of bariatric surgery numbers, 2011–2015 2016 [February 23 2018]*. Available from: https://asmbs.org/resources/estimate-of-bariatric-surgery-numbers.
222. Hanipah ZN, Schauer PR. Surgical treatment of obesity and diabetes. *Gastrointest Endosc Clin N Am* 2017;**27**(2):191–211.
223. Khorgami Z, Shoar S, Andalib A, Aminian A, Brethauer SA, Schauer PR. Trends in utilization of bariatric surgery, 2010–2014: sleeve gastrectomy dominates. *Surg Obes Relat Dis* 2017;**13**(5):774–8.
224. (AHRQ) AfHRaQ. *Statistical brief #23. Bariatric surgery utilization and outcomes in 1998 and 2004*; 2007.
225. Dolan JP, Diggs BS, Sheppard BC, Hunter JG. The national mortality burden and significant factors associated with open and laparoscopic cholecystectomy: 1997-2006. *J Gastrointest Surg* 2009;**13**(12):2292–301.
226. Pedersen AB, Baron JA, Overgaard S, Johnsen SP. Short- and long-term mortality following primary total hip replacement for osteoarthritis: a Danish nationwide epidemiological study. *J Bone Joint Surg (Br)* 2011;**93**(2):172–7.
227. Longitudinal Assessment of Bariatric Surgery C, Flum DR, Belle SH, King WC, Wahed AS, Berk P, et al. Perioperative safety in the longitudinal assessment of bariatric surgery. *N Engl J Med* 2009;**361**(5):445–54.
228. Sjostrom L, Narbro K, Sjostrom CD, Karason K, Larsson B, Wedel H, et al. Effects of bariatric surgery on mortality in Swedish obese subjects. *N Engl J Med* 2007;**357**(8):741–52.
229. Adams TD, Gress RE, Smith SC, Halverson RC, Simper SC, Rosamond WD, et al. Long-term mortality after gastric bypass surgery. *N Engl J Med* 2007;**357**(8):753–61.
230. Angrisani L, Santonicola A, Iovino P, Vitiello A, Zundel N, Buchwald H, et al. Bariatric surgery and endoluminal procedures: IFSO worldwide survey 2014. *Obes Surg* 2017;**27**(9):2279–89.
231. Miras AD, le Roux CW. Mechanisms underlying weight loss after bariatric surgery. *Nat Rev Gastroenterol Hepatol* 2013;**10**(10):575–84.
232. Madsbad S, Dirksen C, Holst JJ. Mechanisms of changes in glucose metabolism and bodyweight after bariatric surgery. *Lancet Diabetes Endocrinol* 2014;**2**(2):152–64.
233. Arble DM, Sandoval DA, Seeley RJ. Mechanisms underlying weight loss and metabolic improvements in rodent models of bariatric surgery. *Diabetologia* 2015;**58**(2):211–20.
234. Mahawar KK, Sharples AJ. Contribution of malabsorption to weight loss after Roux-en-Y gastric bypass: a systematic review. *Obes Surg* 2017;**27**(8):2194–206.
235. O'Brien PE, Brown WA, Dixon JB. Obesity, weight loss and bariatric surgery. *Med J Aust* 2005;**183**(6):310–4.
236. Aarts EO, Dogan K, Koehestanie P, Aufenacker TJ, Janssen IM, Berends FJ. Long-term results after laparoscopic adjustable gastric banding: a mean fourteen year follow-up study. *Surg Obes Relat Dis* 2014;**10**(4):633–40.
237. Marceau P, Hould FS, Simard S, Lebel S, Bourque RA, Potvin M, et al. Biliopancreatic diversion with duodenal switch. *World J Surg* 1998;**22**(9):947–54.
238. Isbell JM, Tamboli RA, Hansen EN, Saliba J, Dunn JP, Phillips SE, et al. The importance of caloric restriction in the early improvements in insulin sensitivity after Roux-en-Y gastric bypass surgery. *Diabetes Care* 2010;**33**(7):1438–42.

239. Kashyap SR, Daud S, Kelly KR, Gastaldelli A, Win H, Brethauer S, et al. Acute effects of gastric bypass versus gastric restrictive surgery on beta-cell function and insulinotropic hormones in severely obese patients with type 2 diabetes. *Int J Obes* 2010;**34**(3):462–71.
240. Bayham BE, Greenway FL, Bellanger DE, O'Neil CE. Early resolution of type 2 diabetes seen after Roux-en-Y gastric bypass and vertical sleeve gastrectomy. *Diabetes Technol Ther* 2012;**14**(1):30–4.
241. Ramon JM, Salvans S, Crous X, Puig S, Goday A, Benaiges D, et al. Effect of Roux-en-Y gastric bypass vs sleeve gastrectomy on glucose and gut hormones: a prospective randomised trial. *J Gastrointest Surg* 2012;**16**(6):1116–22.
242. Buchwald H, Avidor Y, Braunwald E, Jensen MD, Pories W, Fahrbach K, et al. Bariatric surgery: a systematic review and meta-analysis. *JAMA* 2004;**292**(14):1724–37.
243. Hutter MM, Schirmer BD, Jones DB, Ko CY, Cohen ME, Merkow RP, et al. First report from the American College of Surgeons Bariatric Surgery Center Network: laparoscopic sleeve gastrectomy has morbidity and effectiveness positioned between the band and the bypass. *Ann Surg* 2011;**254**(3):410–20. discussion 20–2.
244. Sjostrom L, Lindroos AK, Peltonen M, Torgerson J, Bouchard C, Carlsson B, et al. Lifestyle, diabetes, and cardiovascular risk factors 10 years after bariatric surgery. *N Engl J Med* 2004;**351**(26):2683–93.
245. Sjostrom L, Peltonen M, Jacobson P, Sjostrom CD, Karason K, Wedel H, et al. Bariatric surgery and long-term cardiovascular events. *JAMA* 2012;**307**(1):56–65.
246. Sjostrom L. Review of the key results from the Swedish Obese Subjects (SOS) trial—a prospective controlled intervention study of bariatric surgery. *J Intern Med* 2013;**273**(3):219–34.
247. Schauer PR, Bhatt DL, Kirwan JP, Wolski K, Aminian A, Brethauer SA, et al. Bariatric surgery versus intensive medical therapy for diabetes—5-year outcomes. *N Engl J Med* 2017;**376**(7):641–51.
248. Salinari S, Bertuzzi A, Iaconelli A, Manco M, Mingrone G. Twenty-four hour insulin secretion and beta cell NEFA oxidation in type 2 diabetic, morbidly obese patients before and after bariatric surgery. *Diabetologia* 2008;**51**(7):1276–84.
249. Koves TR, Ussher JR, Noland RC, Slentz D, Mosedale M, Ilkayeva O, et al. Mitochondrial overload and incomplete fatty acid oxidation contribute to skeletal muscle insulin resistance. *Cell Metab* 2008;**7**(1):45–56.
250. Coen PM, Goodpaster BH. Role of intramyocelluar lipids in human health. *Trends Endocrinol Metab* 2012;**23**(8):391–8.
251. Nijhawan S, Richards W, O'Hea MF, Audia JP, Alvarez DF. Bariatric surgery rapidly improves mitochondrial respiration in morbidly obese patients. *Surg Endosc* 2013;**27**(12):4569–73.
252. Vijgen GHEJ, Bouvy ND, Hoeks J, Wijers S, Schrauwen P, van Marken Lichtenbelt WD. Impaired skeletal muscle mitochondrial function in morbidly obese patients is normalized one year after bariatric surgery. *Surg Obes Relat Dis* 2013;**9**(6):936–41.
253. Moehlecke M, Andriatta Blume C, Rheinheimer J, Trindade MRM, Crispim D, Leitao CB. Early reduction of resting energy expenditure and successful weight loss after Roux-en-Y gastric bypass. *Surg Obes Relat Dis* 2017;**13**(2):204–9.
254. Hernandez-Alvarez MI, Chiellini C, Manco M, Naon D, Liesa M, Palacin M, et al. Genes involved in mitochondrial biogenesis/function are induced in response to bilio-pancreatic diversion in morbidly obese individuals with normal glucose tolerance but not in type 2 diabetic patients. *Diabetologia* 2009;**52**(8):1618–27.
255. Dankel SN, Staalesen V, Bjørndal B, Berge RK, Mellgren G, Burri L. Tissue-specific effects of bariatric surgery including mitochondrial function. *J Obes* 2011;**2011**:435245.
256. Dube JJ, Amati F, Stefanovic-Racic M, Toledo FG, Sauers SE, Goodpaster BH. Exercise-induced alterations in intramyocellular lipids and insulin resistance: the athlete's paradox revisited. *Am J Physiol Endocrinol Metab* 2008;**294**(5):E882–8.
257. Coen PM, Menshikova EV, Distefano G, Zheng D, Tanner CJ, Standley RA, et al. Exercise and weight loss improve muscle mitochondrial respiration, lipid partitioning, and insulin sensitivity after gastric bypass surgery. *Diabetes* 2015;**64**(11):3737–50.
258. Woodlief TL, Carnero EA, Standley RA, Distefano G, Anthony SJ, Dubis GS, et al. Dose response of exercise training following roux-en-Y gastric bypass surgery: a randomized trial. *Obesity (Silver Spring)* 2015;**23**(12):2454–61.
259. Carnero EA, Dubis GS, Hames KC, Jakicic JM, Houmard JA, Coen PM, et al. Randomized trial reveals that physical activity and energy expenditure are associated with weight and body composition after RYGB. *Obesity (Silver Spring)* 2017;**25**(7):1206–16.

CHAPTER

# 17

# State of Knowledge and Recent Advances in Prevention and Treatment of Mitochondrial Dysfunction in Obesity and Type 2 Diabetes

*Carles Cantó*[*,†]

[*]Nestlé Institute of Health Sciences S.A., Lausanne, Switzerland [†]École Polytechnique Fédérale de Lausanne, Lausanne, Switzerland

## 1 INTRODUCTION—CAUGHT IN A WEB

Industrialized societies are flooded by information about nutritional facts, be it dietary requirements, meal compositions, or even seasonal recommendations. One would think that such wide knowledge would translate medically into a better prevention of diet-related metabolic complications. Epidemiological data, however, illustrate that, during the past 50 years, these societies have suffered a remarkable increase in the prevalence of chronic metabolic disorders, including obesity, type 2 diabetes mellitus (T2DM), and cardiovascular disease.[1] According to the World Health Organization (https://www.who.int), diabetes alone affects about 60 million people in Europe. Worldwide, high blood glucose and impaired metabolic control account for 3.4 million deaths annually.

Determining the polygenic etiology of diet-linked metabolic disorders has proven to be a great challenge because of the complex interactions between genetic and environmental factors. In addition to the increase of caloric excess, there's a readiness of the people living in industrialized nations to adopt a sedentary lifestyle, with more than 25% of adults not being active at all in their leisure time.[2] The hundreds of thousands of years of natural selection were heavily marked toward a positive selection for fit individuals who could acquire the required elements for survival and avoid danger. Sedentary lifestyles are now so prevalent


https://doi.org/10.1016/B978-0-12-811752-1.00017-1

that "it has become common to refer to exercise as having 'healthy benefits' even though the exercise-trained state is the biologically normal condition."[1] It is a lack of exercise, however, that is abnormal and carries health risks. Exercise remains one of the most effective interventions to prevent and treat the pathophysiology of the metabolic syndrome, in great part by enhancing insulin sensitivity in skeletal muscle.[1]

Given the therapeutic potential of physical activity, there has been a high interest in understanding the molecular basis of its benefits. One of the pioneering breakthroughs in this direction was John Holloszy's finding that exercise stimulated mitochondrial biogenesis in skeletal muscle.[3] Mitochondria are organelles that serve multiple essential cellular functions in eukaryote cells and organisms. The most well-recognized mitochondrial function might be ATP synthesis through oxidative phosphorylation (OXPHOS). Mitochondria, however, also play key roles in fatty acid metabolism, steroidogenesis, $Ca^{2+}$ and ROS signaling, as well as cellular fate decisions.[4] This nodal role of mitochondria at the core of cellular metabolism turns them into a hotspot in the response to metabolic stress. Accordingly, mitochondrial dysfunction has been related to a broad spectrum of diseases, including insulin resistance (IR) and T2DM.[5] However, how much do we understand about the relation between mitochondrial function and IR? And how can we exploit this knowledge for therapeutic purposes?

# 2 WHAT DO WE TALK ABOUT WHEN WE TALK ABOUT MITOCHONDRIAL DYSFUNCTION IN OBESITY AND T2DM?

## 2.1 Mitochondrial Function and Insulin Resistance—Cause or Consequence?

Aging and physical inactivity, which are recognized drivers for IR, are associated with impaired oxidative capacity in skeletal muscle. IR is one of the main morbidities of obesity and a hallmark for T2DM. Therefore, it was logical to think that altered skeletal muscle oxidative capacity could be a feature in obese and T2DM patients. Scattered studies through the last four decades have provided evidence for a reduction in the levels of several mitochondrial marker enzymes in the skeletal muscle from insulin resistant obese and T2DM patients.[6–10] These changes were manifested as a 20%–40% reduction in the enzymatic activities (or protein levels) of enzymes related to the TCA cycle (e.g., citrate synthase), the mitochondrial respiratory chain (e.g., cytochrome C oxidase), or fatty acid oxidation (FAO) (e.g., carnitine palmitoyl transferase). Gene-set enrichment analyses unveiled a coordinated downregulation of multiple genes encoding key enzymes related to oxidative metabolism and mitochondrial function in the skeletal muscle from diabetic patients.[11, 12] This orchestrated repression was linked to a decrease in the activity of the peroxisome proliferator activator receptor gamma coactivator-1 (PGC-1) family of transcriptional coactivators.[11, 12] The exact mechanism by which PGC-1$\alpha/\beta$ or mitochondrial-related genes become downregulated, however, is still unclear.

Although the data point out a clear correlation between IR and lower mitochondrial content, a causal link was missing. A step in this direction was done when mitochondrial function was analyzed in the offspring of T2DM patients. The lean, insulin-resistant offspring of patients with T2DM had a 30% reduction of mitochondrial phosphorylation rates and a 80% higher content of triglycerides in their soleus muscle, as compared with insulin-sensitive control subjects matched for age, height, weight, and activity.[13] These data added weight to the possibility that mitochondrial dysfunction might pave the wave for the development of IR and T2DM.

A causal burden of proof, however, cannot be established from these studies. For example, skeletal muscle mitochondrial content is associated positively with peak oxygen uptake.[14] Although sedentary elderly people with IR might have reductions in mitochondrial content, mitochondrial biogenesis can take place effectively when these individuals are submitted to an exercise training program.[14] Similarly, T2DM patients can increase their whole-body $VO_2$ by about eightfold during exercise,[15] which means an almost 40-fold increase in substrate oxidation rates in muscle. Therefore, reductions in mitochondrial content might reflect altered baseline physical activity in the obese or the elderly, a parameter that is difficult to assess quantitatively in the clinical setting.

A leading hypothesis during the last decade is that decreased mitochondrial function in obese patients could compromise the ability to oxidize fat and prompt the accumulation of lipid secondary species that would interfere with insulin signaling.[16] This brings up an important question: Is the lesser amount of mitochondria compromising FAO needs in obese insulin resistant patients? There is evidence indicating that obese men derive around 43% of their energy from fat during mild exercise, compared to 31% in the lean group.[17] Another study revealed that maximal FAO capacity in leg muscles during cycling exercise is increased in obese patients, in association with higher intramuscular triglyceride levels and higher resting plasma nonesterified fatty acid (NEFA) concentrations.[18] Increased FAO rates in obese and diabetic patients also have been observed in other works (e.g., Ref. 19–21), even though some reports indicate reduced FAO rates in obese and T2DM,[22, 23] which are often balanced by alterations in glucose oxidation.[23] Therefore, although reduced FAO rates might be observed in some obese populations, it is not a universal phenomenon. Consequently, it is unlikely that it acts as the causal event triggering IR. Instead, IR and enhanced FAO rates often coexist in the clinical setting.

Rodents have been largely used to evaluate the influence of mitochondrial function on obesity and IR. The db/db mouse model is one of the top standards for genetically induced obesity, because the db mutation triggers spontaneous hyperphagia, leading to profound IR in skeletal muscle.[24] Mitochondrial respiratory performance in the db/db mice, however, is under the control of tissue-specific mechanisms.[25] This way, mitochondrial respiration was decreased in liver, but increased in glycolytic skeletal muscle, while unaffected in oxidative muscle.[25] Therefore, IR and mitochondrial dysfunction do not go hand in hand in the skeletal muscle of db/db mice.

Diet-induced obesity might represent a more physiological model to address the influence of mitochondrial function in the development of IR and obesity. Mimicking the human scenario, a high-fat diet (HFD) for 8–12 weeks triggers IR in C57Bl/6 mice. In some studies, this has been accompanied by decreased mitochondrial enzymes levels in skeletal muscle.[26, 27] Other labs, however, have found that HFD-induced IR actually happens in parallel to an increase in mitochondrial biogenesis and FAO rates.[28, 29] Understanding the metabolic and molecular aspects behind the effects of HFD could shed light into the conflicting results previously mentioned. In 1963, Randle and colleagues illustrated how metabolites, in addition to hormones, could explain how fat overflow into skeletal muscle could interfere with glucose use.[30] Higher FAO fluxes would generate secondary intermediates, such as citrate, that could allosterically inhibit the glycolytic path, leading to impaired glucose use capacity. This observation has been widely reported and reproduced (see Ref. 31 for review). Accordingly, fasting plasma fatty acid concentrations and intramyocellular lipid content constitute fairly good predictors for IR in sedentary populations.[32–34] Subsequent studies from the Shulman lab using magnetic resonance spectroscopy indicated that the reduced insulin-stimulated glucose use in T2DM patients was largely attributable to defects in glucose transport.[35] This is in line with biochemical observations indicating impaired insulin-stimulated GLUT4 translocation to the plasma membrane in skeletal muscle of diabetic patients.[36]

A number of genetically engineered mouse models of mild and severe reduction in mitochondrial respiratory function display better insulin sensitivity.[37–41] Multiple models of artificially enhanced mitochondrial biogenesis in muscle, however, present signs of IR.[42–44] These results follow up the principles of the Randle cycle, by which defects in FAO are counteracted by enhanced glucose use and vice-versa. Other lines of research indicate that pharmacological strategies aiming to increase mitochondrial biogenesis in mice lead to improvements in glucose management.[45, 46] One has to take into account, however, that these strategies often protect against diet-induced body weight gain, which complicates the interpretation as to whether the protection against IR really comes from the improved mitochondrial function instead of the prevention of obesity per se.

As a whole, the evidence from model organisms and the physiological measurements in the clinical setting do not support the assumption that mitochondrial dysfunction causes IR, even if it could amplify or contribute to its development and complications.

## 2.2 Changes in Mitochondrial Shape—Cause or Consequence?

Electron microscopy measurements in Vastus lateralis muscle from lean, obese nondiabetic volunteers and obese T2DM subjects highlighted how mitochondrial size was reduced ~40% in the last two groups.[9] This suggests that obesity drives a decrease in mitochondrial size. Parallel efforts from the Zorzano lab also detected a fragmented mitochondrial network in the muscle of obese rats and humans.[47] Using a differential mRNA screen, they detected a downregulation of the *Mfn2* gene, which encodes for the Mitofusin 2 (Mfn2) protein, a GTPase enzyme involved in mitochondrial fusion events.[47] Therefore, the reduction in Mfn2 could explain the fragmented mitochondrial architecture observed in the muscle of obese individuals.

Situations of nutrient overload in the absence of a matching energy requirement lead to mitochondrial fission in cultured cells. Pioneering work by the Shirihai lab demonstrated that the mitochondrial network of INS-1 cells became largely fragmented when exposed to lipid-loaded medium.[48] Similar observations have been reported in MEF cells and AML12 hepatocytes.[49] Other works have shown how a nutrient excess based on glucose overload also can lead to mitochondrial fragmentation,[50–52] although this effect might be cell-type specific.[48] When fibroblasts, myocytes, or hepatocytes were submitted to starvation, however, their mitochondria fused and formed elongated networks.[53] The fusion of mitochondria spared them from autophagy and allowed the cell to sustain energy production.[53] Mitochondrial fusion also had an intrinsic bioenergetics effect: Mitochondria displayed more cristae upon fusion and increased dimerization and activity of the ATP synthase complex.[53] This could explain why mitochondria undergo the opposite path (i.e., fission) upon lipid loading or nutrient excess. In this sense, recent experiments demonstrate that mitochondrial fragmentation is a physiological response that increases mitochondrial uncoupling capacity in brown adipocytes.[54] A higher fissioned state might facilitate the access of fatty acids to UCP1, driving its activation.[55] Similar effects of fission on energy dissipation could apply to other tissues.[56]

Transgenic mouse models also support that mitochondrial fission is not per se a mark for mitochondrial dysfunction. Impaired mitochondrial fusion promoted by the deletion of the *Mfn1* gene in the liver (Mfn1-LKO) actually increases hepatic FAO capacity.[49] Mitochondrial respiration in Mfn1 KO MEFs is higher than in WT MEFs upon galactose treatment, which forces cells to rely on oxidative metabolism.[49] In line with this, norepinephrine-induced oxygen consumption in brown adipocytes was impaired by more than 50% when mitochondrial fission was compromised by expressing a dominant negative form of Drp1, a key protein for mitochondrial fission.[54] These observations suggest that mitochondrial fission might enhance FAO capacity as a protective adaptation against lipid overload.

The impact of Mfn2 expression in the muscle from obese subjects might go beyond its influence on mitochondrial fusion processes. Mfn2 is critical for the tethering and functional relationship between the mitochondria and the endoplasmic reticulum (ER),[57] although the exact mechanism remains controversial.[58, 59] Mfn2 plays a crucial role in the lipid and $Ca^{2+}$ transfer between the ER and the mitochondria,[57] and *Mfn2* deletion consistently has been associated with an increase in ER stress markers in most cells and tissues tested to date.[40, 57, 60] In contrast, Mfn1 deficiency in hepatocytes does not lead to ER stress.[49] Mfn2 also has been identified as a key facilitator of the interaction between mitochondria and the lipid droplet in brown adipose tissue (BAT).[40] These roles beyond mitochondrial fusion might explain why Mfn2, but not Mfn1, has been linked to metabolic complications.[47, 61]

Given the lethality of the whole body *Mfn2* knockout mice,[62] a number of tissue-specific knockout models for *Mfn2* have been generated. The first one reported was the specific deletion of *Mfn2* in the liver (Mfn2-LKO).[63] Mfn2-LKO mice exhibit profound abnormalities on glucose control, characterized by fasting hyperglycemia and glucose intolerance even when fed a regular diet.[63] These alterations on glucose management were propelled by excessive mitochondrial ROS production and increased ER stress. In alignment, alleviating ER stress with the molecular chaperone tauroursodeoxycholic acid (TUDCA)[64] was enough to improve insulin signaling in livers from Mfn2-LKO mice.[63] The treatment with N-acetylcysteine (NAC) relieved ER stress and insulin signaling, suggesting that ROS production is a key upstream trigger in the metabolic phenotypes of the Mfn2-LKO mice. The fact that the Mfn1-LKO model does not share these features[49] further suggests that the fragmentation of the mitochondrial network is not cause of metabolic complications in Mfn2-LKO mice.

In a second model, Mfn2 floxed mice were crossed with mice expressing the Cre recombinase under the MEF2C promoter. Mfn2 protein expression in the KO group was markedly reduced (80% decrease) in skeletal muscle, heart, and brain, and a near 50% reduction was detected in adipose tissues, kidney, or liver.[63] These mice display IR on low-fat diets and an exacerbated sensitivity to develop T2DM upon HFD feeding.[63] These effects, however, could derive simply from the defective hepatic Mfn2 levels and more specific models will be required to untangle the effects of Mfn2 in peripheral insulin sensitivity. In this sense, an adipose tissue-specific *Mfn2* knockout mouse (Mfn2-AKO) has been reported recently.[40] The deletion of *Mfn2* in either BAT or white adipose tissue (WAT) led to a heavily compromised lipolytic capacity and blunted Complex I activity in their mitochondria.[40] Surprisingly, mice lacking Mfn2 in adipose tissues were resistant to develop glucose intolerance upon HFD feeding.[40, 65] In fact, insulin sensitivity in Mfn2-defective BAT was enhanced compared to control littermates, in spite of its mitochondrial dysfunction.[40] This was consequent to a glycolytic rewiring in order to maintain thermogenic activity.[40] In the same line, a glycolytic rewiring and protection against IR also was observed in the skeletal muscle of mice lacking optic nerve atrophy 1 (OPA1), the GTPase enzyme mediating inner mitochondrial membrane fusion.[66]

As a whole, these results suggest that mitochondrial fragmentation is characteristic in tissues from obese individuals as a consequence of nutrient excess and lipid overflow. Genetically engineered mouse models, however, suggest that smaller, highly fissioned, mitochondria cannot be causally associated with IR.

## 2.3 Metabolic and Mitochondrial Flexibility

Kelley and colleagues coined the concept of metabolic flexibility, referring to the dynamic changes by which organisms, organs, or single cells adapt fuel use to fuel availability,[67] in particular, the ability to shift from carbohydrate metabolism during feeding times to FAO during fasting. Healthy individuals display a high response to insulin and other hormones in order to dispose nutrients and store them after a meal. T2DM patients are classical examples of metabolic inflexibility, however, failing to adapt fuel preferences to carbohydrate or lipid availability.[67] Longitudinal studies have shown that metabolic inflexibility precedes the development of T2DM even by decades,[68] constituting a hallmark of further metabolic complications.

Changes in substrate use can take place within minutes or hours, a timing that is not enough for the synthesis of new mitochondrial content. These acute changes in mitochondrial function often occur via posttranslational modifications.[69] For example, changes in the acylation status of mitochondrial proteins have been linked to the modulation of their activity (for review, see Ref. 4). The acylation of mitochondrial proteins depends on the accumulation of acetyl-coA and other TCA intermediaries generated during oxidative metabolism.[70] Therefore, intermediate metabolites can influence the flux of substrate oxidation. The deacylation of mitochondrial proteins takes place through the sirtuin family of enzymes, most particularly, SIRT3, SIRT4, and SIRT5.[4] Sirtuins are controlled by $NAD^+$ availability, providing an additional link between metabolism and deacylation activities. SIRT3 deficiency leads to a higher basal protein acetylation in the mitochondria,[71, 72] and makes mice more sensitive to develop mitochondrial dysfunction and metabolic disease upon HFD feeding.[73] The modulation of acylation cues might hold a key to understanding how mitochondria adapt their function to metabolic needs. A clear example for this can be found during situations of fasting and energy stress. The lack of carbohydrate energy sources promotes an increase in $NAD^+$ levels, which, in turn enhances SIRT3 activity.[74] The activation of the energy stress sensor enzyme AMP-activated protein kinase (AMPK) also transcriptionally triggers the expression of SIRT3[75, 76] and enzymes to enhance $NAD^+$ salvaging.[77, 78] This way, SIRT3 levels and activity increase in liver upon prolonged fasting, prompting the deacetylation and activation of enzymes such as the long-chain Acyl-CoA dehydrogenase (LCAD),[79] key for long chain FAO, or the 3-hydroxy,3-methylglutaryl-CoA synthase (HMGCS2),[80] the rate-limiting step in ketone body synthesis.

Mitochondrial architecture also can change quickly in response to nutrients. This is generally achieved by the modulation of Drp1, the key GTPase protein dynamically involved in mitochondrial fission processes.[81] Drp1 generally is localized in the cytosol, but its function is influenced through phosphorylation events triggered by nutrient-responsive and stress-responsive hormones. For example, Drp1 is phosphorylated at S616 by Cdk1 in response to mitotic stimuli, enhancing mitochondrial fission and allowing proper mitochondrial distribution between daughter cells.[82] From a metabolic perspective, the phosphorylation of Drp1 on S637 (S643 in mice, S656 in rats) might be even more relevant. This phosphorylation site is the target of PKA,[83, 84] a key effector engaged by multiple hormones, including glucagon or epinephrine. PKA activity increases during fasting, leading to the phosphorylation of Drp1 at S637, which impairs the ability of Drp1 to bind and oligomerize in the outer mitochondrial membrane (OMM).[53] This allows mitochondrial elongation and favors the oligomerization of ATP synthase,[53] enhancing its efficiency for ATP generation.[85] The outcome of Drp1 phosphorylation at S637, however, might be tissue or cell-type dependent, because Drp1 phosphorylation by PKA promotes fission in brown adipocytes.[54] Energy stress also has been shown to promote the phosphorylation of the mitochondrial fission factor (MFF), a putative receptor for Drp1 in the OMM.[86] This phosphorylation would enhance the ability of MFF to recruit Drp1,[87] leading to mitochondrial fission events and the clearance of daughter mitochondria with deficient respiratory capacity.[87] Altogether, these results illustrate how the acute modulation of mitochondrial architecture has dramatic consequences on cellular bioenergetics, granting metabolic flexibility.

# 3 MITOCHONDRIAL TARGETING FOR INSULIN RESISTANCE

## 3.1 Exercise and Calorie-Restriction Mimetics

The most effective strategies for the prevention and treatment of metabolic complications remain dietary and physical activity interventions. Although physical activity does not necessarily lead to reductions in body weight, it has been correlated to a clear decrease in the risk for metabolic complications.[1, 88] This has raised interest in the generation of exercise- and calorie restriction (CR)-mimetic compounds,[89] that is, natural/synthetic compounds that recapitulate the benefits of these interventions.

The mechanisms behind the benefits of CR or exercise are complex, in great part because they don't have similar effects in all organs.[90] Most efforts have been focused on targeting genes/proteins that provide epistatic effects on CR (or exercise) or whose activity is heavily influenced by these interventions.

SIRT1 and its homologs in lower eukaryotes have been proposed as mediators of the health benefits triggered by CR, because their ablation can compromise the effects of CR on longevity.[91] Whole body SIRT1 overexpression, however, is protective against a number of age-related diseases.[91, 92] SIRT1 is deacetylase enzyme that has a critical influence on metabolism. For example, SIRT1 acts as a modulator of PGC-1α, the master orchestrator of mitochondrial biogenesis. Upon deacetylation by SIRT1, the transcriptional coactivator activity of PGC-1α is enhanced.[93] The fact that SIRT1 activity is controlled by changes in physiological $NAD^+$

availability provides a strong link between intracellular metabolism and transcriptional regulation.[94] Therefore, mice with a physiological overexpression of SIRT1 display better metabolic flexibility.[95]

Multiple quests for SIRT1 activators have been launched, rendering, among others, two widely used compounds: resveratrol and SRT1720. Resveratrol is a natural polyphenol that initially was identified as an agent with cancer chemopreventive activity.[96] SRT1720 later was identified as a SIRT1-activating small molecule that was structurally unrelated to, and 1000-fold more potent than, resveratrol.[97] Both compounds have been shown to prevent HFD-induced body weight gain and IR, as well as the curbing of lifespan associated with obesity in rodents.[45, 46, 97–99] This was attributed to the ability of these compounds to improve mitochondrial function in liver, muscle, and BAT.[45, 46, 97, 99] The mechanism of action of these compounds, however, has been a cause for debate. Although there is data supporting that these compounds might allosterically act on SIRT1 in a substrate-specific fashion,[100] other evidence suggests that SIRT1 activation in response to them might be completely indirect.[101–103] Specifically, it has been proposed that compounds such as resveratrol constitute mild mitochondrial poisons leading to energy stress and AMPK activation.[104] The activation of AMPK can prompt an increase in $NAD^+$ levels, enhancing SIRT1 activity.[78, 105] One of the main arguments favoring that the effects of SIRT1 agonists are not simply because SIRT1 activation comes from genetic SIRT1 gain-of-function models. Although resveratrol or SRT1720 promote an increase in muscle mitochondrial content, mice with either a whole-body[95] or a muscle-specific[106] overexpression of SIRT1 do not show any major change in mitochondrial content or respiratory capacity. Similarly, genetic SIRT1 gain-of-function does not lead to prevention against HFD-induced body weight gain.[95, 107, 108] Therefore, even if SIRT1 activating compounds could lead to direct allosteric SIRT1 activation in vivo, there seems to be a significant burden of off-target effects. In spite of this, genetic SIRT1 activation demonstrates that SIRT1 could be a valuable target to improve insulin sensitivity.[95, 107, 108]

A new branch of sirtuin activators is constituted by $NAD^+$-boosting compounds. They are epitomized by nicotinamide riboside (NR) and nicotinamide mononucleotide (NMN), two salvageable $NAD^+$ precursors.[74] The administration of either compound in rodents consistently has led to increased $NAD^+$ levels and sirtuin activity in most cells and tissues tested.[109–111] They also provided consistent protection, and even reversion, of diet and age-related IR.[110, 111] NR, however, does not increase mitochondrial oxidative capacity in muscle or overall endurance in healthy young rodents,[110, 112] disqualifying it as an exercise mimetic. Nevertheless, it effectively protected from diet or age-related loss in mitochondrial content.[110, 113] Although these compounds hold therapeutic promise, we still fail to understand exactly how they improve insulin sensitivity in mice and whether these effects will be translated to humans, where, at least NR is safe and well tolerated.[114]

AMPK activation is another strategy by which improved mitochondrial function can be achieved. AMPK activation is triggered by muscle contraction and by some, but not all, CR protocols.[115] AMPK is an upstream regulator of PGC-1α, potentially via direct phosphorylation.[116] This phosphorylation might either directly enhance PGC-1α activity as a coactivator[116] or prime PGC-1α for deacetylation by SIRT1.[105] Consequently, AMPK activation constitutes a potent stimulus for mitochondrial biogenesis.[117, 118] AMPK serves as an energy sensor for the cell. Therefore, any drug that lowers ATP levels, increases AMP, or prevents energy substrate use could be an AMPK-activator. A huge group of natural compounds that interfere with mitochondrial function, including resveratrol, indirectly activate AMPK.[104] Several pharma companies have developed different series of selective activators for AMPK.[119–122] Upon activation, AMPK will switch on all possible paths in the cell to obtain energy, including glycolysis, FAO, and autophagy.[118] Conversely, it will switch off energy-consuming pathways that are not vital for cell survival, such as most anabolic paths, cell growth, and division.[118] Administration of systemic AMPK activators improves endurance capacity[122, 123] and glucose management[119–121] in mice and primates. It is unclear whether the effects on mitochondrial biogenesis effects are related to improved glucose management. Mice deficient for AMPK beta subunits, however, remain insulin-sensitive in spite of dramatic impairment in mitochondrial function,[124] suggesting that both aspects do not correlate positively. In spite of the specificity of current drugs and the promising benefits to health, AMPK activation strategies have some pitfalls. First, AMPK triggers protein degradation, which could be a problem in aged, insulin-resistant populations, where muscle mass loss constitutes one of the major problems for mobility and life quality.[125] A second major issue is that AMPK activation can lead to lactic acidosis, in great part because of the high potentiation of carbohydrate catabolism without an evident energy demand.[118] Finally, cardiac hypertrophy has been observed in models with constitute AMPK activation,[119, 126] which is troublesome in populations at risk of cardiovascular disease, such as the obese and the elderly.

A better safety profile for AMPK activators could be achieved by selectively targeting particular AMPK trimer complexes by controlling the amplitude and timing of AMPK activation or by selectively targeting particular tissues. Some of these features were found in what today is the frontline treatment against T2DM: metformin.

## 3.2 Metformin

Metformin is a biguanide whose therapeutic efficacy was discovered before modern target-based drug discovery.[127] The therapeutic effects of metformin on diabetic and insulin-resistant conditions are linked to the capacity of this compound to lower hepatic glucose output. Although its mechanism of action still is largely unresolved, it seems clear that it has a mitochondrial origin. The suppression of gluconeogenesis after metformin treatment is accompanied by inhibition of complex I in the mitochondrial oxidative phosphorylation chain.[128, 129] Therefore, insufficient production of ATP upon metformin treatment could rate-limit the process of hepatic glucose production. The effect of metformin on the cellular levels of adenylates, such as AMP and ATP, is relatively mild, but several studies suggest that it is enough to compromise energy balances. First, metformin treatment results in AMPK activation,[130] caused by alterations in AMP/ATP levels.[104] Second, lactic acidosis has been observed in patients treated with metformin, in line with AMPK activation and defective oxidative capacity.[131] Finally, a strong correlation between ATP levels and gluconeogenic output has been observed in hepatocytes,[132] where even relatively small decreases in ATP content lead to significant effects on glucose production.

The finding that AMPK was activated upon metformin treatment brought some light on its possible mechanism of action. Initial findings supported a role for AMPK by demonstrating that metformin-induced lowering of fasting glycemia in obese mice was blunted in the absence of LKB1, AMPK's upstream kinase.[133] To more accurately demonstrate whether the glucose-lowering effects of metformin were driven by AMPK and not by any of the other

LKB1-dependent kinases, mouse models lacking both the α1 and α2 catalytic subunits of AMPK in hepatocytes were generated.[132] These mice completely lacked AMPK activity in the liver, yet metformin resulted in a glucose-lowering effect comparable with that observed in control mice.[132] Metformin perfectly inhibited glucose production in primary hepatocytes lacking AMPK.[132] Therefore, alterations in cellular bioenergetics might be at the root of metformin effects, yet not by affecting AMPK activity. AICAR blocked hepatic gluconeogenesis in AMPK-deficient hepatocytes,[132] suggesting that AMP-modulated enzymes, other than AMPK, are involved in the process.

The success of metformin is unusual. First, metformin has exceptional safety profiles compared to most natural compounds, including other biguanides.[127] Also, metformin acts in a very tissue-selective fashion, having its predominant effects in the liver. Oral metformin delivery is distributed through the portal vein, which reaches the liver before other peripheral organs.[134] Most importantly, metformin requires the expression of organic cation transporters (e.g., OCT1) in order to be used by cells.[135] OCTs are expressed almost exclusively in the gut, the liver, and the kidney.[136] The fact that metformin has a very weak activity in tissues not expressing OCT1 actually is one of its main advantages. Similar drugs, such as phenformin, that can act on peripheral organs led to dramatic lactic acidosis and have been excluded from pharmaceutical uses.[137] Based on the success of metformin, recent efforts have tried to modify or identify novel protonophore agents specifically targeting the liver[138] or, more globally, the mitochondrial membrane.[139] It is still unclear, however, whether these compounds will reach the privileged pharmacokinetic and pharmacodynamics properties of metformin.

## 3.3 Emerging Approaches—mtUPR and Mitophagy

During the last decade, two emerging fields are focusing on facilitating mitochondrial repair and removal whenever required instead of simply building up mitochondrial content or impairing their function. These new strategies aim at the heart of mitochondrial quality control and, ideally, would facilitate the process without forcing it.

### 3.3.1 *Mitophagy*

Mitochondrial autophagy, also known as mitophagy, is a specialized macroautophagy process that controls mitochondrial number and ensures functional quality.[140] The half-life of mitochondria can differ largely between cell types and tissues.[56] Fission events generally yield heterogeneous daughter mitochondria, one of which is often depolarized.[141] Using mitochondrial tracking-techniques, it was demonstrated that the depolarized daughter mitochondria are six times less likely to rejoin the mitochondrial network through fusion.[141] Rather, they are targeted for degradation by the autophagosome.[141, 142]

In many cultured cell lines, PINK1 and Parkin are the key mediators of mitophagy. PINK1 accumulates in the membrane of depolarized mitochondria, which leads to Parkin recruitment, possibly through the binding to Mfn2.[143] Parkin is a E3 ubiquitin ligase that ubiquitynates OMM proteins, targeting the organelle for degradation.[140] Mutations in Parkin or PINK1 have been identified in Parkinson's disease, characterized by the accumulation of damaged mitochondria.[144, 145] Therefore, strategies aimed to acutely promote fission or to stabilize Parkin in the OMM in a controlled manner could be useful to ensure that damaged mitochondria are removed effectively through autophagy. While fission can be triggered by some nutritional and energy stresses, no drug exists to stabilize PINK1. Recent exciting findings, however, suggest that, although PINK1 is required for mitophagy processes triggered by artificial depolarization, it is dispensable to maintain basal mitophagy rates, even in tissues with high metabolic demand and mitochondrial content.[146] This suggests that mammals might have multiple mitophagy pathways, either redundant or responding to different stimuli. Identifying these additional mechanisms will be important in engineering new quests for pharmacological agents targeting mitophagy.

### 3.3.2 The Mitochondrial Unfolded Protein Response (mtUPR)

A vast majority of the mitochondrial proteome is encoded in the nucleus. This implies that all these proteins need to be imported into the mitochondria, where they need to fold and often assemble with other proteins in stoichiometry. Such a complex regulation leads to a constant proteostatic challenge.[147] To solve this problem, cells contain a number of mitochondrial chaperones aimed at resolving proteostatic problems in the mitochondria. These chaperones are nuclear encoded and their expression is fine-tuned to the degree of mitochondrial proteostatic stress,[148] together constituting what has been coined as the mitochondrial unfolded protein response (mtUPR). An excessive generation of ROS or mitochondrial peptides derived from the cleavage of unfolded proteins can signal to the nuclear genome, stimulating the transcription of nuclear-encoded mitochondrial chaperones, including HSP10 and HSP60.[149] Although the mtUPR path has been mechanistically dissected in worms, this is not yet the case in mammals. It remains unclear how cleaved peptides are transported to the cytosol and whether they require special signatures to trigger the mtUPR. The JNK/c-Jun/CHOP and ATF4 paths also seems to be involved, but the exact mechanisms for their activation and specificity in the response remain undefined.[150, 151]

In spite of the potential obscurities regarding its mechanistic aspects, the regulation of mtUPR seems to be a key mechanism for the therapeutic benefits of a number of compounds for metabolic and age-related diseases. This is the case, for example, of $NAD^+$ boosters, which trigger the mtUPR in a SIRT1 dependent manner.[113, 152, 153] In another study, artificially modifying the stoichiometry of nuclear-encoded versus mitochondrial-encoded subunits of respiratory complexes or ribosomal proteins was proven to be a valid strategy to trigger mtUPR and enhance longevity in worms.[152, 154] Importantly, this was achieved using either antibiotics to reduce mitochondrial translation or agents such as resveratrol or rapamycin, which reduce nuclear-encoded translation. This suggests that the mtUPR might be a common path by which natural compounds operate to promote hormetic responses leading to health benefits.

## 4 CONCLUSIONS AND FUTURE DIRECTIONS

Mitochondria have a strong influence on cellular behavior and fate. Widely known for their contribution to cellular bioenergetics, it is often forgotten that mitochrondria also play key roles in hormones and lipids biosynthesis, $Ca^{2+}$ homeostasis, and apoptotic regulation. The attractive therapeutic potential on targeting mitochondrial function is challenged by the need for selectivity, considering their wide range of actions. One has to wonder what feature in mitochondria should be targeted for a particular indication and keep in mind that mitochondria are heavily specialized in different tissues to achieve different functions. For example, enhancing uncoupling capacity in BAT or WAT has been proposed as a possible treatment for metabolic disorders and to promote body weight loss. Artificially uncoupling mitochondria in other tissues and organs, such as the heart, muscle or brain, however, could be fatal. Therefore, there is a need to have a certain tropism toward the desired target organ/tissue.

There is the widespread notion that metabolic disease is caused by mitochondrial dysfunction, launching quests for compounds to increase mitochondrial content in cells. This strategy has little chance of success in the absence of a matching energy requirement (e.g., physical activity). Similarly, it would be important to combine drugs aimed at increasing mitochondrial biogenesis with others to ensure that mitochondria quality control mechanisms are working effectively. A second point of misconception stems from the fact that observations linking lower mitochondrial content with metabolic disease are mostly correlational, but not causal. In this sense, a number of studies counterintuitively indicate that partially decreasing mitochondrial respiratory capacity can trigger adaptations enhancing glucose use and insulin sensitivity. The threshold between toxicity and hormetic responses, however, is not clearly defined.

Metformin, currently the frontline drug for T2DM, has a number of unique properties in terms of efficacy, bioavailability, and cellular distribution. Finding a compound with such special features would be a radical exception, considering that neither its effects on isolated cells nor its plasma levels would suggest it is strong in vivo efficacy. Aristotle (384–322 BC) said "the secret to humor is surprise"—a phrase that might apply to the quest for new therapeutic compounds.

## Acknowledgements

I would like to thank the Canto lab members for inspiring discussions. CC is funded by the NIHS and by the EU Marie Skłodowska-Curie ITN—ChroMe (H2020-MSCA-ITN-2015-ChroMe—project number 675610). I would like to apologize to all the authors whose relevant work was not cited because of space limitations.

## References

1. Hawley JA, Hargreaves M, Joyner MJ, Zierath JR. Integrative biology of exercise. *Cell* 2014;**159**:738–49.
2. Hawley JA, Holloszy JO. Exercise: it's the real thing! *Nutr Rev* 2009;**67**:172–8.
3. Holloszy JO. Biochemical adaptations in muscle. Effects of exercise on mitochondrial oxygen uptake and respiratory enzyme activity in skeletal muscle. *J Biol Chem* 1967;**242**:2278–82.
4. Kulkarni SS, Canto C. Mitochondrial post-translational modifications and metabolic control: sirtuins and beyond. *Curr Diabetes Rev* 2017;**13**:338–51.
5. Andreux PA, Houtkooper RH, Auwerx J. Pharmacological approaches to restore mitochondrial function. *Nat Rev Drug Discov* 2013;**12**:465–83.
6. Lithell H, Lindgarde F, Hellsing K, Lundqvist G, Nygaard E, Vessby B, Saltin B. Body weight, skeletal muscle morphology, and enzyme activities in relation to fasting serum insulin concentration and glucose tolerance in 48-year-old men. *Diabetes* 1981;**30**:19–25.
7. Vondra K, Rath R, Bass A, Slabochova Z, Teisinger J, Vitek V. Enzyme activities in quadriceps femoris muscle of obese diabetic male patients. *Diabetologia* 1977;**13**:527–9.
8. Simoneau JA, Veerkamp JH, Turcotte LP, Kelley DE. Markers of capacity to utilize fatty acids in human skeletal muscle: relation to insulin resistance and obesity and effects of weight loss. *FASEB J* 1999;**13**:2051–60.
9. Kelley DE, He J, Menshikova EV, Ritov VB. Dysfunction of mitochondria in human skeletal muscle in type 2 diabetes. *Diabetes* 2002;**51**:2944–50.

10. He J, Watkins S, Kelley DE. Skeletal muscle lipid content and oxidative enzyme activity in relation to muscle fiber type in type 2 diabetes and obesity. *Diabetes* 2001;**50**:817–23.
11. Mootha VK, Lindgren CM, Eriksson KF, Subramanian A, Sihag S, Lehar J, Puigserver P, Carlsson E, Ridderstrale M, Laurila E, Houstis N, Daly MJ, Patterson N, Mesirov JP, Golub TR, Tamayo P, Spiegelman B, Lander ES, Hirschhorn JN, Altshuler D, Groop LC. PGC-1alpha-responsive genes involved in oxidative phosphorylation are coordinately downregulated in human diabetes. *Nat Genet* 2003;**34**:267–73.
12. Patti ME, Butte AJ, Crunkhorn S, Cusi K, Berria R, Kashyap S, Miyazaki Y, Kohane I, Costello M, Saccone R, Landaker EJ, Goldfine AB, Mun E, DeFronzo R, Finlayson J, Kahn CR, Mandarino LJ. Coordinated reduction of genes of oxidative metabolism in humans with insulin resistance and diabetes: potential role of PGC1 and NRF1. *Proc Natl Acad Sci U S A* 2003;**100**:8466–71.
13. Petersen KF, Dufour S, Befroy D, Garcia R, Shulman GI. Impaired mitochondrial activity in the insulin-resistant offspring of patients with type 2 diabetes. *N Engl J Med* 2004;**350**:664–71.
14. Broskey NT, Greggio C, Boss A, Boutant M, Dwyer A, Schlueter L, Hans D, Gremion G, Kreis R, Boesch C, Canto C, Amati F. Skeletal muscle mitochondria in the elderly: effects of physical fitness and exercise training. *J Clin Endocrinol Metab* 2014;**99**:1852–61.
15. Larsen S, Ara I, Rabol R, Andersen JL, Boushel R, Dela F, Helge JW. Are substrate use during exercise and mitochondrial respiratory capacity decreased in arm and leg muscle in type 2 diabetes? *Diabetologia* 2009;**52**:1400–8.
16. Petersen KF, Shulman GI. Etiology of insulin resistance. *Am J Med* 2006;**119**:S10–6.
17. Goodpaster BH, Wolfe RR, Kelley DE. Effects of obesity on substrate utilization during exercise. *Obes Res* 2002;**10**:575–84.
18. Ara I, Larsen S, Stallknecht B, Guerra B, Morales-Alamo D, Andersen JL, Ponce-Gonzalez JG, Guadalupe-Grau A, Galbo H, Calbet JA, Helge JW. Normal mitochondrial function and increased fat oxidation capacity in leg and arm muscles in obese humans. *Int J Obes* 2011;**35**:99–108.
19. Golay A, Felber JP, Meyer HU, Curchod B, Maeder E, Jequier E. Study on lipid metabolism in obesity diabetes. *Metab Clin Exp* 1984;**33**:111–6.
20. Boon H, Blaak EE, Saris WH, Keizer HA, Wagenmakers AJ, van Loon LJ. Substrate source utilisation in long-term diagnosed type 2 diabetes patients at rest, and during exercise and subsequent recovery. *Diabetologia* 2007;**50**:103–12.
21. Felber JP, Ferrannini E, Golay A, Meyer HU, Theibaud D, Curchod B, Maeder E, Jequier E, DeFronzo RA. Role of lipid oxidation in pathogenesis of insulin resistance of obesity and type II diabetes. *Diabetes* 1987;**36**:1341–50.
22. Kelley DE, Goodpaster B, Wing RR, Simoneau JA. Skeletal muscle fatty acid metabolism in association with insulin resistance, obesity, and weight loss. *Am J Phys* 1999;**277**:E1130–41.
23. Kelley DE, Simoneau JA. Impaired free fatty acid utilization by skeletal muscle in non-insulin-dependent diabetes mellitus. *J Clin Invest* 1994;**94**:2349–56.
24. Chan TM, Dehaye JP. Hormone regulation of glucose metabolism in the genetically obese-diabetic mouse (db/db): glucose metabolism in the perfused hindquarters of lean and obese mice. *Diabetes* 1981;**30**:211–8.
25. Holmstrom MH, Iglesias-Gutierrez E, Zierath JR, Garcia-Roves PM. Tissue-specific control of mitochondrial respiration in obesity-related insulin resistance and diabetes. *Am J Physiol Endocrinol Metab* 2012;**302**:E731–9.
26. Koves TR, Li P, An J, Akimoto T, Slentz D, Ilkayeva O, Dohm GL, Yan Z, Newgard CB, Muoio DM. Peroxisome proliferator-activated receptor-gamma co-activator 1alpha-mediated metabolic remodeling of skeletal myocytes mimics exercise training and reverses lipid-induced mitochondrial inefficiency. *J Biol Chem* 2005;**280**:33588–98.
27. Sparks LM, Xie H, Koza RA, Mynatt R, Hulver MW, Bray GA, Smith SR. A high-fat diet coordinately downregulates genes required for mitochondrial oxidative phosphorylation in skeletal muscle. *Diabetes* 2005;**54**:1926–33.
28. Turner N, Bruce CR, Beale SM, Hoehn KL, So T, Rolph MS, Cooney GJ. Excess lipid availability increases mitochondrial fatty acid oxidative capacity in muscle: evidence against a role for reduced fatty acid oxidation in lipid-induced insulin resistance in rodents. *Diabetes* 2007;**56**:2085–92.
29. Hancock CR, Han DH, Chen M, Terada S, Yasuda T, Wright DC, Holloszy JO. High-fat diets cause insulin resistance despite an increase in muscle mitochondria. *Proc Natl Acad Sci U S A* 2008;**105**:7815–20.
30. Randle PJ, Garland PB, Hales CN, Newsholme EA. The glucose fatty-acid cycle. Its role in insulin sensitivity and the metabolic disturbances of diabetes mellitus. *Lancet* 1963;**1**:785–9.
31. Hue L, Taegtmeyer H. The Randle cycle revisited: a new head for an old hat. *Am J Physiol Endocrinol Metab* 2009;**297**:E578–91.

32. Perseghin G, Ghosh S, Gerow K, Shulman GI. Metabolic defects in lean nondiabetic offspring of NIDDM parents: a cross-sectional study. *Diabetes* 1997;**46**:1001–9.
33. Krssak M, Falk Petersen K, Dresner A, DiPietro L, Vogel SM, Rothman DL, Roden M, Shulman GI. Intramyocellular lipid concentrations are correlated with insulin sensitivity in humans: a 1H NMR spectroscopy study. *Diabetologia* 1999;**42**:113–6.
34. Perseghin G, Scifo P, De Cobelli F, Pagliato E, Battezzati A, Arcelloni C, Vanzulli A, Testolin G, Pozza G, Del Maschio A, Luzi L. Intramyocellular triglyceride content is a determinant of in vivo insulin resistance in humans: a 1H-13C nuclear magnetic resonance spectroscopy assessment in offspring of type 2 diabetic parents. *Diabetes* 1999;**48**:1600–6.
35. Cline GW, Petersen KF, Krssak M, Shen J, Hundal RS, Trajanoski Z, Inzucchi S, Dresner A, Rothman DL, Shulman GI. Impaired glucose transport as a cause of decreased insulin-stimulated muscle glycogen synthesis in type 2 diabetes. *N Engl J Med* 1999;**341**:240–6.
36. Zierath JR, He L, Guma A, Odegoard Wahlstrom E, Klip A, Wallberg-Henriksson H. Insulin action on glucose transport and plasma membrane GLUT4 content in skeletal muscle from patients with NIDDM. *Diabetologia* 1996;**39**:1180–9.
37. Wredenberg A, Freyer C, Sandstrom ME, Katz A, Wibom R, Westerblad H, Larsson NG. Respiratory chain dysfunction in skeletal muscle does not cause insulin resistance. *Biochem Biophys Res Commun* 2006;**350**:202–7.
38. Pospisilik JA, Knauf C, Joza N, Benit P, Orthofer M, Cani PD, Ebersberger I, Nakashima T, Sarao R, Neely G, Esterbauer H, Kozlov A, Kahn CR, Kroemer G, Rustin P, Burcelin R, Penninger JM. Targeted deletion of AIF decreases mitochondrial oxidative phosphorylation and protects from obesity and diabetes. *Cell* 2007;**131**:476–91.
39. Zechner C, Lai L, Zechner JF, Geng T, Yan Z, Rumsey JW, Collia D, Chen Z, Wozniak DF, Leone TC, Kelly DP. Total skeletal muscle PGC-1 deficiency uncouples mitochondrial derangements from fiber type determination and insulin sensitivity. *Cell Metab* 2010;**12**:633–42.
40. Boutant M, Kulkarni SS, Joffraud M, Ratajczak J, Valera-Alberni M, Combe R, Zorzano A, Canto C. Mfn2 is critical for brown adipose tissue thermogenic function. *EMBO J* 2017;**36**:1543–58.
41. Yang H, Wu JW, Wang SP, Severi I, Sartini L, Frizzell N, Cinti S, Yang G, Mitchell GA. Adipose-specific deficiency of fumarate hydratase in mice protects against obesity, hepatic steatosis and insulin resistance. *Diabetes* 2016.
42. Choi CS, Befroy DE, Codella R, Kim S, Reznick RM, Hwang YJ, Liu ZX, Lee HY, Distefano A, Samuel VT, Zhang D, Cline GW, Handschin C, Lin J, Petersen KF, Spiegelman BM, Shulman GI. Paradoxical effects of increased expression of PGC-1alpha on muscle mitochondrial function and insulin-stimulated muscle glucose metabolism. *Proc Natl Acad Sci U S A* 2008;**105**:19926–31.
43. Finck BN, Bernal-Mizrachi C, Han DH, Coleman T, Sambandam N, LaRiviere LL, Holloszy JO, Semenkovich CF, Kelly DP. A potential link between muscle peroxisome proliferator-activated receptor-alpha signaling and obesity-related diabetes. *Cell Metab* 2005;**1**:133–44.
44. Park SY, Cho YR, Finck BN, Kim HJ, Higashimori T, Hong EG, Lee MK, Danton C, Deshmukh S, Cline GW, Wu JJ, Bennett AM, Rothermel B, Kalinowski A, Russell KS, Kim YB, Kelly DP, Kim JK. Cardiac-specific overexpression of peroxisome proliferator-activated receptor-alpha causes insulin resistance in heart and liver. *Diabetes* 2005;**54**:2514–24.
45. Lagouge M, Argmann C, Gerhart-Hines Z, Meziane H, Lerin C, Daussin F, Messadeq N, Milne J, Lambert P, Elliott P, Geny B, Laakso M, Puigserver P, Auwerx J. Resveratrol improves mitochondrial function and protects against metabolic disease by activating SIRT1 and PGC-1alpha. *Cell* 2006;**127**:1109–22.
46. Feige JN, Lagouge M, Canto C, Strehle A, Houten SM, Milne JC, Lambert PD, Mataki C, Elliott PJ, Auwerx J. Specific SIRT1 activation mimics low energy levels and protects against diet-induced metabolic disorders by enhancing fat oxidation. *Cell Metab* 2008;**8**:347–58.
47. Bach D, Pich S, Soriano FX, Vega N, Baumgartner B, Oriola J, Daugaard JR, Lloberas J, Camps M, Zierath JR, Rabasa-Lhoret R, Wallberg-Henriksson H, Laville M, Palacin M, Vidal H, Rivera F, Brand M, Zorzano A. Mitofusin-2 determines mitochondrial network architecture and mitochondrial metabolism. A novel regulatory mechanism altered in obesity. *J Biol Chem* 2003;**278**:17190–7.
48. Molina AJ, Wikstrom JD, Stiles L, Las G, Mohamed H, Elorza A, Walzer G, Twig G, Katz S, Corkey BE, Shirihai OS. Mitochondrial networking protects beta-cells from nutrient-induced apoptosis. *Diabetes* 2009;**58**:2303–15.
49. Kulkarni SS, Joffraud M, Boutant M, Ratajczak J, Gao AW, Maclachlan C, Hernandez-Alvarez MI, Raymond F, Metairon S, Descombes P, Houtkooper RH, Zorzano A, Canto C. Mfn1 deficiency in the liver protects against diet-induced insulin resistance and enhances the hypoglycemic effect of metformin. *Diabetes* 2016.

50. Yu T, Robotham JL, Yoon Y. Increased production of reactive oxygen species in hyperglycemic conditions requires dynamic change of mitochondrial morphology. *Proc Natl Acad Sci U S A* 2006;**103**:2653–8.
51. Yu T, Jhun BS, Yoon Y. High-glucose stimulation increases reactive oxygen species production through the calcium and mitogen-activated protein kinase-mediated activation of mitochondrial fission. *Antioxid Redox Signal* 2011;**14**:425–37.
52. Theurey P, Tubbs E, Vial G, Jacquemetton J, Bendridi N, Chauvin MA, Alam MR, Le Romancer M, Vidal H, Rieusset J. Mitochondria-associated endoplasmic reticulum membranes allow adaptation of mitochondrial metabolism to glucose availability in the liver. *J Mol Cell Biol* 2016;**8**:129–43.
53. Gomes LC, Di Benedetto G, Scorrano L. During autophagy mitochondria elongate, are spared from degradation and sustain cell viability. *Nat Cell Biol* 2011;**13**:589–98.
54. Wikstrom JD, Mahdaviani K, Liesa M, Sereda SB, Si Y, Las G, Twig G, Petrovic N, Zingaretti C, Graham A, Cinti S, Corkey BE, Cannon B, Nedergaard J, Shirihai OS. Hormone-induced mitochondrial fission is utilized by brown adipocytes as an amplification pathway for energy expenditure. *EMBO J* 2014;**33**:418–36.
55. Garlid KD, Jaburek M, Jezek P, Varecha M. How do uncoupling proteins uncouple? *Biochim Biophys Acta* 2000;**1459**:383–9.
56. Liesa M, Shirihai OS. Mitochondrial dynamics in the regulation of nutrient utilization and energy expenditure. *Cell Metab* 2013;**17**:491–506.
57. de Brito OM, Scorrano L. Mitofusin 2 tethers endoplasmic reticulum to mitochondria. *Nature* 2008;**456**:605–10.
58. Filadi R, Greotti E, Turacchio G, Luini A, Pozzan T, Pizzo P. Mitofusin 2 ablation increases endoplasmic reticulum-mitochondria coupling. *Proc Natl Acad Sci U S A* 2015;**112**:E2174–81.
59. Cosson P, Marchetti A, Ravazzola M, Orci L. Mitofusin-2 independent juxtaposition of endoplasmic reticulum and mitochondria: an ultrastructural study. *PLoS ONE* 2012;**7**:e46293.
60. Schneeberger M, Dietrich MO, Sebastian D, Imbernon M, Castano C, Garcia A, Esteban Y, Gonzalez-Franquesa A, Rodriguez IC, Bortolozzi A, Garcia-Roves PM, Gomis R, Nogueiras R, Horvath TL, Zorzano A, Claret M. Mitofusin 2 in POMC neurons connects ER stress with leptin resistance and energy imbalance. *Cell* 2013;**155**:172–87.
61. Bach D, Naon D, Pich S, Soriano FX, Vega N, Rieusset J, Laville M, Guillet C, Boirie Y, Wallberg-Henriksson H, Manco M, Calvani M, Castagneto M, Palacin M, Mingrone G, Zierath JR, Vidal H, Zorzano A. Expression of Mfn2, the Charcot-Marie-Tooth neuropathy type 2A gene, in human skeletal muscle: effects of type 2 diabetes, obesity, weight loss, and the regulatory role of tumor necrosis factor alpha and interleukin-6. *Diabetes* 2005;**54**:2685–93.
62. Chen H, Detmer SA, Ewald AJ, Griffin EE, Fraser SE, Chan DC. Mitofusins Mfn1 and Mfn2 coordinately regulate mitochondrial fusion and are essential for embryonic development. *J Cell Biol* 2003;**160**:189–200.
63. Sebastian D, Hernandez-Alvarez MI, Segales J, Sorianello E, Munoz JP, Sala D, Waget A, Liesa M, Paz JC, Gopalacharyulu P, Oresic M, Pich S, Burcelin R, Palacin M, Zorzano A. Mitofusin 2 (Mfn2) links mitochondrial and endoplasmic reticulum function with insulin signaling and is essential for normal glucose homeostasis. *Proc Natl Acad Sci U S A* 2012;**109**:5523–8.
64. Ozcan U, Cao Q, Yilmaz E, Lee AH, Iwakoshi NN, Ozdelen E, Tuncman G, Gorgun C, Glimcher LH, Hotamisligil GS. Endoplasmic reticulum stress links obesity, insulin action, and type 2 diabetes. *Science* 2004;**306**:457–61.
65. Mahdaviani K, Benador IY, Su S, Gharakhanian RA, Stiles L, Trudeau KM, Cardamone M, Enriquez-Zarralanga V, Ritou E, Aprahamian T, Oliveira MF, Corkey BE, Perissi V, Liesa M, Shirihai OS. Mfn2 deletion in brown adipose tissue protects from insulin resistance and impairs thermogenesis. *EMBO Rep* 2017;**18**:1123–38.
66. Pereira RO, Tadinada SM, Zasadny FM, Oliveira KJ, Pires KMP, Olvera A, Jeffers J, Souvenir R, McGlauflin R, Seei A, Funari T, Sesaki H, Potthoff MJ, Adams CM, Anderson EJ, Abel ED. OPA1 deficiency promotes secretion of FGF21 from muscle that prevents obesity and insulin resistance. *EMBO J* 2017;**36**:2126–45.
67. Storlien L, Oakes ND, Kelley DE. Metabolic flexibility. *Proc Nutr Soc* 2004;**63**:363–8.
68. Corpeleijn E, Saris WH, Blaak EE. Metabolic flexibility in the development of insulin resistance and type 2 diabetes: effects of lifestyle. *Obes Rev: Off J Int Assoc Study Obes* 2009;**10**:178–93.
69. Gao AW, Canto C, Houtkooper RH. Mitochondrial response to nutrient availability and its role in metabolic disease. *EMBO Mol Med* 2014;**6**:580–9.
70. Wagner GR, Hirschey MD. Nonenzymatic protein acylation as a carbon stress regulated by sirtuin deacylases. *Mol Cell* 2014;**54**:5–16.

71. Lombard DB, Alt FW, Cheng HL, Bunkenborg J, Streeper RS, Mostoslavsky R, Kim J, Yancopoulos G, Valenzuela D, Murphy A, Yang Y, Chen Y, Hirschey MD, Bronson RT, Haigis M, Guarente LP, Farese Jr. RV, Weissman S, Verdin E, Schwer B. Mammalian Sir2 homolog SIRT3 regulates global mitochondrial lysine acetylation. *Mol Cell Biol* 2007;**27**:8807–14.
72. Fernandez-Marcos PJ, Jeninga EH, Canto C, Harach T, de Boer VC, Andreux P, Moullan N, Pirinen E, Yamamoto H, Houten SM, Schoonjans K, Auwerx J. Muscle or liver-specific Sirt3 deficiency induces hyperacetylation of mitochondrial proteins without affecting global metabolic homeostasis. *Sci Rep* 2012;**2**:425.
73. Hirschey MD, Shimazu T, Jing E, Grueter CA, Collins AM, Aouizerat B, Stancakova A, Goetzman E, Lam MM, Schwer B, Stevens RD, Muehlbauer MJ, Kakar S, Bass NM, Kuusisto J, Laakso M, Alt FW, Newgard CB, Farese Jr. RV, Kahn CR, Verdin E. SIRT3 deficiency and mitochondrial protein hyperacetylation accelerate the development of the metabolic syndrome. *Mol Cell* 2011;**44**:177–90.
74. Canto C, Menzies KJ, Auwerx J. NAD(+) Metabolism and the control of energy homeostasis: a balancing act between mitochondria and the nucleus. *Cell Metab* 2015;**22**:31–53.
75. Brandauer J, Andersen MA, Kellezi H, Risis S, Frosig C, Vienberg SG, Treebak JT. AMP-activated protein kinase controls exercise training- and AICAR-induced increases in SIRT3 and MnSOD. *Front Physiol* 2015;**6**:85.
76. Palacios OM, Carmona JJ, Michan S, Chen KY, Manabe Y, Ward 3rd JL, Goodyear LJ, Tong Q. Diet and exercise signals regulate SIRT3 and activate AMPK and PGC-1alpha in skeletal muscle. *Aging* 2009;**1**:771–83.
77. Diguet N, Trammell SAJ, Tannous C, Deloux R, Piquereau J, Mougenot N, Gouge A, Gressette M, Manoury B, Blanc J, Breton M, Decaux JF, Lavery G, Baczko I, Zoll J, Garnier A, Li Z, Brenner C, Mericskay M. Nicotinamide riboside preserves cardiac function in a mouse model of dilated cardiomyopathy. *Circulation* 2017.
78. Fulco M, Cen Y, Zhao P, Hoffman EP, McBurney MW, Sauve AA, Sartorelli V. Glucose restriction inhibits skeletal myoblast differentiation by activating SIRT1 through AMPK-mediated regulation of Nampt. *Dev Cell* 2008;**14**:661–73.
79. Hirschey MD, Shimazu T, Goetzman E, Jing E, Schwer B, Lombard DB, Grueter CA, Harris C, Biddinger S, Ilkayeva OR, Stevens RD, Li Y, Saha AK, Ruderman NB, Bain JR, Newgard CB, Farese, Jr RV, Alt FW, Kahn CR, Verdin E. SIRT3 regulates mitochondrial fatty-acid oxidation by reversible enzyme deacetylation. *Nature* 2010;**464**:121–5.
80. Shimazu T, Hirschey MD, Hua L, Dittenhafer-Reed KE, Schwer B, Lombard DB, Li Y, Bunkenborg J, Alt FW, Denu JM, Jacobson MP, Verdin E. SIRT3 deacetylates mitochondrial 3-hydroxy-3-methylglutaryl CoA synthase 2 and regulates ketone body production. *Cell Metab* 2010;**12**:654–61.
81. Otera H, Ishihara N, Mihara K. New insights into the function and regulation of mitochondrial fission. *Biochim Biophys Acta* 2013;**1833**:1256–68.
82. Taguchi N, Ishihara N, Jofuku A, Oka T, Mihara K. Mitotic phosphorylation of dynamin-related GTPase Drp1 participates in mitochondrial fission. *J Biol Chem* 2007;**282**:11521–9.
83. Chang CR, Blackstone C. Cyclic AMP-dependent protein kinase phosphorylation of Drp1 regulates its GTPase activity and mitochondrial morphology. *J Biol Chem* 2007;**282**:21583–7.
84. Cribbs JT, Strack S. Reversible phosphorylation of Drp1 by cyclic AMP-dependent protein kinase and calcineurin regulates mitochondrial fission and cell death. *EMBO Rep* 2007;**8**:939–44.
85. Strauss M, Hofhaus G, Schroder RR, Kuhlbrandt W. Dimer ribbons of ATP synthase shape the inner mitochondrial membrane. *EMBO J* 2008;**27**:1154–60.
86. Ducommun S, Deak M, Sumpton D, Ford RJ, Nunez Galindo A, Kussmann M, Viollet B, Steinberg GR, Foretz M, Dayon L, Morrice NA, Sakamoto K. Motif affinity and mass spectrometry proteomic approach for the discovery of cellular AMPK targets: identification of mitochondrial fission factor as a new AMPK substrate. *Cell Signal* 2015;**27**:978–88.
87. Toyama EQ, Herzig S, Courchet J, Lewis Jr. TL, Loson OC, Hellberg K, Young NP, Chen H, Polleux F, Chan DC, Shaw RJ. Metabolism. AMP-activated protein kinase mediates mitochondrial fission in response to energy stress. *Science* 2016;**351**:275–81.
88. InterAct C, Ekelund U, Palla L, Brage S, Franks PW, Peters T, Balkau B, Diaz MJ, Huerta JM, Agnoli C, Arriola L, Ardanaz E, Boeing H, Clavel-Chapelon F, Crowe F, Fagherazzi G, Groop L, Fons Johnsen N, Kaaks R, Khaw KT, Key TJ, de Lauzon-Guillain B, May A, Monninkhof E, Navarro C, Nilsson P, Nautrup Ostergaard J, Norat T, Overvad K, Palli D, Panico S, Redondo ML, Ricceri F, Rolandsson O, Romaguera D, Romieu I, Sanchez Perez MJ, Slimani N, Spijkerman A, Teucher B, Tjonneland A, Travier N, Tumino R, Vos W, Vigl M, Sharp S, Langeberg C, Forouhi N, Riboli E, Feskens E, Wareham NJ. Physical activity reduces the risk of incident type 2 diabetes in general and in abdominally lean and obese men and women: the EPIC-InterAct Study. *Diabetologia* 2012;**55**:1944–52.

89. Handschin C. Caloric restriction and exercise "mimetics": ready for prime time? *Pharmacol Res* 2016;**103**:158–66.
90. Boutant M, Kulkarni SS, Joffraud M, Raymond F, Metairon S, Descombes P, Canto C. SIRT1 gain of function does not mimic or enhance the adaptations to intermittent fasting. *Cell Rep* 2016;**14**:2068–75.
91. Chang HC, Guarente L. SIRT1 and other sirtuins in metabolism. *Trends Endocrinol Metab* 2013.
92. Herranz D, Munoz-Martin M, Canamero M, Mulero F, Martinez-Pastor B, Fernandez-Capetillo O, Serrano M. Sirt1 improves healthy ageing and protects from metabolic syndrome-associated cancer. *Nat Commun* 2010;**1**:3.
93. Rodgers JT, Lerin C, Haas W, Gygi SP, Spiegelman BM, Puigserver P. Nutrient control of glucose homeostasis through a complex of PGC-1alpha and SIRT1. *Nature* 2005;**434**:113–8.
94. Canto C, Auwerx J. Targeting sirtuin 1 to improve metabolism: all you need is NAD(+)? *Pharmacol Rev* 2012;**64**:166–87.
95. Boutant M, Joffraud M, Kulkarni SS, Garcia-Casarrubios E, Garcia-Roves PM, Ratajczak J, Fernandez-Marcos PJ, Valverde AM, Serrano M, Canto C. SIRT1 enhances glucose tolerance by potentiating brown adipose tissue function. *Mol Metab* 2015;**4**:118–31.
96. Jang M, Cai L, Udeani GO, Slowing KV, Thomas CF, Beecher CW, Fong HH, Farnsworth NR, Kinghorn AD, Mehta RG, Moon RC, Pezzuto JM. Cancer chemopreventive activity of resveratrol, a natural product derived from grapes. *Science* 1997;**275**:218–20.
97. Milne JC, Lambert PD, Schenk S, Carney DP, Smith JJ, Gagne DJ, Jin L, Boss O, Perni RB, Vu CB, Bemis JE, Xie R, Disch JS, Ng PY, Nunes JJ, Lynch AV, Yang H, Galonek H, Israelian K, Choy W, Iffland A, Lavu S, Medvedik O, Sinclair DA, Olefsky JM, Jirousek MR, Elliott PJ, Westphal CH. Small molecule activators of SIRT1 as therapeutics for the treatment of type 2 diabetes. *Nature* 2007;**450**:712–6.
98. Minor RK, Baur JA, Gomes AP, Ward TM, Csiszar A, Mercken EM, Abdelmohsen K, Shin YK, Canto C, Scheibye-Knudsen M, Krawczyk M, Irusta PM, Martin-Montalvo A, Hubbard BP, Zhang Y, Lehrmann E, White AA, Price NL, Swindell WR, Pearson KJ, Becker KG, Bohr VA, Gorospe M, Egan JM, Talan MI, Auwerx J, Westphal CH, Ellis JL, Ungvari Z, Vlasuk GP, Elliott PJ, Sinclair DA, de Cabo R. SRT1720 improves survival and healthspan of obese mice. *Sci Rep* 2011;**1**:70.
99. Baur JA, Pearson KJ, Price NL, Jamieson HA, Lerin C, Kalra A, Prabhu VV, Allard JS, Lopez-Lluch G, Lewis K, Pistell PJ, Poosala S, Becker KG, Boss O, Gwinn D, Wang M, Ramaswamy S, Fishbein KW, Spencer RG, Lakatta EG, Le Couteur D, Shaw RJ, Navas P, Puigserver P, Ingram DK, de Cabo R, Sinclair DA. Resveratrol improves health and survival of mice on a high-calorie diet. *Nature* 2006;**444**:337–42.
100. Hubbard BP, Gomes AP, Dai H, Li J, Case AW, Considine T, Riera TV, Lee JE, E Y, Lamming DW, Pentelute BL, Schuman ER, Stevens LA, Ling AJ, Armour SM, Michan S, Zhao H, Jiang Y, Sweitzer SM, Blum CA, Disch JS, Ng PY, Howitz KT, Rolo AP, Hamuro Y, Moss J, Perni RB, Ellis JL, Vlasuk GP, Sinclair DA. Evidence for a common mechanism of SIRT1 regulation by allosteric activators. *Science* 2013;**339**:1216–9.
101. Pacholec M, Bleasdale JE, Chrunyk B, Cunningham D, Flynn D, Garofalo RS, Griffith D, Griffor M, Loulakis P, Pabst B, Qiu X, Stockman B, Thanabal V, Varghese A, Ward J, Withka J, Ahn K. SRT1720, SRT2183, SRT1460, and resveratrol are not direct activators of SIRT1. *J Biol Chem* 2010;**285**:8340–51.
102. Um JH, Park SJ, Kang H, Yang S, Foretz M, McBurney MW, Kim MK, Viollet B, Chung JH. AMP-activated protein kinase-deficient mice are resistant to the metabolic effects of resveratrol. *Diabetes* 2010;**59**:554–63.
103. Park SJ, Ahmad F, Um JH, Brown AL, Xu X, Kang H, Ke H, Feng X, Ryall J, Philp A, Schenk S, Kim MK, Sartorelli V, Chung JH. Specific Sirt1 activator-mediated improvement in glucose homeostasis requires Sirt1-independent activation of AMPK. *EBioMedicine* 2017;**18**:128–38.
104. Hawley SA, Ross FA, Chevtzoff C, Green KA, Evans A, Fogarty S, Towler MC, Brown LJ, Ogunbayo OA, Evans AM, Hardie DG. Use of cells expressing gamma subunit variants to identify diverse mechanisms of AMPK activation. *Cell Metab* 2010;**11**:554–65.
105. Cantó C, Gerhart-Hines Z, Feige JN, Lagouge M, Noriega L, Milne JC, Elliott PJ, Puigserver P, Auwerx J. AMPK regulates energy expenditure by modulating NAD+ metabolism and SIRT1 activity. *Nature* 2009;**458**:1056–60.
106. White AT, McCurdy CE, Philp A, Hamilton DL, Johnson CD, Schenk S. Skeletal muscle-specific overexpression of SIRT1 does not enhance whole-body energy expenditure or insulin sensitivity in young mice. *Diabetologia* 2013;**56**:1629–37.
107. Pfluger PT, Herranz D, Velasco-Miguel S, Serrano M, Tschöp MH. Sirt1 protects against high-fat diet-induced metabolic damage. *Proc Natl Acad Sci* 2008;**105**:9793–8.
108. Banks AS, Kon N, Knight C, Matsumoto M, Gutierrez-Juarez R, Rossetti L, Gu W, Accili D. SirT1 gain of function increases energy efficiency and prevents diabetes in mice. *Cell Metab* 2008;**8**:333–41.

109. Ratajczak J, Joffraud M, Trammell SA, Ras R, Canela N, Boutant M, Kulkarni SS, Rodrigues M, Redpath P, Migaud ME, Auwerx J, Yanes O, Brenner C, Canto C. NRK1 controls nicotinamide mononucleotide and nicotinamide riboside metabolism in mammalian cells. *Nat Commun* 2016;**7**:13103.
110. Canto C, Houtkooper RH, Pirinen E, Youn DY, Oosterveer MH, Cen Y, Fernandez-Marcos PJ, Yamamoto H, Andreux PA, Cettour-Rose P, Gademann K, Rinsch C, Schoonjans K, Sauve AA, Auwerx J. The NAD(+) precursor nicotinamide riboside enhances oxidative metabolism and protects against high-fat diet-induced obesity. *Cell Metab* 2012;**15**:838–47.
111. Yoshino J, Mills KF, Yoon MJ, Imai S. Nicotinamide mononucleotide, a key NAD(+) intermediate, treats the pathophysiology of diet- and age-induced diabetes in mice. *Cell Metab* 2011;**14**:528–36.
112. Kourtzidis IA, Stoupas AT, Gioris IS, Veskoukis AS, Margaritelis NV, Tsantarliotou M, Taitzoglou I, Vrabas IS, Paschalis V, Kyparos A, Nikolaidis MG. The NAD(+) precursor nicotinamide riboside decreases exercise performance in rats. *J Int Soc Sports Nutr* 2016;**13**:32.
113. Zhang H, Ryu D, Wu Y, Gariani K, Wang X, Luan P, D'Amico D, Ropelle ER, Lutolf MP, Aebersold R, Schoonjans K, Menzies KJ, Auwerx J. NAD(+) repletion improves mitochondrial and stem cell function and enhances life span in mice. *Science* 2016;**352**:1436–43.
114. Trammell SA, Schmidt MS, Weidemann BJ, Redpath P, Jaksch F, Dellinger RW, Li Z, Abel ED, Migaud ME, Brenner C. Nicotinamide riboside is uniquely and orally bioavailable in mice and humans. *Nat Commun* 2016;**7**:12948.
115. Canto C, Auwerx J. Calorie restriction: is AMPK a key sensor and effector? *Physiology* 2011;**26**:214–24.
116. Jager S, Handschin C, St-Pierre J, Spiegelman BM. AMP-activated protein kinase (AMPK) action in skeletal muscle via direct phosphorylation of PGC-1alpha. *Proc Natl Acad Sci U S A* 2007;**104**:12017–22.
117. Canto C, Auwerx J. AMP-activated protein kinase and its downstream transcriptional pathways. *Cell Mol Life Sci: CMLS* 2010;**67**:3407–23.
118. Hardie DG. AMP-activated/SNF1 protein kinases: conserved guardians of cellular energy. *Nat Rev Mol Cell Biol* 2007;**8**:774–85.
119. Myers RW, Guan HP, Ehrhart J, Petrov A, Prahalada S, Tozzo E, Yang X, Kurtz MM, Trujillo M, Gonzalez Trotter D, Feng D, Xu S, Eiermann G, Holahan MA, Rubins D, Conarello S, Niu X, Souza SC, Miller C, Liu J, Lu K, Feng W, Li Y, Painter RE, Milligan JA, He H, Liu F, Ogawa A, Wisniewski D, Rohm RJ, Wang L, Bunzel M, Qian Y, Zhu W, Wang H, Bennet B, LaFranco Scheuch L, Fernandez GE, Li C, Klimas M, Zhou G, van Heek M, Biftu T, Weber A, Kelley DE, Thornberry N, Erion MD, Kemp DM, Sebhat IK. Systemic pan-AMPK activator MK-8722 improves glucose homeostasis but induces cardiac hypertrophy. *Science* 2017;**357**:507–11.
120. Cool B, Zinker B, Chiou W, Kifle L, Cao N, Perham M, Dickinson R, Adler A, Gagne G, Iyengar R, Zhao G, Marsh K, Kym P, Jung P, Camp HS, Frevert E. Identification and characterization of a small molecule AMPK activator that treats key components of type 2 diabetes and the metabolic syndrome. *Cell Metab* 2006;**3**:403–16.
121. Cokorinos EC, Delmore J, Reyes AR, Albuquerque B, Kjobsted R, Jorgensen NO, Tran JL, Jatkar A, Cialdea K, Esquejo RM, Meissen J, Calabrese MF, Cordes J, Moccia R, Tess D, Salatto CT, Coskran TM, Opsahl AC, Flynn D, Blatnik M, Li W, Kindt E, Foretz M, Viollet B, Ward J, Kurumbail RG, Kalgutkar AS, Wojtaszewski JFP, Cameron KO, Miller RA. Activation of skeletal muscle AMPK promotes glucose disposal and glucose lowering in non-human primates and mice. *Cell Metab* 2017;**25**:1147–59. e1110.
122. Marcinko K, Bujak AL, Lally JS, Ford RJ, Wong TH, Smith BK, Kemp BE, Jenkins Y, Li W, Kinsella TM, Hitoshi Y, Steinberg GR. The AMPK activator R419 improves exercise capacity and skeletal muscle insulin sensitivity in obese mice. *Mol Metab* 2015;**4**:643–51.
123. Narkar VA, Downes M, Yu RT, Embler E, Wang YX, Banayo E, Mihaylova MM, Nelson MC, Zou Y, Juguilon H, Kang H, Shaw RJ, Evans RM. AMPK and PPARdelta agonists are exercise mimetics. *Cell* 2008;**134**:405–15.
124. O'Neill HM, Maarbjerg SJ, Crane JD, Jeppesen J, Jorgensen SB, Schertzer JD, Shyroka O, Kiens B, van Denderen BJ, Tarnopolsky MA, Kemp BE, Richter EA, Steinberg GR. AMP-activated protein kinase (AMPK) beta1beta2 muscle null mice reveal an essential role for AMPK in maintaining mitochondrial content and glucose uptake during exercise. *Proc Natl Acad Sci U S A* 2011;**108**:16092–7.
125. Schiaffino S, Dyar KA, Ciciliot S, Blaauw B, Sandri M. Mechanisms regulating skeletal muscle growth and atrophy. *FEBS J* 2013;**280**:4294–314.
126. Arad M, Benson DW, Perez-Atayde AR, McKenna WJ, Sparks EA, Kanter RJ, McGarry K, Seidman JG, Seidman CE. Constitutively active AMP kinase mutations cause glycogen storage disease mimicking hypertrophic cardiomyopathy. *J Clin Invest* 2002;**109**:357–62.
127. Rena G, Pearson ER, Sakamoto K. Molecular mechanism of action of metformin: old or new insights? *Diabetologia* 2013;**56**:1898–906.

128. Owen MR, Doran E, Halestrap AP. Evidence that metformin exerts its anti-diabetic effects through inhibition of complex 1 of the mitochondrial respiratory chain. *Biochem J* 2000;**348**(Pt 3):607–14.
129. El-Mir MY, Nogueira V, Fontaine E, Averet N, Rigoulet M, Leverve X. Dimethylbiguanide inhibits cell respiration via an indirect effect targeted on the respiratory chain complex I. *J Biol Chem* 2000;**275**:223–8.
130. Zhou G, Myers R, Li Y, Chen Y, Shen X, Fenyk-Melody J, Wu M, Ventre J, Doebber T, Fujii N, Musi N, Hirshman MF, Goodyear LJ, Moller DE. Role of AMP-activated protein kinase in mechanism of metformin action. *J Clin Invest* 2001;**108**:1167–74.
131. Salpeter SR, Greyber E, Pasternak GA, Salpeter EE. Risk of fatal and nonfatal lactic acidosis with metformin use in type 2 diabetes mellitus: systematic review and meta-analysis. *Arch Intern Med* 2003;**163**:2594–602.
132. Foretz M, Hebrard S, Leclerc J, Zarrinpashneh E, Soty M, Mithieux G, Sakamoto K, Andreelli F, Viollet B. Metformin inhibits hepatic gluconeogenesis in mice independently of the LKB1/AMPK pathway via a decrease in hepatic energy state. *J Clin Invest* 2010;**120**:2355–69.
133. Shaw RJ, Lamia KA, Vasquez D, Koo SH, Bardeesy N, Depinho RA, Montminy M, Cantley LC. The kinase LKB1 mediates glucose homeostasis in liver and therapeutic effects of metformin. *Science* 2005;**310**:1642–6.
134. Wilcock C, Bailey CJ. Accumulation of metformin by tissues of the normal and diabetic mouse. *Xenobiotica: Fate Foreign Compounds Biol Syst* 1994;**24**:49–57.
135. Shu Y, Sheardown SA, Brown C, Owen RP, Zhang S, Castro RA, Ianculescu AG, Yue L, Lo JC, Burchard EG, Brett CM, Giacomini KM. Effect of genetic variation in the organic cation transporter 1 (OCT1) on metformin action. *J Clin Invest* 2007;**117**:1422–31.
136. Koepsell H, Lips K, Volk C. Polyspecific organic cation transporters: structure, function, physiological roles, and biopharmaceutical implications. *Pharm Res* 2007;**24**:1227–51.
137. Luft D, Schmulling RM, Eggstein M. Lactic acidosis in biguanide-treated diabetics: a review of 330 cases. *Diabetologia* 1978;**14**:75–87.
138. Perry RJ, Zhang D, Zhang XM, Boyer JL, Shulman GI. Controlled-release mitochondrial protonophore reverses diabetes and steatohepatitis in rats. *Science* 2015;**347**:1253–6.
139. Kenwood BM, Weaver JL, Bajwa A, Poon IK, Byrne FL, Murrow BA, Calderone JA, Huang L, Divakaruni AS, Tomsig JL, Okabe K, Lo RH, Cameron Coleman G, Columbus L, Yan Z, Saucerman JJ, Smith JS, Holmes JW, Lynch KR, Ravichandran KS, Uchiyama S, Santos WL, Rogers GW, Okusa MD, Bayliss DA, Hoehn KL. Identification of a novel mitochondrial uncoupler that does not depolarize the plasma membrane. *Mol Metab* 2014;**3**:114–23.
140. Youle RJ, Narendra DP. Mechanisms of mitophagy. *Nat Rev Mol Cell Biol* 2011;**12**:9–14.
141. Twig G, Elorza A, Molina AJ, Mohamed H, Wikstrom JD, Walzer G, Stiles L, Haigh SE, Katz S, Las G, Alroy J, Wu M, Py BF, Yuan J, Deeney JT, Corkey BE, Shirihai OS. Fission and selective fusion govern mitochondrial segregation and elimination by autophagy. *EMBO J* 2008;**27**:433–46.
142. Narendra D, Tanaka A, Suen DF, Youle RJ. Parkin is recruited selectively to impaired mitochondria and promotes their autophagy. *J Cell Biol* 2008;**183**:795–803.
143. Chen Y, Dorn 2nd GW. PINK1-phosphorylated mitofusin 2 is a Parkin receptor for culling damaged mitochondria. *Science* 2013;**340**:471–5.
144. Lee JY, Nagano Y, Taylor JP, Lim KL, Yao TP. Disease-causing mutations in parkin impair mitochondrial ubiquitination, aggregation, and HDAC6-dependent mitophagy. *J Cell Biol* 2010;**189**:671–9.
145. Matsuda N, Sato S, Shiba K, Okatsu K, Saisho K, Gautier CA, Sou YS, Saiki S, Kawajiri S, Sato F, Kimura M, Komatsu M, Hattori N, Tanaka K. PINK1 stabilized by mitochondrial depolarization recruits Parkin to damaged mitochondria and activates latent Parkin for mitophagy. *J Cell Biol* 2010;**189**:211–21.
146. McWilliams TG, Prescott AR, Montava-Garriga L, Ball G, Singh F, Barini E, Muqit MMK, Brooks SP, Ganley IG. Basal mitophagy occurs independently of PINK1 in mouse tissues of high metabolic demand. *Cell Metab* 2018;**27**:439–49. e435.
147. Quiros PM, Mottis A, Auwerx J. Mitonuclear communication in homeostasis and stress. *Nat Rev Mol Cell Biol* 2016;**17**:213–26.
148. Hartl FU, Bracher A, Hayer-Hartl M. Molecular chaperones in protein folding and proteostasis. *Nature* 2011;**475**:324–32.
149. Haynes CM, Ron D. The mitochondrial UPR - protecting organelle protein homeostasis. *J Cell Sci* 2010;**123**:3849–55.
150. Hill S, Van Remmen H. Mitochondrial stress signaling in longevity: a new role for mitochondrial function in aging. *Redox Biol* 2014;**2**:936–44.

151. Quiros PM, Prado MA, Zamboni N, D'Amico D, Williams RW, Finley D, Gygi SP, Auwerx J. Multi-omics analysis identifies ATF4 as a key regulator of the mitochondrial stress response in mammals. *J Cell Biol* 2017;**216**:2027–45.
152. Mouchiroud L, Houtkooper RH, Moullan N, Katsyuba E, Ryu D, Canto C, Mottis A, Jo YS, Viswanathan M, Schoonjans K, Guarente L, Auwerx J. The NAD(+)/Sirtuin pathway modulates longevity through activation of mitochondrial UPR and FOXO signaling. *Cell* 2013;**154**:430–41.
153. Gariani K, Menzies KJ, Ryu D, Wegner CJ, Wang X, Ropelle ER, Moullan N, Zhang H, Perino A, Lemos V, Kim B, Park YK, Piersigilli A, Pham TX, Yang Y, Ku CS, Koo SI, Fomitchova A, Canto C, Schoonjans K, Sauve AA, Lee JY, Auwerx J. Eliciting the mitochondrial unfolded protein response by nicotinamide adenine dinucleotide repletion reverses fatty liver disease in mice. *Hepatology* 2016;**63**:1190–204.
154. Houtkooper RH, Mouchiroud L, Ryu D, Moullan N, Katsyuba E, Knott G, Williams RW, Auwerx J. Mitonuclear protein imbalance as a conserved longevity mechanism. *Nature* 2013;**497**:451–7.

PART VI

# SYNTHESIS

CHAPTER

# 18

# Mitochondria in Obesity and Type 2 Diabetes: Concluding Review and Research Perspectives

*Béatrice Morio*, Luc Pénicaud†, Michel Rigoulet‡*

*CarMeN Laboratory, Research Unit U. 1060 Inserm/ Lyon1 University/INRA 1397, Lyon Sud Medical School, Oullins, France †STROMALab, Université de Toulouse, EFS, ENVT, Inserm U1031, ERL CNRS 5311, CHU Rangueil, Toulouse, France ‡Institut de Biochimie et Génétique Cellulaires, UMR 5095, CNRS, Université de Bordeaux, Bordeaux, France

## 1 INTRODUCTION

Obesity and type 2 diabetes (T2DM) are metabolic disorders that have emerged as two of the most serious health issues in the world. Both are related to chronic inequality between food intake and energy expenditure, giving rise to energy imbalance over long-term periods of time. This imbalance has major consequences on the metabolic fluxes in the different tissues or organs and at the whole organism level. Within the cell, mitochondria are at the front line to initiate adaptation to these changes and, therefore, are the primary targets affected. Mitochondria are key organelles in which fuel substrates are oxidized to produce ATP from cascades of redox reactions. An important notion to understand the etiology of metabolic diseases, however, is that in all tissues and organs, energy and redox processes are closely interlinked within the mitochondria and govern major cell metabolic adaptations (see review[1]). This chapter is not a summary of the previous ones but has two objectives: to emphasize notions that, in the opinion of the three of us, appear to be most important, and to propose some directions for future research.

## 2 TUNING OF METABOLIC FLUXES IN OBESITY AND T2DM: THE CRUCIAL ROLE OF THE REDOX STATE

Extensive metabolic reorganization occurs in obesity and T2DM in response to excess nutrient intake. This takes place at all levels of the organism, from the interaction between tissues and organs down to the cell.



Mitochondria in Obesity and Type 2 Diabetes
https://doi.org/10.1016/B978-0-12-811752-1.00018-3

Within the cell, mitochondria are the site where coupling between nutrient oxidation and ATP production is the most efficient because of the oxidative phosphorylation (OXPHOS). OXPHOS has a huge ability to recover energy from the reduced coenzymes (e.g., NADH) produced during nutrient oxidation and to couple it to fuel ATP synthesis. This mechanism is the most efficient to maintain the NADH level in the cell, which features the redox potential.

In obesity and T2DM, energy conversion and ATP synthesis have been the most studied factors up to now. In physiopathological situations, however, cells better adjust to a deficit in ATP synthesis than to an insufficient (re)oxidation of reduced cofactors, which leads to mitochondrial metabolic reorganization and oxidative stress.[2–4] When substrate availability exceeds energy needs, the immediate consequence is an alteration of the redox potential, characterized by increased content in reduced coenzymes (NAD(P)H, FADH2) in the cytosol and, most importantly, in the mitochondria. In that condition, the need for ATP supply and NADH oxidation within the mitochondria can be kinetically different. In other words, the need for ATP synthesis can be low, whereas that for NADH oxidation is high. Adaptive processes enable mitochondrial substrates oxidation to increase without altering ATP production, thus wasting as heat the extra redox energy brought by nutrients. These adaptive processes occur early in response to energy imbalance and thus characterize the first step of mitochondrial adaptations in main tissues involved in the pathophysiology of obesity and T2DM (e.g., liver, skeletal muscle, heart, see Chapters 6, 8, and 10). In those tissues, the main consequences are an increased reliance on fatty acid β-oxidation, enhanced mitochondrial respiration rate, decreased ATP/O ratio, and heightened mitochondrial production of reactive oxygen species (mtROS). Mitochondrial adaptive processes, however, differ between tissues. These alterations, observed during the first phase of the onset of the pathologies, can still be present in the longer term, although the energy intake is no longer excessive. This can be explained by changes in metabolic fluxes and to the appearance of metabolic inflexibility. Overall, the observations show that the redox state (i.e., NADH availability), but not ATP production, is the main conductor regulating the metabolic reorganization of cells, tissues, and organs in obesity and T2DM.

Within mitochondria, complex I of the respiratory chain plays a central role in the regulation of substrate oxidation and mtROS production. Complex I is the final common and obligatory step for NADH oxidation because all metabolic pathways leading to NADH production are completed at this level, a location of tight and complex control. When considering mitochondrial oxidation of carbohydrate and lipids, competition occurs between the dehydrogenases of the two specific pathways (pyruvate dehydrogenase and 3-hydroxyacyl-CoA dehydrogenase) as well as those of the Krebs cycle (isocitrate dehydrogenase, α-ketoglutarate dehydrogenase, and malate dehydrogenase). Furthermore, because malate dehydrogenase is also the matricial NADH supplier of the malate–aspartate shuttle, this step constitutes a crossroad among cytosolic and mitochondrial redox states, mitochondrial proton-motive force, pyruvate and fatty acid oxidation.

Given this highly regulated and complex situation, with multiple kinetic links among different and interconnected pathways, at least two opposite and extreme pictures can be considered. In the first, the redox state of the bulk phase of the matricial compartment represents the common intermediate. In this situation, the flux control of the different pathways depends on the capacity of complex I to oxidize NADH and on the specific elasticity and kinetic regulations of each dehydrogenase of the whole system toward the matricial $NADH/NAD^+$

ratio. In the second, the channeling of electron transfers between each dehydrogenase (or some of them) and complex I is involved.[5,6] In this situation, the versatility of the supramolecular organization represents the main controlling factor. Most likely, the actual situation is the result of a combination of these two extreme possibilities in a dynamic compromise, which is continuously adjusted in response to the different physiological or pathological constraints. Nevertheless, because of the near-equilibrium between redox state and protonmotive force at the level of complex I, mitochondrial protonmotive force appears to be the key regulatory factor that determines both the rate and the nature of the substrates used for fueling the cell and the ratio between the production of NADH and FADH2. Moreover, mitochondrial protonmotive force determines complex I mtROS production by the reverse electron flux.

***Future direction:*** *Focus research on the redox state*

*Because the redox state is the main actor that regulates the metabolic reorganization in obesity and T2D, research must continue to decrypt the precise mechanisms involved. For that purpose, there is, more than ever, a need for using and/or developing reliable redox markers in fluids, cells, and cellular compartments, such as mitochondria. These include assessment of redox indicator metabolites (e.g., lactate/pyruvate, β-hydroxybutyrate/acetoacetate, isocitrate/α-cetoglutarat), biophysical methods, and genetically encoded fluorescent probes or proteins associated with real-time imaging, as described in the review from Sies et al.*[1] *For the time being, because each technique has its own limitation, it is strongly recommended to use more than one and cross-compare the results.*

## 3 EMPHASIS ON METABOLIC COMPARTMENTALIZATION, FROM CELL TO MITOCHONDRIA

Cells divide their intracellular space into subcellular compartments, allowing spatiotemporal coordination of various metabolic activities. It is now accepted that compartmentalization of metabolic pathways is essential for cellular functions. It allows the cells to perform different reactions at the same time, so that specific pools of metabolites and distinct signaling pathways are induced simultaneously in different parts of the cell. This is especially important when considering mitochondria, in which intracellular dynamic compartmentalization allows localized and circumscribed production of signals that promote activation of specific signaling pathways involved in various functions. Metabolically, these subcellular compartments favor the optimal activity of specific enzymes, the lesser diffusion of intermediates essential to the committed metabolic pathway, and less futile cycles of metabolites that could be continuously interconverted in reactions operating in opposite directions.

Considering the numerous links and interactions among the different metabolic pathways and the high level of compartmentalization of the cell, it is important to maintain some degree of independence between the interconnected pathways and a high level of coordination. These two strong constraints have led to a dynamic supramolecular organization of the metabolism, as can be seen in all levels of the mitochondria. Mitochondria are important sites of compartmentalization in the cell, interconnected with other organelles, and tightly orchestrated to adapt rapidly and transiently to stresses, such as increased nutrient availability.

*Within the cell,* this is obviously important to manage the dynamics and location of ATP production. For example, in neurons (see Chapters 12 and 14), mitochondria are localized at the most energy-demanding locations, that is, at synaptic terminals for synaptic transmission and axonal branching. Thus, any defect in the transfer, localization, and activity of mitochondria in neurons alter their function, such as nutrient sensing (see Chapter 12), and potentiate or contribute to the onset of neurodegenerative diseases, as observed in obesity and T2DM (see Chapter 14).

*Interactions between mitochondria and other organelles* (e.g. endoplasmic reticulum (ER), peroxisomes) allow the building of specialized and plastic microdomains. Contact points between mitochondria and ER, known as mitochondria-associated ER membranes (MAMs), are now subjected to extensive research in health and diseases (see review[7]). They have been identified as the control tower involved in the synthesis and exchange of specific molecules, such as calcium or phospholipids (see review[8]) as well as in the regulation of mitochondrial morphology, function, and retrograde signaling pathways (see reviews[9, 10]). The dynamics of interaction is tightly regulated by key proteins, such as mitofusins 1 and 2, and by the efficient structuring of exchange bridges between organelles, because we are beginning to understand about the contact points between mitochondria and ER (MAMs) (see review by Vance[10a]).

*At the organelle level,* mitochondria show substantial heterogeneity in morphology and membrane potential within the same cell. Mitochondrial network undergoes coordinated fusion and fission processes to maintain mitochondrial function (see Chapter 1). Recent studies have shown that mitochondrial dynamics is finely tuned according to the balance between cellular energy needs and nutrient availability. Furthermore, rapid and selective transitions between fission and fusion allow reorganization of mitochondrial components and removal of damaged material, thus maintaining a healthy mitochondrial population (i.e., mitochondrial quality control[11]). Recent studies have shown that mitochondrial adaptive processes in response to exposure to excess nutrient result in an increased uncoupled mitochondrial respiration and a decreased ATP production, and fragmentation of the mitochondrial network (see review[12]). This suggests that mitochondrial morphology might adapt to changes in the balance between nutrient availability and ATP needs, and that nutrient overload, and its impact on the redox state, might compromise mitochondrial dynamics and MAM integrity, thus contributing to alteration in mitochondrial metabolism and potentially to the onset of metabolic disorders (see review[12]). This last point is well-detailed in Chapters 12 and 14, which deal with the importance of mitochondrial dynamics in neuronal nutrient sensing and in the progression of neurodegenerative diseases.

*At the supramolecular level,* channeling defines the organization of a metabolic pathway in such a way that metabolites are conveyed from one enzyme to another without dilution in the bulk phase. This enzyme organization has long been recognized in the glycolysis pathway. In mammals, including humans, the binding of glycolytic enzymes to the cell membrane and cytoskeleton has been shown to play an important role in the compartmentation of the glycolytic supply of ATP to the sodium-potassium ATPase.[13] Similarly, fatty acid metabolism is channeled in the cytosol and in the mitochondria.[14, 15] It has been proposed that, after entry into the cell, the fate of fatty acids might be determined by supramolecular organization, which determines their site of activation and their subsequent distribution between main metabolic pathways (i.e., mitochondrial β-oxidation, phospholipids synthesis, cholesterol esterification and so on). The molecular mechanisms of fatty acid channeling, especially in obesity and T2DM, however, still need to be investigated further.

Finally, the organization of the respiratory chain is another example illustrating the metabolic channeling between reduced coenzymes and oxygen. Two alternative models for the arrangement of respiratory chain complexes in the inner membrane have been proposed. In the first model, all components of the respiratory chain diffuse freely and individually in the membrane according to the random collision model (also called liquid state model). Therefore, electron transfer depends on the random and transient interaction between individual respiratory chain complexes and electron carriers, such as quinones and cytochrome C.[16] The second model involves many different stoichiometric assemblies, named supercomplexes, which have been isolated from mitochondria of various origins.[17–19] The solid-state model of organization, in which ordered sequences of redox compounds catalyze electron fluxes, also is supported by kinetics analysis.[20, 21] Evidence obtained by inhibitor titration studies in bovine heart mitochondria are consistent with both models and suggest their coexistence.[22] The crucial function of these respiratory supercomplexes might be the stabilization of each individual complex.[23, 24] Taken together, data argue in favor of a supramolecular organization of the respiratory chain dynamically modulated according to distinct modalities and able to change in response to stimuli, effectors, or kinetic constraints. However, the question of the physiological significance of such organizations and of the limits of their plasticity, especially in obesity and T2DM, still remains.

***Future direction:*** *Focus research on microdomains and molecular organization*

*To decipher the precise mechanisms at the organelle or supramolecular level, reliable sensors are required to precisely assess dynamic alterations in molecular organization and located concentrations and fluxes of production (e.g., ROS, calcium, ATP). These developments should consider specific and sensitive biosensors with an appropriate Kd in order to fully understand the physiological meaning of low levels of the signaling molecules, and should allow combined exploration of signaling molecules (e.g., ROS and calcium, or calcium and ATP). Promising approaches might be feasible because of the fine-tuning of genetically encoded fluorescent probes targeted to specific cellular compartments, such as mitochondria. Combined with real-time recording of the fluorescence, these methods allow high spatial/temporal resolution.*

# 4 MITOCHONDRIAL SENSING AND SIGNALING

Any adaptation leading to changes in mitochondrial bioenergetics can affect three major factors produced by the mitochondria or regulating its function: ROS, calcium, and ATP. Those factors are crucial signal molecules that play key roles in physiology and pathology.

*Mitochondrial redox signaling* is a central actor in the complex cell signaling network in support of cellular differentiation, growth, and maintenance, in most tissues and organs. Mitochondrial oxygen consumption plays a central role in regulating ROS signaling through mtROS generation, especially through the reverse electron transfer in complex I. Under physiological conditions, increase in mtROS is transient (because the antioxidant defenses rapidly quench mtROS before any damage to the cell occurs) and is a key intracellular messenger for rapidly regulating pathways that directly depend on metabolic fluxes (mostly those related to glucose and lipids). There is now extensive evidence supporting an active role for mtROS signaling in physiology and adaptive responses. In that respect, Chapters 7, 9, and 12 describe why and how mtROS production is crucial for adipocyte, pancreatic beta-cell and neuron proliferation, differentiation and/or functioning. Chapter 16 describes the importance of mtROS in improving skeletal muscle insulin signaling in the adaptation to an exercise stimulus.

*Mitochondrial calcium ($Ca^{2+}$) sensing* regulates numerous cellular functions. In the resting state, the cytosolic concentration in $Ca^{2+}$ is very low, contrasting with that observed in the extracellular medium and in certain organelles, such as the ER. This helps to build a highly responsive system, which requires fine regulation because almost all aspects of the cell function are controlled by $Ca^{2+}$. The specific $Ca^{2+}$ sensing mechanisms located at the mitochondria are essential for OXPHOS activity (see Chapter 2). Mitochondrial $Ca^{2+}$ uptake involves tightly orchestrated redox-sensitive interactions between mitochondria and ER at the MAMs' interface (see reviews[25, 26]). Mitochondrial $Ca^{2+}$ sensing emphasis for maintenance of pancreatic β-cells insulin secretion and heart contractibility is discussed in Chapters 9 and 10, respectively.

*Mitochondrial ATP* is involved as an intracellular signal in numerous physiological regulations at the level of different cell types. This is particularly true in pancreatic β-cells and in neuronal cells (see Chapters 9 and 12, respectively). ATP serves as a producer of phosphate and modulates the activity of enzymes and channels, particularly those involved in signal transduction. Another intracellular mechanism involving ATP is that, being the substrate of the adenylate cyclase, it can be transformed into cyclic AMP, a second messenger that triggers calcium signaling by inducing calcium release from intracellular stores. Finally, some cells secrete ATP to communicate with other cells in a process called purinergic signaling.

***Future direction:*** *Open research to new signaling molecules*

*Involvement of major signaling molecules in the pathophysiology of obesity and T2DM has been studied. In addition to ROS, calcium and ATP, however, emphasis should be placed on nitric oxide (NO) and hydrogen sulfide ($H_2S$), which play key roles in mitochondrial sensing and metabolic adaptation. Until recently, NO was virtually impossible to visualize in single cells. The issue has been potentially tackled because of the development of genetically encoded targeted probes (geNOps), which provide a selective, specific and real-time readout of (sub)cellular dynamics. Targeted fluorescent probes for $H_2S$ also are under development.*

## 5 NEED TO STUDY FLUXES MORE THAN STEADY STATE CONCENTRATIONS OF KEY NUTRIENTS AND NUTRIENT-DERIVED METABOLITES

Metabolic profiling allowed investigation of metabolite contents in cells and tissues in physiological and physiopathological situations, highlighting the complexity and the adaptability of the metabolic network. This is beginning to be well understood in various tissues, such as skeletal muscle (see Chapter 6) or kidney (see Chapter 12).

Until now, it has been almost impossible to analyze metabolite content and its kinetics of alteration in small subcellular fractions, although it was obvious that cellular compartments have different metabolomes, regulated by specific and dynamic processes of compartmentalization.[27] In spite of the known importance of organelles compartmentalization, relatively little is known about changes in metabolite content in physiopathological situations, especially within the mitochondria. Overcoming the barrier of compartmentalization is crucial for providing a better understanding of the extent to which compartmentalization of metabolites contributes to physiology and to development of pathologies.[28] This involves developing organelle-specific metabolic profiling in all cell types and experimental models. In that context, picturing mitochondrial metabolome has been made possible because of a new approach developed by Chen et al.[29] The method combines a rapid immunocapture of epitope-tagged mitochondria with a metabolic profiling of metabolites using liquid chromatography and mass spectrometry (LC/MS).

Metabolic adaptations engage dynamic changes in fluxes of key metabolites within and between compartments, from organelles to the cell, tissue, and whole body. To overcome static picturing of metabolomes, a current challenge is to optimize the use of technologies to translate metabolomes into a network of metabolic fluxes (fluxome) from which they are derived.[30] The fluxome provides functional information partaking in the pathology and allows identifying key steps regulating the metabolite fluxes. New modeling approaches allow translating a profile of metabolites into a predicted set of fluxes. Methodologically, computational modeling is integrated with an analytical platform involving linear optimization, continuation and dynamic analyzes, and metabolic control.[30]

***Future direction:*** *Move from metabolomics to fluxomics at all levels of the organism*

*As methodologies improve, allowing better understanding of the alterations in the localization and dynamics of the metabolic reorganizations associated with obesity and T2D, it is important to develop analytical tools and integrated procedures to translate metabolomics into fluxomics. In addition to analytical apparatus and tools, this requires databases, mathematical methods, and metabolic network analysis.*[30]

*New research also is required to develop new approaches and tools for exploring compartment-specific as well as single-cell metabolome, and decipher their alteration in response to stresses. In-depth studies of metabolism might include the quantification of metabolic fluxes between intracellular compartments (flow between mitochondria, cytosol, peroxisome, etc.), up to the level of the single cell, between tissues and organs, up to the whole organism.*

## 6 NECESSITY FOR CHRONOLOGICAL FOLLOW UP OVER LONG PERIODS OF TIME

Numerous studies that explored the involvement of mitochondria in the etiology of pathologies often have compared a healthy situation to a pathological state. These observations, partial and incomplete, first were interpreted as demonstrating a causal link between mitochondrial dysfunction and pathology. However, it is now well accepted that:

- Mitochondria do not adapt linearly to alterations in cell metabolism (mainly redox state),
- Mitochondrial functions (Krebs cycle, respiratory chain, phosphorylation, shuttles, etc.) transiently and differentially adapt in response to alterations in cell metabolism,
- Mitochondrial intrinsic adaptations (increased anaplerosis, fatty acid oxidation, mtROS production, etc.) amply participate to metabolic inflexibility and metabolic disorders,
- Decreased mitochondrial mass and oxidative capacity is not the cause but rather a consequence of combined stresses occurring during the onset of metabolic disorders.

Therefore, kinetic observations are a prerequisite to better understand the complex and numerous mechanisms linking mitochondria to the pathology. Three main phases can be proposed: the early response, the compensatory adaptive process, and the response to toxicity. It is important to conduct studies that take into account the duration and the evolution of the pathology, from the first stages of the metabolic disorders to the occurrence of the symptoms.

Furthermore, one has to take into account that metabolism and metabolic adaptations in response to energy imbalance largely vary according to tissues and organs and contribute differently over time to the onset of metabolic disorders. Particularly, mitochondria are heavily specialized to achieve the metabolic pathways carried out by tissues and organs. Therefore, mechanisms involving mitochondria in the metabolic disorders are specific and largely differ in term of quality, kinetics, and timing (early or late stages of pathology). This supports the importance of developing tissue-specific approaches along with longitudinal follow up.

***Future direction:*** *Necessity for long-term follow up with integrative view*

*Because only a few studies have considered the chronology of the metabolic reorganizations associated with obesity and T2D, it is important to develop follow-up studies with an integrative view. This statement is obvious and essential, however, because we need to understand precisely the chronology of mitochondrial adaptations in each tissue and organ affected during obesity and T2DM.*

# 7 TREATMENTS TARGETING MITOCHONDRIA

Targeting mitochondria in obesity and T2DM is a major challenge. Benefits of lifestyle and surgical interventions on mitochondrial health are multifactorial and complex in their mechanisms (see Chapter 16). On a practical basis, regular physical activity is the best way to prevent excessive weight gain and its associated pathologies. From a purely bioenergetic point of view, beyond the complex beneficial physiological impacts of physical activity, it increases the demand without changing yields (see Chapters 16 and 17).

Some drugs, including xenobiotics such as ethanol and some environmental pollutants, can aggravate progression of metabolic disorders associated with obesity and T2DM through alteration of mitochondrial function. Examples are given in Chapter 15 regarding potentiation of nonalcoholic fatty liver diseases by xenobiotic-induced alteration of liver mitochondria. Few drugs targeting mitochondria have yet to demonstrate their efficiency to improve features of obesity and T2DM (see Chapter 17).

Metformin is the perfect example. For more than a half-century, it has been the most widely prescribed drug to treat T2DM patients. Although still not fully understood, the mechanisms of metformin action in hepatocytes are thought to involve inhibition of complex I[31, 32] and of the mitochondrial redox shuttle glycerophosphate dehydrogenase (G3PDH).[33] Because complex I is less inhibited than G3PDH shuttle, the result is a redirection of the respiratory flow toward complex I, a decrease in glucose synthesis from glycerol (because of inhibition of the G3PDH shuttle) and lactate (because of an increase in cytosolic NADH/NAD+ ratio). Overall, metformin-induced alterations in the redox states (increased in the cytosol and decreased in the mitochondria), of which major outcomes are decreased mitochondrial ATP and ROS production (metformin inhibiting more the reverse flux of electrons than the forward flow at complex I) ([34]); increased glycolysis; and reduced conversion of lactate and glycerol to glucose (thus decreasing the hepatic gluconeogenesis). The ensuing enhanced cytosolic AMP/ATP ratio stimulates AMPK signaling, whereas inhibiting that of cAMP/PKA, thus improving hepatocyte glucose and lipid homeostasis.

One common and important mechanism of metformin's impact on mitochondrial function in tissues and organs other than liver is the drastic inhibition of mtROS production by complex I through the reverse electron flux. Metabolic outcomes, however, can fully differ from those observed in the liver, highlighting the importance of mitochondrial specificities and their adaptation to each tissue function. Although controversial,[35] metformin has been shown to dose-dependently impair in vivo and ex vivo skeletal muscle oxidative capacity in Zucker diabetic fatty rats.[36]

***Future direction:*** *Toward new drug development targeting mitochondria*

*Numerous therapeutic strategies have been put forward, because many potential targets have been identified, from OXPHOS uncoupling to activity of specific enzymes or shuttles, and dynamics of interaction with other organelles (see Chapter 17). $NAD^+$ precursors are also in development to activate mitochondrial metabolism (see Chapter 17). To our understanding, the most interesting treatments are those that can play on the redox state or on the OXPHOS coupling without being an uncoupler (named bypass). In this context, treatments reduce the yield without changing the demand. Works that used respiratory bypass enzymes (NDI1, alternative NADH dehydrogenase, which bypasses complex I inhibition; AOX, alternative oxidase, which bypasses complex III or IV inhibition) in vitro and in genetically modified models have provided evidence of beneficial impacts on cellular metabolism in situations of mitochondrial dysfunction, especially with regards to decreased mtROS production. This could provide new opportunities for preventing or treating obesity and T2DM, however, the work is still at the preliminary stage.*

*Finally, one has to remember that mitochondria are specialized to support each tissue function, and a drug might have opposite effects depending on the tissues or organs. For example, it is well known that uncoupling is an efficient and safe mechanism in the brown adipose tissue but not in the heart, brain, skeletal muscle, etc. (see Chapters 7 and 17). Therefore, targeting mitochondria would be safer and more efficient if tissue-specific.*

## 8 GENERAL CONCLUSION

In physiological and pathological situations, the need for ATP turnover and NADH oxidation flux can be kinetically different. Therefore, when energy substrates far exceed the energy needs, imbalance between ATP synthesis and NADH oxidation leads to an increase in the redox state, which often is linked to an increase in mtROS production and alteration in mitochondrial sensing and signaling processes. Metabolic disorders associated with obesity and T2D result from the fact that the cells adapt better to an ATP synthesis deficiency than to an insufficient (re)oxidation of reduced cofactors. This supports the conclusion that complex I activity and alteration in the redox state are central to the metabolic adaptations in obesity and T2DM. Therefore, any strategy aimed at enhancing NADH oxidation independently from ATP production could be advantageous in preventing metabolic disorders. Because mitochondria are considerably specialized, however, any strategy targeting mitochondrial function has to demonstrate selectivity, given the diversity of its action in the different tissues and organs.

To support the understanding of mitochondrial specificity and involvement in the physiopathology of obesity and T2DM, methodological perspectives point to the need to further develop targeted approaches to subcellular compartments. This will enable exploring signaling pathways and regulatory mechanisms at the microdomain level, taking into account the consequences on subcellular metabolic network reorganization. This can lead to new therapeutic strategies targeting specific pools of compartmentalized enzymes/proteins, such as those involved in the dynamics and function of MAMs.

## Acknowledgments

We would like to thank Eric Fontaine and Yves Gouriou for helpful discussions. This work is dedicated to Pr. Xavier Leverve and Bernard Beaufrère, whose commitment to bioenergetics and clinical researches have inspired the editorial team when building the book.

## References

1. Sies H, Berndt C, Jones DP. Oxidative stress. *Annu Rev Biochem* 2017;**86**:715–48.
2. Bertero E, Maack C. Metabolic remodelling in heart failure. *Nat Rev Cardiol* 2018;**15**:457–70.
3. Nogueira V, Rigoulet M, Piquet MA, Devin A, Fontaine E, Leverve XM. Mitochondrial respiratory chain adjustment to cellular energy demand. *J Biol Chem* 2001;**276**:46104–10.
4. Schöndorf DC, Ivanyuk D, Baden P, Sanchez-Martinez A, De Cicco S, Yu C, Giunta I, Schwarz LK, Di Napoli G, Panagiotakopoulou V, Nestel S, Keatinge M, Pruszak J, Bandmann O, Heimrich B, Gasser T, Whitworth AJ, Deleidi M. The NAD+ precursor nicotinamide riboside rescues mitochondrial defects and neuronal loss in iPSC and fly models of Parkinson's disease. *Cell Rep* 2018;**23**:2976–88.
5. Fukushima T, Decker RV, Anderson WM, Spivey HO. Substrate channeling of NADH and binding of dehydrogenases to complex I. *J Biol Chem* 1989;**264**:16483–8.
6. Ushiroyama T, Fukushima T, Styre JD, Spivey HO. Substrate channeling of NADH in mitochondrial redox processes. *Curr Top Cell Regul* 1992;**33**:291–307.
7. Simmen T, Herrera-Cruz MS. Plastic mitochondria-endoplasmic reticulum (ER) contacts use chaperones and tethers to mould their structure and signaling. *Curr Opin Cell Biol* 2018;**53**:61–9.
8. Van Vliet AR, Verfaillie T, Agostinis P. New functions of mitochondria associated membranes in cellular signaling. *Biochim Biophys Acta* 2014;**1843**:2253–62.
9. Liu Z, Butow RA. Mitochondrial retrograde signaling. *Annu Rev Genet* 2006;**40**:159–85.
10. Picard M, Shirihai OS, Gentil BJ, Burelle Y. Mitochondrial morphology transitions and functions: implications for retrograde signaling? *Am J Phys Regul Integr Comp Phys* 2013;**304**:R393–406.
10a. Vance JE. MAM (mitochondria-associated membranes) in mammalian cells: lipids and beyond. *Biochim Biophys Acta* 2014;**1841**:595–609.
11. Twig G, Elorza A, Molina AJ, Mohamed H, Wikstrom JD, Walzer G, Stiles L, Haigh SE, Katz S, Las G, Alroy J, Wu M, Py BF, Yuan J, Deeney JT, Corkey BE, Shirihai OS. Fission and selective fusion govern mitochondrial segregation and elimination by autophagy. *EMBO J* 2008;**27**:433–46.
12. Liesa M, Shirihai OS. Mitochondrial dynamics in the regulation of nutrient utilization and energy expenditure. *Cell Metab* 2013;**17**:491–506.
13. Okamoto K, Wang W, Rounds J, Chambers EA, Jacobs DO. ATP from glycolysis is required for normal sodium homeostasis in resting fast-twitch rodent skeletal muscle. *Am J Physiol Endocrinol Metab* 2001;**281**:E479–88.
14. Muoio DM, Lewin TM, Wiedmer P, Coleman RA. Acyl-CoAs are functionally channeled in liver: potential role of acyl-CoA synthetase. *Am J Physiol Endocrinol Metab* 2000;**279**:E1366–73.
15. Sumegi B, Srere PA. Binding of the enzymes of fatty acid beta-oxidation and some related enzymes to pig heart inner mitochondrial membrane. *J Biol Chem* 1984;**259**:8748–52.
16. Hackenbrock CR, Chazotte B, Gupte SS. The random collision model and a critical assessment of diffusion and collision in mitochondrial electron transport. *J Bioenerg Biomembr* 1986;**18**:331–68.

17. Eubel H, Jänsch L, Braun HP. New insights into the respiratory chain of plant mitochondria. Supercomplexes and a unique composition of complex II. *Plant Physiol* 2003;**133**:274–86.
18. Schägger H, Pfeiffer K. Supercomplexes in the respiratory chains of yeast and mammalian mitochondria. *EMBO J* 2000;**19**:1777–83.
19. Schägger H, Pfeiffer K. The ratio of oxidative phosphorylation complexes I-V in bovine heart mitochondria and the composition of respiratory chain supercomplexes. *J Biol Chem* 2001;**276**:37861–7.
20. Boumans H, Berden JA, Grivell LA, van Dam K. Metabolic control analysis of the bc1 complex of *Saccharomyces cerevisiae*: effect on cytochrome c oxidase, respiration and growth rate. *Biochem J* 1998;**331**:877–83.
21. Boumans H, Grivell LA, Berden JA. The respiratory chain in yeast behaves as a single functional unit. *J Biol Chem* 1998;**273**:4872–7.
22. Bianchi C, Genova ML, Parenti Castelli G, Lenaz G. The mitochondrial respiratory chain is partially organized in a supercomplex assembly: kinetic evidence using flux control analysis. *J Biol Chem* 2004;**279**:36562–9.
23. Krause F, Scheckhuber CQ, Werner A, Rexroth S, Reifschneider NH, Dencher NA, Osiewacz HD. Supramolecular organization of cytochrome c oxidase- and alternative oxidase-dependent respiratory chains in the filamentous fungus *Podospora anserina*. *J Biol Chem* 2004;**279**:26453–61.
24. Stroh A, Anderka O, Pfeiffer K, Yagi T, Finel M, Ludwig B, Schägger H. Assembly of respiratory complexes I, III, and IV into NADH oxidase supercomplex stabilizes complex I in Paracoccus denitrificans. *J Biol Chem* 2004;**279**:5000–7.
25. Bagur R, Hajnóczky G. Intracellular Ca(2+) sensing: its role in calcium homeostasis and signaling. *Mol Cell* 2017;**66**:780–8.
26. Chernorudskiy AL, Zito E. Regulation of calcium homeostasis by ER redox: a close-up of the ER/mitochondria connection. *J Mol Biol* 2017;**429**:620–32.
27. Pan D, Lindau C, Lagies S, Wiedemann N, Kammerer B. Metabolic profiling of isolated mitochondria and cytoplasm reveals compartment-specific metabolic responses. *Metabolomics* 2018;**14**:59.
28. Van Vranken JG, Rutter J. The whole (cell) is less than the sum of its parts. *Cell* 2016;**166**:1078–9.
29. Chen WW, Freinkman E, Wang T, Birsoy K, Sabatini DM. Absolute quantification of matrix metabolites reveals the dynamics of mitochondrial metabolism. *Cell* 2016;**166**:1324–1337.e11.
30. Cortassa S, Caceres V, Bell LN, O'Rourke B, Paolocci N, Aon MA. From metabolomics to fluxomics: a computational procedure to translate metabolite profiles into metabolic fluxes. *Biophys J* 2015;**108**:163–72.
31. El-Mir MY, Nogueira V, Fontaine E, Avéret N, Rigoulet M, Leverve X. Dimethylbiguanide inhibits cell respiration via an indirect effect targeted on the respiratory chain complex I. *J Biol Chem* 2000;**275**:223–8.
32. Fontaine E. Metformin and respiratory chain complex I: the last piece of the puzzle? *Biochem J* 2014;**463**:e3–5.
33. Madiraju AK, Erion DM, Rahimi Y, Zhang XM, Braddock DT, Albright RA, Prigaro BJ, Wood JL, Bhanot S, MacDonald MJ, Jurczak MJ, Camporez JP, Lee HY, Cline GW, Samuel VT, Kibbey RG, Shulman GI. Metformin suppresses gluconeogenesis by inhibiting mitochondrial glycerophosphate dehydrogenase. *Nature* 2014;**510**:542–6.
34. Batandier C, Guigas B, Detaille D, El-Mir MY, Fontaine E, Rigoulet M, Leverve XM. The ROS production induced by a reverse-electron flux at respiratory-chain complex 1 is hampered by metformin. *J Bioenerg Biomembr* 2006;**38**:33–42.
35. Kane DA, Anderson EJ, Price 3rd JW, Woodlief TL, Lin CT, Bikman BT, Cortright RN, Neufer PD. Metformin selectively attenuates mitochondrial $H_2O_2$ emission without affecting respiratory capacity in skeletal muscle of obese rats. *Free Radic Biol Med* 2010;**49**:1082–7.
36. Wessels B, Ciapaite J, van den Broek NM, Nicolay K, Prompers JJ. Metformin impairs mitochondrial function in skeletal muscle of both lean and diabetic rats in a dose-dependent manner. *PLoS ONE* 2014;**9**:e100525.

# Index

Note: Page numbers followed by *f* indicate figures and *t* indicate tables.

## N

## Q



## R

## X

9780128117521